Ricci Flow *and the* Poincaré Conjecture

Clay Mathematics Monographs
Volume 3

Ricci Flow *and the* Poincaré Conjecture

JOHN MORGAN
GANG TIAN

American Mathematical Society
Clay Mathematics Institute

Clay Mathematics Institute Monograph Series

Editors in chief: S. Donaldson, A. Wiles
Managing editor: J. Carlson
Associate editors:

B. Conrad   I. Daubechies   C. Fefferman
J. Kollár   A. Okounkov   D. Morrison
C. Taubes   P. Ozsváth   K. Smith

2000 *Mathematics Subject Classification.* Primary 53C44, 57M40.

Cover photo: Jules Henri Poincaré, Smithsonian Institution Libraries, Washington, DC.

For additional information and updates on this book, visit
www.ams.org/bookpages/cmim-3

Library of Congress Cataloging-in-Publication Data
Morgan, John W., 1946–
  Ricci flow and the Poincaré conjecture / John Morgan, Gang Tian.
    p. cm. — (Clay mathematics monographs, ISSN 1539-6061 ; v. 3)
  Includes bibliographical references and index.
  ISBN 978-0-8218-4328-4 (alk. paper)
  1. Ricci flow. 2. Poincaré conjecture. I. Tian, G. II. Title. III. Series.

QA670.M67   2007
516.3′62—dc22
                                                      2007062016

    Copying and reprinting. Individual readers of this publication, and nonprofit libraries acting for them, are permitted to make fair use of the material, such as to copy a chapter for use in teaching or research. Permission is granted to quote brief passages from this publication in reviews, provided the customary acknowledgment of the source is given.
    Republication, systematic copying, or multiple reproduction of any material in this publication is permitted only under license from the American Mathematical Society. Requests for such permission should be addressed to the Acquisitions Department, American Mathematical Society, 201 Charles Street, Providence, Rhode Island 02904-2294, USA. Requests can also be made by e-mail to reprint-permission@ams.org.

© 2007 by the authors. All rights reserved.
Published by the American Mathematical Society, Providence, RI,
for the Clay Mathematics Institute, Cambridge, MA.
Printed in the United States of America.

  ∞ The paper used in this book is acid-free and falls within the guidelines
established to ensure permanence and durability.
Visit the AMS home page at http://www.ams.org/
Visit the Clay Mathematics Institute home page at http://www.claymath.org/

10 9 8 7 6 5 4 3 2 1   12 11 10 09 08 07

Contents

Introduction   ix
1. Overview of Perelman's argument   x
2. Background material from Riemannian geometry   xvi
3. Background material from Ricci flow   xix
4. Perelman's advances   xxv
5. The standard solution and the surgery process   xxxi
6. Extending Ricci flows with surgery   xxxiv
7. Finite-time extinction   xxxvii
8. Acknowledgements   xl
9. List of related papers   xlii

Part 1. Background from Riemannian Geometry and Ricci flow   1

Chapter 1. Preliminaries from Riemannian geometry   3
1. Riemannian metrics and the Levi-Civita connection   3
2. Curvature of a Riemannian manifold   5
3. Geodesics and the exponential map   10
4. Computations in Gaussian normal coordinates   16
5. Basic curvature comparison results   18
6. Local volume and the injectivity radius   19

Chapter 2. Manifolds of non-negative curvature   21
1. Busemann functions   21
2. Comparison results in non-negative curvature   23
3. The soul theorem   24
4. Ends of a manifold   27
5. The splitting theorem   28
6. ϵ-necks   30
7. Forward difference quotients   33

Chapter 3. Basics of Ricci flow   35
1. The definition of Ricci flow   35
2. Some exact solutions to the Ricci flow   36
3. Local existence and uniqueness   39
4. Evolution of curvatures   41

5.	Curvature evolution in an evolving orthonormal frame	42
6.	Variation of distance under Ricci flow	45
7.	Shi's derivative estimates	50
8.	Generalized Ricci flows	59

Chapter 4. The maximum principle — 63
1. Maximum principle for scalar curvature — 63
2. The maximum principle for tensors — 65
3. Applications of the maximum principle — 67
4. The strong maximum principle for curvature — 69
5. Pinching toward positive curvature — 75

Chapter 5. Convergence results for Ricci flow — 83
1. Geometric convergence of Riemannian manifolds — 83
2. Geometric convergence of Ricci flows — 90
3. Gromov-Hausdorff convergence — 92
4. Blow-up limits — 99
5. Splitting limits at infinity — 100

Part 2. Perelman's length function and its applications — 103

Chapter 6. A comparison geometry approach to the Ricci flow — 105
1. $\mathcal{L}$-length and $\mathcal{L}$-geodesics — 105
2. The $\mathcal{L}$-exponential map and its first-order properties — 112
3. Minimizing $\mathcal{L}$-geodesics and the injectivity domain — 116
4. Second-order differential inequalities for $\widetilde{L}^{\overline{\tau}}$ and $L_x^{\overline{\tau}}$ — 119
5. Reduced length — 129
6. Local Lipschitz estimates for l_x — 133
7. Reduced volume — 140

Chapter 7. Complete Ricci flows of bounded curvature — 149
1. The functions L_x and l_x — 149
2. A bound for $\min l_x^\tau$ — 152
3. Reduced volume — 164

Chapter 8. Non-collapsed results — 169
1. A non-collapsing result for generalized Ricci flows — 169
2. Application to compact Ricci flows — 176

Chapter 9. κ-non-collapsed ancient solutions — 179
1. Preliminaries — 179
2. The asymptotic gradient shrinking soliton for κ-solutions — 183
3. Splitting results at infinity — 203
4. Classification of gradient shrinking solitons — 206
5. Universal κ — 220
6. Asymptotic volume — 221

7.	Compactness of the space of 3-dimensional κ-solutions	225
8.	Qualitative description of κ-solutions	230

Chapter 10. Bounded curvature at bounded distance 245
1. Pinching toward positive: the definitions 245
2. The statement of the theorem 245
3. The incomplete geometric limit 247
4. Cone limits near the end $\mathcal{E}$ for rescalings of U_∞ 255
5. Comparison of the two types of limits 263
6. The final contradiction 265

Chapter 11. Geometric limits of generalized Ricci flows 267
1. A smooth blow-up limit defined for a small time 267
2. Long-time blow-up limits 271
3. Incomplete smooth limits at singular times 279
4. Existence of strong δ-necks sufficiently deep in a 2ϵ-horn 287

Chapter 12. The standard solution 293
1. The initial metric 293
2. Standard Ricci flows: The statement 295
3. Existence of a standard flow 296
4. Completeness, positive curvature, and asymptotic behavior 297
5. Standard solutions are rotationally symmetric 300
6. Uniqueness 306
7. Solution of the harmonic map flow 308
8. Completion of the proof of uniqueness 322
9. Some corollaries 325

Part 3. Ricci flow with surgery 329

Chapter 13. Surgery on a δ-neck 331
1. Notation and the statement of the result 331
2. Preliminary computations 334
3. The proof of Theorem 13.2 339
4. Other properties of the result of surgery 341

Chapter 14. Ricci Flow with surgery: the definition 343
1. Surgery space-time 343
2. The generalized Ricci flow equation 348

Chapter 15. Controlled Ricci flows with surgery 353
1. Gluing together evolving necks 353
2. Topological consequences of Assumptions (1) – (7) 356
3. Further conditions on surgery 359
4. The process of surgery 361
5. Statements about the existence of Ricci flow with surgery 362

6. Outline of the proof of Theorem 15.9 — 365

Chapter 16. Proof of non-collapsing — 367
1. The statement of the non-collapsing result — 367
2. The proof of non-collapsing when $R(x) = r^{-2}$ with $r \leq r_{i+1}$ — 368
3. Minimizing $\mathcal{L}$-geodesics exist when $R(x) \leq r_{i+1}^{-2}$: the statement — 368
4. Evolution of neighborhoods of surgery caps — 369
5. A length estimate — 375
6. Completion of the proof of Proposition 16.1 — 391

Chapter 17. Completion of the proof of Theorem 15.9 — 395
1. Proof of the strong canonical neighborhood assumption — 395
2. Surgery times don't accumulate — 408

Part 4. Completion of the proof of the Poincaré Conjecture — 413

Chapter 18. Finite-time extinction — 415
1. The result — 415
2. Disappearance of components with non-trivial π_2 — 420
3. Components with non-trivial π_3 — 429
4. First steps in the proof of Proposition 18.18 — 432

Chapter 19. Completion of the Proof of Proposition 18.24 — 437
1. Curve-shrinking — 437
2. Basic estimates for curve-shrinking — 441
3. Ramp solutions in $M \times S^1$ — 445
4. Approximating the original family Γ — 449
5. The case of a single $c \in S^2$ — 453
6. The completion of the proof of Proposition 18.24 — 461
7. Proof of Lemma 19.31: annuli of small area — 464
8. Proof of the first inequality in Lemma 19.24 — 481

Appendix. 3-manifolds covered by canonical neighborhoods — 497
1. Shortening curves — 497
2. The geometry of an ϵ-neck — 497
3. Overlapping ϵ-necks — 502
4. Regions covered by ϵ-necks and (C, ϵ)-caps — 504
5. Subsets of the union of cores of (C, ϵ)-caps and ϵ-necks. — 508

Bibliography — 515

Index — 519

Introduction

In this book we present a complete and detailed proof of

The Poincaré Conjecture: every closed, smooth, simply connected 3-manifold is diffeomorphic[1] to S^3.

This conjecture was formulated by Henri Poincaré [**58**] in 1904 and has remained open until the recent work of Perelman. The arguments we give here are a detailed version of those that appear in Perelman's three preprints [**53, 55, 54**]. Perelman's arguments rest on a foundation built by Richard Hamilton with his study of the Ricci flow equation for Riemannian metrics. Indeed, Hamilton believed that Ricci flow could be used to establish the Poincaré Conjecture and more general topological classification results in dimension 3, and laid out a program to accomplish this. The difficulty was to deal with singularities in the Ricci flow. Perelman's breakthrough was to understand the qualitative nature of the singularities sufficiently to allow him to prove the Poincaré Conjecture (and Theorem 0.1 below which implies the Poincaré Conjecture). For a detailed history of the Poincaré Conjecture, see Milnor's survey article [**50**].

A class of examples closely related to the 3-sphere are *the 3-dimensional spherical space-forms*, i.e., the quotients of S^3 by free, linear actions of finite subgroups of the orthogonal group $O(4)$. There is a generalization of the Poincaré Conjecture, called the **3-dimensional spherical space-form conjecture**, which conjectures that any closed 3-manifold with finite fundamental group is diffeomorphic to a 3-dimensional spherical space-form. Clearly, a special case of the 3-dimensional spherical space-form conjecture is the Poincaré Conjecture.

As indicated in Remark 1.4 of [**54**], the arguments we present here not only prove the Poincaré Conjecture, they prove the 3-dimensional space-form conjecture. In fact, the purpose of this book is to prove the following more general theorem.

---

[1] Every topological 3-manifold admits a differentiable structure and every homeomorphism between smooth 3-manifolds can be approximated by a diffeomorphism. Thus, classification results about topological 3-manifolds up to homeomorphism and about smooth 3-manifolds up to diffeomorphism are equivalent. In this book 'manifold' means 'smooth manifold.'

THEOREM 0.1. *Let M be a closed, connected 3-manifold and suppose that the fundamental group of M is a free product of finite groups and infinite cyclic groups. Then M is diffeomorphic to a connected sum of spherical space-forms, copies of $S^2 \times S^1$, and copies of the unique (up to diffeomorphism) non-orientable 2-sphere bundle over S^1.*

This immediately implies an affirmative resolution of the Poincaré Conjecture and of the 3-dimensional spherical space-form conjecture.

COROLLARY 0.2. *(a) A closed, simply connected 3-manifold is diffeomorphic to S^3. (b) A closed 3-manifold with finite fundamental group is diffeomorphic to a 3-dimensional spherical space-form.*

Before launching into a more detailed description of the contents of this book, one remark on the style of the exposition is in order. Because of the importance and visibility of the results discussed here, and because of the number of incorrect claims of proofs of these results in the past, we felt that it behooved us to work out and present the arguments in great detail. Our goal was to make the arguments clear and convincing and also to make them more easily accessible to a wider audience. As a result, experts may find some of the points are overly elaborated.

1. Overview of Perelman's argument

In dimensions less than or equal to 3, any Riemannian metric of constant Ricci curvature has constant sectional curvature. Classical results in Riemannian geometry show that the universal cover of a closed manifold of constant positive curvature is diffeomorphic to the sphere and that the fundamental group is identified with a finite subgroup of the orthogonal group acting linearly and freely on the universal cover. Thus, one can approach the Poincaré Conjecture and the more general 3-dimensional spherical space-form problem by asking the following question. Making the appropriate fundamental group assumptions on 3-manifold M, how does one establish the existence of a metric of constant Ricci curvature on M? The essential ingredient in producing such a metric is the Ricci flow equation introduced by Richard Hamilton in [**29**]:

$$\frac{\partial g(t)}{\partial t} = -2\mathrm{Ric}(g(t)),$$

where $\mathrm{Ric}(g(t))$ is the Ricci curvature of the metric $g(t)$. The fixed points (up to rescaling) of this equation are the Riemannian metrics of constant Ricci curvature. For a general introduction to the subject of the Ricci flow see Hamilton's survey paper [**34**], the book by Chow-Knopf [**13**], or the book by Chow, Lu, and Ni [**14**]. The Ricci flow equation is a (weakly) parabolic partial differential flow equation for Riemannian metrics on a smooth manifold. Following Hamilton, one defines a Ricci flow to be a family of

1. OVERVIEW OF PERELMAN'S ARGUMENT

Riemannian metrics $g(t)$ on a fixed smooth manifold, parameterized by t in some interval, satisfying this equation. One considers t as time and studies the equation as an initial value problem: Beginning with any Riemannian manifold (M, g_0) find a Ricci flow with (M, g_0) as initial metric; that is to say find a one-parameter family $(M, g(t))$ of Riemannian manifolds with $g(0) = g_0$ satisfying the Ricci flow equation. This equation is valid in all dimensions but we concentrate here on dimension 3. In a sentence, the method of proof is to begin with any Riemannian metric on the given smooth 3-manifold and flow it using the Ricci flow equation to obtain the constant curvature metric for which one is searching. There are two examples where things work in exactly this way, both due to Hamilton. (i) If the initial metric has positive Ricci curvature, Hamilton proved over twenty years ago, [**29**], that under the Ricci flow the manifold shrinks to a point in finite time, that is to say, there is a finite-time singularity, and, as we approach the singular time, the diameter of the manifold tends to zero and the curvature blows up at every point. Hamilton went on to show that, in this case, rescaling by a time-dependent function so that the diameter is constant produces a one-parameter family of metrics converging smoothly to a metric of constant positive curvature. (ii) At the other extreme, in [**36**] Hamilton showed that if the Ricci flow exists for all time and if there is an appropriate curvature bound together with another geometric bound, then as $t \to \infty$, after rescaling to have a fixed diameter, the metric converges to a metric of constant negative curvature.

The results in the general case are much more complicated to formulate and much more difficult to establish. While Hamilton established that the Ricci flow equation has short-term existence properties, i.e., one can define $g(t)$ for t in some interval $[0, T)$ where T depends on the initial metric, it turns out that if the topology of the manifold is sufficiently complicated, say it is a non-trivial connected sum, then no matter what the initial metric is one must encounter finite-time singularities, forced by the topology. More seriously, even if the manifold has simple topology, beginning with an arbitrary metric one expects to (and cannot rule out the possibility that one will) encounter finite-time singularities in the Ricci flow. These singularities, unlike in the case of positive Ricci curvature, occur along proper subsets of the manifold, not the entire manifold. Thus, to derive the topological consequences stated above, it is not sufficient in general to stop the analysis the first time a singularity arises in the Ricci flow. One is led to study a more general evolution process called *Ricci flow with surgery*, first introduced by Hamilton in the context of four-manifolds, [**35**]. This evolution process is still parameterized by an interval in time, so that for each t in the interval of definition there is a compact Riemannian 3-manifold M_t. But there is a discrete set of times at which the manifolds and metrics undergo topological and metric discontinuities (surgeries). In each of the complementary intervals to the singular times, the evolution is the usual Ricci flow, though, because of the surgeries, the topological type of the manifold M_t changes as t moves from one complementary interval to the next. From an analytic point of view, the surgeries at the discontinuity times are introduced in order to 'cut away' a neighborhood of the singularities as they develop and insert by hand, in place of the 'cut away' regions, geometrically nice regions. This allows one to continue the Ricci flow (or more precisely, restart the Ricci flow with the new metric constructed at the discontinuity time). Of course, the surgery process also changes the topology. To be able to say anything useful topologically about such a process, one needs results about Ricci flow, and one also needs to control both the topology and the geometry of the surgery process at the singular times. For example, it is crucial for the topological applications that we do surgery along 2-spheres rather than surfaces of higher genus. Surgery along 2-spheres produces the connected sum decomposition, which is well-understood topologically, while, for example, Dehn surgeries along tori can completely destroy the topology, changing any 3-manifold into any other.

The change in topology turns out to be completely understandable and amazingly, the surgery processes produce exactly the topological operations needed to cut the manifold into pieces on which the Ricci flow can produce the metrics sufficiently controlled so that the topology can be recognized.

The bulk of this book (Chapters 1-17 and the Appendix) concerns the establishment of the following long-time existence result for Ricci flow with surgery.

THEOREM 0.3. *Let (M, g_0) be a closed Riemannian 3-manifold. Suppose that there is no embedded, locally separating $\mathbb{R}P^2$ contained[2] in M. Then there is a Ricci flow with surgery defined for all $t \in [0, \infty)$ with initial metric (M, g_0). The set of discontinuity times for this Ricci flow with surgery is a discrete subset of $[0, \infty)$. The topological change in the 3-manifold as one crosses a surgery time is a connected sum decomposition together with removal of connected components, each of which is diffeomorphic to one of $S^2 \times S^1$, $\mathbb{R}P^3 \# \mathbb{R}P^3$, the non-orientable 2-sphere bundle over S^1, or a manifold admitting a metric of constant positive curvature.*

While Theorem 0.3 is central for all applications of Ricci flow to the topology of three-dimensional manifolds, the argument for the 3-manifolds described in Theorem 0.1 is simplified, and avoids all references to the nature of the flow as time goes to infinity, because of the following finite-time extinction result.

---

[2]I.e., no embedded $\mathbb{R}P^2$ in M with trivial normal bundle. Clearly, all orientable manifolds satisfy this condition.

THEOREM 0.4. *Let M be a closed 3-manifold whose fundamental group is a free product of finite groups and infinite cyclic groups[3]. Let g_0 be any Riemannian metric on M. Then M admits no locally separating $\mathbb{R}P^2$, so that there is a Ricci flow with surgery defined for all positive time with (M, g_0) as initial metric as described in Theorem 0.3. This Ricci flow with surgery becomes extinct after some time $T < \infty$, in the sense that the manifolds M_t are empty for all $t \geq T$.*

This result is established in Chapter 18 following the argument given by Perelman in [**54**], see also [**15**].

We immediately deduce Theorem 0.1 from Theorems 0.3 and 0.4 as follows: Let M be a 3-manifold satisfying the hypothesis of Theorem 0.1. Then there is a finite sequence $M = M_0, M_1, \ldots, M_k = \emptyset$ such that for each i, $1 \leq i \leq k$, M_i is obtained from M_{i-1} by a connected sum decomposition or M_i is obtained from M_{i-1} by removing a component diffeomorphic to one of $S^2 \times S^1$, $\mathbb{R}P^3 \# \mathbb{R}P^3$, a non-orientable 2-sphere bundle over S^1, or a 3-dimensional spherical space-form. Clearly, it follows by downward induction on i that each connected component of M_i is diffeomorphic to a connected sum of 3-dimensional spherical space-forms, copies of $S^2 \times S^1$, and copies of the non-orientable 2-sphere bundle over S^1. In particular, $M = M_0$ has this form. Since M is connected by hypothesis, this proves the theorem. In fact, this argument proves the following:

COROLLARY 0.5. *Let (M_0, g_0) be a connected Riemannian manifold with no locally separating $\mathbb{R}P^2$. Let $(\mathcal{M}, G)$ be a Ricci flow with surgery defined for $0 \leq t < \infty$ with (M_0, g_0) as initial manifold. Then the following four conditions are equivalent:*

(1) *$(\mathcal{M}, G)$ becomes extinct after a finite time, i.e., $M_T = \emptyset$ for all T sufficiently large,*
(2) *M_0 is diffeomorphic to a connected sum of three-dimensional spherical space-forms and S^2-bundles over S^1,*
(3) *the fundamental group of M_0 is a free product of finite groups and infinite cyclic groups,*
(4) *no prime[4] factor of M_0 is acyclic, i.e., every prime factor of M_0 has either non-trivial π_2 or non-trivial π_3.*

PROOF. Repeated application of Theorem 0.3 shows that (1) implies (2). The implication (2) implies (3) is immediate from van Kampen's theorem.

---

[3]In [**54**] Perelman states the result for 3-manifolds without prime factors that are acyclic. It is a standard exercise in 3-manifold topology to show that Perelman's condition is equivalent to the group theory hypothesis stated here; see Corollary 0.5.

[4]A three-manifold P is prime if every separating 2-sphere in P bounds a three-ball in P. Equivalently, P is prime if it admits no non-trivial connected sum decomposition. Every closed three-manifold decomposes as a connected sum of prime factors with the decomposition being unique up to diffeomorphism of the factors and the order of the factors.

The fact that (3) implies (1) is Theorem 0.4. This shows that (1), (2) and (3) are all equivalent. Since three-dimensional spherical space-forms and S^2-bundles over S^1 are easily seen to be prime, (2) implies (4). Thus, it remains only to see that (4) implies (3). We consider a manifold M satisfying (4), a prime factor P of M, and universal covering $\widetilde{P}$ of P. First suppose that $\pi_2(P) = \pi_2(\widetilde{P})$ is trivial. Then, by hypothesis $\pi_3(P) = \pi_3(\widetilde{P})$ is non-trivial. By the Hurewicz theorem this means that $H_3(\widetilde{P})$ is non-trivial, and hence that $\widetilde{P}$ is a compact, simply connected three-manifold. It follows that $\pi_1(P)$ is finite. Now suppose that $\pi_2(P)$ is non-trivial. Then P is not diffeomorphic to $\mathbb{R}P^3$. Since P is prime and contains no locally separating $\mathbb{R}P^2$, it follows that P contains no embedded $\mathbb{R}P^2$. Then by the sphere theorem there is an embedded 2-sphere in P that is homotopically non-trivial. Since P is prime, this sphere cannot separate, so cutting P open along it produces a connected manifold P_0 with two boundary 2-spheres. Since P_0 is prime, it follows that P_0 is diffeomorphic to $S^2 \times I$ and hence P is diffeomorphic to a 2-sphere bundle over the circle. □

REMARK 0.6. (i) The use of the sphere theorem is unnecessary in the above argument for what we actually prove is that if every prime factor of M has non-trivial π_2 or non-trivial π_3, then the Ricci flow with surgery with (M, g_0) as initial metric becomes extinct after a finite time. In fact, the sphere theorem for closed 3-manifolds follows from the results here.
(ii) If the initial manifold is simpler then all the time-slices are simpler: If $(\mathcal{M}, G)$ is a Ricci flow with surgery whose initial manifold is prime, then every time-slice is a disjoint union of connected components, all but at most one being diffeomorphic to a 3-sphere and if there is one not diffeomorphic to a 3-sphere, then it is diffeomorphic to the initial manifold. If the initial manifold is a simply connected manifold M_0, then every component of every time-slice M_T must be simply connected, and thus *a posteriori* every time-slice is a disjoint union of manifolds diffeomorphic to the 3-sphere. Similarly, if the initial manifold has finite fundamental group, then every connected component of every time-slice is either simply connected or has the same fundamental group as the initial manifold.
(iii) The conclusion of this result is a natural generalization of Hamilton's conclusion in analyzing the Ricci flow on manifolds of positive Ricci curvature in [**29**]. Namely, under appropriate hypotheses, during the evolution process of Ricci flow with surgery the manifold breaks into components each of which disappears in finite time. As a component disappears at some finite time, the metric on that component is well enough controlled to show that the disappearing component admits a non-flat, homogeneous Riemannian metric of non-negative sectional curvature, i.e., a metric locally isometric to either a round S^3 or to a product of a round S^2 with the usual metric on $\mathbb{R}$. The existence of such a metric on a component immediately gives the

topological conclusion of Theorem 0.1 for that component, i.e., that it is diffeomorphic to a 3-dimensional spherical space-form, to $S^2 \times S^1$ to a non-orientable 2-sphere bundle over S^1, or to $\mathbb{R}P^3 \# \mathbb{R}P^3$. The biggest difference between these two results is that Hamilton's hypothesis is geometric (positive Ricci curvature) whereas Perelman's is homotopy theoretic (information about the fundamental group).

(iv) It is also worth pointing out that it follows from Corollary 0.5 that the manifolds that satisfy the four equivalent conditions in that corollary are exactly the closed, connected, 3-manifolds that admit a Riemannian metric of positive scalar curvature, cf, [**62**] and [**26**].

One can use Ricci flow in a more general study of 3-manifolds than the one we carry out here. There is a conjecture due to Thurston, see [**69**], known as Thurston's Geometrization Conjecture or simply as the Geometrization Conjecture for 3-manifolds. It conjectures that every 3-manifold without locally separating $\mathbb{R}P^2$'s (in particular every orientable 3-manifold) is a connected sum of prime 3-manifolds each of which admits a decomposition along incompressible[5] tori into pieces that admit locally homogeneous geometries of finite volume. Modulo questions about cofinite-volume lattices in $SL_2(\mathbb{C})$, proving this conjecture leads to a complete classification of 3-manifolds without locally separating $\mathbb{R}P^2$'s, and in particular to a complete classification of all orientable 3-manifolds. (See Peter Scott's survey article [**63**].) By passing to the orientation double cover and working equivariantly, these results can be extended to all 3-manifolds.

Perelman in [**55**] has stated results which imply a positive resolution of Thurston's Geometrization conjecture. Perelman's proposed proof of Thurston's Geometrization Conjecture relies in an essential way on Theorem 0.3, namely the existence of Ricci flow with surgery for all positive time. But it also involves a further analysis of the limits of these Ricci flows as time goes to infinity. This further analysis involves analytic arguments which are exposed in Sections 6 and 7 of Perelman's second paper ([**55**]), following earlier work of Hamilton ([**36**]) in a simpler case of bounded curvature. They also involve a result (Theorem 7.4 from [**55**]) from the theory of manifolds with curvature locally bounded below that are collapsed, related to results of Shioya-Yamaguchi [**67**]. The Shioya-Yamaguchi results in turn rely on an earlier, unpublished work of Perelman proving the so-called 'Stability Theorem.' Recently, Kapovich, [**43**] has put a preprint on the archive giving a proof of the stability result. We have been examining another approach, one suggested by Perelman in [**55**], avoiding the stability theorem, cf, [**44**] and [**51**]. It is our view that the collapsing results needed for the Geometrization Conjecture are in place, but that before a definitive statement that the Geometrization Conjecture has been resolved can be made these

---

[5]I.e., embedded by a map that is injective on π_1.

arguments must be subjected to the same close scrutiny that the arguments proving the Poincaré Conjecture have received. This process is underway.

In this book we do not attempt to explicate any of the results beyond Theorem 0.3 described in the previous paragraph that are needed for the Geometrization Conjecture. Rather, we content ourselves with presenting a proof of Theorem 0.1 above which, as we have indicated, concerns initial Riemannian manifolds for which the Ricci flow with surgery becomes extinct after finite time. We are currently preparing a detailed proof, along the lines suggested by Perelman, of the further results that will complete the proof of the Geometrization Conjecture.

As should be clear from the above overview, Perelman's argument did not arise in a vacuum. Firstly, it resides in a context provided by the general theory of Riemannian manifolds. In particular, various notions of convergence of sequences of manifolds play a crucial role. The most important is geometric convergence (smooth convergence on compact subsets). Even more importantly, Perelman's argument resides in the context of the theory of the Ricci flow equation, introduced by Richard Hamilton and extensively studied by him and others. Perelman makes use of almost every previously established result for 3-dimensional Ricci flows. One exception is Hamilton's proposed classification results for 3-dimensional singularities. These are replaced by Perelman's strong qualitative description of singularity development for Ricci flows on compact 3-manifolds.

The first five chapters of the book review the necessary background material from these two subjects. Chapters 6 through 11 then explain Perelman's advances. In Chapter 12 we introduce the standard solution, which is the manifold constructed by hand that one 'glues in' in doing surgery. Chapters 13 through 17 describe in great detail the surgery process and prove the main analytic and topological estimates that are needed to show that one can continue the process for all positive time. At the end of Chapter 17 we have established Theorem 0.3. Chapter 18 and 19 discuss the finite-time extinction result. Lastly, there is an appendix on some topological results that were needed in the surgery analysis in Chapters 13-17.

2. Background material from Riemannian geometry

2.1. Volume and injectivity radius. One important general concept that is used throughout is the notion of a manifold being non-collapsed at a point. Suppose that we have a point x in a complete Riemannian n-manifold. Then we say that the manifold is κ-*non-collapsed* at x provided that the following holds: For any r such that the norm of the Riemann curvature tensor, $|\mathrm{Rm}|$, is $\leq r^{-2}$ at all points of the metric ball, $B(x,r)$, of radius r centered at x, we have $\mathrm{Vol}\, B(x,r) \geq \kappa r^n$. There is a relationship between this notion and the injectivity radius of M at x. Namely, if $|\mathrm{Rm}| \leq r^{-2}$ on $B(x,r)$ and if $B(x,r)$ is κ-non-collapsed then the injectivity radius of M

2. BACKGROUND MATERIAL FROM RIEMANNIAN GEOMETRY

at x is greater than or equal to a positive constant that depends only on r and κ. The advantage of working with the volume non-collapsing condition is that, unlike for the injectivity radius, there is a simple equation for the evolution of volume under Ricci flow.

Another important general result is the Bishop-Gromov volume comparison result that says that if the Ricci curvature of a complete Riemannian n-manifold M is bounded below by a constant $(n-1)K$, then for any $x \in M$ the ratio of the volume of $B(x,r)$ to the volume of the ball of radius r in the space of constant curvature K is a non-increasing function whose limit as $r \to 0$ is 1.

All of these basic facts from Riemannian geometry are reviewed in the first chapter.

2.2. Manifolds of non-negative curvature. For reasons that should be clear from the above description and in any event will become much clearer shortly, manifolds of non-negative curvature play an extremely important role in the analysis of Ricci flows with surgery. We need several general results about them. The first is the soul theorem for manifolds of non-negative sectional curvature. A *soul* is a compact, totally geodesic submanifold. The entire manifold is diffeomorphic to the total space of a vector bundle over any of its souls. If a non-compact n-manifold has positive sectional curvature, then any soul for it is a point, and in particular, the manifold is diffeomorphic to Euclidean space. In addition, the distance function f from a soul has the property that, for every $t > 0$, the pre-image $f^{-1}(t)$ is homeomorphic to an $(n-1)$-sphere and the pre-image under this distance function of any non-degenerate interval $I \subset \mathbb{R}^+$ is homeomorphic to $S^{n-1} \times I$.

Another important result is the splitting theorem, which says that, if a complete manifold of non-negative sectional curvature has a geodesic line (an isometric copy of $\mathbb{R}$) that is distance minimizing between every pair of its points, then that manifold is a metric product of a manifold of one lower dimension and $\mathbb{R}$. In particular, if a complete n-manifold of non-negative sectional curvature has two ends, then it is a metric product $N^{n-1} \times \mathbb{R}$ where N^{n-1} is a compact manifold.

Also, we need some of the elementary comparison results from Toponogov theory. These compare ordinary triangles in the Euclidean plane with triangles in a manifold of non-negative sectional curvature whose sides are minimizing geodesics in that manifold.

2.3. Canonical neighborhoods. Much of the analysis of the geometry of Ricci flows revolves around the notion of canonical neighborhoods. Fix some $\epsilon > 0$ sufficiently small. There are two types of non-compact canonical neighborhoods: ϵ-necks and ϵ-caps. An ϵ-neck in a Riemannian 3-manifold (M, g) centered at a point $x \in M$ is a submanifold $N \subset M$ and a diffeomorphism $\psi\colon S^2\times(-\epsilon^{-1},\epsilon^{-1})\to N$ such that $x\in\psi(S^2\times\{0\})$ and such that the pullback of the rescaled metric, $\psi^*(R(x)g)$, is within ϵ in the $C^{[1/\epsilon]}$-topology of the product of the round metric of scalar curvature 1 on S^2 with the usual metric on the interval $(-\epsilon^{-1},\epsilon^{-1})$. (Throughout, $R(x)$ denotes the scalar curvature of (M,g) at the point x.) An ϵ-cap is a non-compact submanifold $\mathcal{C}\subset M$ with the property that a neighborhood N of infinity in $\mathcal{C}$ is an ϵ-neck, such that every point of N is the center of an ϵ-neck in M, and such that the *core*, $\mathcal{C}\setminus\overline{N}$, of the ϵ-cap is diffeomorphic to either a 3-ball or a punctured $\mathbb{R}P^3$. It will also be important to consider ϵ-caps that, after rescaling to make $R(x)=1$ for some point x in the cap, have bounded geometry (bounded diameter, bounded ratio of the curvatures at any two points, and bounded volume). If C represents the bound for these quantities, then we call the cap a (C,ϵ)-cap. See FIG. 1. An ϵ-tube in M is a submanifold of M diffeomorphic to $S^2\times(0,1)$ which is a union of ϵ-necks and with the property that each point of the ϵ-tube is the center of an ϵ-neck in M.

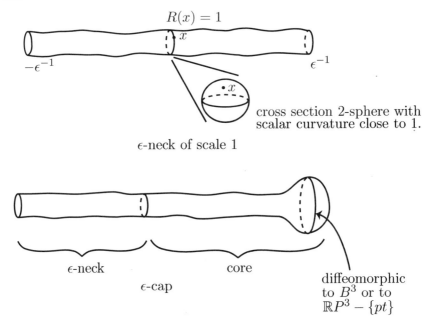

FIGURE 1. Canonical neighborhoods.

There are two other types of canonical neighborhoods in 3-manifolds – (i) a C-component and (ii) an ϵ-round component. The C-component is a compact, connected Riemannian manifold of positive sectional curvature diffeomorphic to either S^3 or $\mathbb{R}P^3$ with the property that rescaling the metric by $R(x)$ for any x in the component produces a Riemannian manifold whose diameter is at most C, whose sectional curvature at any point and in any 2-plane direction is between C^{-1} and C, and whose volume is between C^{-1}

and C. An ϵ-round component is a component on which the metric rescaled by $R(x)$ for any x in the component is within ϵ in the $C^{[1/\epsilon]}$-topology of a round metric of scalar curvature 1.

As we shall see, the singularities at time T of a 3-dimensional Ricci flow are contained in subsets that are unions of canonical neighborhoods with respect to the metrics at nearby, earlier times $t' < T$. Thus, we need to understand the topology of manifolds that are unions of ϵ-tubes and ϵ-caps. The fundamental observation is that, provided that ϵ is sufficiently small, when two ϵ-necks intersect (in more than a small neighborhood of the boundaries) their product structures almost line up, so that the two ϵ-necks can be glued together to form a manifold fibered by S^2's. Using this idea we show that, for $\epsilon > 0$ sufficiently small, if a connected manifold is a union of ϵ-tubes and ϵ-caps, then it is diffeomorphic to $\mathbb{R}^3$, $S^2 \times \mathbb{R}$, S^3, $S^2 \times S^1$, $\mathbb{R}P^3 \# \mathbb{R}P^3$, the total space of a line bundle over $\mathbb{R}P^2$, or the non-orientable 2-sphere bundle over S^1. This topological result is proved in the appendix at the end of the book. **We shall fix $\epsilon > 0$ sufficiently small so that these results hold.**

There is one result relating canonical neighborhoods and manifolds of positive curvature of which we make repeated use: Any complete 3-manifold of positive curvature does not admit ϵ-necks of arbitrarily high curvature. In particular, if M is a complete Riemannian 3-manifold with the property that every point of scalar curvature greater than r_0^{-2} has a canonical neighborhood, then M has bounded curvature. This turns out to be of central importance and is used repeatedly.

All of these basic facts about Riemannian manifolds of non-negative curvature are recalled in the second chapter.

3. Background material from Ricci flow

Hamilton [**29**] introduced the Ricci flow equation,

$$\frac{\partial g(t)}{\partial t} = -2\mathrm{Ric}(g(t)).$$

This is an evolution equation for a one-parameter family of Riemannian metrics $g(t)$ on a smooth manifold M. The Ricci flow equation is weakly parabolic and is strictly parabolic modulo the 'gauge group', which is the group of diffeomorphisms of the underlying smooth manifold. One should view this equation as a non-linear, tensor version of the heat equation. From it, one can derive the evolution equation for the Riemannian metric tensor, the Ricci tensor, and the scalar curvature function. These are all parabolic equations. For example, the evolution equation for scalar curvature $R(x, t)$ is

(0.1) $$\frac{\partial R}{\partial t}(x, t) = \triangle R(x, t) + 2|\mathrm{Ric}(x, t)|^2,$$

illustrating the similarity with the heat equation. (Here $\triangle$ is the Laplacian with non-positive spectrum.)

3.1. First results. Of course, the first results we need are uniqueness and short-time existence for solutions to the Ricci flow equation for compact manifolds. These results were proved by Hamilton ([**29**]) using the Nash-Moser inverse function theorem, ([**28**]). These results are standard for strictly parabolic equations. By now there is a fairly standard method for working 'modulo' the gauge group (the group of diffeomorphisms) and hence arriving at a strictly parabolic situation where the classical existence, uniqueness and smoothness results apply. The method for the Ricci flow equation goes under the name of 'DeTurck's trick.'

There is also a result that allows us to patch together local solutions $(U, g(t))$, $a \leq t \leq b$, and $(U, h(t))$, $b \leq t \leq c$, to form a smooth solution defined on the interval $a \leq t \leq c$ provided that $g(b) = h(b)$.

Given a Ricci flow $(M, g(t))$ we can always translate, replacing t by $t+t_0$ for some fixed t_0, to produce a new Ricci flow. We can also rescale by any positive constant Q by setting $h(t) = Qg(Q^{-1}t)$ to produce a new Ricci flow.

3.2. Gradient shrinking solitons. Suppose that (M, g) is a complete Riemannian manifold, and suppose that there is a constant $\lambda > 0$ with the property that

$$\mathrm{Ric}(g) = \lambda g.$$

In this case, it is easy to see that there is a Ricci flow given by

$$g(t) = (1 - 2\lambda t)g.$$

In particular, all the metrics in this flow differ by a constant factor depending on time and the metric is a decreasing function of time. These are called *shrinking solitons*. Examples are compact manifolds of constant positive Ricci curvature.

There is a closely related, but more general, class of examples: the *gradient shrinking solitons*. Suppose that (M, g) is a complete Riemannian manifold, and suppose that there is a constant $\lambda > 0$ and a function $f \colon M \to \mathbb{R}$ satisfying

$$\mathrm{Ric}(g) = \lambda g - \mathrm{Hess}^g f.$$

In this case, there is a Ricci flow which is a shrinking family after we pull back by the one-parameter family of diffeomorphisms generated by the time-dependent vector field $\frac{1}{1-2\lambda t}\nabla_g f$. An example of a gradient shrinking soliton is the manifold $S^2 \times \mathbb{R}$ with the family of metrics being the product of the shrinking family of round metrics on S^2 and the constant family of standard metrics on $\mathbb{R}$. The function f is $s^2/4$ where s is the Euclidean parameter on $\mathbb{R}$.

3.3. Controlling higher derivatives of curvature. Now let us discuss the smoothness results for geometric limits. The general result along these lines is Shi's theorem, see [65, 66]. Again, this is a standard type of result for parabolic equations. Of course, the situation here is complicated somewhat by the existence of the gauge group. Roughly, Shi's theorem says the following. Let us denote by $B(x, t_0, r)$ the metric ball in $(M, g(t_0))$ centered at x and of radius r. If we can control the norm of the Riemann curvature tensor on a backward neighborhood of the form $B(x, t_0, r) \times [0, t_0]$, then for each $k > 0$ we can control the kth covariant derivative of the curvature on $B(x, t_0, r/2^k) \times [0, t_0]$ by a constant over $t^{k/2}$. This result has many important consequences in our study because it tells us that geometric limits are smooth limits. Maybe the first result to highlight is the fact (established earlier by Hamilton) that if $(M, g(t))$ is a Ricci flow defined on $0 \le t < T < \infty$, and if the Riemann curvature is uniformly bounded for the entire flow, then the Ricci flow extends past time T.

In the third chapter this material is reviewed and, where necessary, slight variants of results and arguments in the literature are presented.

3.4. Generalized Ricci flows. Because we cannot restrict our attention to Ricci flows, but rather must consider more general objects, Ricci flows with surgery, it is important to establish the basic analytic results and estimates in a context more general than that of Ricci flow. We choose to do this in the context of generalized Ricci flows.

A generalized 3-dimensional Ricci flow consists of a smooth manifold $\mathcal{M}$ of dimension 4 (the space-time) together with a time function $\mathbf{t} \colon \mathcal{M} \to \mathbb{R}$ and a smooth vector field χ. These are required to satisfy:

(1) Each $x \in \mathcal{M}$ has a neighborhood of the form $U \times J$, where U is an open subset in $\mathbb{R}^3$ and $J \subset \mathbb{R}$ is an interval, in which $\mathbf{t}$ is the projection onto J and χ is the unit vector field tangent to the one-dimensional foliation $\{u\} \times J$ pointing in the direction of increasing $\mathbf{t}$. We call $\mathbf{t}^{-1}(t)$ the t time-slice. It is a smooth 3-manifold.
(2) The image $\mathbf{t}(\mathcal{M})$ is a connected interval I in $\mathbb{R}$, possibly infinite. The boundary of $\mathcal{M}$ is the pre-image under $\mathbf{t}$ of the boundary of I.
(3) The level sets $\mathbf{t}^{-1}(t)$ form a codimension-one foliation of $\mathcal{M}$, called the horizontal foliation, with the boundary components of $\mathcal{M}$ being leaves.
(4) There is a metric G on the horizontal distribution, i.e., the distribution tangent to the level sets of $\mathbf{t}$. This metric induces a Riemannian metric on each t time-slice, varying smoothly as we vary the time-slice. We define the curvature of G at a point $x \in \mathcal{M}$ to be the curvature of the Riemannian metric induced by G on the time-slice M_t at x.

(5) Because of the first property the integral curves of χ preserve the horizontal foliation and hence the horizontal distribution. Thus, we can take the Lie derivative of G along χ. The Ricci flow equation is then

$$\mathcal{L}_\chi(G) = -2\mathrm{Ric}(G).$$

Locally in space-time the horizontal metric is simply a smoothly varying family of Riemannian metrics on a fixed smooth manifold and the evolution equation is the ordinary Ricci flow equation. This means that the usual formulas for the evolution of the curvatures as well as much of the analytic analysis of Ricci flows still hold in this generalized context. In the end, a Ricci flow with surgery is a more singular type of space-time, but it will have an open dense subset which is a generalized Ricci flow, and all the analytic estimates take place in this open subset.

The notion of canonical neighborhoods make sense in the context of generalized Ricci flows. There is also the notion of a strong ϵ-neck. Consider an embedding $\psi\colon \left(S^2 \times (-\epsilon^{-1}, \epsilon^{-1})\right) \times (-1, 0]$ into space-time such that the time function pulls back to the projection onto $(-1, 0]$ and the vector field χ pulls back to $\partial/\partial t$. If there is such an embedding into an appropriately shifted and rescaled version of the original generalized Ricci flow so that the pull-back of the rescaled horizontal metric is within ϵ in the $C^{[1/\epsilon]}$-topology of the product of the shrinking family of round S^2's with the Euclidean metric on $(-\epsilon^{-1}, \epsilon^{-1})$, then we say that ψ is a strong ϵ-neck in the generalized Ricci flow.

3.5. The maximum principle. The Ricci flow equation satisfies various forms of the maximum principle. The fourth chapter explains this principle, which is due to Hamilton (see Section 4 of [**34**]), and derives many of its consequences, which are also due to Hamilton (cf. [**36**]). This principle and its consequences are at the core of all the detailed results about the nature of the flow. We illustrate the idea by considering the case of the scalar curvature. A standard scalar maximum principle argument applied to Equation (0.1) proves that the minimum of the scalar curvature is a non-decreasing function of time. In addition, it shows that if the minimum of scalar curvature at time 0 is positive then we have

$$R_{\min}(t) \geq R_{\min}(0) \left(\frac{1}{1 - \frac{2t}{n} R_{\min}(0)} \right),$$

and thus the equation develops a singularity at or before time $n/(2R_{\min}(0))$.

While the above result about the scalar curvature is important and is used repeatedly, the most significant uses of the maximum principle involve the tensor version, established by Hamilton, which applies for example to the Ricci tensor and the full curvature tensor. These have given the most significant understanding of the Ricci flows, and they form the core of the arguments that Perelman uses in his application of Ricci flow to 3-dimensional topology. Here are the main results established by Hamilton:

(1) For 3-dimensional flows, if the Ricci curvature is positive, then the family of metrics becomes singular at finite time and as the family becomes singular, the metric becomes closer and closer to round; see [**29**].
(2) For 3-dimensional flows, as the scalar curvature goes to $+\infty$ the ratio of the absolute value of any negative eigenvalue of the Riemann curvature to the largest positive eigenvalue goes to zero; see [**36**]. This condition is called *pinched toward positive curvature*.
(3) Motivated by a Harnack inequality for the heat equation established by Li-Yau [**48**], Hamilton established a Harnack inequality for the curvature tensor under the Ricci flow for complete manifolds $(M, g(t))$ with bounded, non-negative curvature operator; see [**32**]. In the applications to three dimensions, we shall need the following consequence for the scalar curvature: Suppose that $(M, g(t))$ is a Ricci flow defined for all $t \in [T_0, T_1]$ of complete manifolds of non-negative curvature operator with bounded curvature. Then

$$\frac{\partial R}{\partial t}(x,t) + \frac{R(x,t)}{t - T_0} \geq 0.$$

In particular, if $(M, g(t))$ is an ancient solution (i.e., defined for all $t \leq 0$) of bounded, non-negative curvature, then $\partial R(x,t)/\partial t \geq 0$.
(4) If a complete 3-dimensional Ricci flow $(M, g(t))$, $0 \leq t \leq T$, has non-negative curvature, if $g(0)$ is not flat, and if there is at least one point (x, T) such that the Riemann curvature tensor of $g(T)$ has a flat direction in $\wedge^2 TM_x$, then M has a cover $\widetilde{M}$ so that for each $t > 0$ the Riemannian manifold $(\widetilde{M}, g(t))$ splits as a Riemannian product of a surface of positive curvature and a Euclidean line. Furthermore, the flow on the cover $\widetilde{M}$ is the product of a 2-dimensional flow and the trivial one-dimensional Ricci flow on the line; see Sections 8 and 9 of [**30**].
(5) In particular, there is no Ricci flow $(U, g(t))$ with non-negative curvature tensor defined for $0 \leq t \leq T$ with $T > 0$, such that $(U, g(T))$ is isometric to an open subset in a non-flat, 3-dimensional metric cone.

3.6. Geometric limits. In the fifth chapter we discuss geometric limits of Riemannian manifolds and of Ricci flows. Let us review the history of these ideas. The first results about geometric limits of Riemannian manifolds go back to Cheeger in his thesis in 1967; see [**6**]. Here Cheeger obtained topological results. In [**25**] Gromov proposed that geometric limits should exist in the Lipschitz topology and suggested a result along these lines, which also was known to Cheeger. In [**23**], Greene-Wu gave a rigorous proof of the compactness theorem suggested by Gromov and also enhanced the convergence to be $C^{1,\alpha}$-convergence by using harmonic coordinates; see also [**56**]. Assuming that all the derivatives of curvature are bounded, one can apply elliptic theory to the expression of curvature in harmonic coordinates and deduce C^∞-convergence. These ideas lead to various types of compactness results that go under the name Cheeger-Gromov compactness for Riemannian manifolds. Hamilton in [**33**] extended these results to Ricci flows. We shall use the compactness results for both Riemannian manifolds and for Ricci flows. In a different direction, geometric limits were extended to the non-smooth context by Gromov in [**25**] where he introduced a weaker topology, called the Gromov-Hausdorff topology and proved a compactness theorem.

Recall that a sequence of based Riemannian manifolds (M_n, g_n, x_n) is said to *converge geometrically* to a based, complete Riemannian manifold $(M_\infty, g_\infty, x_\infty)$ if there is a sequence of open subsets $U_n \subset M_\infty$ with compact closures, with $x_\infty \in U_1 \subset \overline{U}_1 \subset U_2 \subset \overline{U}_2 \subset U_3 \subset \cdots$ with $\cup_n U_n = M_\infty$, and embeddings $\varphi_n \colon U_n \to M_n$ sending x_∞ to x_n so that the pullback metrics, $\varphi_n^* g_n$, converge uniformly on compact subsets of M_∞ in the C^∞-topology to g_∞. Notice that the topological type of the limit can be different from the topological type of the manifolds in the sequence. There is a similar notion of geometric convergence for a sequence of based Ricci flows.

Certainly, one of the most important consequences of Shi's results, cited above, is that, in concert with Cheeger-Gromov compactness, it allows us to form smooth geometric limits of sequences of based Ricci flows. We have the following result of Hamilton's; see [**33**]:

THEOREM 0.7. *Let $(M_n, g_n(t), (x_n, 0))$ be a sequence of based Ricci flows defined for $t \in (-T, 0]$ with the $(M_n, g_n(t))$ being complete. Suppose that:*
  (1) *There is $r > 0$ and $\kappa > 0$ such that for every n the metric ball $B(x_n, 0, r) \subset (M_n, g_n(0))$ is κ-non-collapsed.*
  (2) *For each $A < \infty$ there is $C = C(A) < \infty$ such that the Riemann curvature on $B(x_n, 0, A) \times (-T, 0]$ is bounded by C.*

Then after passing to a subsequence there is a geometric limit which is a based Ricci flow $(M_\infty, g_\infty(t), (x_\infty, 0))$ defined for $t \in (-T, 0]$.

To emphasize, the two conditions that we must check in order to extract a geometric limit of a subsequence based at points at time zero are: (i) uniform non-collapsing at the base point in the time zero metric, and (ii) for each $A < \infty$ uniformly bounded curvature for the restriction of the flow to the metric balls of radius A centered at the base points.

Most steps in Perelman's argument require invoking this result in order to form limits of appropriate sequences of Ricci flows, often rescaled to make the scalar curvatures at the base point equal to 1. If, before rescaling, the scalar curvature at the base points goes to infinity as we move through the sequence, then the resulting limit of the rescaled flows has non-negative sectional curvature. This is a consequence of the fact that the sectional curvatures of the manifolds in the sequence are uniformly pinched toward positive. It is for exactly this reason that non-negative curvature plays such an important role in the study of singularity development in 3-dimensional Ricci flows.

4. Perelman's advances

So far we have been discussing the results that were known before Perelman's work. They concern almost exclusively Ricci flow (though Hamilton in [**35**] had introduced the notion of surgery and proved that surgery can be performed preserving the condition that the curvature is pinched toward positive, as in (2) above). Perelman extended in two essential ways the analysis of Ricci flow – one involves the introduction of a new analytic functional, *the reduced length*, which is the tool by which he establishes the needed non-collapsing results, and the other is a delicate combination of geometric limit ideas and consequences of the maximum principle together with the non-collapsing results in order to establish bounded curvature at bounded distance results. These are used to prove in an inductive way the existence of canonical neighborhoods, which is a crucial ingredient in proving that it is possible to do surgery iteratively, creating a flow defined for all positive time.

While it is easiest to formulate and consider these techniques in the case of Ricci flow, in the end one needs them in the more general context of Ricci flow with surgery since we inductively repeat the surgery process, and in order to know at each step that we can perform surgery we need to apply these results to the previously constructed Ricci flow with surgery. We have chosen to present these new ideas only once – in the context of generalized Ricci flows – so that we can derive the needed consequences in all the relevant contexts from this one source.

4.1. The reduced length function. In Chapter 6 we come to the first of Perelman's major contributions. Let us first describe it in the context of an ordinary 3-dimensional Ricci flow, but viewing the Ricci flow as a horizontal metric on a space-time which is the manifold $M \times I$, where I is the interval of definition of the flow. Suppose that $I = [0, T)$ and fix $(x, t) \in M \times (0, T)$. We consider paths $\gamma(\tau)$, $0 \leq \tau \leq \overline{\tau}$, in space-time with the property that for every $\tau \leq \overline{\tau}$ we have $\gamma(\tau) \in M \times \{t - \tau\}$ and $\gamma(0) = x$. These paths are said to be *parameterized by backward time*. See FIG. 2. The $\mathcal{L}$-*length* of such a path is given by

$$\mathcal{L}(\gamma) = \int_0^{\overline{\tau}} \sqrt{\tau} \left(R(\gamma(\tau)) + |\gamma'(\tau)|^2 \right) d\tau,$$

where the derivative on γ refers to the spatial derivative. There is also the closely related *reduced length*

$$\ell(\gamma) = \frac{\mathcal{L}(\gamma)}{2\sqrt{\overline{\tau}}}.$$

There is a theory for the functional $\mathcal{L}$ analogous to the theory for the usual energy function[6]. In particular, there is the notion of an $\mathcal{L}$-geodesic, and the reduced length as a function on space-time $\ell_{(x,t)} \colon M \times [0,t) \to \mathbb{R}$. One establishes a crucial monotonicity for this reduced length along $\mathcal{L}$-geodesics. Then one defines the *reduced volume*

$$\widetilde{V}_{(x,t)}(U \times \{\overline{t}\}) = \int_{U \times \{\overline{t}\}} \overline{\tau}^{-3/2} e^{-\ell_{(x,t)}(q,\overline{\tau})} d\mathrm{vol}_{g(\overline{\tau})}(q),$$

where, as before $\overline{\tau} = t - \overline{t}$. Because of the monotonicity of $\ell_{(x,t)}$ along $\mathcal{L}$-geodesics, the reduced volume is also non-increasing under the flow (forward in $\overline{\tau}$ and hence backward in time) of open subsets along $\mathcal{L}$-geodesics. This is the fundamental tool which is used to establish non-collapsing results which in turn are essential in proving the existence of geometric limits.

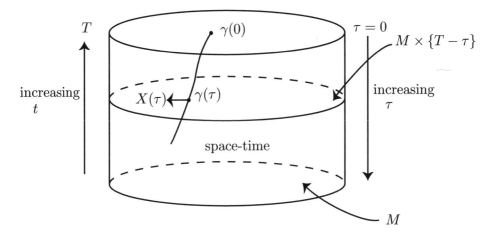

FIGURE 2. Curves in space-time parameterized by τ.

The definitions and the analysis of the reduced length function and the reduced volume as well as the monotonicity results are valid in the context of the generalized Ricci flow. The only twist to be aware of is that in the more general context one cannot always extend $\mathcal{L}$-geodesics; they may run 'off the edge' of space-time. Thus, the reduced length function and reduced volume cannot be defined globally, but only on appropriate open subsets of a time-slice (those reachable by minimizing $\mathcal{L}$-geodesics). But as long as

---

[6] Even though this functional is called a length, the presence of the $|\gamma'(\tau)|^2$ in the integrand means that it behaves more like the usual energy functional for paths in a Riemannian manifold.

one can flow an open set U of a time-slice along minimizing $\mathcal{L}$-geodesics in the direction of decreasing $\overline{\tau}$, the reduced volumes of the resulting family of open sets form a monotone non-increasing function of $\overline{\tau}$. This turns out to be sufficient to extend the non-collapsing results to Ricci flow with surgery, provided that we are careful in how we choose the parameters that go into the definition of the surgery process.

4.2. Application to non-collapsing results. As we indicated in the previous paragraph, one of the main applications of the reduced length function is to prove non-collapsing results for 3-dimensional Ricci flows with surgery. In order to make this argument work, one takes a weaker notion of κ-non-collapsed by making a stronger curvature bound assumption: one considers points (x,t) and constants r with the property that $|\text{Rm}| \leq r^{-2}$ on $P(x,t,r,-r^2) = B(x,t,r) \times (t-r^2, t]$. The κ-non-collapsing condition applies to these balls and says that $\text{Vol}(B(x,t,r)) \geq \kappa r^3$. The basic idea in proving non-collapsing is to use the fact that, as we flow forward in time via minimizing $\mathcal{L}$-geodesics, the reduced volume is a non-decreasing function. Hence, a lower bound of the reduced volume of an open set at an earlier time implies the same lower bound for the corresponding open subset at a later time. This is contrasted with direct computations (related to the heat kernel in $\mathbb{R}^3$) that say if the manifold is highly collapsed near (x,t) (i.e., satisfies the curvature bound above but is not κ-non-collapsed for some small κ) then the reduced volume $\widetilde{V}_{(x,t)}$ is small at times close to t. Thus, to show that the manifold is non-collapsed at (x,t) we need only find an open subset at an earlier time that is reachable by minimizing $\mathcal{L}$-geodesics and that has a reduced volume bounded away from zero.

One case where it is easy to do this is when we have a Ricci flow of compact manifolds or of complete manifolds of non-negative curvature. Hence, these manifolds are non-collapsed at all points with a non-collapsing constant that depends only on the geometry of the initial metric of the Ricci flow. Non-collapsing results are crucial and are used repeatedly in dealing with Ricci flows with surgery in Chapters 10 – 17, for these give one of the two conditions required in order to take geometric limits.

4.3. Application to ancient κ-non-collapsed solutions. There is another important application of the length function, which is to the study of non-collapsed, ancient solutions in dimension 3. In the case that the generalized Ricci flow is an ordinary Ricci flow either on a compact manifold or on a complete manifold (with bounded curvatures) one can say much more about the reduced length function and the reduced volume. Fix a point (x_0, t_0) in space-time. First of all, one shows that every point (x,t) with $t < t_0$ is reachable by a minimizing $\mathcal{L}$-geodesic and thus that the reduced length is defined as a function on all points of space at all times $t < t_0$. It turns out to be a locally Lipschitz function in both space and time and hence its gradient and its time derivative exist as L^2-functions and satisfy important differential inequalities in the weak sense.

These results apply to a class of Ricci flows called κ-solutions, where κ is a positive constant. By definition a κ-solution is a Ricci flow defined for all $t \in (-\infty, 0]$, each time slice is a non-flat, complete 3-manifold of non-negative, bounded curvature and each time-slice is κ-non-collapsed. The differential inequalities for the reduced length from any point $(x, 0)$ imply that, for any $t < 0$, the minimum value of $\ell_{(x,0)}(y, t)$ for all $y \in M$ is at most $3/2$. Furthermore, again using the differential inequalities for the reduced length function, one shows that for any sequence $t_n \to -\infty$, and any points (y_n, t_n) at which the reduced length function is bounded above by $3/2$, there is a subsequence of based Riemannian manifolds, $(M, \frac{1}{|t_n|}g(t_n), y_n)$, with a geometric limit, and this limit is a gradient shrinking soliton. This gradient shrinking soliton is called an *asymptotic soliton* for the original κ-solution, see FIG. 3.

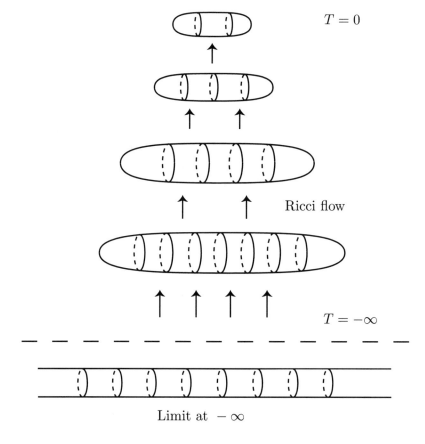

FIGURE 3. The asymptotic soliton.

The point is that there are only two types of 3-dimensional gradient shrinking solitons – (i) those finitely covered by a family of shrinking round

3-spheres and (ii) those finitely covered by a family of shrinking round cylinders $S^2 \times \mathbb{R}$. If a κ-solution has a gradient shrinking soliton of the first type then it is in fact isomorphic to its gradient shrinking soliton. More interesting is the case when the κ-solution has a gradient shrinking soliton which is of the second type. If the κ-solution does not have strictly positive curvature, then it is isomorphic to its gradient shrinking soliton. Furthermore, there is a constant $C_1 < \infty$ depending on ϵ (which remember is taken sufficiently small) such that a κ-solution of strictly positive curvature either is a C_1-component, or is a union of cores of (C_1, ϵ)-caps and points that are the center points of ϵ-necks.

In order to prove the above results (for example the uniformity of C_1 as above over all κ-solutions) one needs the following result:

THEOREM 0.8. *The space of based κ-solutions, based at points with scalar curvature zero in the zero time-slice is compact.*

This result does not generalize to ancient solutions that are not non-collapsed because, in order to prove compactness, one has to take limits of subsequences, and in doing this the non-collapsing hypothesis is essential. See Hamilton's work [34] for more on general ancient solutions (i.e., those that are not necessarily non-collapsed).

Since $\epsilon > 0$ is sufficiently small so that all the results from the appendix about manifolds covered by ϵ-necks and ϵ-caps hold, the above results about gradient shrinking solitons lead to a rough qualitative description of all κ-solutions. There are those which do not have strictly positive curvature. These are gradient shrinking solitons, either an evolving family of round 2-spheres times $\mathbb{R}$ or the quotient of this family by an involution. Non-compact κ-solutions of strictly positive curvature are diffeomorphic to $\mathbb{R}^3$ and are the union of an ϵ-tube and a core of a (C_1, ϵ)-cap. The compact ones of strictly positive curvature are of two types. The first type are positive, constant curvature shrinking solitons. Solutions of the second type are diffeomorphic to either S^3 or $\mathbb{R}P^3$. Each time-slice of a κ-solution of the second type either is of uniformly bounded geometry (curvature, diameter, and volume) when rescaled so that the scalar curvature at a point is 1, or admits an ϵ-tube whose complement is either a disjoint union of the cores of two (C_1, ϵ)-caps.

This gives a rough qualitative understanding of κ-solutions. Either they are round, or they are finitely covered by the product of a round surface and a line, or they are a union of ϵ-tubes and cores of (C_1, ϵ)-caps, or they are diffeomorphic to S^3 or $\mathbb{R}P^3$ and have bounded geometry (again after rescaling so that there is a point of scalar curvature 1). This is the source of canonical neighborhoods for Ricci flows: the point is that this qualitative result remains true for any point x in a Ricci flow that has an appropriate size neighborhood within ϵ in the $C^{[1/\epsilon]}$-topology of a neighborhood in a κ-solution. For example, if we have a sequence of based generalized flows $(\mathcal{M}_n, G_n, x_n)$ converging to a based κ-solution, then for all n sufficiently large x will have a canonical neighborhood, one that is either an ϵ-neck centered at that point, a (C_1, ϵ)-cap whose core contains the point, a C_1-component, or an ϵ-round component.

4.4. Bounded curvature at bounded distance. Perelman's other major breakthrough is his result establishing bounded curvature at bounded distance for blow-up limits of generalized Ricci flows. As we have alluded to several times, many steps in the argument require taking (smooth) geometric limits of a sequence of based generalized flows about points of curvature tending to infinity. To study such a sequence we rescale each term in the sequence so that its curvature at the base point becomes 1. Nevertheless, in taking such limits we face the problem that even though the curvature at the point we are focusing on (the points we take as base points) was originally large and has been rescaled to be 1, there may be other points in the same time-slice of much larger curvature, which, even after the rescalings, can tend to infinity. If these points are at uniformly bounded (rescaled) distance from the base points, then they would preclude the existence of a smooth geometric limit of the based, rescaled flows. In his arguments, Hamilton avoided this problem by always focusing on points of maximal curvature (or almost maximal curvature). That method will not work in this case. The way to deal with this possible problem is to show that a generalized Ricci flow satisfying appropriate conditions satisfies the following. For each $A < \infty$ there are constants $Q_0 = Q_0(A) < \infty$ and $Q(A) < \infty$ such that any point x in such a generalized flow for which the scalar curvature $R(x) \geq Q_0$ and for any y in the same time-slice as x with $d(x,y) < AR(x)^{-1/2}$ satisfies $R(y)/R(x) < Q(A)$. As we shall see, this and the non-collapsing result are the fundamental tools that allow Perelman to study neighborhoods of points of sufficiently large curvature by taking smooth limits of rescaled flows, so essential in studying the prolongation of Ricci flows with surgery.

The basic idea in proving this result is to assume the contrary and take an incomplete geometric limit of the rescaled flows based at the counterexample points. The existence of points at bounded distance with unbounded, rescaled curvature means that there is a point at infinity at finite distance from the base point where the curvature blows up. A neighborhood of this point at infinity is cone-like in a manifold of non-negative curvature. This contradicts Hamilton's maximum principle result (5) in Section 3.5) that the result of a Ricci flow of manifolds of non-negative curvature is never an open subset of a cone. (We know that any 'blow-up limit' like this has non-negative curvature because of the curvature pinching result.) This contradiction establishes the result.

5. The standard solution and the surgery process

Now we are ready to discuss 3-dimensional Ricci flows with surgery.

5.1. The standard solution. In preparing the way for defining the surgery process, we must construct a metric on the 3-ball that we shall glue in when we perform surgery. This we do in Chapter 12. We fix a non-negatively curved, rotationally symmetric metric on $\mathbb{R}^3$ that is isometric near infinity to $S^2 \times [0, \infty)$ where the metric on S^2 is the round metric of scalar curvature 1, and outside this region has positive sectional curvature, see FIG. 4. Any such metric will suffice for the gluing process, and we fix one and call it the *standard metric*. It is important to understand Ricci flow with the standard metric as initial metric. Because of the special nature of this metric (the rotational symmetry and the asymptotic nature at infinity), it is fairly elementary to show that there is a unique solution of bounded curvature on each time-slice to the Ricci flow equation with the standard metric as the initial metric; this flow is defined for $0 \leq t < 1$; and for any $T < 1$ outside of a compact subset $X(T)$ the restriction of the flow to $[0, T]$ is close to the evolving round cylinder. Using the length function, one shows that the Ricci flow is non-collapsed, and that the bounded curvature and bounded distance result applies to it. This allows one to prove that every point (x, t) in this flow has one of the following types of neighborhoods:

(1) (x, t) is contained in the core of a (C_2, ϵ)-cap, where $C_2 < \infty$ is a given universal constant depending only on ϵ.
(2) (x, t) is the center of a strong ϵ-neck.
(3) (x, t) is the center of an evolving ϵ-neck whose initial slice is at time zero.

These form the second source of models for canonical neighborhoods in a Ricci flow with surgery. Thus, we shall set $C = C(\epsilon) = \max(C_1(\epsilon), C_2(\epsilon))$ and we shall find (C, ϵ)-canonical neighborhoods in Ricci flows with surgery.

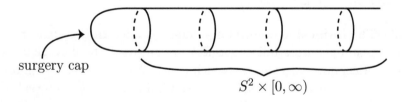

FIGURE 4. The standard metric.

5.2. Ricci flows with surgery. Now it is time to introduce the notion of a Ricci flow with surgery. To do this we formulate an appropriate notion of 4-dimensional space-time that allows for the surgery operations. We define *space-time* to be a 4-dimensional Hausdorff singular space with a time function **t** with the property that each time-slice is a compact, smooth

3-manifold, but level sets at different times are not necessarily diffeomorphic. Generically space-time is a smooth 4-manifold, but there are *exposed regions* at a discrete set of times. Near a point in the exposed region space-time is a 4-manifold with boundary. The singular points of space-time are the boundaries of the exposed regions. Near these, space-time is modeled on the product of $\mathbb{R}^2$ with the square $(-1,1) \times (-1,1)$, the latter having a topology in which the open sets are, in addition to the usual open sets, open subsets of $(0,1) \times [0,1)$, see FIG. 5. There is a natural notion of smooth functions on space-time. These are smooth in the usual sense on the open subset of non-singular points. Near the singular points, and in the local coordinates described above, they are required to be pull-backs from smooth functions on $\mathbb{R}^2 \times (-1,1) \times (-1,1)$ under the natural map. Space-time is equipped with a smooth vector field χ with $\chi(\mathbf{t}) = 1$.

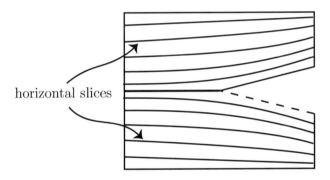

FIGURE 5. Model for singularities in space-time.

A Ricci flow with surgery is a smooth horizontal metric G on a space-time with the property that the restriction of G, $\mathbf{t}$ and χ to the open subset of smooth points forms a generalized Ricci flow. We call this the *associated generalized Ricci flow* for the Ricci flow with surgery.

5.3. The inductive conditions necessary for doing surgery. With all this preliminary work out of the way, we are ready to show that one can construct Ricci flow with surgery which is precisely controlled both topologically and metrically. This result is proved inductively, one interval of time after another, and it is important to keep track of various properties as we go along to ensure that we can continue to do surgery. Here we discuss the conditions we verify at each step.

Fix $\epsilon > 0$ sufficiently small and let $C = \max(C_1, C_2) < \infty$, where C_1 is the constant associated to ϵ for κ-solutions and C_2 is the constant associated to ϵ for the standard solution. We say that a point x in a generalized Ricci flow has a (C,ϵ)-canonical neighborhood if one of the following holds:

(1) x is contained in a C-component of its time-slice.

(2) x is contained in a connected component of its time-slice that is within ϵ of round in the $C^{[1/\epsilon]}$-topology.
(3) x is contained in the core of a (C, ϵ)-cap.
(4) x is the center of a strong ϵ-neck.

We shall study Ricci flows with surgery defined for $0 \le t < T < \infty$ whose associated generalized Ricci flows satisfy the following properties:

(1) The initial metric is normalized, meaning that for the metric at time zero the norm of the Riemann curvature is bounded above by 1 and the volume of any ball of radius 1 is at least half the volume of the unit ball in Euclidean space.
(2) The curvature of the flow is pinched toward positive.
(3) There is $\kappa > 0$ so that the associated generalized Ricci flow is κ-non-collapsed on scales at most ϵ, in the sense that we require only that balls of radius $r \le \epsilon$ be κ-non-collapsed.
(4) There is $r_0 > 0$ such that any point of space-time at which the scalar curvature is $\ge r_0^{-2}$ has a (C, ϵ)-canonical neighborhood.

The main result is that, having a Ricci flow with surgery defined on some time interval satisfying these conditions, it is possible to extend it to a longer time interval in such a way that it still satisfies the same conditions, possibly allowing the constants κ and r_0 defining these conditions to get closer to zero, but keeping them bounded away from 0 on each compact time interval. We repeat this construction inductively. It is easy to see that there is a bound on the number of surgeries in each compact time interval. Thus, in the end, we create a Ricci flow with surgery defined for all positive time. As far as we know, it may be the case that in the entire flow, defined for all positive time, there are infinitely many surgeries.

5.4. Surgery. Let us describe how we extend a Ricci flow with surgery satisfying all the conditions listed above and becoming singular at time $T < \infty$. Fix $T^- < T$ so that there are no surgery times in the interval $[T^-, T)$. Then we can use the Ricci flow to identify all the time-slices M_t for $t \in [T^-, T)$, and hence view this part of the Ricci flow with surgery as an ordinary Ricci flow. Because of the canonical neighborhood assumption, there is an open subset $\Omega \subset M_{T^-}$ on which the curvature stays bounded as $t \to T$. Hence, by Shi's results, there is a limiting metric at time T on Ω. Furthermore, the scalar curvature is a proper function, bounded below, from Ω to $\mathbb{R}$, and each end of Ω is an ϵ-tube where the cross-sectional area of the 2-spheres goes to zero as we go to the end of the tube. We call such tubes ϵ-*horns*. We are interested in ϵ-horns whose boundary is contained in the part of Ω where the scalar curvature is bounded above by some fixed finite constant ρ^{-2}. We call this region Ω_ρ. Using the bounded curvature at bounded distance result, and using the non-collapsing hypothesis, one shows that given any $\delta > 0$ there is $h = h(\delta, \rho, r_0)$ such that for any ϵ-horn

$\mathcal{H}$ whose boundary lies in Ω_ρ and for any $x \in \mathcal{H}$ with $R(x) \geq h^{-2}$, the point x is the center of a strong δ-neck.

Now we are ready to describe the surgery procedure. It depends on our choice of standard solution on $\mathbb{R}^3$ and on a choice of $\delta > 0$ sufficiently small. For each ϵ-horn in Ω whose boundary is contained in Ω_ρ, fix a point of curvature $(h(\delta, \rho, r_0))^{-2}$ and fix a strong δ-neck centered at this point. Then we cut the ϵ-horn open along the central 2-sphere S of this neck and remove the end of the ϵ-horn that is cut off by S. Then we glue in a ball of a fixed radius around the tip from the standard solution, after scaling the metric on this ball by $(h(\delta, \rho, r_0))^2$. To glue these two metrics together we must use a partition of unity near the 2-spheres that are matched. There is also a delicate point that we first bend in the metrics slightly so as to achieve positive curvature near where we are gluing. This is an idea due to Hamilton, and it is needed in order to show that the condition of curvature pinching toward positive is preserved. In addition, we remove all components of Ω that do not contain any points of Ω_ρ.

This operation produces a new compact 3-manifold. One continues the Ricci flow with surgery by letting this Riemannian manifold at time T evolve under the Ricci flow. See FIG. 6.

5.5. Topological effect of surgery.
Looking at the situation just before the surgery time, we see a finite number of disjoint submanifolds, each diffeomorphic to either $S^2 \times I$ or the 3-ball, where the curvature is large. In addition there may be entire components of where the scalar curvature is large. The effect of 2-sphere surgery is to do a finite number of ordinary topological surgeries along 2-spheres in the $S^2 \times I$. This simply effects a partial connected-sum decomposition and may introduce new components diffeomorphic to S^3. We also remove entire components, but these are covered by ϵ-necks and ϵ-caps so that they have standard topology (each one is diffeomorphic to S^3, $\mathbb{R}P^3$, $\mathbb{R}P^3 \# \mathbb{R}P^3$, $S^2 \times S^1$, or the non-orientable 2-sphere bundle over S^1). Also, we remove C-components and ϵ-round components (each of these is either diffeomorphic to S^3 or $\mathbb{R}P^3$ or admits a metric of constant positive curvature). Thus, the topological effect of surgery is to do a finite number of ordinary 2-sphere topological surgeries and to remove a finite number of topologically standard components.

6. Extending Ricci flows with surgery

We consider Ricci flows with surgery that are defined on the time interval $0 \leq t < T$, with $T < \infty$, and that satisfy four conditions. These conditions are: (i) normalized initial metric, (ii) curvature pinched toward positive, (iii) all points of scalar curvature $\geq r^{-2}$ have canonical neighborhoods, and (iv) the flow is κ-non-collapsed on scales $\leq \epsilon$. The crux of the argument is to show that it is possible to extend to such a Ricci flow with surgery to a Ricci flow with surgery defined for all $t \in [0, T')$ for some $T' > T$, keeping

6. EXTENDING RICCI FLOWS WITH SURGERY

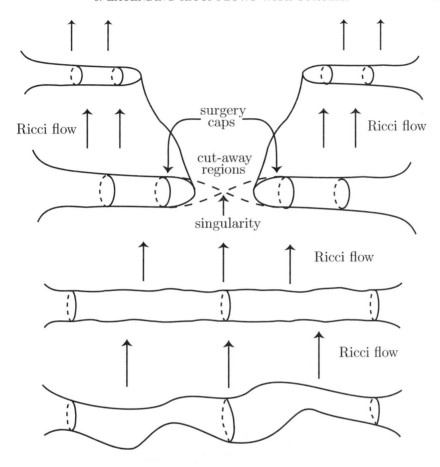

FIGURE 6. Surgery.

these four conditions satisfied (possibly with different constants $r' < r$ and $\kappa' < \kappa$). In order to do this we need to choose the surgery parameter $\delta > 0$ sufficiently small. There is also the issue of whether the surgery times can accumulate.

Of course, the initial metric does not change as we extend surgery so that the condition that the normalized initial metric is clearly preserved as we extend surgery. As we have already remarked, Hamilton had proved earlier that one can do surgery in such a way as to preserve the condition that the curvature is pinched toward positive. The other two conditions require more work, and, as we indicated above, the constants may decay to zero as we extend the Ricci flow with surgery.

If we have all the conditions for the Ricci flow with surgery up to time T, then the analysis of the open subset on which the curvature remains bounded holds, and given $\delta > 0$ sufficiently small, we do surgery on the central S^2 of a strong δ-neck in each ϵ-horn meeting Ω_ρ. In addition we remove entirely all components that do not contain points of Ω_ρ. We then glue in the cap from the standard solution. This gives us a new compact 3-manifold and we restart the flow from this manifold.

The κ-non-collapsed result is extended to the new part of the Ricci flow with surgery using the fact that it holds at times previous to T. To establish this extension one uses $\mathcal{L}$-geodesics in the associated generalized Ricci flow and reduced volume as indicated before. In order to get this argument to work, one must require $\delta > 0$ to be sufficiently small; how small is determined by r_0.

The other thing that we must establish is the existence of canonical neighborhoods for all points of sufficiently large scalar curvature. Here the argument is by contradiction. We consider all Ricci flows with surgery that satisfy all four conditions on $[0, T)$ and we suppose that we can find a sequence of such containing points (automatically at times $T' > T$) of arbitrarily large curvature where there are not canonical neighborhoods. In fact, we take the points at the first such time violating this condition. We base our flows at these points. Now we consider rescaled versions of the generalized flows so that the curvature at these base points is rescaled to 1. We are in a position to apply the bounded curvature and bounded distance results to this sequence, and of course the κ-non-collapsing results which have already been established. There are two possibilities. The first is that the rescaled sequence converges to an ancient solution. This ancient solution has non-negative curvature by the pinching hypothesis. General results about 3-manifolds of non-negative curvature imply that it also has bounded curvature. It is κ-non-collapsed. Thus, in this case the limit is a κ-solution. This produces the required canonical neighborhoods for the base points of the tail of the sequence modeled on the canonical neighborhoods of points in a κ-solution. This contradicts the assumption that none of these points has a canonical neighborhood.

The other possibility is that one can take a partial smooth limit but that this limit does not extend all the way back to $-\infty$. The only way this can happen is if there are surgery caps that prevent extending the limit back to $-\infty$. This means that the base points in our sequence are all within a fixed distance and time (after the rescaling) of a surgery region. But in this case results from the nature of the standard solution show that if we have taken $\delta > 0$ sufficiently small, then the base points have canonical neighborhoods modeled on the canonical neighborhoods in the standard solution, again contradicting our assumption that none of the base points has a canonical neighborhood. In order to show that our base points have neighborhoods near those of the standard solution, one appeals to a geometric limit argument as $\delta \to 0$. This argument uses the uniqueness of the Ricci flow for the standard solution. (Actually, Bruce Kleiner pointed out to us that one only needs a compactness result for the space of all Ricci flows with the standard metric as initial metric, not uniqueness, and the compactness result can be proved by the same arguments that prove the compactness of the space of κ-solutions.)

Interestingly enough, in order to establish the uniqueness of the Ricci flow for the standard solution, as well as to prove that this flow is defined for time $[0, 1)$ and to prove that at infinity it is asymptotic to an evolving cylinder, requires the same results – non-collapsing and the bounded curvature at bounded distance that we invoked above. For this reason, we order the material described here as follows. First, we introduce generalized Ricci flows, and then introduce the length function in this context and establish the basic monotonicity results. Then we have a chapter on stronger results for the length function in the case of complete manifolds with bounded curvature. At this point we are in a position to prove the needed results about the Ricci flow from the standard solution. Then we are ready to define the surgery process and prove the inductive non-collapsing results and the existence of canonical neighborhoods.

In this way, one establishes the existence of canonical neighborhoods. Hence, one can continue to do surgery, producing a Ricci flow with surgery defined for all positive time. Since these arguments are inductive, it turns out that the constants in the non-collapsing and in the canonical neighborhood statements decay in a predetermined rate as time goes to infinity.

Lastly, there is the issue of ruling out the possibility that the surgery times accumulate. The idea here is very simple: Under Ricci flow during an elapsed time T, volume increases at most by a multiplicative factor which is a fixed exponential of the time T. Under each surgery there is a removal of at least a fixed positive amount of volume depending on the surgery scale h, which in turns depends on δ and r_0. Since both δ and r_0 are bounded away from zero on each finite interval, there can be at most finitely many surgeries in each finite interval. This argument allows for the possibility, noted in Section 5.3, that in the entire flow all the way to infinity there are infinitely many surgeries. It is still unknown whether that possibility ever happens.

This completes our outline of the proof of Theorem 0.3.

7. Finite-time extinction

The last topic we discuss is the proof of the finite-time extinction for Ricci flows with initial metrics satisfying the hypothesis of Theorem 0.4.

As we present it, the finite extinction result has two steps. The first step is to show that there is $T < \infty$ (depending on the initial metric) such that for all $t \geq T$, all connected components of the t-time-slice M_t have trivial π_2. First, an easy topological argument shows that only finitely many of the 2-sphere surgeries in a Ricci flow with surgery can be along homotopically non-trivial 2-spheres. Thus, after some time T_0 all 2-sphere surgeries are along homotopically trivial 2-spheres. Such a surgery does not affect π_2.

Thus, after time T_0, the only way that π_2 can change is by removal of components with non-trivial π_2. (An examination of the topological types of components that are removed shows that there are only two types of such components with non-trivial π_2: 2-sphere bundles over S^1 and $\mathbb{R}P^3\#\mathbb{R}P^3$.) We suppose that at every $t \geq T_0$ there is a component of M_t with non-trivial π_2. Then we can find a connected open subset $\mathcal{X}$ of $\mathbf{t}^{-1}([T_0,\infty))$ with the property that for each $t \geq T_0$ the intersection $\mathcal{X}(t) = \mathcal{X} \cap M_t$ is a component of M_t with non-trivial π_2. We define a function $W_2 \colon [T_0,\infty) \to \mathbb{R}$ associated with such an $\mathcal{X}$. The value $W_2(t)$ is the minimal area of all homotopically non-trivial 2-spheres mapping into $\mathcal{X}(t)$. This minimal area $W_2(t)$ is realized by a harmonic map of S^2 into $\mathcal{X}(t)$. The function W_2 varies continuously under Ricci flow and at a surgery is lower semi-continuous. Furthermore, using an idea that goes back to Hamilton (who applied it to minimal disks) one shows that the forward difference quotient of the minimal area satisfies

$$\frac{dW_2(t)}{dt} \leq -4\pi + \frac{3}{(4t+1)}W_2(t).$$

(Here, the explicit form of the bound for the forward difference quotient depends on the way we have chosen to normalize initial metric and also on Hamilton's curvature pinching result.)

But any function $W_2(t)$ with these properties and defined for all $t > T_0$, becomes negative at some finite T_1 (depending on the initial value). This is absurd since $W_2(t)$ is the minimum of positive quantities. This contradiction shows that such a path of components with non-trivial π_2 cannot exist for all $t \geq T_0$. In fact, it even gives a computable upper bound on how long such a component $\mathcal{X}$, with every time-slice having non-trivial π_2, can exist in terms of the minimal area of a homotopically non-trivial 2-sphere mapping into $\mathcal{X}(T_0)$. It follows that there is $T < \infty$ with the property that every component of M_T has trivial π_2. This condition then persists for all $t \geq T$.

Three remarks are in order. This argument showing that eventually every component of the time-slice t has trivial π_2 is not necessary for the topological application (Theorem 0.4), or indeed, for any other topological application. The reason is the sphere theorem (see [**39**]), which says that if $\pi_2(M)$ is non-trivial then either M is diffeomorphic to an S^2 bundle over S^1 or M has a non-trivial connected sum decomposition. Thus, we can establish results for all 3-manifolds if we can establish them for 3-manifolds with $\pi_2 = 0$. Secondly, the reason for giving this argument is that it is pleasing to see Ricci flow with surgery implementing the connected sum decomposition required for geometrization of 3-manifolds. Also, this argument is a simpler version of the one that we use to deal with components with non-trivial π_3. Lastly, these results on Ricci flow do not use the sphere theorem so that establishing the cutting into pieces with trivial π_2 allows us to give a different proof of this result (though admittedly one using much deeper ideas).

Let us now fix $T < \infty$ such that for all $t \geq T$ all the time-slices M_t have trivial π_2. There is a simple topological consequence of this and our assumption on the initial manifold. If M is a compact 3-manifold whose fundamental group is either a non-trivial free product or an infinite cyclic group, then M admits a homotopically non-trivial embedded 2-sphere. Since we began with a manifold M_0 whose fundamental group is a free product of finite groups and infinite cyclic groups, it follows that for $t \geq T$ every component of M_t has finite fundamental group. Fix $t \geq T$. Then each component of M_t has a finite cover that is simply connected, and thus, by an elementary argument in algebraic topology, each component of M_t has non-trivial π_3. The second step in the finite-time extinction argument is to use a non-trivial element in this group analogously to the way we used homotopically non-trivial 2-spheres to show that eventually the manifolds have trivial π_2.

There are two approaches to this second step: the first is due to Perelman in [**54**] and the other due to Colding-Minicozzi in [**15**]. In their approach Colding-Minicozzi associate to a non-trivial element in $\pi_3(M)$ a non-trivial element in $\pi_1(\mathrm{Maps}(S^2, M))$. This element is represented by a one-parameter family of 2-spheres (starting and ending at the constant map) representing a non-trivial element $\xi \in \pi_3(M_0)$. They define the *width* of this homotopy class by $W(\xi, t)$ by associating to each representative the maximal energy of the 2-spheres in the family and then minimizing over all representatives of the homotopy class. Using results of Jost [**42**], they show that this function satisfies the same forward difference inequality that W_2 satisfies (and has the same continuity property under Ricci flow and the same semi-continuity under surgery). Since $W(\xi, t)$ is always ≥ 0 if it is defined, this forward difference quotient inequality implies that the manifolds M_t must eventually become empty.

While this approach seemed completely natural to us, and while we believe that it works, we found the technical details daunting[7] (because one is forced to consider critical points of index 1 of the energy functional rather than minima). For this reason we chose to follow Perelman's approach. He represents a non-trivial element in $\pi_3(M)$ as a non-trivial element in $\xi \in \pi_2(\Lambda M, *)$ where ΛM is the free loop space of M. He then associates to a family $\Gamma \colon S^2 \to \Lambda M$ of homotopically trivial loops an invariant $W(\Gamma)$ which is the maximum of the areas of minimal spanning disks for the loops $\Gamma(c)$ as c ranges over S^2. The invariant of a non-trivial homotopy class ξ is then the infimum over all representatives Γ for ξ of $W(\Gamma)$. As before, this function is continuous under Ricci flow and is lower semi-continuous under

---

[7]Colding and Minicozzi tell us they plan to give an expanded version of their argument with a more detailed proof.

surgery (unless the surgery removes the component in question). It also satisfies a forward difference quotient

$$\frac{dW(\xi)}{dt} \leq -2\pi + \frac{3}{4t+1}W(\zeta).$$

The reason for the term -2π instead of -4π which occurs in the other cases is that we are working with minimal 2-disks instead of minimal 2-spheres. Once this forward difference quotient estimate (and the continuity) have been established the argument finishes in the same way as the other cases: a function W with the properties we have just established cannot be non-negative for all positive time. This means the component in question, and indeed all components at later time derived from it, must disappear in finite time. Hence, under the hypothesis on the fundamental group in Theorem 0.4 the entire manifold must disappear at finite time.

Because this approach uses only minima for the energy or area functional, one does not have to deal with higher index critical points. But one is forced to face other difficulties though – namely boundary issues. Here, one must prescribe the deformation of the family of boundary curves before computing the forward difference quotient of the energy. The obvious choice is the curve-shrinking flow (see [**2**]). Unfortunately, this flow can only be defined when the curve in question is immersed and even in this case the curve-shrinking flow can develop singularities even if the Ricci flow does not. Following Perelman, or indeed [**2**], one uses the device of taking the product with a small circle and using loops, called *ramps*, that go around that circle once. In this context the curve-shrinking flow remains regular as long as the Ricci flow does. One then projects this flow to a flow of families of 2-spheres in the free loop space of the time-slices of the original Ricci flow. Taking the length of the circle sufficiently small yields the boundary deformation needed to establish the forward difference quotient result. This requires a compactness result which holds under local total curvature bounds. This compactness result holds outside a subset of time-interval of small total measure, which is sufficient for the argument. At the very end of the argument we need an elementary but complicated result on annuli, which we could not find in the literature. For more details on these points see Chapter 18.

8. Acknowledgements

In sorting out Perelman's arguments we have had the aid of many people. First of all, during his tour of the United States in the Spring of 2003, Perelman spent enormous amounts of time explaining his ideas and techniques both in public lectures and in private sessions to us and to many others. Since his return to Russia, he has also freely responded to our questions by e-mail. Richard Hamilton has also given unstintingly of his time, explaining to us his results and his ideas about Perelman's results. We have benefitted tremendously from the work of Bruce Kleiner and John Lott. They produced a lengthy set of notes filling in the details in Perelman's arguments [45]. We have referred to those notes countless times as we came to grips with Perelman's ideas. In late August and early September of 2004, Kleiner, Lott and the two of us participated in a workshop at Princeton University, supported by the Clay Mathematics Institute, going through Perelman's second paper (the one on Ricci flow with surgery) in detail. This workshop played a significant role in convincing us that Perelman's arguments were complete and correct and also in convincing us to write this book. We thank all the participants of this workshop and especially Guo-Feng Wei, Peng Lu, Yu Ding, and X.-C. Rong who, together with Kleiner and Lott, made significant contributions to the workshop. Before, during, and after this workshop, we have benefitted from private conversations too numerous to count with Bruce Kleiner and John Lott.

Several of the analytic points were worked out in detail by others, and we have freely adapted their work. Rugang Ye wrote a set of notes proving the Lipschitz properties of the length function and proving the fact that the cut locus for the reduced length function is of measure zero. Closely related to this, Ye gave detailed arguments proving the requisite limiting results for the length function required to establish the existence of a gradient shrinking soliton. None of these points was directly addressed by Perelman, and it seemed to us that they needed careful explanation. Also, the proof of the uniqueness of the standard solution was established jointly by the second author and Peng Lu, and the proof of the refined version of Shi's theorem where one has control on a certain number of derivatives at time zero was also shown to us by Peng Lu. We also benefitted from Ben Chow's expertise and vast knowledge of the theory of Ricci flow especially during the ClayMath/MSRI Summer School on Ricci Flow in 2005. We had several very helpful conversations with Tom Mrowka and also with Robert Bryant, especially about annuli of small area.

The second author gave courses at Princeton University and ran seminars on this material at MIT and Princeton. Natasa Sesum and Xiao-Dong Wang, see [64], wrote notes for the seminars at MIT, and Edward Fan and Alex Subotic took notes for the seminars and courses at Princeton, and they produced preliminary manuscripts. We have borrowed freely from these manuscripts, and it is a pleasure to thank each of them for their efforts in the early stages of this project.

The authors thank all the referees for their time and effort which helped us to improve the presentation. In particular, we wish to thank Cliff Taubes who carefully read the entire manuscript and gave us innumerable comments and suggestions for clarifications and improvements, and Arthur Greenspoon who carefully copy-edited the book. We also thank Terry Tao for much helpful feedback which improved the exposition. We also thank Colin Rourke for his comments on the finite-time extinction argument.

We thank Lori Lejeune for drawing the figures and John Etnyre for technical help in inserting these figures into the manuscript.

Lastly, during this work we were both generously supported by the Clay Mathematical Institute, and it is a pleasure to thank the Clay Mathematics Institute for its support and to thank its staff, especially Vida Salahi, for their help in preparing this manuscript. We also thank the National Science Foundation for its continued support and the second author thanks the Jim Simons Foundation for its support during the period that he was a faculty member at MIT.

9. List of related papers

For the readers' convenience we gather here references to all the closely related articles.

First and foremost are Perelman's three preprints, [**53**], [**55**], and [**54**]. The first of these introduces the main techniques in the case of Ricci flow, the second discusses the extension of these techniques to Ricci flow with surgery, and the last gives the short-cut to the Poincaré Conjecture and the 3-dimensional spherical space-form conjecture, avoiding the study of the limits as time goes to infinity and collapsing space arguments. There are the detailed notes by Bruce Kleiner and John Lott, [**45**], which greatly expand and clarify Perelman's arguments from the first two preprints. There is also a note on Perelman's second paper by Yu Ding [**17**]. There is the article by Colding-Minicozzi [**15**], which gives their alternate approach to the material in Perelman's third preprint. Collapsing space arguments needed for the full geometrization conjecture are discussed in Shioya-Yamaguchi [**67**]. Lastly, after we had submitted a preliminary version of this manuscript for refereeing, H.-D. Cao and X.-P. Zhu published an article on the Poincaré Conjecture and Thurston's Geometrization Conjecture; see [**5**].

Part 1

Background from Riemannian Geometry and Ricci flow

CHAPTER 1

Preliminaries from Riemannian geometry

In this chapter we will recall some basic facts in Riemannian geometry. For more details we refer the reader to [18] and [57]. Throughout, we always adopt Einstein's summation convention on repeated indices and 'manifold' means a paracompact, Hausdorff, smooth manifold.

1. Riemannian metrics and the Levi-Civita connection

Let M be a manifold and let p be a point of M. Then TM denotes the tangent bundle of M and T_pM is the tangent space at p. Similarly, T^*M denotes the cotangent bundle of M and T_p^*M is the cotangent space at p. For any vector bundle $\mathcal{V}$ over M we denote by $\Gamma(\mathcal{V})$ the vector space of smooth sections of $\mathcal{V}$.

DEFINITION 1.1. Let M be an n-dimensional manifold. A *Riemannian metric* g on M is a smooth section of $T^*M \otimes T^*M$ defining a positive definite symmetric bilinear form on T_pM for each $p \in M$. In local coordinates $(x^1, \ldots, x^n)$, one has a natural local basis $\{\partial_1, \ldots, \partial_n\}$ for TM, where $\partial_i = \frac{\partial}{\partial x^i}$. The metric tensor $g = g_{ij} dx^i \otimes dx^j$ is represented by a smooth matrix-valued function
$$g_{ij} = g(\partial_i, \partial_j).$$
The pair (M, g) is a *Riemannian manifold*. We denote by (g^{ij}) the inverse of the matrix (g_{ij}).

Using a partition of unity, one can easily see that any manifold admits a Riemannian metric. A Riemannian metric on M allows us to measure lengths of smooth paths in M and hence to define a distance function by setting $d(p, q)$ equal to the infimum of the lengths of smooth paths from p to q. This makes M a metric space. For a point p in a Riemannian manifold (M, g) and for $r > 0$ we denote the metric ball of radius r centered at p in M by $B(p, r)$ or by $B_g(p, r)$ if the metric needs specifying or emphasizing. It is defined by
$$B(p, r) = \{q \in M \mid d(p, q) < r\}.$$

THEOREM 1.2. *Given a Riemannian metric g on M, there uniquely exists a torsion-free connection on TM making g parallel, i.e., there is a unique*

$\mathbb{R}$-*linear mapping* $\nabla \colon \Gamma(TM) \to \Gamma(T^*M \otimes TM)$ *satisfying the Leibnitz formula*
$$\nabla(fX) = df \otimes X + f\nabla X,$$
and the following two additional conditions for all vector fields X and Y:
- *(g orthogonal)* $d(g(X,Y)) = g(\nabla X, Y) + g(X, \nabla Y)$;
- *(Torsion-free)* $\nabla_X Y - \nabla_Y X - [X,Y] = 0$ *(where, as is customary, we denote $\nabla Y(X)$ by $\nabla_X Y$).*

We call the above connection the *Levi-Civita connection* of the metric and ∇X the *covariant derivative* of X. On a Riemannian manifold we always use the Levi-Civita connection.

In local coordinates $(x^1, \ldots, x^n)$ the Levi-Civita connection ∇ is given by the $\nabla_{\partial_i}(\partial_j) = \Gamma_{ij}^k \partial_k$, where the Christoffel symbols Γ_{ij}^k are the smooth functions

(1.1) $$\Gamma_{ij}^k = \frac{1}{2} g^{kl} (\partial_i g_{lj} + \partial_j g_{il} - \partial_l g_{ij}).$$

Note that the above two additional conditions for the Levi-Civita connection ∇ correspond respectively to
- $\Gamma_{ij}^k = \Gamma_{ji}^k$,
- $\partial_k g_{ij} = g_{lj} \Gamma_{ki}^l + g_{il} \Gamma_{kj}^l$.

The covariant derivative extends to all tensors. In the special case of a function f we have $\nabla(f) = df$. Note that there is a possible confusion between this and the notation in the literature, since one often sees ∇f written for the gradient of f, which is the vector field dual to df. We always use ∇f to mean df, and we will denote the gradient of f by $(\nabla f)^*$.

The covariant derivative allows us to define the Hessian of a smooth function at any point, not just a critical point. Let f be a smooth real-valued function on M. We define the Hessian of f, denoted $\mathrm{Hess}(f)$, as follows:

(1.2) $$\mathrm{Hess}(f)(X,Y) = X(Y(f)) - \nabla_X Y(f).$$

LEMMA 1.3. *The Hessian is a contravariant, symmetric two-tensor, i.e., for vector fields X and Y we have*
$$\mathrm{Hess}(f)(X,Y) = \mathrm{Hess}(f)(Y,X)$$
and
$$\mathrm{Hess}(f)(\phi X, \psi Y) = \phi\psi \mathrm{Hess}(f)(X,Y)$$
for all smooth functions ϕ, ψ. Other formulas for the Hessian are
$$\mathrm{Hess}(f)(X,Y) = \langle \nabla_X(\nabla f), Y \rangle = \nabla_X(\nabla_Y(f)) = \nabla^2 f(X,Y).$$
Also, in local coordinates we have
$$\mathrm{Hess}(f)_{ij} = \partial_i \partial_j f - (\partial_k f) \Gamma_{ij}^k.$$

PROOF. The proof of symmetry is direct from the torsion-free assumption:
$$\text{Hess}(f)(X,Y) - \text{Hess}(f)(Y,X) = [X,Y](f) - (\nabla_X Y - \nabla_Y X)(f) = 0.$$
The fact that $\text{Hess}(f)$ is a tensor is also established by direct computation. The equivalence of the various formulas is also immediate:
$$\begin{align}(1.3)\quad \langle \nabla_X(\nabla f), Y \rangle &= X(\langle \nabla f, Y \rangle) - \langle \nabla f, \nabla_X Y \rangle \\ &= X(Y(f)) - \nabla_X Y(f) = \text{Hess}(f)(X,Y).\end{align}$$
Since $df = (\partial_r f)dx^r$ and $\nabla(dx^k) = -\Gamma^k_{ij}dx^i \otimes dx^j$, it follows that
$$\nabla(df) = \left(\partial_i \partial_j f - (\partial_k f)\Gamma^k_{ij}\right) dx^i \otimes dx^j.$$
It is direct from the definition that
$$\text{Hess}(f)_{ij} = \text{Hess}(f)(\partial_i, \partial_j) = \partial_i \partial_j f - (\partial_k f)\Gamma^k_{ij}.$$
□

When the metric that we are using to define the Hessian is not clear from the context, we introduce it into the notation and write $\text{Hess}_g(f)$ to denote the Hessian of f with respect to the metric g.

The Laplacian $\triangle f$ is defined as the trace of the Hessian: That is to say, in local coordinates near p we have
$$\triangle f(p) = \sum_{ij} g^{ij} \text{Hess}(f)(\partial_i, \partial_j).$$
Thus, if $\{X_i\}$ is an orthonormal basis for $T_p M$ then
$$(1.4)\qquad \triangle f(p) = \sum_i \text{Hess}(f)(X_i, X_i).$$
Notice that this is the form of the Laplacian that is non-negative at a local minimum, and consequently has a non-positive spectrum.

2. Curvature of a Riemannian manifold

For the rest of this chapter (M,g) is a Riemannian manifold.

DEFINITION 1.4. The *Riemann curvature tensor* of M is the $(1,3)$-tensor on M,
$$\mathcal{R}(X,Y)Z = \nabla^2_{X,Y}Z - \nabla^2_{Y,X}Z = \nabla_X \nabla_Y Z - \nabla_Y \nabla_X Z - \nabla_{[X,Y]}Z,$$
where $\nabla^2_{X,Y}Z = \nabla_X \nabla_Y Z - \nabla_{\nabla_X Y}Z$.

In local coordinates the curvature tensor can be represented as
$$\mathcal{R}(\partial_i, \partial_j)\partial_k = R_{ij}{}^l{}_k \partial_l,$$
where
$$R_{ij}{}^l{}_k = \partial_i \Gamma^l_{jk} - \partial_j \Gamma^l_{ik} + \Gamma^s_{jk}\Gamma^l_{is} - \Gamma^s_{ik}\Gamma^l_{js}.$$

Using the metric tensor g, we can change $\mathcal{R}$ to a $(0,4)$-tensor as follows:

$$\mathcal{R}(X,Y,Z,W) = g(\mathcal{R}(X,Y)W, Z).$$

(Notice the change of order in the last two variables.) Notice that we use the same symbol and the same name for both the $(1,3)$ tensor and the $(0,4)$ tensor; which one we are dealing with in a given context is indicated by the index structure or the variables to which the tensor is applied. In local coordinates, the Riemann curvature tensor can be represented as

$$\begin{aligned}\mathcal{R}(\partial_i, \partial_j, \partial_k, \partial_l) &= R_{ijkl} \\ &= g_{ks} R_{ij}{}^s{}_l \\ &= g_{ks}(\partial_i \Gamma^s_{jl} - \partial_j \Gamma^s_{il} + \Gamma^t_{jl}\Gamma^s_{it} - \Gamma^t_{il}\Gamma^s_{jt}).\end{aligned}$$

One can easily verify the following:

CLAIM 1.5. *The Riemann curvature tensor $\mathcal{R}$ satisfies the following properties:*
- *(Symmetry) $R_{ijkl} = -R_{jikl}$, $R_{ijkl} = -R_{ijlk}$, $R_{ijkl} = R_{klij}$.*
- *(1st Bianchi identity) The sum of R_{ijkl} over the cyclic permutation of any three indices vanishes.*
- *(2nd Bianchi identity) $R_{ijkl,h} + R_{ijlh,k} + R_{ijhk,l} = 0$, where*

$$R_{ijkl,h} = (\nabla_{\partial_h} \mathcal{R})_{ijkl}.$$

There are many important related curvatures.

DEFINITION 1.6. The *sectional curvature* of a 2-plane $P \subset T_pM$ is defined as

$$K(P) = \mathcal{R}(X, Y, X, Y),$$

where $\{X, Y\}$ is an orthonormal basis of P. We say that (M, g) has *positive sectional curvature* (resp., *negative sectional curvature*) if $K(P) > 0$ (resp., $K(P) < 0$) for every 2-plane P. There are analogous notions of non-negative and non-positive sectional curvature.

In local coordinates, suppose that $X = X^i \partial_i$ and $Y = Y^i \partial_i$. Then we have

$$K(P) = R_{ijkl} X^i Y^j X^k Y^l.$$

A Riemannian manifold is said to have *constant sectional curvature* if $K(P)$ is the same for all $p \in M$ and all 2-planes $P \subset T_pM$. One can show that a manifold (M, g) has constant sectional curvature λ if and only if

$$R_{ijkl} = \lambda(g_{ik}g_{jl} - g_{il}g_{jk}).$$

Of course, the sphere of radius r in $\mathbb{R}^n$ has constant sectional curvature $1/r^2$, $\mathbb{R}^n$ with the Euclidean metric has constant sectional curvature 0, and the

hyperbolic space $\mathbb{H}^n$, which in the Poincaré model is given by the unit disk with the metric
$$\frac{4(dx_1^2 + \cdots + dx_n^2)}{(1-|x|^2)^2},$$
or in the upper half-space model with coordinates $(x^1, \ldots, x^n)$ is given by
$$\frac{ds^2}{(x^n)^2},$$
has constant sectional curvature -1. In all three cases we denote the constant curvature metric by g_{st}.

DEFINITION 1.7. Using the metric, one can replace the Riemann curvature tensor $\mathcal{R}$ by a symmetric bilinear form Rm on $\wedge^2 TM$. In local coordinates let $\varphi = \varphi^{ij} \partial_i \wedge \partial_j$ and $\psi = \psi^{kl} \partial_k \wedge \partial_l$ be local sections of $\wedge^2 TM$. The formula for Rm is
$$\text{Rm}(\varphi, \psi) = R_{ijkl} \varphi^{ij} \psi^{kl}.$$
We call Rm the *curvature operator*. We say (M, g) has *positive curvature operator* if $\text{Rm}(\varphi, \varphi) > 0$ for any non-zero 2-form $\varphi = \varphi^{ij} \partial_i \wedge \partial_j$ and has *non-negative curvature operator* if $\text{Rm}(\varphi, \varphi) \geq 0$ for any $\varphi \in \wedge^2 TM$.

If the curvature operator is a positive (resp., non-negative) operator then the manifold is positively (resp., non-negatively) curved.

DEFINITION 1.8. The *Ricci curvature tensor*, denoted Ric or Ric_g when it is necessary to specify the metric, is a symmetric contravariant two-tensor. In local coordinates it is defined by
$$\text{Ric}(X, Y) = g^{kl} R(X, \partial_k, Y, \partial_l).$$
The value of this tensor at $p \in M$ is given by $\sum_{i=1}^n R(X(p), e_i, Y(p), e_i)$ where $\{e_1, \ldots, e_n\}$ is an orthonormal basis of $T_p M$. Clearly Ric is a symmetric bilinear form on TM, given in local coordinates by
$$\text{Ric} = \text{Ric}_{ij} dx^i \otimes dx^j,$$
where $\text{Ric}_{ij} = \text{Ric}(\partial_i, \partial_j)$. The *scalar curvature* is defined by:
$$R = R_g = \text{tr}_g \text{Ric} = g^{ij} \text{Ric}_{ij}.$$
We will say that Ric $\geq k$ (or $\leq k$) if all the eigenvalues of Ric are $\geq k$ (or $\leq k$).

Clearly, the curvatures are natural in the sense that if $F \colon N \to M$ is a diffeomorphism and if g is a Riemannian metric on M, then F^*g is a Riemannian metric on N and we have $\text{Rm}(F^*g) = F^*(\text{Rm}(g))$, $\text{Ric}(F^*g) = F^*(\text{Ric}(g))$, and $R(F^*g) = F^*(R(g))$.

2.1. Consequences of the Bianchi identities.

There is one consequence of the second Bianchi identity that will be important later. For any contravariant two-tensor ω on M (such as Ric or Hess(f)) we define the contravariant one-tensor div(ω) as follows: For any vector field X we set

$$\mathrm{div}(\omega)(X) = \nabla^*\omega(X) = g^{rs}\nabla_r(\omega)(X, \partial_s).$$

LEMMA 1.9.
$$dR = 2\mathrm{div}(\mathrm{Ric}) = 2\nabla^*\mathrm{Ric}.$$

For a proof see Proposition 6 of Chapter 2 on page 40 of [57].

We shall also need a formula relating the connection Laplacian on contravariant one-tensors with the Ricci curvature. Recall that for a smooth function f, we defined the symmetric two-tensor $\nabla^2 f$ by

$$\nabla^2 f(X, Y) = \nabla_X \nabla_Y(f) - \nabla_{\nabla_X(Y)}(f) = \mathrm{Hess}(f)(X, Y),$$

and then defined the Laplacian

$$\triangle f = \mathrm{tr}\nabla^2 f = g^{ij}(\nabla^2 f)_{ij}.$$

These operators extend to tensors of any rank. Suppose that ω is a contravariant tensor of rank k. Then we define $\nabla^2 \omega$ to be a contravariant tensor of rank $k+2$ given by

$$\nabla^2\omega(\cdot, X, Y) = (\nabla_X \nabla_Y \omega)(\cdot) - \nabla_{\nabla_X(Y)}\omega(\cdot).$$

This expression is not symmetric in the vector fields X, Y but the commutator is given by evaluating the curvature operator $\mathcal{R}(X, Y)$ on ω. We define the connection Laplacian on the tensor ω to be

$$\triangle\omega = g^{ij}\nabla^2(\omega)(\partial_i, \partial_j).$$

Direct computation gives the standard Bochner formula relating these Laplacians with the Ricci curvature; see for example Proposition 4.36 on page 168 of [22].

LEMMA 1.10. *Let f be a smooth function on a Riemannian manifold. Then we have the following formula for contravariant one-tensors:*

$$\triangle df = d\triangle f + \mathrm{Ric}((\nabla f)^*, \cdot).$$

2.2. First examples.

The most homogeneous Riemannian manifolds are those of constant sectional curvature. These are easy to classify; see Corollary 10 of Chapter 5 on page 147 of [57].

THEOREM 1.11. (**Uniformization Theorem**) *If (M^n, g) is a complete, simply-connected Riemannian manifold of constant sectional curvature λ, then:*

(1) *If $\lambda = 0$, then M^n is isometric to Euclidean n-space.*
(2) *If $\lambda > 0$ there is a diffeomorphism $\phi \colon M \to S^n$ such that $g = \lambda^{-1}\phi^*(g_{\mathrm{st}})$ where g_{st} is the usual metric on the unit sphere in $\mathbb{R}^{n+1}$.*

(3) If $\lambda < 0$ there is a diffeomorphism $\phi\colon M \to \mathbb{H}^n$ such that $g = |\lambda|^{-1}\phi^*(g_{\text{st}})$ where g_{st} is the Poincaré metric of constant curvature -1 on $\mathbb{H}^n$.

Of course, if (M^n, g) is a complete manifold of constant sectional curvature, then its universal covering satisfies the hypothesis of the theorem and hence is one of $S^n, \mathbb{R}^n$, or $\mathbb{H}^n$, up to a constant scale factor. This implies that (M, g) is isometric to a quotient of one of these simply connected spaces of constant curvature by the free action of a discrete group of isometries. Such a Riemannian manifold is called a *space-form*.

DEFINITION 1.12. The Riemannian manifold (M, g) is said to be an *Einstein manifold with Einstein constant* λ if $\text{Ric}(g) = \lambda g$.

EXAMPLE 1.13. Let M be an n-dimensional manifold with n being either 2 or 3. If (M, g) is Einstein with Einstein constant λ, one can easily show that M has constant sectional curvature $\frac{\lambda}{n-1}$, so that in fact M is a space-form.

2.3. Cones. Another class of examples that will play an important role in our study of the Ricci flow is that of cones.

DEFINITION 1.14. Let (N, g) be a Riemannian manifold. We define the *open cone* over (N, g) to be the manifold $N \times (0, \infty)$ with the metric $\widetilde{g}$ defined as follows: For any $(x, s) \in N \times (0, \infty)$ we have
$$\widetilde{g}(x, s) = s^2 g(x) + ds^2.$$

Fix local coordinates $(x^1, \ldots, x^n)$ on N. Let Γ_{ij}^k; $1 \leq i, j, k \leq n$, be the Christoffel symbols for the Levi-Civita connection on N. Set $x^0 = s$. In the local coordinates $(x^0, x^1, \ldots, x^n)$ for the cone we have the Christoffel symbols $\widetilde{\Gamma}_{ij}^k$, $0 \leq i, j, k \leq n$, for $\widetilde{g}$. The relation between the metrics gives the following relations between the two sets of Christoffel symbols:

$$\widetilde{\Gamma}_{ij}^k = \Gamma_{ij}^k; \quad 1 \leq i, j, k \leq n,$$
$$\widetilde{\Gamma}_{ij}^0 = -sg_{ij}; \quad 1 \leq i, j \leq n,$$
$$\widetilde{\Gamma}_{i0}^j = \widetilde{\Gamma}_{0i}^j = s^{-1}\delta_i^j; \quad 1 \leq i, j \leq n,$$
$$\widetilde{\Gamma}_{i0}^0 = 0; \quad 0 \leq i \leq n,$$
$$\widetilde{\Gamma}_{00}^i = 0; \quad 0 \leq i \leq n.$$

Denote by $\mathcal{R}_g$ the curvature tensor for g and by $\mathcal{R}_{\widetilde{g}}$ the curvature tensor for $\widetilde{g}$. Then the above formulas lead directly to:

$$\mathcal{R}_{\widetilde{g}}(\partial_i, \partial_j)(\partial_0) = 0; \quad 0 \leq i, j \leq n,$$
$$\mathcal{R}_{\widetilde{g}}(\partial_i, \partial_j)(\partial_i) = \mathcal{R}_g(\partial_i, \partial_j)(\partial_i) + g_{ii}\partial_j - g_{ji}\partial_i \quad 1 \leq i, j \leq n.$$

This allows us to compute the Riemann curvatures of the cone in terms of those of N.

PROPOSITION 1.15. *Let N be a Riemannian manifold of dimension $n-1$. Fix $(x,s) \in c(N) = N \times (0, \infty)$. With respect to the coordinates $(x^0, \ldots, x^n)$ the curvature operator $\mathrm{Rm}_{\widetilde{g}}(p,s)$ of the cone decomposes as*

$$\begin{pmatrix} 0 & 0 \\ s^2(\mathrm{Rm}_g(p) - \wedge^2 g(p)) & 0 \end{pmatrix},$$

where $\wedge^2 g(p)$ is the symmetric form on $\wedge^2 T_p N$ induced by g.

COROLLARY 1.16. *For any $p \in N$ let $\lambda_1, \ldots, \lambda_{(n-1)(n-2)/2}$ be the eigenvalues of $\mathrm{Rm}_g(p)$. Then for any $s > 0$ there are $(n-1)$ zero eigenvalues of $\mathrm{Rm}_{\widetilde{g}}(p,s)$. The other $(n-1)(n-2)/2$ eigenvalues of $\mathrm{Rm}_{\widetilde{g}}(p,s)$ are $s^{-2}(\lambda_i - 1)$.*

PROOF. Clearly from Proposition 1.15, we see that under the orthogonal decomposition $\wedge^2 T_{(p,s)} c(N) = \wedge^2 T_p N \oplus T_p N$ the second subspace is contained in the null space of $\mathrm{Rm}_{\widetilde{g}}(p,s)$, and hence contributes $(n-1)$ zero eigenvalues. Likewise, from this proposition we see that the eigenvalues of the restriction of $\mathrm{Rm}_{\widetilde{g}}(p,s)$ to the subspace $\wedge^2 T_p N$ are given by $s^{-4}(s^2(\lambda_i - 1)) = s^{-2}(\lambda_i - 1)$. □

3. Geodesics and the exponential map

Here we review standard material about geodesics, Jacobi fields, and the exponential map.

3.1. Geodesics and the energy functional.

DEFINITION 1.17. Let I be an open interval. A smooth curve $\gamma \colon I \to M$ is called a *geodesic* if $\nabla_{\dot\gamma} \dot\gamma = 0$.

In local coordinates, we write $\gamma(t) = (x^1(t), \ldots, x^n(t))$ and this equation becomes

$$0 = \nabla_{\dot\gamma} \dot\gamma(t) = \left(\sum_k \left(\ddot x^k(t) + \dot x^i(t) \dot x^j(t) \Gamma^k_{ij}(\gamma(t)) \right) \partial_k \right).$$

This is a system of 2nd order ODE's. The local existence, uniqueness and smoothness of a geodesic through any point $p \in M$ with initial velocity vector $v \in T_p M$ follow from the classical ODE theory. Given any two points in a complete manifold, a standard limiting argument shows that there is a rectifiable curve of minimal length between these points. Any such curve is a geodesic. We call geodesics that minimize the length between their endpoints *minimizing geodesics*.

We have the classical theorem showing that on a complete manifold all geodesics are defined for all time (see Theorem 16 of Chapter 5 on p. 137 of [**57**]).

3. GEODESICS AND THE EXPONENTIAL MAP

THEOREM 1.18. *(Hopf-Rinow) If (M, g) is complete as a metric space, then every geodesic extends to a geodesic defined for all time.*

Geodesics are critical points of the energy functional. Let (M, g) be a complete Riemannian manifold. Consider the space of C^1-paths in M parameterized by the unit interval. On this space we have the energy functional

$$E(\gamma) = \frac{1}{2} \int_0^1 \langle \gamma'(t), \gamma'(t) \rangle dt.$$

Suppose that we have a one-parameter family of paths parameterized by $[0, 1]$, all having the same initial point p and the same final point q. By this we mean that we have a surface $\tilde{\gamma}(t, u)$ with the property that for each u the path $\gamma_u = \tilde{\gamma}(\cdot, u)$ is a path from p to q parameterized by $[0, 1]$. Let $\tilde{X} = \partial \tilde{\gamma}/\partial t$ and $\tilde{Y} = \partial \tilde{\gamma}/\partial u$ be the corresponding vector fields along the surface swept out by $\tilde{\gamma}$, and denote by X and Y the restriction of these vector fields along γ_0. We compute

$$\frac{dE(\gamma_u)}{du}\Big|_{u=0} = \left(\int_0^1 \langle \nabla_{\tilde{Y}} \tilde{X}, \tilde{X} \rangle dt \right)\Big|_{u=0}$$
$$= \left(\int_0^1 \langle \nabla_{\tilde{X}} \tilde{Y}, \tilde{X} \rangle dt \right)\Big|_{u=0}$$
$$= -\left(\int_0^1 \langle \nabla_{\tilde{X}} \tilde{X}, \tilde{Y} \rangle dt \right)\Big|_{u=0} = -\int_0^1 \langle \nabla_X X, Y \rangle,$$

where the first equality in the last line comes from integration by parts and the fact that $\tilde{Y}$ vanishes at the endpoints. Given any vector field Y along γ_0 there is a one-parameter family $\tilde{\gamma}(t, u)$ of paths from p to q with $\tilde{\gamma}(t, 0) = \gamma_0$ and with $\tilde{Y}(t, 0) = Y$. Thus, from the above expression we see that γ_0 is a critical point for the energy functional on the space of paths from p to q parameterized by the interval $[0, 1]$ if and only if γ_0 is a geodesic.

Notice that it follows immediately from the geodesic equation that the length of a tangent vector along a geodesic is constant. Thus, if a geodesic is parameterized by $[0, 1]$ we have

$$E(\gamma) = \frac{1}{2} L(\gamma)^2.$$

It is immediate from the Cauchy-Schwarz inequality that for any curve μ parameterized by $[0, 1]$ we have

$$E(\mu) \geq \frac{1}{2} L(\mu)^2$$

with equality if and only if $|\mu'|$ is constant. In particular, a curve parameterized by $[0, 1]$ minimizes distance between its endpoints if it is a minimum for the energy functional on all paths parameterized by $[0, 1]$ with the given endpoints.

3.2. Families of geodesics and Jacobi fields.

Consider a family of geodesics $\widetilde{\gamma}(u,t) = \gamma_u(t)$ parameterized by the interval $[0,1]$ with $\gamma_u(0) = p$ for all u. Here, unlike the discussion above, we allow $\gamma_u(1)$ to vary with u. As before define vector fields along the surface swept out by $\widetilde{\gamma}$: $\widetilde{X} = \partial\widetilde{\gamma}/\partial t$ and let $\widetilde{Y} = \partial\widetilde{\gamma}/\partial u$. We denote by X and Y the restriction of these vector fields to the geodesic $\gamma_0 = \gamma$. Since each γ_u is a geodesic, we have $\nabla_{\widetilde{X}}\widetilde{X} = 0$. Differentiating this equation in the $\widetilde{Y}$-direction yields $\nabla_{\widetilde{Y}}\nabla_{\widetilde{X}}\widetilde{X} = 0$. Interchanging the order of differentiation, using $\nabla_{\widetilde{X}}\widetilde{Y} = \nabla_{\widetilde{Y}}\widetilde{X}$, and then restricting to γ, we get the *Jacobi equation*:

$$\nabla_X \nabla_X Y + \mathcal{R}(Y,X)X = 0.$$

Notice that the left-hand side of the equation depends only on the value of Y along γ, not on the entire family. We denote the left-hand side of this equation by $\mathrm{Jac}(Y)$, so that the Jacobi equation now reads

$$\mathrm{Jac}(Y) = 0.$$

The fact that all the geodesics begin at the same point at time 0 means that $Y(0) = 0$. A vector field Y along a geodesic γ is said to be a *Jacobi field* if it satisfies this equation and vanishes at the initial point p. A Jacobi field is determined by its first derivative at p, i.e., by $\nabla_X Y(0)$. We have just seen that this is the equation describing, to first order, variations of γ by a family of geodesics with the same starting point.

Jacobi fields are also determined by the energy functional. Consider the space of paths parameterized by $[0,1]$ starting at a given point p but free to end anywhere in the manifold. Let γ be a geodesic (parameterized by $[0,1]$) from p to q. Associated to any one-parameter family $\widetilde{\gamma}(t,u)$ of paths parameterized by $[0,1]$ starting at p we associate the second derivative of the energy at $u = 0$. Straightforward computation gives

$$\left.\frac{d^2 E(\gamma_u)}{du^2}\right|_{u=0} = \langle \nabla_X Y(1), Y(1)\rangle + \langle X(1), \nabla_Y \widetilde{Y}(1,0)\rangle - \int_0^1 \langle \mathrm{Jac}(Y), Y\rangle dt.$$

Notice that the first term is a boundary term from the integration by parts, and it depends not just on the value of Y (i.e., on $\widetilde{Y}$ restricted to γ) but also on the first-order variation of $\widetilde{Y}$ in the Y direction. There is the associated bilinear form that comes from two-parameter families $\widetilde{\gamma}(t, u_1, u_2)$ whose value at $u_1 = u_1 = 0$ is γ. It is

$$\left.\frac{d^2 E}{du_1 du_2}\right|_{0,0} = \langle \nabla_X Y_1(1), Y_2(1)\rangle + \langle X(1), \nabla_{Y_1}\widetilde{Y}_2(1,0)\rangle - \int_0^1 \langle \mathrm{Jac}(Y_1), Y_2\rangle dt.$$

Notice that restricting to the space of vector fields that vanish at both endpoints, the second derivatives depend only on Y_1 and Y_2 and the formula is

$$\left.\frac{d^2 E}{du_1 du_2}\right|_{0,0} = -\int_0^1 \langle \mathrm{Jac}(Y_1), Y_2\rangle dt,$$

so that this expression is symmetric in Y_1 and Y_2. The associated quadratic form on the space of vector fields along γ vanishing at both endpoints

$$-\int_0^1 \langle \text{Jac}(Y), Y\rangle dt$$

is the second derivative of the energy function at γ for any one-parameter family whose value at 0 is γ and whose first variation is given by Y.

3.3. Minimal geodesics.

DEFINITION 1.19. *Let γ be a geodesic beginning at $p \in M$. For any $t > 0$ we say that $q = \gamma(t)$ is a* conjugate point along γ *if there is a non-zero Jacobi field along γ vanishing at $\gamma(t)$.*

PROPOSITION 1.20. *Suppose that $\gamma\colon [0,1] \to M$ is a minimal geodesic. Then for any $t < 1$ the restriction of γ to $[0,t]$ is the unique minimal geodesic between its endpoints and there are no conjugate points on $\gamma([0,1))$, i.e., there is no non-zero Jacobi field along γ vanishing at any $t \in [0,1)$.*

We shall sketch the proof. For more details see Proposition 19 and Lemma 14 of Chapter 5 on pp. 139 and 140 of [**57**].

PROOF. (Sketch) Fix $0 < t_0 < 1$. Suppose that there were a different geodesic $\mu\colon [0, t_0] \to M$ from $\gamma(0)$ to $\gamma(t_0)$, whose length was at most that of $\gamma|_{[0,t_0]}$. The fact that μ and $\gamma|_{[0,t_0]}$ are distinct means that $\mu'(t_0) \neq \gamma'(t_0)$. Then the curve formed by concatenating μ with $\gamma|_{[t_0,1]}$ is a curve from $\gamma(0)$ to $\gamma(1)$ whose length is at most that of γ. But this concatenated curve is not smooth at $\mu(t_0)$, and hence it is not a geodesic, and in particular there is a curve with shorter length (a minimal geodesic) between these points. This is contrary to our assumption that γ was minimal.

To establish that there are no conjugate points at $\gamma(t_0)$ for $t_0 < 1$ we need the following claim.

CLAIM 1.21. *Suppose that γ is a minimal geodesic and Y is a field vanishing at both endpoints. Let $\widetilde{\gamma}(t,u)$ be any one-parameter family of curves parameterized by $[0,1]$, with $\gamma_0 = \gamma$ and with $\gamma_u(0) = \gamma_0(0)$ for all u. Suppose that the first-order variation of $\widetilde{\gamma}$ at $u = 0$ is given by Y. Then*

$$\left.\frac{d^2 E(\gamma_u)}{du^2}\right|_{u=0} = 0$$

if and only if Y is a Jacobi field.

PROOF. Suppose that $\widetilde{\gamma}(u,t)$ is a one-parameter family of curves from $\gamma(0)$ to $\gamma(1)$ with $\gamma_0 = \gamma$ and Y is the first-order variation of this family along γ. Since γ is a minimal geodesic we have

$$-\int_0^1 \langle \text{Jac}(Y), Y\rangle dt = \left.\frac{d^2 E(\gamma_u)}{du^2}\right|_{u=0} \geq 0.$$

The associated symmetric bilinear form

$$B_\gamma(Y_1, Y_2) = -\int_\gamma \langle \text{Jac}(Y_1), Y_2 \rangle dt$$

is symmetric when Y_1 and Y_2 are constrained to vanish at both endpoints. Since the associated quadratic form is non-negative, we see by the usual argument for symmetric bilinear forms that $B_\gamma(Y, Y) = 0$ if and only if $B_\gamma(Y, \cdot) = 0$ as a linear functional on the space of vector fields along γ vanishing at point endpoints. This of course occurs if and only if $\text{Jac}(Y) = 0$. □

Now let us use this claim to show that there are no conjugate points on $\gamma|_{(0,1)}$. If for some $t_0 < 1$, $\gamma(t_0)$ is a conjugate point along γ, then there is a non-zero Jacobi field $Y(t)$ along γ with $Y(t_0) = 0$. Notice that since Y is non-trivial, $\nabla_X Y(t_0) \neq 0$. Extend $Y(t)$ to a vector field $\hat{Y}$ along all of γ by setting it equal to 0 on $\gamma|_{[t_0, 1]}$. Since the restriction of Y to $\gamma([0, t_0])$ is a Jacobi field vanishing at both ends and since $\gamma|_{[0,t_0]}$ is a minimal geodesic, the second-order variation of length of $\gamma|_{[0,t_0]}$ in the Y-direction is zero. It follows that the second-order variation of length along $\hat{Y}$ vanishes. But $\hat{Y}$ is not smooth (at $\gamma(t_0)$) and hence it is not a Jacobi field along γ. This contradicts the fact discussed in the previous paragraph that for minimal geodesics the null space of the quadratic form is exactly the space of Jacobi fields. □

3.4. The exponential mapping.

DEFINITION 1.22. For any $p \in M$, we can define the *exponential map* at p, $\exp_p$. It is defined on an open neighborhood O_p of the origin in T_pM and is defined by $\exp_p(v) = \gamma_v(1)$, the endpoint of the unique geodesic $\gamma_v \colon [0, 1] \to M$ starting from p with initial velocity vector v. We always take $O_p \subset T_pM$ to be the maximal domain on which $\exp_p$ is defined, so that O_p is a star-shaped open neighborhood of $0 \in T_pM$. By the Hopf-Rinow Theorem, if M is complete, then the exponential map is defined on all of T_pM.

By the inverse function theorem there exists $r_0 = r_0(p, M) > 0$, such that the restriction of $\exp_p$ to the ball $B_{g|T_pM}(0, r_0)$ in T_pM is a diffeomorphism onto $B_g(p, r_0)$. Fix g-orthonormal linear coordinates on T_pM. Transferring these coordinates via $\exp_p$ to coordinates on $B(p, r_0)$ gives us *Gaussian normal coordinates* on $B(p, r_0) \subset M$.

Suppose now that M is complete, and fix a point $p \in M$. For every $q \in M$, there is a length-minimizing path from p to q. When parameterized at constant speed equal to its length, this path is a geodesic with domain interval $[0, 1]$. Consequently, $\exp_p \colon T_pM \to M$ is onto. The differential of the exponential mapping is given by Jacobi fields: Let $\gamma \colon [0, 1] \to M$ be a geodesic from p to q, and let $X \in T_pM$ be $\gamma'(0)$. Then the exponential mapping at p is a smooth map from $T_p(M) \to M$ sending X to q. Fix $Z \in T_pM$. Then there is a unique Jacobi field Y_Z along γ with $\nabla_X Y_Z(0) = Z$. The association $Z \mapsto Y_Z(1) \in T_qM$ is a linear map from $T_p(M) \to T_qM$. Under the natural identification of T_pM with the tangent plane to T_pM at the point Z, this linear mapping is the differential of $\exp_p \colon T_pM \to M$ at the point $X \in T_pM$.

COROLLARY 1.23. *Suppose that γ is a minimal geodesic parameterized by $[0,1]$ starting at p. Let $X(0) = \gamma'(0) \in T_pM$. Then for each $t_0 < 1$ the restriction $\gamma|_{[0,t_0]}$ is a minimal geodesic and $\exp_p \colon T_pM \to M$ is a local diffeomorphism near $t_0 X(0)$.*

PROOF. Of course, $\exp_p(t_0 X(0)) = \gamma(t_0)$. According to the previous discussion, the kernel of the differential of the exponential mapping at $t_0 X(0)$ is identified with the space of Jacobi fields along γ vanishing at $\gamma(t_0)$. According to Proposition 1.20 the only such Jacobi field is the trivial one. Hence, the differential of $\exp_p$ at $t_0 X(0)$ is an isomorphism, completing the proof. □

DEFINITION 1.24. There is an open neighborhood $U_p \subset T_pM$ of 0 consisting of all $v \in T_pM$ for which: (i) γ_v is the unique minimal geodesic from p to $\gamma_v(1)$, and (ii) $\exp_p$ is a local diffeomorphism at v. We set $\mathcal{C}_p \subset M$ equal to $M \setminus \exp_p(U_p)$. Then $\mathcal{C}_p$ is called the *cut locus from p*. It is a closed subset of measure 0.

It follows from Corollary 1.23 that $U \subset T_pM$ is a star-shaped open neighborhood of $0 \in T_pM$.

PROPOSITION 1.25. *The map*
$$\exp_p \colon U_p \to M \setminus \mathcal{C}_p$$
is a diffeomorphism.

For a proof see p. 139 of [**57**].

DEFINITION 1.26. The *injectivity radius* $\text{inj}_M(p)$ of M at p is the supremum of the $r > 0$ for which the restriction of $\exp_p \colon T_pM \to M$ to the ball $B(0,r)$ of radius r in T_pM is a diffeomorphism into M. Clearly, $\text{inj}_M(p)$ is the distance in T_pM from 0 to the frontier of U_p. It is also the distance in M from p to the cut locus $\mathcal{C}_p$.

Suppose that $\text{inj}_M(p) = r$. There are two possibilities: Either there is a broken, closed geodesic through p, broken only at p, of length $2r$, or there is a geodesic γ of length r emanating from p whose endpoint is a conjugate point along γ. The first case happens when the exponential mapping is not one-to-one of the closed ball of radius r in T_pM, and the second happens when there is a tangent vector in T_pM of length r at which $\exp_p$ is not a local diffeomorphism.

4. Computations in Gaussian normal coordinates

In this section we compute the metric and the Laplacian (on functions) in local Gaussian coordinates. A direct computation shows that in Gaussian normal coordinates on a metric ball about $p \in M$ the metric takes the form

$$(1.5) \qquad g_{ij}(x) = \delta_{ij} + \frac{1}{3}R_{iklj}x^k x^l + \frac{1}{6}R_{iklj,s}x^k x^l x^s$$
$$+ \left(\frac{1}{20}R_{iklj,st} + \frac{2}{45}\sum_m R_{iklm}R_{jstm}\right)x^k x^l x^s x^t + O(r^5),$$

where r is the distance from p. (See, for example Proposition 3.1 on page 41 of [**60**], with the understanding that, with the conventions there, the quantity R_{ijkl} there differs by sign from ours.)

Let γ be a geodesic in M emanating from p in the direction v. Choose local coordinates $\theta^1, \ldots, \theta^{n-1}$ on the unit sphere in T_pM in a neighborhood of $v/|v|$. Then $(r, \theta^1, \ldots, \theta^{n-1})$ are local coordinates at any point of the ray emanating from the origin in the v direction (except at p). Transferring these via $\exp_p$ produces local coordinates $(r, \theta^1, \ldots, \theta^{n-1})$ along γ. Using Gauss's lemma (Lemma 12 of Chapter 5 on p. 133 of [**57**]), we can write the metric locally as

$$g = dr^2 + r^2 h_{ij}(r, \theta)d\theta^i \otimes d\theta^j.$$

Then the volume form

$$dV = \sqrt{\det(g_{ij})} dr \wedge d\theta^1 \wedge \cdots \wedge d\theta^{n-1}$$
$$= r^{n-1}\sqrt{\det(h_{ij})} dr \wedge d\theta^1 \wedge \cdots \wedge d\theta^{n-1}.$$

LEMMA 1.27. *The Laplacian operator acting on scalar functions on M is given in local coordinates by*

$$\triangle = \frac{1}{\sqrt{\det(g)}}\partial_i\left(g^{ij}\sqrt{\det(g)}\partial_j\right).$$

PROOF. Let us compute the derivative at a point p. We have

$$\frac{1}{\sqrt{\det(g)}}\partial_i\left(g^{ij}\sqrt{\det(g)}\partial_j\right)f = g^{ij}\partial_i\partial_j f + \partial_i g^{ij}\partial_j f + \frac{1}{2}g^{ij}\partial_i Tr(\widetilde{g})\partial_j f,$$

where $\widetilde{g} = g(p)^{-1}g$. On the other hand from the definition of the Laplacian, Equation (1.4), and Equation (1.3) we have

$$\triangle f = g^{ij}\text{Hess}(f)(\partial_i, \partial_j) = g^{ij}\left(\partial_i\partial_j(f) - \nabla_{\partial_i}\partial_j f\right) = g^{ij}\partial_i\partial_j f - g^{ij}\Gamma^k_{ij}\partial_k f.$$

Thus, to prove the claim it suffices to show that

$$g^{ij}\Gamma^k_{ij} = -(\partial_i g^{ik} + \frac{1}{2}g^{ik}\text{Tr}(\partial_i\widetilde{g})).$$

From the definition of the Christoffel symbols we have

$$g^{ij}\Gamma^k_{ij} = \frac{1}{2}g^{ij}g^{kl}(\partial_i g_{jl} + \partial_j g_{il} - \partial_l g_{ij}).$$

Of course, $g^{ij}\partial_i g_{jl} = -\partial_i g^{ij} g_{jl}$, so that $g^{ij}g^{kl}\partial_i g_{jl} = -\partial_i g^{ik}$. It follows by symmetry that $g^{ij}g^{jl}\partial_j g_{il} = -\partial_i g^{ik}$. The last term is clearly $-\frac{1}{2}g^{ik}\text{Tr}(\partial_i \widetilde{g})$. □

Using Gaussian local coordinates near p, we have

$$\triangle r = \frac{1}{r^{n-1}\sqrt{\det(h)}}\partial_r\left(r^{n-1}\sqrt{\det(h)}\right)$$
$$= \frac{n-1}{r} + \partial_r \log\left(\sqrt{\det(h)}\right).$$

From this one computes directly that

$$\triangle r = \frac{n-1}{r} - \frac{r}{3}\text{Ric}(v,v) + O(r^2),$$

where $v = \dot{r}(0)$, cf, p.265-268 of [**57**]. So

$$\triangle r \leq \frac{n-1}{r} \quad \text{when} \quad r \ll 1 \text{ and Ric} > 0.$$

This local computation has the following global analogue.

EXERCISE 1.28. (E.Calabi, 1958) Let $f(x) = d(p,x)$ be the distance function from p. If (M,g) has Ric ≥ 0, then

$$\triangle f \leq \frac{n-1}{f}$$

in the sense of distributions.

[Compare [**57**], p. 284 Lemma 42].

REMARK 1.29. The statement that $\triangle f \leq \frac{n-1}{f}$ *in the sense of distributions* (or equivalently *in the weak sense*) means that for any non-negative test function ϕ, that is to say for any compactly supported C^∞-function ϕ, we have

$$\int_M f\triangle\phi\, d\text{vol} \leq \int_M \left(\frac{n-1}{f}\right)\phi\, d\text{vol}.$$

Since the triangle inequality implies that $|f(x) - f(y)| \leq d(x,y)$, it follows that f is Lipschitz, and hence that the restriction of ∇f to any compact subset of M is an L^2 one-form. Integration by parts then shows that

$$\int_M f\triangle\phi\, d\text{vol} = -\int_M \langle\nabla f, \nabla\phi\rangle\, d\text{vol}.$$

Since $|\nabla f| = 1$ and $\triangle f$ is the mean curvature of the geodesic sphere $\partial B(x,r)$, $\mathrm{Ric}(v,v)$ measures the difference of the mean curvature between the standard Euclidean sphere and the geodesic sphere in the direction v. Another important geometric object is the shape operator associated to f, denoted S. By definition it is the Hessian of f; i.e., $S = \nabla^2 f = \mathrm{Hess}(f)$.

5. Basic curvature comparison results

In this section we will recall some of the basic curvature comparison results in Riemannian geometry. The reader can refer to [**57**], Section 1 of Chapter 9 for details.

We fix a point $p \in M$. For any real number $k \geq 0$ let H_k^n denote the simply connected, complete Riemannian n-manifold of constant sectional curvature $-k$. Fix a point $q_k \in H_k^n$, and consider the exponential map $\exp_{q_k}\colon T_{q_k}(H_k^n) \to H_k^n$. This map is a global diffeomorphism. Let us consider the pullback, $\tilde{h}_k$, of the Riemannian metric on H_k^n to $T_{q_k} H_k^n$. A formula for this tensor is easily given in polar coordinates on $T_{q_k}(H_k^n)$ in terms of the following function.

DEFINITION 1.30. We define a function sn_k as follows:

$$\mathrm{sn}_k(r) = \begin{cases} r & \text{if } k = 0, \\ \frac{1}{\sqrt{k}} \sinh(\sqrt{k} r) & \text{if } k > 0. \end{cases}$$

The function $\mathrm{sn}_k(r)$ is the solution to the equation

$$\varphi'' - k\varphi = 0,$$
$$\varphi(0) = 0,$$
$$\varphi'(0) = 1.$$

We define $\mathrm{ct}_k(r) = \mathrm{sn}_k'(r)/\sqrt{k}\mathrm{sn}_k(r)$.

Now we can compare manifolds of varying sectional curvature with those of constant curvature.

THEOREM 1.31. *(Sectional Curvature Comparison) Fix $k \geq 0$. Let (M, g) be a Riemannian manifold with the property that $-k \leq K(P)$ for every 2-plane P in TM. Fix a minimizing geodesic $\gamma\colon [0, r_0) \to M$ parameterized at unit speed with $\gamma(0) = p$. Impose Gaussian polar coordinates $(r, \theta^1, \ldots, \theta^{n-1})$ on a neighborhood of γ so that $g = dr^2 + g_{ij}\theta^i \otimes \theta^j$. Then for all $0 < r < r_0$ we have*

$$(g_{ij}(r,\theta))_{1 \leq i,j \leq n-1} \leq \mathrm{sn}_k^2(r),$$

and the shape operator associated to the distance function from p, f, satisfies

$$(S_{ij}(r,\theta))_{1 \leq i,j \leq n-1} \leq \sqrt{k}\mathrm{ct}_k(r).$$

There is also an analogous result for a positive upper bound to the sectional curvature, but in fact all we shall need is the local diffeomorphism property of the exponential mapping.

LEMMA 1.32. *Fix $K \geq 0$. If $|\mathrm{Rm}(x)| \leq K$ for all $x \in B(p, \pi/\sqrt{K})$, then $\exp_p$ is a local diffeomorphism from the ball $B(0, \pi/\sqrt{K})$ in T_pM to the ball $B(p, \pi/\sqrt{K})$ in M.*

There is a crucial comparison result for volume which involves the Ricci curvature.

THEOREM 1.33. **(Ricci curvature comparison)** *Fix $k \geq 0$. Assume that (M, g) satisfies $\mathrm{Ric} \geq -(n-1)k$. Let $\gamma \colon [0, r_0) \to M$ be a minimal geodesic of unit speed. Then for any $r < r_0$ at $\gamma(r)$ we have*
$$\sqrt{\det g(r, \theta)} \leq \mathrm{sn}_k^{n-1}(r)$$
and
$$\mathrm{Tr}(S)(r, \theta) \leq (n-1)\frac{\mathrm{sn}_k'(r)}{\mathrm{sn}_k(r)}.$$

Note that the inequality in Remark 1.29 follows from this theorem.

The comparison result in Theorem 1.33 holds out to every radius, a fact that will be used repeatedly in our arguments. This result evolved over the period 1964-1980 and now is referred to as the Bishop-Gromov inequality; see Proposition 4.1 of [**11**]

THEOREM 1.34. *(Relative Volume Comparison, Bishop-Gromov 1964-1980) Suppose (M, g) is a Riemannian manifold. Fix a point $p \in M$, and suppose that $B(p, R)$ has compact closure in M. Suppose that for some $k \geq 0$ we have $\mathrm{Ric} \geq -(n-1)k$ on $B(p, R)$. Recall that H_k^n is the simply connected, complete manifold of constant curvature $-k$ and $q_k \in H_k^n$ is a point. Then*
$$\frac{\mathrm{Vol}\, B(p, r)}{\mathrm{Vol}\, B_{H_k^n} B(q_k, r)}$$
is a non-increasing function of r for $r < R$, whose limit as $r \to 0$ is 1. In particular, if the Ricci curvature of (M, g) is ≥ 0 on $B(p, R)$, then $\mathrm{Vol}\, B(p, r)/r^n$ is a non-increasing function of r for $r < R$.

6. Local volume and the injectivity radius

As the following results show, in the presence of bounded curvature the volume of a ball $B(p, r)$ in M is bounded away from zero if and only if the injectivity radius of M at p is bounded away from zero.

PROPOSITION 1.35. *Fix an integer $n > 0$. For every $\epsilon > 0$ there is $\delta > 0$ depending on n and ϵ such that the following holds. Suppose that (M^n, g) is a complete Riemannian manifold of dimension n and that $p \in M$. Suppose that $|\mathrm{Rm}(x)| \leq r^{-2}$ for all $x \in B(p, r)$. If the injectivity radius of M at p is at least ϵr, then $\mathrm{Vol}(B(p, r)) \geq \delta r^n$.*

PROOF. Suppose that $|\text{Rm}(x)| \leq r^{-2}$ for all $x \in B(p,r)$. Replacing g by $r^2 g$ allows us to assume that $r = 1$. Without loss of generality we can assume that $\epsilon \leq 1$. The map $\exp_p$ is a diffeomorphism on the ball $B(0, \epsilon)$ in the tangent space, and by Theorem 1.31 the volume of $B(p, \epsilon)$ is at least that of the ball of radius ϵ in the n-sphere of radius 1. This gives a lower bound to the volume of $B(p, \epsilon)$, and a fortiori to $B(p, 1)$, in terms of n and ϵ. □

We shall normally work with volume, which behaves nicely under Ricci flow, but in order to take limits we need to bound the injectivity radius away from zero. Thus, the more important, indeed crucial, result for our purposes is the converse to the previous proposition; see Theorem 4.3, especially Inequality (4.22), on page 46 of [11], or see Theorem 5.8 on page 96 of [7].

THEOREM 1.36. *Fix an integer $n > 0$. For every $\epsilon > 0$ there is $\delta > 0$ depending on n and ϵ such that the following holds. Suppose that (M^n, g) is a complete Riemannian manifold of dimension n and that $p \in M$. Suppose that $|\text{Rm}(x)| \leq r^{-2}$ for all $x \in B(p, r)$. If $\text{Vol}\, B(p, r) \geq \epsilon r^n$ then the injectivity radius of M at p is at least δr.*

CHAPTER 2

Manifolds of non-negative curvature

In studying singularity development in 3-dimensional Ricci flows one forms blow-up limits. By this we mean the following. One considers a sequence of points x_k in the flow converging to the singularity. It will be the case that $R(x_k)$ tends to ∞ as k tends to ∞. We form a sequence of based Riemannian manifolds labeled by k, where the kth Riemannian manifold is obtained by taking the time-slice of x_k, rescaling its metric by $R(x_k)$, and then taking x_k as the base point. This creates a sequence with the property that for each member of the sequence the scalar curvature at the base point is 1. Because of a pinching result of Hamilton's (see Chapter 4), if there is a geometric limit of this sequence, or of any subsequence of it, then that limit is non-negatively curved. Hence, it is important to understand the basic properties of Riemannian manifolds of non-negative curvature in order to study singularity development. In this chapter we review the properties that we shall need. We suppose that M is non-compact and of positive (resp., non-negative) curvature. The key to understanding these manifolds is the Busemann function associated to a minimizing geodesic ray.

1. Busemann functions

A geodesic ray $\lambda \colon [0, \infty) \to M$ is said to be *minimizing* if the restriction of λ to every compact subinterval of $[0, \infty)$ is a length-minimizing geodesic arc, i.e., a geodesic arc whose length is equal to the distance between its endpoints. Likewise, a geodesic line $\lambda \colon (-\infty, \infty) \to M$ is said to be *minimizing* if its restriction to every compact sub-interval of $\mathbb{R}$ is a length minimizing geodesic arc.

Clearly, if a sequence of minimizing geodesic arcs λ_k converges to a geodesic arc, then the limiting geodesic arc is also minimizing. More generally, if λ_k is a sequence of length minimizing geodesic arcs whose initial points converge and whose lengths go to infinity, then, after passing to a subsequence, there is a limit which is a minimizing geodesic ray. (The existence of a limit of a subsequence is a consequence of the fact that a geodesic ray is determined by its initial point and its initial tangent direction.) Similarly, if I_k is an sequence of compact intervals with the property that every compact subset of $\mathbb{R}$ is contained in I_k for all sufficiently large k, if for each k the map $\lambda_k \colon I_k \to M$ is a minimizing geodesic arc, and if $\lim_{k \to \infty} \lambda_k(0)$ exists, then, after passing to a subsequence there is a limit which is a minimizing geodesic line. Using these facts one establishes the following elementary lemma.

LEMMA 2.1. *Suppose that M is a complete, connected, non-compact Riemannian manifold and let p be a point of M. Then M has a minimizing geodesic ray emanating from p. If M has more than one end, then it has a minimizing line.*

DEFINITION 2.2. Suppose that $\lambda\colon [0,\infty) \to M$ is a minimizing geodesic ray with initial point p. For each $t \geq 0$ we consider $B_{\lambda,t}(x) = d(\lambda(t), x) - t$. This is a family of functions satisfying $|B_{\lambda,t}(x) - B_{\lambda,t}(y)| \leq d(x,y)$. Since λ is a minimizing geodesic, $B_{\lambda,t}(p) = 0$ for all t. It follows that $B_{\lambda,t}(x) \geq -d(x,p)$ for all $x \in M$. Thus, the family of functions $B_{\lambda,t}$ is pointwise bounded below. The triangle inequality shows that for each $x \in M$ the function $B_{\lambda,t}(x)$ is a non-increasing function of t. It follows that, for each $x \in M$, $\lim_{t \to \infty} B_{\lambda,t}(x)$ exists. We denote this limit by $B_\lambda(x)$. This is the *Busemann function* for λ.

Clearly, $B_\lambda(x) \geq -d(x, \lambda(0))$. By equicontinuity $B_\lambda(x)$ is a continuous function of x and in fact a Lipschitz function satisfying $|B_\lambda(x) - B_\lambda(y)| \leq d(x,y)$ for all $x, y \in X$. Clearly $B_\lambda(\lambda(s)) = -s$ for all $s \geq 0$. Since B_λ is Lipschitz, ∇B_λ is well-defined as an L^2-vector field.

PROPOSITION 2.3. *Suppose that M is complete and of non-negative Ricci curvature. Then, for any minimizing geodesic ray λ, the Busemann function B_λ satisfies $\Delta B_\lambda \leq 0$ in the weak sense.*

PROOF. First notice that since B_λ is Lipschitz, ∇B_λ is an L^2-vector field on M. That is to say, $B_\lambda \in W^{1,2}_{\text{loc}}$, i.e., B_λ locally has one derivative in L^2. Hence, there is a sequence of C^∞-functions f_n converging to B_λ in $W^{1,2}_{\text{loc}}$. Let φ be a test function (i.e., a compactly supported C^∞-function). Integrating by parts yields

$$-\int_M \langle \nabla f_n, \nabla \varphi \rangle d\text{vol} = \int_M f_n \Delta \varphi \, d\text{vol}.$$

Using the fact that f_n converges to B_λ in $W^{1,2}_{\text{loc}}$ and taking limits yields

$$-\int_M \langle \nabla B_\lambda, \nabla \varphi \rangle d\text{vol} = \int_M B_\lambda \Delta \varphi \, d\text{vol}.$$

Thus, to prove the proposition we need only show that if φ is a non-negative test function, then

$$-\int_M \langle \nabla B_\lambda, \nabla \varphi \rangle d\text{vol} \leq 0.$$

For a proof of this see Proposition 1.1 and its proof on pp. 7 and 8 in [**61**]. □

2. Comparison results in non-negative curvature

Let us review some elementary comparison results for manifolds of non-negative curvature. These form the basis for Toponogov theory, [**70**]. For any pair of points x, y in a complete Riemannian manifold, s_{xy} denotes a minimizing geodesic from x to y. We set $|s_{xy}| = d(x,y)$ and call it the *length* of the side. A *triangle* in a Riemannian manifold consists of three vertices a, b, c and three sides s_{ab}, s_{ac}, s_{bc}. We denote by $\angle_a$ the angle of the triangle at a, i.e., the angle at a between the geodesic rays s_{ab} and s_{ac}.

THEOREM 2.4. (**Length comparison**) *Let (M, g) be a manifold of non-negative curvature. Suppose that $\triangle = \triangle(a, b, c)$ is a triangle in M and let $\triangle' = \triangle(a', b', c')$ be a Euclidean triangle.*

(1) *Suppose that the corresponding sides of $\triangle$ and $\triangle'$ have the same lengths. Then the angle at each vertex of $\triangle'$ is no larger than the corresponding angle of $\triangle$. Furthermore, for any α and β less than $|s_{ab}|$ and $|s_{ac}|$ respectively, let x, resp. x', be the point on s_{ab}, resp. $s_{a'b'}$, at distance α from a, resp. a', and let y, resp. y', be the point on s_{ac}, resp. $s_{a'c'}$, at distance β from a, resp. a'. Then $d(x, y) \geq d(x', y')$.*

(2) *Suppose that $|s_{ab}| = |s_{a'b'}|$, that $|s_{ac}| = |s_{a'c'}|$ and that $\angle_a = \angle_{a'}$. Then $|s_{b'c'}| \geq |s_{bc}|$.*

See FIG. 1. For a proof of this result see Theorem 4.2 on page 161 of [**60**], or Theorem 2.2 on page 42 of [**7**].

One corollary is a monotonicity result. Suppose that $\triangle(a, b, c)$ is a triangle in a complete manifold of non-negative curvature. Define a function $EA(u, v)$ defined for $0 \leq u \leq |s_{ab}|$ and $0 \leq v \leq |s_{ac}|$ as follows. For u and v in the indicated ranges, let $x(u)$ be the point on s_{ab} at distance u from a and let $y(v)$ be the point of s_{ac} at distance v from a. Let $EA(u, v)$ be the angle at a' of the Euclidean triangle with side lengths $|s_{a'b'}| = u$, $|s_{a'c'}| = v$ and $|s_{b'c'}| = d(x(u), y(v))$.

COROLLARY 2.5. *Under the assumptions of the previous theorem, the function $EA(u, v)$ is a monotone non-increasing function of each variable u and v when the other variable is held fixed.*

Suppose that α, β, γ are three geodesics emanating from a point p in a Riemannian manifold. Let $\angle_p(\alpha, \beta)$, $\angle_p(\beta, \gamma)$ and $\angle_p(\alpha, \gamma)$ be the angles of these geodesics at p as measured by the Riemannian metric. Then of course

$$\angle_p(\alpha, \beta) + \angle_p(\beta, \gamma) + \angle_p(\alpha, \gamma) \leq 2\pi$$

since this inequality holds for the angles between straight lines in Euclidean n-space. There is a second corollary of Theorem 2.4 which gives an analogous result for the associated Euclidean angles.

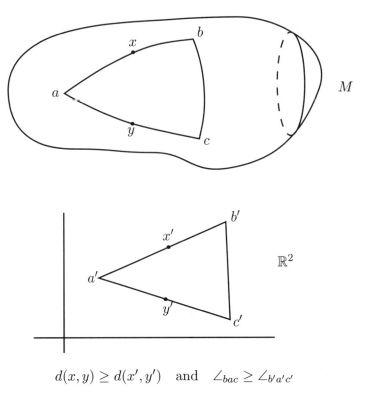

$$d(x,y) \geq d(x',y') \quad \text{and} \quad \angle_{bac} \geq \angle_{b'a'c'}$$

FIGURE 1. Toponogov comparison.

COROLLARY 2.6. *Let (M,g) be a complete Riemannian manifold of non-negative curvature. Let p,a,b,c be four points in M and let α, β, γ be minimizing geodesic arcs from the point p to a,b,c respectively. Let $T(apb)$, $T(bpc)$ and $T(cpa)$ be the triangles in M made out of these minimizing geodesics and minimizing geodesics between a,b,c. Let $T(a'p'b')$, $T(b'p'c')$ and $T(c'p'a')$ be planar triangles with the same side lengths. Then*

$$\angle_{p'} T(a'p'b') + \angle_{p'} T(b'p'c') + \angle_{p'} T(c'p'a') \leq 2\pi.$$

PROOF. Consider the sum of these angles as the geodesic arcs in M are shortened without changing their direction. By the first property of Theorem 2.4 the sum of the angles of these triangles is a monotone decreasing function of the lengths. Of course, the limit as the lengths all go to zero is the corresponding Euclidean angle. The result is now clear. □

3. The soul theorem

A subset X of a Riemannian manifold (M,g) is said to be *totally convex* if every geodesic segment with endpoints in X is contained in X. Thus, a point p in M is totally convex if and only if there is no broken geodesic arc in M broken exactly at x.

THEOREM 2.7. *(Cheeger-Gromoll, see [8] and [10])* Suppose that (M,g) is a connected, complete, non-compact Riemannian manifold of non-negative sectional curvature. Then M contains a soul $S \subset M$. By definition a soul is a compact, totally geodesic, totally convex submanifold (automatically of positive codimension). Furthermore, M is diffeomorphic to the total space of the normal bundle of the S in M. If (M,g) has positive curvature, then any soul for it is a point, and consequently M is diffeomorphic to $\mathbb{R}^n$.

REMARK 2.8. We only use the soul theorem for manifolds with positive curvature and the fact that any soul of such a manifold is a point. A proof of this result first appears in [24].

The rest of this section is devoted to a sketch of the proof of this result. Our discussion follows closely that in [57] starting on p. 349. We shall need more information about complete, non-compact manifolds of non-negative curvature, so we review a little of their theory as we sketch the proof of the soul theorem.

LEMMA 2.9. *Let (M,g) be a complete, non-compact Riemannian manifold of non-negative sectional curvature and let $p \in M$. For every $\epsilon > 0$ there is a compact subset $K = K(p,\epsilon) \subset M$ such that for all points $q \notin K$, if γ and μ are minimizing geodesics from p to q, then the angle that γ and μ make at q is less than ϵ.*

See FIG. 2.

PROOF. The proof is by contradiction. Fix $0 < \epsilon < 1$ sufficiently small so that $\cos(\epsilon/2) < 1 - \epsilon^2/12$. Suppose that there is a sequence of points q_n tending to infinity such that for each n there are minimizing geodesics γ_n and μ_n from p to q_n making angle at least ϵ at q_n. For each n let $d_n = d(p, q_n)$. By passing to a subsequence we can suppose that for all n and m the cosine of the angle at p between γ_n and γ_m at least $1 - \epsilon^2/24$, and the cosine of the angle at p between μ_n and μ_m is at least $1 - \epsilon^2/24$. We can also assume that for all $n \geq 1$ we have $d_{n+1} \geq (100/\epsilon^2)d_n$. Let $\delta_n = d(q_n, q_{n+1})$. Applying the first Toponogov property at p, we see that $\delta_n^2 \leq d_n^2 + d_{n+1}^2 - 2d_n d_{n+1}(1 - \epsilon^2/24)$. Applying the same property at q_n we have
$$d_{n+1}^2 \leq d_n^2 + \delta_n^2 - 2d_n \delta_n \cos(\theta),$$
where $\theta \leq \pi$ is the angle at q_n between γ_n and a minimal geodesic joining q_n to q_{n+1}. Thus,
$$\cos(\theta) \leq \frac{d_n - d_{n+1}(1 - \epsilon^2/24)}{\delta_n}.$$
By the triangle inequality (and the fact that $\epsilon < 1$) we have $\delta_n \geq (99/\epsilon)d_n$ and $\delta_n \geq d_{n+1}(1 - (\epsilon^2/100))$. Thus,
$$\cos(\theta) \leq \epsilon^2/99 - (1 - \epsilon^2/24)/(1 - (\epsilon^2/100)) < -(1 - \epsilon^2/12).$$

This implies that $\cos(\pi - \theta) > (1 - \epsilon^2/12)$, which implies that $\pi - \theta < \epsilon/2$. That is to say, the angle at q_n between γ_n and a shortest geodesic from q_n to q_{n+1} is between $\pi - \epsilon/2$ and π. By symmetry, the same is true for the angle between μ_n and the same shortest geodesic from q_n to q_{n+1}. Thus, the angle between γ_n and μ_n at q_n is less than ϵ, contradicting our assumption. $\square$

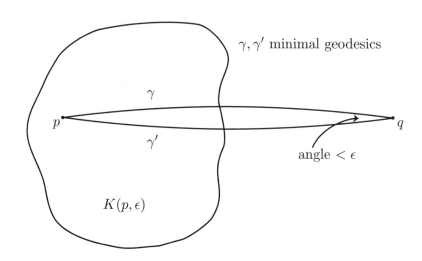

FIGURE 2. Shallow angles of minimal geodesics.

COROLLARY 2.10. *Fix (M, g) a complete, non-compact manifold of non-negative sectional curvature. Let $p \in M$ and define a function $f\colon M \to \mathbb{R}$ by $f(q) = d(p, q)$. Then there is $R < \infty$ such that for $R \leq s < s'$ we have:*
  (1) *$f^{-1}([s, s'])$ is homeomorphic to $f^{-1}(s) \times [s, s']$ and in particular the level sets $f^{-1}(s)$ and $f^{-1}(s')$ are homeomorphic;*
  (2) *$f^{-1}([s, \infty))$ is homeomorphic to $f^{-1}(s) \times [s, \infty)$.*

PROOF. Given (M, g) and $p \in M$ as in the statement of the corollary, choose a constant $R < \infty$ such that any two minimal geodesics from p to a point q with $d(p, q) \geq R/2$ make an angle at most $\pi/6$ at q. Now following [**57**] p. 335, it is possible to find a smooth unit vector field X on $U = M - \overline{B(p, R/2)}$ with the property that $f(\cdot) = d(p, \cdot)$ is increasing along any integral curve for X at a rate bounded below by $\cos(\pi/3)$. In particular, for any $s \geq R$ each integral curve of X crosses the level set $f^{-1}(s)$ in a single point. Using this vector field we see that for any $s, s' > R$, the pre-image $f^{-1}([s, s'])$ is homeomorphic to $f^{-1}(s) \times [s, s']$ and that the end $f^{-1}([s, \infty))$ is homeomorphic to $f^{-1}(s) \times [s, \infty)$. $\square$

In a complete, non-compact n-manifold of positive curvature any soul is a point. While the proof of this result uses the same ideas as discussed above, we shall not give a proof. Rather we refer the reader to Theorem 84 of [57] on p. 349. A soul has the property that if two minimal geodesics emanate from p and end at the same point $q \neq p$, then the angle that they make at q is less than $\pi/2$. Also, of course, the exponential mapping is a diffeomorphism sufficiently close to the soul. Applying the above lemma and a standard compactness argument, we see that in fact there is $\epsilon > 0$ such that all such pairs of minimal geodesics from p ending at the same point make an angle less than $\pi/2 - \epsilon$ at that point. Hence, in this case there is a vector field X on all of M vanishing only at the soul, and agreeing with the gradient of the distance function near the soul, so that the distance function from p is strictly increasing to infinity along each flow line of X (except the fixed point). Using X one establishes that M is diffeomorphic to $\mathbb{R}^n$. It also follows that all the level surfaces $f^{-1}(s)$ for $s > 0$ are homeomorphic to S^{n-1} and for $0 < s < s'$ the preimage $f^{-1}([s, s'])$ is homeomorphic to $S^{n-1} \times [s, s']$.

There is an analogue of this result for the distance function from any point, not just a soul.

COROLLARY 2.11. *Let (M, g) be a complete, non-compact Riemannian n-manifold of positive curvature. Then for any point $p \in M$ there is a constant $R = R(p)$ such that for any $s < s'$ with $R \leq s$ both $f^{-1}(s, s')$ and $f^{-1}(s, \infty)$ are homotopy equivalent to S^{n-1}.*

PROOF. Given (M, g) and p, fix $R < \infty$ sufficiently large so that Corollary 2.10 holds. Since M is diffeomorphic to $\mathbb{R}^n$ it has only one end and hence the level sets $f^{-1}(s)$ for $s \geq R$ are connected. Given any compact subset $K \subset M$ there is a larger compact set B (a ball) such that $M \setminus B$ has trivial fundamental group and trivial homology groups H_i for $i < n - 1$. Hence for any subset $Z \subset M \setminus B$, the inclusion of $Z \to M \setminus K$ induces the trivial map on π_1 and on H_i for $i < n - 1$. Clearly, for any $R \leq s < b$ the inclusion $f^{-1}(b, \infty) \to f^{-1}(s, \infty)$ is a homotopy equivalence. Thus, it must be the case that $f^{-1}(b, \infty)$ has trivial fundamental group and H_i for $i < n - 1$. Hence, the same is true for $f^{-1}(s, \infty)$ for any $s \geq R$. Lastly, since $f^{-1}(s, \infty)$ is connected and simply connected (hence orientable) and has two ends, it follows by the non-compact form of Poincaré duality that $H_{n-1}(f^{-1}(s, \infty)) \cong \mathbb{Z}$. Hence, by the Hurewicz theorem $f^{-1}(s, \infty)$ is homotopy equivalent to S^{n-1} for any $s \geq R$. Of course, it is also true for $R \leq s \leq s'$ that $f^{-1}(s, s')$ is homotopy equivalent to S^{n-1}. □

4. Ends of a manifold

Let us review the basic notions about ends of a manifold.

DEFINITION 2.12. Let M be a connected manifold. Consider the inverse system of spaces indexed by the compact, codimension-0 submanifolds $K \subset M$, where the space associated to K is the finite set $\pi_0(M \setminus K)$ with the discrete topology. The inverse limit of this inverse system is the *space of ends* of M. It is a compact space. An *end* of M is a point of the space of ends. An end $\mathcal{E}$ determines a complementary component of each compact, codimension-0 submanifold $K \subset M$, called a *neighborhood* of the end. Conversely, by definition these neighborhoods are cofinal in the set of all neighborhoods of the end. A sequence $\{x_n\}$ in M *converges to the end* $\mathcal{E}$ if it is eventually in every neighborhood of the end. In fact, what we are doing is defining a topology on the union of M and its space of ends that makes this union a compact, connected Hausdorff space which is a compactification of M.

A proper map between topological manifolds induces a map on the space of ends, and in fact induces a map on the compactifications sending the subspace of ends of the compactification of the domain to the subspace of ends of the compactification of the range.

We say that a path $\gamma\colon [a,b) \to M$ is a path *to the end* $\mathcal{E}$ if it is a proper map and it sends the end $\{b\}$ of $[a,b)$ to the end $\mathcal{E}$ of M. This condition is equivalent to saying that given a neighborhood U of $\mathcal{E}$ there is a neighborhood of the end $\{b\}$ of $[a,b)$ that maps to U.

Now suppose that M has a Riemannian metric g. Then we can distinguish between ends at finite and infinite distance. An end is at *finite distance* if there is a rectifiable path of finite length to the end. Otherwise, the end is at infinite distance. If an end is at finite distance we have the notion of the distance from a point $x \in M$ to the end. It is the infimum of the lengths of rectifiable paths from x to the end. This distance is always positive. Also, notice that the Riemannian manifold is complete if and only if has no end at finite distance.

5. The splitting theorem

In this section we give a proof of the following theorem which is originally due to Cheeger-Gromoll [**9**]. The weaker version giving the same conclusion under the stronger hypothesis of non-negative sectional curvature (which is in fact all we need in this work) was proved earlier by Toponogov, see [**70**].

THEOREM 2.13. *Suppose that M is complete, of non-negative Ricci curvature and suppose that M has at least two ends. Then M is isometric to a product $N \times \mathbb{R}$ where N is a compact manifold.*

PROOF. We begin the proof by establishing a result of independent interest, which was formulated as the main theorem in [**9**].

LEMMA 2.14. *Any complete Riemannian manifold X of non-negative Ricci curvature containing a minimizing line is isometric to a product $N \times \mathbb{R}$ for some Riemannian manifold N.*

PROOF. Given a minimizing line $\lambda\colon \mathbb{R} \to X$, define $\lambda_\pm \colon [0,\infty) \to X$ by $\lambda_+(t) = \lambda(t)$ and $\lambda_-(t) = \lambda(-t)$. Then we have the Busemann functions $B_+ = B_{\lambda_+}$ and $B_- = B_{\lambda_-}$. Proposition 2.3 applies to both B_+ and B_- and shows that $\Delta(B_+ + B_-) \leq 0$. On the other hand, using the fact that λ is distance minimizing, we see that for any $s, t > 0$ and for any $x \in M$ we have $d(x, \lambda(t)) + d(x, \lambda(-s)) \geq s + t$, and hence $B_+(x) + B_-(x) \geq 0$. Clearly, $B_+(x) + B_-(x) = 0$ for any x in the image of λ. Thus, the function $B_+ + B_-$ is everywhere ≥ 0, vanishes at least at one point and satisfies $\Delta(B_+ + B_-) \leq 0$ in the weak sense. This is exactly the set-up for the maximum principle, cf. [**57**], p. 279.

THEOREM 2.15. **(The Maximum Principle)** *Let f be a real-valued continuous function on a connected Riemannian manifold with $\Delta f \geq 0$ in the weak sense. Then f is locally constant near any local maximum. In particular, if f achieves its maximum then it is a constant.*

Applying this result to $-(B_+ + B_-)$, we see that $B_+ + B_- = 0$, so that $B_- = -B_+$. It now follows that $\Delta B_+ = 0$ in the weak sense. By standard elliptic regularity results this implies that B_+ is a smooth harmonic function.

Next, we show that for all $x \in M$ we have $|\nabla B_+(x)| = 1$. Fix $x \in M$. Take a sequence t_n tending to infinity and consider minimizing geodesics $\mu_{+,n}$ from x to $\lambda_+(t_n)$. By passing to a subsequence we can assume that there is a limit as $n \to \infty$. This limit is a minimizing geodesic ray μ_+ from x, which we think of as being 'asymptotic at infinity' to λ_+. Similarly, we construct a minimizing geodesic ray μ_- from x asymptotic at infinity to λ_+. Since μ_+ is a minimal geodesic ray, it follows that for any t the restriction $\mu_+|_{[0,t]}$ is the unique length minimizing geodesic from x to $\mu_+(t)$ and that $\mu_+(t)$ is not a conjugate point along μ_+. It follows by symmetry that x is not a conjugate point along the reversed geodesic $-\mu_+|_{[0,t]}$ and hence that $x \in U_{\mu_+(t)}$. This means that the function $d(\mu_+(t), \cdot)$ is smooth at x with gradient equal to the unit tangent vector in the negative direction at x to μ_+, and consequently that $B_{\mu_+,t}$ is smooth at x. Symmetrically, for any $t > 0$ the function $B_{\mu_-,t}$ is smooth at x with the opposite gradient. Notice that these gradients have norm 1. We have

$$B_{\mu_+,t} + B_+(x) \geq B_+ = -B_- \geq -(B_{\mu_-,t} + B_-(x)).$$

Of course, $B_{\mu_+,t}(x) = 0$ and $B_{\mu_-,t}(x) = 0$, so that

$$B_{\mu_+,t}(x) + B_+(x) = -(B_{\mu_-,t}(x) + B_-(x)).$$

This squeezes B_+ between two smooth functions with the same value and same gradient at x and hence shows that B_+ is C^1 at x and $|\nabla B_+(x)|$ is of norm 1.

Thus, B defines a smooth Riemannian submersion from $M \to \mathbb{R}$ which implies that M is isometric to a product of the fiber over the origin with $\mathbb{R}$. □

This result together with Lemma 2.1 shows that if M satisfies the hypothesis of the theorem, then it can be written as a Riemannian product $M = N \times \mathbb{R}$. Since M has at least two ends, it follows immediately that N is compact. This completes the proof of the theorem. □

6. ϵ-necks

Certain types of (incomplete) Riemannian manifolds play an especially important role in our analysis. The purpose of this section is to introduce these manifolds and use them to prove one essential result in Riemannian geometry.

For all of the following definitions we fix $0 < \epsilon < 1/2$. Set k equal to the greatest integer less than or equal to ϵ^{-1}. In particular, $k \geq 2$.

DEFINITION 2.16. Suppose that we have a fixed metric g_0 on a manifold M and an open submanifold $X \subset M$. We say that another metric g on X is *within ϵ of $g_0|_X$ in the $C^{[1/\epsilon]}$-topology* if, setting $k = [1/\epsilon]$ we have

$$(2.1) \qquad \sup_{x \in X} \left(|g(x) - g_0(x)|^2_{g_0} + \sum_{\ell=1}^{k} |\nabla^\ell_{g_0} g(x)|^2_{g_0} \right) < \epsilon^2,$$

where the covariant derivative $\nabla^\ell_{g_0}$ is the Levi-Civita connection of g_0 and norms are the pointwise g_0-norms on

$$\mathrm{Sym}^2 T^*M \otimes \underbrace{T^*M \otimes \cdots \otimes T^*M}_{\ell-\text{times}}.$$

More generally, given two smooth families of metrics $g(t)$ and $g_0(t)$ on M defined for t in some interval I we say that the family $g(t)|_X$ is within ϵ of the family $g_0(t)|_X$ in the $C^{[1/\epsilon]}$-topology if we have

$$\sup_{(x,t) \in X \times I} \left(|g(x,t) - g_0(x,t)|^2_{g_0(t)} + \sum_{\ell=1}^{k} \left|\nabla^\ell_{g_0} g(x,t)\right|^2_{g_0} \right) < \epsilon^2.$$

REMARK 2.17. Notice that if we view a one-parameter family of metrics $g(t)$ as a curve in the space of metrics on X with the $C^{[1/\epsilon]}$-topology then this is the statement that the two paths are pointwise within ϵ of each other. It says nothing about the derivatives of the paths, or equivalently about the time derivatives of the metrics and of their covariant derivatives. We will always be considering paths of metrics satisfying the Ricci flow equation. In this context two one-parameter families of metrics that are close in the C^{2k}-topology exactly when the rth time derivatives of the sth-covariant derivatives are close for all r, s with $s + 2r \leq 2k$.

The first object of interest is one that, up to scale, is close to a long, round cylinder.

DEFINITION 2.18. Let (N, g) be a Riemannian manifold and $x \in N$ a point. Then *an ϵ-neck structure on (N, g) centered at x* consists of a diffeomorphism

$$\varphi \colon S^2 \times (-\epsilon^{-1}, \epsilon^{-1}) \to N,$$

with $x \in \varphi(S^2 \times \{0\})$, such that the metric $R(x)\varphi^*g$ is within ϵ in the $C^{[1/\epsilon]}$-topology of the product of the usual Euclidean metric on the open interval with the metric of constant Gaussian curvature $1/2$ on S^2. We also use the terminology N *is an ϵ-neck centered at x*. The image under φ of the family of submanifolds $S^2 \times \{t\}$ is called the *family of 2-spheres of the ϵ-neck*. The submanifold $\varphi(S^2 \times \{0\})$ is called *the central 2-sphere* of the ϵ-neck structure. We denote by $s_N \colon N \to \mathbb{R}$ the composition $p_2 \circ \varphi^{-1}$, where p_2 is the projection of $S^2 \times (-\epsilon^{-1}, \epsilon^{-1})$ to the second factor. There is also the vector field $\partial/\partial s_N$ on N which is φ_* of the standard vector field in the interval-direction of the product. We also use the terminology of the *plus* and *minus* end of the ϵ-neck in the obvious sense. The opposite (or reversed) ϵ-neck structure is the one obtained by composing the structure map with $\mathrm{Id}_{S^2} \times -1$. We define the *positive half of the neck* to be the region $s_N^{-1}(0, \epsilon^{-1})$ and the *negative half* to be the region $s_N^{-1}(-\epsilon^{-1}, 0)$. For any other fraction, e.g., the left-hand three-quarters, the right-hand one-quarter, there are analogous notions, all measured with respect to $s_N \colon N \to (-\epsilon^{-1}, \epsilon^{-1})$. We also use the terminology the middle one-half, or middle one-third of the ϵ-neck; again these regions have their obvious meaning when measured via s_N.

An *ϵ-neck* in a Riemannian manifold X is a codimension-zero submanifold N and an ϵ-structure on N centered at some point $x \in N$.

The *scale* of an ϵ-neck N centered at x is $R(x)^{-1/2}$. The scale of N is denoted r_N. Intuitively, this is a measure of the radius of the cross-sectional S^2 in the neck. In fact, the extrinsic diameter of any S^2 factor in the neck is close to $\sqrt{2}\pi r_N$. See FIG. 1 in the introduction.

Here is the result that will be so important in our later arguments.

PROPOSITION 2.19. *The following holds for any $\epsilon > 0$ sufficiently small. Let (M, g) be a complete, positively curved Riemannian 3-manifold. Then (M, g) does not contain ϵ-necks of arbitrarily small scale.*

PROOF. The result is obvious if M is compact, so we assume that M is non-compact. Let $p \in M$ be a soul for M (Theorem 2.7), and let f be the distance function from p. Then $f^{-1}(s)$ is connected for all $s > 0$.

LEMMA 2.20. *Suppose that $\epsilon > 0$ is sufficiently small that Lemma A.10 from the appendix holds. Let (M, g) be a non-compact 3-manifold of positive curvature and let $p \in M$ be a soul for it. Then for any ϵ-neck N disjoint from p the central 2-sphere of N separates the soul from the end of the manifold. In particular, if two ϵ-necks N_1 and N_2 in M are disjoint from* *each other and from p, then the central 2-spheres of N_1 and N_2 are the boundary components of a region in M diffeomorphic to $S^2 \times I$.*

PROOF. Let N be an ϵ-neck disjoint from p. By Lemma A.10 for any point z in the middle third of N, the boundary of the metric ball $B(p, d(z,p))$ is a topological 2-sphere in N isotopic in N to the central 2-sphere of N. Hence, the central 2-sphere separates the soul from the end of M. The second statement follows immediately by applying this to N_1 and N_2. □

Let N_1 and N_2 be disjoint ϵ-necks, each disjoint from the soul. By the previous lemma, the central 2-spheres S_1 and S_2 of these necks are smoothly isotopic to each other and they are the boundary components of a region diffeomorphic to $S^2 \times I$. Reversing the indices if necessary we can assume that N_2 is closer to ∞ than N_1, i.e., further from the soul. Reversing the directions of the necks if necessary, we can arrange that for $i = 1, 2$ the function s_{N_i} is increasing as we go away from the soul. We define C^∞-functions ψ_i on N_i, functions depending only on s_{N_i}, as follows. The function ψ_1 is 0 on the negative side of the middle third of N_1 and increases to be identically 1 on the positive side of the middle third. The function ψ_2 is 1 on the negative side of the middle third of N_2 and decreases to be 0 on the positive side. We extend ψ_1, ψ_2 to a function ψ defined on all of M by requiring that it be identically 1 on the region X between N_1 and N_2 and to be identically 0 on $M \setminus (N_1 \cup X \cup N_2)$.

Let λ be a geodesic ray from the soul of M to infinity, and B_λ its Busemann function. Let N be any ϵ-neck disjoint from the soul, with s_N direction chosen so that it points away from the soul. At any point of the middle third of N where B_λ is smooth, ∇B_λ is a unit vector in the direction of the unique minimal geodesic ray from the end of λ to this point. Invoking Lemma A.4 from the appendix we see that at such points ∇B_λ is close to $-R(x)^{1/2}\partial/\partial s_N$, where $x \in N$ is the center of the ϵ-neck. Since ∇B_λ is L^2 its non-smooth points have measure zero and hence, the restriction of ∇B_λ to the middle third of N is close in the L^2-sense to $-R(x)^{1/2}\partial/\partial s_N$.

Applying this to N_1 and N_2 we see that

$$(2.2) \qquad \int_M \langle \nabla B_\lambda, \nabla \psi \rangle d\mathrm{vol} = \left(\alpha_2 R(x_2)^{-1} - \alpha_1 R(x_1)^{-1} \right) \mathrm{Vol}_{h_0}(S^2)),$$

where $h(0)$ is the round metric of scalar curvature 1 and where each of α_1 and α_2 limits to 1 as ϵ goes to 0. Since $\psi \geq 0$, Proposition 2.3 tells us that the left-hand side of Equation (2.2) must be ≥ 0. This shows that, provided that ϵ is sufficiently small, $R(x_2)$ is bounded above by $2R(x_1)$. This completes the proof of the proposition. □

COROLLARY 2.21. *Fix $\epsilon > 0$ sufficiently small so that Lemma A.10 holds. Then there is a constant $C < \infty$ depending on ϵ such that the following holds. Suppose that M is a non-compact 3-manifold of positive sectional curvature.*

Suppose that N is an ϵ-neck in M centered at a point x and disjoint from a soul p of M. Then for any ϵ-neck N' that is separated from p by N with center x' we have $R(x') \leq CR(x)$.

7. Forward difference quotients

Let us review quickly some standard material on forward difference quotients.

Let $f\colon [a,b] \to \mathbb{R}$ be a continuous function on an interval. We say that the *forward difference quotient of f at a point $t \in [a,b)$*, denoted $\frac{df}{dt}(t)$, is *less than c* provided that

$$\overline{\lim}_{\Delta t \to 0^+} \frac{f(t + \Delta t) - f(t)}{\Delta t} \leq c.$$

We say that it is greater than or equal to c' if

$$c' \leq \underline{\lim}_{\Delta t \to 0^+} \frac{f(t + \Delta t) - f(t)}{\Delta t}.$$

Standard comparison arguments show:

LEMMA 2.22. *Suppose that $f\colon [a,b] \to \mathbb{R}$ is a continuous function. Suppose that ψ is a C^1-function on $[a,b] \times \mathbb{R}$ and suppose that $\frac{df}{dt}(t) \leq \psi(t,f(t))$ for every $t \in [a,b)$ in the sense of forward difference quotients. Suppose also that there is a function $G(t)$ defined on $[a,b]$ that satisfies the differential equation $G'(t) = \psi(t, G(t))$ and has $f(a) \leq G(a)$. Then $f(t) \leq G(t)$ for all $t \in [a,b]$.*

The application we shall make of these results is the following.

PROPOSITION 2.23. *Let M be a smooth manifold with a smooth vector field χ and a smooth function $\mathbf{t}\colon M \to [a,b]$ with $\chi(\mathbf{t}) = 1$. Suppose also that $F\colon M \to \mathbb{R}$ is a smooth function with the properties:*

(1) *for each $t_0 \in [a,b]$ the restriction of F to the level set $\mathbf{t}^{-1}(t_0)$ achieves its maximum, and*
(2) *the subset $\mathcal{Z}$ of M consisting of all x for which $F(x) \geq F(y)$ for all $y \in \mathbf{t}^{-1}(\mathbf{t}(x))$ is a compact set.*

Suppose also that at each $x \in \mathcal{Z}$ we have $\chi(F(x)) \leq \psi(\mathbf{t}(x), F(x))$. Set $F_{\max}(t) = \max_{x \in \mathbf{t}^{-1}(t)} F(x)$. Then $F_{\max}(t)$ is a continuous function and

$$\frac{dF_{\max}}{dt}(t) \leq \psi(t, F_{\max}(t))$$

in the sense of forward difference quotients. Suppose that $G(t)$ satisfies the differential equation

$$G'(t) = \psi(t, G(t))$$

and has initial condition $F_{\max}(a) \leq G(a)$. Then for all $t \in [a,b]$ we have

$$F_{\max}(t) \leq G(t).$$

PROOF. Under the given hypothesis it is a standard and easy exercise to establish the statement about the forward difference quotient of $F_{\max}$. The second statement then is an immediate corollary of the previous result. □

CHAPTER 3

Basics of Ricci flow

In this chapter we introduce the Ricci flow equation due to R. Hamilton [29]. For the basic material on the Ricci flow equation see [13].

1. The definition of Ricci flow

DEFINITION 3.1. The *Ricci flow equation* is the following evolution equation for a Riemannian metric:

$$\frac{\partial g}{\partial t} = -2\text{Ric}(g). \tag{3.1}$$

A solution to this equation (or a *Ricci flow*) is a one-parameter family of metrics $g(t)$, parameterized by t in a non-degenerate interval I, on a smooth manifold M satisfying Equation (3.1). If I has an initial point t_0, then $(M, g(t_0))$ is called *the initial condition of* or *the initial metric for* the Ricci flow (or of the solution).

Let us give a quick indication of what the Ricci flow equation means. In harmonic coordinates $(x^1, \ldots, x^n)$ about p, that is to say coordinates where $\triangle x^i = 0$ for all i, we have

$$\text{Ric}_{ij} = \text{Ric}(\frac{\partial}{\partial x^i}, \frac{\partial}{\partial x^j}) = -\frac{1}{2}\triangle g_{ij} + Q_{ij}(g^{-1}, \partial g)$$

where Q is a quadratic form in g^{-1} and ∂g, and so in particular is a lower order term in the derivatives of g. See Lemma 3.32 on page 92 of [13]. So, in these coordinates, the Ricci flow equation is actually a heat equation for the Riemannian metric

$$\frac{\partial}{\partial t} g = \triangle g + 2Q(g^{-1}, \partial g).$$

DEFINITION 3.2. We introduce some notation that will be used throughout. Given a Ricci flow $(M^n, g(t))$ defined for t contained in an interval I, then the *space-time* for this flow is $M \times I$. The t *time-slice* of space-time is the Riemannian manifold $M \times \{t\}$ with the Riemannian metric $g(t)$. Let $\mathcal{H}T(M \times I)$ be the *horizontal tangent bundle* of space-time, i.e., the bundle of tangent vectors to the time-slices. It is a smooth, rank-n subbundle of the tangent bundle of space-time. The evolving metric $g(t)$ is then a smooth section of $\text{Sym}^2 \mathcal{H}T^*(M \times I)$. We denote points of space-time as pairs (p, t).

Given (p,t) and any $r > 0$ we denote by $B(p,t,r)$ the metric ball of radius r centered at (p,t) in the t time-slice. For any $\Delta t > 0$ for which $[t - \Delta t, t] \subset I$, we define the *backwards parabolic neighborhood* $P(x,t,r,-\Delta t)$ to be the product $B(x,t,r) \times [t - \Delta t, t]$ in space-time. Notice that the intersection of $P(x,t,r,-\Delta t)$ with a time-slice other than the t time-slice need not be a metric ball in that time-slice. There is the corresponding notion of a forward parabolic neighborhood $P(x,t,r,\Delta t)$ provided that $[t, t+\Delta t] \subset I$.

2. Some exact solutions to the Ricci flow

2.1. Einstein manifolds. Let g_0 be an Einstein metric: $\mathrm{Ric}(g_0) = \lambda g_0$, where λ is a constant. Then for any positive constant c, setting $g = cg_0$ we have $\mathrm{Ric}(g) = \mathrm{Ric}(g_0) = \lambda g_0 = \frac{\lambda}{c} g$. Using this we can construct solutions to the Ricci flow equation as follows. Consider $g(t) = u(t)g_0$. If this one-parameter family of metrics is a solution of the Ricci flow, then

$$\frac{\partial g}{\partial t} = u'(t) g_0$$
$$= -2\mathrm{Ric}(u(t)g_0)$$
$$= -2\mathrm{Ric}(g_0)$$
$$= -2\lambda g_0.$$

So $u'(t) = -2\lambda$, and hence $u(t) = 1 - 2\lambda t$. Thus $g(t) = (1 - 2\lambda t)g_0$ is a solution of the Ricci flow. The cases $\lambda > 0, \lambda = 0$, and $\lambda < 0$ correspond to *shrinking*, *steady* and *expanding* solutions. Notice that in the shrinking case the solution exists for $t \in [0, \frac{1}{2\lambda})$ and goes singular at $t = \frac{1}{2\lambda}$.

EXAMPLE 3.3. The standard metric on each of $S^n, \mathbb{R}^n$, and $\mathbb{H}^n$ is Einstein. Ricci flow is contracting on S^n, constant on $\mathbb{R}^n$, and expanding on $\mathbb{H}^n$. The Ricci flow on S^n has a finite-time singularity where the diameter of the manifold goes to zero and the curvature goes uniformly to $+\infty$. The Ricci flow on $\mathbb{H}^n$ exists for all $t \geq 0$ and as t goes to infinity the distance between any pair of points grows without bound and the curvature goes uniformly to zero.

EXAMPLE 3.4. $\mathbb{C}P^n$ equipped with the Fubini-Study metric, which is induced from the standard metric of S^{2n+1} under the Hopf fibration with the fibers of great circles, is Einstein.

EXAMPLE 3.5. Let h_0 be the round metric on S^2 with constant Gaussian curvature $1/2$. Set $h(t) = (1-t)h_0$. Then the flow

$$(S^2, h(t)), \quad -\infty < t < 1,$$

is a Ricci flow. We also have the product of this flow with the trivial flow on the line: $(S^2 \times \mathbb{R}, h(t) \times ds^2)$, $-\infty < t < 1$. This is called the *standard shrinking round cylinder*.

The standard shrinking round cylinder is a model for evolving ϵ-necks. In Chapter 1 we introduced the notion of an ϵ-neck. In the case of flows, in order to take smooth geometric limits, it is important to have a stronger version of this notion. In this stronger notion, the neck not only exists in one time-slice but it exists backwards in the flow for an appropriate amount of time and is close to the standard shrinking round cylinder on the entire time interval. The existence of evolving necks is exploited when we study limits of Ricci flows.

DEFINITION 3.6. Let $(M, g(t))$ be a Ricci flow. An *evolving ϵ-neck centered at* (x, t_0) *and defined for rescaled time* t_1 is an ϵ-neck

$$\varphi \colon S^2 \times (-\epsilon^{-1}, \epsilon^{-1}) \xrightarrow{\cong} N \subset (M, g(t))$$

centered at (x, t_0) with the property that pull-back via φ of the family of metrics $R(x, t_0)g(t')|_N$, $-t_1 < t' \le 0$, where $t_1 = R(x, t_0)^{-1}(t - t_0)$, is within ϵ in the $C^{[1/\epsilon]}$-topology of the product of the standard metric on the interval with evolving round metric on S^2 with scalar curvature $1/(1-t')$ at time t'. A *strong ϵ-neck centered at* (x, t_0) in a Ricci flow is an evolving ϵ-neck centered at (x, t_0) and defined for rescaled time 1, see FIG. 1.

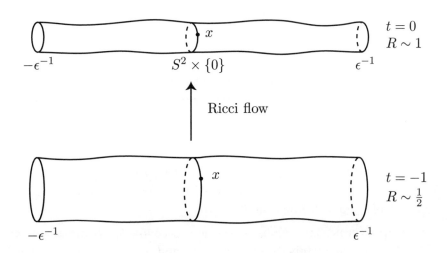

FIGURE 1. Strong ϵ-neck of scale 1.

2.2. Solitons. A *Ricci soliton* is a Ricci flow $(M, g(t))$, $0 \le t < T \le \infty$, with the property that for each $t \in [0, T)$ there is a diffeomorphism $\varphi_t \colon M \to M$ and a constant $\sigma(t)$ such that $\sigma(t)\varphi_t^* g(0) = g(t)$. That is to say, in a Ricci soliton all the Riemannian manifolds $(M, g(t))$ are isometric up to a scale factor that is allowed to vary with t. The soliton is said to be *shrinking* if $\sigma'(t) < 0$ for all t. One way to generate Ricci solitons is the following: Suppose that we have a vector field X on M and a constant λ and a metric $g(0)$ such that

$$\text{(3.2)} \qquad -\text{Ric}(g(0)) = \frac{1}{2}\mathcal{L}_X g(0) - \lambda g(0).$$

We set $T = \infty$ if $\lambda \leq 0$ and equal to $(2\lambda)^{-1}$ if $\lambda > 0$. Then, for all $t \in [0, T)$ we define a function

$$\sigma(t) = 1 - 2\lambda t,$$

and a vector field

$$Y_t(x) = \frac{X(x)}{\sigma(t)}.$$

Then we define φ_t as the one-parameter family of diffeomorphisms generated by the time-dependent vector fields Y_t.

CLAIM 3.7. *The flow $(M, g(t))$, $0 \leq t < T$, where $g(t) = \sigma(t)\varphi_t^* g(0)$, is a soliton. It is a shrinking soliton if $\lambda > 0$.*

PROOF. We check that this flow satisfies the Ricci flow equation; from that, the result follows immediately. We have

$$\begin{aligned}
\frac{\partial g(t)}{\partial t} &= \sigma'(t)\varphi_t^* g(0) + \sigma(t)\varphi_t^* \mathcal{L}_{Y(t)} g(0) \\
&= \varphi_t^*(-2\lambda + \mathcal{L}_X)g(0) \\
&= \varphi_t^*(-2\text{Ric}(g(0))) = -2\text{Ric}(\varphi_t^*(g(0))).
\end{aligned}$$

Since $\text{Ric}(\alpha g) = \text{Ric}(g)$ for any $\alpha > 0$, it follows that

$$\frac{\partial g(t)}{\partial t} = -2\text{Ric}(g(t)).$$

$\square$

There is one class of shrinking solitons which are of special importance to us. These are the gradient shrinking solitons.

DEFINITION 3.8. A shrinking soliton $(M, g(t))$, $0 \leq t < T$, is said to be a *gradient shrinking soliton* if the vector field X in Equation (3.2) is the gradient of a smooth function f on M.

PROPOSITION 3.9. *Suppose we have a complete Riemannian manifold $(M, g(0))$, a smooth function $f \colon M \to \mathbb{R}$, and a constant $\lambda > 0$ such that*

$$\text{(3.3)} \qquad -\text{Ric}(g(0)) = \text{Hess}(f) - \lambda g(0).$$

Then there is $T > 0$ and a gradient shrinking soliton $(M, g(t))$ defined for $0 \leq t < T$.

PROOF. Since

$$\mathcal{L}_{\nabla f} g(0) = 2\text{Hess}(f),$$

Equation (3.3) is the soliton equation, Equation (3.2), with the vector field X being the gradient vector field ∇f. It is a shrinking soliton by assumption since $\lambda > 0$. $\square$

DEFINITION 3.10. In this case we say that $(M, g(0))$ and $f\colon M \to \mathbb{R}$ *generate* a gradient shrinking soliton.

3. Local existence and uniqueness

The following is the first basic result in the theory – local existence and uniqueness for Ricci flow in the case of compact manifolds.

THEOREM 3.11. **(Hamilton, cf. [29].)** *Let (M, g_0) be a compact Riemannian manifold of dimension n. Then there is a $T > 0$ depending on (M, g_0) and a Ricci flow $(M, g(t))$, $0 \le t < T$, with $g(0) = g_0$. Furthermore, if we have Ricci flows with initial conditions (M, g_0) at time 0 defined respectively on time intervals I and I', then these flows agree on $I \cap I'$.*

We remark that the Ricci flow is a weakly parabolic system where degeneracy comes from the gauge invariance of the equation under diffeomorphisms. Therefore the short-time existence does not come from general theory. R. Hamilton's original proof of the short-time existence was involved and used the Nash-Moser inverse function theorem, [28]. Soon after, DeTurck [16] substantially simplified the short-time existence proof by breaking the diffeomorphism invariance of the equation. For the reader's convenience, and also because in establishing the uniqueness for Ricci flows from the standard solution in Section 6 of Chapter 12 we use a version of this idea in the non-compact case, we sketch DeTurck's argument.

PROOF. Let's sketch the proof due to DeTurck [16], cf, Section 3 of Chapter 3 starting on page 78 of [13] for more details. First, we compute the first variation at a Riemannian metric g of minus twice the Ricci curvature tensor in the direction h:

$$\delta_g(-2\mathrm{Ric})(h) = \triangle h - \mathrm{Sym}(\nabla V) + S$$

where:

(1) V is the one-form given by

$$V_k = \frac{1}{2} g^{pq}(\nabla_p h_{qk} + \nabla_q h_{pk} - \nabla_k h_{pq}),$$

(2) $\mathrm{Sym}(\nabla V)$ is the symmetric two-tensor obtained by symmetrizing the covariant derivative of V, and
(3) S is a symmetric two-tensor constructed from the inverse of the metric, the Riemann curvature tensor and h, but involves no derivatives of h.

Now let g_0 be the initial metric. For any metric g we define a one-form $\hat{W}$ by taking the trace, with respect to g, of the matrix-valued one-form that is the difference of the connections of g and g_0. Now we form a second-order operator of g by setting

$$P(g) = \mathcal{L}_{\hat{W}} g,$$

the Lie derivative of g with respect to the vector field W dual to $\hat{W}$. Thus, in local coordinates we have $P(g)_{ij} = \nabla_i \hat{W}_j + \nabla_j \hat{W}_i$. The linearization at g of the second-order operator P in the direction h is symmetric and is given by

$$\delta_g P(h) = \mathrm{Sym}(\nabla V) + T$$

where T is a first-order operator in h. Thus, defining $Q = -2\mathrm{Ric} + P$ we have

$$\delta_g(Q)(h) = \triangle h + U$$

where U is a first-order operator in h. Now we introduce the Ricci-DeTurck flow

$$(3.4) \qquad \frac{\partial g}{\partial t} = -2\mathrm{Ric}(g) + P.$$

The computations above show that the Ricci-DeTurck flow is strictly parabolic. Thus, Equation (3.4) has a short-time solution $\overline{g}(t)$ with $\overline{g}(0) = g_0$ by the standard PDE theory. Given this solution $\overline{g}(t)$ we define the time-dependent vector field $W(t) = W(\overline{g}(t), g_0)$ as above. Let ϕ_t be a one-parameter family of diffeomorphisms, with $\phi_0 = \mathrm{Id}$, generated by this time-dependent vector field, i.e.,

$$\frac{\partial \phi_t}{\partial t} = W(t).$$

Then, direct computation shows that $g(t) = \phi_t^* \overline{g}(t)$ solves the Ricci flow equation. □

In performing surgery at time T, we will have an open submanifold Ω of the compact manifold with the following property. As t approaches T from below, the metrics $g(t)|_\Omega$ converge smoothly to a limiting metric $g(T)$ on Ω. We will 'cut away' the rest of the manifold $M \setminus \Omega$ where the metrics are not converging and glue in a piece E coming from the standard solution to form a new compact manifold M'. Then we extend the Riemannian metric $g(T)$ on Ω to one defined on $M' = \Omega \cup E$. The resulting Riemannian manifold forms the initial manifold at time T for continuing the Ricci flow $\widetilde{g}(t)$ on an interval $T \leq t < T'$. It is important to know that the two Ricci flows $(\Omega, g(t))$, $t \leq T$ and $(\Omega, \widetilde{g}(t))$, $T \leq t < T'$ glue together to make a smooth solution spanning across the surgery time T. That this is true is a consequence of the following elementary result.

PROPOSITION 3.12. *Suppose that $(U, g(t))$, $a \leq t < b$, is a Ricci flow and suppose that there is a Riemannian metric $g(b)$ on U such that as $t \to b$ the metrics $g(t)$ converge in the C^∞-topology, uniformly on compact subsets, to $g(b)$. Suppose also that $(U, g(t))$, $b \leq t < c$, is a Ricci flow. Then the one-parameter family of metrics $g(t)$, $a \leq t < c$, is a C^∞-family and is a solution to the Ricci flow equation on the entire interval $[a, c)$.*

4. Evolution of curvatures

Let us fix a set $(x^1, \ldots, x^n)$ of local coordinates. The Ricci flow equation, written in local coordinates as

$$\frac{\partial g_{ij}}{\partial t} = -2\mathrm{Ric}_{ij},$$

implies a heat equation for the Riemann curvature tensor R_{ijkl} which we now derive. Various second-order derivatives of the curvature tensor are likely to differ by terms quadratic in the curvature tensors. To this end we introduce the tensor

$$B_{ijkl} = g^{pr}g^{qs}R_{ipjq}R_{krls}.$$

Note that we have the obvious symmetries

$$B_{ijkl} = B_{jilk} = B_{klij},$$

but the other symmetries of the curvature tensor R_{ijkl} may fail to hold for B_{ijkl}.

THEOREM 3.13. *The curvature tensor R_{ijkl}, the Ricci curvature Ric_{ij}, the scalar curvature R, and the volume form $\mathrm{dvol}(x,t)$ satisfy the following evolution equations under Ricci flow:*

$$\frac{\partial R_{ijkl}}{\partial t} = \triangle R_{ijkl} + 2(B_{ijkl} - B_{ijlk} - B_{iljk} + B_{ikjl})$$
(3.5)
$$\quad - g^{pq}(R_{pjkl}\mathrm{Ric}_{qi} + R_{ipkl}\mathrm{Ric}_{qj} + R_{ijpl}\mathrm{Ric}_{qk} + R_{ijkp}\mathrm{Ric}_{ql}),$$

(3.6) $\quad \dfrac{\partial}{\partial t}\mathrm{Ric}_{jk} = \triangle \mathrm{Ric}_{jk} + 2g^{pq}g^{rs}R_{pjkr}\mathrm{Ric}_{qs} - 2g^{pq}\mathrm{Ric}_{jp}\mathrm{Ric}_{qk},$

(3.7) $\quad \dfrac{\partial}{\partial t}R = \triangle R + 2|\mathrm{Ric}|^2,$

(3.8)
$$\frac{\partial}{\partial t}\mathrm{dvol}(x,t) = -R(x,t)\mathrm{dvol}(x,t).$$

These equations are contained in Lemma 6.15 on page 179, Lemma 6.9 on page 176, Lemma 6.7 on page 176, and Equation (6.5) on page 175 of [13], respectively.

Let us derive some consequences of these evolution equations. The first result is obvious from the Ricci flow equation and will be used implicitly throughout the paper.

LEMMA 3.14. *Suppose that $(M, g(t))$, $a < t < b$ is a Ricci flow of non-negative Ricci curvature with M a connected manifold. Then for any points $x, y \in M$ the function $d_{g(t)}(x,y)$ is a non-increasing function of t.*

PROOF. The Ricci flow equation tells us that non-negative Ricci curvature implies that $\partial g/\partial t \leq 0$. Hence, the length of any tangent vector in M, and consequently the length of any path in M, is a non-increasing function of

t. Since the distance between points is the infimum over all rectifiable paths from x to y of the length of the path, this function is also a non-increasing function of t. □

LEMMA 3.15. *Suppose that $(M, g(t))$, $0 \leq t \leq T$, is a Ricci flow and $|\mathrm{Rm}(x,t)| \leq K$ for all $x \in M$ and all $t \in [0, T]$. Then there are constants A, A' depending on K, T and the dimension such that:*

(1) *For any non-zero tangent vector $v \in T_x M$ and any $t \leq T$ we have*
$$A^{-1}\langle v, v\rangle_{g(0)} \leq \langle v, v\rangle_{g(t)} \leq A\langle v, v\rangle_{g(0)}.$$

(2) *For any open subset $U \subset M$ and any $t \leq T$ we have*
$$(A')^{-1}\mathrm{Vol}_0(U) \leq \mathrm{Vol}_t(U) \leq A'\mathrm{Vol}_0(U).$$

PROOF. The Ricci flow equation yields
$$\frac{d}{dt}\left(\langle v, v\rangle_{g(t)}\right) = -2\mathrm{Ric}(v, v).$$

The bound on the Riemann curvature gives a bound on Ric. Integrating yields the result. The second statement is proved analogously using Equation (3.8). □

5. Curvature evolution in an evolving orthonormal frame

It is often best to study the evolution of the representative of the tensor in an orthonormal frame F. Let $(M, g(t))$, $0 \leq t < T$, be a Ricci flow, and suppose that $\mathcal{F}$ is a frame on an open subset $U \subset M$ consisting of vector fields $\{F_1, F_2, \ldots, F_n\}$ on U that are $g(0)$-orthonormal at every point. Since the metric evolves by the Ricci flow, to keep the frame orthonormal we must evolve it by an equation involving Ricci curvature. We evolve this local frame according to the formula

(3.9) $$\frac{\partial F_a}{\partial t} = \mathrm{Ric}(F_a, \cdot)^*,$$

i.e., assuming that in local coordinates $(x^1, \ldots, x^n)$, we have
$$F_a = F_a^i \frac{\partial}{\partial x_i},$$

then the evolution equation is
$$\frac{\partial F_a^i}{\partial t} = g^{ij}\mathrm{Ric}_{jk}F_a^k.$$

Since this is a linear ODE, there are unique solutions for all times $t \in [0, T)$.

The next remark to make is that this frame remains orthonormal:

CLAIM 3.16. *Suppose that $\mathcal{F}(0) = \{F_a\}_a$ is a local $g(0)$-orthonormal frame, and suppose that $\mathcal{F}(t)$ evolves according to Equation (3.9). Then for all $t \in [0, T)$ the frame $\mathcal{F}(t)$ is a local $g(t)$-orthonormal frame.*

5. CURVATURE EVOLUTION IN AN EVOLVING ORTHONORMAL FRAME

PROOF.
$$\frac{\partial}{\partial t}\langle F_a(t), F_b(t)\rangle_{g(t)} = \langle \frac{\partial F_a}{\partial t}, F_b\rangle + \langle F_b, \frac{\partial F_b}{\partial t}\rangle + \frac{\partial g}{\partial t}(F_a, F_b)$$
$$= \mathrm{Ric}(F_a, F_b) + \mathrm{Ric}(F_b, F_a) - 2\mathrm{Ric}(F_a, F_b) = 0.$$

□

Notice that if $\mathcal{F}'(0) = \{F'_a\}_a$ is another frame related to $\mathcal{F}(0)$ by, say,
$$F'_a = A^b_a F_b,$$
then
$$F_a(t) = A^b_a F_b(t).$$
This means that the evolution of frames actually defines a bundle automorphism
$$\Phi\colon TM|_U \times [0, T) \to TM|_U \times [0, T)$$
covering the identity map of $U \times [0, T)$ which is independent of the choice of initial frame and is the identity at time $t = 0$. Of course, since the resulting bundle automorphism is independent of the initial frame, it globalizes to produce a bundle isomorphism
$$\Phi\colon TM \times [0, T) \to TM \times [0, T)$$
covering the identity on $M \times [0, T)$. We view this as an evolving identification Φ_t of TM with itself. This identification is the identity at $t = 0$. The content of Claim 3.16 is:

COROLLARY 3.17.
$$\Phi^*_t(g(t)) = g(0).$$

Returning to the local situation of the orthonormal frame $\mathcal{F}$, we set $\mathcal{F}^* = \{F^1, \ldots, F^n\}$ equal to the dual coframe to $\{F_1, \ldots, F_n\}$. In this coframe the Riemann curvature tensor is given by $R_{abcd}F^a F^b F^c F^d$ where

(3.10)
$$R_{abcd} = R_{ijkl}F^i_a F^j_b F^k_c F^l_d.$$

One advantage of working in the evolving frame is that the evolution equation for the Riemann curvature tensor simplifies:

LEMMA 3.18. *Suppose that the orthonormal frame $\mathcal{F}(t)$ evolves by Formula (3.9). Then we have the evolution equation*
$$\frac{\partial R_{abcd}}{\partial t} = \Delta R_{abcd} + 2(B_{abcd} + B_{acbd} - B_{abdc} - B_{adbc}),$$
where $B_{abcd} = \sum_{e,f} R_{aebf} R_{cedf}$.

PROOF. For a proof see Theorem 2.1 in [**32**]. □

Of course, the other way to describe all of this is to consider the four-tensor $\Phi_t^*(\mathcal{R}_{g(t)}) = R_{abcd}F^a F^b F^c F^d$ on M. Since Φ_t is a bundle map but not a bundle map induced by a diffeomorphism, even though the pullback of the metric $\Phi_t^* g(t)$ is constant, it is not the case that the pullback of the curvature $\Phi_t^* \mathcal{R}_{g(t)}$ is constant. The next proposition gives the evolution equation for the pullback of the Riemann curvature tensor.

It simplifies the notation somewhat to work directly with a basis of $\wedge^2 TM$. We chose an orthonormal basis

$$\{\varphi^1, \ldots, \varphi^{\frac{n(n-1)}{2}}\},$$

of $\wedge^2 T_p^* M$ where we have

$$\varphi^\alpha(F_a, F_b) = \varphi^\alpha_{ab}$$

and write the curvature tensor in this basis as $\mathcal{T} = (\mathcal{T}_{\alpha\beta})$ so that

(3.11) $$R_{abcd} = \mathcal{T}_{\alpha\beta} \varphi^\alpha_{ab} \varphi^\beta_{cd}.$$

PROPOSITION 3.19. *The evolution of the curvature operator* $\mathcal{T}(t) = \Phi_t^* \mathrm{Rm}(g(t))$ *is given by*

$$\frac{\partial \mathcal{T}_{\alpha\beta}}{\partial t} = \triangle \mathcal{T}_{\alpha\beta} + \mathcal{T}^2_{\alpha\beta} + \mathcal{T}^\sharp_{\alpha\beta},$$

where $\mathcal{T}^2_{\alpha\beta} = \mathcal{T}_{\alpha\gamma}\mathcal{T}_{\gamma\beta}$ *is the operator square;* $\mathcal{T}^\sharp_{\alpha\beta} = c_{\alpha\gamma\zeta} c_{\beta\delta\eta} \mathcal{T}_{\gamma\delta}\mathcal{T}_{\zeta\eta}$ *is the Lie algebra square; and* $c_{\alpha\beta\gamma} = \langle [\varphi^\alpha, \varphi^\beta], \varphi^\gamma \rangle$ *are the structure constants of the Lie algebra* $\mathrm{so}(n)$ *relative to the basis* $\{\varphi^\alpha\}$. *The structure constants* $c_{\alpha\beta\gamma}$ *are fully antisymmetric in the three indices.*

PROOF. We work in local coordinates that are orthonormal at the point. By the first Bianchi identity

$$R_{abcd} + R_{acdb} + R_{adbc} = 0,$$

we get

$$\sum_{e,f} R_{abef} R_{cdef} = \sum_{e,f} (-R_{aefb} - R_{afbe})(-R_{cefd} - R_{cfde})$$

$$= \sum_{e,f} 2R_{aebf} R_{cedf} - 2R_{aebf} R_{cfde}$$

$$= 2(B_{abcd} - B_{adbc}).$$

Note that

$$\sum_{e,f} R_{abef} R_{cdef} = \sum_{e,f} \mathcal{T}_{\alpha\beta} \varphi^\alpha_{ab} \varphi^\beta_{ef} \mathcal{T}_{\gamma\lambda} \varphi^\gamma_{cd} \varphi^\lambda_{ef}$$

$$= \mathcal{T}_{\alpha\beta} \varphi^\alpha_{ab} \mathcal{T}_{\gamma\lambda} \varphi^\gamma_{cd} \delta^{\beta\lambda}$$

$$= \mathcal{T}^2_{\alpha\beta} \varphi^\alpha_{ab} \varphi^\beta_{cd}.$$

Also,
$$2(B_{acbd} - B_{adbc}) = 2\sum_{e,f}(R_{aecf}R_{bedf} - R_{aedf}R_{becf})$$
$$= 2\sum_{e,f}(\mathcal{T}_{\alpha\beta}\varphi^\alpha_{ae}\varphi^\beta_{cf}\mathcal{T}_{\gamma\lambda}\varphi^\gamma_{be}\varphi^\lambda_{df} - \mathcal{T}_{\alpha\beta}\varphi^\alpha_{ae}\varphi^\beta_{df}\mathcal{T}_{\gamma\lambda}\varphi^\gamma_{be}\varphi^\lambda_{cf})$$
$$= 2\sum_{e,f}\mathcal{T}_{\alpha\beta}\mathcal{T}_{\gamma\lambda}\varphi^\alpha_{ae}\varphi^\gamma_{be}(\varphi^\beta_{cf}\varphi^\lambda_{df} - \varphi^\beta_{df}\varphi^\lambda_{cf})$$
$$= 2\sum_{e}\mathcal{T}_{\alpha\beta}\mathcal{T}_{\gamma\lambda}\varphi^\alpha_{ae}\varphi^\gamma_{be}[\varphi^\beta,\varphi^\lambda]_{cd}$$
$$= \sum_{e}\mathcal{T}_{\alpha\beta}\mathcal{T}_{\gamma\lambda}[\varphi^\beta,\varphi^\lambda]_{cd}(\varphi^\alpha_{ae}\varphi^\gamma_{be} - \varphi^\alpha_{be}\varphi^\gamma_{ae})$$
$$= \mathcal{T}_{\alpha\beta}\mathcal{T}_{\gamma\delta}[\varphi^\beta,\varphi^\lambda]_{cd}[\varphi^\alpha,\varphi^\gamma]_{ab}$$
$$= \mathcal{T}^\sharp_{\alpha\beta}\varphi^\alpha_{ab}\varphi^\beta_{cd}.$$

So we can rewrite the equation for the evolution of the curvature tensor given in Lemma 3.18 as
$$\frac{\partial R_{abcd}}{\partial t} = \triangle R_{abcd} + \mathcal{T}^2_{\alpha\beta}\varphi^\alpha_{ab}\varphi^\beta_{cd} + \mathcal{T}^\sharp_{\alpha\beta}\varphi^\alpha_{ab}\varphi^\beta_{cd},$$
or equivalently as
$$\frac{\partial \mathcal{T}_{\alpha\beta}}{\partial t} = \triangle \mathcal{T}_{\alpha\beta} + \mathcal{T}^2_{\alpha\beta} + \mathcal{T}^\sharp_{\alpha\beta}.$$
We abbreviate the last equation as
$$\frac{\partial \mathcal{T}}{\partial t} = \triangle \mathcal{T} + \mathcal{T}^2 + \mathcal{T}^\sharp.$$

□

REMARK 3.20. Notice that neither $\mathcal{T}^2$ nor $\mathcal{T}^\sharp$ satisfies the Bianchi identity, but their sum does.

6. Variation of distance under Ricci flow

There is one result that we will use several times in the arguments to follow. Since it is an elementary result (though the proof is somewhat involved), we have chosen to include it here.

PROPOSITION 3.21. *Let $t_0 \in \mathbb{R}$ and let $(M, g(t))$ be a Ricci flow defined for t in an interval containing t_0 with $(M, g(t))$ complete for every t in this interval. Fix a constant $K < \infty$. Let x_0, x_1 be two points of M and let $r_0 > 0$ such that $d_{t_0}(x_0, x_1) \geq 2r_0$. Suppose that $\mathrm{Ric}(x, t_0) \leq (n-1)K$ for all $x \in B(x_0, r_0, t_0) \cup B(x_1, r_0, t_0)$. Then*
$$\left.\frac{d(d_t(x_0, x_1))}{dt}\right|_{t=t_0} \geq -2(n-1)\left(\frac{2}{3}Kr_0 + r_0^{-1}\right).$$

If the distance function $d_t(x_0, x_1)$ is not a differentiable function of t at $t = t_0$, then this inequality is understood as an inequality for the forward difference quotient.

REMARK 3.22. Of course, if the distance function is differentiable at $t = t_0$, then the derivative statement is equivalent to the forward difference quotient statement. Thus, in the proof of this result we shall always work with the forward difference quotients.

PROOF. The first step in the proof is to replace the distance function by the length of minimal geodesics. The following is standard.

CLAIM 3.23. *Suppose that for every minimal $g(t_0)$-geodesic γ from x_0 to x_1 the function $\ell_t(\gamma)$ which is the $g(t)$-length of γ satisfies*

$$\frac{d(\ell_t(\gamma))}{dt}\bigg|_{t=t_0} \geq C.$$

Then

$$\frac{d(d_t(x_0, x_1))}{dt}\bigg|_{t=t_0} \geq C,$$

where, as in the statement of the proposition, if the distance function is not differentiable at t_0, then the inequality in the conclusion is interpreted by replacing the derivative on the left-hand side with the liminf *of the forward difference quotients of $d_t(x_0, x_1)$ at t_0.*

The second step in the proof is to estimate the time derivative of a minimal geodesic under the hypothesis of the proposition.

CLAIM 3.24. *Assuming the hypothesis of the proposition, for any minimal $g(t_0)$-geodesic γ from x_0 to x_1, we have*

$$\frac{d(\ell_t(\gamma))}{dt}\bigg|_{t=t_0} \geq -2(n-1)\left(\frac{2}{3}Kr_0 + r_0^{-1}\right).$$

PROOF. Fix a minimal $g(t_0)$-geodesic $\gamma(u)$ from x_0 to x_1, parameterized by arc length. We set $d = d_{t_0}(x_0, x_1)$, we set $X(u) = \gamma'(u)$, and we take tangent vectors $Y_1, \ldots, Y_{n-1}$ in $T_{x_0}M$ which together with $X(0) = \gamma'(0)$ form an orthonormal basis. We let $Y_i(u)$ be the parallel translation of Y_i along γ. Define $f: [0, d] \to [0, 1]$ by:

$$f(u) = \begin{cases} u/r_0, & 0 \leq u \leq r_0, \\ 1, & r_0 \leq u \leq d - r_0, \\ (d-u)/r_0, & d - u \leq r_0 \leq d, \end{cases}$$

and define

$$\widetilde{Y}_i(u) = f(u)Y_i(u).$$

See FIG. 2. For $1 \leq i \leq n-1$, let $s''_{\widetilde{Y}_i}(\gamma)$ be the second variation of the $g(t_0)$-length of γ along $\widetilde{Y}_i$. Since γ is a minimal $g(t_0)$-geodesic, for all i we have

(3.12) $$s''_{\widetilde{Y}_i}(\gamma) \geq 0.$$

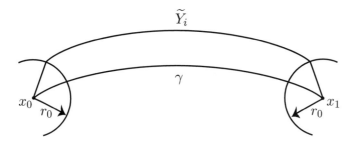

FIGURE 2. $\widetilde{Y}_i$ along γ.

Let us now compute $s''_{\widetilde{Y}_i}(\gamma)$ by taking a two-parameter family $\gamma(u,s)$ such that the curve $\gamma(u,0)$ is the original minimal geodesic and $\frac{\partial}{\partial s}(\gamma(u,s))|_{s=0} = \widetilde{Y}_i(u)$. We denote by $X(u,s)$ the image $D\gamma_{(u,s)}(\partial/\partial u)$ and by $\widetilde{Y}_i(u,s)$ the image $D\gamma_{(u,s)}(\partial/\partial s)$. We wish to compute

(3.13)
$$\begin{aligned}
s''_{\widetilde{Y}_i}(\gamma) &= \frac{d^2}{ds^2}\left(\int_0^d \sqrt{X(u,s), X(u,s)}\, du\right)\Big|_{s=0} \\
&= \frac{d}{ds}\left(\int_0^d \langle X(u,s), X(u,s)\rangle^{-1/2}\langle X(u,s), \nabla_{\widetilde{Y}_i} X(u,s)\rangle du\right)\Big|_{s=0} \\
&= \int_0^d -\langle X(u,0), X(u,0)\rangle^{-3/2}\langle X(u,0), \nabla_{\widetilde{Y}_i} X(u,0)\rangle^2 du \\
&\quad + \int_0^d \frac{\langle \nabla_{\widetilde{Y}_i} X(u,0), \nabla_{\widetilde{Y}_i} X(u,0)\rangle + \langle X(u,0), \nabla_{\widetilde{Y}_i}\nabla_{\widetilde{Y}_i} X(u,0)\rangle}{\langle X(u,0), X(u,0)\rangle^{1/2}} du.
\end{aligned}$$

Using the fact that X and $\widetilde{Y}_i$ commute (since they are the coordinate partial derivatives of a map of a surface into M) and using the fact that $Y_i(u)$ is parallel along γ, meaning that $\nabla_X(Y_i)(u) = 0$, we see that $\nabla_{\widetilde{Y}_i} X(u,0) = \nabla_X \widetilde{Y}_i(u,0) = f'(u) Y_i(u)$. By construction $\langle Y_i(u), X(u,0)\rangle = 0$. It follows that

$$\langle \nabla_{\widetilde{Y}_i} X(u,0), X(u,0)\rangle = \langle \nabla_X(\widetilde{Y}_i)(u,0), X(u,0)\rangle = \langle f'(u)Y_i(u), X(u,0)\rangle = 0.$$

Also, $\langle X(u,0), X(u,0)\rangle = 1$, and by construction $\langle Y_i(u,0), Y_i(u,0)\rangle = 1$. Thus, restricting to $s=0$ and, for simplicity of notation, leaving the variable

u implicit, Equation (3.13) simplifies to

$$(3.14) \quad s''_{\widetilde{Y}_i}(\gamma) = \int_0^d \left((f')^2 \langle Y_i, Y_i \rangle + \langle \nabla_{\widetilde{Y}_i} \nabla_X \widetilde{Y}_i, X \rangle \right) du$$

$$= \int_0^d \left(\langle R(\widetilde{Y}_i, X)\widetilde{Y}_i, X \rangle - \langle \nabla_X \nabla_{\widetilde{Y}_i} \widetilde{Y}_i, X \rangle + (f')^2 \right) du.$$

We have

$$\langle \nabla_X \nabla_{\widetilde{Y}_i} \widetilde{Y}_i, X \rangle = \frac{d}{du} \langle \nabla_{\widetilde{Y}_i} \widetilde{Y}_i, X \rangle - \langle \nabla_{\widetilde{Y}_i} \widetilde{Y}_i, \nabla_X X \rangle = \frac{d}{du} \langle \nabla_{\widetilde{Y}_i} \widetilde{Y}_i, X \rangle,$$

where the last equality is a consequence of the geodesic equation, $\nabla_X X = 0$. It follows that

$$\int_0^d \langle \nabla_X \nabla_{\widetilde{Y}_i} \widetilde{Y}_i, X \rangle du = \int_0^d \frac{d}{du} \langle \nabla_{\widetilde{Y}_i} \widetilde{Y}_i, X \rangle = 0,$$

where the last equality is a consequence of the fact that $\widetilde{Y}_i$ vanishes at the end points.

Consequently, plugging these into Equation (3.14) we have

$$(3.15) \quad s''_{\widetilde{Y}_i}(\gamma) = \int_0^d \left(\langle R(\widetilde{Y}_i, X)\widetilde{Y}_i(u,0), X(u,0) \rangle + (f'(u))^2 \right) du.$$

Of course, it is immediate from the definition that $f'(u)^2 = 1/r_0^2$ for $u \in [0, r_0]$ and for $u \in [d - r_0, d]$ and is zero otherwise. Also, from the definition of the vector fields Y_i we have

$$\sum_{i=1}^{n-1} \langle R(Y_i, X)Y_i(u), X(u) \rangle = -\text{Ric}_{g(t_0)}(X(u), X(u)),$$

so that

$$\sum_{i=1}^{n-1} \langle R(\widetilde{Y}_i, X)\widetilde{Y}_i(u), X(u) \rangle = -f^2(u)\text{Ric}_{g(t_0)}(X(u), X(u)).$$

Hence, summing Equalities (3.15) for $i = 1, \ldots, n-1$ and using Equation (3.12) gives

$$0 \le \sum_{i=1}^{n-1} s''_{\widetilde{Y}_i}(\gamma) = \int_0^{r_0} \left[\frac{u^2}{r_0^2} \left(-\text{Ric}_{g(t_0)}(X(u), X(u)) \right) + \frac{n-1}{r_0^2} \right] du$$

$$+ \int_{r_0}^{d-r_0} -\text{Ric}_{g(t_0)}(X(u), X(u)) du$$

$$+ \int_{d-r_0}^{d} \left[\frac{(d-u)^2}{r_0^2} \left(-\text{Ric}_{g(t_0)}(X(u), X(u)) \right) + \frac{n-1}{r_0^2} \right] du.$$

Rearranging the terms yields

$$0 \leq -\int_0^d \mathrm{Ric}_{g(t_0)}(X(u), X(u))du$$
$$+ \int_0^{r_0} \left[\left(1 - \frac{u^2}{r_0^2}\right)\left(\mathrm{Ric}_{g(t_0)}(X(u), X(u))\right) + \frac{n-1}{r_0^2}\right] du$$
$$+ \int_{d-r_0}^d \left[\left(1 - \frac{(d-u)^2}{r_0^2}\right)\left(\mathrm{Ric}_{g(t_0)}(X(u), X(u))\right) + \frac{n-1}{r_0^2}\right] du.$$

Since

$$\frac{d(\ell_t(\gamma))}{dt}\bigg|_{t=t_0} = \frac{d}{dt}\left[\left(\int_0^d \sqrt{\langle X(u), X(u)\rangle}dt\right)^{1/2}\right]\bigg|_{t=t_0}$$
$$= -\int_0^d \mathrm{Ric}_{g(t_0)}(X(u), X(u))du,$$

we have

$$\frac{d(\ell_t(\gamma))}{dt}\bigg|_{t=t_0} \geq -\left\{\int_0^{r_0}\left[\left(1 - \frac{u^2}{r_0^2}\right)\left(\mathrm{Ric}_{g(t_0)}(X(u), X(u))\right) + \frac{n-1}{r_0^2}\right] du \right.$$
$$\left. + \int_{d-r_0}^d \left[\left(1 - \frac{(d-u)^2}{r_0^2}\right)\left(\mathrm{Ric}_{g(t_0)}(X(u), X(u))\right) + \frac{n-1}{r_0^2}\right] du\right\}.$$

Now, since $|X(u)| = 1$, by the hypothesis of the proposition we have the estimate $\mathrm{Ric}_{g(t_0)}(X(u), X(u)) \leq (n-1)K$ on the regions of integration on the right-hand side of the above inequality. Thus,

$$\frac{d(\ell_t(\gamma))}{dt}\bigg|_{t=t_0} \geq -2(n-1)\left(\frac{2}{3}r_0 K + r_0^{-1}\right).$$

This completes the proof of Claim 3.24. □

Claims 3.23 and 3.24 together prove the proposition. □

COROLLARY 3.25. *Let $t_0 \in \mathbb{R}$ and let $(M, g(t))$ be a Ricci flow defined for t in an interval containing t_0 and with $(M, g(t))$ complete for every t in this interval. Fix a constant $K < \infty$. Suppose that $\mathrm{Ric}(x, t_0) \leq (n-1)K$ for all $x \in M$. Then for any points $x_0, x_1 \in M$ we have*

$$\frac{d(d_t(x_0, x_1))}{dt}\bigg|_{t=t_0} \geq -4(n-1)\sqrt{\frac{2K}{3}}$$

in the sense of forward difference quotients.

PROOF. There are two cases to consider: (i): $d_{t_0}(x_0, x_1) \geq \sqrt{\frac{6}{K}}$ and (ii): $d_{t_0}(x_0, x_1) < \sqrt{\frac{6}{K}}$. In Case (i) we take $r_0 = \sqrt{3/2K}$ in Proposition 3.21, and we conclude that the liminf at t_0 of the difference quotients for $d_t(x_0, x_1)$ is at most $-4(n-1)\sqrt{\frac{2K}{3}}$. In Case (ii) w let $\gamma(u)$ be any minimal $g(t_0)$-geodesic from x_0 to x_1 parameterized by arc length. Since

$$\frac{d}{dt}(\ell_t(\gamma))|_{t=t_0} = -\int_\gamma \mathrm{Ric}_{g(t_0)}(\gamma'(u),\gamma'(u))du,$$

we see that

$$\frac{d}{dt}(\ell_t(\gamma))|_{t=t_0} \geq -(n-1)K\sqrt{6/K} = -(n-1)\sqrt{6K}.$$

By Claim 3.23, this implies that the liminf of the forward difference quotient of $d_t(x_0, x_1)$ at $t = t_0$ is at least $-(n-1)\sqrt{6K} \geq -4(n-1)\sqrt{2K/3}$. □

COROLLARY 3.26. *Let $(M, g(t))$, $a \leq t \leq b$, be a Ricci flow with $(M, g(t))$ complete for every $t \in [0, T)$. Fix a positive function $K(t)$, and suppose that $\mathrm{Ric}_{g(t)}(x,t) \leq (n-1)K(t)$ for all $x \in M$ and all $t \in [a, b]$. Let x_0, x_1 be two points of M. Then*

$$d_a(x_0, x_1) \leq d_b(x_0, x_1) + 4(n-1)\int_a^b \sqrt{\frac{2K(t)}{3}}dt.$$

PROOF. By Corollary 3.25 we have

$$\frac{d}{dt}d_t(x_0, x_1)|_{t=t'} \geq -4(n-1)\sqrt{\frac{2K(t')}{3}} \tag{3.16}$$

in the sense of forward difference quotients. Thus, this result is an immediate consequence of Lemma 2.22. □

7. Shi's derivative estimates

The last 'elementary' result we discuss is Shi's result controlling all derivatives in terms of a bound on curvature. This is a consequence of the parabolic nature of the Ricci flow equation. More precisely, we can control all derivatives of the curvature tensor at a point $p \in M$ and at a time t provided that we have an upper bound for the curvature on an entire backward parabolic neighborhood of (p, t) in space-time. The estimates become weaker as the parabolic neighborhood shrinks, either in the space direction or the time direction.

Recall that for any $K < \infty$ if (M, g) is a Riemannian manifold with $|\mathrm{Rm}| \leq K$ and if for some $r \leq \pi/\sqrt{K}$ the metric ball $B(p, r)$ has compact closure in M, then the exponential mapping $\exp_p$ is defined on the ball $B(0, r)$ of radius r centered at the origin of T_pM and $\exp_p \colon B(0, r) \to M$ is a local diffeomorphism onto $B(p, r)$.

The first of Shi's derivative estimates controls the first derivative of Rm.

THEOREM 3.27. *There is a constant $C = C(n)$, depending only on the dimension n, such that the following holds for every $K < \infty$, for every $T > 0$, and for every $r > 0$. Suppose that $(U, g(t))$, $0 \leq t \leq T$, is an n-dimensional Ricci flow with $|\mathrm{Rm}(x, t)| \leq K$ for all $x \in U$ and $t \in [0, T]$.*

Suppose that $p \in U$ has the property that $B(p, 0, r)$ has compact closure in U. Then

$$|\nabla \mathrm{Rm}(p,t)| \leq CK \left(\frac{1}{r^2} + \frac{1}{t} + K \right)^{1/2}.$$

For a proof of this result see Chapter 6.2, starting on page 212, of [**14**].

We also need higher derivative estimates. These are also due to Shi, but they take a slightly different form. (See Theorem 6.9 on page 210 of [**14**].)

THEOREM 3.28. (**Shi's Derivative Estimates**) *Fix the dimension n of the Ricci flows under consideration. Let $K < \infty$ and $\alpha > 0$ be positive constants. Then for each non-negative integer k and each $r > 0$ there is a constant $C_k = C_k(K, \alpha, r, n)$ such that the following holds. Let $(U, g(t))$, $0 \leq t \leq T$, be a Ricci flow with $T \leq \alpha/K$. Fix $p \in U$ and suppose that the metric ball $B(p, 0, r)$ has compact closure in U. If*

$$|\mathrm{Rm}(x,t)| \leq K \quad \text{for all} \quad (x,t) \in P(x, 0, r, T),$$

then

$$|\nabla^k (\mathrm{Rm}(y,t))| \leq \frac{C_k}{t^{k/2}}$$

for all $y \in B(p, 0, r/2)$ and all $t \in (0, T]$.

For a proof of this result see Chapter 6.2 of [**14**] where these estimates are proved for the first and second derivatives of Rm. The proofs of the higher derivatives follow similarly. Below, we shall prove a stronger form of this result including the proof for all derivatives.

We shall need a stronger version of this result, a version which is well-known but for which there seems to be no good reference. The stronger version takes as hypothesis C^k-bounds on the initial conditions and produces a better bound on the derivatives of the curvature at later times. The argument is basically the same as that of the result cited above, but since there is no good reference for it we include the proof, which was shown to us by Peng Lu.

THEOREM 3.29. *Fix the dimension n of the Ricci flows under consideration. Let $K < \infty$ and $\alpha > 0$ be given positive constants. Fix an integer $l \geq 0$. Then for each integer $k \geq 0$ and for each $r > 0$ there is a constant $C'_{k,l} = C'_{k,l}(K, \alpha, r, n)$ such that the following holds. Let $(U, g(t))$, $0 \leq t \leq T$, be a Ricci flow with $T \leq \alpha/K$. Fix $p \in U$ and suppose that the metric ball $B(p, 0, r)$ has compact closure in U. Suppose that*

$$|\mathrm{Rm}(x,t)| \leq K \quad \text{for all } x \in U \text{ and all } t \in [0,T],$$

$$\left|\nabla^\beta \mathrm{Rm}(x,0)\right| \leq K \quad \text{for all } x \in U \quad \text{and all} \quad \beta \leq l.$$

Then

$$\left|\nabla^k \mathrm{Rm}(y,t)\right| \leq \frac{C'_{k,l}}{t^{\max\{k-l,0\}/2}}$$

for all $y \in B(p, 0, r/2)$ and all $t \in (0, T]$. In particular if $k \leq l$, then for $y \in B(p, 0, r/2)$ and $t \in (0, T]$ we have

$$\left|\nabla^k \operatorname{Rm}(y, t)\right| \leq C'_{k,l}.$$

REMARK 3.30. Clearly, the case $l = 0$ of Theorem 3.29 is Shi's theorem (Theorem 3.27).

Theorem 3.29 leads immediately to the following:

COROLLARY 3.31. *Suppose that $(M, g(t))$, $0 \leq t \leq T$, is a Ricci flow with $(M, g(t))$ being complete and with $T < \infty$. Suppose that $\operatorname{Rm}(x, 0)$ is bounded in the C^∞-topology independent of $x \in M$ and suppose that $|\operatorname{Rm}(x,t)|$ is bounded independent of $x \in M$ and $t \in [0, T]$. Then the operator $\operatorname{Rm}(x, t)$ is bounded in the C^∞-topology independent of $(x, t) \in M \times [0, T]$.*

For a proof of Theorem 3.28 see [**65, 66**]. We give the proof of a stronger result, Theorem 3.29.

PROOF. The first remark is that establishing Theorem 3.29 for one value of r immediately gives it for all $r' \geq 2r$. The reason is that for such r' any point $y \in B(p, 0, r'/2)$ has the property that $B(y, 0, r) \subset B(p, 0, r')$ so that a curvature bound on $B(p, 0, r')$ will imply one on $B(y, 0, r)$ and hence by the result for r will imply the higher derivative bounds at y.

Thus, without loss of generality we can suppose that $r \leq \pi/2\sqrt{K}$. We shall assume this from now on in the proof. Since $B(p, 0, r)$ has compact closure in M, for some $r < r' < \pi/\sqrt{K}$ the ball $B(p, 0, r')$ also has compact closure in M. This means that the exponential mapping from the ball of radius r' in T_pM is a local diffeomorphism onto $B(p, 0, r')$.

The proof is by induction: We assume that we have established the result for $k = 0, \ldots, m$, and then we shall establish it for $k = m+1$. The inductive hypothesis tells us that there are constants A_j, $0 \leq j \leq m$, depending on (l, K, α, r, n) such that for all $(x, t) \in B(p, 0, r/2) \times (0, T]$ we have

(3.17) $$\left|\nabla^j \operatorname{Rm}(x, t)\right| \leq A_j t^{-\max\{j-l, 0\}/2}.$$

Applying the inductive result to $B(y, 0, r/2)$ with $y \in B(p, 0, r/2)$ we see that, replacing the A_j by the constants associated with $(l, K, \alpha, r/2, n)$, we have the same inequality for all $y \in B(x, 0, 3r/4)$.

We fix a constant $C \geq \max(4A_m^2, 1)$ and consider

$$F_m(x, t) = \left(C + t^{\max\{m-l, 0\}} |\nabla^m \operatorname{Rm}(x, t)|^2\right) t^{\max\{m+1-l, 0\}} \left|\nabla^{m+1} \operatorname{Rm}(x, t)\right|^2.$$

Notice that bounding F_m above by a constant $(C'_{m+1, l})^2$ will yield

$$|\nabla^{m+1} \operatorname{Rm}(x, t)|^2 \leq \frac{(C'_{m+1, l})^2}{t^{\max\{m+1-l, 0\}}},$$

and hence will complete the proof of the result.

Bounding F_m above (assuming the inductive hypothesis) is what is accomplished in the rest of this proof. The main calculation is the proof of the following claim under the inductive hypothesis.

CLAIM 3.32. *With F_m as defined above and with $C \geq \max(4A_m^2, 1)$, there are constants c_1 and C_0, C_1 depending on C as well as $K, \alpha, A_1, \ldots, A_m$ for which the following holds on $B(p, 0, 3r/4) \times (0, T]$:*

$$\left(\frac{\partial}{\partial t} - \Delta\right) F_m(x,t) \leq -\frac{c_1 \left(F_m(x,t) - C_0\right)^2}{t^{s\{\max\{m-l+1,0\}\}}} + \frac{C_1}{t^{s\{\max\{m-l+1,0\}\}}},$$

where

$$s\{n\} = \begin{cases} +1, & \text{if } n > 0, \\ 0, & \text{if } n = 0, \\ -1, & \text{if } n < 0. \end{cases}$$

Let us assume this claim and use it to prove Theorem 3.29. We fix $C = \max\{4A_m^2, 1\}$, and consider the resulting function F_m. The constants c_1, C_0, C_1 from Claim 3.32 depend only on K, α, and $A_1, \ldots, A_m$. Since $r \leq \pi/2\sqrt{K}$, and $B(p, 0, r)$ has compact closure in U, there is some $r' > r$ so that the exponential mapping $\exp_p \colon B(0, r') \to U$ is a local diffeomorphism onto $B(p, 0, r')$. Pulling back by the exponential map, we replace the Ricci flow on U by a Ricci flow on $B(0, r')$ in $T_p M$. Clearly, it suffices to establish the estimates in the statement of the proposition for $B(0, r/2)$. This remark allows us to assume that the exponential mapping is a diffeomorphism onto $B(p, 0, r)$. Bounded curvature then comes into play in the following crucial proposition, which goes back to Shi. The function given in the next proposition allows us to localize the computation in the ball $B(p, 0, r)$.

PROPOSITION 3.33. *Fix constants $0 < \alpha$ and the dimension n. Then there is a constant $C_2' = C_2'(\alpha, n)$ and for each $r > 0$ and $K < \infty$ there is a constant $C_2 = C_2(K, \alpha, r, n)$ such that the following holds. Suppose that $(U, g(t))$, $0 \leq t \leq T$, is an n-dimensional Ricci flow with $T \leq \alpha/K$. Suppose that $p \in U$ and that $B(p, 0, r)$ has compact closure in U and that the exponential mapping from the ball of radius r in $T_p U$ to $B(p, 0, r)$ is a diffeomorphism. Suppose that $|\mathrm{Rm}(x, 0)| \leq K$ for all $x \in B(p, 0, r)$. There is a smooth function $\eta \colon B(p, 0, r) \to [0, 1]$ satisfying the following for all $t \in [0, T]$:*

(1) *η has compact support in $B(p, 0, r/2)$.*
(2) *The restriction of η to $B(p, 0, r/4)$ is identically 1.*
(3) *$|\Delta_{g(t)} \eta| \leq C_2(K, \alpha, r, n)$.*
(4) *$\frac{|\nabla \eta|_{g(t)}^2}{\eta} \leq \frac{C_2'(\alpha, n)}{r^2}$.*

For a proof of this result see Lemma 6.62 on page 225 of [**14**].

We can apply this proposition to our situation, because we are assuming that $r \le \pi/2\sqrt{K}$ so that the exponential mapping is a local diffeomorphism onto $B(p, 0, r)$ and we have pulled the Ricci flow back to the ball in the tangent space.

Fix any $y \subset B(p, 0, r/2)$ and choose η as in the previous proposition for the constants $C_2(\alpha, n)$ and $C_2'(K, \alpha, r/4, n)$. Notice that $B(y, 0, r/4) \subset B(p, 0, 3r/4)$ so that the conclusion of Claim 3.32 holds for every (z, t) with $z \in B(y, 0, r/4)$ and $t \in [0, T]$. We shall show that the restriction of ηF_m to $P(y, 0, r/4, T)$ is bounded by a constant that depends only on $K, \alpha, r, n, A_1, \ldots, A_m$. It will then follow immediately that the restriction of F_m to $P(y, 0, r/8, T)$ is bounded by the same constant. In particular, the values of $F_m(y, t)$ are bounded by the same constant for all $y \in B(p, 0, r/2)$ and $t \in [0, T]$.

Consider a point $(x, t) \in B(y, 0, r/2) \times [0, T]$ where ηF_m achieves its maximum; such a point exists since the ball $B(y, 0, r/2) \subset B(p, 0, r)$, and hence $B(y, 0, r/2)$ has compact closure in U. If $t = 0$, then ηF_m is bounded by $(C + K^2)K^2$ which is a constant depending only on K and A_m. This, of course, immediately implies the result. Thus we can assume that the maximum is achieved at some $t > 0$. When $s\{\max\{m + 1 - l, 0\}\} = 0$, according to Claim 3.32, we have

$$\left(\frac{\partial}{\partial t} - \Delta\right) F_m \le -c_1 (F_m - C_0)^2 + C_1.$$

We compute

$$\left(\frac{\partial}{\partial t} - \Delta\right)(\eta F_m) \le \eta\left(-c_1 (F_{m-C_0})^2 + C_1\right) - \Delta\eta \cdot F_m - 2\nabla\eta \cdot \nabla F_m.$$

Since (x, t) is a maximum point for ηF_m and since $t > 0$, a simple maximum principle argument shows that

$$\left(\frac{\partial}{\partial t} - \Delta\right)\eta F_m(x, t) \ge 0.$$

Hence, in this case we conclude that

$$0 \le \left(\frac{\partial}{\partial t} - \Delta\right)(\eta(x) F_m(x, t)) \le \eta(x)\left(-c_1 (F_m(x, t) - C_0)^2 + C_1\right)$$
$$- \Delta\eta(x) \cdot F_m(x, t) - 2\nabla\eta(x) \cdot \nabla F_m(x, t).$$

Hence,

$$c_1 \eta(F_m(x, t) - C_0)^2 \le \eta(x)C_1 - \Delta\eta(x) \cdot F_m(x, t) - 2\nabla\eta(x) \cdot \nabla F_m(x, t).$$

Since we are proving that F_m is bounded, we are free to argue by contradiction and assume that $F_m(x, t) \ge 2C_0$, implying that $F_m(x, t) - C_0 \ge$

$F_m(x,t)/2$. Using this inequality yields

$$\eta(x)(F_m(x,t) - C_0) \leq \frac{2\eta C_1}{c_1 F_m(x,t)} - \frac{2\Delta\eta(x)}{c_1} - \frac{4}{c_1 F_m(x,t)} \nabla\eta(x) \cdot \nabla F_m(x,t)$$

$$\leq \frac{\eta C_1}{c_1 C_0} - \frac{2\Delta\eta(x)}{c_1} - \frac{4}{c_1 F_m(x,t)} \nabla\eta(x) \cdot \nabla F_m(x,t).$$

Since (x,t) is a maximum for ηF_m we have

$$0 = \nabla(\eta(x) F_m(x,t)) = \nabla\eta(x) F_m(x,t) + \eta(x) \nabla F_m(x,t),$$

so that

$$\frac{\nabla\eta(x)}{\eta(x)} = -\frac{\nabla F_m(x,t)}{F_m(x,t)}.$$

Plugging this in gives

$$\eta(x) F_m(x,t) \leq \frac{C_1}{c_1 C_0} - \frac{2\Delta\eta(x)}{c_1} + 4\frac{|\nabla\eta(x)|^2}{c_1 \eta(x)} + \eta C_0.$$

Of course, the gradient and Laplacian of η are taken at the point (x,t). Thus, because of the properties of η given in Proposition 3.33, it immediately follows that $\eta F_m(x,t)$ is bounded by a constant depending only on $K, n, \alpha, r, c_1, C_0, C_1$, and as we have already seen, c_1, C_0, C_1 depend only on $K, \alpha, A_1, \ldots, A_m$.

Now suppose that $s\{\max\{m - l + 1, 0\}\} = 1$. Again we compute the evolution inequality for ηF_m. The result is

$$\left(\frac{\partial}{\partial t} - \Delta\right)(\eta F_m) \leq \eta\left(-\frac{c_1}{t}(F_m - C_0)^2 + \frac{C_1}{t}\right) - \Delta\eta \cdot F_m - 2\nabla\eta \cdot \nabla F_m.$$

Thus, using the maximum principle as before, we have

$$\left(\frac{\partial}{\partial t} - \Delta\right)\eta F_m(x,t) \geq 0.$$

Hence,

$$\frac{\eta(x) c_1 (F_m(x,t) - C_0)^2}{t} \leq \frac{\eta(x) C_1}{t} - \Delta\eta(x) F_m(x,t) - 2\nabla\eta(x) \cdot \nabla F_m(x,t).$$

Using the assumption that $F_m(x,t) \geq 2C_0$ as before, and rewriting the last term as before, we have

$$\eta F_m(x,t) \leq \frac{\eta(x) C_1}{c_1 C_0} - \frac{2t\Delta\eta(x)}{c_1} + \frac{4t|\nabla\eta(x)|^2}{c_1 \eta(x)} + \eta C_0.$$

The right-hand side is bounded by a constant depending only on $K, n, \alpha, r, c_1, C_0$, and C_1. We conclude that in all cases ηF_m is bounded by a constant depending only on $K, n, \alpha, r, c_1, C_0, C_1$, and hence on $K, n, \alpha, r, A_1, \ldots, A_m$.

This proves that for any $y \in B(p, 0, r/2)$, the value $\eta F_m(x,t)$ is bounded by a constant A_{m+1} depending only on $(m+1, l, K, n, \alpha, r)$ for all $(x,t) \in B(y, 0, r/2) \times [0, T]$. Since $\eta(y) = 1$, for all $0 \leq t \leq T$ we have

$$t^{\max\{m+1-l,0\}} |\nabla^{m+1} \mathrm{Rm}(y,t)|^2 \leq F_m(y,t) = \eta(y) F_m(y,t) \leq A_{m+1}.$$

This completes the inductive proof that the result holds for $k = m + 1$ and hence establishes Theorem 3.29, modulo the proof of Claim 3.32. $\square$

Now we turn to the proof of Claim 3.32.

PROOF. In this argument we fix $(x,t) \in B(p,0,3r/4) \times (0,T]$ and we drop (x,t) from the notation. Recall that by Equations (7.4a) and (7.4b) on p. 229 of [**13**] we have

$$(3.18) \quad \frac{\partial}{\partial t} \left|\nabla^\ell \operatorname{Rm}\right|^2 \leq \Delta \left|\nabla^\ell \operatorname{Rm}\right|^2 - 2\left|\nabla^{\ell+1} \operatorname{Rm}\right|^2$$
$$+ \sum_{i=0}^{\ell} c_{\ell,j} \left|\nabla^i \operatorname{Rm}\right| \left|\nabla^{\ell-i} \operatorname{Rm}\right| \left|\nabla^\ell \operatorname{Rm}\right|,$$

where the constants $c_{\ell,j}$ depend only on ℓ and j.

Hence, setting $m_l = \max\{m + 1 - l, 0\}$ and denoting $c_{m+1,i}$ by $\widetilde{c}_i$, we have

$$(3.19) \quad \frac{\partial}{\partial t}\left(t^{m_l}\left|\nabla^{m+1}\operatorname{Rm}\right|^2\right)$$
$$\leq \Delta\left(t^{m_l}\left|\nabla^{m+1}\operatorname{Rm}\right|^2\right) - 2t^{m_l}\left|\nabla^{m+2}\operatorname{Rm}\right|^2$$
$$+ t^{m_l}\sum_{i=0}^{m+1} \widetilde{c}_i \left|\nabla^i\operatorname{Rm}\right|\left|\nabla^{m+1-i}\operatorname{Rm}\right|\left|\nabla^{m+1}\operatorname{Rm}\right| + m_l t^{m_l-1}\left|\nabla^{m+1}\operatorname{Rm}\right|^2$$
$$\leq \Delta\left(t^{m_l}\left|\nabla^{m+1}\operatorname{Rm}\right|^2\right) - 2t^{m_l}\left|\nabla^{m+2}\operatorname{Rm}\right|^2$$
$$+ (\widetilde{c}_0 + \widetilde{c}_{m+1})t^{m_l}\left|\operatorname{Rm}\right|\left|\nabla^{m+1}\operatorname{Rm}\right|^2$$
$$+ t^{m_l}\sum_{i=1}^{m} \widetilde{c}_i \left|\nabla^i\operatorname{Rm}\right|\left|\nabla^{m+1-i}\operatorname{Rm}\right|\left|\nabla^{m+1}\operatorname{Rm}\right| + m_l t^{m_l-1}\left|\nabla^{m+1}\operatorname{Rm}\right|^2.$$

Using the inductive hypothesis, Inequality (3.17), there is a constant $A < \infty$ depending only on $K, \alpha, A_1, \ldots, A_m$ such that

$$\sum_{i=1}^{m} \widetilde{c}_i \left|\nabla^i\operatorname{Rm}\right|\left|\nabla^{m+1-i}\operatorname{Rm}\right| \leq At^{-m_l/2}.$$

Also, let $c = \widetilde{c}_0 + \widetilde{c}_{m+1}$ and define a new constant B by

$$B = c(\alpha + K) + m_l.$$

Then, since $t \leq T \leq \alpha/K$ and $m_l \geq 0$, we have

$$((\widetilde{c}_0 + \widetilde{c}_{m+1})t\left|\operatorname{Rm}\right| + m_l)t^{m_l-1} \leq \frac{Bt^{m_l}}{t^{s(m_l)}}.$$

Putting this together allows us to rewrite Inequality (3.19) as

$$\frac{\partial}{\partial t}\left(t^{m_l}\left|\nabla^{m+1}\operatorname{Rm}\right|^2\right) \leq \Delta\left(t^{m_l}\left|\nabla^{m+1}\operatorname{Rm}\right|^2\right) - 2t^{m_l}\left|\nabla^{m+2}\operatorname{Rm}\right|^2$$
$$+ At^{m_l/2}\left|\nabla^{m+1}\operatorname{Rm}\right|$$
$$+ (ct\left|\operatorname{Rm}\right| + m_l)\,t^{m_l-1}\left|\nabla^{m+1}\operatorname{Rm}\right|^2$$
$$\leq \Delta\left(t^{m_l}\left|\nabla^{m+1}\operatorname{Rm}\right|^2\right) - 2t^{m_l}\left|\nabla^{m+2}\operatorname{Rm}\right|^2$$
$$+ \frac{B}{t^{s(m_l)}}t^{m_l}\left|\nabla^{m+1}\operatorname{Rm}\right|^2 + At^{m_l/2}\left|\nabla^{m+1}\operatorname{Rm}\right|.$$

Completing the square gives

$$\frac{\partial}{\partial t}\left(t^{m_l}\left|\nabla^{m+1}\operatorname{Rm}\right|^2\right) \leq \Delta\left(t^{m_l}\left|\nabla^{m+1}\operatorname{Rm}\right|^2\right) - 2t^{m_l}\left|\nabla^{m+2}\operatorname{Rm}\right|^2$$
$$+ (B+1)t^{m_l-s(m_l)}\left|\nabla^{m+1}\operatorname{Rm}\right|^2 + \frac{A^2}{4}t^{s(m_l)}.$$

Let $\hat{m}_l = \max\{m-l, 0\}$. From (3.18) and the induction hypothesis, there is a constant D, depending on $K, \alpha, A_1, \ldots, A_m$ such that

$$\frac{\partial}{\partial t}\left(t^{\hat{m}_l}\left|\nabla^m\operatorname{Rm}\right|^2\right) \leq \Delta\left(t^{\hat{m}_l}\left|\nabla^m\operatorname{Rm}\right|^2\right) - 2t^{\hat{m}_l}\left|\nabla^{m+1}\operatorname{Rm}\right|^2$$
$$+ \hat{m}_l t^{\hat{m}_l-1}\left|\nabla^m\operatorname{Rm}\right|^2 + D.$$

Now, defining new constants $\widetilde{B} = B+1$ and $\widetilde{A} = A^2/4$ we have

$$\left(\frac{\partial}{\partial t}-\Delta\right)F_m = \left(\frac{\partial}{\partial t}-\Delta\right)\left[\left(C+t^{\hat{m}_l}\left|\nabla^m\operatorname{Rm}\right|^2\right)t^{m_l}\left|\nabla^{m+1}\operatorname{Rm}\right|^2\right] \leq$$
$$\left(C+t^{\hat{m}_l}\left|\nabla^m\operatorname{Rm}\right|^2\right)\left(\widetilde{A}t^{s(m_l)} - 2t^{m_l}\left|\nabla^{m+2}\operatorname{Rm}\right|^2 + \frac{\widetilde{B}t^{m_l}\left|\nabla^{m+1}\operatorname{Rm}\right|^2}{t^{s\{m_l\}}}\right)$$
$$+ \left(-2t^{\hat{m}_l}\left|\nabla^{m+1}\operatorname{Rm}\right|^2 + \hat{m}_l t^{\hat{m}_l-1}\left|\nabla^m\operatorname{Rm}\right|^2 + D\right)t^{m_l}\left|\nabla^{m+1}\operatorname{Rm}\right|^2$$
$$- 2t^{\hat{m}_l+m_l}\nabla\left(\left|\nabla^m\operatorname{Rm}\right|^2\right)\cdot\nabla\left(\left|\nabla^{m+1}\operatorname{Rm}\right|^2\right).$$

Since $C \geq 4t^{\hat{m}_l}\left|\nabla^m\operatorname{Rm}\right|^2$, this implies

(3.20) $\left(\frac{\partial}{\partial t}-\Delta\right)F_m \leq -10t^{\hat{m}_l+m_l}\left|\nabla^m\operatorname{Rm}\right|^2\left|\nabla^{m+2}\operatorname{Rm}\right|^2$
$$- 8t^{\hat{m}_l+m_l}\left|\nabla^m\operatorname{Rm}\right|\left|\nabla^{m+1}\operatorname{Rm}\right|^2\left|\nabla^{m+2}\operatorname{Rm}\right| - 2t^{\hat{m}_l+m_l}\left|\nabla^{m+1}\operatorname{Rm}\right|^4$$
$$+ \left(C+t^{\hat{m}_l}\left|\nabla^m\operatorname{Rm}\right|^2\right)\left(\widetilde{B}t^{m_l-s(m_l)}\left|\nabla^{m+1}\operatorname{Rm}\right|^2 + \widetilde{A}t^{s(m_l)}\right)$$
$$+ \left(\hat{m}_l t^{\hat{m}_l-1}\left|\nabla^m\operatorname{Rm}\right|^2 + D\right)t^{m_l}\left|\nabla^{m+1}\operatorname{Rm}\right|^2.$$

Now we can write the first three terms on the right-hand side of Inequality (3.20) as $-t^{\hat{m}_l+m_l}$ times

$$(3.21) \quad \left(\sqrt{10}\,|\nabla^{m+2}\operatorname{Rm}|\,|\nabla^m\operatorname{Rm}| + \frac{4}{\sqrt{10}}\,|\nabla^{m+1}\operatorname{Rm}|^2\right)^2 + \frac{2}{5}\,|\nabla^{m+1}\operatorname{Rm}|^4.$$

In addition we have

$$(3.22) \quad C + t^{\hat{m}_l}\,|\nabla^m\operatorname{Rm}|^2 \leq C + A_m^2.$$

Let us set $\widetilde{D} = \max(\alpha/K, 1)D$. If $\hat{m}_l = 0$, then

$$(3.23) \quad \hat{m}_l t^{\hat{m}_l - 1}\,|\nabla^m\operatorname{Rm}|^2 + D = D \leq \frac{\widetilde{D}}{t^{s(m_l)}} = \hat{m}_l A_m^2 + D \leq \frac{\hat{m}_l A_m^2 + \widetilde{D}}{t^{s(m_l)}}.$$

On the other hand, if $\hat{m}_l > 0$, then $s(\hat{m}_l) = s(m_l) = 1$ and hence

$$\hat{m}_l t^{\hat{m}_l - 1}\,|\nabla^m\operatorname{Rm}|^2 + D \leq \frac{1}{t^{s(m_l)}}\hat{m}_l A_m^2 + D \leq \frac{\hat{m}_l A_m^2 + \widetilde{D}}{t^{s(m_l)}}.$$

Since $\hat{m}_l = m_l - s(m_l)$, Inequalities (3.21), (3.22), and (3.23) allow us to rewrite Inequality (3.20) as

$$\left(\frac{\partial}{\partial t} - \Delta\right) F_m \leq -\frac{2}{5t^{s(m_l)}} t^{2m_l}\,|\nabla^{m+1}\operatorname{Rm}|^4$$

$$+ (C + A_m^2)\left(\frac{\widetilde{B}t^{m_l}}{t^{s(m_l)}}\,|\nabla^{m+1}\operatorname{Rm}|^2 + \widetilde{A}t^{s(m_l)}\right) + \frac{\hat{m}_l A_m^2 + \widetilde{D}}{t^{s(m_l)}} t^{m_l}\,|\nabla^{m+1}\operatorname{Rm}|^2.$$

Setting

$$B' = (C + A_m^2)\widetilde{B} + (\hat{m}_l A_m^2 + \widetilde{D}),$$

and $A' = \widetilde{A}(C + A_m^2)$ we have

$$\left(\frac{\partial}{\partial t} - \Delta\right) F_m \leq -\frac{2}{5t^{s(m_l)}}\left(t^{m_l}\,|\nabla^{m+1}\operatorname{Rm}|^2\right)^2$$

$$+ \frac{B'}{t^{s(m_l)}} t^{m_l}\,|\nabla^{m+1}\operatorname{Rm}|^2 + A't^{s(m_l)}.$$

We rewrite this as

$$\left(\frac{\partial}{\partial t} - \Delta\right) F_m \leq -\frac{2}{5t^{s(m_l)}}\left(t^{m_l}\,|\nabla^{m+1}\operatorname{Rm}|^2 - \frac{5B'}{4}\right)^2$$

$$+ \frac{5(B')^2}{8t^{s(m_l)}} + A't^{s(m_l)},$$

and hence

$$\left(\frac{\partial}{\partial t} - \Delta\right) F_m \leq -\frac{2}{5t^{s(m_l)}}\left(t^{m_l}\,|\nabla^{m+1}\operatorname{Rm}|^2 - B''\right)^2 + \frac{A''}{t^{s(m_l)}}$$

where the constants B'' and A'' are defined by $B'' = 5B'/4$ and

$$A'' = (\max\{\alpha/K, 1\})^2 + 5(B')^2/8.$$

(Recall that $t \leq T \leq \alpha/K$.) Let
$$Y = (C + t^{\hat{m}_l} |\nabla^m \operatorname{Rm}|^2).$$
(Notice that Y is not a constant.) Of course, by definition
$$F_m = Y t^{m_l} |\nabla^{m+1} \operatorname{Rm}|^2.$$
Then the previous inequality becomes
$$\left(\frac{\partial}{\partial t} - \Delta\right) F_m \leq -\frac{2}{5 t^{s(m_l)} Y^2} \left(Y t^{m_l} |\nabla^{m+1} \operatorname{Rm}|^2 - B'' Y\right)^2 + \frac{A''}{t^{s(m_l)}}.$$
Since $C \leq Y \leq 5C/4$ we have
$$\left(\frac{\partial}{\partial t} - \Delta\right) F_m \leq -\frac{32}{125 t^{s(m_l)} C^2} \left(F_m - B'' Y\right)^2 + \frac{A''}{t^{s(m_l)}}.$$
At any point where $F_m \geq 5CB''/4$, the last inequality gives
$$\left(\frac{\partial}{\partial t} - \Delta\right) F_m \leq -\frac{32}{125 t^{s(m_l)} C^2} \left(F_m - 5CB''/4\right)^2 + \frac{A''}{t^{s(m_l)}}.$$
At any point where $F_m \leq 5CB''/4$, since $F_m \geq 0$ and $0 \leq B'' Y \leq 5CB''/4$, we have $(F_m - B'' Y)^2 \leq 25 C^2 (B'')^2 / 16$, so that
$$-\frac{32}{125 t^{s(m_l)} C^2} \left(F_m - 5CB''/4\right)^2 \geq -2(B'')^2 / 5 t^{s(m_l)}.$$
Thus, in this case we have
$$\left(\frac{\partial}{\partial t} - \Delta\right) F_m \leq \frac{A''}{t^{s(m_l)}} \leq -\frac{32}{125 t^{s(m_l)} C^2} \left(F_m - 5CB''/4\right)^2 + \frac{A'' + 2(B'')^2/5}{t^{s(m_l)}}.$$
These two cases together prove Claim 3.32. $\square$

8. Generalized Ricci flows

In this section we introduce a generalization of the Ricci flow equation. The generalization does not involve changing the PDE that gives the flow. Rather it allows for the global topology of space-time to be different from a product.

8.1. Space-time. There are two basic ways to view an n-dimensional Ricci flow: (i) as a one-parameter family of metrics $g(t)$ on a fixed smooth n-dimensional manifold M, and (ii) as a partial metric (in the horizontal directions) on the $(n+1)$-dimensional manifold $M \times I$. We call the latter $(n+1)$-dimensional manifold *space-time* and the horizontal slices are the *time-slices*. In defining the generalized Ricci flow, it is the second approach that we generalize.

DEFINITION 3.34. By *space-time* we mean a smooth $(n+1)$-dimensional manifold $\mathcal{M}$ (possibly with boundary), equipped with a smooth function $\mathbf{t} \colon \mathcal{M} \to \mathbb{R}$, called *time* and a smooth vector field χ subject to the following axioms:

(1) The image of **t** is an interval I (possibly infinite) and the boundary of $\mathcal{M}$ is the preimage under **t** of ∂I.
(2) For each $x \in \mathcal{M}$ there is an open neighborhood $U \subset \mathcal{M}$ of x and a diffeomorphism $f\colon V \times J \to U$, where V is an open subset in $\mathbb{R}^n$ and J is an interval with the property that (i) **t** is the composition of f^{-1} followed by the projection onto the interval J and (ii) χ is the image under f of the unit vector field in the positive direction tangent to the foliation by the lines $\{v\} \times J$ of $V \times J$.

Notice that it follows that $\chi(\mathbf{t}) = 1$.

DEFINITION 3.35. The *time-slices* of space-time are the level sets **t**. These form a foliation of $\mathcal{M}$ of codimension 1. For each $t \in I$ we denote by $M_t \subset \mathcal{M}$ the t time-slice, that is to say $\mathbf{t}^{-1}(t)$. Notice that each boundary component of $\mathcal{M}$ is contained in a single time-slice. The *horizontal distribution* $\mathcal{H}T\mathcal{M}$ is the distribution tangent to this foliation. A *horizontal metric* on space-time is a smoothly varying positive definite inner product on $\mathcal{H}T\mathcal{M}$.

Notice that a horizontal metric on space-time induces an ordinary Riemannian metric on each time-slice. Conversely, given a Riemannian metric on each time-slice M_t, the condition that they fit together to form a horizontal metric on space-time is that they vary smoothly on space-time. We define the curvature of a horizontal metric G to be the section of the dual of the symmetric square of $\wedge^2 \mathcal{H}T\mathcal{M}$ whose value at each point x with $\mathbf{t}(x) = t$ is the usual Riemann curvature tensor of the induced metric on M_t at the point x. This is a smooth section of $\operatorname{Sym}^2(\wedge^2 \mathcal{H}T^*\mathcal{M})$. The Ricci curvature and the scalar curvature of a horizontal metric are given in the usual way from its Riemann curvature. The Ricci curvature is a smooth section of $\operatorname{Sym}^2(\mathcal{H}T^*\mathcal{M})$ while the scalar curvature is a smooth function on $\mathcal{M}$.

8.2. The generalized Ricci flow equation. Because of the second condition in the definition of space-time, the vector field χ preserves the horizontal foliation and hence the horizontal distribution. Thus, we can form the Lie derivative of a horizontal metric with respect to χ.

DEFINITION 3.36. An *n-dimensional generalized Ricci flow* consists of a space-time $\mathcal{M}$ that is $(n+1)$-dimensional and a horizontal metric G satisfying the generalized Ricci flow equation:
$$\mathcal{L}_\chi(G) = -2\operatorname{Ric}(G).$$

REMARK 3.37. Let $(\mathcal{M}, G)$ be a generalized Ricci flow and let $x \in \mathcal{M}$. Pulling G back to the local coordinates $V \times J$ defined near any point gives a one-parameter family of metrics $(V, g(t))$, $t \in J$, satisfying the usual Ricci flow equation. It follows that all the usual evolution formulas for Riemann curvature, Ricci curvature, and scalar curvature hold in this more general context.

Of course, any ordinary Ricci flow is a generalized Ricci flow where space-time is a product $M \times I$ with time being the projection to I and χ being the unit vector field in the positive I-direction.

8.3. More definitions for generalized Ricci flows.

DEFINITION 3.38. Let $\mathcal{M}$ be a space-time. Given a space C and an interval $I \subset \mathbb{R}$ we say that an embedding $C \times I \to \mathcal{M}$ is *compatible with the time and the vector field* if: (i) the restriction of $\mathbf{t}$ to the image agrees with the projection onto the second factor and (ii) for each $c \in C$ the image of $\{c\} \times I$ is the integral curve for the vector field χ. If in addition C is a subset of M_t we require that $t \in I$ and that the map $C \times \{t\} \to M_t$ be the identity. Clearly, by the uniqueness of integral curves for vector fields, two such embeddings agree on their common interval of definition, so that, given $C \subset M_t$ there is a maximal interval I_C containing t such that such an embedding, compatible with time and the vector field, is defined on $C \times I$. In the special case when $C = \{x\}$ for a point $x \in M_t$ we say that such an embedding is *the flow line* through x. The embedding of the maximal interval through x compatible with time and the vector field χ is called *the domain of definition* of the flow line through x. For a more general subset $C \subset M_t$ there is an embedding $C \times I$ compatible with time and the vector field χ if and only if for every $x \in C$, I is contained in the domain of definition of the flow line through x.

DEFINITION 3.39. We say that t is a *regular time* if there is $\epsilon > 0$ and a diffeomorphism $M_t \times (t - \epsilon, t + \epsilon) \to \mathbf{t}^{-1}((t - \epsilon, t + \epsilon))$ compatible with time and the vector field. A time is *singular* if it is not regular. Notice that if all times are regular, then space-time is a product $M_t \times I$ with $\mathbf{t}$ and χ coming from the second factor. If the image $\mathbf{t}(\mathcal{M})$ is an interval I bounded below, then the *initial time* for the flow is the greatest lower bound for I. If I includes $(-\infty, A]$ for some A, then the initial time for the generalized Ricci flow is $-\infty$.

DEFINITION 3.40. Suppose that $(\mathcal{M}, G)$ is a generalized Ricci flow and that $Q > 0$ is a positive constant. Then we can define a new generalized Ricci flow by setting $G' = QG$, $\mathbf{t}' = Q\mathbf{t}$ and $\chi' = Q^{-1}\chi$. It is easy to see that the result still satisfies the generalized Ricci flow equation. We denote this new generalized Ricci flow by $(Q\mathcal{M}, QG)$ where the changes in $\mathbf{t}$ and χ are denoted by the factor of Q in front of $\mathcal{M}$.

It is also possible to translate a generalized solution $(\mathcal{M}, G)$ by replacing the time function $\mathbf{t}$ by $\mathbf{t}' = \mathbf{t} + a$ for any constant a, leaving G and χ unchanged.

DEFINITION 3.41. Let $(\mathcal{M}, G)$ be a generalized Ricci flow and let x be a point of space-time. Set $t = \mathbf{t}(x)$. For any $r > 0$ we define $B(x, t, r) \subset M_t$

to be the metric ball of radius r centered at x in the Riemannian manifold $(M_t, g(t))$. For any $\Delta t > 0$ we say that $P(x, t, r, \Delta t)$, respectively, $P(x, r, t, -\Delta t)$, *exists* in $\mathcal{M}$ if there is an embedding $B(x, t, r) \times [t, t + \Delta t]$, respectively, $B(x, t, r) \times [t - \Delta t, t]$, into $\mathcal{M}$ compatible with time and the vector field. When this embedding exists, its image is defined to be the *forward parabolic neighborhood* $P(x, t, r, \Delta t)$, respectively the *backward parabolic neighborhood* $P(x, t, r, -\Delta t)$. See FIG. 3.

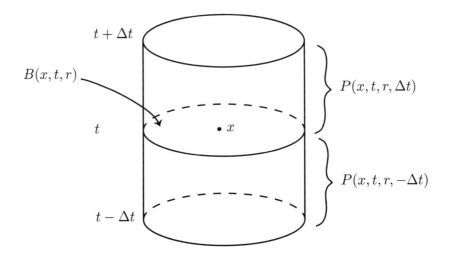

FIGURE 3. Parabolic neighborhoods.

CHAPTER 4

The maximum principle

Recall that the maximum principle for the heat equation says that if h is a solution to the heat equation

$$\frac{\partial h}{\partial t} = \Delta h$$

on a compact manifold and if $h(x,0) \geq 0$ for all $x \in M$, then $h(x,t) \geq 0$ for all (x,t). In this chapter we discuss analogues of this result for the scalar curvature, the Ricci curvature, and the sectional curvature under Ricci flow. Of course, in all three cases we are working with quasi-linear versions of the heat equation so it is important to control the lower order (non-linear) terms and in particular show that at zero curvature they have the appropriate sign. Also, in the latter two cases we are working with tensors rather than with scalars and hence we require a tensor version of the maximum principle, which was established by Hamilton in [35].

As further applications of these results beyond just establishing non-negativity, we indicate Hamilton's result that if the initial conditions have positive Ricci curvature, then the solution becomes singular at finite time and as it does it becomes round (pinching to round). We also give Hamilton's result showing that at points where the scalar curvature is sufficiently large the curvature is pinched toward positive. This result is crucial for understanding singularity development. As a last application, we give Hamilton's Harnack inequality for Ricci flows of non-negative curvature.

The maximum principle is used here in two different ways. The first assumes non-negativity of something (e.g., a curvature) at time zero and uses the maximum principle to establish non-negativity of this quantity at all future times. The second assumes non-negativity of something at all times and positivity at one point, and then uses the maximum principle to establish positivity at all points and all later times. In the latter application one compares the solution with a solution to the linear heat equation where such a property is known classically to hold.

1. Maximum principle for scalar curvature

Let us begin with the easiest evolution equation, that for the scalar curvature, where the argument uses only the (non-linear) version of the maximum principle. This result is valid in all dimensions:

PROPOSITION 4.1. *Let $(M, g(t))$, $0 \le t < T$, be a Ricci flow with M a compact n-dimensional manifold. Denote by $R_{\min}(t)$ the minimum value of the scalar curvature of $(M, g(t))$. Then:*

- *$R_{\min}(t)$ is a non-decreasing function of t.*
- *If $R_{\min}(0) \ge 0$, then*

$$R_{\min}(t) \ge R_{\min}(0) \left(\frac{1}{1 - \frac{2t}{n} R_{\min}(0)} \right),$$

in particular,

$$T \le \frac{n}{2 R_{\min}(0)}.$$

- *If $R_{\min}(0) < 0$, then*

$$R_{\min}(t) \ge -\frac{n |R_{\min}(0)|}{2t |R_{\min}(0)| + n}.$$

PROOF. According to Equation (3.7), the evolution equation for R is

$$\frac{\partial}{\partial t} R(x, t) = \Delta R(x, t) + 2|\mathrm{Ric}(x, t)|^2.$$

Since M is compact, the function $R_{\min}(t)$ is continuous but may not be C^1 at points where the minimum of the scalar curvature is achieved at more than one point.

The first thing to notice is the following:

CLAIM 4.2. *If $R(x, t) = R_{\min}(t)$, then $(\partial R / \partial t)(x, t) \ge \frac{2}{n} R^2(x, t)$.*

PROOF. This is immediate from the evolution equation for R, the fact that if $R(x, t) = R_{\min}(t)$, then $\Delta R(x, t) \ge 0$, and the fact that R is the trace of Ric which implies by the Cauchy-Schwarz inequality that $|R|^2 \le n |\mathrm{Ric}|^2$. □

Now it follows that:

CLAIM 4.3.
$$\frac{d}{dt}(R_{\min}(t)) \ge \frac{2}{n} R_{\min}^2(t),$$

where, at times t where $R_{\min}(t)$ is not smooth, this inequality is interpreted as an inequality for the forward difference quotients.

PROOF. This is immediate from the first statement in Proposition 2.23. □

It follows immediately from Claim 4.3 and Lemma 2.22 that $R_{\min}(t)$ is a non-decreasing function of t. This establishes the first item and also the second item in the case when $R_{\min}(0) = 0$.

Suppose that $R_{\min}(0) \ne 0$. Consider the function

$$S(t) = \frac{-1}{R_{\min}(t)} - \frac{2t}{n} + \frac{1}{R_{\min}(0)}.$$

Clearly, $S(0) = 0$ and $S'(t) \geq 0$ (in the sense of forward difference quotients), so that by Lemma 2.22 we have $S(t) \geq 0$ for all t. This means that

$$\frac{1}{R_{\min}(t)} \leq \frac{1}{R_{\min}(0)} - \frac{2t}{n} \tag{4.1}$$

provided that $R_{\min}$ is not ever zero on the interval $[0, t]$. If $R_{\min}(0) > 0$, then by the first item, $R_{\min}(t) > 0$ for all t for which the flow is defined, and the inequality in the second item of the proposition is immediate from Equation (4.1). The third inequality in the proposition also follows easily from Equation (4.1) when $R_{\min}(t) < 0$. But if $R_{\min}(t) \geq 0$, then the third item is obvious. □

2. The maximum principle for tensors

For the applications to the Ricci curvature and the curvature tensor we need a version of the maximum principle for tensors that is due to Hamilton; see [**30**].

Suppose that V is a finite-dimensional real vector space and $Z \subset V$ is a closed convex set. For each z in the frontier of Z we define the *tangent cone to Z at z*, denoted $T_z Z$, to be the intersection of all closed half-spaces H of V such that $z \in \partial H$ and $Z \subset H$. For $z \in \mathrm{int}\, Z$ we define $T_z Z = V$. Notice that $v \notin T_z Z$ if and only if there is an affine linear function ℓ vanishing at z non-positive on Z and positive on v.

DEFINITION 4.4. Let Z be a closed convex subset of a finite-dimensional real vector space V. We say that a smooth vector field ψ defined on an open neighborhood U of Z in V *preserves Z* if for every $z \in Z$ we have $\psi(z) \in T_z Z$.

It is an easy exercise to show the following; see Lemma 4.1 on page 183 of [**30**]:

LEMMA 4.5. *Let Z be a closed convex subset in a finite-dimensional real vector space V. Let ψ be a smooth vector field defined on an open neighborhood of Z in V. Then ψ preserves Z if and only if every integral curve $\gamma \colon [0, a) \to V$ for ψ with $\gamma(0) \in Z$ has $\gamma(t) \in Z$ for all $t \in [0, a)$. Said more informally, ψ preserves Z if and only if every integral curve for ψ that starts in Z remains in Z.*

2.1. The global version. The maximum principle for tensors generalizes this to tensor flows evolving by parabolic equations. First we introduce a generalization of the notion of a vector field preserving a closed convex set to the context of vector bundles.

DEFINITION 4.6. Let $\pi \colon \mathcal{V} \to M$ be a vector bundle and let $\mathcal{Z} \subset \mathcal{V}$ be a closed subset. We say that $\mathcal{Z}$ is *convex* if for every $x \in M$ the fiber Z_x of $\mathcal{Z}$ over x is a convex subset of the vector space fiber V_x of $\mathcal{V}$ over x. Let ψ

be a fiberwise vector field on an open neighborhood $\mathcal{U}$ of $\mathcal{Z}$ in $\mathcal{V}$. We say that ψ *preserves* $\mathcal{Z}$ if for each $x \in M$ the restriction of ψ to the fiber U_x of $\mathcal{U}$ over x preserves Z_x.

The following global version of the maximum principle for tensors is Theorem 4.2 of [**30**].

THEOREM 4.7. (**The maximum principle for tensors**) *Let (M, g) be a compact Riemannian manifold. Let $\mathcal{V} \to M$ be a tensor bundle and let $\mathcal{Z} \subset \mathcal{V}$ be a closed, convex subset invariant under the parallel translation induced by the Levi-Civita connection. Suppose that ψ is a fiberwise vector field defined on an open neighborhood of $\mathcal{Z}$ in $\mathcal{V}$ that preserves $\mathcal{Z}$. Suppose that $\mathcal{T}(x, t)$, $0 \le t \le T$, is a one-parameter family of sections of $\mathcal{V}$ that evolves according to the parabolic equation*

$$\frac{\partial \mathcal{T}}{\partial t} = \Delta \mathcal{T} + \psi(\mathcal{T}).$$

If $\mathcal{T}(x, 0)$ is contained in $\mathcal{Z}$ for all $x \in M$, then $\mathcal{T}(x, t)$ is contained in $\mathcal{Z}$ for all $x \in M$ and for all $0 \le t \le T$.

For a proof we refer the reader to Theorem 4.3 and its proof (and the related Theorem 4.2 and its proof) in [**30**].

There is a slight improvement of this result where the convex set $\mathcal{Z}$ is allowed to vary with t. It is proved by the same argument; see Theorem 4.8 on page 101 of [**13**].

THEOREM 4.8. *Let (M, g) be a compact Riemannian manifold. Let $\mathcal{V} \to M$ be a tensor bundle and let $\mathcal{Z} \subset \mathcal{V} \times [0, T]$ be a closed subset with the property that for each $t \in [0, T]$ the time-slice $\mathcal{Z}(t)$ is a convex subset of $\mathcal{V} \times \{t\}$ invariant under the parallel translation induced by the Levi-Civita connection. Suppose that ψ is a fiberwise vector field defined on an open neighborhood of $\mathcal{Z}$ in $\mathcal{V} \times [0, T]$ that preserves the family $\mathcal{Z}(t)$, in the sense that any integral curve $\gamma(t)$, $t_0 \le t \le t_1$, for ψ with the property that $\gamma(t_0) \in \mathcal{Z}(t_0)$ has $\gamma(t) \in \mathcal{Z}(t)$ for every $t \in [t_0, t_1]$. Suppose that $\mathcal{T}(x, t)$, $0 \le t \le T$, is a one-parameter family of sections of $\mathcal{V}$ that evolves according to the parabolic equation*

$$\frac{\partial \mathcal{T}}{\partial t} = \Delta \mathcal{T} + \psi(\mathcal{T}).$$

If $\mathcal{T}(x, 0)$ is contained in $\mathcal{Z}(0)$ for all $x \in M$, then $\mathcal{T}(x, t)$ is contained in $\mathcal{Z}(t)$ for all $x \in M$ and for all $0 \le t \le T$.

2.2. The local version. Here is the local result. It is proved by the same argument as given in the proof of Theorem 4.3 in [**30**].

THEOREM 4.9. *Let (M, g) be a Riemannian manifold. Let $\overline{U} \subset M$ be a compact, smooth, connected, codimension-0 submanifold. Let $\mathcal{V} \to M$ be a tensor bundle and let $\mathcal{Z} \subset \mathcal{V}$ be a closed, convex subset. Suppose that ψ is a* fiberwise vector field defined on an open neighborhood of $\mathcal{Z}$ in $\mathcal{V}$ preserving $\mathcal{Z}$. Suppose that $\mathcal{Z}$ is invariant under the parallel translation induced by the Levi-Civita connection. Suppose that $\mathcal{T}(x,t)$, $0 \leq t \leq T$, is a one-parameter family of sections of $\mathcal{V}$ that evolves according to the parabolic equation

$$\frac{\partial \mathcal{T}}{\partial t} = \Delta \mathcal{T} + \psi(\mathcal{T}).$$

If $\mathcal{T}(x,0)$ is contained in $\mathcal{Z}$ for all $x \in \overline{U}$ and if $\mathcal{T}(x,t) \in \mathcal{Z}$ for all $x \in \partial \overline{U}$ and all $0 \leq t \leq T$, then $\mathcal{T}(x,t)$ is contained in $\mathcal{Z}$ for all $x \in \overline{U}$ and all $0 \leq t \leq T$.

3. Applications of the maximum principle

Now let us give some applications of these results to Riemann and Ricci curvature. In order to do this we first need to specialize the above general maximum principles for tensors to the situation of the curvature.

3.1. Ricci flows with normalized initial conditions. As we have already seen, the Ricci flow equation is invariant under multiplying space and time by the same scale. This means that there can be no absolute constants in the results about Ricci surgery. To break this gauge symmetry and make the constants absolute, we impose scale fixing (or rather scale bounding) conditions on the initial metrics of the flows that we shall consider. The following definition makes precise the exact conditions that we shall use.

DEFINITION 4.10. We say that a Ricci flow $(M, g(t))$ has *normalized initial conditions* if 0 is the initial time for the flow and if the compact Riemannian manifold $(M^n, g(0))$ satisfies:
 (1) $|\text{Rm}(x,0)| \leq 1$ for all $x \in M$.
 (2) Let ω_n be the volume of the ball of radius 1 in n-dimensional Euclidean space. Then $\text{Vol}(B(x,0,r)) \geq (\omega_n/2)r^n$ for any $p \in M$ and any $r \leq 1$.

We also use the terminology $(M, g(0))$ *is normalized* to indicate that it satisfies these two conditions.

The evolution equation for the Riemann curvature and a standard maximum principle argument show that if $(M, g(0))$ has an upper bound on the Riemann curvature and a lower bound on the volume of balls of a fixed radius, then the flow has Riemann curvature bounded above and volumes of balls bounded below on a fixed time interval. Here is the result in the context of normalized initial condition.

PROPOSITION 4.11. *There is $\kappa_0 > 0$ depending only on the dimension n such that the following holds. Let $(M^n, g(t))$, $0 \leq t \leq T$, be a Ricci flow with bounded curvature, with each $(M, g(t))$ being complete, and with normalized initial conditions. Then $|\text{Rm}(x,t)| \leq 2$ for all $x \in M$ and all*

$t \in [0, \min(T, 2^{-4})]$. Furthermore, for any $t \in [0, \min(T, 2^{-4})]$ and any $x \in M$ and any $r \leq 1$ we have $\operatorname{Vol} B(x, t, r) \geq \kappa_0 r^n$.

PROOF. The bound on the Riemann curvature follows directly from Lemma 6.1 on page 207 of [**14**] and the definition of normalized initial conditions. Once we know that the Riemann curvature is bounded by 2 on $[0, 2^{-4}]$, there is an $0 < r_0$ depending on n such that for every $x \in M$ and every $r \leq r_0$ we have $B(x, 0, r_0 r) \subset B(x, t, r) \subset B(x, 0, 1)$. Also, from the bound on the Riemann curvature and the evolution equation for volume given in Equation (3.7), we see that there is $A < \infty$ such that $\operatorname{Vol}_t(B(x, 0, s)) \geq A^{-1} \operatorname{Vol}_0(B(x, 0, s))$. Putting this together we see that

$$\operatorname{Vol}_t(B(x, t, r)) \geq A^{-1}(\omega_n/2) r_0^n r^n.$$

This proves the result. □

3.2. Extending flows. There is one other consequence that will be important for us. For a reference see [**14**] Theorem 6.3 on page 208.

PROPOSITION 4.12. *Let $(M, g(t))$, $0 \leq t < T < \infty$, be a Ricci flow with M a compact manifold. Then either the flow extends to an interval $[0, T')$ for some $T' > T$ or $|\operatorname{Rm}|$ is unbounded on $M \times [0, T)$.*

3.3. Non-negative curvature is preserved. We need to consider the tensor versions of the maximum principle when the tensor in question is the Riemann or Ricci curvature and the evolution equation is that induced by the Ricci flow. This part of the discussion is valid in dimension 3 only. We begin by evaluating the expressions in Equation (3.19) in the 3-dimensional case. Fix a symmetric bilinear form $\mathcal{S}$ on a 3-dimensional real vector space V with a positive definite inner product. The inner product determines an identification of $\wedge^2 V$ with V^*. Hence, $\wedge^2 \mathcal{S}^*$ is identified with a symmetric automorphism of V, denoted by $\mathcal{S}^\sharp$.

LEMMA 4.13. *Let (M, g) be a Riemannian 3-manifold. Let $\mathcal{T}$ be the curvature operator written with respect to the evolving frame as in Proposition 3.19. Then the evolution equation given in Proposition 3.19 is:*

$$\frac{\partial \mathcal{T}}{\partial t} = \Delta \mathcal{T} + \psi(\mathcal{T})$$

where

$$\psi(\mathcal{T}) = \mathcal{T}^2 + \mathcal{T}^\sharp.$$

In particular, in an orthonormal basis in which

$$\mathcal{T} = \begin{pmatrix} \lambda & 0 & 0 \\ 0 & \mu & 0 \\ 0 & 0 & \nu \end{pmatrix}$$

with $\lambda \geq \mu \geq \nu$, the vector field is given by

$$\psi(\mathcal{T}) = \mathcal{T}^2 + \mathcal{T}^\sharp = \begin{pmatrix} \lambda^2 + \mu\nu & 0 & 0 \\ 0 & \mu^2 + \lambda\nu & 0 \\ 0 & 0 & \nu^2 + \lambda\mu \end{pmatrix}.$$

COROLLARY 4.14. *Let $(M, g(t))$, $0 \leq t \leq T$, be a Ricci flow with M a compact, connected 3-manifold. Suppose that $\mathrm{Rm}(x, 0) \geq 0$ for all $x \in M$. Then $\mathrm{Rm}(x, t) \geq 0$ for all $x \in M$ and all $t \in [0, T]$.*

PROOF. Let $\nu_x \colon \mathrm{Sym}^2(\wedge^2 T_x^* M) \to \mathbb{R}$ associate to each endomorphism its smallest eigenvalue. Then $\nu_x(\mathcal{T})$ is the minimum over all lines in $\wedge^2 T_x M$ of the trace of the restriction of $\mathcal{T}$ to that line. As a minimum of linear functions, ν_x is a convex function. In particular, $Z_x = \nu_x^{-1}([0, \infty))$ is a convex subset. We let $\mathcal{Z}$ be the union over all x of Z_x. Clearly, $\mathcal{Z}$ is a closed convex subset of the tensor bundle. Since parallel translation is orthogonal, $\mathcal{Z}$ is invariant under parallel translation. The expressions in Lemma 4.13 show that if $\mathcal{T}$ is an endomorphism of $\wedge^2 T_x^* M$ with $\nu(\mathcal{T}) \geq 0$, then the symmetric matrix $\psi(\mathcal{T})$ is non-negative. This implies that ν_x is non-decreasing in the direction $\psi(\mathcal{T})$ at the point $\mathcal{T}$. That is to say, for each $x \in M$, the vector field $\psi(\mathcal{T})$ preserves the set $\{\nu_x^{-1}([c, \infty))\}$ for any $c \geq 0$. The hypothesis that $\mathrm{Rm}(x, 0) \geq 0$ means that $\mathrm{Rm}(x, 0) \in \mathcal{Z}$ for all $x \in M$. Applying Theorem 4.7 proves the result. □

COROLLARY 4.15. *Suppose that $(M, g(t))$, $0 \leq t \leq T$, is a Ricci flow with M a compact, connected 3-manifold with $\mathrm{Ric}(x, 0) \geq 0$ for all $x \in M$. Then $\mathrm{Ric}(x, t) \geq 0$ for all $t > 0$.*

PROOF. The statement that $\mathrm{Ric}(x, t) \geq 0$ is equivalent to the statement that for every two-plane in $\wedge^2 T_x M$ the trace of the Riemann curvature operator on this plane is ≥ 0. For $\mathcal{T} \in \mathrm{Sym}^2(\wedge^2 T_x^* M)$, we define $s(\mathcal{T})$ as the minimum over all two-planes P in $\wedge^2 TM$ of the trace of $\mathcal{T}$ on P. The restriction s_x of s to the fiber over x is the minimum of a collection of linear functions and hence is convex. Thus, the subset $\mathcal{S} = s^{-1}([0, \infty))$ is convex. Clearly, s is preserved by orthogonal isomorphisms, so $\mathcal{S}$ is invariant under parallel translation. Let $\lambda \geq \mu \geq \nu$ be the eigenvalues of $\mathcal{T}$. According to Lemma 4.13 the derivative of s_x at $\mathcal{T}$ in the $\psi(\mathcal{T})$-direction is $(\mu^2 + \lambda\nu) + (\nu^2 + \lambda\mu) = (\mu^2 + \nu^2) + \lambda(\mu + \nu)$. The condition that $s(\mathcal{T}) \geq 0$ is the condition that $\nu + \mu \geq 0$, and hence $\mu \geq 0$, implying that $\lambda \geq 0$. Thus, if $s(\mathcal{T}) \geq 0$, it is also the case that the derivative of s_x in the $\psi(\mathcal{T})$-direction is non-negative. This implies that ψ preserves $\mathcal{S}$. Applying Theorem 4.7 gives the result. □

4. The strong maximum principle for curvature

First let us state the strong maximum principle for the heat equation.

THEOREM 4.16. *Let $\overline{U}$ be a compact, connected manifold, possibly with boundary. Let $h(x,t)$, $0 \leq t \leq T$, be a solution to the heat equation*

$$\frac{\partial h(x,t)}{\partial t} = \Delta h(x,t).$$

Suppose that h has Dirichlet boundary conditions in the sense that $h(x,t) = 0$ for all $(x,t) \in \partial \overline{U} \times [0,T]$. If $h(x,0) \geq 0$ for all $x \in \overline{U}$, then $h(x,t) \geq 0$ for all $(x,t) \in \overline{U} \times [0,T]$. If, in addition, there is $y \in \overline{U}$ with $h(y,0) > 0$, then $h(x,t) > 0$ for all $(x,t) \in \mathrm{int}(\overline{U}) \times (0,T]$.

We shall use this strong maximum principle to establish an analogous result for the curvature tensors. The hypotheses are in some ways more restrictive – they are set up to apply to the Riemann and Ricci curvature.

PROPOSITION 4.17. *Let (M,g) be a Riemannian manifold and let $\mathcal{V}$ be a tensor bundle. Suppose that $\overline{U}$ is a compact, connected, smooth submanifold of M of codimension 0. Consider a one-parameter family of sections $\mathcal{T}(x,t)$ defined for $0 \leq t \leq T$, of $\mathcal{V}$. Suppose that $\mathcal{T}$ evolves according to the equation*

$$\frac{\partial \mathcal{T}}{\partial t} = \Delta \mathcal{T} + \psi(\mathcal{T})$$

for some smooth, fiberwise vector field $\psi(\mathcal{T})$ defined on $\mathcal{V}$. Suppose that $s \colon \mathcal{V} \to \mathbb{R}$ is a function satisfying the following properties:
  (1) *For each $x \in M$ the restriction s_x to the fiber $\mathcal{V}_x$ of $\mathcal{V}$ over x is a convex function.*
  (2) *For any A satisfying $s_x(A) \geq 0$ the vector $\psi(A)$ is contained in the tangent cone of the convex set $\{y | s_x(y) \geq s_x(A)\}$ at the point A.*
  (3) *s is invariant under parallel translation.*

Suppose that $s(\mathcal{T}(x,0)) \geq 0$ for all $x \in \overline{U}$ and that $s(\mathcal{T}(x,t)) \geq 0$ for all $x \in \partial \overline{U}$ and all $t \in [0,T]$. Suppose also that there is $x_0 \in \mathrm{int}(\overline{U})$ with $s(\mathcal{T}(x_0,0)) > 0$. Then $s(\mathcal{T}(x,t)) > 0$ for all $(x,t) \in \mathrm{int}(\overline{U}) \times (0,T]$.

PROOF. Let $h \colon \overline{U} \times \{0\} \to \mathbb{R}$ be a smooth function with $h(x,0) = 0$ for all $x \in \partial \overline{U}$ and with $s(\mathcal{T}(x,0)) \geq h(x,0) \geq 0$ for all $x \in \overline{U}$. We choose h so that $h(x_0,0) > 0$. Let $h(x,t)$, $0 \leq t < \infty$, be the solution to the heat equation on $\overline{U}$,

$$\frac{\partial h}{\partial t} = \Delta h,$$

with Dirichlet boundary conditions $h(x,t) = 0$ for all $x \in \partial \overline{U}$ and all $t \geq 0$ and with the given initial conditions.

Consider the tensor bundle $\mathcal{V} \oplus \mathbb{R}$ over M. We define

$$Z_x = \left\{ (\mathcal{T},h) \in V_x \oplus \mathbb{R} \middle| s_x(\mathcal{T}) \geq h \geq 0 \right\}.$$

The union over all $x \in M$ of the Z_x defines a closed convex subset $\mathcal{Z} \subset \mathcal{V} \oplus \mathbb{R}$ which is invariant under parallel translation since s is. We consider the family of sections $(\mathcal{T}(x,t), h(x,t))$, $0 \leq t \leq T$, of $\mathcal{V} \oplus \mathbb{R}$. These evolve by

$$\frac{d(\mathcal{T}(x,t), h(x,t))}{dt} = (\Delta \mathcal{T}(x,t), \Delta h(x,t)) + \widetilde{\psi}(\mathcal{T}(x,t), h(x,t))$$

where $\widetilde{\psi}(\mathcal{T}, h) = (\psi(\mathcal{T}), 0)$. Clearly, by our hypotheses, the vector field $\widetilde{\psi}$ preserves the convex set $\mathcal{Z}$. Applying the local version of the maximum principle (Theorem 4.9), we conclude that $\mathcal{T}(x,t) \geq h(x,t)$ for all $(x,t) \in \overline{U} \times [0,T]$.

The result then follows immediately from Theorem 4.16. $\square$

4.1. Applications of the strong maximum principle. We have the following applications of the strong maximum principle.

THEOREM 4.18. *Let $(U, g(t))$, $0 \leq t \leq T$, be a 3-dimensional Ricci flow with non-negative sectional curvature with U connected but not necessarily complete and with $T > 0$. If $R(p, T) = 0$ for some $p \in U$, then $(U, g(t))$ is flat for every $t \in [0, T]$.*

PROOF. We suppose that there is $p \in U$ with $R(p, T) = 0$. Since all the metrics in the flow are of non-negative sectional curvature, if the flow does not consist entirely of flat manifolds, then there is $(q, t) \in U \times [0, T]$ with $R(q, t) > 0$. Clearly, by continuity, we can assume $t < T$. By restricting to the time interval $[t, T]$ and shifting by $-t$ we can arrange that $t = 0$. Let V be a compact, connected smooth submanifold with boundary whose interior contains q and p. Let $h(y, 0)$ be a smooth non-negative function with support in V, positive at q, such that $R(y, 0) \geq h(y, 0)$ for all $y \in V$. Let $h(y, t)$ be the solution to the heat equation on $V \times [0, T]$ that vanishes on ∂V. Of course, $h(y, T) > 0$ for all $y \in \text{int}(V)$. Also, from Equation (3.7) we have

$$\frac{\partial}{\partial t}(R - h) = \Delta(R - h) + 2|\text{Ric}|^2,$$

so that $(R - h)(y, 0) \geq 0$ on $(V \times \{0\}) \cup (\partial V \times [0, T])$. It follows from the maximum principle that $(R - h) \geq 0$ on all of $V \times [0, T]$. In particular, $R(p, T) \geq h(p, T) > 0$. This is a contradiction, establishing the theorem. $\square$

COROLLARY 4.19. *Fix $T > 0$. Suppose that $(U, g(t))$, $0 \leq t \leq T$, is a Ricci flow such that for each t, the Riemannian manifold $(U, g(t))$ is a (not necessarily complete) connected, 3-manifold of non-negative sectional curvature. Suppose that $(U, g(0))$ is not flat and that for some $p \in M$ the Ricci curvature at (p, T) has a zero eigenvalue. Then for each $t \in (0, T]$ the Riemannian manifold $(U, g(t))$ splits locally as a product of a surface of positive curvature and a line, and under this local splitting the flow is locally the product of a Ricci flow on the surface and the trivial flow on the line.*

PROOF. First notice that it follows from Theorem 4.18 that because $(U, g(0))$ is not flat, we have $R(y, t) > 0$ for every $(y, t) \in U \times (0, T]$.

We consider the function s on $\mathrm{Sym}^2(\wedge^2 T_y^* U)$ that associates to each endomorphism the sum of the smallest two eigenvalues. Then s_y is the minimum of the traces on 2-dimensional subsets in $\wedge^2 T_y U$. Thus, s is a convex function, and the subset $\mathcal{S} = s^{-1}([0, \infty))$ is a convex subset. Clearly, this subset is invariant under parallel translation. By the computations in the proof of Corollary 4.15 it is invariant under the vector field $\psi(\mathcal{T})$. The hypothesis of the corollary tells us that $s(p, T) = 0$. Suppose that $s(q, t) > 0$ for some $(q, t) \in U \times [0, T]$. Of course, by continuity we can take $t < T$. Shift the time parameter so that $t = 0$, and fix a compact connected, codimension-0 submanifold V containing p, q in its interior. Then by Theorem 4.17 $s(y, T) > 0$ for all $y \in \mathrm{int}(V)$ and in particular $s(p, T) > 0$. This is a contradiction, and we conclude that $s(q, t) = 0$ for all $(q, t) \in U \times [0, T]$.

Since we have already established that each $R(y, t) > 0$ for all $(y, t) \in U \times (0, T]$, so that $\mathrm{Rm}(y, t)$ is not identically zero, this means that for all $y \in U$ and all $t \in (0, T]$, the null space of the operator $\mathrm{Rm}(y, t)$ is a 2-dimensional subspace of $\wedge^2 T_y U$. This 2-dimensional subspace is dual to a line in $T_x M$. Thus, we have a one-dimensional distribution (a line bundle in the tangent bundle) $\mathcal{D}$ in $U \times (0, T]$ with the property that the sectional curvature $\mathrm{Rm}(y, t)$ vanishes on any 2-plane containing the line $\mathcal{D}(y, t)$. The fact that the sectional curvature of $g(t)$ vanishes on all two-planes in $T_y M$ containing $\mathcal{D}(y, t)$ means that its eigenvalues are $\{\lambda, 0, 0\}$ where $\lambda > 0$ is the sectional curvature of the $g(t)$-orthogonal 2-plane to $\mathcal{D}(y, t)$. Hence $\mathcal{R}(V(y, t), \cdot, \cdot, \cdot) = 0$.

Locally in space and time, there is a unique (up to sign) vector field $V(y, t)$ that generates $\mathcal{D}$ and satisfies $|V(y, t)|^2_{g(t)} = 1$. We wish to show that this local vector field is invariant under parallel translation and time translation; cf. Lemma 8.2 in [30]. Fix a point $x \in M$, a direction X at x, and a time t. Let $\widetilde{V}(y, t)$ be a parallel extension of $V(x, t)$ along a curve C passing through x in the X-direction, and let $\widetilde{W}(y, t)$ be an arbitrary parallel vector field along C. Since the sectional curvature is non-negative, we have $\mathcal{R}(\widetilde{V}, \widetilde{W}, \widetilde{V}, \widetilde{W})(y) \geq 0$ for all $y \in C$; furthermore, this expression vanishes at x. Hence, its first variation vanishes at x. That is to say

$$\nabla\left(\mathcal{R}(\widetilde{V}, \widetilde{W}, \widetilde{V}, \widetilde{W})\right)(x, t) = (\nabla \mathcal{R})(\widetilde{V}, \widetilde{W}, \widetilde{V}, \widetilde{W})$$

vanishes at (x, t). Since this is true for all $\widetilde{W}$, it follows that the null space of the quadratic form $\nabla \mathcal{R}(x, t)$ contains the null space of $\mathcal{R}(x, t)$, and thus

$$(\nabla \mathcal{R})(V(x, t), \cdot, \cdot, \cdot) = 0.$$

Now let us consider three parallel vector fields $\widetilde{W}_1, \widetilde{W}_2$, and $\widetilde{W}_3$ along C. We compute $0 = \nabla_X\left(\mathcal{R}(V(y, t), \widetilde{W}_1(y, t), \widetilde{W}_2(y, t), \widetilde{W}_3(y, t))\right)$. (Notice that while the $\widetilde{W}_i$ are parallel along C, $V(y, t)$ is defined to be the vector field spanning $\mathcal{D}(y, t)$ rather than a parallel extension of $V(x, t)$.) Given the

above result we find that
$$0 = 2\mathcal{R}(\nabla_X V(x,t), \widetilde{W}_1(x,t), \widetilde{W}_2(x,t), \widetilde{W}_3(x,t)).$$
Since this is true for all triples of vector fields $\widetilde{W}_i(x,t)$, it follows that $\nabla_X V(x,t)$ is a real multiple of $V(x,t)$. But since $|V(y,t)|^2_{g(t)} = 1$, we see that $\nabla_X V(x,t)$ is orthogonal to $V(x,t)$. We conclude that $\nabla_X V(x,t) = 0$. Since x and X are general, this shows that the local vector field $V(x,t)$ is invariant under the parallel translation associated to the metric $g(t)$.

It follows that locally $(M, g(t))$ is a Riemannian product of a surface of positive curvature with a line. Under this product decomposition, the curvature is the pullback of the curvature of the surface. Hence, by Equation (3.5), under Ricci flow on the 3-manifold, the time derivative of the curvature at time t also decomposes as the pullback of the time derivative of the curvature of the surface under Ricci flow on the surface. In particular, $(\partial \mathcal{R}/\partial t)(V, \cdot, \cdot, \cdot) = 0$. It now follows easily that $\partial V(x,t)/\partial t = 0$.

This completes the proof that the unit vector field in the direction $\mathcal{D}(x,t)$ is invariant under parallel translation and under time translation. Thus, there is a local Riemannian splitting of the 3-manifold into a surface and a line, and this splitting is invariant under the Ricci flow. This completes the proof of the corollary. □

In the complete case, this local product decomposition globalizes in some cover; see Lemma 9.1 in [**30**].

COROLLARY 4.20. *Suppose that $(M, g(t))$, $0 \leq t \leq T$, is a Ricci flow of complete, connected Riemannian 3-manifolds with $\mathrm{Rm}(x,t) \geq 0$ for all (x,t) and with $T > 0$. Suppose that $(M, g(0))$ is not flat and that for some $x \in M$ the endomorphism $\mathrm{Rm}(x,T)$ has a zero eigenvalue. Then M has a cover $\widetilde{M}$ such that, denoting the induced family of metrics on this cover by $\widetilde{g}(t)$, we have that $(\widetilde{M}, \widetilde{g}(t))$ splits as a product*
$$(N, h(t)) \times (\mathbb{R}, ds^2)$$
where $(N, h(t))$ is a surface of positive curvature for all $0 < t \leq T$. The Ricci flow is a product of the Ricci flow $(N, h(t))$, $0 \leq t \leq T$, with the trivial flow on $\mathbb{R}$.

REMARK 4.21. Notice that there are only four possibilities for the cover required by the corollary. It can be trivial, or a normal $\mathbb{Z}$-cover, or it can be a two-sheeted cover or a normal infinite dihedral group cover. In the first two cases, there is a unit vector field on M parallel under $g(t)$ for all t spanning the null direction of Ric. In the last two cases, there is no such vector field, only a non-orientable line field.

Let (N, g) be a Riemannian manifold. Recall from Definition 1.14 that the open cone on (N, g) is the space $N \times (0, \infty)$ with the Riemannian metric

$\widetilde{g}(x,s) = s^2 g(x) + ds^2$. An extremely important result for us is that open pieces in non-flat cones cannot arise as the result of Ricci flow with non-negative curvature.

PROPOSITION 4.22. *Suppose that $(U, g(t))$, $0 \le t \le T$, is a 3-dimensional Ricci flow with non-negative sectional curvature, with U being connected but not necessarily complete and $T > 0$. Suppose that $(U, g(T))$ is isometric to a non-empty open subset of a cone over a Riemannian manifold. Then $(U, g(t))$ is flat for every $t \in [0, T]$.*

PROOF. If $(U, g(T))$ is flat, then by Theorem 4.18 for every $t \in [0, T]$ the Riemannian manifold $(U, g(t))$ is flat.

We must rule out the possibility that $(U, g(T))$ is non-flat. Suppose that $(U, g(T))$ is an open subset in a non-flat cone. According to Proposition 1.15, for each $x \in U$ the Riemann curvature tensor of $(U, g(T))$ at x has a 2-dimensional null space in $\wedge^2 T_x U$. Since we are assuming that $(U, g(T))$ is not flat, the third eigenvalue of the Riemann curvature tensor is not identically zero. Restricting to a smaller open subset if necessary, we can assume that the third eigenvalue is never zero. By the computations in Proposition 1.15, the non-zero eigenvalue is not constant, and in fact it scales by s^{-2} in the terminology of that proposition, as we move along the cone lines. Of course, the 2-dimensional null-space for the Riemann curvature tensor at each point is equivalent to a line field in the tangent bundle of the manifold. Clearly, that line field is the line field along the cone lines. Corollary 4.19 says that since the Riemann curvature of $(U, g(T))$ has a 2-dimensional null-space in $\wedge^2 T_x U$ at every point $x \in U$, the Riemannian manifold $(U, g(T))$ locally splits as a Riemannian product of a line with a surface of positive curvature, and the 2-dimensional null-space for the Riemann curvature tensor is equivalent to the line field in the direction of the second factor. Along these lines the non-zero eigenvalue of the curvature is constant. This is a contradiction and establishes the result. □

We also have Hamilton's result (Theorem 15.1 in [**29**]) that compact 3-manifolds of non-negative Ricci curvature become round under Ricci flow:

THEOREM 4.23. *Suppose that $(M, g(t))$, $0 \le t < T$, is a Ricci flow with M being a compact 3-dimensional manifold. If $\mathrm{Ric}(x, 0) \ge 0$ for all $x \in M$, then either $\mathrm{Ric}(x, t) > 0$ for all $(x, t) \in M \times (0, T)$ or $\mathrm{Ric}(x, t) = 0$ for all $(x, t) \in M \times [0, T)$. Suppose that $\mathrm{Ric}(x, t) > 0$ for some (x, t) and that the flow is maximal in the sense that there is no $T' > T$ and an extension of the given flow to a flow defined on the time interval $[0, T')$. For each (x, t), let $\lambda(x, t)$, resp. $\nu(x, t)$, denote the largest, resp. smallest, eigenvalue of $\mathrm{Rm}(x, t)$ on $\wedge^2 T_x M$. Then as t tends to T the Riemannian manifolds*

$(M, g(t))$ are becoming round in the sense that
$$\lim_{t \to T} \frac{\max_{x \in M} \lambda(x,t)}{\min_{x \in M} \nu(x,t)} = 1.$$

Furthermore, for any $x \in M$ the largest eigenvalue $\lambda(x,t)$ tends to ∞ as t tends to T, and rescaling $(M, g(t))$ by $\lambda(x,t)$ produces a family of Riemannian manifolds converging smoothly as t goes to T to a compact round manifold. In particular, the underlying smooth manifold supports a Riemannian metric of constant positive curvature so that the manifold is diffeomorphic to a 3-dimensional spherical space-form.

Hamilton's proof in [29] uses the maximum principle and Shi's derivative estimates.

4.2. Solitons of positive curvature. One nice application of this pinching result is the following theorem.

THEOREM 4.24. *Let (M, g) be a compact 3-dimensional soliton of positive Ricci curvature. Then (M, g) is round. In particular, (M, g) is the quotient of S^3 with a round metric by a finite subgroup of $O(4)$ acting freely; that is to say, M is a 3-dimensional spherical space-form.*

PROOF. Let $(M, g(t))$, $0 \leq t < T$, be the maximal Ricci flow with initial manifold (M, g). Since $\operatorname{Ric}(x, 0) > 0$ for all $x \in M$, it follows from Theorem 4.23 that $T < \infty$ and that as t tends to T the metrics $g(t)$ converge smoothly to a round metric. Since all the manifolds $(M, g(t))$ are isometric up to diffeomorphism and a constant conformal factor, this implies that all the $g(t)$ are of constant positive curvature.

The last statement is a standard consequence of the fact that the manifold has constant positive curvature. □

REMARK 4.25. After we give a stronger pinching result in the next section, we shall improve this result, replacing the positive Ricci curvature assumption by the *a priori* weaker assumption that the soliton is a shrinking soliton.

5. Pinching toward positive curvature

As the last application of the maximum principle for tensors we give a theorem due to R. Hamilton (Theorem 4.1 in [36]) and T. Ivey [41] which shows that, in dimension 3, as the scalar curvature gets large, the sectional curvatures pinch toward positive. Of course, if the sectional curvatures are non-negative, then the results in the previous section apply. Here, we are considering the case when the sectional curvature is not everywhere positive. The pinching result says roughly the following: At points where the Riemann curvature tensor has a negative eigenvalue, the smallest (thus negative) eigenvalue of the Riemann curvature tensor divided by the largest eigenvalue limits to zero as the scalar curvature grows. This result is central in the analysis of singularity development in finite time for a 3-dimensional Ricci flow.

THEOREM 4.26. *Let $(M, g(t))$, $0 \le t < T$, be a Ricci flow with M a compact 3-manifold. Assume that for every $x \in M$ the smallest eigenvalue of* $\operatorname{Rm}(x, 0)$, *denoted* $\nu(x, 0)$, *is at least* -1. *Set* $X(x, t) = \max(-\nu(x, t), 0)$. *Then we have:*

(1) $R(x, t) \ge \frac{-6}{4t+1}$, *and*
(2) *for all (x, t) for which $0 < X(x, t)$,*

$$R(x, t) \ge 2X(x, t)\left(\log X(x, t) + \log(1 + t) - 3\right).$$

For any fixed t, the limit as X goes to 0 from above of

$$X(\log(X) + \log(1 + t) - 3)$$

is zero, so that it is natural to interpret this expression to be zero when $X = 0$. Of course, when $X(x, t) = 0$ all the eigenvalues of $\operatorname{Rm}(x, t)$ are non-negative so that $R(x, t) \ge 0$ as well. Thus, with this interpretation of the expression in part 2 of the theorem, it remains valid even when $X(x, t) = 0$.

REMARK 4.27. This theorem tells us, among other things, that as the scalar curvature goes to infinity, the absolute values of all the negative eigenvalues (if any) of Rm are arbitrarily small with respect to the scalar curvature.

The proof we give below follows Hamilton's original proof in [**36**] very closely.

PROOF. First note that by Proposition 4.1, if $R_{\min}(0) \ge 0$, then the same is true for $R_{\min}(t)$ for every $t > 0$ and thus the first inequality stated in the theorem is clearly true. If $R_{\min}(0) < 0$, the first inequality stated in the theorem follows easily from the last inequality in Proposition 4.1.

We turn now to the second inequality in the statement of the theorem. Consider the tensor bundle $\mathcal{V} = \operatorname{Sym}^2(\wedge^2 T^* M)$. Then the curvature operator, written in the evolving frame $\mathcal{T}(x, t)$, is a one-parameter family of smooth sections of this bundle, evolving by

$$\frac{\partial \mathcal{T}}{\partial t} = \Delta \mathcal{T} + \psi(\mathcal{T}).$$

We consider two subsets of $\mathcal{V}$. There are two solutions to

$$x(\log(x) + (\log(1 + t) - 3) = -3/(1 + t).$$

5. PINCHING TOWARD POSITIVE CURVATURE

One is $x = 1/(1+t)$; let $\xi(t) > 1/(1+t)$ be the other. We set $S(\mathcal{T}) = \text{tr}(\mathcal{T})$, so that $R = 2S$, and we set $X(\mathcal{T}) = \max(-\nu(\mathcal{T}), 0)$. Define

$$\mathcal{Z}_1(t) = \{\mathcal{T} \in \mathcal{V} | S(\mathcal{T}) \geq -\frac{3}{(1+t)}\},$$

$$\mathcal{Z}_2(t) = \{\mathcal{T} \in \mathcal{V} | S(\mathcal{T}) \geq f_t(X(\mathcal{T})), \quad \text{if} \quad X(\mathcal{T}) \geq \xi(t)\},$$

where $f_t(x) = x(\log x + \log(1+t) - 3)$. Then we define

$$\mathcal{Z}(t) = \mathcal{Z}_1(t) \cap \mathcal{Z}_2(t).$$

CLAIM 4.28. *For each $x \in M$ and each $t \geq 0$, the fiber $Z(x,t)$ of $\mathcal{Z}(t)$ over x is a convex subset of $\text{Sym}^2(\wedge^2 T^*M)$.*

PROOF. First consider the function $f_t(x) = x(\log(x) + \log(1+t) - 3)$ on the interval $[\xi(t), \infty)$. Direct computation shows that $f'(x) > 0$ and $f''(x) > 0$ on this interval. Hence, for every $t \geq 0$ the region $\mathcal{C}(t)$ in the S-X plane, defined by $S \geq -3/(1+t)$ and in addition by $S \geq f_t(X)$ when $X \geq \xi(t)$, is convex and has the property that if $(S, X) \in \mathcal{C}(t)$ then so is (S, X') for all $X' \leq X$. (See FIG. 1). By definition an element $\mathcal{T} \in \mathcal{V}$ is contained in $\mathcal{Z}(t)$ if and only if $(S(\mathcal{T}), X(\mathcal{T})) \in \mathcal{C}(t)$. Now fix $t \geq 0$ and suppose that $\mathcal{T}_1$ and $\mathcal{T}_2$ are elements of $\text{Sym}^2(\wedge^2 T^*M_x)$ such that setting $S_i = \text{tr}(\mathcal{T}_i)$ and $X_i = X(\mathcal{T}_i)$ we have $(S_i, X_i) \in \mathcal{C}(t)$ for $i = 1, 2$. Then we consider $\mathcal{T} = s\mathcal{T}_1 + (1-s)\mathcal{T}_2$ for some $s \in [0,1]$. Let $S = \text{tr}(\mathcal{T})$ and $X = X(\mathcal{T})$. Since $\mathcal{C}(t)$ is convex, we know that $(sS_1 + (1-s)S_2, sX_1 + (1-s)X_2) \in \mathcal{C}(t)$, so that $\mathcal{T} \in \mathcal{Z}(t)$. Clearly, $S = sS_1 + (1-s)S_2$, so that we conclude that $(S, (sX_1 + (1-s)X_2)) \in \mathcal{C}(t)$. But since ν is a convex function, X is a concave function, i.e., $X \leq sX_1 + (1-s)X_2$. Hence $(S, X) \in \mathcal{C}(t)$. □

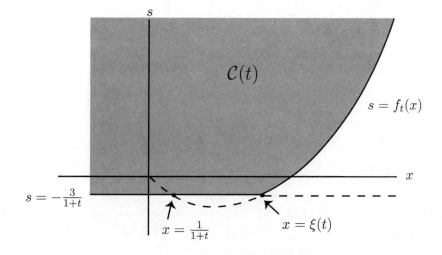

FIGURE 1. Curvature convex set.

CLAIM 4.29. *$\mathcal{T}(x, 0) \in Z(x, 0)$ for all $x \in M$.*

PROOF. Note that by the hypothesis of the theorem the trace of $\mathrm{Rm}(x,0)$ is at least -3 so $(S(x,0), X(x,0)) \in \mathcal{C}(0)$ for all $x \in M$. On the other hand, if $0 < X(x,0)$, then since $X(x,0) \le 1$ we have $S(x,0) \ge -3X(x,0) \ge X(\log X - 3)$. This completes the proof that $\mathcal{T}(x,0) \in \mathcal{C}(0)$ for all $x \in M$. □

CLAIM 4.30. *The vector field $\psi(\mathcal{T}) = \mathcal{T}^2 + \mathcal{T}^\sharp$ preserves the family $\mathcal{Z}(t)$ of convex sets.*

PROOF. We denote the eigenvalues of $\mathrm{Rm}(x,t)$ by $\lambda(x,t) \ge \mu(x,t) \ge \nu(x,t)$. Fix $x \in M$ and let $\gamma(t)$, $t_0 \le t \le T$, be an integral curve for ψ with $\gamma(t_0) \in Z(x,t_0)$. We wish to show that $\gamma(t) \in Z(x,t)$ for all $t \in [t_0, T]$. The function $S(t) = S(\gamma(t))$ satisfies

$$\frac{dS}{dt} = \lambda^2 + \mu^2 + \nu^2 + \lambda\mu + \lambda\nu + \mu\nu = \frac{1}{2}\left((\lambda+\mu)^2 + (\lambda+\nu)^2 + (\mu+\nu)^2\right).$$

By Cauchy-Schwarz we have

$$(\lambda+\mu)^2 + (\lambda+\nu)^2 + (\mu+\nu)^2) \ge \frac{4S^2}{3} \ge \frac{2S^2}{3}.$$

Since $\gamma(t_0) \in Z(x,t_0)$ we have $S(t_0) \ge -3/(1+t_0)$. It then follows that

(4.2) $\qquad\qquad S(t) \ge -3/(1+t) \quad \text{for all} \quad t \ge t_0.$

Now let us consider the evolution of $X(t) = X(\gamma(t))$. Assume that we are at a point t for which $X(t) > 0$. For this computation we set $Y = -\mu$.

$$\frac{dX}{dt} = -\frac{d\nu}{dt} = -\nu^2 - \mu\lambda = -X^2 + Y\lambda,$$
$$\frac{dS}{dt} = \frac{d(\nu+\mu+\lambda)}{dt} = \nu^2 + \mu^2 + \lambda^2 + \mu\lambda + \nu\lambda + \nu\mu$$
$$= X^2 + Y^2 + \lambda^2 + XY - \lambda(X+Y).$$

Putting this together yields

(4.3) $\qquad\qquad X\frac{dS}{dt} - (S+X)\frac{dX}{dt} = X^3 + I,$

where $I = XY^2 + \lambda Y(Y-X) + \lambda^2(X-Y)$.

CLAIM 4.31. *$I \ge 0$.*

PROOF. First we consider the case when $Y \le 0$. This means that $\mu \ge 0$ and hence that $\lambda \ge 0$. Since by definition $X \ge 0$, we have $X \ge Y$. This immediately gives $I \ge 0$. Now let us consider the case when $Y > 0$ which means that $\nu \le \mu < 0$. In this case, we have

$$I = Y^3 + (X-Y)(\lambda^2 - \lambda Y + Y^2) > 0$$

since $X \ge Y$ and $\lambda^2 - \lambda Y + Y^2 = (\lambda - \frac{Y}{2})^2 + \frac{3Y^2}{4} > 0$. □

The above claim and Equation (4.3) immediately imply that

$$X\frac{dS}{dt} - (S+X)\frac{dX}{dt} \geq X^3. \tag{4.4}$$

Set $W = \frac{S}{X} - \log X$, then rewriting Equation (4.4) in terms of W gives

$$\frac{dW}{dt} \geq X. \tag{4.5}$$

Now suppose that $\gamma(t) \not\in Z(x,t)$ for some $t \in [t_0, T]$. Let $t_1 < T$ be maximal subject to the condition that $\gamma(t) \in Z(x,t)$ for all $t_0 \leq t \leq t_1$. Of course, $\gamma(t_1) \in \partial Z(x,t_1)$ which implies that $(S(t_1), X(t_1)) \in \partial \mathcal{C}(t_1)$. There are two possibilities: either $S(t_1) = -3/(1+t_1)$ and $X(t_1) < \xi(t_1)$ or $X(t_1) \geq \xi(t_1) > 1/(1+t_1)$ and $S(t_1) = f_{t_1}(X(t_1))$. But Equation (4.2) implies that $S(t) \geq -3/(1+t)$ for all t. Hence, if the first case holds then $\gamma(t) \in Z(x,t)$ for t in some interval $[t_0, t_1']$ with $t_1' > t_1$. This contradicts the maximality of t_1. Thus, it must be the case that $X(t_1) \geq \xi(t_1)$. But then $X(t) > \frac{1}{1+t}$ for all t sufficiently close to t_1. Hence, by Equation (4.5) we have

$$\frac{dW}{dt}(t) \geq X(t) > \frac{1}{1+t},$$

for all t sufficiently close to t_1. Also, since $S(t_1) = f_{t_1}(X(t_1))$, we have $W(t_1) = \log(1+t_1) - 3$. It follows immediately that $W(t) \geq \log(1+t) - 3$ for all $t > t_1$ sufficiently close to t_1. This proves that $S(t) \geq f_t(X(t))$ for all $t \geq t_1$ sufficiently close to t_1, again contradicting the maximality of t_1.

This contradiction proves that ψ preserves the family $\mathcal{Z}(t)$. □

By Theorem 4.8, the previous three claims imply that $\mathcal{T}(x,t) \in \mathcal{Z}(t)$ for all $x \in M$ and all $t \in [0,T)$. That is to say, $S(x,t) \geq -3/(1+t)$ and $S(x,t) \geq f_t(X(x,t))$ whenever $X(x,t) \geq \xi(t)$. For $X \in [1/(1+t), \xi(t)]$ we have $f_t(X) \leq -3/(1+t)$, and thus in fact $S(x,t) \geq f_t(X(x,t))$ as long as $X(x,t) \geq 1/(1+t)$. On the other hand, if $0 < X(x,t) \leq 1/(1+t)$ then $f_t(X(x,t)) < -3X(x,t) \leq S(x,t)$. On the other hand, since $X(x,t)$ is the negative of the smallest eigenvalue of $\mathcal{T}(x,t)$ and $S(x,t)$ is the trace of this matrix, we have $S(x,t) \geq -3X(x,t)$. Thus, $S(x,t) \geq f_t(X(x,t))$ in this case as well. This completes the proof of Theorem 4.26. □

Actually, the proof establishes a stronger result which we shall need.

THEOREM 4.32. *Fix $a \geq 0$. Let $(M, g(t))$, $a \leq t < T$, be a Ricci flow with M a compact 3-manifold. Suppose the eigenvalues of $\mathrm{Rm}(x,t)$ are $\lambda(x,t) \geq \mu(x,t) \geq \nu(x,t)$ and set $X(x,t) = \max(-\nu(x,t), 0)$. Assume that for every $x \in M$ we have*

$$R(x,a) \geq \frac{-6}{4a+1}$$

and

$$R(x,a) \geq 2X(x,a)\left(\log X(x,a) + \log(1+a) - 3\right),$$

where the second inequality holds whenever $X(x,a) > 0$. Then for all $a \leq t < T$ and all $x \in M$ we have:

(4.6) $\quad R(x,t) \geq \dfrac{-6}{4t+1}.$

(4.7) $\quad R(x,t) \geq 2X(x,t)\left(\log X(x,t) + \log(1+t) - 3\right)\quad$ if $X(x,t) > 0$.

Once again it is natural to interpret the right-hand side of the second inequality to be zero when $X(x,t) = 0$. With this convention, the second inequality remains true even when $X(x,t) = 0$.

COROLLARY 4.33. *Fix $a \geq 0$. Suppose that $(M, g(t))$, $a \leq t < T$, is a Ricci flow with M a compact 3-manifold, and suppose that the two hypotheses of the previous theorem hold. Then there is a continuous function ϕ such that for all $R_0 < \infty$, if $R(x,t) \leq R_0$ then $|\mathrm{Rm}(x,t)| \leq \phi(R_0)$.*

PROOF. Fix $R_0 \geq e^4$ sufficiently large, and suppose that $R(x,t) \leq R_0$. If $X(x,t) = 0$, then $|\mathrm{Rm}(x,t)| \leq R(x,t)/2$. If $X(x,t) > 0$, then by Theorem 4.32 it is bounded by R_0. Thus, $\lambda(x,t) \leq 3R_0$. Thus, we have an upper bound on $\lambda(x,t)$ and a lower bound on $\nu(x,t)$ in terms of R_0. □

This theorem leads to a definition.

DEFINITION 4.34. Let $(\mathcal{M}, G)$ be a generalized Ricci flow whose domain of definition is contained in $[0, \infty)$. Then we say that $(\mathcal{M}, G)$ has *curvature pinched toward positive* if for every $x \in \mathcal{M}$ the following two conditions hold:

$$R(x) \geq \dfrac{-6}{4\mathbf{t}(x) + 1},$$
$$R(x) \geq 2X(x)\left(\log X(x) + \log(1 + \mathbf{t}(x)) - 3\right) \quad \text{if } X(x) > 0,$$

where, as in the statement of Theorem 4.26, $X(x)$ is the maximum of zero and the negative of the smallest eigenvalue of $\mathrm{Rm}(x)$.

The content of Theorem 4.32 is that if $(M, g(t))$, $0 \leq a \leq t < T$, is a Ricci flow with M a compact 3-manifold and if the curvature of $(M, g(a))$ is pinched toward positive, then the same is true for the entire flow.

5.1. Application of the pinching result. As an application of this pinching toward positive curvature result we establish a strengthening of Theorem 4.24.

THEOREM 4.35. *Let (M, g) be a compact 3-dimensional shrinking soliton, i.e., there is a Ricci flow $(M, g(t))$, $0 \leq t < T$, so that for each $t \in [0, T)$ there is a constant $c(t)$ with $\lim_{t \to T} c(t) = 0$ and with the property that there is an isometry from $(M, g(t))$ to $(M, c(t)g)$. Then (M, g) is round.*

PROOF. By rescaling we can assume that for all $x \in M$ all the eigenvalues of $\mathrm{Rm}(x, 0)$ have absolute value ≤ 1. This implies that $(M, g(0))$ satisfies the hypothesis of Theorem 4.26. Our first goal is to show that

$\mathrm{Rm}(x,0) \geq 0$ for all $x \in M$. Suppose that this is not true; then there is a point x with $X(x,0) > 0$. Consider $A = X(x,0)/R(x,0)$. For each $t < T$ let $x_t \in M$ be the image of x under the isometry from $(M, g(0))$ to $(M, c(t)g(t))$. Then $X(x_t, t) = c^{-1}(t) X(x, 0)$ and $X(x_t, t)/R(x_t, t) = A$. Since $c(t)$ tends to 0 as t approaches T, this contradicts Theorem 4.26. Now, according to Theorem 4.18 either all $(M, g(t))$ are flat or $\mathrm{Rm}(x,t) > 0$ for all $(x,t) \in M \times (0, T)$. But if the $(M, g(t))$ are all flat, then the flow is trivial and hence the diameters of the $(M, g(t))$ do not go to zero as t approaches T, contradicting the hypothesis. Hence, $\mathrm{Rm}(x,t) > 0$ for all $(x,t) \in M \times (0, T)$. According to Theorem 4.23 this means that as the singularity develops the metrics are converging to round. By the shrinking soliton hypothesis, this implies that all the metrics $(M, g(t))$, $0 < t < T$, are in fact round. Of course, it then follows that (M, g) is round. □

The following more general result was first given by T. Ivey [**41**].

THEOREM 4.36. *Any 3-dimensional compact Ricci soliton g_0 is Einstein.*

Since we do not need this result, we do not include a proof.

5.2. The Harnack inequality. The last consequence of the maximum principle that we need is Hamilton's version of the Harnack inequality for Ricci flows, see Theorem 1.1 and Corollary 1.2 of [**32**].

THEOREM 4.37. *Suppose that $(M, g(t))$ is a Ricci flow defined for (T_0, T_1) with $(M, g(t))$ a complete manifold of non-negative curvature operator with bounded curvature for each $t \in (T_0, T_1)$. Then for any time-dependent vector field $\chi(x,t)$ on M we have:*

$$\frac{\partial R(x,t)}{\partial t} + \frac{R(x,t)}{t - T_0} + 2\langle \chi(x,t), \nabla R(x,t) \rangle + 2\mathrm{Ric}(x,t)(\chi(x,t), \chi(x,t)) \geq 0.$$

In particular, we have

$$\frac{\partial R(x,t)}{\partial t} + \frac{R(x,t)}{t - T_0} \geq 0.$$

REMARK 4.38. Notice that the second result follows from the first by taking $\chi = 0$.

COROLLARY 4.39. *If $(M, g(t))$ is a Ricci flow defined for $-\infty < t \leq 0$ with $(M, g(t))$ a complete manifold of bounded, non-negative curvature operator for each t, then*

$$\frac{\partial R(x,t)}{\partial t} \geq 0.$$

PROOF. Apply the above theorem with $\chi(x,t) = 0$ for all (x,t) and for a sequence of $T_0 \to -\infty$. □

The above is the differential form of Hamilton's Harnack inequality. There is also the integrated version, also due to Hamilton; see Corollary 1.3 of [**32**].

THEOREM 4.40. *Suppose that $(M, g(t))$ is a Ricci flow defined for $t_1 \leq t \leq t_2$ with $(M, y(l))$ a complete manifold of non-negative, bounded curvature operator for all $t \in [t_1, t_2]$. Let x_1 and x_2 be two points of M. Then*

$$\log\left(\frac{R(x_2, t_2)}{R(x_1, t_1)}\right) \geq -\frac{d_{t_1}^2(x_2, x_1)}{2(t_2 - t_1)}.$$

PROOF. Apply the differential form of the Harnack inequality to

$$\chi = -\nabla(\log R)/2 = -\nabla R/2R,$$

and divide by R. The result is

$$R^{-1}(\partial R/\partial t) - |\nabla(\log R)|^2 + \frac{\operatorname{Ric}(\nabla(\log R), \nabla(\log R))}{2R} \geq 0.$$

Since $\operatorname{Ric}(A, A)/R \leq |A|^2$, it follows that

$$\frac{\partial}{\partial t}(\log R) - \frac{|\nabla(\log R)|^2}{2} \geq 0.$$

Let d be the $g(t_1)$-distance from x_1 to x_2 and let $\gamma\colon [t_1, t_2] \to M$ be a $g(t_1)$-geodesic from x_1 to x_2, parameterized at speed $d/(t_2 - t_1)$. Then let $\mu(t) = (\gamma(t), t)$ be a path in space-time. We compute

$$\log\left(\frac{R(x_2, t_2)}{R(x_1, t_1)}\right) = \int_{t_1}^{t_2} \frac{d}{dt}\log(R(\mu(t))dt$$

$$= \int_{t_1}^{t_2} \frac{\frac{\partial R}{\partial t}(\mu(t))}{R(\mu(t))} + \langle \nabla(\log R)(\mu(t)), \frac{d\mu}{dt}(\mu(t))\rangle dt$$

$$\geq \int_{t_1}^{t_2} \frac{1}{2}|\nabla(\log R)(\mu(t))|^2 - |\nabla(\log R)(\mu(t))| \cdot \left|\frac{d\gamma}{dt}\right| dt$$

$$\geq -\frac{1}{2}\int_{t_1}^{t_2} \left|\frac{d\gamma}{dt}\right|^2 dt,$$

where the last inequality comes by completing the square. Since $\operatorname{Ric}(x, t) \geq 0$, we have $|d\gamma/dt|_{g(t)} \leq |d\gamma/dt|_{g(t_1)}$, and thus

$$\log\left(\frac{R(x_2, t_2)}{R(x_1, t_1)}\right) \geq -\frac{1}{2}\int_{t_1}^{t_2} \left|\frac{d\gamma}{dt}\right|^2_{g(t_1)} dt.$$

Since γ is a $g(t_1)$-geodesic, this latter expression is

$$-\frac{d_{g(t_1)}^2(x_1, x_2)}{2(t_2 - t_1)}.$$

□

CHAPTER 5

Convergence results for Ricci flow

The most obvious notion of smooth convergence of Riemannian manifolds is the C^∞-version of Cheeger-Gromov compactness: We have a sequence of Riemannian metrics g_n on a fixed smooth manifold M converging uniformly on compact subsets of M in the C^∞-topology to a limit metric g_∞. There is also a version of this compactness for based, complete Riemannian manifolds. The most common starts with a sequence of based, complete Riemannian manifolds (M_n, g_n, x_n), typically of unbounded diameter. Then a geometric limit is a based, complete $(M_\infty, g_\infty, x_\infty)$ so that for every $R < \infty$ the metric balls $B(x_n, R) \subset M_n$ converge uniformly in the C^∞-topology to the metric ball $B(x_\infty, R) \subset M_\infty$. This allows the topology to change – even if all the M_n are diffeomorphic, M_∞ can have a different topological type; for example the M_n could all be compact and M_∞ could be non-compact.

But we also need to be able to deal with incomplete limits. In the case of incomplete limits, the basic idea remains the same, but it requires some care to give a definition of a geometric limit that makes it unique up to canonical isometry. One must somehow impose conditions that imply that the limit eventually fills up most of each of the manifolds in the sequence.

1. Geometric convergence of Riemannian manifolds

Above we referred to filling up 'most' of the manifold. The measure of most of the manifold is in terms of the δ-regular points as defined below.

DEFINITION 5.1. Let (U, g) be a Riemannian manifold. Let $\delta > 0$ be given. We say that $p \in U$ is a δ-regular point if for every $r' < \delta$ the metric ball $B(p, r')$ has compact closure in U. Equivalently, p is δ-regular if the exponential mapping at p is defined on the open ball of radius δ centered at the origin in $T_p U$, i.e., if each geodesic ray emanating from p extends to a geodesic defined on $[0, \delta)$. We denote by $\text{Reg}_\delta(U, g)$ the subset of δ-regular points in (U, g). For any $x \in \text{Reg}_\delta(U, g)$ we denote by $\text{Reg}_\delta(U, g, x)$ the connected component of $\text{Reg}_\delta(U, g)$ containing x.

Intuitively, the δ-regular points of (U, g) are at distance at least δ from the boundary on U.

LEMMA 5.2. $\text{Reg}_\delta(U, g)$ *is a closed subset of* U.

PROOF. Suppose that p_n converges to p as n tends to ∞ and suppose that $p_n \in \text{Reg}_\delta(U, g)$ for all n. Fix $r' < \delta$ and consider the ball $B(p, r')$. For all n sufficiently large, this ball is contained in $B(p_n, (\delta + r')/2)$, and hence has compact closure. □

Now we are ready for the basic definition of geometric convergence of Riemannian manifolds.

DEFINITION 5.3. For each k let (U_k, g_k, x_k) be a based, connected Riemannian manifold. A *geometric limit* of the sequence $\{U_k, g_k, x_k\}_{k=0}^\infty$ is a based, connected Riemannian manifold $(U_\infty, g_\infty, x_\infty)$ with the extra data:

(1) An increasing sequence $V_k \subset U_\infty$ of connected open subsets of U_∞ whose union is U_∞ and which satisfy the following for all k:
   (a) the closure $\overline{V}_k$ is compact,
   (b) $\overline{V}_k \subset V_{k+1}$,
   (c) V_k contains x_∞.
(2) For each $k \geq 0$ a smooth embedding $\varphi_k \colon (V_k, x_\infty) \to (U_k, x_k)$ with the properties that:
   (a) $\lim_{k \to \infty} \varphi_k^* g_k = g_\infty$, where the limit is in the uniform C^∞-topology on compact subsets of U_∞.
   (b) For any $\delta > 0$ and any $R < \infty$ for all k sufficiently large, $x_k \in \text{Reg}_\delta(U_k, g_k)$ and for any $\ell \geq k$ the image $\varphi_\ell(V_k)$ contains $B(x_\ell, R) \cap \text{Reg}_\delta(U_\ell, g_\ell, x_\ell)$.

We also say that the sequence *converges geometrically to* $(U_\infty, g_\infty, x_\infty)$ if there exist (V_k, φ_k) as required in the above definition. We also say that $(U_\infty, g_\infty, x_\infty)$ is the *geometric limit* of the sequence.

More generally, given $(U_\infty, g_\infty, x_\infty)$, a sequence of open subsets and $\{V_k\}_{k=1}^\infty$ satisfying (1) above, and smooth maps $\varphi_k \colon V_k \to U_k$ satisfying (2a) above, we say that $(U_\infty, g_\infty, x_\infty)$ is a *partial geometric limit* of the sequence.

REMARK 5.4. Conditions (1) and (2a) in the definition above also appear in the definition in the case of complete limits. It is Condition (2b) that is extra in this incomplete case. It says that once k is sufficiently large then the image $\varphi_\ell(V_k)$ contains all points satisfying two conditions: they are at most a given bounded distance from x_ℓ, and also they are at least a fixed distance from the boundary of U_ℓ.

Notice that if the (U_k, g_k) have uniformly bounded volume by, say, V, then any geometric limit has volume $\leq V$.

LEMMA 5.5. *The geometric limit of a sequence (U_k, g_k, x_k) is unique up to based isometry.*

PROOF. Suppose that $(U_\infty, g_\infty, x_\infty)$ and $(U'_\infty, g'_\infty, x'_\infty)$ are geometric limits. Let $\{V_k, \varphi_k\}$ and $\{V'_k, \varphi'_k\}$ be the sequences of open subsets and maps as required by the definition of the limit.

Fix k. Since V_k is connected and has compact closure, there are $R < \infty$ and $\delta > 0$ such that $V_k \subset B(x_\infty, R) \cap \operatorname{Reg}_\delta(U_\infty, g_\infty, x_\infty)$. Let x be contained in the closure of V_k. Then by the triangle inequality the closed ball $\overline{B(x, \delta/3)}$ is contained in $B(x_\infty, R+\delta) \cap \operatorname{Reg}_{\delta/2}(U_\infty, g_\infty, x_\infty)$. Since the union of these closed balls as x ranges over $\overline{V}_k$ is a compact set, for all ℓ sufficiently large, the restriction of $\varphi_\ell^* g_\ell$ to the union of these balls is close to the restriction of g_∞ to the same subset. In particular, for all ℓ sufficiently large and any $x \in V_k$ we see that $\varphi_\ell(B(x, \delta/3))$ contains $B(\varphi_\ell(x), \delta/4)$. Thus, for all ℓ sufficiently large $\varphi_\ell(V_k) \subset B(x_\ell, R+2\delta) \cap \operatorname{Reg}_{\delta/4}(U_\ell, g_\ell, x_\ell)$. This implies that, for given k, for all ℓ sufficiently large $\varphi_\ell(V_k) \subset \varphi'_\ell(V'_\ell)$. Of course, $(\varphi'_\ell)^{-1} \circ \varphi_\ell(x_\infty) = x'_\infty$. Fix k and pass to a subsequence of ℓ, such that as $\ell \to \infty$, the compositions $(\varphi'_\ell)^{-1} \circ (\varphi_\ell|_{V_k}) : V_k \to U'_\infty$ converge to a base-point preserving isometric embedding of V_k into U'_∞. Clearly, as we pass from k to $k' > k$ and take a further subsequence of ℓ these limiting isometric embeddings are compatible. Their union is then a base-point preserving isometric embedding of U_∞ into U'_∞.

The last thing we need to see is that the embedding of U_∞ into U'_∞ constructed in the previous paragraph is onto. For each n we have $\overline{V}'_n \subset V'_{n+1}$. Since $\overline{V}'_n$ is compact and connected, it follows that there are $R < \infty$ and $\delta > 0$ (depending on n) such that $\overline{V}'_n \subset B(x'_\infty, R) \cap \operatorname{Reg}_\delta(V_{n+1}, g'_\infty, x'_\infty)$. Since V'_{n+1} has compact closure in U'_∞, as ℓ tends to ∞ the metrics $(\varphi'_\ell)^* g_\ell$ converge uniformly on V_{n+1} to $g'_\infty|_{V_{n+1}}$. This means that there are $R' < \infty$ and $\delta' > 0$ (depending on n) such that for all ℓ sufficiently large, $\varphi'_\ell(V_n) \subset B(x_k, R') \cap \operatorname{Reg}_{\delta'}(U_\ell, g_\ell, x_\ell)$. This implies that for all k sufficiently large and any $\ell \geq k$ the image $\varphi'_\ell(V'_n)$ is contained in the image of $\varphi_\ell(V_k)$. Hence, for all k sufficiently large and any $\ell \geq k$ we have $V'_n \subset (\varphi'_\ell)^{-1}(\varphi_\ell(V_k))$. Hence, the isometric embedding $U_\infty \to U'_\infty$ constructed above contains V'_n. Since this is true for every n, it follows that this isometric embedding is in fact an isometry $U_\infty \to U'_\infty$. $\square$

Here is the basic existence result.

THEOREM 5.6. *Suppose that $\{(U_k, g_k, x_k)\}_{k=1}^\infty$ is a sequence of based, connected, n-dimensional Riemannian manifolds. In addition, suppose the following:*

(1) *There is $\delta > 0$ such that $x_k \in \operatorname{Reg}_\delta(U_k, g_k)$ for all k.*
(2) *For each $R < \infty$ and $\delta > 0$ there is a constant $V(R, \delta) < \infty$ such that $\operatorname{Vol}(B(x_k, R) \cap \operatorname{Reg}_\delta(U_k, x_k)) \leq V(R, \delta)$ for all k sufficiently large.*
(3) *For each non-negative integer ℓ, each $\delta > 0$, and each $R < \infty$, there is a constant $C(\ell, \delta, R)$ such that for every k sufficiently large we have*

$$|\nabla^\ell \operatorname{Rm}(g_k)| \leq C(\ell, \delta, R)$$

on all of $B(x_k, R) \cap \text{Reg}_\delta(U_k, g_k)$.

(4) For every $R < \infty$ there are $r_0 > 0$ and $\kappa > 0$ such that for every k sufficiently large, for every $\delta \leq r_0$ and every $x \in B(x_k, R) \cap \text{Reg}_\delta(U_k, g_k, x_k)$ the volume of the metric ball $B(x, \delta) \subset U_k$ is at least $\kappa \delta^n$.

Then, after passing to a subsequence, there exists a based Riemannian manifold $(U_\infty, g_\infty, x_\infty)$ that is a geometric limit of the sequence $\{(U_k, g_k, x_k)\}_{k=1}^\infty$.

Before giving the proof of this result, we begin with a standard lemma.

LEMMA 5.7. *Suppose that we have a sequence of n-dimensional balls (B_k, h_k) of radius r in Riemannian n-manifolds. Suppose that for each ℓ there is a constant $C(\ell)$ such that for every k, we have $|\nabla^\ell \text{Rm}(h_k)| \leq C(\ell)$ throughout B_k. Suppose also that for each n the exponential mapping from the tangent space at the center of B_k induces a diffeomorphism from a ball in the tangent space onto B_k. Then choosing an isometric identification of the tangent spaces at the central points of the B_k with $\mathbb{R}^n$ and pulling back the metrics h_k via the exponential mapping to metrics $\widetilde{h}_k$ on the ball B of radius r in $\mathbb{R}^n$ gives us a family of metrics on B that, after passing to a subsequence, converge in the C^∞-topology, uniformly on compact subsets of B, to a limit.*

The basic point in proving this lemma is to 'find the right gauge,' which in this case means find local coordinates so that the metric tensor is controlled by the curvature. The correct local coordinates are the Gaussian coordinates centered at the center of the ball.

PROOF. (of the theorem). Fix $R < \infty$ and $\delta > 0$. Let

$$X(\delta, R) = B(x_k, R) \cap \text{Reg}_{2\delta}(U_k, g_k, x_k).$$

From the non-collapsing assumption and the curvature bound assumption if follows from Theorem 1.36 that there is a uniform positive lower bound (independent of k) to the injectivity radius of every point in $X(\delta, R)$. Fix $0 < \delta' \leq \min(r_0, \delta/2)$ much less than this injectivity radius. We also choose $\delta' > 0$ sufficiently small so that any ball of radius $2\delta'$ in $B(x_k, R + \delta) \cap \text{Reg}_\delta(U_k, g_k, x_k)$ is geodesically convex. (This is possible because of the curvature bound.) We cover $X(\delta, R)$ by balls $B'_1, \ldots, B'_N$ of radii $\delta'/2$ centered at points of $X(\delta, R)$ with the property that the sub-balls of radius $\delta'/4$ are disjoint. We denote by $B'_i \subset B_i \subset \widetilde{B}_i$ the metric balls with the same center and radii $\delta'/2$, δ', and $2\delta'$ respectively. Notice that each of the balls $\widetilde{B}_i$ is contained in $B(x_k, R + \delta) \cap \text{Reg}_\delta(U_k, g_k, x_k)$. Because $\delta' \leq r_0$, because $\text{Vol}\, B(x_k, R + \delta)$ is bounded independent of k, and because the concentric balls of radius $\delta'/4$ are disjoint, there is a uniform bound (independent of k) to the number of such balls. Passing to a subsequence we can assume that the number of balls in these coverings is the same for all k. We number them $\widetilde{B}_1, \ldots, \widetilde{B}_N$. Next, using the exponential mapping at the central point, identify each of these balls with the ball of radius $2\delta'$ in $\mathbb{R}^n$. By passing to a further subsequence we can arrange that the metrics on each $\widetilde{B}_i$ converge uniformly. (This uses the fact that the concentric balls of radius $2\delta \geq 4\delta'$ are embedded in the U_k by the exponential mapping.) Now we pass to a further subsequence so that the distance between the centers of the balls converges, and so that for any pair $\widetilde{B}_i$ and $\widetilde{B}_j$ for which the limiting distance between their centers is less than $4\delta'$, the overlap functions in the U_k also converge. The limits of the overlap functions defines a limiting equivalence relation on $\coprod_i \widetilde{B}_i$.

This allows us to form a limit manifold $\widehat{U}_\infty$. It is the quotient of the disjoint union of the $\widetilde{B}_i$ with the limit metrics under the limit equivalence relation. We set $(U_\infty(\delta, R), g_\infty(\delta, R), x_\infty(\delta, R))$ equal to the submanifold of $\widehat{U}_\infty$ that is the union of the sub-balls $B_i \subset \widetilde{B}_i$ of radii δ'. A standard argument using partitions of unity and the geodesic convexity of the balls $\widetilde{B}_i$ shows that, for all k sufficiently large, there are smooth embeddings $\varphi_k(\delta, R)\colon U_\infty(\delta, R) \to B(x_k, R + \delta) \cap \operatorname{Reg}_\delta(U_k, g_k, x_k)$ sending $x_\infty(\delta, R)$ to x_k and converging as $k \to \infty$, uniformly in the C^∞-topology on each B_i, to the identity. Furthermore, the images of each of these maps contains $B(x_k, R) \cap \operatorname{Reg}_{2\delta}(U_k, g_k, x_k)$; compare [6]. Also, the pull-backs under these embeddings of the metrics g_k converge uniformly to $g_\infty(\delta, R)$.

Repeat the process with R replaced by $2R$ and $\delta = \delta_1$ replaced by $\delta_2 \leq \delta_1/2$. This produces
$$(U_\infty(\delta_2, 2R), g_\infty(\delta_2, 2R), x_\infty(\delta_2, 2R))$$
and, for all k sufficiently large, embeddings $\varphi_k(\delta_2, 2R)$ of this manifold into
$$B(x_k, 2R + \delta_2) \cap \operatorname{Reg}_{\delta_2}(U_k, g_k, x_k).$$
Hence, the image of these embeddings contains the images of the original embeddings. The compositions $(\varphi_k(\delta_2, 2R))^{-1} \circ \varphi_k(\delta, R)$ converge to an isometric embedding
$$(U_\infty(\delta, R), g_\infty(\delta, R), x_\infty(\delta, R)) \to (U_\infty(\delta_2, 2R), g_\infty(\delta_2, 2R), x_\infty(\delta_2, 2R)).$$

Repeating this construction infinitely often produces a based Riemannian manifold $(U_\infty, g_\infty, x_\infty)$ which is written as an increasing union of open subsets $V_k = U_\infty(\delta_k, 2^k R)$, where the δ_k tend to zero as k tends to ∞. For each k the open subset V_k has compact closure contained in V_{k+1}. By taking a subsequence of the original sequence we have maps $\varphi_k\colon V_k \to U_k$ so that (2a) in the definition of geometric limits holds. Condition (2b) clearly holds by construction. □

Now let us turn to complete Riemannian manifolds, where the result is the C^∞-version of the classical Cheeger-Gromov compactness.

LEMMA 5.8. *Suppose that (U_k, g_k, x_k) is a sequence of based Riemannian manifolds and that there is a partial geometric limit $(U_\infty, g_\infty, x_\infty)$ that is a complete Riemannian manifold. Then this partial geometric limit is a geometric limit.*

PROOF. Since the balls $B(x_\infty, R)$ have compact closure in U_∞ and since
$$\mathrm{Reg}_\delta(U_\infty, g_\infty, x_\infty) = U_\infty$$
for every $\delta > 0$, it is easy to see that the extra condition, (2b), in Definition 5.3 is automatic in this case. □

Now as an immediate corollary of Theorem 5.6 we have the following.

THEOREM 5.9. *Let $\{(M_k, g_k, x_k)\}_{k=1}^\infty$ be a sequence of connected, based Riemannian manifolds. Suppose that:*
  (1) *For every $A < \infty$ the ball $B(x_k, A)$ has compact closure in M_k for all k sufficiently large.*
  (2) *For each integer $\ell \geq 0$ and each $A < \infty$ there is a constant $C = C(\ell, A)$ such that for each $y_k \in B(x_k, A)$ we have*
$$\left|\nabla^\ell \mathrm{Rm}(g_k)(y_k)\right| \leq C$$
  for all k sufficiently large.
  (3) *Suppose also that there is a constant $\delta > 0$ with $\mathrm{inj}_{(M_k, g_k)}(x_k) \geq \delta$ for all k sufficiently large.*

Then after passing to a subsequence there is a geometric limit which is a complete Riemannian manifold.

PROOF. By the curvature bounds, it follows from the Bishop-Gromov theorem (Theorem 1.34) that for each $A < \infty$ there is a uniform bound to the volumes of the balls $B(x_k, A)$ for all k sufficiently large. It also follows from the same result that the uniform lower bound on the injectivity radius at the central point implies that for each $A < \infty$ there is a uniform lower bound for the injectivity radius on the entire ball $B(x_k, A)$, again for k sufficiently large. Given these two facts, it follows immediately from Theorem 5.6 that there is a geometric limit.

Since, for every $A < \infty$, the $B(x_k, A)$ have compact closure in M_k for all k sufficiently large, it follows that for every $A < \infty$ the ball $B(x_\infty, A)$ has compact closure in M_∞. This means that (M_∞, g_∞) is complete. □

COROLLARY 5.10. *Suppose that $\{(M_k, g_k, x_k)\}_{k=1}^\infty$ is a sequence of based, connected Riemannian manifolds. Suppose that the first two conditions in Theorem 5.9 hold and that there are constants $\kappa > 0$ and $\delta > 0$ such that $\mathrm{Vol}_{g_k} B(x_k, \delta) \geq \kappa \delta^n$ for all k. Then after passing to a subsequence there is a geometric limit which is a complete Riemannian manifold.*

PROOF. Let $A = \max(\delta^{-2}, C(0,\delta))$, where $C(0,\delta)$ is the constant given in the second condition in Theorem 5.9. Rescale, replacing the Riemannian metric g_k by Ag_k. Of course, the first condition of Theorem 5.9 still holds as does the second with different constants, and we have $|\text{Rm}_{Ag_k}(y_k)| \leq 1$ for all $y_k \in B_{Ag_k}(x_k, \sqrt{A}\delta)$. Also, $\text{Vol}\, B_{Ag_k}(x_k, \sqrt{A}\delta) \geq \kappa(\sqrt{A}\delta)^n$. Thus, by the Bishop-Gromov inequality (Theorem 1.34), we have $\text{Vol}_{Ag_k} B(x_k, 1) \geq \kappa/\Omega$ where
$$\Omega = \frac{V(\sqrt{A}\delta)}{(\sqrt{A}\delta)^n V(1)},$$
and $V(a)$ is the volume of the ball of radius a in hyperbolic n-space (the simply connected n-manifold of constant curvature -1). Since $\sqrt{A}\delta \geq 1$, this proves that, for all k sufficiently large, the absolute values of the sectional curvatures of Ag_k on $B_{Ag_k}(x_k, 1)$ are bounded by 1 and that $\text{Vol}_{Ag_k} B(x_k, 1)$ is bounded below by a positive constant independent of k. According to Theorem 1.36 these conditions imply that there is $r > 0$, such that for all k sufficiently large, the injectivity radius of (M_k, Ag_k) at x_k is at least r. Hence, for all k sufficiently large, the injectivity radius at x_k of (M_k, g_k) is bounded below by $r/\sqrt{A}$. This means that the original sequence of manifolds satisfies the third condition in Theorem 5.9. Invoking this theorem gives the result. □

1.1. Geometric convergence of manifolds in the case of Ricci flow. As the next theorem shows, because of Shi's theorem, it is much easier to establish the geometric convergence manifolds in the context of Ricci flows than in general.

THEOREM 5.11. *Suppose that $(\mathcal{M}_k, G_k, x_k)$ is a sequence of based, generalized n-dimensional Ricci flows with $\mathbf{t}(x_k) = 0$. Let (M_k, g_k) be the 0 time-slice of $(\mathcal{M}_k, G_k)$. Suppose that for each $A < \infty$ there are constants $C(A) < \infty$ and $\delta(A) > 0$ such that for all k sufficiently large the following hold:*

(1) *the ball $B(x_k, 0, A)$ has compact closure in M_k,*
(2) *there is an embedding $B(x_k, 0, A) \times (-\delta(A), 0] \to \mathcal{M}_k$ compatible with the time function and with the vector field,*
(3) *$|\text{Rm}| \leq C(A)$ on the image of the embedding in the item (2), and*
(4) *there is $r_0 > 0$ and $\kappa > 0$ such that $\text{Vol}\, B(x_k, 0, r_0) \geq \kappa r_0^n$ for all k sufficiently large.*

Then after passing to a subsequence there is a geometric limit $(M_\infty, g_\infty, x_\infty)$ of the 0 time-slices (M_k, g_k, x_k). This limit is a complete Riemannian manifold.

PROOF. The first condition in Theorem 5.9 holds by our first assumption. It is immediate from Shi's theorem (Theorem 3.28) that the second condition of Theorem 5.9 holds. The result is then immediate from Corollary 5.10. □

2. Geometric convergence of Ricci flows

In this section we extend this notion of geometric convergence for based Riemannian manifolds in the obvious way to geometric convergence of based Ricci flows. Then we give Hamilton's theorem about the existence of such geometric limits.

DEFINITION 5.12. Let $\{(\mathcal{M}_k, G_k, x_k)\}_{k=1}^{\infty}$ be a sequence of based, generalized Ricci flows. We suppose that $\mathbf{t}(x_k) = 0$ for all k and we denote by (M_k, g_k) the time-slice of $(\mathcal{M}_k, G_k)$. For some $0 < T \leq \infty$, we say that a based Ricci flow $(M_\infty, g_\infty(t), (x_\infty, 0))$ defined for $t \in (-T, 0]$ is a *partial geometric limit Ricci flow* if:

(1) There are open subsets $x_\infty \in V_1 \subset V_2 \subset \cdots \subset M_\infty$ satisfying (1) of Definition 5.3 with M_∞ in place of U_∞,
(2) there is a sequence $0 < t_1 < t_2 < \cdots$ with $\lim_{k \to \infty} t_k = T$,
(3) and maps
$$\widetilde{\varphi}_k \colon V_k \times [-t_k, 0] \to \mathcal{M}_k$$
compatible with time and the vector field

such that the sequence of horizontal families of metrics $\widetilde{\varphi}_k^* G_k$ converges uniformly on compact subsets of $M_\infty \times (-T, 0]$ in the C^∞-topology to the horizontal family of metrics $g_\infty(t)$ on $M_\infty \times (-T, 0]$.

Notice that the restriction to the 0 time-slices of a partial geometric limit of generalized Ricci flows is a partial geometric limit of the 0 time-slices.

DEFINITION 5.13. For $0 < T \leq \infty$, if $(M_\infty, g_\infty(t), x_\infty)$, $-T < t \leq 0$, is a partial geometric limit Ricci flow of the based, generalized Ricci flows $(\mathcal{M}_k, G_k, x_k)$ and if $(M_\infty, g_\infty(0), x_\infty)$ is a geometric limit of the 0 time-slices, then we say that the partial geometric limit is a *geometric limit Ricci flow defined on the time interval* $(-T, 0]$.

Again Shi's theorem, together with a computation of Hamilton, allows us to form geometric limits of generalized Ricci flows. We have the following result due originally to Hamilton [**33**].

PROPOSITION 5.14. *Fix constants* $-\infty \leq T' \leq 0 \leq T \leq \infty$ *and suppose that* $T' < T$. *Let* $\{(\mathcal{M}_k, G_k, x_k)\}_{k=1}^{\infty}$ *be a sequence of based, generalized Ricci flows. Suppose that* $\mathbf{t}(x_k) = 0$ *for all* k, *and denote by* (M_k, g_k) *the 0 time-slice of* $(\mathcal{M}_k, G_k)$. *Suppose that there is a partial geometric limit* $(M_\infty, g_\infty, x_\infty)$ *for the* $(\mathcal{M}_k, g_k, x_k)$ *with open subsets* $\{V_k \subset M_\infty\}$ *and maps* $\varphi_k \colon V_k \to M_k$ *as in Definition 5.3. Suppose that for every compact subset* $K \subset M_\infty$ *and every compact interval* $I \subset (T', T)$ *containing* 0, *for all* k *sufficiently large, there is an embedding* $\widetilde{\varphi}_k(K, I) \colon K \times I \to \mathcal{M}_k$ *compatible* with time and the vector field and extending the map φ_k on the 0 time-slice. Suppose in addition that for every k sufficiently large there is a uniform bound (independent of k) to the norm of Riemann curvature on the image of $\widetilde{\varphi}_k(K, I)$. Then after passing to a subsequence the flows $\widetilde{\varphi}_k^* G_k$ converge to a partial geometric limit Ricci flow $g_\infty(t)$ defined for $t \in (T', T)$.

PROOF. Suppose that we have a partial geometric limit of the time-zero slices as stated in the proposition. Fix a compact subset $K \subset M_\infty$ and a compact sub-interval $I \subset (T', T)$. For all k sufficiently large we have embeddings $\widetilde{\varphi}_k(K, I)$ as stated. We consider the flows $g_k(K, I)(t)$ on $K \times I$ defined by pulling back the horizontal metrics G_k under the maps $\widetilde{\varphi}_k(K, I)$. These of course satisfy the Ricci flow equation on $K \times I$. Furthermore, by assumption the flows $g_k(K, I)(t)$ have uniformly bounded curvature. Then under these hypothesis, Shi's theorem can be used to show that the curvatures of the $g_k(K, I)$ are uniformly bounded C^∞-topology. The basic computation done by Hamilton in [**33**] shows that after passing to a further subsequence, the Ricci flows $g_k(K, I)$ converge uniformly in the C^∞-topology to a limit flow on $K \times I$. A standard diagonalization argument allows us to pass to a further subsequence so that the pullbacks $\widetilde{\varphi}_k^* G_k$ converge uniformly in the C^∞-topology on every compact subset of $M_\infty \times (T', T)$. Of course, the limit satisfies the Ricci flow equation. □

This 'local' result leads immediately to the following result for complete limits.

THEOREM 5.15. *Fix* $-\infty \leq T' \leq 0 \leq T \leq \infty$ *with* $T' < T$. *Let* $\{(\mathcal{M}_k, G_k, x_k)\}_{k=1}^\infty$ *be a sequence of based, generalized Ricci flows. Suppose that* $\mathbf{t}(x_k) = 0$ *for all* k, *and denote by* (M_k, g_k) *the 0 time-slice of* $(\mathcal{M}_k, G_k)$. *Suppose that for each* $A < \infty$ *and each compact interval* $I \subset (T', T)$ *containing 0 there is a constant* $C(A, I)$ *such that the following hold for all* k *sufficiently large:*
  (1) *the ball* $B_{g_k}(x_k, 0, A)$ *has compact closure in* M_k,
  (2) *there is an embedding* $B_{g_k}(x_k, 0, A) \times I \to \mathcal{M}_k$ *compatible with time and with the vector field,*
  (3) *the norms of the Riemann curvature of* G_k *on the image of the embedding in the previous item are bounded by* $C(A, I)$, *and*
  (4) *there is* $r_0 > 0$ *and* $\kappa > 0$ *with* $\mathrm{Vol}\, B(x_k, 0, r_0) \geq \kappa r_0^n$ *for all* k *sufficiently large.*

Then after passing to a subsequence there is a flow $(\mathcal{M}_\infty, g_\infty(t), (x_\infty, 0))$ *which is the geometric limit. It is a solution to the Ricci flow equation defined for* $t \in (T', T)$. *For every* $t \in (T', T)$ *the Riemannian manifold* $(M_\infty, g_\infty(t))$ *is complete.*

PROOF. By Theorem 5.11 there is a geometric limit $(M_\infty, g_\infty(0))$ of the 0 time-slices, and the limit is a complete Riemannian manifold. Then by

Proposition 5.14 there is a geometric limit flow defined on the time interval (T', T). Since for every $t \in (T', T)$ there is a compact interval I containing 0 and t, it follows that the Riemann curvature of the limit is bounded on $M_\infty \times I$. This means that the metrics $g_\infty(0)$ and $g_\infty(t)$ are commensurable with each other. Since $g_\infty(0)$ is complete so is $g_\infty(t)$. □

COROLLARY 5.16. *Suppose that $(U, g(t))$, $0 \leq t < T < \infty$, is a Ricci flow. Suppose that $|\mathrm{Rm}(x, t)|$ is bounded independent of $(x, t) \in U \times [0, T)$. Then for any open subset $V \subset U$ with compact closure in U, there is an extension of the Ricci flow $(V, g(t)|_V)$ past time T.*

PROOF. Take a sequence $t_n \to T$ and consider the sequence of Riemannian manifolds $(V, g(t_n))$. By Shi's theorem and the fact that V has compact closure in U, the restriction of this sequence of metrics to V has uniformly bounded curvature derivatives. Hence, this sequence has a convergent subsequence with limit (V, g_∞), where the convergence is uniform in the C^∞-topology. Now by Hamilton's result [**33**] it follows that, passing to a further subsequence, the flows $(V, g(T + t - t_n), (p, 0))$ converge to a flow $(V, g_\infty(t), (p, 0))$ defined on $(0, T]$. Clearly, for any $0 < t < T$ we have $g_\infty(t) = g(t)$. That is to say, we have extended the original Ricci flow smoothly to time T. Once we have done this, we extend it to a Ricci flow on $[T, T_1)$ for some $T_1 > T$ using the local existence results. The extension to $[T, T_1)$ fits together smoothly with the flow on $[0, T]$ by Proposition 3.12. □

3. Gromov-Hausdorff convergence

Let us begin with the notion of the Gromov-Hausdorff distance between based metric spaces of finite diameter. Let Z be a metric space. We define the Hausdorff distance between subsets of Z as follows: $d_H^Z(X, Y)$ is the infimum of all $\delta \geq 0$ such that X is contained in the δ-neighborhood of Y and Y is contained in the δ-neighborhood of X. For metric spaces X and Y we define the Gromov-Hausdorff distance between them, denoted $D_{GH}(X, Y)$, to be the infimum over all metric spaces Z and isometric embeddings $f \colon X \to Z$ and $g \colon Y \to Z$ of the Hausdorff distance between $f(X)$ and $g(Y)$. For pointed metric spaces (X, x) and (Y, y) of finite diameter, we define the Gromov-Hausdorff distance between them, denoted $D_{GH}((X, x), (Y, y))$, to be the infimum of $D_H^Z(f(X), g(Y))$ over all triples $((Z, z), f, g)$ where (Z, z) is a pointed metric space and $f \colon (X, x) \to (Z, z)$ and $g \colon (Y, y) \to (Z, z)$ are base-point preserving isometries.

To see that D_{GH} is a distance function we must establish the triangle inequality. For this it is convenient to introduce δ-nets in metric spaces.

DEFINITION 5.17. A δ-net in (X, x) is a subset L of X containing x whose δ-neighborhood covers X and for which there is some $\delta' > 0$ with $d(\ell_1, \ell_2) \geq \delta'$ for all $\ell_1 \neq \ell_2$ in L.

Clearly, the Gromov-Hausdorff distance from a based metric space (X, x) to a δ-net (L, x) contained in it is at most δ. Furthermore, for every $\delta > 0$ the based space (X, x) has a δ-net: Consider subsets $L \subset X$ containing x with the property that the $\delta/2$-balls centered at the points of L are disjoint. Any maximal such subset (with respect to the inclusion relation) is a δ-net in X.

LEMMA 5.18. *The Gromov-Hausdorff distance satisfies the triangle inequality.*

PROOF. Suppose $D_{GH}((X,x),(Y,y)) = a$ and $D_{GH}((Y,y),(Z,z)) = b$. Fix any $\delta > 0$. Then there is a metric d_1 on $X \vee Y$ such that d_1 extends the metrics on X, Y and the $(a+\delta)$-neighborhood of X is all of $X \vee Y$ as is the $(a+\delta)$-neighborhood of Y. Similarly, there is a metric d_2 on $Y \vee Z$ with the analogous properties (with b replacing a). Take a δ-net $(L, y) \subset (Y, y)$, and define
$$d(x', z') = \inf_{\ell \in L} d(x', \ell) + d(\ell, z').$$
We claim that $d(x', z') > 0$ unless $x' = z'$ is the common base point. The reason is that if $\inf_{\ell \in L} d(x', \ell) = 0$, then by the triangle inequality, any sequence of $\ell_n \in L$ with $d(x', \ell_n)$ converging to zero is a Cauchy sequence, and hence is eventually constant. This means that for all n sufficiently large, $x' = \ell_n \in L \cap X$ and hence x' is the common base point. Similarly for z'.

A straightforward computation shows that the function d above, together with the given metrics on X and Z, define a metric on $X \vee Z$ with the property that the $(a + b + 3\delta)$-neighborhood of X is all of $X \vee Z$ and likewise for Z. Since we can do this for any $\delta > 0$, we conclude that $D_{GH}((X, x), (Z, z)) \leq a + b$. □

Thus, the Gromov-Hausdorff distance is a pseudo-metric. In fact, the restriction of the Gromov-Hausdorff distance to complete metric spaces of bounded diameter is a metric. We shall not establish this result, though we prove below closely related results about the uniqueness of Gromov-Hausdorff limits.

DEFINITION 5.19. We say that a sequence of based metric spaces (X_k, x_k) of uniformly bounded diameter *converges in the Gromov-Hausdorff sense* to a based metric space (Y, y) of finite diameter if
$$\lim_{k \to \infty} D_{GH}((X_k, x_k), (Y, y)) = 0.$$
Thus, a based metric space (X, x) of bounded diameter is the limit of a sequence of δ_n-nets $L_n \subset X$ provided that $\delta_n \to 0$ as $n \to \infty$.

EXAMPLE 5.20. A sequence of compact n-manifolds of diameter tending to zero has a point as Gromov-Hausdorff limit.

DEFINITION 5.21. Suppose that $\{(X_k, x_k)\}_k$ converges in the Gromov-Hausdorff sense to (Y, y). Then a *realization sequence* for this convergence is a sequence of triples $((Z_k, z_k), f_k, g_k)$ where, for each k, the pair (Z_k, z_k) is a based metric space,

$$f_k \colon (X_k, x_k) \to (Z_k, z_k) \quad \text{and} \quad g_k \colon (Y, y) \to (Z_k, z_k)$$

are isometric embeddings and $D_{GH}(f_k(X_k), g_k(Y)) \to 0$ as $k \to \infty$. Given a realization sequence for the convergence, we say that a sequence $\ell_k \in X_k$ converges to $\ell \in Y$ (relative to the given realization sequence) if $d(f_k(\ell_k), g_k(\ell))$ tends to 0 as $i \to \infty$.

Notice that, with a different realization sequence for the convergence, a sequence $\ell_k \in X_k$ can converge to a different point of Y. Also notice that, given a realization sequence for the convergence, every $y \in Y$ is the limit of some sequence $x_k \in X_k$, a sequence $x_k \in X_k$ has at most one limit in Y, and if Y is compact then every sequence $x_k \in X_k$ has a subsequence converging to a point of Y. Lastly, notice that under any realization sequence for the convergence, the base points x_k converge to the base point y.

LEMMA 5.22. *Let (X_k, x_k) be a sequence of metric spaces whose diameters are uniformly bounded. Then the (X_k, x_k) converge in the Gromov-Hausdorff sense to (X, x) if and only if the following holds for every $\delta > 0$. For every δ-net $L \subset X$, for every $\eta > 0$, and for every k sufficiently large, there is a $(\delta + \eta)$-net $L_k \subset X_k$ and a bijection $L_k \to L$ sending x_k to x so that the push forward of the metric on L_k induced from that of X_k is $(1 + \eta)$-bi-Lipschitz equivalent to the metric on L induced from X.*

For a proof see Proposition 3.5 on page 36 of [**25**].

LEMMA 5.23. *Let (X_k, x_k) be a sequence of based metric spaces whose diameters are uniformly bounded. Suppose that (Y, y) and (Y', y') are limits in the Gromov-Hausdorff sense of this sequence and each of Y and Y' are compact. Then (Y, y) is isometric to (Y', y').*

PROOF. By the triangle inequality for Gromov-Hausdorff distance, it follows from the hypothesis of the lemma that $D_{GH}((Y, y), (Y', y')) = 0$. Fix $\delta > 0$. Since $D_{GH}((Y, y), (Y', y')) = 0$, for any $n > 0$ and finite $(1/n)$-net $L_n \subset Y$ containing y there is an embedding $\varphi_n \colon L_n \to Y'$ sending y to y' such that the image is a $2/n$-net in Y' and such that the map from L_n to its image is a $(1 + \delta)$-bi-Lipschitz homeomorphism. Clearly, we can suppose that in addition the L_n are nested: $L_n \subset L_{n+1} \subset \cdots$. Since Y' is compact and L_n is finite, and we can pass to a subsequence so that $\lim_{k \to \infty} \varphi_k|_{L_n}$ converges to a map $\psi_n \colon L_n \to Y'$ which is a $(1+\delta)$-bi-Lipschitz map onto its image which is a $2/n$ net in Y'. By a standard diagonalization argument, we can arrange that $\psi_{n+1}|_{L_n} = \psi_n$ for all n. The $\{\psi_n\}$ then define an embedding $\cup_n L_n \to Y'$ that is a $(1 + \delta)$-bi-Lipschitz map onto its image which is a dense subset of Y'. Clearly, using the compactness of Y' this map extends to a $(1+\delta)$-bi-Lipschitz embedding $\psi_\delta \colon (Y, y) \to (Y', y')$ onto a dense subset of Y'. Since Y is also compact, this image is in fact all of Y'. That is to say, ψ_δ is a $(1+\delta)$-bi-Lipschitz homeomorphism $(Y, y) \to (Y', y')$. Now perform this construction for a sequence of $\delta_n \to 0$ and $(1 + \delta_n)$-bi-Lipschitz homeomorphisms $\psi_{\delta_n} \colon (Y, y) \to (Y', y')$. These form an equicontinuous family so that by passing to a subsequence we can extract a limit $\psi \colon (Y, y) \to (Y', y')$. Clearly, this limit is an isometry. $\square$

Now let us consider the more general case of spaces of not necessarily bounded diameter. It turns out that the above definition is too restrictive when applied to such spaces. Rather one takes:

DEFINITION 5.24. For based metric spaces (X_k, x_k) (not necessarily of finite diameter) to *converge in the Gromov-Hausdorff sense to a based metric space (Y, y)* means that for each $r > 0$ there is a sequence $\delta_k \to 0$ such that the sequence of balls $B(x_k, r + \delta_k)$ in (X_k, x_k) converges in the Gromov-Hausdorff sense to the ball $B(y, r)$ in Y.

Thus, a sequence of cylinders $S^{n-1} \times \mathbb{R}$ with any base points and with the radii of the cylinders going to zero has the real line as Gromov-Hausdorff limit.

LEMMA 5.25. *Let (X_k, x_k) be a sequence of locally compact metric spaces. Suppose that (Y, y) and (Y', y') are complete, locally compact, based metric spaces that are limits of the sequence in the Gromov-Hausdorff sense. Then there is an isometry $(Y, y) \to (Y', y')$.*

PROOF. We show that for each $r < \infty$ there is an isometry between the closed balls $\overline{B(y, r)}$ and $\overline{B(y', r)}$. By the local compactness and completeness, these closed balls are compact. Each is the limit in the Gromov-Hausdorff sense of a sequence $B(x_k, r + \delta_k)$ for some $\delta_k \to 0$ as $k \to \infty$. Thus, invoking the previous lemma we see that these closed balls are isometric. We take a sequence $r_n \to \infty$ and isometries $\varphi_n \colon (B(y, r_n), y) \to (B(y', r_n), y')$. By a standard diagonalization argument, we pass to a subsequence such that for each $r < \infty$ the sequence $\varphi_n|_{B(y,r)}$ of isometries converges to an isometry $\varphi_r \colon B(y, r) \to B(y', r)$. These then fit together to define a global isometry $\varphi \colon (Y, y) \to (Y', y')$. $\square$

If follows from this that if a sequence of points $\ell_k \in X_k$ converges to $\ell \in Y$ under one realization sequence for the convergence and to $\ell' \in Y$ under another, then there is an isometry of (Y, y) to itself carrying ℓ to ℓ'.

EXAMPLE 5.26. Let (M_n, g_n, x_n) be a sequence of based Riemannian manifolds converging geometrically to $(M_\infty, g_\infty, x_\infty)$. Then the sequence also converges in the Gromov-Hausdorff sense to the same limit.

3.1. Pre-compactnes.
There is a fundamental compactness result due to Gromov. We begin with a definition.

DEFINITION 5.27. A *length space* is a connected metric space (X, d) such that for any two points x, y there is a rectifiable arc γ with endpoints x and y and with the length of γ equal to $d(x, y)$.

For any based metric space (X, x) and constants $\delta > 0$ and $R < \infty$, let $N(\delta, R, X)$ be the maximal number of disjoint δ-balls in X that can be contained in $B(x, R)$.

THEOREM 5.28. *Suppose that (X_k, x_k) is a sequence of based length spaces. Then there is a based length space (X, x) that is the limit in the Gromov-Hausdorff sense of a subsequence of the (X_k, x_k) if for every $\delta > 0$ and $R < \infty$ there is an $N < \infty$ such that $N(\delta, R, X_k) \leq N$ for all k. On the other hand, if the sequence (X_k, x_k) has a Gromov-Hausdorff limit, then for every $\delta > 0$ and $R < \infty$ the $N(\delta, R, X_k)$ are bounded independent of k.*

For a proof of this result see Proposition 5.2 on page 63 of [**25**].

3.2. The Tits cone.
Let (M, g) be a complete, non-compact Riemannian manifold of non-negative sectional curvature. Fix a point $p \in M$, and let γ and μ be minimal geodesic rays emanating from p. For each $r > 0$ let $\gamma(r)$ and $\mu(r)$ be the points along these geodesic rays at distance r from p. Then by part 1 of Theorem 2.4 we see that

$$\ell(\gamma, \mu, r) = \frac{d(\gamma(r), \mu(r))}{r}$$

is a non-increasing function of r. Hence, there is a limit $\ell(\gamma, \mu) \geq 0$ of $\ell(\gamma, \mu, r)$ as $r \to \infty$. We define the *angle at infinity* between γ and μ, $0 \leq \theta_\infty(\gamma, \mu) \leq \pi$, to be the angle at b of the Euclidean triangle a, b, c with side lengths $|ab| = |bc| = 1$ and $|bc| = \ell(\gamma, \mu)$, see FIG. 1. If ν is a third geodesic ray emanating from p, then clearly, $\theta_\infty(\gamma, \mu) + \theta_\infty(\mu, \nu) \geq \theta_\infty(\gamma, \nu)$.

DEFINITION 5.29. Now we define a metric space whose underlying space is the quotient space of the equivalence classes of minimal geodesic rays emanating from p, with two rays equivalent if and only if the angle at infinity between them is zero. The pseudo-distance function θ_∞ descends to a metric on this space. This space is a length space [**4**]. Notice that the distance between any two points in this metric space is at most π. We denote this space by $S_\infty(M, p)$.

CLAIM 5.30. *$S_\infty(M, p)$ is a compact space.*

PROOF. Let $\{[\gamma_n]\}_n$ be a sequence of points in $S_\infty(M, p)$. We show that there is a subsequence with a limit point. By passing to a subsequence we can arrange that the unit tangent vectors to the γ_n at p converge to a unit tangent vector τ, say. Fix $d < \infty$, and let x_n be the point of γ_n at distance

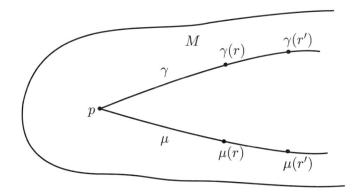

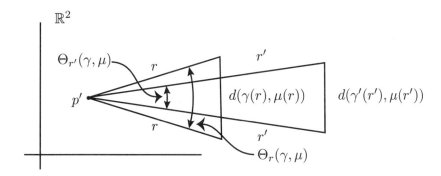

FIGURE 1. Angles at infinity.

d from p. Then by passing to a subsequence we can arrange that the x_n converge to a point x. The minimizing geodesic segments $[p, x_n]$ on γ_n then converge to a minimizing geodesic segment connecting p to x. Performing this construction for a sequence of d tending to infinity, and then taking a diagonal subsequence, produces a minimizing geodesic ray γ from p whose class is the limit of a subsequence of the $\{[\gamma_n]\}$. □

We define the *Tits cone* of M at p, denoted $\mathcal{T}(M, p)$, to be the cone over $S_\infty(M, p)$, i.e., the quotient of the space $S_\infty(M, p) \times [0, \infty)$ where all points $(x, 0)$ are identified (becoming the cone point). The cone metric on this space is given as follows: Let (x_1, a_1) and (x_2, a_2) be points of $S_\infty(M, p) \times [0, \infty)$. Then the distance between their images in the cone is determined by

$$d^2([x_1, a_1], [x_2, a_2]) = a_1^2 + a_2^2 - 2a_1 a_2 \cos(\theta_\infty(x_1, x_2)).$$

It is an easy exercise to show that the Tits cone of M at p is in fact independent of the choice of p. From the previous claim, it follows that the Tits cone of M is locally compact and complete.

PROPOSITION 5.31. *Let (M, g) be a non-negatively curved, complete, non-compact Riemannian manifold of dimension k. Fix a point $p \in M$ and let $\{x_n\}_{n=1}^{\infty}$ be a sequence tending to infinity in M. Let $\lambda_n = d^2(p, x_n)$ and consider the sequence of based Riemannian manifolds (M, g_n, p), where $g_n = \lambda_n^{-1} g$. Then there is a subsequence converging in the Gromov-Hausdorff sense. Any Gromov-Hausdorff limit (X, g_∞, x_∞) of a subsequence (X, g_∞) is isometric to the Tits cone $\mathcal{T}(M, p)$ with base point the cone point.*

PROOF. Let c be the cone point of $\mathcal{T}(M, p)$, and denote by d the distance function on $\mathcal{T}(M, p)$. Consider the ball $B(c, R) \subset \mathcal{T}(M, p)$. Since $S_\infty(M, p)$ is the metric completion of the quotient space of minimal geodesic rays emanating from p, for any $\delta > 0$ there is a δ-net $L \subset B(c, R)$ consisting of the cone point together with points of the form $([\gamma], t)$ where γ is a minimal geodesic ray emanating from p and $t > 0$. We define a map from $\psi_n \colon L \to (M, g_n)$ by sending the cone point to p and sending $([\gamma], t)$ to the point at g_n-distance t from p along γ. Clearly, $\psi_n(L)$ is contained in $B_{g_n}(p, R)$. From the second item of Theorem 2.4 and the monotonicity of angles it follows that the map $\psi_n \colon L \to (M, g_n)$ is a distance non-decreasing map; i.e., $\psi_n^*(g_n | \psi_n(L)) \geq d|_L$. On the other hand, by the monotonicity, $\psi_{n+1}^*(g_{n+1} | \psi_{n+1}(L)) \leq \psi_n^*(g_n | \psi_n(L))$ and this non-increasing sequence of metrics converges to $d|_L$. This proves that for any $\delta > 0$ for all n sufficiently large, the embedding ψ_n is a $(1 + \delta)$-bi-Lipschitz homeomorphism.

It remains to show that for any $\eta > 0$ the images $\psi_n(L)$ are eventually $(\delta + \eta)$-nets in $B_{g_n}(p, R)$. Suppose not. Then after passing to a subsequence, for each n we have a point $x_n \in B_{g_n}(p, R)$ whose distance from $\psi_n(L)$ is at least $\delta + \eta$. In particular, $d_{g_n}(x_n, p) \geq \delta$. Consider a sequence of minimal geodesic rays μ_n connecting p to the x_n. Since the g-length of μ_n is at least $n\delta$, by passing to a further subsequence, we can arrange that the μ_n converge to a minimal geodesic ray γ emanating from p. By passing to a further subsequence if necessary, we arrange that $d_{g_n}(x_n, p)$ converges to $r > 0$. Now consider the points $\widetilde{x}_n$ on γ at g-distance $\sqrt{\lambda_n} r$ from p. Clearly, from the second item of Theorem 2.4 and the fact that the angle at p between the μ_n and μ tends to zero as $n \to \infty$ we have $d_{g_n}(x_n, \widetilde{x}_n) \to 0$ as $n \to \infty$. Hence, it suffices to show that for all n sufficiently large, $\widetilde{x}_n$ is within δ of $\psi_n(L)$ to obtain a contradiction. Consider the point $z = ([\mu], r) \in \mathcal{T}(M, p)$. There is a point $\ell = ([\gamma], t') \in L$ within distance δ of z in the metric d. Let $\widetilde{y}_n \in M$ be the point in M at g-distance $\sqrt{\lambda_n} t'$ along γ. Of course, $\widetilde{y}_n = \psi_n(\ell)$. Then $d_{g_n}(\widetilde{x}_n, \widetilde{y}_n) \to d(\ell, z) < \delta$. Hence, for all n sufficiently large, $d_{g_n}(\widetilde{x}_n, \widetilde{y}_n) < \delta$. This proves that for all n sufficiently large $\widetilde{x}_n$ is within δ of $\psi_n(L)$ and hence for all n sufficiently large x_n is within $\delta + \eta$ of $\psi_n(L)$.

We have established that for every positive δ, η and every $R < \infty$ there is a finite δ-net L in $(\mathcal{T}(M, p), c)$ and, for all n sufficiently large, an

4. Blow-up limits

Here we introduce a type of geometric limit. These were originally introduced and studied by Hamilton in [**34**], where, among other things, he showed that 3-dimensional blow-up limits have non-negative sectional curvature. We shall use repeatedly blow-up limits and the positive curvature result in the arguments in the later sections.

DEFINITION 5.32. Let $(\mathcal{M}_k, G_k, x_k)$ be a sequence of based, generalized Ricci flows. We suppose that $\mathbf{t}(x_k) = 0$ for all n. We set Q_k equal to $R(x_k)$. We denote by $(Q_k\mathcal{M}_k, Q_k G_k, x_k)$ the family of generalized flows that have been rescaled so that $R_{Q_k G_k}(x_k) = 1$. Suppose that $\lim_{k\to\infty} Q_k = \infty$ and that after passing to a subsequence there is a geometric limit of the sequence $(Q_k\mathcal{M}_k, Q_k G_k, x_k)$ which is a Ricci flow defined for $-T < t \leq 0$. Then we call this limit a *blow-up limit* of the original based sequence. In the same fashion, if there is a geometric limit for a subsequence of the zero time-slices of the $(Q_k\mathcal{M}_k, Q_k G_k, x_k)$, then we call this limit the blow-up limit of the 0 time-slices.

The significance of the condition that the generalized Ricci flows have curvature pinched toward positive is that, as Hamilton originally established in [**34**], the latter condition implies that any blow-up limit has non-negative curvature.

THEOREM 5.33. *Let $(\mathcal{M}_k, G_k, x_k)$ be a sequence of generalized Ricci flows of dimension 3, each of which has time interval of definition contained in $[0, \infty)$ and each of which has curvature pinched toward positive. Suppose that $Q_k = R(x_k)$ tends to infinity as k tends to infinity. Let $t_k = \mathbf{t}(x_k)$ and let $(\mathcal{M}'_k, G'_k, x_k)$ be the result of shifting time by $-t_k$ so that $\mathbf{t}'(x_k) = 0$. Then any blow-up limit of the sequence $(\mathcal{M}_k, G'_k, x_k)$ has non-negative Riemann curvature. Similarly, any blow-up limit of the zero time-slices of this sequence has non-negative curvature.*

PROOF. Let us consider the case of the geometric limit of the zero time-slice first. Let $(M_\infty, g_\infty(0), x_\infty)$ be a blow-up limit of the zero time-slices in the sequence. Let $V_k \subset M_\infty$ and $\varphi_k \colon V_k \to (M_k)_0$ be as in the definition of the geometric limit. Let $y \in M_\infty$ be a point and let $\lambda(y) \geq \mu(y) \geq \nu(y)$ be the eigenvalues of the Riemann curvature operator for g_∞ at y. Let $\{y_k\}$ be a sequence in $Q_k\mathcal{M}'_k$ converging to y, in the sense that $y_k = \varphi_k(y)$ for all k

sufficiently large. Then

$$\lambda(y) = \lim_{n\to\infty} Q_k^{-1}\lambda(y_k),$$
$$\mu(y) = \lim_{n\to\infty} Q_k^{-1}\mu(y_k),$$
$$\nu(y) = \lim_{n\to\infty} Q_k^{-1}\nu(y_k).$$

Since by Equation (4.6) we have $R(y_k) \geq -6$ for all k and since by hypothesis Q_k tends to infinity as n does, it follows that $R(y) \geq 0$. Thus if $\lambda(y) = 0$, then $\text{Rm}(y) = 0$ and the result is established at y. Hence, we may assume that $\lambda(y) > 0$, which means that $\lambda(y_k)$ tends to infinity as k does. If $\nu(y_k)$ remains bounded below as k tends to infinity, then $Q_k^{-1}\nu(y_k)$ converges to a limit which is ≥ 0, and consequently $Q_k^{-1}\mu(y_k) \geq Q_k^{-1}\nu(y_k)$ has a non-negative limit. Thus, in this case the Riemann curvature of g_∞ at y is non-negative. On the other hand, if $\nu(y_k)$ goes to $-\infty$ as k does, then according to Equation (4.7) the ratio of $X(y_k)/R(y_k)$ goes to zero. Since $Q_k^{-1}R(y_k)$ converges to the finite limit $R(y)$, the product $Q_k^{-1}X(y_k)$ converges to zero as k goes to infinity. This means that $\nu(y) = 0$ and consequently that $\mu(y) \geq 0$. Thus, once again we have non-negative curvature for g_∞ at y.

The argument in the case of a geometric limit flow is identical. □

COROLLARY 5.34. *Suppose that $(M_k, g_k(t))$ is a sequence of Ricci flows each of which has time interval of definition contained in $[0, \infty)$ with each M_k being a compact 3-manifold. Suppose further that, for each k, we have $|\text{Rm}(p_k, 0)| \leq 1$ for all $p_k \in M_k$. Then any blow-up limit of this sequence of Ricci flows has non-negative curvature.*

PROOF. According to Theorem 4.26 the hypotheses imply that for every k the Ricci flow $(M_k, g_k(t))$ has curvature pinched toward positive. From this, the corollary follows immediately from the previous theorem. □

5. Splitting limits at infinity

In our later arguments we shall need a splitting result at infinity in the non-negative curvature case. Assuming that a geometric limit exists, the splitting result is quite elementary. For this reason we present it here, though it will not be used until Chapter 9.

The main result of this section gives a condition under which a geometric limit automatically splits off a line; see FIG. 2.

THEOREM 5.35. *Let (M, g) be a complete, connected manifold of non-negative sectional curvature. Let $\{x_n\}$ be a sequence of points going off to infinity, and suppose that we can find scaling factors $\lambda_n > 0$ such that the based Riemannian manifolds $(M, \lambda_n g, x_n)$ have a geometric limit $(M_\infty, g_\infty, x_\infty)$. Suppose that there is a point $p \in M$ such that $\lambda_n d^2(p, x_n) \to \infty$ as $n \to \infty$. Then, after passing to a subsequence, minimizing geodesic arcs γ_n from x_n to p converge to a minimizing geodesic ray in M_∞. This minimizing geodesic* ray is part of a minimizing geodesic line ℓ in M_∞. In particular, there is a Riemannian product decomposition $M_\infty = N \times \mathbb{R}$ with the property that ℓ is $\{x\} \times \mathbb{R}$ for some $x \in N$.

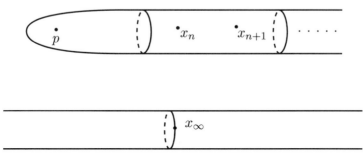

limit as n tends to infinity

FIGURE 2. Splitting at infinity.

PROOF. Let d_n be the distance from p to x_n. Consider minimizing geodesic arcs γ_n from p to x_n. By passing to a subsequence we can assume that tangent directions at p of these arcs converge. Hence, for every $0 < \delta < 1$ there is N such that for all $n, m \geq N$ the angle between γ_n and γ_m at p is less than δ. For any n we can choose $m(n)$ such that $d_{m(n)} \geq d_n(1+1/\delta)$. Let μ_n be a minimizing geodesic from x_n to $x_{m(n)}$. Now applying the Toponogov comparison (first part of Theorem 2.4) and the usual law of cosines in Euclidean space, we see that the distance d from x_n to $x_{m(n)}$ satisfies

$$d_{m(n)} - d_n \leq d \leq \sqrt{d_n^2 + d_{m(n)}^2 - 2d_n d_{m(n)} \cos(\delta)}.$$

Let $\theta_n = \angle_{x'_n}$ of the Euclidean triangle $\triangle(x'_n, p', x'_{m(n)})$ with $|s_{x'_n p'}| = d_n$, $|s_{x'_n x'_{m(n)}}| = d$ and $|s_{p' x'_{m(n)}}| = d_{m(n)}$. Then for any $\alpha < d_n$ and $\beta < d$ let x and y be the points on $s_{x_n p}$ and on $s_{x_n x_{m(n)}}$ at distances α and β respectively from x_n. Given this, according to the Toponogov comparison result (first part of Theorem 2.4), we have

$$d(x, y) \geq \sqrt{\alpha^2 + \beta^2 - 2\alpha\beta\cos(\theta_n)}.$$

The angle θ_n satisfies:

$$d_n^2 + d^2 - 2d_n d \cos(\theta_n) = d_{m(n)}^2.$$

Thus,
$$\cos(\theta_n) = \frac{d_n^2 + d^2 - d_{m(n)}^2}{2d_n d}$$
$$\leq \frac{2d_n^2 - 2d_n d_{m(n)}\cos(\delta)}{2d_n d}$$
$$= \frac{d_n}{d} - \frac{d_{m(n)}}{d}\cos(\delta)$$
$$\leq \delta - (1-\delta)\cos(\delta).$$

Since $\delta \to 0$ as $n \to \infty$, it follows that given any $\delta > 0$, for all n sufficiently large, $1 + \cos(\theta_n) < \delta$.

We are assuming the based Riemannian manifolds $\{(M, \lambda_n g, x_n)\}_{n=1}^\infty$ converge to a geometric limit $(M_\infty, g_\infty, x_\infty)$, and that $d_{\lambda_n g_n}(p, x_n) \to \infty$ as $n \to \infty$, so that the lengths of the γ_n tend to infinity in the metrics $\lambda_n g_n$. This also means that the lengths of μ_n, measured in the metrics $\lambda_n g_n$, tend to infinity. Thus, by passing to a subsequence we can assume that each of these families, $\{\gamma_n\}$ and $\{\mu_n\}$, of minimizing geodesic arcs converges to a minimizing geodesic arc, which we denote $\widetilde{\gamma}$ and $\widetilde{\mu}$, respectively, in M_∞ emanating from x_∞. The above computation shows that the angle between these arcs is π and hence that their union is a geodesic, say ℓ. The same computation shows that ℓ is minimizing.

The existence of the minimizing geodesic line ℓ together with the fact that the sectional curvatures of the limit are ≥ 0 implies by Lemma 2.14 that the limit manifold is a Riemannian product $N \times \mathbb{R}$ in such a way that ℓ is of the form $\{x\} \times \mathbb{R}$ for some $x \in N$. $\square$

Part 2

Perelman's length function and its applications

CHAPTER 6

A comparison geometry approach to the Ricci flow

In this section we discuss Perelman's notions, introduced in [53], of the $\mathcal{L}$-length in the context of generalized Ricci flows. This is a functional defined on paths in space-time parameterized by backward time, denoted τ. The $\mathcal{L}$-length is the analogue in this context of the energy for paths in a Riemannian manifold. We derive the associated Euler-Lagrange equation for the $\mathcal{L}$-length; the critical paths are then $\mathcal{L}$-geodesics. Using $\mathcal{L}$-geodesics we define the L-exponential mapping. We derive the $\mathcal{L}$-Jacobi equation and relate $\mathcal{L}$-Jacobi fields to the differential of the $\mathcal{L}$-exponential mapping. There exists an analogue of the interior of the cut locus. It is the open subset, depending on the parameter τ, of the tangent space of initial vectors for $\mathcal{L}$-geodesics which are minimizing out to time τ and at which the $\mathcal{L}$-geodesic map is a local diffeomorphism at time τ. The difference between this situation and that of geodesics in a Riemannian manifold is that there is such an open set in the tangent space for each positive τ. The analogue of the fact that, for ordinary geodesics, the interior of the cut locus in the tangent space is star-shaped from the origin is that the open set of 'good' initial conditions at τ is contained in the open subset of 'good' initial conditions at time τ' for any $\tau' < \tau$. All of these results are local and are established in the context of generalized Ricci flows. In the next section we consider the case of ordinary Ricci flows, where we are able to extend our results over the entire manifold.

There are two applications of this theory in our study. In Chapter 8 we use the theory of $\mathcal{L}$-geodesics and the associated notion of reduced volume to establish non-collapsing results. These are crucial when we wish to take blow-up limits in studying singularities in Ricci flows and Ricci flows with surgery. The second application will be in Chapter 9 to κ-solutions (ancient, κ-non-collapsed solutions of bounded non-negative curvature). Here the second-order inequalities on the length function that we establish in this section are used to prove the existence of an asymptotic soliton for any κ-solution. This asymptotic soliton is important for giving qualitative results on κ-solutions.

1. $\mathcal{L}$-length and $\mathcal{L}$-geodesics

The running assumption throughout this section is that we have an n-dimensional generalized Ricci flow $(\mathcal{M}, G)$. In particular, the space-time

$\mathcal{M}$ is a smooth manifold of dimension $n+1$ whose boundary lies at the initial and final times (if they exist). Recall that its tangent bundle naturally decomposes as the direct sum of the sub-line bundle spanned by the vector field χ and the horizontal tangent bundle, denoted $\mathcal{H}T\mathcal{M}$. We also fix a time T in the time interval of definition of the flow distinct from the initial time.

DEFINITION 6.1. Let $0 \leq \tau_1 < \tau_2$ be given and let $\gamma\colon [\tau_1, \tau_2] \to \mathcal{M}$ be a continuous map. We say that γ is *parameterized by backward time* provided that $\gamma(\tau) \in M_{T-\tau}$ for all $\tau \in [\tau_1, \tau_2]$.

Throughout this section the paths γ that we consider shall be parameterized by backward time. We begin with the definition of $\mathcal{L}$-length of such a path.

DEFINITION 6.2. Let $\gamma\colon [\tau_1, \tau_2] \to \mathcal{M}$, $0 \leq \tau_1 < \tau_2$, be a C^1-path parameterized by backward time. We define $X_\gamma(\tau)$ to be the horizontal projection of the tangent vector $d\gamma(\tau)/d\tau$, so that $d\gamma/d\tau = -\chi + X_\gamma(\tau)$ with $X_\gamma(\tau) \in \mathcal{H}T\mathcal{M}$. We define the *$\mathcal{L}$-length* of γ to be

$$\mathcal{L}(\gamma) = \int_{\tau_1}^{\tau_2} \sqrt{\tau} \left(R(\gamma(\tau)) + |X_\gamma(\tau)|^2 \right) d\tau,$$

where the norm of $X_\gamma(\tau)$ is measured using the metric $G_{T-\tau}$ on $\mathcal{H}T\mathcal{M}$. When γ is clear from the context, we write X for X_γ; see FIG. 2 from the Introduction.

With a view toward better understanding the properties of the paths that are critical points of this functional, the so-called $\mathcal{L}$-geodesics, especially near $\tau = 0$, it is helpful to introduce a convenient reparameterization. We set $s = \sqrt{\tau}$. We use the notation $A(s)$ to denote the horizontal component of the derivative of γ with respect to the variable s. One sees immediately by the chain rule that

(6.1) $$A(s^2) = 2sX(s^2) \quad \text{or} \quad A(\tau) = 2\sqrt{\tau}X(\tau).$$

With respect to the variable s, the $\mathcal{L}$-functional is

(6.2) $$\mathcal{L}(\gamma) = \int_{\sqrt{\tau_1}}^{\sqrt{\tau_2}} \left(\frac{1}{2}|A(s)|^2 + 2R(\gamma(s))s^2 \right) ds.$$

Let's consider the simplest example.

EXAMPLE 6.3. Suppose that our generalized Ricci flow is a constant family of Euclidean metrics on $\mathbb{R}^n \times [0, T]$. That is to say, $g(t) = g_0$ is the usual Euclidean metric. Then we have $R(\gamma(\tau)) \equiv 0$. Using the change of variables $s = \sqrt{\tau}$, we have

$$\mathcal{L}(\gamma) = \frac{1}{2} \int_{\sqrt{\tau_1}}^{\sqrt{\tau_2}} |A(s)|^2 \, ds,$$

1. $\mathcal{L}$-LENGTH AND $\mathcal{L}$-GEODESICS

which is the standard energy functional in Riemannian geometry for the path $\gamma(s)$. The minimizers for this functional are the maps $s \mapsto (\alpha(s), T - s^2)$ where $\alpha(s)$ is a straight line in $\mathbb{R}^n$ parameterized at constant speed. Written in the τ variables the minimizers are

$$\gamma(\tau) = (x + \sqrt{\tau} v, T - \tau),$$

straight lines parameterized at speed varying linearly with $\sqrt{\tau}$.

1.1. $\mathcal{L}$-geodesics.

LEMMA 6.4. *The Euler-Lagrange equation for critical paths for the $\mathcal{L}$-length is*

(6.3) $$\nabla_X X - \frac{1}{2}\nabla R + \frac{1}{2\tau}X + 2\mathrm{Ric}(X, \cdot)^* = 0.$$

REMARK 6.5. $\mathrm{Ric}(X, \cdot)$ is a horizontal one-form along γ and its dual $\mathrm{Ric}(X, \cdot)^*$ is a horizontal tangent vector field along γ.

PROOF. First, let us suppose that the generalized Ricci flow is an ordinary Ricci flow $(M, g(t))$. Let $\gamma_u(\tau) = \gamma(\tau, u)$ be a family of curves parameterized by backward time. Let

$$\widetilde{Y}(\tau, u) = \frac{\partial \gamma}{\partial u}.$$

Then $\widetilde{X}(\tau, u) = X_{\gamma_u}(\tau, u)$ and $\widetilde{Y}(\tau, u)$ are the coordinate vector fields along the surface obtained by taking the projection of $\gamma(\tau, u)$ into M. Thus, $[\widetilde{X}, \widetilde{Y}] = 0$. We denote by X and Y the restrictions of $\widetilde{X}$ and $\widetilde{Y}$, respectively to γ_0. We have

$$\frac{d}{du}\mathcal{L}(\gamma_u)\big|_{u=0} = \frac{d}{du}\left(\int_{\tau_1}^{\tau_2} \sqrt{\tau}(R(\gamma_u(\tau)) + |\widetilde{X}(\tau, u)|^2)d\tau\right)\bigg|_{u=0}$$

$$= \int_{\tau_1}^{\tau_2} \sqrt{\tau}(\langle \nabla R, Y\rangle + 2\langle (\nabla_Y \widetilde{X})|_{u=0}, X\rangle)d\tau.$$

On the other hand, since $\partial g/\partial \tau = 2\mathrm{Ric}$ and since $[\widetilde{X}, \widetilde{Y}] = 0$, we have

$$2\frac{d}{d\tau}(\sqrt{\tau}\langle Y, X\rangle_{g(T-\tau)}) = \frac{1}{\sqrt{\tau}}\langle Y, X\rangle + 2\sqrt{\tau}\langle \nabla_X Y, X\rangle + 2\sqrt{\tau}\langle Y, \nabla_X X\rangle$$

$$+ 4\sqrt{\tau}\mathrm{Ric}(Y, X)$$

$$= \frac{1}{\sqrt{\tau}}\langle Y, X\rangle + 2\sqrt{\tau}\langle (\nabla_Y \widetilde{X})|_{u=0}, X\rangle + 2\sqrt{\tau}\langle Y, \nabla_X X\rangle$$

$$+ 4\sqrt{\tau}\mathrm{Ric}(Y, X).$$

Using this we obtain

(6.4)
$$\frac{d}{du}\mathcal{L}(\cdot\gamma_u)\big|_{u=0} = \int_{\tau_1}^{\tau_2}\Big(2\frac{d}{d\tau}\big[(\sqrt{\tau})\langle Y, X\rangle\big] - \frac{1}{\sqrt{\tau}}\langle Y, X\rangle$$
$$+ \sqrt{\tau}(\langle \nabla R, Y\rangle - 2\langle Y, \nabla_X X\rangle - 4\mathrm{Ric}(X, Y))\Big)d\tau$$
$$= 2\sqrt{\tau}\langle Y, X\rangle\big|_{\tau_1}^{\tau_2}$$
$$+ \int_{\tau_1}^{\tau_2}\sqrt{\tau}\langle Y, (\nabla R - \frac{1}{\tau}X - 2\nabla_X X - 4\mathrm{Ric}(X, \cdot)^*)\rangle d\tau.$$

Now we drop the assumption that the generalized Ricci flow is an ordinary Ricci flow. Still we can partition the interval $[\tau_1, \tau_2]$ into finitely many sub-intervals with the property that the restriction of γ_0 to each of the sub-intervals is contained in a patch of space-time on which the generalized Ricci flow is isomorphic to an ordinary Ricci flow. The above argument then applies to each of the sub-intervals. Adding up Equation (6.4) over these sub-intervals shows that the same equation for the first variation of length for the entire family γ_u holds.

We consider a variation $\gamma(\tau, u)$ with fixed endpoints, so that $Y(\tau_1) = Y(\tau_2) = 0$. Thus, the condition that γ be a critical path for the $\mathcal{L}$-length is that the integral expression vanish for all variations Y satisfying $Y(\tau_1) = Y(\tau_2) = 0$. Equation (6.4) holds for all such Y if and only if γ satisfies Equation (6.3). □

REMARK 6.6. In the Euler-Lagrange equation, ∇R is the horizontal gradient, and the equation is an equation of horizontal vector fields along γ.

DEFINITION 6.7. Let γ be a curve parameterized by backward time. Then γ is an $\mathcal{L}$-*geodesic* if it is a critical point of the $\mathcal{L}$-length. Equation (6.3) is the $\mathcal{L}$-*geodesic equation*.

Written with respect to the variable $s = \sqrt{\tau}$ the $\mathcal{L}$-geodesic equation becomes

(6.5) $$\nabla_{A(s)}A(s) - 2s^2\nabla R + 4s\mathrm{Ric}(A(s), \cdot)^* = 0.$$

Notice that in this form the ODE is regular even at $s = 0$.

LEMMA 6.8. *Let* $\gamma\colon [0, \tau_2] \to \mathcal{M}$ *be an $\mathcal{L}$-geodesic. Then* $\lim_{\tau\to 0}\sqrt{\tau}X_\gamma(\tau)$ *exists. The $\mathcal{L}$-geodesic γ is completely determined by this limit (and by τ_2).*

PROOF. Since the ODE in Equation (6.5) is non-singular even at zero, it follows that $A(s)$ is a smooth function of s in a neighborhood of $s = 0$. The lemma follows easily by the change of variables formula, $A(\tau) = 2\sqrt{\tau}X_\gamma(\tau)$. □

DEFINITION 6.9. An $\mathcal{L}$-geodesic is said to be *minimizing* if there is no curve parameterized by backward time with the same endpoints and with smaller $\mathcal{L}$-length.

1.2. The $\mathcal{L}$-Jacobi equation. Consider a family $\gamma(\tau, u)$ of $\mathcal{L}$-geodesics parameterized by u and defined on $[\tau_1, \tau_2]$ with $0 \leq \tau_1 < \tau_2$. Let $Y(\tau)$ be the horizontal vector field along γ defined by

$$Y(\tau) = \frac{\partial}{\partial u}\gamma(\tau, u)|_{u=0}.$$

LEMMA 6.10. $Y(\tau)$ *satisfies the $\mathcal{L}$-Jacobi equation:*

$$(6.6) \quad \nabla_X \nabla_X Y + \mathcal{R}(Y, X)X - \frac{1}{2}\nabla_Y(\nabla R)$$
$$+ \frac{1}{2\tau}\nabla_X Y + 2(\nabla_Y \mathrm{Ric})(X, \cdot)^* + 2\mathrm{Ric}(\nabla_X Y, \cdot)^* = 0.$$

This is a second-order linear equation for Y. Supposing that $\tau_1 > 0$, there is a unique horizontal vector field Y along γ solving this equation, vanishing at τ_1 with a given first-order derivative along γ at τ_1. Similarly, there is a unique solution Y to this equation, vanishing at τ_2 and with a given first-order derivative at τ_2.

PROOF. Given a family $\gamma(\tau, u)$ of $\mathcal{L}$-geodesics, then from Lemma 6.4 we have

$$\nabla_{\widetilde{X}} \widetilde{X} = \frac{1}{2}\nabla R(\gamma) - \frac{1}{2\tau}\widetilde{X} - 2\mathrm{Ric}(\widetilde{X}, \cdot)^*.$$

Differentiating this equation in the u-direction along the curve $u = 0$ yields

$$\nabla_Y \nabla_{\widetilde{X}} \widetilde{X}|_{u=0} = \frac{1}{2}\nabla_Y(\nabla R) - \frac{1}{2\tau}\nabla_Y(\widetilde{X})|_{u=0} - 2\nabla_Y(\mathrm{Ric}(\widetilde{X}, \cdot))^*|_{u=0}.$$

Of course, we have

$$\nabla_Y(\mathrm{Ric}(\widetilde{X}, \cdot)^*)|_{u=0} = (\nabla_Y \mathrm{Ric})(X, \cdot)^* + \mathrm{Ric}(\nabla_Y \widetilde{X}|_{u=0}, \cdot)^*.$$

Plugging this in, interchanging the orders of differentiation on the left-hand side, using $\nabla_{\widetilde{Y}} \widetilde{X} = \nabla_{\widetilde{X}} \widetilde{Y}$, and restricting to $u = 0$, yields the equation given in the statement of the lemma. This equation is a regular, second-order linear equation for all $\tau > 0$, and hence is determined by specifying the value and first derivative at any $\tau > 0$. $\square$

Equation (6.6) is obtained by applying ∇_Y to Equation (6.3) and exchanging orders of differentiation. The result, Equation (6.6), is a second-order differential equation for Y that makes no reference to an extension of $\gamma(\tau)$ to an entire family of curves.

DEFINITION 6.11. A field $Y(\tau)$ along an $\mathcal{L}$-geodesic is called an *$\mathcal{L}$-Jacobi field* if it satisfies the $\mathcal{L}$-Jacobi equation, Equation (6.6), and if it vanishes at τ_1. For any horizontal vector field Y along γ we denote by $\mathrm{Jac}(Y)$ the expression on the left-hand side of Equation (6.6).

In fact, there is a similar result even for $\tau_1 = 0$.

LEMMA 6.12. *Let γ be an $\mathcal{L}$-geodesic defined on $[0, \tau_2]$ and let $Y(\tau)$ be an $\mathcal{L}$-Jacobi field along γ. Then*
$$\lim_{\tau \to 0} \sqrt{\tau} \nabla_X Y$$
exists. Furthermore, $Y(\tau)$ is completely determined by this limit.

PROOF. We use the variable $s = \sqrt{\tau}$, and let $A(s)$ be the horizontal component of $d\gamma/ds$. Then differentiating the $\mathcal{L}$-geodesic equation written with respect to this variable we see
$$\nabla_A \nabla_A Y = -\mathcal{R}(Y, A)A + 2s^2 \nabla_Y (\nabla R) - 4s(\nabla_Y \mathrm{Ric})(A, \cdot) - 4s\mathrm{Ric}(\nabla_A Y, \cdot).$$
Hence, for each tangent vector Z, there is a unique solution to this equation with the two initial conditions $Y(0) = 0$ and $\nabla_A Y(0) = Z$.

On the other hand, from Equation (6.1) we have $\nabla_X(Y) = \frac{1}{2\sqrt{\tau}} \nabla_A(Y)$, so that
$$\sqrt{\tau} \nabla_X(Y) = \frac{1}{2} \nabla_A(Y).$$
□

1.3. Second order variation of $\mathcal{L}$. We shall need the relationship of the $\mathcal{L}$-Jacobi equation to the second-order variation of $\mathcal{L}$. This is given in the next proposition.

PROPOSITION 6.13. *Suppose that γ is a minimizing $\mathcal{L}$-geodesic. Then, for any vector field Y along γ, vanishing at both endpoints, and any family γ_u of curves parameterized by backward time with $\gamma_0 = \gamma$ and with the u-derivative of the family at $u = 0$ being the vector field Y along γ, we have*
$$\frac{d^2}{du^2} \mathcal{L}(\gamma_u)|_{u=0} = -\int_{\tau_1}^{\tau_2} 2\sqrt{\tau} \langle \mathrm{Jac}(Y), Y \rangle d\tau.$$
This quantity vanishes if and only if Y is an $\mathcal{L}$-Jacobi field.

Let us begin the proof of this proposition with the essential computation.

LEMMA 6.14. *Let γ be an $\mathcal{L}$-geodesic defined on $[\tau_1, \tau_2]$, and let Y_1 and Y_2 be horizontal vector fields along γ vanishing at τ_1. Suppose that γ_{u_1, u_2} is any family of curves parameterized by backward time with the property that $\gamma_{0,0} = \gamma$ and the derivative of the family in the u_i-direction at $u_1 = u_2 = 0$ is Y_i. Let $\widetilde{Y}_i$ be the image of $\partial/\partial u_i$ under γ_{u_1, u_2} and let $\widetilde{X}$ be the image of the horizontal projection of $\partial/\partial \tau$ under this same map, so that the restrictions of these three vector fields to the curve $\gamma_{0,0} = \gamma$ are Y_1, Y_2 and X respectively. Then we have*
$$\frac{\partial}{\partial u_1} \frac{\partial}{\partial u_2} \mathcal{L}(\gamma_{u_1,u_2})|_{u_1=u_2=0} = 2\sqrt{\tau_2} Y_1(\tau_2) \langle \widetilde{Y}_2(\tau_2, u_1, 0), \widetilde{X}(\tau_2, u_1, 0) \rangle |_{u_1=0}$$
$$- \int_{\tau_1}^{\tau_2} 2\sqrt{\tau} \langle \mathrm{Jac}(Y_1), Y_2 \rangle d\tau.$$

PROOF. According to Equation (6.4) we have

$$\frac{\partial}{\partial u_2}\mathcal{L}(\gamma)(u_1, u_2) = 2\sqrt{\tau_2}\langle \widetilde{Y}_2(\tau_2, u_1, u_2), \widetilde{X}(\tau_2, u_1, u_2)\rangle$$
$$- \int_{\tau_1}^{\tau_2} 2\sqrt{\tau}\langle EL(\widetilde{X}(\tau, u_1, u_2)), \widetilde{Y}_2(\tau, u_1, u_2)\rangle d\tau,$$

where $EL(\widetilde{X}(\tau, u_1, u_2))$ is the Euler-Lagrange expression for geodesics, i.e., the left-hand side of Equation (6.3). Differentiating again yields:

(6.7) $\quad \dfrac{\partial}{\partial u_1}\dfrac{\partial}{\partial u_2}\mathcal{L}(\gamma_{u_1, u_2})|_{0,0} = 2\sqrt{\tau_2}Y_1(\tau_2)\langle \widetilde{Y}_2(\tau_2, u_1, 0), \widetilde{X}(\tau_2, u_1, 0)\rangle|_{u_1=0}$
$$- \int_{\tau_1}^{\tau_2} 2\sqrt{\tau}\left(\langle \nabla_{Y_1} EL(\widetilde{X}), Y_2\rangle + \langle EL(X), \nabla_{Y_1}\widetilde{Y}_2\rangle\right)(\tau, 0, 0)d\tau.$$

Since $\gamma_{0,0} = \gamma$ is a geodesic, the second term in the integrand vanishes, and since $[\widetilde{X}, \widetilde{Y}_1] = 0$, we have $\nabla_{Y_1} EL(\widetilde{X}(\tau, 0, 0)) = \text{Jac}(Y_1)(\tau)$. This proves the lemma. $\square$

REMARK 6.15. Let $\gamma(\tau, u)$ be a family of curves as above with $\gamma(\tau, 0)$ being an $\mathcal{L}$-geodesic. It follows from Lemma 6.14 and the remark after the introduction of the $\mathcal{L}$-Jacobi equation that the second-order variation of length at $u = 0$ of this family is determined by the vector field $Y(\tau) = \partial\gamma/\partial u$ along $\gamma(\cdot, 0)$ and by the second-order information about the curve $\gamma(\bar{\tau}, u)$ at $u = 0$.

COROLLARY 6.16. *Let γ be an $\mathcal{L}$-geodesic and let Y_1, Y_2 be vector fields along γ vanishing at τ_1. Suppose $Y_1(\tau_2) = Y_2(\tau_2) = 0$. Then the bilinear pairing*

$$- \int_{\tau_1}^{\tau_2} 2\sqrt{\tau}\langle \text{Jac}(Y_1), Y_2\rangle d\tau$$

is a symmetric function of Y_1 and Y_2.

PROOF. Given Y_1 and Y_2 along γ we construct a two-parameter family of curves parameterized by backward time as follows. Let $\gamma(\tau, u_1)$ be the value at u_1 of the geodesic through $\gamma(\tau)$ with tangent vector $Y_1(\tau)$. This defines a family of curves parameterized by backward time, the family being parameterized by u_1 sufficiently close to 0. We extend Y_1 and X to vector fields on this entire family by defining them to be $\partial/\partial u_1$ and the horizontal projection of $\partial/\partial \tau$, respectively. Now we extend the vector field Y_2 along γ to a vector field on this entire one-parameter family of curves. We do this so that $Y_2(\tau_2, u_1) = Y_1(\tau_2, u_1)$. Now given this extension $Y_2(\tau, u_1)$ we define a two-parameter family of curves parameterized by backward time by setting $\gamma(\tau, u_1, u_2)$ equal to the value at u_2 of the geodesic through $\gamma(\tau, u_1)$ in the direction $Y_2(\tau, u_1)$. We then extend Y_1, Y_2, and X over this entire family by letting them be $\partial/\partial u_1, \partial/\partial u_2$, and the horizontal projection of $\partial/\partial \tau$, respectively. Applying Lemma 6.14 and using the fact that $Y_i(\overline\tau) = 0$ we conclude that

$$\frac{\partial}{\partial u_1}\frac{\partial}{\partial u_2}\mathcal{L}(\gamma)|_{u_1=u_2=0} = -\int_{\tau_1}^{\tau_2} 2\sqrt{\tau}\langle \operatorname{Jac}(Y_1), Y_2\rangle d\tau$$

and symmetrically that

$$\frac{\partial}{\partial u_2}\frac{\partial}{\partial u_1}\mathcal{L}(\gamma)|_{u_1=u_2=0} = -\int_{\tau_1}^{\tau_2} 2\sqrt{\tau}\langle \operatorname{Jac}(Y_2), Y_1\rangle d\tau.$$

Since the second cross-partials are equal, the corollary follows. □

Now we are in a position to establish Proposition 6.13.

PROOF. (Of Proposition 6.13) From the equation in Lemma 6.14, the equality of the second variation of $\mathcal{L}$-length at $u = 0$ and the integral is immediate from the fact that $Y(\tau_2) = 0$. It follows immediately that, if Y is an $\mathcal{L}$-Jacobi field vanishing at τ_2, then the second variation of the length vanishes at $u = 0$. Conversely, suppose given a family γ_u with $\gamma_0 = \gamma$ with the property that the second variation of length vanishes at $u = 0$, and that the vector field $Y = (\partial\gamma/\partial u)|_{u=0}$ along γ vanishes at the end points. It follows that the integral also vanishes. Since γ is a minimizing $\mathcal{L}$-geodesic, for any variation W, vanishing at the endpoints, the first variation of the length vanishes and the second variation of length is non-negative. That is to say,

$$-\int_{\tau_1}^{\tau_2} 2\sqrt{\tau}\langle \operatorname{Jac}(W), W\rangle d\tau \geq 0$$

for all vector fields W along γ vanishing at the endpoints. Hence, the restriction to the space of vector fields along γ vanishing at the endpoints of the symmetric bilinear form

$$B(Y_1, Y_2) = -\int_{\tau_1}^{\tau_2} 2\sqrt{\tau}\langle \operatorname{Jac}(Y_1), (Y_2)\rangle d\tau,$$

is positive semi-definite. Since $B(Y, Y) = 0$, it follows that $B(Y, \cdot) = 0$; that is to say, $\operatorname{Jac}(Y) = 0$. □

2. The $\mathcal{L}$-exponential map and its first-order properties

We use $\mathcal{L}$-geodesics in order to define the $\mathcal{L}$-exponential map.

For this section we fix $\tau_1 \geq 0$ and a point $x \in \mathcal{M}$ with $\mathbf{t}(x) = T - \tau_1$. We suppose that $T - \tau_1$ is greater than the initial time of the generalized Ricci flow. Then, for every $Z \in T_x M_{T-\tau_1}$, there is a maximal $\mathcal{L}$-geodesic, denoted γ_Z, defined on some positive τ-interval, with $\gamma_Z(\tau_1) = x$ and with $\sqrt{\tau_1}X(\tau_1) = Z$. (In the case $\tau_1 = 0$, this equation is interpreted to mean $\lim_{\tau\to 0}\sqrt{\tau}X(\tau) = Z$.)

2. THE $\mathcal{L}$-EXPONENTIAL MAP AND ITS FIRST-ORDER PROPERTIES

DEFINITION 6.17. We define *the domain of definition of* $\mathcal{L}\exp_x$, denoted $\mathcal{D}_x$, to be the subset of $T_x M_{T-\tau_1} \times (\tau_1, \infty)$ consisting of all pairs (Z, τ) for which $\tau > \tau_1$ is in the maximal domain of definition of γ_Z. Then we define $\mathcal{L}\exp_x \colon \mathcal{D}_x \to \mathcal{M}$ by setting $\mathcal{L}\exp_x(Z, \tau) = \gamma_Z(\tau)$ for all $(Z, \tau) \in \mathcal{D}_x$. (See FIG. 1.) We define the map $\widetilde{L}\colon \mathcal{D}_x \to \mathbb{R}$ by $\widetilde{L}(Z, \tau) = \mathcal{L}\left(\gamma_Z|_{[\tau_1, \tau]}\right)$. Lastly, for any $\tau > \tau_1$ we denote by $\mathcal{L}\exp_x^\tau$ the restriction of $\mathcal{L}\exp_x$ to the slice

$$\mathcal{D}_x^\tau = \mathcal{D}_x \cap (T_x M_{T-\tau_1} \times \{\tau\}),$$

which is *the domain of definition of* $\mathcal{L}\exp_x^\tau$. We also denote by $\widetilde{L}^\tau$ the restriction of $\widetilde{L}$ to this slice. We will implicitly identify $\mathcal{D}_x^\tau$ with a subset of $T_x M_{T-\tau_1}$.

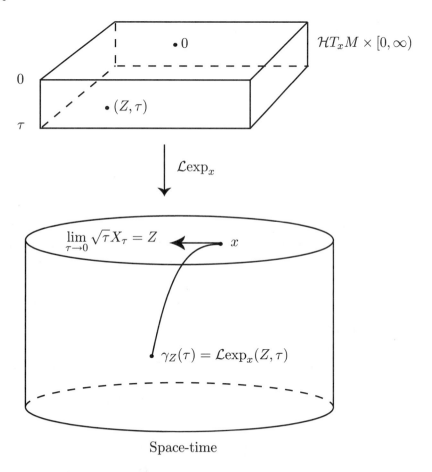

FIGURE 1. The map $\mathcal{L}\exp_x$.

LEMMA 6.18. $\mathcal{D}_x$ *is an open subset of* $T_x M_{T-\tau_1} \times (\tau_1, \infty)$; *its intersection with each line* $\{Z\} \times (\tau_1, \infty)$ *is a non-empty interval whose closure contains* τ_1. *Furthermore,* $\mathcal{L}\exp_x \colon \mathcal{D}_x \to \mathcal{M}$ *is a smooth map, and* $\widetilde{L}$ *is a smooth function.*

PROOF. The tangent vector in space-time of the $\mathcal{L}$-geodesic γ is the vector field $-\chi + X_\gamma(\tau)$ along γ, where $X_\gamma(\tau)$ satisfies Equation (6.3). As above, in the case $\tau_1 = 0$, it is convenient to replace the independent variable τ by $s = \sqrt{\tau}$, so that the ODE becomes Equation (6.5) which is regular at 0. With this change, the lemma then follows immediately by the usual results on existence, uniqueness and C^∞-variation with parameters of ODE's. $\square$

2.1. The differential of $\mathcal{L}\exp$. Now we compute the differential of $\mathcal{L}\exp$.

LEMMA 6.19. *Let $Z \in \mathcal{D}_x^{\bar{\tau}} \subset T_x M_{T-\tau_1}$. The differential of $\mathcal{L}\exp_x^{\bar{\tau}}$ at the point Z is given as follows: For each $W \in T_x(M_{T-\tau_1})$ there is a unique $\mathcal{L}$-Jacobi field $Y_W(\tau)$ along γ_Z with the property that $\sqrt{\tau_1} Y_W(\tau_1) = 0$ and $\sqrt{\tau_1} \nabla_X(Y_W)(\tau_1) = W$. We have*

$$d_Z \mathcal{L}\exp_x^{\bar{\tau}}(W) = Y_W(\bar{\tau}).$$

Again, in case $\tau_1 = 0$, both of the conditions on Y_W are interpreted as the limits as $\tau \to 0$.

PROOF. Let $Z(u)$ be a curve in $\mathcal{D}_x^{\bar{\tau}}$ with $Z(0) = Z$. Let γ_u be the $\mathcal{L}$-geodesic starting at x with $\sqrt{\tau_1} X_{\gamma_u}(\tau_1) = Z(u)$. Then, clearly,

$$d_Z \mathcal{L}\exp_x^{\bar{\tau}}\left(\frac{dZ}{du}(0)\right) = \frac{\partial}{\partial u}\left(\gamma_u(\bar{\tau})\right)|_{u=0}.$$

On the other hand, the vector field $Y(\tau) = (\partial \gamma_u(\tau)/\partial u)|_{u=0}$ is an $\mathcal{L}$-Jacobi field along γ_Z. Thus, to complete the proof in the case when $\tau_1 > 0$ we need only see that $\nabla_X \widetilde{Y}(\tau_1) = \nabla_Y \widetilde{X}(\tau_1)$. This is clear since, as we have already seen, $[\widetilde{X}, \widetilde{Y}] = 0$.

When $\tau_1 = 0$, we complete the argument using the following claim.

CLAIM 6.20. *If $\tau_1 = 0$, then*

$$\frac{\partial}{\partial u}\left(\lim_{\tau \to 0} \sqrt{\tau} X(\tau, u)\right)|_{u=0} = \lim_{\tau \to 0} \sqrt{\tau} \frac{d}{d\tau} Y(\tau).$$

PROOF. This follows immediately by changing variables, replacing τ by $s = \sqrt{\tau}$. $\square$

This completes the proof of Lemma 6.19. $\square$

2.2. Positivity of the second variation. Suppose that γ is a minimizing $\mathcal{L}$-geodesic. Then variations of γ fixing the endpoints give curves whose $\mathcal{L}$-length is no less than that of γ. In fact, there is a second-order version of this inequality which we shall need later.

COROLLARY 6.21. *Let $Z \in T_x M_{T-\tau_1}$. Suppose that the associated $\mathcal{L}$-geodesic γ_Z minimizes $\mathcal{L}$-length between its endpoints, x and $\gamma_Z(\bar{\tau})$, and*

2. THE $\mathcal{L}$-EXPONENTIAL MAP AND ITS FIRST-ORDER PROPERTIES

that $d_Z \mathcal{L}\mathrm{exp}_x^{\overline{\tau}}$ is an isomorphism. Then for any family γ_u of curves parameterized by backward time with $Y = (\partial\gamma/\partial u)|_{u=0}$ vanishing at both endpoints, we have

$$\frac{d^2}{du^2}\mathcal{L}(\gamma_u)|_{u=0} \geq 0,$$

with equality if and only if $Y = 0$.

PROOF. According to Proposition 6.13 the second variation in the Y-direction is non-negative and vanishes if and only if Y is an $\mathcal{L}$-Jacobi field. But since $d_Z \mathcal{L}\mathrm{exp}_x^{\overline{\tau}}$ is a diffeomorphism, by Lemma 6.19 there are no non-zero $\mathcal{L}$-Jacobi fields vanishing at both endpoints of γ_Z. $\square$

2.3. The gradient of $\widetilde{L}^\tau$. Recall that $\widetilde{L}^\tau$ is the map from $\mathcal{D}_x^\tau$ to $\mathbb{R}$ that assigns to each Z the $\mathcal{L}$-length of $\gamma_Z|_{[\tau_1,\tau]}$. We compute its gradient.

LEMMA 6.22. *Suppose that* $Z \in \mathcal{D}_x^\tau$. *Then for any* $\widetilde{Y} \in T_x M_{T-\tau_1} = T_Z(\mathcal{D}_x^\tau)$ *we have*

$$\langle \nabla \widetilde{L}^\tau, \widetilde{Y} \rangle = 2\sqrt{\tau}\langle X(\tau), d_Z\left(\mathcal{L}\mathrm{exp}_x^\tau\right)(\widetilde{Y})\rangle.$$

PROOF. Since $\mathcal{D}_x^\tau$ is an open subset of $T_x(M_{T-\tau_1})$, it follows that for any $\widetilde{Y} \in T_x(M_{T-\tau_1})$ there is a one-parameter family $\gamma_u(\tau') = \gamma(\tau', u)$ of $\mathcal{L}$-geodesics, defined for $\tau_1 \leq \tau' \leq \tau$, starting at x with $\gamma(\cdot, 0) = \gamma_Z$ and with $\frac{\partial}{\partial u}\left(\sqrt{\tau_1}X(\tau_1)\right) = \widetilde{Y}$. (Again, when $\tau_1 = 0$, this equation is interpreted to mean $\frac{\partial}{\partial u}\lim_{\tau' \to 0}(\sqrt{\tau'}X(\tau', u)) = \widetilde{Y}$.) Let $Y(\tau') = \frac{\partial}{\partial u}(\gamma(\tau', u))|_{u=0}$ be the corresponding $\mathcal{L}$-Jacobi field along γ_Z. Since $\gamma(\tau_1, u) = x$ for all u, we have $Y(\tau_1) = 0$. Since $\gamma(\cdot, u)$ is an $\mathcal{L}$-geodesic for all u, according to Equation (6.4), and in the case $\tau_1 = 0$, using the fact that $\sqrt{\tau}X(\tau')$ approaches a finite limit as $\tau \to 0$, we have

$$\frac{d}{du}\mathcal{L}(\gamma_u)|_{u=0} = 2\sqrt{\tau}\langle X(\tau), Y(\tau)\rangle.$$

By Lemma 6.19 we have $Y(\tau) = d_Z \mathcal{L}\mathrm{exp}_x^\tau(\widetilde{Y})$. Thus,

$$\langle \nabla \widetilde{L}^\tau, \widetilde{Y}\rangle = \frac{d}{du}\mathcal{L}(\gamma_u)|_{u=0} = 2\sqrt{\tau}\langle X(\tau), Y(\tau)\rangle = 2\sqrt{\tau}\langle X(\tau), d_Z(\mathcal{L}\mathrm{exp}_x^\tau)(\widetilde{Y})\rangle.$$

$\square$

2.4. Local diffeomorphism near the initial τ. Now let us use the nature of the $\mathcal{L}$-Jacobi equation to study $\mathcal{L}\mathrm{exp}_x$ for $\tau > \tau_1$ but τ sufficiently close to τ_1.

LEMMA 6.23. *For any* x *in* $\mathcal{M}$ *with* $\mathbf{t}(x) = T - \tau_1$ *and any* $Z \in T_x M_{t-\tau_1}$, *there is* $\delta > 0$ *such that for any* τ *with* $\tau_1 < \tau < \tau_1 + \delta$ *the map* $\mathcal{L}\mathrm{exp}_x^\tau$ *is a local diffeomorphism from a neighborhood of* Z *in* $T_x M_{T-\tau_1}$ *to* $M_{T-\tau}$.

PROOF. Fix x and Z as in the statement of the lemma. To establish the result it suffices to prove that there is $\delta > 0$ such that $d_Z \mathcal{L}\exp_x^\tau$ is an isomorphism for all $\tau_1 < \tau < \tau_1 + \delta$. By Lemma 6.19 it is enough to find a $\delta > 0$ such that any $\mathcal{L}$-Jacobi field Y along γ_Z with $\sqrt{\tau_1}\nabla_X Y(\tau_1) \neq 0$ does not vanish on the interval $(\tau_1, \tau_1 + \delta)$. Because the $\mathcal{L}$-Jacobi equation is linear, it suffices to consider the case of $\mathcal{L}$-Jacobi fields with $|\nabla_X Y(\tau_1)| = 1$. The space of such fields is identified with the unit sphere in $T_x M_{T-\tau_1}$. Let us consider first the case when $\tau_1 \neq 0$. Then for any such tangent vector $\nabla_X Y(\tau_1) \neq 0$. Since $Y(\tau_1) = 0$, it follows that $Y(\tau) \neq 0$ in some interval $(\tau_1, \tau_1 + \delta)$, where δ can depend on Y. Using compactness of the unit sphere in the tangent space, we see that there is $\delta > 0$ independent of Y so that the above holds.

In case when $\tau_1 = 0$, it is convenient to shift to the $s = \sqrt{\tau}$ parameterization. Then the geodesic equation and the $\mathcal{L}$-Jacobi equation are non-singular at the origin. Also, letting $A = d\gamma_Z/ds$ we have $\nabla_A Y = 2\lim_{\tau \to 0} \sqrt{\tau}\nabla_X Y$. In these variables, the argument for $\tau_1 = 0$ is the same as the one above for $\tau_1 > 0$. □

REMARK 6.24. When $\tau_1 > 0$ it is possible to consider the $\mathcal{L}\exp_x^\tau$ defined for $0 < \tau < \tau_1$. In this case, the curves are moving backward in τ and hence are moving forward with respect to the time parameter **t**. Two comments are in order. First of all, for $\tau < \tau_1$, the gradient of $\widetilde{L}_x^\tau$ is $-2\sqrt{\tau}X(\tau)$. The reason for the sign reversal is that the length is given by the integral from τ to τ_1 and hence its derivative with respect to τ is the negative of the integrand. The second thing to remark is that Lemma 6.23 is true for $\tau < \tau_1$ with τ sufficiently close to τ_1.

3. Minimizing $\mathcal{L}$-geodesics and the injectivity domain

Now we discuss the analogue of the interior of the cut locus for the usual exponential map of a Riemannian manifold. In this section we keep the assumption that $x \in \mathcal{M}$ with $\mathbf{t}(x) = T - \tau_1$ for some $\tau_1 \geq 0$.

DEFINITION 6.25. The *injectivity set* $\widetilde{\mathcal{U}}_x \subset \mathcal{D}_x \subset (T_x M_{T-\tau_1} \times (\tau_1, \infty))$ is the subset of all $(Z, \tau) \in \mathcal{D}_x$ with the following properties:

(1) The map $\mathcal{L}\exp_x^\tau$ is a local diffeomorphism near Z from $T_x(M_{T-\tau_1})$ to $M_{T-\tau}$.
(2) There is a neighborhood $\mathcal{Z}$ of Z in $\mathcal{D}_x^\tau$ such that for every $Z' \in \mathcal{Z}$ the $\mathcal{L}$-geodesic $\gamma_{Z'}|_{[\tau_1,\tau]}$ is the unique minimizing path parameterized by backward time for the $\mathcal{L}$-length. That is to say, the $\mathcal{L}$-length of $\gamma_{Z'}|_{[\tau_1,\tau]}$ is less than the $\mathcal{L}$-length of any other path parameterized by backward time between the same endpoints.

For any $\tau > \tau_1$, we set $\widetilde{\mathcal{U}}_x(\tau) \subset T_x M_{T-\tau_1}$ equal to the slice of $\widetilde{\mathcal{U}}_x$ at τ, i.e., $\widetilde{\mathcal{U}}_x(\tau)$ is determined by the equation

$$\widetilde{\mathcal{U}}_x(\tau) \times \{\tau\} = \widetilde{\mathcal{U}}_x \cap (T_x M_{T-\tau_1} \times \{\tau\}).$$

It is clear from the definition that $\widetilde{\mathcal{U}}_x \subset \mathcal{D}_x$ is an open subset and hence $\widetilde{\mathcal{U}}_x$ is an open subset of $T_x M_{T-\tau_1} \times (\tau_1, \infty)$. Of course, this implies that $\widetilde{\mathcal{U}}_x(\tau)$ is an open subset of $\mathcal{D}_x^\tau$ for every $\tau > \tau_1$.

DEFINITION 6.26. We set $\mathcal{U}_x \subset \mathcal{M}$ equal to $\mathcal{L}\exp_x(\widetilde{\mathcal{U}}_x)$. We call this subset of $\mathcal{M}$ the *injectivity domain (of x)*. For any $\tau > \tau_1$ we set $\mathcal{U}_x(\tau) = \mathcal{U}_x \cap M_{T-\tau}$.

By definition, for any $(Z, \tau) \in \widetilde{\mathcal{U}}_x$ with $\mathcal{L}\exp_x(Z, \tau) = q$, the $\mathcal{L}$-geodesic $\gamma_Z|_{[\tau_1, \tau]}$ is a minimizing $\mathcal{L}$-geodesic to q. In particular, for every $q \in \mathcal{U}_x$ there is a minimizing $\mathcal{L}$-geodesic from x to q.

DEFINITION 6.27. The function $L_x \colon \mathcal{U}_x \to \mathbb{R}$ assigns to each q in $\mathcal{U}_x$ the length of any minimizing $\mathcal{L}$-geodesic from x to q. For any $\tau > \tau_1$, we denote by L_x^τ the restriction of L_x to the $T - \tau$ time-slice of $\mathcal{U}_x$, i.e., the restriction of L_x to $\mathcal{U}_x(\tau)$.

This brings us to the analogue of the fact that in Riemannian geometry the restriction to the interior of the cut locus of the exponential mapping is a diffeomorphism onto an open subset of the manifold.

PROPOSITION 6.28. *The map*

$$\mathcal{L}\exp_x \colon \widetilde{\mathcal{U}}_x \to \mathcal{M}$$

is a diffeomorphism onto the open subset $\mathcal{U}_x$ of $\mathcal{M}$. The function $L \colon \mathcal{U}_x \to \mathbb{R}$ that associates to each $q \in \mathcal{U}_x$ the length of the unique minimizing $\mathcal{L}$-geodesic from x to q is a smooth function and

$$L_x \circ \mathcal{L}\exp_x|_{\widetilde{\mathcal{U}}_x} = \widetilde{L}|_{\widetilde{\mathcal{U}}_x}.$$

PROOF. We consider the differential of $\mathcal{L}\exp_x$ at any $(Z, \tau) \in \widetilde{\mathcal{U}}_x$. By construction the restriction of this differential to $T_x M_{T-\tau_1}$ is a local isomorphism onto $\mathcal{HTM}$ at the image point. On the other hand, the differential of $\mathcal{L}\exp_x$ in the τ direction is $\gamma_Z'(\tau)$, whose 'vertical' component is $-\chi$. By the inverse function theorem this shows that $\mathcal{L}\exp_x$ is a local diffeomorphism at (Z, τ), and its image is an open subset of $\mathcal{M}$. The uniqueness in Condition 2 of Definition 6.25 immediately implies that the restriction of $\mathcal{L}\exp_x$ to $\widetilde{\mathcal{U}}_x$ is one-to-one, and hence that it is a global diffeomorphism onto its image $\mathcal{U}_x$.

Since for every $(Z, \tau) \in \widetilde{\mathcal{U}}_x$ the $\mathcal{L}$-geodesic $\gamma_Z|_{[\tau_1, \tau]}$ is $\mathcal{L}$-minimizing, we see that $L_x \circ \mathcal{L}\exp_x|_{\widetilde{\mathcal{U}}_x} = \widetilde{L}|_{\widetilde{\mathcal{U}}_x}$ and that $L_x \colon \mathcal{U}_x \to \mathbb{R}$ is a smooth function. □

According to Lemma 6.22 we have:

COROLLARY 6.29. *At any $q \in \mathcal{U}_x(\tau)$ we have*
$$\nabla L_x^\tau(q) = 2\sqrt{\tau} X(\tau)$$
where $X(\tau)$ is the horizontal component of $\gamma'(\tau)$, and where γ is the unique minimizing $\mathcal{L}$-geodesic connecting x to q. (See FIG. *2 in the Introduction.)*

At the level of generality that we are working (arbitrary generalized Ricci flows) there is no result analogous to the fact in Riemannian geometry that the image under the exponential mapping of the interior of the cut locus is an open dense subset of the manifold. There is an analogue in the special case of Ricci flows on compact manifolds or on complete manifolds of bounded curvature. These will be discussed in Chapter 7.

3.1. Monotonicity of the $\widetilde{\mathcal{U}}_x(\tau)$ with respect to τ. Next, we have the analogue of the fact in Riemannian geometry that the cut locus is star-shaped.

PROPOSITION 6.30. *Let $\overline{\tau}' > \overline{\tau}$. Then $\widetilde{\mathcal{U}}_x(\overline{\tau}') \subset \widetilde{\mathcal{U}}_x(\overline{\tau}) \subset T_x M_{T-\tau_1}$.*

PROOF. For $Z \in \widetilde{\mathcal{U}}_x(\overline{\tau}')$, we shall show that: (i) the $\mathcal{L}$-geodesic $\gamma_{Z'}|_{[\tau_1,\overline{\tau}]}$ is the unique minimizing $\mathcal{L}$-geodesic from x to $\gamma_Z(\overline{\tau})$, and (ii) the differential $d_Z \mathcal{L}\exp_x^{\overline{\tau}}$ is an isomorphism. Given these two conditions, it follows from the definition that $\widetilde{\mathcal{U}}_x(\overline{\tau}')$ is contained in $\widetilde{\mathcal{U}}_x(\overline{\tau})$.

We will show that the $\mathcal{L}$-geodesic $\gamma_Z|_{[\tau_1,\overline{\tau}]}$ is the unique minimizing $\mathcal{L}$- geodesic to its endpoint. If there is an $\mathcal{L}$-geodesic γ_1, distinct from $\gamma_Z|_{[\tau_1,\overline{\tau}]}$, from x to $\gamma_Z(\overline{\tau})$ whose $\mathcal{L}$-length is at most that of $\gamma_Z|_{[\tau_1,\overline{\tau}]}$, then there is a broken path $\gamma_1 * \gamma_Z|_{[\overline{\tau},\overline{\tau}']}$ parameterized by backward time whose $\mathcal{L}$-length is at most that of γ_Z. Since this latter path is not smooth, its $\mathcal{L}$-length cannot be the minimum, which is a contradiction.

Now suppose that $d_Z \mathcal{L}\exp_x^{\overline{\tau}}$ is not an isomorphism. The argument is similar to the one above, using a non-zero $\mathcal{L}$-Jacobi field vanishing at both endpoints rather than another geodesic. Let τ_2' be the first τ for which $d_Z \mathcal{L}\exp_x^\tau$ is not an isomorphism. According to Lemma 6.23, $\tau_1 < \tau_2' \leq \overline{\tau}$. Since $\mathcal{L}\exp_x^{\tau_2'}$ is not a local diffeomorphism at (Z, τ_2'), by Lemma 6.19 there is a non-zero $\mathcal{L}$-Jacobi field Y along $\gamma_Z|_{[\tau_1,\tau_2']}$ vanishing at both ends. Since $\gamma_Z|_{[\tau_1,\tau_2']}$ is $\mathcal{L}$-minimizing, according to Proposition 6.13, the second variation of the length of $\gamma_Z|_{[\tau_1,\tau_2']}$ in the Y-direction vanishes, in the sense that if $\gamma(u,\tau)$ is any one-parameter family of paths parameterized by backward time from x to $\gamma_Z(\tau_2')$ with $(\partial\gamma/\partial u)|_{u=0} = Y$, then
$$\frac{\partial^2 \mathcal{L}(\gamma_u)}{\partial u^2}\bigg|_{u=0} = 0.$$

Extend Y to a horizontal vector field $\widehat{Y}$ along γ_Z by setting $\widehat{Y}(\tau) = 0$ for all $\tau \in [\tau_2', \overline{\tau}]$. Of course, the extended horizontal vector field $\widehat{Y}$ is not C^2 at τ_2' since Y, being a non-zero $\mathcal{L}$-Jacobi field, does not vanish to second order there. This is the first-order variation of the family $\hat{\gamma}(u,\tau)$ that agrees with $\gamma(u,\tau)$ for all $\tau \leq \tau_2'$ and has $\hat{\gamma}(u,\tau) = \gamma_Z(\tau)$ for all $\tau \in [\tau_2', \overline{\tau}]$. Of course, the second-order variation of this extended family at $u = 0$ agrees with the second-order variation of the original family at $u = 0$, and hence vanishes. But according to Proposition 6.13 this means that $\widehat{Y}$ is an $\mathcal{L}$-Jacobi field, which is absurd since it is not a C^2-vector field. $\square$

We shall also need a closely related result.

PROPOSITION 6.31. *Let γ be a minimizing $\mathcal{L}$-geodesic defined for $[\tau_1, \overline{\tau}]$. Fix $0 \leq \tau_1 < \tau_2 < \overline{\tau}$, and set $q_2 = \gamma(\tau_2)$, and $Z_2 = \sqrt{\tau_2} X_\gamma(\tau_2)$. Then, the map $\mathcal{L}\exp_{q_2}$ is a diffeomorphism from a neighborhood of $\{Z_2\} \times (\tau_2, \overline{\tau}]$ in $T_q \mathcal{M}_{T-\tau_2} \times (\tau_2, \infty)$ onto a neighborhood of the image of $\gamma|_{(\tau_2, \overline{\tau}]}$.*

PROOF. It suffices to show that the differential of $\mathcal{L}\exp_{q_2}^\tau$ is an isomorphism for all $\tau \in (\tau_2, \overline{\tau}]$. If this is not the case, then there is a $\tau' \in (\tau_2, \overline{\tau}]$ and a non-zero $\mathcal{L}$-Jacobi field Y along $\gamma_Z|_{[\tau_2, \tau']}$ vanishing at both ends. We extend Y to a horizontal vector field $\widehat{Y}$ along all of $\gamma_Z|_{[\tau_1, \tau']}$ by setting it equal to zero on $[\tau_1, \tau_2]$. Since Y is an $\mathcal{L}$-Jacobi field, the second-order variation of $\mathcal{L}$-length in the direction of Y is zero, and consequently the second-order variation of the length of $\gamma_Z|_{[\tau_1, \tau']}$ vanishes. Hence by Proposition 6.13 it must be the case that $\widehat{Y}$ is an $\mathcal{L}$-Jacobi field. This is impossible since $\widehat{Y}$ is not smooth at τ'. $\square$

We finish this section with a computation of the τ-derivative of L_x.

LEMMA 6.32. *Suppose that $q \in \mathcal{U}_x$ with $\mathbf{t}(q) = T - \overline{\tau}$ for some $\overline{\tau} > \tau_1$. Let $\gamma : [\tau_1, \overline{\tau}] \to \mathcal{M}$ be the unique minimizing $\mathcal{L}$-geodesic from x to q. Then we have*

$$(6.8) \qquad \frac{\partial L_x}{\partial \tau}(q) = 2\sqrt{\overline{\tau}} R(q) - \sqrt{\overline{\tau}} \left(R(q) + |X(\overline{\tau})|^2 \right).$$

PROOF. By definition and the Fundamental Theorem of Calculus, we have

$$\frac{d}{d\tau} L_x(\gamma(\tau)) = \sqrt{\tau} \left(R(\gamma(\tau)) + |X(\tau)|^2 \right).$$

On the other hand since $\gamma'(\tau) = -\partial/\partial t + X(\tau)$ the chain rule implies

$$\frac{d}{d\tau} L_x(\gamma(\tau)) = \langle \nabla L_x, X(\tau) \rangle + \frac{\partial L_x}{\partial \tau}(\gamma(\tau)),$$

so that

$$\frac{\partial L_x}{\partial \tau}(\gamma(\tau)) = \sqrt{\tau} \left(R(\gamma(\tau)) + |X(\tau)|^2 \right) - \langle \nabla L_x, X(\tau) \rangle.$$

Now using Corollary 6.29, and rearranging the terms gives the result. $\square$

4. Second-order differential inequalities for $\widetilde{L}^{\overline{\tau}}$ and $L_x^{\overline{\tau}}$

Throughout this section we fix $x \in \mathcal{M}$ with $x \in \mathcal{M}_{T-\tau_1}$.

4.1. The second variation formula for $\widetilde{L}^{\overline{\tau}}$.
Our goal here is to compute the second variation of $\widetilde{L}^{\overline{\tau}}$ in the direction of a horizontal vector field $Y(\tau)$ along an $\mathcal{L}$-geodesic γ. Here is the main result of this subsection.

PROPOSITION 6.33. *Fix $0 \le \tau_1 < \overline{\tau}$. Let γ be an $\mathcal{L}$-geodesic defined on $[\tau_1, \overline{\tau}]$ and let $\gamma_u = \widetilde{\gamma}(\tau, u)$ be a smooth family of curves parameterized by backward time with $\gamma_0 = \gamma$. Let $\widetilde{Y}(\tau, u)$ be $\partial \widetilde{\gamma}/\partial u$ and let $\widetilde{X}$ be the horizontal component of $\partial \widetilde{\gamma}/\partial \tau$. These are horizontal vector fields along the image of $\widetilde{\gamma}$. We set Y and X equal to the restrictions of $\widetilde{Y}$ and $\widetilde{X}$, respectively, to γ. We assume that $Y(\tau_1) = 0$. Then*

$$\frac{d^2}{du^2}\left(\mathcal{L}(\gamma_u)\right)|_{u=0} = 2\sqrt{\overline{\tau}}\langle \nabla_{Y(\overline{\tau})}\widetilde{Y}(\overline{\tau},u)|_{u=0}, X(\overline{\tau})\rangle + \int_{\tau_1}^{\overline{\tau}} \sqrt{\tau}\Big[\text{Hess}(R)(Y,Y)$$
$$+ 2\langle \mathcal{R}(Y,X)Y, X\rangle - 4(\nabla_Y \text{Ric})(X,Y)$$
$$+ 2(\nabla_X \text{Ric})(Y,Y) + 2|\nabla_X Y|^2\Big]d\tau.$$

As we shall see, this is simply a rewriting of the equation in Lemma 6.14 in the special case when $u_1 = u_2$.

We begin the proof of this result with the following computation.

CLAIM 6.34. *Let $\gamma(\tau)$ be a curve parameterized by backward time. Let Y be a horizontal vector field along γ and let X be the horizontal component of $\partial \widetilde{\gamma}/\partial \tau$. Then*

$$\frac{\partial}{\partial \tau}\langle \nabla_X Y, Y\rangle = \langle \nabla_X Y, \nabla_X Y\rangle + \langle \nabla_X \nabla_X Y, Y\rangle$$
$$+ 2\text{Ric}(\nabla_X Y, Y)) + (\nabla_X \text{Ric})(Y,Y)).$$

PROOF. We can break $\frac{\partial}{\partial \tau}\langle \nabla_X Y, Y\rangle$ into two parts: the first assumes that the metric is constant and the second deals with the variation with τ of the metric. The first contribution is the usual formula

$$\frac{\partial}{\partial \tau}\langle \nabla_X Y, Y\rangle_{G(T-\tau_0)} = \langle \nabla_X Y, \nabla_X Y\rangle_{G(T-\tau_0)} + \langle \nabla_X \nabla_X Y, Y\rangle_{G(T-\tau_0)}.$$

This gives us the first two terms of the right-hand side of the equation in the claim.

We show that the last two terms in that equation come from differentiating the metric with respect to τ. To do this recall that in local coordinates, writing the metric $G(T-\tau)$ as g_{ij}, we have

$$\langle \nabla_X Y, Y\rangle = g_{ij}\big(X^k \partial_k Y^i + \Gamma^i_{kl} X^k Y^l\big)Y^j.$$

There are two contributions coming from differentiating the metric with respect to τ. The first is when we differentiate g_{ij}. This leads to

$$2\text{Ric}_{ij}\big(X^k \partial_k Y^i + \Gamma^i_{kl} X^k Y^l\big)Y^j = 2\text{Ric}(\nabla_X Y, Y).$$

The other contribution is from differentiating the Christoffel symbols. This yields
$$g_{ij}\frac{\partial \Gamma^i_{kl}}{\partial \tau}X^k Y^l Y^j.$$
Differentiating the formula $\Gamma^i_{kl} = \frac{1}{2}g^{si}(\partial_k g_{sl} + \partial_l g_{sk} - \partial_s g_{kl})$ leads to
$$\begin{aligned}g_{ij}\frac{\partial \Gamma^i_{kl}}{\partial \tau} &= -2\mathrm{Ric}_{ij}\Gamma^i_{kl} + g_{ij}g^{si}(\partial_k \mathrm{Ric}_{sl} + \partial_l \mathrm{Ric}_{sk} - \partial_s \mathrm{Ric}_{kl})\\ &= -2\mathrm{Ric}_{ij}\Gamma^i_{kl} + \partial_k \mathrm{Ric}_{jl} + \partial_l \mathrm{Ric}_{jk} - \partial_j \mathrm{Ric}_{kl}.\end{aligned}$$
Thus, we have
$$\begin{aligned}g_{ij}\frac{\partial \Gamma^i_{kl}}{\partial \tau}X^k Y^l Y^j &= \left(-2\mathrm{Ric}_{ij}\Gamma^i_{kl} + \partial_k \mathrm{Ric}_{jl}\right)X^k Y^l Y^j\\ &= (\nabla_X \mathrm{Ric})(Y,Y).\end{aligned}$$
This completes the proof of the claim. □

Now we return to the proof of the second variational formula in Proposition 6.33.

PROOF. According to Lemma 6.14 we have
$$\frac{d^2}{du^2}\mathcal{L}|_0 = 2\sqrt{\overline{\tau}}Y(\overline{\tau})(\langle \widetilde{Y}(\overline{\tau},u),\widetilde{X}(\overline{\tau},u)\rangle)|_0 - \int_{\tau_1}^{\tau_2} 2\sqrt{\tau}\langle \mathrm{Jac}(Y),Y\rangle d\tau.$$
We plug in Equation 6.6 for $\mathrm{Jac}(Y)$ and this makes the integrand
$$\sqrt{\tau}\langle \nabla_Y(\nabla R), Y\rangle + 2\sqrt{\tau}\langle \mathcal{R}(Y,X)Y,X\rangle - 2\sqrt{\tau}\langle \nabla_X \nabla_X Y, Y\rangle$$
$$- \frac{1}{\sqrt{\tau}}\langle \nabla_X Y, Y\rangle - 4\sqrt{\tau}(\nabla_Y \mathrm{Ric})(X,Y) - 4\sqrt{\tau}\mathrm{Ric}(\nabla_X Y, Y).$$
The first term is $\sqrt{\tau}\mathrm{Hess}(R)(Y,Y)$. Let us deal with the third and fourth terms. According to the previous claim, we have
$$\frac{\partial}{\partial \tau}(2\sqrt{\tau}\langle \nabla_X Y, Y\rangle) = \frac{1}{\sqrt{\tau}}\langle \nabla_X Y, Y\rangle + 2\sqrt{\tau}\langle \nabla_X \nabla_X Y, Y\rangle$$
$$+ 2\sqrt{\tau}\langle \nabla_X Y, \nabla_X Y\rangle + 4\sqrt{\tau}\mathrm{Ric}(\nabla_X Y, Y) + 2\sqrt{\tau}(\nabla_X \mathrm{Ric})(Y,Y).$$
This allows us to replace the two terms under consideration by
$$-\frac{\partial}{\partial t}(2\sqrt{\tau}\langle \nabla_X Y, Y\rangle) + 2\sqrt{\tau}\langle \nabla_X Y, \nabla_X Y\rangle$$
$$+ 4\sqrt{\tau}\mathrm{Ric}(\nabla_X Y, Y) + 2\sqrt{\tau}(\nabla_X \mathrm{Ric})(Y,Y).$$
Integrating the total derivative out of the integrand and canceling terms leaves the integrand as
$$\sqrt{\tau}\mathrm{Hess}(R)(Y,Y) + 2\sqrt{\tau}\langle \mathcal{R}(Y,X)Y,X\rangle$$
$$+ 2\sqrt{\tau}|\nabla_X Y|^2 - 4\sqrt{\tau}(\nabla_Y \mathrm{Ric})(X,Y) + 2\sqrt{\tau}(\nabla_X \mathrm{Ric})(Y,Y),$$

and makes the boundary term (the one in front of the integral) equal to

$$2\sqrt{\bar\tau}\bigl(Y(\bar\tau)\langle \widetilde{Y}(\bar\tau,u),\widetilde{X}(\bar\tau,u)\rangle|_0 - \langle\nabla_X Y(\bar\tau),Y(\bar\tau)\rangle\bigr)$$
$$= 2\sqrt{\bar\tau}\langle X(\bar\tau),\nabla_Y \overset{\approx}{Y}(\bar\tau,u)|_0\rangle.$$

This completes the proof of the proposition. □

4.2. Inequalities for the Hessian of $L_x^{\bar\tau}$. Now we shall specialize the type of vector fields along γ. This will allow us to give an inequality for the Hessian of $\mathcal{L}$ involving the integral of the vector field along γ. These lead to inequalities for the Hessian of $L_x^{\bar\tau}$. The main result of this section is Proposition 6.37 below. In the end we are interested in the case when the $\tau_1 = 0$. In this case the formulas simplify. The reason for working here in the full generality of all τ_1 is in order to establish differential inequalities at points not in the injectivity domain. As in the case of the theory of geodesics, the method is to establish weak inequalities at these points by working with smooth barrier functions. In the geodesic case the barriers are constructed by moving the initial point out the geodesic a small amount. Here the analogue is to move the initial point of an $\mathcal{L}$-geodesic from $\tau_1 = 0$ to a small positive τ_1. Thus, the case of general τ_1 is needed so that we can establish the differential inequalities for the barrier functions that yield the weak inequalities at non-smooth points.

DEFINITION 6.35. Let $q \in \mathcal{U}_x(\bar\tau)$ and let $\gamma\colon [\tau_1,\bar\tau] \to \mathcal{M}$ be the unique minimizing $\mathcal{L}$-geodesic from x to q. We say that a horizontal vector field $Y(\tau)$ along γ is *adapted* if it solves the following ODE on $[\tau_1,\bar\tau]$:

$$(6.9) \qquad \nabla_X Y(\tau) = -\mathrm{Ric}(Y(\tau),\cdot)^* + \frac{1}{2\sqrt{\tau}(\sqrt{\tau}-\sqrt{\tau_1})}Y(\tau).$$

Direct computation shows the following:

LEMMA 6.36. *Suppose that $Y(\tau)$ is an adapted vector field along γ. Then*

$$(6.10) \qquad \frac{d}{d\tau}\langle Y(\tau),Y(\tau)\rangle = 2\mathrm{Ric}(Y(\tau),Y(\tau)) + 2\langle \nabla_X Y(\tau),Y(\tau)\rangle$$
$$= \frac{1}{\sqrt{\tau}(\sqrt{\tau}-\sqrt{\tau_1})}\langle Y(\tau),Y(\tau)\rangle.$$

It follows that

$$|Y(\tau)|^2 = C\frac{(\sqrt{\tau}-\sqrt{\tau_1})^2}{(\sqrt{\bar\tau}-\sqrt{\tau_1})^2},$$

where $C = |Y(\bar\tau)|^2$.

Now we come to the main result of this subsection, which is an extremely important inequality for the Hessian of $L_x^{\bar\tau}$.

4. SECOND-ORDER DIFFERENTIAL INEQUALITIES FOR $\widetilde{L}^{\overline{\tau}}$ AND $L_x^{\overline{\tau}}$

PROPOSITION 6.37. *Suppose that $q \in \mathcal{U}_x(\overline{\tau})$, that $Z \in \widetilde{\mathcal{U}}_x(\overline{\tau})$ is the pre-image of q, and that γ_Z is the $\mathcal{L}$-geodesic to q determined by Z. Suppose that $Y(\tau)$ is an adapted vector field along γ_Z. Then*

$$(6.11) \quad \operatorname{Hess}(L_x^{\overline{\tau}})(Y(\overline{\tau}), Y(\overline{\tau})) \leq \left(\frac{|Y(\overline{\tau})|^2}{\sqrt{\overline{\tau}} - \sqrt{\tau_1}}\right) - 2\sqrt{\overline{\tau}}\operatorname{Ric}(Y(\overline{\tau}), Y(\overline{\tau}))$$
$$- \int_{\tau_1}^{\overline{\tau}} \sqrt{\tau} H(X, Y) d\tau,$$

where

$$(6.12) \quad \begin{aligned} H(X, Y) = &-\operatorname{Hess}(R)(Y, Y) - 2\langle \mathcal{R}(Y, X)Y, X\rangle \\ &- 4(\nabla_X \operatorname{Ric})(Y, Y) + 4(\nabla_Y \operatorname{Ric})(Y, X) \\ &- 2\frac{\partial \operatorname{Ric}}{\partial \tau}(Y, Y) + 2|\operatorname{Ric}(Y, \cdot)|^2 - \frac{1}{\tau}\operatorname{Ric}(Y, Y). \end{aligned}$$

We have equality in Equation (6.11) if and only if the adapted vector field Y is also an $\mathcal{L}$-Jacobi field.

REMARK 6.38. In spite of the notation, $H(X, Y)$ is a purely quadratic function of the vector field Y along γ_Z.

We begin the proof of this proposition with three elementary lemmas. The first is an immediate consequence of the definition of $\widetilde{\mathcal{U}}_x(\overline{\tau})$.

LEMMA 6.39. *Suppose that $q \in \mathcal{U}_x(\overline{\tau})$ and that $\gamma \colon [\tau_1, \overline{\tau}] \to M$ is the minimizing $\mathcal{L}$-geodesic from x to q. Then for every tangent vector $Y(\overline{\tau}) \in T_q M_{T-\overline{\tau}}$ there is a one-parameter family of $\mathcal{L}$-geodesics $\widetilde{\gamma}(\tau, u)$ defined on $[\tau_1, \overline{\tau}]$ with $\widetilde{\gamma}(0, u) = x$ for all u, with $\widetilde{\gamma}(\tau, 0) = \gamma(\tau)$ and $\partial\widetilde{\gamma}(\overline{\tau}, 0)/\partial u = Y(\overline{\tau})$. Also, for every $Z \in T_x M_{T-\tau_1}$ there is a family of $\mathcal{L}$-geodesics $\widetilde{\gamma}(\tau, u)$ such that $\gamma(0, u) = x$ for all u, $\widetilde{\gamma}(\tau, 0) = \gamma(\tau)$ and such that, setting $Y(\tau) = \frac{\partial}{\partial u}\widetilde{\gamma}_u(\tau)|_0$, we have*

$$\nabla_{\sqrt{\tau_1} X(\tau_1)} Y(\tau_1) = Z.$$

LEMMA 6.40. *Let γ be a minimizing $\mathcal{L}$-geodesic from x, and let $Y(\tau)$ be an $\mathcal{L}$-Jacobi field along γ. Then*

$$2\sqrt{\overline{\tau}}\langle \nabla_X Y(\overline{\tau}), Y(\overline{\tau})\rangle = \operatorname{Hess}(L_x^{\overline{\tau}})(Y(\overline{\tau}), Y(\overline{\tau})).$$

PROOF. Let $\gamma(\tau, u)$ be a one-parameter family of $\mathcal{L}$-geodesics emanating from x with $\gamma(u, 0)$ being the $\mathcal{L}$-geodesic in the statement of the lemma and with $\frac{\partial}{\partial u}\gamma(\tau, 0) = Y(\tau)$. We have the extensions of $X(\tau)$ and $Y(\tau)$ to vector fields $\widetilde{X}(\tau, u)$ and $\widetilde{Y}(\tau, u)$ defined at $\gamma(\tau, u)$ for all τ and u. Of course,

$$2\sqrt{\overline{\tau}}\langle \nabla_Y \widetilde{X}(\overline{\tau}, u)|_0, Y(\overline{\tau})\rangle$$
$$= Y(\langle 2\sqrt{\overline{\tau}}\widetilde{X}(\overline{\tau}, u), \widetilde{Y}(\overline{\tau}, u)\rangle)|_0 - \langle 2\sqrt{\overline{\tau}}X(\overline{\tau}), \nabla_Y \widetilde{Y}(\overline{\tau}, u)|_0\rangle.$$

Then by Corollary 6.29 we have

$$2\sqrt{\overline{\tau}}\langle \nabla_Y \widetilde{X}(\overline{\tau},u)|_0, Y(\overline{\tau})\rangle = Y(\overline{\tau})(\langle \nabla L_x^{\overline{\tau}}, \widetilde{Y}(\overline{\tau},u)\rangle)|_0 - \langle \nabla L_x^{\overline{\tau}}, \nabla_{Y(\overline{\tau})}\widetilde{Y}(\overline{\tau},u)|_0\rangle$$
$$= Y(\overline{\tau})(\widetilde{Y}(\overline{\tau},u)L_x^{\overline{\tau}})|_0 - \nabla_{Y(\overline{\tau})}(\widetilde{Y}(\overline{\tau},u)|_0)(L_x^{\overline{\tau}})$$
$$= \operatorname{Hess}(L_x^{\overline{\tau}})(Y(\overline{\tau}), Y(\overline{\tau})).$$

□

Now suppose that we have a horizontal vector field that is both adapted and $\mathcal{L}$-Jacobi. We get:

LEMMA 6.41. *Suppose that $q \in \mathcal{U}_x(\overline{\tau})$, that $Z \in \widetilde{\mathcal{U}}_x(\overline{\tau})$ is the pre-image of q, and that γ_Z is the $\mathcal{L}$-geodesic to q determined by Z. Suppose further that $Y(\tau)$ is a horizontal vector field along γ that is both adapted and an $\mathcal{L}$-Jacobi field. Then, we have*

$$\frac{1}{2\sqrt{\overline{\tau}}(\sqrt{\overline{\tau}} - \sqrt{\tau_1})}|Y(\overline{\tau})|^2 = \frac{1}{2\sqrt{\overline{\tau}}}\operatorname{Hess}(L_x^{\overline{\tau}})(Y(\overline{\tau}), Y(\overline{\tau})) + \operatorname{Ric}(Y(\overline{\tau}), Y(\overline{\tau})).$$

PROOF. From the definition of an adapted vector field $Y(\tau)$ we have

$$\operatorname{Ric}(Y(\tau), Y(\tau)) + \langle \nabla_X Y(\tau), Y(\tau)\rangle = \frac{1}{2\sqrt{\tau}(\sqrt{\tau} - \sqrt{\tau_1})}\langle Y(\tau), Y(\tau)\rangle.$$

Since $Y(\tau)$ is an $\mathcal{L}$-Jacobi field, according to Lemma 6.40 we have

$$\langle \nabla_X Y(\overline{\tau}), Y(\overline{\tau})\rangle = \frac{1}{2\sqrt{\overline{\tau}}}\operatorname{Hess}(L_x^{\overline{\tau}})(Y(\overline{\tau}), Y(\overline{\tau})).$$

Putting these together gives the result. □

Now we are ready to begin the proof of Proposition 6.37.

PROOF. Let $\widetilde{\gamma}(\tau, u)$ be a family of curves with $\gamma(\tau, 0) = \gamma_Z$ and with $\frac{\partial}{\partial u}\gamma(\tau, u) = \widetilde{Y}(\tau, u)$. We denote by Y the horizontal vector field which is the restriction of $\widetilde{Y}$ to $\gamma_0 = \gamma_Z$. We set $q(u) = \widetilde{\gamma}(\overline{\tau}, u)$. By restricting to a smaller neighborhood of 0 in the u-direction, we can assume that $q(u) \in \mathcal{U}_x(\overline{\tau})$ for all u. Then $\mathcal{L}(\widetilde{\gamma}_u) \geq L_x^{\overline{\tau}}(q(u))$. Of course, $L_x^{\overline{\tau}}(q(0)) = \mathcal{L}(\gamma_Z)$. This implies that

$$\frac{d}{du}L_x^{\overline{\tau}}(q(u))\big|_0 = \frac{d}{du}\mathcal{L}(\gamma_u)\big|_0,$$

and

$$Y(\overline{\tau})(\widetilde{Y}(\overline{\tau},u)(L_x^{\overline{\tau}}))|_0 = \frac{d^2}{du^2}L_x^{\overline{\tau}}(q(u))\big|_0 \leq \frac{d^2}{du^2}\mathcal{L}(\gamma_u)\big|_0.$$

Recall that $\nabla L_x^{\overline{\tau}}(q) = 2\sqrt{\overline{\tau}}X(\overline{\tau})$, so that

$$\nabla_{Y(\overline{\tau})}\widetilde{Y}(\overline{\tau},u)|_0(L_x^{\overline{\tau}}) = \langle \nabla_{Y(\overline{\tau})}\widetilde{Y}(\overline{\tau},u)|_0, \nabla L^{\overline{\tau}}\rangle = 2\sqrt{\overline{\tau}}\langle \nabla_{Y(\overline{\tau})}\widetilde{Y}(\overline{\tau},u)|_0, X(\overline{\tau})\rangle.$$

Thus, by Proposition 6.33, and using the fact that $Y(\tau_1) = 0$, we have

$$\operatorname{Hess}(L^{\overline{\tau}})(Y(\overline{\tau}), Y(\overline{\tau})) = Y(\overline{\tau})\left(\widetilde{Y}(\overline{\tau}, u)(L_x^{\overline{\tau}})\right)|_0 - \nabla_{Y(\overline{\tau})}\widetilde{Y}(\overline{\tau}, u)|_0(L_x^{\overline{\tau}})$$

$$\leq \frac{d^2}{du^2}\mathcal{L}(\gamma_u) - 2\sqrt{\overline{\tau}}\langle \nabla_{Y(\overline{\tau})}\widetilde{Y}(\overline{\tau}, u)|_0, X(\overline{\tau})\rangle$$

$$= \int_{\tau_1}^{\overline{\tau}} \sqrt{\tau}\Big[\operatorname{Hess}(R)(Y, Y) + 2\langle \mathcal{R}(Y, X)Y, X\rangle$$

$$\quad - 4(\nabla_Y\operatorname{Ric})(X, Y) + 2(\nabla_X\operatorname{Ric})(Y, Y)$$

$$\quad + 2|\nabla_X Y|^2\Big]d\tau.$$

Equation (6.9) and the fact that $|Y(\tau)|^2 = |Y(\overline{\tau})|^2 \frac{(\sqrt{\tau} - \sqrt{\tau_1})^2}{(\sqrt{\overline{\tau}} - \sqrt{\tau_1})^2}$, give

$$\operatorname{Hess}(L^{\overline{\tau}})(Y(\overline{\tau}), Y(\overline{\tau}))$$

$$\leq \int_{\tau_1}^{\overline{\tau}} \sqrt{\tau}\Big[\operatorname{Hess}(R)(Y, Y) + 2\langle \mathcal{R}(Y, X)Y, X\rangle - 4(\nabla_Y\operatorname{Ric})(X, Y)$$

$$\quad + 2(\nabla_X\operatorname{Ric})(Y, Y) + 2|\operatorname{Ric}(Y, \cdot)|^2\Big]d\tau$$

$$+ \int_{\tau_1}^{\overline{\tau}} \left[\frac{|Y(\overline{\tau})|^2}{2\sqrt{\tau}(\sqrt{\overline{\tau}} - \sqrt{\tau_1})^2} - \frac{2}{(\sqrt{\overline{\tau}} - \sqrt{\tau_1})}\operatorname{Ric}(Y, Y)\right]d\tau.$$

Using the definition of $H(X, Y)$ given in Equation (6.12), allows us to write

$$\operatorname{Hess}(L^{\overline{\tau}})(Y(\overline{\tau}), Y(\overline{\tau}))$$

$$\leq -\int_{\tau_1}^{\overline{\tau}} \sqrt{\tau}H(X, Y)d\tau$$

$$+ \int_{\tau_1}^{\overline{\tau}} \Big[\sqrt{\tau}\big(-2(\nabla_X\operatorname{Ric})(Y, Y) - 2\frac{\partial \operatorname{Ric}}{\partial \tau}(Y, Y) + 4|\operatorname{Ric}(Y, \cdot)|^2\big)$$

$$+ \frac{|Y(\overline{\tau})|^2}{2\sqrt{\tau}(\sqrt{\overline{\tau}} - \sqrt{\tau_1})^2} - \left(\frac{2}{(\sqrt{\overline{\tau}} - \sqrt{\tau_1})} + \frac{1}{\sqrt{\tau}}\right)\operatorname{Ric}(Y, Y)\Big]d\tau.$$

To simplify further, we compute, using Equation (6.9),

$$\frac{d}{d\tau}\left(\operatorname{Ric}(Y(\tau), Y(\tau))\right) = \frac{\partial \operatorname{Ric}}{\partial \tau}(Y, Y) + 2\operatorname{Ric}(\nabla_X Y, Y) + (\nabla_X\operatorname{Ric})(Y, Y)$$

$$= \frac{\partial \operatorname{Ric}}{\partial \tau}(Y, Y) + (\nabla_X\operatorname{Ric})(Y, Y)$$

$$\quad + \frac{1}{\sqrt{\tau}(\sqrt{\overline{\tau}} - \sqrt{\tau_1})}\operatorname{Ric}(Y, Y) - 2|\operatorname{Ric}(Y, \cdot)|^2.$$

Consequently, we have

$$\frac{d\left(2\sqrt{\tau}\mathrm{Ric}(Y(\tau),Y(\tau))\right)}{d\tau}$$
$$= 2\sqrt{\tau}\left(\frac{\partial \mathrm{Ric}}{\partial \tau}(Y,Y) + (\nabla_X \mathrm{Ric})(Y,Y) - 2|\mathrm{Ric}(Y,\cdot)|^2\right)$$
$$+ \left(\frac{2}{(\sqrt{\tau}-\sqrt{\tau_1})} + \frac{1}{\sqrt{\tau}}\right)\mathrm{Ric}(Y,Y).$$

Using this, and the fact that $Y(\tau_1) = 0$, gives

(6.13) $\mathrm{Hess}(L_x^{\overline{\tau}})(Y(\overline{\tau}), Y(\overline{\tau}))$
$$\leq -\int_{\tau_1}^{\overline{\tau}} \left(\sqrt{\tau}H(X,Y) - \frac{d}{d\tau}\left(2\sqrt{\tau}\mathrm{Ric}(Y,Y)\right) - \frac{|Y(\tau)|^2}{2\sqrt{\tau}(\sqrt{\tau}-\sqrt{\tau_1})^2}\right)d\tau$$
$$= \frac{|Y(\overline{\tau})|^2}{\sqrt{\overline{\tau}}-\sqrt{\tau_1}} - 2\sqrt{\overline{\tau}}\mathrm{Ric}(Y(\overline{\tau}),Y(\overline{\tau})) - \int_{\tau_1}^{\overline{\tau}} \sqrt{\tau}H(X,Y)d\tau.$$

This proves Inequality (6.11). Now we examine when equality holds in this expression. Given an adapted vector field $Y(\tau)$ along γ, let $\mu(v)$ be a geodesic through $\gamma(\overline{\tau}, 0)$ with tangent vector $Y(\overline{\tau})$. Then there is a one-parameter family $\mu(\tau, v)$ of minimizing $\mathcal{L}$-geodesics with the property that $\mu(\overline{\tau}, v) = \mu(v)$. Let $\widetilde{Y}'(\tau, v)$ be $\partial\mu(\tau, v)/\partial v$. It is an $\mathcal{L}$-Jacobi field with $\widetilde{Y}'(\overline{\tau}, 0) = Y(\overline{\tau})$. Since $L_x \circ \mathcal{L}\exp_x = \widetilde{L}$, we see that

$$\frac{d^2}{dv^2}\mathcal{L}(\mu_v)|_{v=0} = \frac{d^2}{du^2}L_x^{\overline{\tau}}(\mu(u))|_0.$$

Hence, the assumption that we have equality in (6.11) implies that

$$\frac{d^2}{dv^2}\mathcal{L}(\mu_v)|_{v=0} = \frac{d^2}{du^2}\mathcal{L}(\widetilde{\gamma}_u)|_0.$$

Now we extend this one-parameter family to a two-parameter family $\mu(\tau, u, v)$ so that $\partial\mu(\tau, 0, 0)/\partial v = \widetilde{Y}'$ and $\partial\mu(\tau, 0, 0)/\partial u = Y(\tau)$. Let w be the variable $u - v$, and let $\widetilde{W}$ be the tangent vector in this coordinate direction, so that $\widetilde{W} = \widetilde{Y} - \widetilde{Y}'$. We denote by W the restriction of $\widetilde{W}$ to $\gamma_{0,0} = \gamma_Z$. By Remark 6.15 the second partial derivative of the length of this family in the u-direction at $u = v = 0$ agrees with the second derivative of the length of the original family $\widetilde{\gamma}$ in the u-direction.

CLAIM 6.42.
$$\frac{\partial}{\partial v}\frac{\partial}{\partial w}\mathcal{L}(\mu)|_{u=v=0} = \frac{\partial}{\partial w}\frac{\partial}{\partial v}\mathcal{L}(\mu)|_{v=w=0} = 0.$$

PROOF. Of course, the second cross-partials are equal. By Lemma 6.14 we have
$$\frac{\partial}{\partial v}\frac{\partial}{\partial w}\mathcal{L}(\mu)|_{v=w=0} = 2\sqrt{\overline{\tau}}\widetilde{Y}'(\overline{\tau})\langle\widetilde{W}(\overline{\tau}), X(\overline{\tau})\rangle - \int_{\tau_1}^{\overline{\tau}} 2\sqrt{\tau}\langle\mathrm{Jac}(\widetilde{Y}'), W\rangle d\tau.$$

Since $W(\bar{\tau}) = 0$ and since $\nabla_{\tilde{Y}'}(\widetilde{W}) = \nabla_W(\tilde{Y}')$, we see that the boundary term in the above expression vanishes. The integral vanishes since $\tilde{Y}'$ is an $\mathcal{L}$-Jacobi field. $\square$

If Inequality (6.11) is an equality, then
$$\frac{\partial^2}{\partial v^2}\mathcal{L}(\mu)|_{u=v=0} = \frac{\partial^2}{\partial u^2}\mathcal{L}(\mu)|_{u=v=0}.$$
We write $\partial/\partial u = \partial/\partial v + \partial/\partial w$. Expanding the right-hand side and canceling the common terms gives
$$0 = \left(\frac{\partial}{\partial v}\frac{\partial}{\partial w} + \frac{\partial}{\partial w}\frac{\partial}{\partial v} + \frac{\partial^2}{\partial w^2}\right)\mathcal{L}(\mu)|_{u=v=0}.$$
The previous claim tells us that the first two terms on the right-hand side of this equation vanish, and hence we conclude
$$\frac{\partial^2}{\partial w^2}\mathcal{L}(\mu)|_{u=v=0} = 0.$$
Since W vanishes at both endpoints this implies, according to Proposition 6.13, that $\widetilde{W}(\tau, 0, 0) = 0$ for all τ, or in other words $Y(\tau) = \tilde{Y}'(\tau, 0, 0)$ for all τ. Of course by construction $\tilde{Y}'(\tau, 0, 0)$ is an $\mathcal{L}$-Jacobi field. This shows that equality holds only if the adapted vector field $Y(\tau)$ is also an $\mathcal{L}$-Jacobi field.

Conversely, if the adapted vector field $Y(\tau)$ is also an $\mathcal{L}$-Jacobi field, then inequality between the second variations at the beginning of the proof is an equality. In the rest of the argument we dealt only with equalities. Hence, in this case Inequality (6.11) is an equality.

This shows that we have equality in (6.11) if and only if the adapted vector field $Y(\tau)$ is also an $\mathcal{L}$-Jacobi field. $\square$

4.3. Inequalities for $\triangle L_x^{\bar{\tau}}$. The inequalities for the Hessian of $L_x^{\bar{\tau}}$ lead to inequalities for $\triangle L_x^{\bar{\tau}}$ which we establish in this section. Here is the main result.

PROPOSITION 6.43. *Suppose that $q \in \mathcal{U}_x(\bar{\tau})$, that $Z \in \widetilde{\mathcal{U}}_x(\bar{\tau})$ is the preimage of q and that γ_Z is the $\mathcal{L}$-geodesic determined by Z. Then*

(6.14) $$\triangle L_x^{\bar{\tau}}(q) \leq \frac{n}{\sqrt{\bar{\tau}} - \sqrt{\tau_1}} - 2\sqrt{\bar{\tau}}R(q) - \frac{1}{(\sqrt{\bar{\tau}} - \sqrt{\tau_1})^2}\mathcal{K}_{\tau_1}^{\bar{\tau}}(\gamma_Z),$$

where, for any path γ parameterized by backward time on the interval $[\tau_1, \bar{\tau}]$ taking value x at $\tau = \tau_1$ we define
$$\mathcal{K}_{\tau_1}^{\bar{\tau}}(\gamma) = \int_{\tau_1}^{\bar{\tau}} \sqrt{\tau}(\sqrt{\tau} - \sqrt{\tau_1})^2 H(X)d\tau,$$

with

(6.15) $$H(X) = -\frac{\partial R}{\partial \tau} - \frac{1}{\tau}R - 2\langle \nabla R, X\rangle + 2\text{Ric}(X, X),$$

128   6. A COMPARISON GEOMETRY APPROACH TO THE RICCI FLOW

where X is the horizontal projection of $\gamma'(\tau)$. Furthermore, Inequality (6.14) is an equality if and only if for every $Y \in T_q(M_{T-\bar{\tau}})$ the unique adapted vector field $Y(\tau)$ along γ satisfying $Y(\bar{\tau}) = Y$ is an $\mathcal{L}$-Jacobi field. In this case

$$\mathrm{Ric} + \frac{1}{2\sqrt{\bar{\tau}}}\mathrm{Hess}(L_x^{\bar{\tau}}) = \frac{1}{2\sqrt{\bar{\tau}}(\sqrt{\bar{\tau}} - \sqrt{\tau_1})}G(T - \bar{\tau}).$$

PROOF. Choose an orthonormal basis $\{Y_\alpha\}$ for $T_q(M_{T-\bar{\tau}})$. For each α, extend $\{Y_\alpha\}$ to an adapted vector field along the $\mathcal{L}$-geodesic γ_Z by solving

$$\nabla_X Y_\alpha = \frac{1}{2\sqrt{\tau}(\sqrt{\tau} - \sqrt{\tau_1})}Y_\alpha - \mathrm{Ric}(Y_\alpha, \cdot)^*.$$

As in Equation (6.10), we have

$$\frac{d}{d\tau}\langle Y_\alpha, Y_\beta \rangle = \langle \nabla_X Y_\alpha, Y_\beta \rangle + \langle \nabla_X Y_\beta, Y_\alpha \rangle + 2\mathrm{Ric}(Y_\alpha, Y_\beta)$$

$$= \frac{1}{\sqrt{\tau}(\sqrt{\tau} - \sqrt{\tau_1})}\langle Y_\alpha, Y_\beta \rangle.$$

By integrating we get

$$\langle Y_\alpha, Y_\beta \rangle(\tau) = \frac{(\sqrt{\tau} - \sqrt{\tau_1})^2}{(\sqrt{\bar{\tau}} - \sqrt{\tau_1})^2}\delta_{\alpha\beta}.$$

To simplify the notation we set

$$I(\tau) = \frac{\sqrt{\bar{\tau}} - \sqrt{\tau_1}}{\sqrt{\tau} - \sqrt{\tau_1}}$$

and $W_\alpha(\tau) = I(\tau)Y_\alpha(\tau)$. Then $\{W_\alpha(\tau)\}_\alpha$ form an orthonormal basis at τ. Consequently, summing Inequality (6.13) over α gives

(6.16) $$\triangle L_x^{\bar{\tau}}(q) \leq \frac{n}{\sqrt{\bar{\tau}} - \sqrt{\tau_1}} - 2\sqrt{\bar{\tau}}R(q) - \sum_\alpha \int_{\tau_1}^{\bar{\tau}} \sqrt{\tau}H(X, Y_\alpha)d\tau.$$

To establish Inequality (6.14) it remains to prove the following claim.

CLAIM 6.44.

$$\sum_\alpha H(X, Y_\alpha) = \frac{(\sqrt{\tau} - \sqrt{\tau_1})^2}{(\sqrt{\bar{\tau}} - \sqrt{\tau_1})^2}H(X).$$

PROOF. To prove the claim we sum Equation (6.12) giving

$$I^2(\tau)\sum_\alpha H(X, Y_\alpha) = \sum_\alpha H(X, W_\alpha)$$

$$= -\triangle R + 2\mathrm{Ric}(X, X) + 4\sum_\alpha (\nabla_{W_\alpha}\mathrm{Ric})(W_\alpha, X)$$

$$- 2\sum_\alpha \mathrm{Ric}_\tau(W_\alpha, W_\alpha) + 2|\mathrm{Ric}|^2 - \frac{1}{\tau}R - 4\langle \nabla R, X \rangle.$$

Taking the trace of the second Bianchi identity, we get

$$\sum_{\alpha}(\nabla_{W_\alpha}\text{Ric})(W_\alpha, X) = \frac{1}{2}\langle \nabla R, X\rangle.$$

In addition by Equation (3.7), recalling that $\partial R/\partial \tau = -\partial R/\partial t$, we have

$$\frac{\partial R}{\partial \tau} = -\triangle R - 2|\text{Ric}|^2.$$

On the other hand,

$$\frac{\partial R}{\partial \tau} = \partial(g^{ij}R_{ij})/\partial \tau = -2|\text{Ric}|^2 + \sum_{\alpha}\frac{\partial \text{Ric}}{\partial \tau}(W_\alpha, W_\alpha),$$

and so $\sum_\alpha \frac{\partial \text{Ric}}{\partial \tau}(W_\alpha, W_\alpha) = -\triangle R$. Putting all this together gives

$$I^2(\tau)\sum_\alpha H(X, Y_\alpha) = H(X).$$

□

Clearly, Inequality (6.14) follows immediately from Inequality (6.16) and the claim. The last statement of Proposition 6.43 follows directly from the last statement of Proposition 6.37 and Lemma 6.41. This completes the proof of Proposition 6.43. □

5. Reduced length

We introduce the reduced length function both on the tangent space and on space-time. The reason that the reduced length l_x is easier to work with is that it is scale invariant when $\tau_1 = 0$. Throughout this section we fix $x \in \mathcal{M}$ with $\mathbf{t}(x) = T - \tau_1$. We shall always suppose that $T - \tau_1$ is greater than the initial time of the generalized Ricci flow.

5.1. The reduced length function l_x on space-time.

DEFINITION 6.45. We define the $\mathcal{L}$-reduced length (from x)

$$l_x : \mathcal{U}_x \to \mathbb{R}$$

by setting

$$l_x(q) = \frac{L_x(q)}{2\sqrt{\tau}},$$

where $\tau = T - \mathbf{t}(q)$. We denote by l_x^τ the restriction of l_x to the slice $\mathcal{U}_x(\tau)$.

In order to understand the differential inequalities that l_x satisfies, we first need to introduce a quantity closely related to the function $\mathcal{K}_{\tau_1}^{\overline{\tau}}$ defined in Proposition 6.43.

DEFINITION 6.46. For any $\mathcal{L}$-geodesic γ parameterized by $[\tau_1, \overline{\tau}]$ we define
$$K_{\tau_1}^{\overline{\tau}}(\gamma) = \int_{\tau_1}^{\overline{\tau}} \tau^{3/2} H(X) d\tau.$$
In the special case when $\tau_1 = 0$ we denote this integral by $K^{\overline{\tau}}(\gamma)$.

The following is immediate from the definitions.

LEMMA 6.47. *For any $\mathcal{L}$-geodesic γ defined on $[0, \overline{\tau}]$ both $K_{\tau_1}^{\overline{\tau}}(\gamma)$ and $\mathcal{K}_{\tau_1}^{\overline{\tau}}(\gamma)$ are continuous in τ_1 and at $\tau_1 = 0$ they take the same value. Also,*
$$\left(\frac{\tau_1}{\overline{\tau}}\right)^{3/2} \left(R(\gamma(\tau_1)) + |X(\tau_1)|^2\right)$$
is continuous for all $\tau_1 > 0$ and has limit 0 as $\tau_1 \to 0$. Here, as always, $X(\tau_1)$ is the horizontal component of γ' at $\tau = \tau_1$.

LEMMA 6.48. *Let $q \in \mathcal{U}_x(\overline{\tau})$, let $Z \in \widetilde{\mathcal{U}}_x$ be the pre-image of q and let γ_Z be the $\mathcal{L}$-geodesic determined by Z. Then we have*

(6.17) $\overline{\tau}^{-\frac{3}{2}} K_{\tau_1}^{\overline{\tau}}(\gamma_Z) = \frac{l_x(q)}{\overline{\tau}} - (R(q) + |X(\overline{\tau})|^2) + \left(\frac{\tau_1}{\overline{\tau}}\right)^{3/2} \left(R(x) + |X(\tau_1)|^2\right).$

In the case when $\tau_1 = 0$, the last term on the right-hand side of Equation (6.17) vanishes.

PROOF. Using the $\mathcal{L}$-geodesic equation and the definition of H we have
$$\frac{d}{d\tau}(R(\gamma_Z(\tau)) + |X(\tau)|^2) = \frac{\partial R}{\partial \tau}(\gamma_Z(\tau)) + \langle \nabla R(\gamma_Z(\tau)), X(\tau) \rangle$$
$$+ 2\langle \nabla_X X(\tau), X(\tau) \rangle + 2\mathrm{Ric}(X(\tau), X(\tau))$$
$$= \frac{\partial R}{\partial \tau}(\gamma_Z(\tau)) + 2X(\tau)(R) - \frac{1}{\tau}|X(\tau)|^2$$
$$- 2\mathrm{Ric}(X(\tau), X(\tau))$$
$$= -H(X(\tau)) - \frac{1}{\tau}(R(\gamma_Z(\tau)) + |X(\tau)|^2).$$
Thus
$$\frac{d}{d\tau}(\tau^{\frac{3}{2}}(R(\gamma_Z(\tau)) + |X(\tau)|^2)) = \frac{1}{2}\sqrt{\tau}(R(\gamma_Z(\tau)) + |X(\tau)|^2) - \tau^{\frac{3}{2}} H(X(\tau)).$$
Integration from τ_1 to $\overline{\tau}$ gives
$$\overline{\tau}^{3/2}\left(R(q)) + |X(\overline{\tau})|^2\right) - \tau_1^{3/2}(R(x) + |X(\tau_1)|^2) = \frac{L_x^{\overline{\tau}}(q)}{2} - K_{\tau_1}^{\overline{\tau}}(\gamma_Z),$$
which is equivalent to Equation (6.17). In the case when $\tau_1 = 0$, the last term on the right-hand side vanishes since
$$\lim_{\tau \to 0} \tau^{3/2} |X(\tau)|^2 = 0.$$
$\square$

5. REDUCED LENGTH

Now we come to the most general of the differential inequalities for l_x that will be so important in what follows. Whenever the expression $\left(\frac{\tau_1}{\tau}\right)^{3/2}\left(R(x)+|X(\tau_1)|^2\right)$ appears in a formula, it is interpreted to be zero in the case when $\tau_1 = 0$.

LEMMA 6.49. *For any $q \in \mathcal{U}_x(\overline{\tau})$, let $Z \in \widetilde{\mathcal{U}}_x(\overline{\tau})$ be the pre-image of q and let γ_Z be the $\mathcal{L}$-geodesic determined by Z. Then we have*

$$\frac{\partial l_x}{\partial \tau}(q) = R(q) - \frac{l_x(q)}{\overline{\tau}} + \frac{K_{\tau_1}^{\overline{\tau}}(\gamma_Z)}{2\overline{\tau}^{3/2}} - \frac{1}{2}\left(\frac{\tau_1}{\overline{\tau}}\right)^{3/2}\left(R(x)+|X(\tau_1)|^2\right),$$

$$|\nabla l_x^{\overline{\tau}}(q)|^2 = |X(\overline{\tau})|^2$$
$$= \frac{l_x^{\overline{\tau}}(q)}{\overline{\tau}} - \frac{K_{\tau_1}^{\overline{\tau}}(\gamma_Z)}{\overline{\tau}^{3/2}} - R(q) + \left(\frac{\tau_1}{\overline{\tau}}\right)^{3/2}\left(R(x)+|X(\tau_1)|^2\right),$$

$$\triangle l_x^{\overline{\tau}}(q) = \frac{1}{2\sqrt{\overline{\tau}}}\triangle L_x^{\overline{\tau}}(q) \le \frac{n}{2\sqrt{\overline{\tau}}(\sqrt{\overline{\tau}}-\sqrt{\tau_1})} - R(q) - \frac{K_{\tau_1}^{\overline{\tau}}(\gamma_Z)}{2\sqrt{\overline{\tau}}(\sqrt{\overline{\tau}}-\sqrt{\tau_1})^2}.$$

PROOF. It follows immediately from Equation (6.8) that

$$\frac{\partial l_x}{\partial \tau} = R - \frac{1}{2}(R+|X|^2) - \frac{l_x}{2\tau}.$$

Using Equation (6.17) this gives the first equality stated in the lemma. It follows immediately from Corollary 6.29 that $\nabla l_x^{\tau} = X(\tau)$ and hence $|\nabla l_x^{\tau}|^2 = |X(\tau)|^2$. From this and Equation (6.17) the second equation follows. The last inequality is immediate from Proposition 6.43.

When $\tau_1 = 0$, the last terms on the right-hand sides of the first two equations vanish, since the last term on the right-hand side of Equation (6.17) vanishes in this case. □

When $\tau_1 = 0$, which is the case of main interest, all these formulas simplify and we get:

THEOREM 6.50. *Suppose that $x \in M_T$ so that $\tau_1 = 0$. For any $q \in \mathcal{U}_x(\overline{\tau})$, let $Z \in \widetilde{\mathcal{U}}_x(\overline{\tau})$ be the pre-image of q and let γ_Z be the $\mathcal{L}$-geodesic determined by Z. As usual, let $X(\tau)$ be the horizontal projection of $\gamma_Z'(\tau)$. Then we have*

$$\frac{\partial l_x}{\partial \tau}(q) = R(q) - \frac{l_x(q)}{\overline{\tau}} + \frac{K^{\overline{\tau}}(\gamma_Z)}{2\overline{\tau}^{3/2}},$$

$$|\nabla l_x^{\overline{\tau}}(q)|^2 = |X(\overline{\tau})|^2 = \frac{l_x^{\overline{\tau}}(q)}{\overline{\tau}} - \frac{K^{\overline{\tau}}(\gamma_Z)}{\overline{\tau}^{3/2}} - R(q),$$

$$\triangle l_x^{\overline{\tau}}(q) = \frac{1}{2\sqrt{\overline{\tau}}}\triangle L_x^{\overline{\tau}}(q) \le \frac{n}{2\overline{\tau}} - R(q) - \frac{K^{\overline{\tau}}(\gamma_Z)}{2\overline{\tau}^{3/2}}.$$

PROOF. This is immediate from the formulas in the previous lemma. □

Now let us reformulate the differential inequalities in Theorem 6.50 in a way that will be useful later.

COROLLARY 6.51. *Suppose that $x \in M_T$ so that $\tau_1 = 0$. Then for $q \in \mathcal{U}_x^{\overline{\tau}}$ we have*

$$\frac{\partial l_x}{\partial \tau}(q) + \triangle l_x^{\overline{\tau}}(q) - \frac{(n/2) - l_x^{\overline{\tau}}(q)}{\overline{\tau}} \leq 0,$$

$$\frac{\partial l_x}{\partial \tau}(q) - \triangle l_x^{\overline{\tau}}(q) + |\nabla l_x^{\overline{\tau}}(q)|^2 - R(q) + \frac{n}{2\overline{\tau}} \geq 0,$$

$$2\triangle l_x^{\overline{\tau}}(q) - |\nabla l_x^{\overline{\tau}}(q)|^2 + R(q) + \frac{l_x^{\overline{\tau}}(q) - n}{\overline{\tau}} \leq 0.$$

In fact, setting

$$\delta = \frac{n}{2\overline{\tau}} - R(q) - \frac{K^{\overline{\tau}}(\gamma_Z)}{2\overline{\tau}^{3/2}} - \triangle l_x^{\overline{\tau}}(q),$$

then $\delta \geq 0$ and

$$\frac{\partial l_x}{\partial \tau}(q) - \triangle l_x^{\overline{\tau}}(q) + |\nabla l_x^{\overline{\tau}}(q)|^2 - R(q) + \frac{n}{2\overline{\tau}} = \delta,$$

$$2\triangle l_x^{\overline{\tau}}(q) - |\nabla l_x^{\overline{\tau}}(q)|^2 + R(q) + \frac{l_x^{\overline{\tau}}(q) - n}{\overline{\tau}} = -2\delta.$$

5.2. The tangential version $\widetilde{l}$ of the reduced length function.
For any path $\gamma \colon [\tau_1, \overline{\tau}] \to (\mathcal{M}, G)$ parameterized by backward time we define

$$l(\gamma) = \frac{1}{2\sqrt{\overline{\tau}}} \mathcal{L}(\gamma).$$

This leads immediately to a reduced length on $\widetilde{\mathcal{U}}_x$.

DEFINITION 6.52. We define $\widetilde{l} \colon \widetilde{\mathcal{U}}_x \to \mathbb{R}$ by

$$\widetilde{l}(Z, \tau) = \frac{\widetilde{L}(Z, \tau)}{2\sqrt{\tau}} = l(\gamma_Z|_{[\tau_1, \overline{\tau}]}).$$

At first glance it may appear that the computations of the gradient and τ-derivatives for l_x pass immediately to those for $\widetilde{l}$. For the spatial derivative this is correct, but for the τ-derivative it is not true. As the computation below shows, the τ-derivatives of $\widetilde{l}$ and l_x do not agree under the identification $\mathcal{L}\exp_x$. The reason is that this identification does not line up the τ-vector field in the domain with $-\partial/\partial t$ in the range. So it is an entirely different computation with a different answer.

LEMMA 6.53.
$$\frac{\partial \widetilde{l}(Z, \tau)}{\partial \tau} = \frac{1}{2}\left(R(\gamma_Z(\tau)) + |X(\tau)|^2\right) - \frac{\widetilde{l}(Z, \tau)}{2\tau}.$$

PROOF. By the Fundamental Theorem of Calculus

$$\frac{\partial}{\partial \tau} \widetilde{L}(Z, \tau) = \sqrt{\tau}\left(R(\gamma_z(\tau)) + |X(\tau)|^2\right).$$

Thus,
$$\frac{\partial}{\partial \tau}\tilde{l}(Z,\tau) = \frac{1}{2}\left(R(\gamma_z(\tau)) + |X(\tau)|^2\right) - \frac{\tilde{l}(Z,\tau)}{2\tau}.$$

□

COROLLARY 6.54. *Suppose that $x \in M_T$ so that $\tau_1 = 0$. Then*
$$\frac{\partial}{\partial \tau}\tilde{l}(Z,\tau) = -\frac{K^\tau(\gamma_Z)}{2\tau^{\frac{3}{2}}}.$$

PROOF. This is immediate from Lemma 6.53 and Lemma 6.48 (after the latter is rewritten using $\tilde{L}$ instead of L_x). □

6. Local Lipschitz estimates for l_x

It is important for the applications to have results on the Lipschitz properties of l_x, or equivalently L_x. Of course, these are the analogues of the fact that in Riemannian geometry the distance function from a point is Lipschitz. The proof of the Lipschitz property given here is based on the exposition in [**72**]. In this section, we fix $x \in M_{T-\tau_1} \subset \mathcal{M}$.

6.1. Statement and corollaries.

DEFINITION 6.55. Let $(\mathcal{M}, G)$ be a generalized Ricci flow and let $x \in M_{T-\tau_1} \subset \mathcal{M}$. The reduced length function l_x is defined on the subset of $\mathcal{M}$ consisting of all points $y \in \mathcal{M}$ for which there is a minimizing $\mathcal{L}$-geodesic from x to y. The value $l_x(y)$ is the quotient of $\mathcal{L}$-length of any such minimizing $\mathcal{L}$-geodesic divided by $2\sqrt{\tau}$.

Here is the main result of this subsection.

PROPOSITION 6.56. *Let $(\mathcal{M}, G)$ be a generalized Ricci flow and let $x \in M_{T-\tau_1} \subset \mathcal{M}$. Let $\epsilon > 0$ be given and let $A \subset \mathcal{M} \cap \mathbf{t}^{-1}(-\infty, T - \tau_1 + \epsilon)$. Suppose that there is a subset $F \subset \mathcal{M}$ on which $|\mathrm{Ric}|$ and $|\nabla R|$ are bounded and a neighborhood $\nu(A)$ of A contained in F with the property that for every point $z \in \nu(A)$ there is a minimizing $\mathcal{L}$-geodesic from x to z contained in F. Then l_x is defined on all of $\nu(A)$. Furthermore, there is a smaller neighborhood $\nu_0(A) \subset \nu(A)$ of A on which l_x is a locally Lipschitz function with respect to the Riemannian metric, denoted $\widehat{G}$, on $\mathcal{M}$ which is defined as the orthogonal sum of the Riemannian metric G on $\mathcal{HTM}$ and the metric dt^2 on the tangent line spanned by χ.*

COROLLARY 6.57. *With A and $\nu_0(A)$ as in Proposition 6.56, the restriction of l_x to $\nu_0(A) \cap M_{T-\overline{\tau}}$ is a locally Lipschitz function with respect to the metric $G_{T-\overline{\tau}}$.*

6.2. The proof of Proposition 6.56. Proposition 6.56 follows from a much more precise, though more complicated to state, result. In order to state this more technical result we introduce the following definition.

DEFINITION 6.58. Let $y \in \mathcal{M}$ with $\mathbf{t}(y) = t$ and suppose that for some $\epsilon > 0$ there is an embedding $\iota \colon B(y,t,r) \times (t-\epsilon, t+\epsilon) \to \mathcal{M}$ that is compatible with time and the vector field. Then we denote by $\widetilde{P}(y, r, \epsilon) \subset \mathcal{M}$ the image of ι. Whenever we introduce $\widetilde{P}(y, r, \epsilon) \subset \mathcal{M}$ implicitly we are asserting that such an embedding exists.

For $A \subset \mathcal{M}$, if $\widetilde{P}(a, \epsilon, \epsilon) \subset \mathcal{M}$ exists for every $a \in A$, then we denote by $\nu_\epsilon(A)$ the union over all $a \in A$ of $\widetilde{P}(a, \epsilon, \epsilon)$.

Now we are ready for the more precise, technical result.

PROPOSITION 6.59. *Given constants $\epsilon > 0$, $\overline{\tau}_0 < \infty$, $l_0 < \infty$, and $C_0 < \infty$, there are constants $C < \infty$ and $0 < \delta < \epsilon$ depending only on the given constants such that the following holds. Let $(\mathcal{M}, G)$ be a generalized Ricci flow and let $x \in \mathcal{M}$ be a point with $\mathbf{t}(x) = T - \tau_1$. Let $y \in \mathcal{M}$ be a point with $\mathbf{t}(y) = t = T - \overline{\tau}$ where $\tau_1 + \epsilon \leq \overline{\tau} \leq \overline{\tau}_0$. Suppose that there is a minimizing $\mathcal{L}$-geodesic γ from x to y with $l(\gamma) \leq l_0$. Suppose that the ball $B(y, t, \epsilon)$ has compact closure in $\mathcal{M}_t$ and that $\widetilde{P}(y, \epsilon, \epsilon) \subset \mathcal{M}$ exists and that the sectional curvatures of the restriction of G to this submanifold are bounded by C_0. Lastly, suppose that for every point of the form $z \in \widetilde{P}(y, \delta, \delta)$ there is a minimizing $\mathcal{L}$-geodesic from x to z with $|\text{Ric}|$ and $|\nabla R|$ bounded by C_0 along this geodesic. Then for all $(b, t') \in B(y, t, \delta) \times (t - \delta, t + \delta)$ we have*

$$|l_x(y) - l_x(\iota(b, t'))| \leq C\sqrt{d_t(y, b)^2 + |t - t'|^2}.$$

Before proving Proposition 6.59, let us show how it implies Proposition 6.56.

PROOF. (that Proposition 6.59 implies Proposition 6.56) Suppose given $\epsilon > 0$, A, $\nu(A)$ and F as in the statement of Proposition 6.56. For each $y \in A$ there is $0 < \epsilon' < \epsilon$ and a neighborhood $\nu'(y)$ with (i) the closure $\overline{\nu}'(y)$ of $\nu'(y)$ being a compact subset of $\nu(A)$ and (ii) for each $z \in \overline{\nu}'(y)$ the parabolic neighborhood $\widetilde{P}(z, \epsilon', \epsilon')$ exists and has compact closure in $\nu(A)$. It follows that for every $z \in \overline{\nu}'(A)$, Rm_G is bounded on $\widetilde{P}(z, \epsilon', \epsilon')$ and every point of $\widetilde{P}(z, \epsilon', \epsilon')$ is connected to x by a minimizing $\mathcal{L}$-geodesic in F. Thus, Proposition 6.59, with ϵ replaced by ϵ', applies to z. In particular, l_x is continuous at z, and hence is continuous on all of $\overline{\nu}'(y)$. Thus, l_x is bounded on $\overline{\nu}'(y)$. Since we have uniform bounds for the curvature on $\widetilde{P}(z, \epsilon', \epsilon')$ according to Proposition 6.59 there are constants $C < \infty$ and $0 < \delta < \epsilon'$ such that for any $z \in \overline{\nu}'(y)$ and any $z' \in \widetilde{P}(z, \delta, \delta)$, we have

$$|l_x(z) - l_x(z')| \leq C|z - z'|_{G(\mathbf{t}(z)) + dt^2}.$$

Since we have a uniform bound for the curvature on $\widetilde{P}(z, \epsilon', \epsilon')$ independent of $z \in \nu'(y)$, the metrics $\widehat{G} = G + dt^2$ and $G(\mathbf{t}(z)) + dt^2$ are uniformly comparable on all of $\widetilde{P}(z, \delta, \delta)$. It follows that there is a constant $C' < \infty$ such that for all $z \in \nu'(y)$ and all $z' \in \widetilde{P}(z, \delta, \delta)$ we have

$$|l_x(z) - l_x(z')| \leq C'|Z - z'|_{\widehat{G}}.$$

We set $\nu_0(A) = \cup_{y \in A} \nu'(y)$. This is an open neighborhood of A contained in $\nu(A)$ on which l_x is locally Lipschitz with respect to the metric $\widehat{G}$. □

Now we turn to the proof of Proposition 6.59. We begin with several preliminary results.

LEMMA 6.60. *Suppose that γ is an $\mathcal{L}$-geodesic defined on $[\tau_1, \overline{\tau}]$, and suppose that for all $\tau \in [\tau_1, \overline{\tau}]$ we have $|\nabla R(\gamma(\tau))| \leq C_0$ and $|\mathrm{Ric}(\gamma(\tau))| \leq C_0$. Then*

$$\max_\tau \left(\sqrt{\tau}|X_\gamma(\tau)|\right) \leq C_1 \min_\tau \left(\sqrt{\tau}|X_\gamma(\tau)|\right) + \frac{(C_1 - 1)}{2}\sqrt{\overline{\tau}},$$

where $C_1 = e^{2C_0 \overline{\tau}}$.

PROOF. The geodesic equation in terms of the variable s, Equation (6.5), gives

$$\begin{aligned}
\frac{d|\gamma'(s)|^2}{ds} &= 2\langle \nabla_{\gamma'(s)} \gamma'(s), \gamma'(s) \rangle + 4s\mathrm{Ric}(\gamma'(s), \gamma'(s)) \\
(6.18) \quad &= 4s^2 \langle \nabla R, \gamma'(s) \rangle - 4s\mathrm{Ric}(\gamma'(s), \gamma'(s)).
\end{aligned}$$

Thus, by our assumption on $|\nabla R|$ and $|\mathrm{Ric}|$ along γ, we have

$$\left|\frac{d|\gamma'(s)|^2}{ds}\right| \leq 4C_0 s^2 |\gamma'(s)| + 4C_0 s |\gamma'(s)|^2.$$

It follows that

$$\left|\frac{d|\gamma'(s)|}{ds}\right| \leq 2C_0 s^2 + 2C_0 s |\gamma'(s)| \leq 2C_0 \overline{\tau} + 2C_0 \sqrt{\overline{\tau}} |\gamma'(s)|,$$

and hence that

$$-2C_0 \sqrt{\overline{\tau}} ds \leq \frac{d|\gamma'(s)|}{\sqrt{\overline{\tau}} + |\gamma'(s)|} \leq 2C_0 \sqrt{\overline{\tau}} ds.$$

Suppose that $s_0 < s_1$. Integrating from s_0 to s_1 gives

$$|\gamma'(s_1)| \leq C|\gamma'(s_0)| + (C - 1)\sqrt{\overline{\tau}},$$
$$|\gamma'(s_0)| \leq C|\gamma'(s_1)| + (C - 1)\sqrt{\overline{\tau}}.$$

where

$$C = e^{2C_0 \sqrt{\overline{\tau}}(s_1 - s_0)}.$$

Since $\sqrt{\tau} X_\gamma(\tau) = \frac{1}{2}\gamma'(s)$, this completes the proof of the lemma. □

136   6. A COMPARISON GEOMETRY APPROACH TO THE RICCI FLOW

COROLLARY 6.61. *Given $\bar{\tau}_0 < \infty$, $C_0 < \infty$, $\epsilon > 0$, and $l_0 < \infty$, there is a constant C_2 depending only on C_0, l_0, ϵ and $\bar{\tau}_0$ such that the following holds. Let γ be an $\mathcal{L}$-geodesic defined on $[\tau_1, \bar{\tau}]$ with $\tau_1 + \epsilon \leq \bar{\tau} \leq \bar{\tau}_0$ and with $|\nabla R(\gamma(\tau))| \leq C_0$ and $|\mathrm{Ric}(\gamma(\tau))| \leq C_0$ for all $\tau \in [\tau_1, \bar{\tau}]$. Suppose also that $l(\gamma) \leq l_0$. Then, we have*

$$\max_\tau \left(\sqrt{\tau} |X_\gamma(\tau)| \right) \leq C_2.$$

PROOF. From the definition $\mathcal{L}(\gamma) = \int_{\sqrt{\tau_1}}^{\sqrt{\bar{\tau}}} (2s^2 R + \frac{1}{2}|\gamma'(s)|^2) ds$. Because of the bound on $|\mathrm{Ric}|$ (which implies that $|R| \leq 3C_0$) we have

$$\frac{1}{2} \int_{\sqrt{\tau_1}}^{\sqrt{\bar{\tau}}} |\gamma'(s)|^2 ds \leq \mathcal{L}(\gamma) + 2C_0 \bar{\tau}^{3/2}.$$

Thus,

$$(\sqrt{\bar{\tau}} - \sqrt{\tau_1}) \min(|\gamma'(s)|^2) \leq 2\mathcal{L}(\gamma) + 4C_0 \bar{\tau}^{3/2}.$$

The bounds $\tau_1 + \epsilon \leq \bar{\tau} \leq \bar{\tau}_0$, then imply that $\min |\gamma'(s)|^2 \leq C''$ for some C'' depending on C_0, l_0, ϵ, and $\bar{\tau}_0$. Since $\sqrt{\tau} X_\gamma(\tau) = \frac{1}{2} \gamma'(s)$, we have

$$\min_\tau \left(\sqrt{\tau} |X_\gamma(\tau)| \right) \leq C'$$

for some constant C' depending only on C_0, l_0, ϵ and $\bar{\tau}_0$. The result is now immediate from Lemma 6.60. □

Now we are ready to show that, for z sufficiently close to y, the reduced length $l_x(z)$ is bounded above by a constant depending on the curvature bounds, on $l_x(y)$, and on the distance in space-time from z to y.

LEMMA 6.62. *Given constants $\epsilon > 0$, $\bar{\tau}_0 < \infty$, $C_0 < \infty$, and $l_0 < \infty$, there are $C_3 < \infty$ and $0 < \delta_2 \leq \epsilon/4$ depending only on the given constants such that the following holds. Let $y \in \mathcal{M}$ be a point with $\mathbf{t}(y) = t_0 = T - \bar{\tau}$ where $\tau_1 + \epsilon \leq \bar{\tau} \leq \bar{\tau}_0$. Suppose that there is a minimizing $\mathcal{L}$-geodesic γ from x to y with $l_x(\gamma) \leq l_0$. Suppose that $|\nabla R|$ and $|\mathrm{Ric}|$ are bounded by C_0 along γ. Suppose also that the ball $B(y, t_0, \epsilon)$ has compact closure in M_{t_0} and that there is an embedding*

$$\iota \colon B(y, t_0, \epsilon) \times (t_0 - \epsilon, t_0 + \epsilon) \xrightarrow{\cong} \widetilde{P}(y, \epsilon, \epsilon) \subset \mathcal{M}$$

compatible with time and the vector field so that the sectional curvatures of the restriction of G to the image of this embedding are bounded by C_0. Then for any point $b \in B(y, t_0, \delta_2)$ and for any $t' \in (t_0 - \delta_2, t_0 + \delta_2)$ there is a curve γ_1 from x to the point $z = \iota(b, t')$, parameterized by backward time, such that

$$l(\gamma_1) \leq l(\gamma) + C_3 \sqrt{d_{t_0}(y, b)^2 + |t_0 - t'|^2}.$$

PROOF. Let C_2 be the constant depending on C_0, l_0, ϵ, and $\bar{\tau}_0$ from Corollary 6.61, and set

$$C' = \frac{\sqrt{2}}{\sqrt{\epsilon}} C_2.$$

Since $\bar{\tau} \geq \epsilon$, it follows that $\bar{\tau} - \epsilon/2 \geq \epsilon/2$, so that by Corollary 6.61 we have $|X_\gamma(\tau)| \leq C'$ for all $\tau \in [\bar{\tau} - \epsilon/2, \bar{\tau}]$. Set $0 < \delta_0$ sufficiently small (how small depends only on C_0) such that for all $(z,t) \in \widetilde{P}(y, \epsilon, \delta_0)$ we have

$$\frac{1}{2} g(z,t) \leq g(z, t_0) \leq 2g(z,t),$$

and define

$$\delta_2 = \min\left(\frac{\epsilon}{8}, \frac{\epsilon}{8C'}, \frac{\delta_0}{4}\right).$$

Let $b \in B(y, t_0, \delta_2)$ and $t' \in (t_0 - \delta_2, t_0 + \delta_2)$ be given. Set

$$\alpha = \sqrt{d_{t_0}(y,b)^2 + |t_0 - t'|^2},$$

set $t_1 = t_0 - 2\alpha$, and let $\tau_1 = T - t_1$. Notice that $\alpha < \sqrt{2}\delta_2 < \epsilon/4$, so that the norm of the Ricci curvature is bounded by C' on $\iota(B(y, t_0, \epsilon) \times (t_1, t_0 + 2\alpha))$.

CLAIM 6.63. $\gamma(\tau_1) \in \widetilde{P}(y, \epsilon, \epsilon)$ and writing $\gamma(\tau_1) = \iota(c, t_1)$ we have $d_{t_0}(c, b) \leq (4C' + 1)\alpha$.

PROOF. Since $|X_\gamma(\tau)| \leq C'$ for all $\tau \in [\bar{\tau} - 2\alpha, \bar{\tau}]$, and $\delta_2 \leq \delta_0/4$, it follows that $2\alpha \leq \delta_0$ and hence that $|X_\gamma(\tau)|_{g(t_0)} \leq 2C'$ for all $\tau \in [\bar{\tau} - 2\alpha, \bar{\tau}]$. Since $\gamma(\bar{\tau}) = y$, this implies that

$$d_{t_0}(y, c) \leq 4C'\alpha.$$

The claim then follows from the triangle inequality. $\square$

Now let $\bar{\mu} \colon [\bar{\tau} - 2\alpha, T - t'] \to B(y, t_0, \epsilon)$ be a shortest $g(t_0)$-geodesic from c to b, parameterized at constant $g(t_0)$-speed, and let μ be the path parameterized by backward time defined by

$$\mu(\tau) = \iota(\bar{\mu}(\tau), T - \tau)$$

for all $\tau \in [\bar{\tau} - 2\alpha, T - t']$. Then the concatenation $\gamma_1 = \gamma|_{[\tau_1, \bar{\tau} - 2\alpha]} * \mu$ is a path parameterized by backward time from x to $\iota(b, t')$.

CLAIM 6.64. There is a constant C'_1 depending only on C_0, C', and $\bar{\tau}_0$ such that

$$l(\gamma_1) \leq l(\gamma|_{[\tau_1, \bar{\tau} - 2\alpha]}) + C'_1 \alpha$$

PROOF. First notice that since $\bar{\tau} = T - t_0$ and $|t' - t_0| \leq \alpha$ we have $(T - t') - (\bar{\tau} - 2\alpha) = 2\alpha + (t' - t_0) \geq \alpha$. According to Claim 6.63 this implies that the $g(t_0)$-speed of μ is at most $(4C' + 1)$, and hence that $|X_\mu(\tau)|_{g(T-\tau)} \leq 8C' + 2$ for all $\tau \in [\bar{\tau} - 2\alpha, T - t']$. Consequently, $R + |X_\mu|^2$ is bounded above along μ by a constant $\widetilde{C}$ depending only on C' and C_0. This implies that $\mathcal{L}(\mu) \leq \widetilde{C}\alpha\sqrt{T - t'}$. Of course, $T - t' \leq \bar{\tau} + \epsilon < 2\bar{\tau} \leq 2\bar{\tau}_0$. This completes the proof of the claim. $\square$

On the other hand, since $R \geq -3C_0$ in $P(y, \epsilon, \epsilon)$ and $|X|^2 \geq 0$, we see that
$$\mathcal{L}(\gamma|_{[\tau_1, \bar{\tau}-2\alpha]}) \leq \mathcal{L}(\gamma) + 6C_0\alpha\sqrt{\bar{\tau}_0}.$$
Together with the previous claim this establishes Lemma 6.62. □

This is a one-sided inequality which says that the nearby values of l_x are bounded above in terms of $l_x(y)$, the curvature bounds, and the distance in space-time from y. In order to complete the proof of Proposition 6.59 we must establish inequalities in the opposite direction. This requires reversing the roles of the points.

PROOF. (of Proposition 6.59) Let δ_2 and C_3 be the constants as in Lemma 6.62 associated to $\epsilon/2$, $\bar{\tau}_0$, C_0, and l_0. We shall choose $C \geq C_3$ and $\delta \leq \delta_2$ so that by Lemma 6.62 we will automatically have
$$l_x(\iota(b,t')) \leq l_x(y) + C_3\sqrt{d_{t_0}(y,b)^2 + |t_0-t'|^2}$$
$$\leq l_x(y) + C\sqrt{d_{t_0}(y,b)^2 + |t_0-t'|^2}$$
for all $\iota(b,t') \in P(y,\delta,\delta)$. It remains to choose C and δ so that
$$l_x(y) \leq l_x(\iota(b,t')) + C\sqrt{d_{t_0}(y,b)^2 + |t_0-t'|^2}.$$
Let δ_2' and C_3' be the constants given by Lemma 6.62 for the following set of input constants: $C_0' = C_0$, $\bar{\tau}_0$ replaced by $\bar{\tau}_0' = \bar{\tau}_0 + \epsilon/2$, and l_0 replaced by $l_0' = l_0 + \sqrt{2}C_3\delta_2$, and ϵ replaced by $\epsilon' = \epsilon/4$. Then set $C = \max(2C_3', C_3)$.
Let $z = \iota(b,t') \in \widetilde{P}(y,\delta,\delta)$.

CLAIM 6.65. *For δ sufficiently small (how small depends on δ_2 and δ_2') we have $B(z,t',\epsilon/4) \subset B(y,t_0,\epsilon)$.*

PROOF. Since $|t_0 - t'| < \delta \leq \delta_2$, and by construction $\delta_2 < \delta_0$, it follows that for any $c \in B(y,t_0,\epsilon)$ we have $d_{t'}(b,c) \leq 2d_{t_0}(b,c)$. Since $d_{t_0}(y,b) < \delta \leq \epsilon/4$, the result is immediate from the triangle inequality. □

By the above and the fact that $\delta \leq \epsilon/4$, the sectional curvatures on $\widetilde{P}(z,\epsilon/4,\epsilon/4)$ are bounded by C_0. By Lemma 6.62 there is a curve parameterized by backward time from x to z whose l-length is at most l_0'. Thus the l-length of any minimizing $\mathcal{L}$-geodesic from x to z is at most l_0'. By assumption we have a minimizing $\mathcal{L}$-geodesic with the property that $|\text{Ric}|$ and $|\nabla R|$ are bounded by C_0 along the $\mathcal{L}$-geodesic.

Of course, $t_0 - \delta < t' < t_0 + \delta$, so that $\tau_1 + \epsilon/2 < T - t' \leq \bar{\tau}_0 + \epsilon/4$. This means that Lemma 6.62 applies to show that for very $w = \iota(c,t) \in \widetilde{P}(z, \delta_2', \delta_2')$, we have
$$l_x(w) \leq l_x(z) + C_3'\sqrt{d_{t'}(b,c)^2 + |t-t'|^2}.$$

The proof is then completed by showing the following:

CLAIM 6.66. $y \in \widetilde{P}(z, \delta_2', \delta_2')$.

PROOF. By construction $|t' - t_0| < \delta \leq \delta_2'$. Also, $d_{t_0}(y, b) < \delta \leq \delta_2'/2$. Since $d_{t_0} \leq 2d_{t'}$, we have $d_{t'}(y, b) < \delta_2'$ the claim is then immediate. □

It follows immediately that
$$l_x(y) \leq l_x(z) + C_3'\sqrt{d_{t'}(b, y)^2 + |t_0 - t'|^2}$$
$$\leq l_x(z) + 2C_3'\sqrt{d_{t_0}(b, y)^2 + |t_0 - t'|^2}$$
$$\leq l_x(z) + C\sqrt{d_{t_0}(b, y)^2 + |t_0 - t'|^2}.$$

This completes the proof of Proposition 6.59. □

COROLLARY 6.67. *Let $(\mathcal{M}, G)$ be a generalized Ricci flow and let $x \in \mathcal{M}$ with $\mathbf{t}(x) = T - \tau_1$. Let $A \subset \mathcal{M} \cap \mathbf{t}^{-1}(-\infty, T - \tau_1)$ be a subset whose intersection with each time-slice M_t is measurable. Suppose that there is a subset $F \subset \mathcal{M}$ such that $|\nabla R|$ and $|\mathrm{Ric}|$ are bounded on F and such that every minimizing $\mathcal{L}$ geodesic from x to any point in a neighborhood, $\nu(A)$, of A is contained in F. Then for each $\tau \in (\tau_1, \overline{\tau}]$ the intersection of A with $\mathcal{U}_x(\tau)$ is an open subset of full measure in $A \cap M_{T-\tau}$.*

PROOF. Since $\mathcal{U}_x(\tau)$ is an open subset of $M_{T-\tau}$, the complement of $\nu(A) \cap \mathcal{U}_x(\tau)$ in $\nu(A) \cap M_{T-\tau}$ is a closed subset of $\nu(A) \cap M_{T-\tau}$. Since there is a minimizing $\mathcal{L}$-geodesic to every point of $\nu(A) \cap M_{T-\tau}$, the $\mathcal{L}$-exponential map $\mathcal{L}\exp_x^\tau$ is onto $\nu(A) \cap M_{T-\tau}$.

CLAIM 6.68. *The complement of $\nu(A) \cap \mathcal{U}_x(\tau)$ in $\nu(A)$ is contained in the union of two sets: The first is the set of points z where there is more than one minimizing $\mathcal{L}$-geodesic from x ending at z and, if Z is the initial condition for any minimizing $\mathcal{L}$-geodesic to z, then the differential of $\mathcal{L}\exp_x^\tau$ at any Z is an isomorphism. The second is the intersection of the set of critical values of $\mathcal{L}\exp^\tau$ with $\nu(A) \cap M_{T-\tau}$.*

PROOF. Suppose that $q \in \nu(A) \cap M_{T-\tau}$ is not contained in $\mathcal{U}_x$. Let γ_Z be a minimal $\mathcal{L}$-geodesic from x to q. If the differential of $\mathcal{L}\exp_x$ is not an isomorphism at Z, then q is contained in the second set given in the claim. Thus, we can assume that the differential of $\mathcal{L}\exp_x$ at Z, and hence $\mathcal{L}\exp_x$ identifies a neighborhood $\widetilde{V}$ of Z in $\mathcal{H}T_z\mathcal{M}$ with a neighborhood $V \subset \nu(A)$ of q in $M_{T-\tau}$. Suppose that there is no neighborhood $\widetilde{V}' \subset \widetilde{V}$ of Z so that the $\mathcal{L}$-geodesics are unique minimal $\mathcal{L}$-geodesics to their endpoints in M_{T_τ}. Then there is a sequence of minimizing $\mathcal{L}$-geodesics γ_n whose endpoints converge to q, but so that no γ_n has initial condition contained in $\widetilde{V}'$. By hypothesis all of these geodesics are contained in F and hence $|\mathrm{Ric}|$ and $|\nabla R|$ are uniformly bounded on these geodesics. Also, by the continuity of $\mathcal{L}$, the $\mathcal{L}$-lengths of γ_n are uniformly bounded as n tends to infinity. By Corollary 6.61 we see that the initial conditions $Z_n = \sqrt{\tau_1}X_{\gamma_n}(\tau_1)$ (meaning the limit as $\tau \to 0$ of these quantities in the case when $\tau_1 = 0$) are of uniformly bounded norm. Hence, passing to a subsequence we can arrange that the Z_n converge to some Z_∞. The tangent vector Z_∞ is the initial condition of an $\mathcal{L}$-geodesic γ_∞. Since the γ_n are minimizing $\mathcal{L}$-geodesics to a sequence of points q_n converging to q, by continuity it follows that γ_∞ is a minimizing $\mathcal{L}$-geodesic to q. Since none of the Z_n is contained in $\widetilde{V}'$, it follows that $Z_\infty \neq Z$. This is a contradiction, showing that throughout some neighborhood $\widetilde{V}'$ of Z the $\mathcal{L}$-geodesics are unique minimizing $\mathcal{L}$-geodesics and completing the proof of the claim. □

According to the next claim, the first subset given in Claim 6.68 is contained in the set of points of $\nu(A) \cap M_{T-\tau}$ where L_x^τ is non-differentiable. Since L_x^τ is a locally Lipschitz function on $\nu(A)$, this subset is of measure zero in $\nu(A)$; see Rademacher's Theorem on p. 81 of [20]. The second set is of measure zero by Sard's theorem. This proves, modulo the next claim, that $\mathcal{U}_x(\tau) \cap A$ is full measure in $A \cap M_{T-\tau}$.

CLAIM 6.69. *Let $z \in M_{T-\tau}$. Suppose that there is a neighborhood of z in $M_{T-\tau}$ with the property that every point of the neighborhood is the endpoint of a minimizing $\mathcal{L}$-geodesic from x, so that L_x^τ is defined on this neighborhood of z. Suppose that there are two distinct, minimizing $\mathcal{L}$-geodesics γ_{Z_1} and γ_{Z_2} from x ending at z with the property that the differential of $\mathcal{L}\exp^\tau$ is an isomorphism at both Z_1 and Z_2. Then the function L_x^τ is non-differentiable at z.*

PROOF. Suppose that $\gamma_{Z_0}|_{[0,\tau]}$ is an $\mathcal{L}$-minimal $\mathcal{L}$-geodesic and that $d_{Z_0}\mathcal{L}\exp_x^\tau$ is an isomorphism. Then use $\mathcal{L}\exp_x^\tau$ to identify a neighborhood of $Z_0 \in T_xM$ with a neighborhood of z in $M_{T-\tau}$, and push the function $\widetilde{\mathcal{L}}_x^\tau$ on this neighborhood of Z_0 down to a function L_{Z_0} on a neighborhood in $M_{T-\tau}$ of z. According to Lemma 6.22 the resulting function L_{Z_0} is smooth and its gradient at z is $2\sqrt{\tau}X(\tau)$. Now suppose that there is a second $\mathcal{L}$-minimizing $\mathcal{L}$-geodesic to z with initial condition $Z_1 \neq Z_0$ and with $d_{Z_1}\mathcal{L}\exp_x^\tau$ being an isomorphism. Then near z the function L_x^τ is less than or equal to the minimum of two smooth functions L_{Z_0} and L_{Z_1}. We have $L_{Z_0}(z) = L_{Z_1}(z) = L_x^\tau(z)$, and furthermore, L_{Z_0} and L_{Z_1} have distinct gradients at z. It follows that L_x^τ is not smooth at z. □

This completes the proof of Corollary 6.67. □

7. Reduced volume

Here, we assume that $x \in M_T \subset \mathcal{M}$, so that $\tau_1 = 0$ in this subsection.

DEFINITION 6.70. *Let $A \subset \mathcal{U}_x(\tau)$ be a measurable subset of $M_{T-\tau}$. The $\mathcal{L}$-reduced volume of A from x (or the reduced volume for short) is defined to be*
$$\widetilde{V}_x(A) = \int_A \tau^{-\frac{n}{2}} \exp(-l_x(q)) dq$$
where dq is the volume element of the metric $G(T-\tau)$.

LEMMA 6.71. *Let $A \subset \mathcal{U}_x(\tau)$ be a measurable subset. Define $\widetilde{A} \subset \widetilde{\mathcal{U}}_x(\tau)$ to be the pre-image under $\mathcal{L}\exp_x^\tau$ of A. Then*

$$\widetilde{V}_x(A) = \int_{\widetilde{A}} \tau^{-\frac{n}{2}} \exp(-\widetilde{l}(Z,\tau)) \mathcal{J}(Z,\tau) dZ,$$

where dZ is the usual Euclidean volume element and $\mathcal{J}(Z,\tau)$ is the Jacobian determinant of $\mathcal{L}\exp_x^\tau$ at $Z \in T_x M_T$.

PROOF. This is simply the change of variables formula for integration. □

Before we can study the reduced volume we must study the function that appears as the integrand in its definition. To understand the limit as $\tau \to 0$ requires a rescaling argument.

7.1. Rescaling. Fix $Q > 0$. We rescale to form $(Q\mathcal{M}, QG)$ and then we shift the time by $T - QT$ so that the time-slice M_T in the original flow is the T time-slice of the new flow. We call the result $(\mathcal{M}', G')$. Recall that $\tau = T - \mathbf{t}$ is the parameter for $\mathcal{L}$-geodesics in $(\mathcal{M}, G)$ The corresponding parameter in the rescaled flow $(\mathcal{M}', G')$ is $\tau' = T - \mathbf{t}' = Q\tau$. We denote by $\mathcal{L}'\exp_x$ the $\mathcal{L}$-exponential map from x in $(\mathcal{M}', G')$, and by l'_x the reduced length function for this Ricci flow. The associated function on the tangent space is denoted $\widetilde{l}'$.

LEMMA 6.72. *Let $(\mathcal{M}, G)$ be a generalized Ricci flow and let $x \in M_T \subset \mathcal{M}$. Fix $Q > 0$ and let $(\mathcal{M}', G')$ be the Q scaling and shifting of $(\mathcal{M}, G)$ as described in the previous paragraph. Let $\iota \colon \mathcal{M} \to \mathcal{M}'$ be the identity map. Suppose that $\gamma \colon [0, \overline{\tau}] \to \mathcal{M}$ is a path parameterized by backward time with $\gamma(0) = x$. Let $\beta \colon [0, Q\overline{\tau}] \to Q\mathcal{M}$ be defined by*

$$\beta(\tau') = \iota(\gamma(\tau'/Q)).$$

Then $\beta(0) = x$ and β is parameterized by backward time in $(\mathcal{M}', G')$, and $\mathcal{L}(\beta) = \sqrt{Q}\mathcal{L}(\gamma)$. Furthermore, β is an $\mathcal{L}$-geodesic if and only if γ is. In this case, if $Z = \lim_{\tau \to 0} \sqrt{\tau} X_\gamma(\tau)$ then $\sqrt{Q^{-1}} Z = \lim_{\tau' \to 0} \sqrt{\tau'} X_\beta(\tau')$

REMARK 6.73. Notice that $|Z|_G^2 = |\sqrt{Q^{-1}} Z|_{G'}^2$.

PROOF. It is clear that $\beta(0) = x$ and that β is parameterized by backward time in $(\mathcal{M}', G')$. Because of the scaling of space and time by Q, we have $R_{G'} = R_G/Q$ and $X_\beta(\tau') = d\iota(X_\gamma(\tau))/Q$, and hence $|X_\beta(\tau')|_{G'}^2 = \frac{1}{Q}|X_\gamma(\tau)|_G^2$. A direct change of variables in the integral then shows that

$$\mathcal{L}(\beta) = \sqrt{Q}\mathcal{L}(\gamma).$$

It follows that β is an $\mathcal{L}$-geodesic if and only if γ is. The last statement follows directly. □

Immediately from the definitions we see the following:

COROLLARY 6.74. *With notation as above, and with the substitution $\tau' = Q\tau$, for any $Z \in \mathcal{H}T_xM$ and any $\tau > 0$ we have*

$$\mathcal{L}'\exp_x(\sqrt{Q^{-1}}Z,\tau') = \iota(\mathcal{L}\exp_x(Z,\tau))$$

and

$$\widetilde{l}'(\sqrt{Q^{-1}}Z,\tau') = \widetilde{l}(Z,\tau),$$

whenever these are defined.

7.2. The integrand in the reduced volume integral. Now we turn our attention to the integrand (over $\widetilde{\mathcal{U}}_x(\tau)$) in the reduced volume integral. Namely, set

$$f(\tau) = \tau^{-n/2} e^{-\widetilde{l}(Z,\tau)} \mathcal{J}(Z,\tau),$$

where $\mathcal{J}(Z,\tau)$ is the Jacobian determinant of $\mathcal{L}\exp_x^\tau$ at the point $Z \in \widetilde{U}_x(\tau) \subset T_xM_T$. We wish to see that this quantity is invariant under the rescaling.

LEMMA 6.75. *With the notation as above let $\mathcal{J}'(Z,\tau')$ denote the Jacobian determinant of $\mathcal{L}'\exp_x$. Then, with the substitution $\tau' = Q\tau$, we have*

$$(\tau')^{-n/2} e^{-\widetilde{l}'(\sqrt{Q^{-1}}Z,\tau')} \mathcal{J}'(\sqrt{Q^{-1}}Z,\tau') = \tau^{-n/2} e^{-\widetilde{l}(Z,\tau)} \mathcal{J}(Z,\tau).$$

PROOF. It follows from the first equation in Corollary 6.74 that

$$J(\iota)\mathcal{J}(Z,\tau) = J(\sqrt{Q^{-1}})\mathcal{J}'(\sqrt{Q^{-1}}Z,\tau'),$$

where $J(\iota)$ is the Jacobian determinant of ι at $\mathcal{L}\exp_x(Z,\tau)$ and $J(\sqrt{Q^{-1}})$ is the Jacobian determinant of multiplication by $\sqrt{Q^{-1}}$ as a map from T_xM_T to itself, where the domain has the metric G and the range has metric $G' = QG$. Clearly, with these conventions, we have $J(\iota) = Q^{n/2}$ and $J(\sqrt{Q^{-1}}) = 1$. Hence, we conclude

$$Q^{n/2}\mathcal{J}(Z,\tau) = \mathcal{J}'(\sqrt{Q^{-1}}Z,\tau').$$

Letting γ be the $\mathcal{L}$-geodesic in $(\mathcal{M}, G)$ with initial condition Z and β the $\mathcal{L}$-geodesic in $(\mathcal{M}', G')$ with initial condition $\sqrt{Q^{-1}}Z$, by Lemma 6.72 we have $\gamma(\tau) = \beta(\tau')$. From Corollary 6.74 and the definition of the reduced length, we get

$$\widetilde{l}'(\sqrt{Q^{-1}}Z,\tau') = \widetilde{l}(\gamma,\tau).$$

Plugging these in gives the result. □

Let us evaluate $f(\tau)$ in the case of $\mathbb{R}^n$ with the Ricci flow being the constant family of Euclidean metrics.

EXAMPLE 6.76. Let the Ricci flow be the constant family of standard metrics on $\mathbb{R}^n$. Fix $x = (p,T) \in \mathbb{R}^n \times (-\infty,\infty)$. Then

$$\mathcal{L}\exp_x(Z,\tau) = (p + 2\sqrt{\tau}Z, T - \tau).$$

In particular, the Jacobian determinant of $\mathcal{L}\exp^{\tau}_{(x,T)}$ is constant and equal to $2^n \tau^{n/2}$. The $\tilde{l}$-length of the $\mathcal{L}$-geodesic $\gamma_Z(\tau) = (p + 2\sqrt{\tau}Z, T - \tau)$, $0 \leq \tau \leq \overline{\tau}$, is $|Z|^2$.

Putting these computations together gives the following.

CLAIM 6.77. *In the case of the constant flow on Euclidean space we have*
$$f(\tau) = \tau^{-n/2} e^{-\tilde{l}(Z,\tau)} \mathcal{J}(Z,\tau) = 2^n e^{-|Z|^2}.$$

This computation has consequences for all Ricci flows.

PROPOSITION 6.78. *Let $(\mathcal{M}, G)$ be a generalized Ricci flow and let $x \in M_T \subset \mathcal{M}$. Then, for any $A < \infty$, there is $\delta > 0$ such that the map $\mathcal{L}\exp_x$ is defined on $B(0, A) \times (0, \delta)$, where $B(0, A)$ is the ball of radius A centered at the origin in $T_x M_T$. Moreover, the map $\mathcal{L}\exp_x$ defines a diffeomorphism of $B(0, A) \times (0, \delta)$ onto an open subset of $\mathcal{M}$. Furthermore,*
$$\lim_{\tau \to 0} \tau^{-n/2} e^{-\tilde{l}(Z,\tau)} \mathcal{J}(Z,\tau) = 2^n e^{-|Z|^2},$$
where the convergence is uniform on each compact subset of $T_x M_T$.

PROOF. First notice that since T is greater than the initial time of $\mathcal{M}$, there is $\epsilon > 0$, and an embedding $\rho \colon B(x, T, \epsilon) \times [T - \epsilon, T] \to \mathcal{M}$ compatible with time and the vector field. By taking $\epsilon > 0$ smaller if necessary, we can assume that the image of ρ has compact closure in $\mathcal{M}$. By compactness every higher partial derivative (both spatial and temporal) of the metric is bounded on the image of ρ.

Now take a sequence of positive constants τ_k tending to 0 as $k \to \infty$, and set $Q_k = \tau_k^{-1}$. We let $(\mathcal{M}_k, G_k)$ be the Q_k-rescaling and shifting of $(\mathcal{M}, G)$ as described at the beginning of this section. The rescaled version of ρ is an embedding
$$\rho_k \colon B_{G_k}(x, T, \sqrt{Q_k}\epsilon) \times [T - Q_k \epsilon, T] \to \mathcal{M}_k$$
compatible with the time function $\mathbf{t}_k$ and the vector field. Furthermore, uniformly on the image of ρ_k, every higher partial derivative of the metric is bounded by a constant that goes to zero with k. Thus, the generalized Ricci flows $(\mathcal{M}_k, G_k)$ based at x converge geometrically to the constant family of Euclidean metrics on $\mathbb{R}^n$. Since the ODE given in Equation (6.5) is regular even at 0, this implies that the $\mathcal{L}$-exponential maps for these flows converge uniformly on the balls of finite radius centered at the origin of the tangent spaces at x to the $\mathcal{L}$-exponential map of $\mathbb{R}^n$ at the origin. Of course, if $Z \in T_x M_T$ is an initial condition for an $\mathcal{L}$-geodesic in $(\mathcal{M}, G)$, then $\sqrt{Q_k^{-1}} Z$ is the initial condition for the corresponding $\mathcal{L}$-geodesic in $(\mathcal{M}_k, G_k)$. But $|Z|_G = |\sqrt{Q_k^{-1}} Z|_{G_k}$, so that if $Z \in B_G(0, A)$ then $\sqrt{Q_k^{-1}} Z \in B_{G_k}(0, A)$. In particular, we see that, for any $A < \infty$ and for all k sufficiently large, the $\mathcal{L}$-geodesics are defined on $B_{G_k}(0, A) \times (0, 1]$ and the image is contained in

the image of ρ_k. Rescaling shows that for any $A < \infty$ there is k for which the $\mathcal{L}$-exponential map is defined on $B_G(0, A) \times (0, \tau_k]$ and has image contained in ρ.

Let $Z \subset B_G(0, A)$, and let γ be the $\mathcal{L}$-geodesic with $\lim_{\tau \to 0} \sqrt{\tau} X_\gamma(\tau) = Z$. Let γ_k be the corresponding $\mathcal{L}$-geodesic in $(\mathcal{M}_k, G_k)$. Then

$$\lim_{\tau \to 0} \sqrt{\tau} X_{\gamma_k}(\tau) = \sqrt{\tau_k} Z = Z_k.$$

Of course, $|Z_k|^2_{G_k} = |Z|^2_G$, meaning that, for every k, Z_k is contained in the ball $B_{G_k}(0, A) \subset T_x M_T$. Hence, by passing to a subsequence we can assume that, in the geometric limit, the $\sqrt{\tau_k} Z$ converge to a tangent vector Z' in the ball of radius A centered at the origin in the tangent space to Euclidean space. Of course $|Z'|^2 = |Z|^2_G$. By Claim 6.77, this means that we have

$$\lim_{k \to \infty} 1^{-n/2} e^{-\widetilde{l}_k(\sqrt{Q_k^{-1}} Z, 1)} \mathcal{J}_k(\sqrt{Q_k^{-1}} Z, 1) = 2^n e^{-|Z|^2},$$

with $\mathcal{J}_k$ the Jacobian determinant of the $\mathcal{L}$-exponential map for $(\mathcal{M}_k, G_k)$. Of course, since $\tau_k = Q_k^{-1}$, by Lemma 6.75 we have

$$1^{-n/2} e^{-\widetilde{l}_k(\sqrt{Q_k^{-1}} Z, 1)} \mathcal{J}_k(\sqrt{Q_k^{-1}} Z, 1) = \tau_k^{-n/2} e^{-\widetilde{l}(Z, \tau_k)} \mathcal{J}(Z, \tau_k).$$

This establishes the limiting result.

Since the geometric limits are uniform on balls of finite radius centered at the origin in the tangent space, the above limit also is uniform over each of these balls. □

COROLLARY 6.79. *Let $(\mathcal{M}, G)$ be a generalized Ricci flow whose sectional curvatures are bounded. For any $x \in M_T$ and any $R < \infty$ for all $\tau > 0$ sufficiently small, the ball of radius R centered at the origin in $T_x M_T$ is contained in $\widetilde{\mathcal{U}}_x(\tau)$.*

PROOF. According to the last result, given $R < \infty$, for all $\delta > 0$ sufficiently small the ball of radius R centered at the origin in $T_x M_T$ is contained in $\mathcal{D}_x^\delta$, in the domain of definition of $\mathcal{L}\exp_x^\delta$ as given in Definition 6.17, and $\mathcal{L}\exp_x$ is a diffeomorphism on this subset. We shall show that if $\delta > 0$ is sufficiently small, then the resulting $\mathcal{L}$-geodesic γ is the unique minimizing $\mathcal{L}$-geodesic. If not then there must be another, distinct $\mathcal{L}$-geodesic to this point whose $\mathcal{L}$-length is no greater than that of γ. According to Lemma 6.60 there is a constant C_1 depending on the curvature bound and on δ such that if Z is an initial condition for an $\mathcal{L}$-geodesic, then for all $\tau \in (0, \delta)$ we have

$$C_1^{-1}\left(|Z| - \frac{(C_1 - 1)}{2}\sqrt{\delta}\right) \leq \sqrt{\tau}|X(\tau)| \leq C_1|Z| + \frac{(C_1 - 1)}{2}\sqrt{\delta}.$$

From the formula given in Lemma 6.60 for C_1, it follows that, fixing the bound of the curvature and its derivatives, $C_1 \to 1$ as $\delta \to 0$. Thus, with a given curvature bound, for δ sufficiently small, $\sqrt{\tau}|X(\tau)|$ is almost a constant along $\mathcal{L}$-geodesics. Hence, the integral of $\sqrt{\tau}|X(\tau)|^2$ is approximately

$2\sqrt{\delta}|Z|^2$. On the other hand, the absolute value of the integral of $\sqrt{\tau}R(\gamma(\tau))$ is at most $2C_0\delta^{3/2}/3$ where C_0 is an upper bound for the absolute value of the scalar curvature.

Given $R < \infty$, choose $\delta > 0$ sufficiently small such that $\mathcal{L}\exp_x$ is a diffeomorphism on the ball of radius $9R$ centered at the origin and such that the following estimate holds: The $\mathcal{L}$-length of an $\mathcal{L}$-geodesic defined on $[0,\delta]$ with initial condition Z is between $\sqrt{\delta}|Z|^2$ and $3\sqrt{\delta}|Z|^2$. To ensure the latter estimate we need only take δ sufficiently small, given the curvature bounds and the dimension. Hence, for these δ no $\mathcal{L}$-geodesic with initial condition outside the ball of radius $9R$ centered at the origin in T_xM_T can be as short as any $\mathcal{L}$-geodesic with initial condition in the ball of radius R centered at the same point. This means that the $\mathcal{L}$-geodesics defined on $[0,\delta]$ with initial condition $|Z|$ with $|Z| < R$ are unique minimizing $\mathcal{L}$-geodesics. □

7.3. Monotonicity of reduced volume. Now we are ready to state and prove our main result concerning the reduced volume.

THEOREM 6.80. *Fix $x \in M_T \subset \mathcal{M}$. Let $A \subset \mathcal{U}_x \subset \mathcal{M}$ be an open subset. We suppose that for any $0 < \tau \leq \overline{\tau}$ and any $y \in A_\tau = A \cap M_{T-\tau}$ the minimizing $\mathcal{L}$-geodesic from x to y is contained in $A \cup \{x\}$. Then $\widetilde{V}_x(A_\tau)$ is a non-increasing function of τ for all $0 < \tau \leq \overline{\tau}$.*

PROOF. Fix $\tau_0 \in (0,\overline{\tau}]$. To prove the theorem we shall show that for any $0 < \tau < \tau_0$ we have $\widetilde{V}_x(A_\tau) \geq \widetilde{V}_x(A_{\tau_0})$. Let $\widetilde{A}_{\tau_0} \subset \widetilde{\mathcal{U}}_x(\tau_0)$ be the pre-image under $\mathcal{L}\exp_x^{\tau_0}$ of A_{τ_0}. For each $0 < \tau \leq \tau_0$ we set
$$A_{\tau,\tau_0} = \mathcal{L}\exp_x^\tau(\widetilde{A}_{\tau_0}) \subset M_{T-\tau}.$$
It follows from the assumption on A that $A_{\tau,\tau_0} \subset A_\tau$, so that $\widetilde{V}_x(A_{\tau,\tau_0}) \leq \widetilde{V}_x(A_\tau)$. Thus, it suffices to show that for all $0 < \tau \leq \tau_0$ we have
$$\widetilde{V}_x(A_{\tau,\tau_0}) \geq \widetilde{V}_x(\tau_0).$$
Since
$$\widetilde{V}_x(A_{\tau,\tau_0}) = \int_{\widetilde{A}_{\tau_0}} \tau^{-\frac{n}{2}}\exp(-\widetilde{l}(Z,\tau))\mathcal{J}(Z,\tau)dZ,$$
the theorem follows from:

PROPOSITION 6.81. *For each $Z \in \widetilde{\mathcal{U}}_x(\overline{\tau}) \subset T_xM_T$ the function*
$$f(Z,\tau) = \tau^{-\frac{n}{2}}e^{-\widetilde{l}(Z,\tau)}\mathcal{J}(Z,\tau)$$
is a non-increasing function of τ on the interval $(0,\overline{\tau}]$ with $\lim_{\tau \to 0}f(Z,\tau) = 2^ne^{-|Z|^2}$, the limit being uniform on any compact subset of T_xM_T.

PROOF. First, we analyze the Jacobian $\mathcal{J}(Z,\tau)$. We know that $\mathcal{L}\exp_x^\tau$ is smooth in a neighborhood of Z. Choose a basis $\{\partial_\alpha\}$ for T_xM_T such that ∂_α pushes forward under the differential at Z of $\mathcal{L}\exp_x^\tau$ to an orthonormal basis $\{Y_\alpha\}$ for $M_{T-\tau}$ at $\gamma_Z(\tau)$. Notice that, letting τ' range from 0 to τ and taking the push-forward of the ∂_α under the differential at Z of $\mathcal{L}\exp_x^{\tau'}$ produces a basis of $\mathcal{L}$-Jacobi fields $\{Y_\alpha(\tau')\}$ along γ_Z. With this understood, we have:

$$\frac{\partial}{\partial \tau}\ln \mathcal{J}|_\tau = \frac{d}{d\tau}\ln(\sqrt{\det(\langle Y_\alpha, Y_\beta\rangle)})$$
$$= \frac{1}{2}\Big(\frac{d}{d\tau}\sum_\alpha |Y_\alpha|^2\Big)\Big|_\tau.$$

By Lemma 6.40 and by Proposition 6.37 (recall that $\tau_1 = 0$) we have

$$\frac{1}{2}\frac{d}{d\tau}|Y_\alpha(\tau)|^2 = \frac{1}{2\sqrt{\tau}}\mathrm{Hess}(L)(Y_\alpha, Y_\alpha) + \mathrm{Ric}(Y_\alpha, Y_\alpha)$$
(6.19)
$$\leq \frac{1}{2\tau} - \frac{1}{2\sqrt{\tau}}\int_0^\tau \sqrt{\tau'}H(X, \widetilde{Y}_\alpha(\tau'))d\tau',$$

where $\widetilde{Y}_\alpha(\tau')$ is the adapted vector field along γ with $\widetilde{Y}(\tau) = Y_\alpha(\tau)$. Summing over α as in the proof of Proposition 6.43 and Claim 6.44 yields

$$\frac{\partial}{\partial\tau}\ln \mathcal{J}(Z,\tau)|_\tau \leq \frac{n}{2\tau} - \frac{1}{2\sqrt{\tau}}\sum_\alpha \int_0^\tau \sqrt{\tau'}H(X, \widetilde{Y}_{\alpha(\tau')})d\tau'$$
(6.20)
$$= \frac{n}{2\tau} - \frac{1}{2}\tau^{-\frac{3}{2}}K^\tau(\gamma_Z).$$

On $\widetilde{\mathcal{U}}_x(\tau)$ the expression $\tau^{-\frac{n}{2}}e^{-\widetilde{l}(Z,\tau)}\mathcal{J}(Z,\tau)$ is positive, and so we have

$$\frac{\partial}{\partial\tau}\ln\Big(\tau^{-\frac{n}{2}}e^{-\widetilde{l}(Z,\tau)}\mathcal{J}(Z,\tau)\Big) \leq \Big(-\frac{n}{2\tau} - \frac{d\widetilde{l}}{d\tau} + \frac{n}{2\tau} - \frac{1}{2}\tau^{-\frac{3}{2}}K^\tau(\gamma_Z)\Big).$$

Corollary 6.54 says that the right-hand side of the previous inequality is zero. Hence, we conclude

(6.21)
$$\frac{d}{d\tau}\Big(\tau^{-\frac{n}{2}}e^{-\widetilde{l}(X,\tau)}\mathcal{J}(X,\tau)\Big) \leq 0.$$

This proves the inequality given in the statement of the proposition. The limit statement as $\tau \to 0$ is contained in Proposition 6.78. □

As we have already seen, this proposition implies Theorem 6.80, and hence the proof of this theorem is complete. □

Notice that we have established the following:

COROLLARY 6.82. *For any measurable subset $A \subset \mathcal{U}_x(\tau)$ the reduced volume $\widetilde{V}_x(A)$ is at most $(4\pi)^{n/2}$.*

PROOF. Let $\widetilde{A} \subset \widetilde{\mathcal{U}}_x(\tau)$ be the pre-image of A. We have seen that

$$\widetilde{V}_x(A) = \int_A \tau^{-n/2}e^{-l(q,\tau)}dq = \int_{\widetilde{A}} \tau^{-n/2}e^{-\widetilde{l}(Z,\tau)}\mathcal{J}(Z,\tau)dz.$$

By Theorem 6.80 we see that $\tau^{-n/2}e^{-\widetilde{l}(Z,\tau)}\mathcal{J}(Z,\tau)$ is a non-increasing function of τ whose limit as $\tau \to 0$ is the restriction of $2^n e^{-|Z|^2}$ to $\widetilde{A}$. The result is immediate from Lebesgue dominated convergence. □

CHAPTER 7

Complete Ricci flows of bounded curvature

In this chapter we establish strong results for $\mathcal{L}\exp_x$ in the case of ordinary Ricci flow on complete n-manifolds with appropriate curvature bounds. In particular, for these flows we show that there is a minimizing $\mathcal{L}$-geodesic to every point. This means that l_x is everywhere defined. We extend the differential inequalities for l_x established in Section 4 of Chapter 6 at the 'smooth points' to weak inequalities (i.e., inequalities in the distributional sense) valid on the whole manifold. Using this we prove an upper bound for the minimum of l_x^τ.

Let us begin with a definition that captures the necessary curvature bound for these results.

DEFINITION 7.1. Let $(M, g(t))$, $a \leq t \leq b$, be a Ricci flow. We say that the flow is *complete of bounded curvature* if for each $t \in [a,b]$ the Riemannian manifold $(M, g(t))$ is complete and if there is $C < \infty$ such that $|\mathrm{Rm}|(p,t) \leq C$ for all $p \in M$ and all $t \in [a,b]$. Let I be an interval and let $(M, g(t))$, $t \in I$, be a Ricci flow. Then we say that the flow is *complete with curvature locally bounded in time* if for each compact subinterval $J \subset I$ the restriction of the flow to $(M, g(t))$, $t \in J$, is complete of bounded curvature.

1. The functions L_x and l_x

Throughout Chapter 7 we have a Ricci flow $(M, g(t))$, $0 \leq t \leq T < \infty$, and we set $\tau = T - t$. All the results of the last chapter apply in this context, but in fact in this context there are much stronger results, which we develop here.

1.1. Existence of $\mathcal{L}$-geodesics. We assume here that $(M, g(t))$, $0 \leq t \leq T < \infty$, is a Ricci flow which is complete of bounded curvature. In Shi's Theorem (Theorem 3.28) we take K equal to the bound of the norm of the Riemannian curvature on $M \times [0, T]$, we take $\alpha = 1$, and we take $t_0 = T$. It follows from Theorem 3.28 that there is a constant $C(K,T)$ such that $|\nabla R(x,t)| \leq C/t^{1/2}$. Thus, for any $\epsilon > 0$ we have a uniform bound for $|\nabla R|$ on $M \times [\epsilon, T]$. Also, because of the uniform bound for the Riemann curvature and the fact that $T < \infty$, there is a constant C, depending on the curvature bound and T such that

(7.1) $$C^{-1} g(x,t) \leq g(x,0) \leq C g(x,t)$$

for all $(x,t) \in M \times [0,T]$.

LEMMA 7.2. *Assume that M is connected. Given $p_1, p_2 \in M$ and $0 \leq \tau_1 < \tau_2 \leq T$, there is a minimizing $\mathcal{L}$-geodesic: $\gamma\colon [\tau_1, \tau_2] \to M \times [0,T]$ connecting (p_1, τ_1) to (p_2, τ_2).*

PROOF. For any curve γ parameterized by backward time, we set $\overline{\gamma}$ equal to the path in M that is the image under projection of γ. We set $A(s) = \overline{\gamma}'(s)$. Define

$$c((p_1, \tau_1), (p_2, \tau_2)) = \inf\{\mathcal{L}(\gamma) | \gamma\colon [\tau_1, \tau_2] \to M, \overline{\gamma}(\tau_1) = p_1, \overline{\gamma}(\tau_2) = p_2\}.$$

From Equation (6.2) we see that the infimum exists since, by assumption, the curvature is uniformly bounded (below). Furthermore, for a minimizing sequence γ_i, we have $\int_{s_1}^{s_2} |A_i(s)|^2 \, ds \leq C_0$, for some constant C_0, where $s_i = \sqrt{\tau_i}$ for $i = 1, 2$. It follows from this and the inequality in Equation (7.1) that there is a constant $C_1 < \infty$ such that for all i we have

$$\int_{s_1}^{s_2} |A_i|_{g(0)}^2 \, d\tau \leq C_1.$$

Therefore the sequence $\{\gamma_i\}$ is uniformly continuous with respect to the metric $g(0)$; by Cauchy-Schwarz we have

$$\left|\overline{\gamma}_i(s) - \overline{\gamma}_i(s')\right|_{g(0)} \leq \int_{s'}^{s} |A_i|_{g(0)} \, ds \leq \sqrt{C_1}\sqrt{s - s'}.$$

By the uniform continuity, we see that a subsequence of the γ_i converges uniformly pointwise to a continuous curve γ parameterized by s, the square root backward time. By passing to a subsequence we can arrange that the γ_i converge weakly in $H^{2,1}$. Of course, the limit in $H^{2,1}$ is represented by the continuous limit γ. That is to say, after passing to a subsequence, the γ_i converge uniformly and weakly in $H^{2,1}$ to a continuous curve γ. Let $A(s)$ be the L^2-derivative of γ. Weak convergence in $H^{2,1}$ implies that $\int_{s'}^{s} |A(s)|^2 ds \leq \lim_{i \to \infty} \int_{s'}^{s} |A_i(s)|^2 ds$, so that $\mathcal{L}(\gamma) \leq \lim_{i \to \infty} \mathcal{L}(\gamma_i)$. This means that γ minimizes the $\mathcal{L}$-length. Being a minimizer of $\mathcal{L}$-length, γ satisfies the Euler-Lagrange equation and is smooth by the regularity theorem of differential equations. This then is the required minimizing $\mathcal{L}$-geodesic from (p_1, τ_1) to (p_2, τ_2). □

Let us now show that it is always possible to uniquely extend $\mathcal{L}$-geodesics up to time T.

LEMMA 7.3. *Fix $0 \leq \tau_1 < \tau_2 < T$ and let $\gamma\colon [\tau_1, \tau_2] \to M \times [0, T]$ be an $\mathcal{L}$-geodesic. Then γ extends uniquely to an $\mathcal{L}$-geodesic $\gamma\colon [0, T) \to M \times (0, T]$.*

PROOF. We work with the parameter $s = \sqrt{\tau}$. According to Equation (6.5), we have

$$\nabla_{\gamma'(s)} \gamma'(s) = 2s^2 \nabla R - 4s \operatorname{Ric}(\gamma'(s), \cdot).$$

This is an everywhere non-singular ODE. Since the manifolds $(M, g(t))$ are complete and their metrics are uniformly related as in Inequality (7.1), to show that the solution is defined on the entire interval $s \in [0, \sqrt{T})$ we need only show that there is a uniform bound to the length, or equivalently the energy of γ of any compact subinterval of $[0, T)$ on which it is defined. Fix $\epsilon > 0$. It follows immediately from Lemma 6.60, and the fact that the quantities R, $|\nabla R|$ and $|\mathrm{Rm}|$ are bounded on $M \times [\epsilon, T]$, that there is a bound on $\max |\gamma'(s)|$ in terms of $|\gamma'(\tau_1)|$, for all $s \in [0, \sqrt{T - \epsilon}]$ for which γ is defined. Since $(M, g(0))$ is complete, this, together with a standard extension result for second-order ODEs, implies that γ extends uniquely to the entire interval $[0, \sqrt{T - \epsilon}]$. Changing the variable from s to $\tau = s^2$ shows that the $\mathcal{L}$-geodesic extends uniquely to the entire interval $[0, T - \epsilon]$. Since this is true for every $\epsilon > 0$, this completes the proof. □

Let $p \in M$ and set $x = (p, T) \in M \times [0, T]$. Recall that from Definition 6.25 for every $\tau > 0$, the injectivity set $\widetilde{\mathcal{U}}_x(\tau) \subset T_p M$ consists of all $Z \in T_p M$ for which (i) the $\mathcal{L}$-geodesic $\gamma_Z|_{[0,\tau]}$ is the unique minimizing $\mathcal{L}$-geodesic from x to its endpoint, (ii) the differential of $\mathcal{L}\exp_x^\tau$ is an isomorphism at Z, and (iii) for all Z' sufficiently close to Z the $\mathcal{L}$-geodesic $\gamma_{Z'}|_{[0,\tau]}$ is the unique minimizing $\mathcal{L}$-geodesic to its endpoint. The image of $\widetilde{\mathcal{U}}_x(\tau)$ is denoted $\mathcal{U}_x(\tau) \subset M$.

The existence of minimizing $\mathcal{L}$-geodesics from x to every point of $M \times (0, T)$ means that the functions L_x and l_x are defined on all of $M \times (0, T)$. This leads to:

DEFINITION 7.4. Suppose that $(M, g(t))$, $0 \leq t \leq T < \infty$, is a Ricci flow, complete of bounded curvature. We define the function $L_x \colon M \times [0, T) \to \mathbb{R}$ by assigning to each (q, t) the length of any $\mathcal{L}$-minimizing $\mathcal{L}$-geodesic from x to $y = (q, t) \in M \times [0, T)$. Clearly, the restriction of this function to $\mathcal{U}_x$ agrees with the smooth function L_x given in Definition 6.26. We define $L_x^\tau \colon M \to \mathbb{R}$ to be the restriction of L_x to $M \times \{T - \tau\}$. Of course, the restriction of L_x^τ to $\mathcal{U}_x(\tau)$ agrees with the smooth function L_x^τ defined in the last chapter. We define $l_x \colon M \times [0, T) \to \mathbb{R}$ by $l_x(y) = L_x(y)/2\sqrt{\tau}$, where, as always $\tau = T - t$, and we define $l_x^\tau(q) = l_x(q, T - \tau)$.

1.2. Results about l_x and $\mathcal{U}_x(\tau)$. Now we come to our main result about the nature of $\mathcal{U}_x(\tau)$ and the function l_x in the context of Ricci flows which are complete and of bounded curvature.

PROPOSITION 7.5. *Let $(M, g(t))$, $0 \leq t \leq T < \infty$, be a Ricci flow that is complete and of bounded curvature. Let $p \in M$, let $x = (p, T) \in M \times [0, T]$, and let $\tau \in (0, T)$.*

(1) *L_x and l_x are locally Lipschitz functions on $M \times (0, T)$.*
(2) *$\mathcal{L}\exp_x^\tau$ is a diffeomorphism from $\widetilde{\mathcal{U}}_x(\tau)$ onto an open subset $\mathcal{U}_x(\tau)$ of M.*

(3) *The complement of $\mathcal{U}_x(\tau)$ in M is a closed subset of M of zero Lebesgue measure.*
(4) *For every $\tau < \tau' < T$ we have*
$$\mathcal{U}_x(\tau') \subset \mathcal{U}_x(\tau).$$

PROOF. By Shi's Theorem (Theorem 3.28) the curvature bound on $M \times [0,T]$ implies that for each $\epsilon > 0$ there is a bound for $|\nabla R|$ on $M \times (\epsilon, T]$. Thus, Proposition 6.59 shows that L_x is a locally Lipschitz function on $M \times (\epsilon, T)$. Since this is true for every $\epsilon > 0$, L_x is a locally Lipschitz function on $M \times (0, T)$. Of course, the same is true for l_x. The second statement is contained in Proposition 6.28, and the last one is contained in Proposition 6.30. It remains to prove the third statement, namely that the complement of $\mathcal{U}_x(\tau)$ is closed nowhere dense. This follows immediately from Corollary 6.67 since $|\text{Ric}|$ and $|\nabla R|$ are bounded on $F = M \times [T-\tau, T]$. □

COROLLARY 7.6. *The function l_x is a continuous function on $M \times (0,T)$ and is smooth on the complement of a closed subset $\mathcal{C}$ that has the property that its intersection with each $M \times \{t\}$ is of zero Lebesgue measure in $M \times \{t\}$. For each $\tau \in (0,T)$ the gradient ∇l_x^τ is then a smooth vector field on the complement of $\mathcal{C} \cap M_{T_\tau}$, and it is a locally essentially bounded vector field in the following sense. For each $q \in M$ there is a neighborhood $V \subset M$ of q such that the restriction of $|\nabla l_x^\tau|$ to $V \setminus (V \cap \mathcal{C})$ is a bounded smooth function. Similarly, $\partial l_x / \partial t$ is an essentially bounded smooth vector field on $M \times (0, T)$.*

2. A bound for $\min l_x^\tau$

We continue to assume that we have a Ricci flow $(M, g(t))$, $0 \leq t \leq T < \infty$, complete and of bounded curvature and a point $x = (p, T) \in M \times [0, T]$. Our purpose here is to extend the first differential inequality given in Corollary 6.51 to a differential inequality in the weak or distributional sense for l_x valid on all of $M \times (0,T)$. We then use this to establish that $\min_{q \in M} l_x^\tau(q) \leq n/2$ for all $0 < \tau < T$.

In establishing inequalities in the non-smooth case the notion of a support function or a barrier function is often convenient.

DEFINITION 7.7. Let P be a smooth manifold and let $f \colon P \to \mathbb{R}$ be a continuous function. An *upper barrier* for f at $p \in P$ is a smooth function φ defined on a neighborhood of p in P, say U, satisfying $\varphi(p) = f(p)$ and $\varphi(u) \geq f(u)$ for all $u \in U$, see FIG. 1.

PROPOSITION 7.8. *Let $(M, g(t))$, $0 \leq t \leq T < \infty$, be an n-dimensional Ricci flow, complete of bounded curvature. Let $x = (p, T) \in M \times [0, T]$, and for any $(q,t) \in M \times [0,T]$, set $\tau = T - t$. Then for any (q,t), with $0 < t < T$, we have*
$$\frac{\partial l_x}{\partial \tau}(q,\tau) + \triangle l_x(q,\tau) \leq \frac{(n/2) - l_x(q,\tau)}{\tau}$$

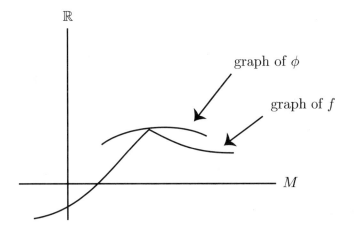

FIGURE 1. Upper barrier.

in the barrier sense. This means that for each $\epsilon > 0$ there is a neighborhood U of (q,t) in $M \times [0,T]$ and an upper barrier φ for l_x at this point defined on U satisfying

$$\frac{\partial \varphi}{\partial \tau}(q,\tau) + \triangle \varphi(q,\tau) \le \frac{(n/2) - l_x(q,\tau)}{\tau} + \epsilon.$$

REMARK 7.9. The operator $\triangle$ in the above statement is the horizontal Laplacian, i.e., the Laplacian of the restriction of the indicated function to the slice $M \times \{t = T - \tau\}$ as defined using the metric $g(T - \tau)$ on this slice.

PROOF. If $(q, T - \tau) \in \mathcal{U}_x$, then l_x is smooth near $(q, T - \tau)$, and the result is immediate from the first inequality in Corollary 6.51.

Now consider a general point $(q, t = T - \tau)$ with $0 < t < T$. According to Lemma 7.2 there is a minimizing $\mathcal{L}$-geodesic γ from $x = (p, T)$ to $(q, t = T - \tau)$. Let γ be any minimizing $\mathcal{L}$-geodesic between these points. Fix $0 < \tau_1 < \tau$ let $q_1 = \gamma(\tau_1)$ and set $t_1 = T - \tau_1$. Even though q_1 is contained in the t_1 time-slice, we keep $\tau = T - t$ so that paths beginning at q_1 are parameterized by intervals in the τ-line of the form $[\tau_1, \tau']$ for some $\tau' < T$. Consider $\mathcal{L}\exp_{(q_1,t_1)} \colon T_{q_1}M \times (\tau_1, T) \to M \times (0, t_1)$. According to Proposition 6.31 there is a neighborhood $\widetilde{V}$ of $\{\sqrt{\tau_1}X_\gamma(\tau_1)\} \times (\tau_1, \tau]$ which is mapped diffeomorphically by $\mathcal{L}\exp_{(q_1,t_1)} \colon T_{q_1}M \times (\tau_1, \tau) \to M \times (0, t_1)$ onto a neighborhood V of $\gamma((\tau_1, \tau))$. (Of course, the neighborhood V depends on τ_1.) Let $L_{(q_1,t_1)}$ be the length function on V obtained by taking the $\mathcal{L}$-lengths of geodesics parameterized by points of $\widetilde{V}$. Let $\varphi_{\tau_1} \colon V \to \mathbb{R}$ be defined by

$$\varphi_{\tau_1}(q', \tau') = \frac{1}{2\sqrt{\tau'}}\left(\mathcal{L}(\gamma|_{[0,\tau_1]}) + L_{(q_1,t_1)}(q', T - \tau')\right).$$

Clearly, φ_{τ_1} is an upper barrier for l_x at (q,τ). According to Lemma 6.49 we have

$$\frac{\partial \varphi_{\tau_1}}{d\tau}(q,\tau) + \triangle \varphi_{\tau_1}(q,\tau) \leq \frac{n}{2\sqrt{\tau}(\sqrt{\tau} - \sqrt{\tau_1})} - \frac{\varphi_{\tau_1}(q,\tau)}{\tau} + \frac{1}{2\tau^{3/2}} \mathcal{L}(\gamma|_{[0,\tau_1]})$$
$$+ \frac{K_{\tau_1}^\tau(\gamma)}{2\tau^{3/2}} - \frac{\mathcal{K}_{\tau_1}^\tau(\gamma)}{2\sqrt{\tau}(\sqrt{\tau} - \sqrt{\tau_1})^2}$$
$$- \frac{1}{2}\left(\frac{\tau_1}{\tau}\right)^{3/2} \left(R(q_1,t_1) + |X(\tau_1)|^2\right).$$

By Lemma 6.47, it follows easily that

$$\lim_{\tau_1 \to 0^+} \frac{\partial \varphi_{\tau_1}}{\partial \tau}(q,t) + \triangle \varphi_{\tau_1}(q,t) \leq \frac{(n/2) - l_x(q,t)}{\tau}.$$

This establishes the result. □

THEOREM 7.10. *Suppose that $(M, g(t))$, $0 \leq t \leq T < \infty$, is an n-dimensional Ricci flow, complete of bounded curvature. Then for any $x = (p, T) \in M \times [0, T]$ and for every $0 < \tau < T$ there is a point $q_\tau \in M$ such that $l_x(q_\tau, \tau) \leq \frac{n}{2}$.*

PROOF. We set $l_{\min}(\tau) = \inf_{q \in M} l_x(q, \tau)$. (We are not excluding the possibility that this infimum is $-\infty$.) To prove this corollary we first need to establish the following claim.

CLAIM 7.11. *For every $\tau \in (0, T)$ the function $l_x(\cdot, \tau)$ achieves its minimum. Furthermore, for every compact interval $I \subset (0, T)$ the subset of $(q, \tau) \in M \times I$ for which $l_x(q, \tau) = l_{\min}(\tau)$ is a compact set.*

First, let us assume this claim and use it to prove the theorem. We set $l_{\min}(\tau) = \min_{q \in M} l_x(q, \tau)$. (This minimum exists by the first statement in the claim.) From the compactness result in the claim, it follows (see for example Proposition 2.23) that $l_{\min}(\tau)$ is a continuous function of τ.

Suppose that $l_x(\cdot, \tau)$ achieves its minimum at q. Then by the previous result, for any $\epsilon > 0$ there is an upper barrier φ for l_x at (q, τ) defined on an open subset U of $(q, \tau) \in M \times (0, T)$ and satisfying

$$\frac{d\varphi}{d\tau}(q,\tau) + \triangle \varphi(q,\tau) \leq \frac{(n/2) - l_x(q,t)}{\tau} + \epsilon.$$

Since $l_x(q, \tau) = l_{\min}(\tau)$, it follows that $\varphi(q', \tau) \geq \varphi(q, \tau)$ for all $(q', \tau) \in U \cap M_{T-\tau}$. This means that $\triangle \varphi(q, \tau) \geq 0$, and we conclude that

$$\frac{d\varphi}{d\tau}(q,\tau) \leq \frac{(n/2) - l_{\min}(\tau)}{\tau} + \epsilon.$$

Since φ is an upper barrier for l_x at (q, τ) it follows immediately that

$$\operatorname{limsup}_{\tau' \to \tau^+} \frac{l_x(q, \tau') - l_x(q, \tau)}{\tau' - \tau} \leq \frac{(n/2) - l_x(q, \tau)}{\tau} + \epsilon.$$

Since this is true for every $\epsilon > 0$, we see that

$$\limsup_{\tau' \to \tau^+} \frac{l_x(q, \tau') - l_x(q, \tau)}{\tau' - \tau} \leq \frac{(n/2) - l_x(q, \tau)}{\tau}.$$

Since $l_{\min}(\tau) = l_x(q, \tau)$, the same inequality holds for the forward difference quotient of $l_{\min}$ at τ. That is to say, we have

$$\limsup_{\tau' \to \tau^+} \frac{l_{\min}(\tau') - l_{\min}(\tau)}{\tau' - \tau} \leq \frac{(n/2) - l_{\min}(\tau)}{\tau}.$$

The preceding equation implies that if $l_{\min}(\tau) \leq n/2$ then $l_{\min}(\tau') \leq n/2$ for every $\tau' \geq \tau$. On the other hand $\lim_{\tau \to 0} l_{\min}(\tau) = 0$. Then reason for this is that the path $\tau' \mapsto (P, T - \tau')$ for $\tau' \in [0, \tau]$ has $\mathcal{L}$-length $O(\tau^{3/2})$ as $\tau \to 0$. It follows that $l_{\min}(\tau) < n/2$ when τ is small.

To complete the proof of Theorem 7.10, it remains to prove Claim 7.11.

PROOF. In the case when M is compact, the claim is obvious. We consider the case when M is complete and the flow has bounded curvature. Since the curvature on $M \times [0, T]$ is bounded, according to Inequality (7.1) there is a constant C such that for all $t, t' \in [0, T]$ we have

$$C^{-1} g(t') \leq g(t) \leq C g(t').$$

For any compact interval $I \subset (0, T)$, there is $l_0 < \infty$ such that $l_{\min}(\tau) \leq l_0$ for all $\tau \in I$. According to Corollary 6.61, for every $\overline{\tau} \in I$ and all $\mathcal{L}$-geodesics from x to points $(q, T - \overline{\tau})$ of lengths at most $2|l_0|$ there is an upper bound, say C_2, to $|\sqrt{\tau} X_\gamma(\tau)|$. Thus, $|X_\gamma(\tau)| \leq \frac{C_2}{\sqrt{\tau}}$, and hence

$$|X_\gamma(\tau)|_{g(T)} \leq C_2 \sqrt{C}/\sqrt{\tau}$$

for these geodesics. Thus,

$$\int_0^{\overline{\tau}} |X_\gamma(\tau)|_{g(T)} d\tau \leq 2\sqrt{\overline{\tau}} C_2 \sqrt{C}.$$

This shows that there is $A < \infty$ such that for each $\overline{\tau} \in I$ and for any $\mathcal{L}$-geodesic γ defined on $[0, \overline{\tau}]$ of length at most $2|l_0|$ the following holds. Letting $q \in M$ be such that $\gamma(\overline{\tau}) = (q, T - \tau)$, the point q lies in $B_T(p, A)$. This implies that the endpoints of all such $\mathcal{L}$-geodesics lie in a fixed compact subset of M independent of $\overline{\tau} \in I$ and the geodesic. Since the set of (q, τ) where $l_x(q, \tau) = l_{\min}(\tau)$ is clearly a closed set, it follows that the subset of $M \times I$ of all $(q, \tau) \in M \times I$ for which $l_x(q, \tau) = l_{\min}(\tau)$ is compact. The last thing to show is that for every $\tau \in I$ the function $l_x(\cdot, \tau)$ achieves its minimum. Fix $\tau \in I$ and let q_n be a minimizing sequence for $l_x(\cdot, \tau)$. We have already established that the q_n are contained in a compact subset of M, and hence we can assume that they converge to a limit $q \in M$. Clearly, by the continuity of l_x we have $l_x(q, \tau) = \lim_{n \to \infty} l_x(q_n, \tau) = \inf_{q' \in M} l_x(q', \tau)$, so that $l_x(\cdot, \tau)$ achieves its minimum at q. $\square$

Having established the claim, we have now completed the proof of Theorem 7.10. □

Actually, the proof given here also shows the following, which will be useful later.

COROLLARY 7.12. *Suppose that $(\mathcal{M}, G)$ is a generalized n-dimensional Ricci flow and that $x \in \mathcal{M}$ is given and set $t_0 = \mathbf{t}(x)$. We suppose that there is an open subset $U \subset \mathbf{t}^{-1}(-\infty, t_0)$ with the following properties:*

(1) *For every $y \in U$ there is a minimizing $\mathcal{L}$-geodesic from x to y.*
(2) *There are $r > 0$ and $\Delta t > 0$ such that the backward parabolic neighborhood $P(x, t_0, r, -\Delta t)$ of x exists in $\mathcal{M}$ and has the property that $P \cap \mathbf{t}^{-1}(-\infty, t_0)$ is contained in U.*
(3) *For each compact interval (including the case of degenerate intervals consisting of a single point) $I \subset (-\infty, t_0)$ the subset of points $y \in \mathbf{t}^{-1}(I) \cap U$ for which $\mathcal{L}(y) = \inf_{z \in \mathbf{t}^{-1}(\mathbf{t}(y)) \cap U} \mathcal{L}(z)$ is compact and non-empty.*

Then for every $t < t_0$ the minimum of the restriction of l_x to the time-slice $\mathbf{t}^{-1}(t) \cap U$ is at most $n/2$.

2.1. Extension of the other inequalities in Corollary 6.51.

The material in this subsection is adapted from [**72**]. It captures (in a weaker way) the fact that, in the case of geodesics on a Riemannian manifold, the interior of the cut locus in $T_x M$ is star-shaped from the origin.

THEOREM 7.13. *Let $(M, g(t))$, $0 \leq t \leq T < \infty$, be a Ricci flow, complete and of bounded curvature, and let $x = (p, T) \in M \times [0, T]$. The last two inequalities in Corollary 6.51, namely*

$$\frac{\partial l_x}{\partial \tau} + |\nabla l_x^\tau|^2 - R + \frac{n}{2\tau} - \triangle l_x^\tau \geq 0,$$

$$-|\nabla l_x^\tau|^2 + R + \frac{l_x^\tau - n}{\tau} + 2\triangle l_x^\tau \leq 0$$

hold in the weak or distributional sense on all of $M \times \{\tau\}$ for all $\tau > 0$. This means that for any $\tau > 0$ and for any non-negative, compactly supported, smooth function $\phi(q)$ on M we have the following two inequalities:

$$\int_{M \times \{\tau\}} \left[\phi \cdot \left(\frac{\partial l_x}{\partial \tau} + |\nabla l_x^\tau|^2 - R + \frac{n}{2\tau} \right) - l_x^\tau \triangle \phi \right] \, d\mathrm{vol}(g(t)) \geq 0,$$

$$\int_{M \times \{\tau\}} \left[\phi \cdot \left(-|\nabla l_x^\tau|^2 + R + \frac{l_x^\tau - n}{\tau} \right) + 2 l_x^\tau \triangle \phi \right] \, d\mathrm{vol}(g(t)) \leq 0.$$

Furthermore, equality holds in either of these weak inequalities for all functions ϕ as above and all τ if and only if it holds in both. In that case l_x is a smooth function on space-time and the equalities hold in the usual smooth sense.

2. A BOUND FOR $\min l_x^\tau$

REMARK 7.14. The terms in these inequalities are interpreted in the following way: First of all, ∇l_x^τ and $\triangle l_x^\tau$ are computed using only the spatial derivatives (i.e., they are horizontal differential operators). Secondly, since l_x is a locally Lipschitz function defined on all of $M \times (0, T)$, we have seen that $\partial l_x / \partial t$ and $|\nabla l_x^\tau|^2$ are continuous functions on the open subset $\mathcal{U}_x(\tau)$ of full measure in $M \times \{\tau\}$ and furthermore, that they are locally bounded on all of $M \times \{\tau\}$ in the sense that for any $q \in M$ there is a neighborhood V of q such that the restriction of $|\nabla l_x^\tau|^2$ to $V \cap \mathcal{U}_x(\tau)$ is bounded. This means that $\partial l_x / \partial t$ and $|\nabla l_x^\tau|^2$ are elements of $L_{\text{loc}}^\infty(M)$ and hence can be integrated against any smooth function with compact support. In particular, they are distributions.

Since ∇l_x^τ is a smooth, locally bounded vector field on an open subset of full measure, for any compactly supported test function ϕ, integration by parts yields

$$\int \triangle \phi \cdot l_x^\tau \, d\text{vol}(g(t)) = -\int \langle \nabla \phi, \nabla l_x^\tau \rangle \, d\text{vol}(g(t)).$$

Thus, formulas in Theorem 7.13 can also be taken to mean:

$$\int_{M \times \{\tau\}} \left[\phi \cdot \left(\frac{\partial l_x}{\partial \tau} + |\nabla l_x^\tau|^2 - R + \frac{n}{2\tau} \right) + \langle \nabla l_x^\tau, \nabla \phi \rangle \right] d\text{vol}(g(t)) \geq 0,$$

$$\int_{M \times \{\tau\}} \left[\phi \cdot \left(-|\nabla l_x^\tau|^2 + R + \frac{l_x^\tau - n}{\tau} \right) - 2\langle \nabla l_x^\tau, \nabla \phi \rangle \right] d\text{vol}(g(t)) \leq 0.$$

The rest of this subsection is devoted to the proof of these inequalities. We fix $(M, g(t))$, $0 \leq t \leq T < \infty$, as in the statement of the theorem. We fix x and denote by L and l the functions L_x and l_x. We also fix τ, and we denote by L^τ and l^τ the restrictions of L and l to the slice $M \times \{T - \tau\}$. We begin with a lemma.

LEMMA 7.15. *There is a continuous function $C \colon M \times (0, T) \to \mathbb{R}$ such that for each point $(q, t) \in M \times (0, T)$, setting $\tau = T - t$, the following holds. There is an upper barrier $\varphi_{(q,t)}$ for L_x^τ at the point q defined on a neighborhood $U_{(q,t)}$ of q in M satisfying $|\nabla \varphi_{(q,t)}(q)| \leq C(q, t)$ and*

$$\text{Hess}(\varphi)(v, v) \leq C(q, t)|v|^2$$

for all tangent vectors $v \in T_q M$.

PROOF. By Proposition 7.5, L is a locally Lipschitz function on $M \times (0, T)$, and in particular is continuous. The bound $C(q, t)$ will depend only on the bounds on curvature and its first two derivatives and on the function $L(q, t)$. Fix (q, t) and let γ be a minimizing $\mathcal{L}$-geodesic from x to (q, t). (The existence of such a minimizing geodesic is established in Lemma 7.2.) Fix $\tau_1 > 0$, with $\tau_1 < (T-t)/2$, let $t_1 = T - \tau_1$, and let $q_1 = \gamma(\tau_1)$. Consider $\varphi_{(q,t)} = \mathcal{L}(\gamma_{[0,\tau_1]}) + L_{(q_1,t_1)}^\tau$. This is an upper barrier for L_x^τ at q

defined in some neighborhood $V \subset M$ of q. Clearly, $\nabla \varphi_{(q,t)} = \nabla L^\tau_{(q_1,t_1)}$ and $\mathrm{Hess}(\varphi_{(q,t)}) = \mathrm{Hess}(L^\tau_{(q_1,t_1)})$.

According to Corollary 6.29 we have $\nabla L^\tau_{(q_1,t_1)}(q) = 2\sqrt{\tau} X_\gamma(\tau)$. On the other hand, by Corollary 6.61 there is a bound on $\sqrt{\tau}|X_\gamma(\tau)|$ depending only on the bounds on curvature and its first derivatives, on τ and τ_1 and on $l_x(q,\tau)$. Of course, by Shi's theorem (Theorem 3.27) for every $\epsilon > 0$ the norms of the first derivatives of curvature on $M \times [\epsilon, T]$ are bounded in terms of ϵ and the bounds on curvature. This proves that $|\nabla \varphi_{(q,t)}(q)|$ is bounded by a continuous function $C(q,t)$ defined on all of $M \times (0,T)$.

Now consider Inequality (6.11) for γ at $\overline{\tau} = \tau$. It is clear that the first two terms on the right-hand side are bounded by $C|Y(\tau)|^2$, where C depends on the curvature bound and on $T - t$. We consider the last term, $\int_{\tau_1}^\tau \sqrt{\tau'} H(X,Y) d\tau'$. We claim that this integral is also bounded by $C'|Y(\tau)|^2$ where C' depends on the bounds on curvature and its first and second derivatives along γ_1 and on $T - t$. We consider $\tau' \in [\tau_1, \tau]$. Of course, $\sqrt{\tau'}|X(\tau')|$ is bounded on this interval. Also,

$$|Y(\tau')| = \left(\frac{\sqrt{\tau'} - \sqrt{\tau_1}}{\sqrt{\tau} - \sqrt{\tau_1}}\right)|Y(\tau)| \le \frac{\sqrt{\tau'}}{\sqrt{\tau}}|Y(\tau)|.$$

Hence $|Y(\tau')|/\sqrt{\tau'}$ and $|Y(\tau')||X(\tau')|$ are bounded in terms of $T - t$, $|Y(\tau)|$, and the bound on $\sqrt{\tau'}|X(\tau')|$ along the $\mathcal{L}$-geodesic. It then follows from Equation (6.12) that $H(X,Y)$ is bounded along the $\mathcal{L}$-geodesic by $C|Y(\overline{\tau})|^2$ where the constant C depends on $T - t$ and the bounds on curvature and its first two derivatives. $\square$

Of course, if $(q,t) \in \mathcal{U}_x(\tau)$, then this argument shows that the Hessian of L^τ_x is bounded near (q,t).

At this point in the proof of Theorem 7.13 we wish to employ arguments using convexity. To carry these out we find it convenient to work with a Euclidean metric and usual convexity rather than the given metric $g(t)$ and convexity measured using $g(t)$-geodesics. In order to switch to a Euclidean metric we must find one that well approximates $g(t)$. The following is straightforward to prove.

CLAIM 7.16. *For each point $(q,t) \in M \times (0,T)$ there is an open metric ball $B_{(q,t)}$ centered at q in $(M, g(t))$ which is the diffeomorphic image of a ball $\widetilde{B} \subset T_q M$ under the exponential map for $g(t)$ centered at q such that the following hold:*

(1) *$B_{(q,t)} \subset U_{(q,t)}$ so that the upper barrier $\varphi_{(q,t)}$ from Lemma 7.15 is defined on all of $B_{(q,t)}$.*
(2) *The constants $C(z,t)$ of Lemma 7.15 satisfy $C(z,t) \le 2C(q,t)$ for all $z \in B_{(q,t)}$.*

(3) The push-forward, h, under the exponential mapping of the Euclidean metric on T_qM satisfies
$$h/2 \le g \le 2h.$$

(4) The Christoffel symbols Γ_{ij}^k for the metric $g(t)$ written using normal coordinates are bounded in absolute value by $1/(8n^3C(q,t))$ where n is the dimension of M.

Instead of working in the given metric $g(t)$ on $B_{(q,t)}$ we shall use the Euclidean metric h as in the above claim. For any function f on $B_{(q,t)}$ we denote by $\text{Hess}(f)$ the Hessian of f with respect to the metric $g(t)$ and by $\text{Hess}^h(f)$ the Hessian of f with respect to the metric h. By Formula (1.2), for any $z \in B_{(q,t)}$ and any $v \in T_zM$, we have

$$\text{Hess}(\varphi_{(z,t)})(v,v) = \text{Hess}^h(\varphi_{(z,t)})(v,v) - \sum_{i,j,k} v^i v^j \Gamma_{ij}^k \frac{\partial \varphi_{(z,t)}}{\partial x^k}.$$

Thus, it follows from the above assumptions on the Γ_{ij}^k and the bound on $|\nabla \varphi_{(z,t)}|$ that for all $z \in B_{(q,t)}$ we have

$$(7.2) \qquad \left| \text{Hess}(\varphi_{(z,t)})(v,v) - \text{Hess}^h(\varphi_{(z,t)})(v,v) \right| \le \frac{1}{4}|v|_h^2,$$

and hence for every $z \in B_{(q,t)}$ we have

$$\text{Hess}^h(\varphi_{(z,t)})(v,v) \le 2C(q,t)|v|_g^2 + \frac{|v|_h^2}{4} \le \left(4C(q,t) + \frac{1}{4}\right)|v|_h^2.$$

This means:

CLAIM 7.17. *For each $(q,t) \in M \times (0,T)$ there is a smooth function*
$$\psi_{(q,t)} \colon B_{(q,t)} \to \mathbb{R}$$
with the property that at each $z \in B_{(q,t)}$ there is an upper barrier $b_{(z,t)}$ for $L^\tau + \psi_{(q,t)}$ at z with
$$\text{Hess}^h(b_{(z,t)})(v,v) \le -3|v|_h^2/2$$
for all $v \in T_zM$.

PROOF. Set
$$\psi_{(q,t)} = -(2C(q,t) + 1)d_h^2(q,\cdot).$$
Then for any $z \in B_{(q,t)}$ the function $b_{(z,t)} = \varphi_{(z,t)} + \psi_{(q,t)}$ is an upper barrier for $L^\tau + \psi_{(q,t)}$ at z. Clearly, for all $v \in T_zM$ we have
$$\text{Hess}^h(b_{(z,t)})(v,v) = \text{Hess}^h(\varphi_{(z,t)})(v,v) + \text{Hess}^h(\psi_{(q,t)})(v,v) \le -3|v|_h^2/2.$$
□

This implies that if $\alpha\colon [a,b] \to B_{(q,t)}$ is any Euclidean straight-line segment in $B_{(q,t)}$ parameterized by Euclidean arc length and if $z = \alpha(s)$ for some $s \in (a,b)$, then
$$(b_{(z,t)} \circ \alpha)''(s) \leq -3/2.$$

CLAIM 7.18. *Suppose that $\beta\colon [-a,a] \to \mathbb{R}$ is a continuous function and that at each $s \in (-a,a)$ there is an upper barrier $\hat{b}_s$ for β at s with $\hat{b}_s'' \leq -3/2$. Then*
$$\frac{\beta(a)+\beta(-a)}{2} \leq \beta(0) - \frac{3}{4}a^2.$$

PROOF. Fix $c < 3/4$ and define a continuous function
$$A(s) = \frac{(\beta(-s)+\beta(s))}{2} + cs^2 - \beta(0)$$
for $s \in [0,a]$. Clearly, $A(0) = 0$. Using the upper barrier at 0 we see that for $s > 0$ sufficiently small, $A(s) < 0$. For any $s \in (0,a)$ there is an upper barrier $c_s = (\hat{b}_s + \hat{b}_{-s})/2 + cs^2 - \beta(0)$ for $A(s)$ at s, and $c_s''(t) \leq 2c - 3/2 < 0$. By the maximum principle this implies that A has no local minimum in $(0,a)$, and consequently that it is a non-increasing function of s on this interval. That is to say, $A(s) < 0$ for all $s \in (0,a)$ and hence $A(a) \leq 0$, i.e., $(\beta(a)+\beta(-a))/2 + ca^2 \leq \beta(0)$. Since this is true for every $c < 3/4$, the result follows. □

Now applying this to Euclidean intervals in $B_{(q,t)}$ we conclude:

COROLLARY 7.19. *For any $(q,t) \in M \times (0,T)$, the function*
$$\beta_{(q,t)} = L^\tau + \psi_{(q,t)}\colon B_{(q,t)} \to \mathbb{R}$$
is uniformly strictly convex with respect to h. In fact, let $\alpha\colon [a,b] \to B_{(q,t)}$ be a Euclidean geodesic arc. Let y,z be the endpoints of α, let w be its midpoint, and let $|\alpha|$ denote the length of this arc (all defined using the Euclidean metric). We have
$$\beta_{(q,t)}(w) \geq \frac{(\beta_{(q,t)}(y) + \beta_{(q,t)}(z))}{2} + \frac{3}{16}|\alpha|^2.$$

What follows is a simple interpolation result (see [23]). For each $q \in M$ we let $B'_{(q,t)} \subset B_{(q,t)}$ be a smaller ball centered at q, so that $B'_{(q,t)}$ has compact closure in $B_{(q,t)}$.

CLAIM 7.20. *Fix $(q,t) \in M \times (0,T)$, and let $\beta_{(q,t)}\colon B_{(q,t)} \to \mathbb{R}$ be as above. Let $S \subset M$ be the singular locus of L^τ, i.e., $S = M \setminus \mathcal{U}_x(\tau)$. Set $S_{(q,t)} = B_{(q,t)} \cap S$. Of course, $\beta_{(q,t)}$ is smooth on $B_{(q,t)} \setminus S_{(q,t)}$. Then there is a sequence of smooth functions $\{f_k\colon B'_{(q,t)} \to \mathbb{R}\}_{k=1}^\infty$ with the following properties:*

(1) *As $k \to \infty$ the functions f_k converge uniformly to $\beta_{(q,t)}$ on $B'_{(q,t)}$.*

(2) For any $\epsilon > 0$ sufficiently small, let $\nu_\epsilon(S_{(q,t)})$ be the ϵ-neighborhood (with respect to the Euclidean metric) in $B_{(q,t)}$ of $S_{(q,t)} \cap B_{(q,t)}$. Then, as $k \to \infty$ the restrictions of f_k to $B'_{(q,t)} \setminus \left(B'_{(q,t)} \cap \nu_\epsilon(S_{(q,t)})\right)$ converge uniformly in the C^∞-topology to the restriction of $\beta_{(q,t)}$ to this subset.

(3) For each k, and for any $z \in B'_{(q,t)}$ and any $v \in T_z M$ we have
$$\mathrm{Hess}(f_k)(v,v) \leq -|v|^2_{g(t)}/2.$$
That is to say, f_k is strictly convex with respect to the metric $g(t)$.

PROOF. Fix $\epsilon > 0$ sufficiently small so that for any $z \in B'_{(q,t)}$ the Euclidean ϵ-ball centered at z is contained in $B_{(q,t)}$. Let B_0 be the ball of radius ϵ centered at the origin in $\mathbb{R}^n$ and let $\xi \colon B_0 \to \mathbb{R}$ be a non-negative C^∞-function with compact support and with $\int_{B_0} \xi d\mathrm{vol}_h = 1$. We define
$$\beta^\epsilon_{(q,t)}(z) = \int_{B_0} \xi(y) \beta_{(q,t)}(z+y) dy,$$
for all $z \in B'_{(q,t)}$. It is clear that for each $\epsilon > 0$ sufficiently small, the function $\beta^\epsilon_{(q,t)} \colon B'_{(q,t)} \to \mathbb{R}$ is C^∞ and that as $\epsilon \to 0$ the $\beta^\epsilon_{(q,t)}$ converge uniformly on $B'_{(q,t)}$ to $\beta_{(q,t)}$. It is also clear that for every $\epsilon > 0$ sufficiently small, the conclusion of Corollary 7.19 holds for $\beta^\epsilon_{(q,t)}$ and for each Euclidean straight-line segment α in $B'_{(q,t)}$. This implies that $\mathrm{Hess}^h(\beta^\epsilon_{(q,t)})(v,v) \leq -3|v|^2_h/2$, and hence that by Inequality (7.2) that
$$\mathrm{Hess}(\beta^\epsilon_{(q,t)})(v,v) \leq -|v|^2_h = -|v|^2_{g(t)}/2.$$
This means that $\beta^\epsilon_{(q,t)}$ is convex with respect to $g(t)$. Now take a sequence $\epsilon_k \to 0$ and let $f_k = \beta^{\epsilon_k}_{(q,t)}$. Lastly, it is a standard fact that f_k converge uniformly in the C^∞-topology to $\beta_{(q,t)}$ on any subset of $B'_{(q,t)}$ whose closure is disjoint from $S_{(q,t)}$. □

DEFINITION 7.21. For any continuous function ψ that is defined on $B'_{(q,t)} \setminus (S_{(q,t)} \cap B'_{(q,t)})$ we define
$$\int_{(B'_{(q,t)})^*} \psi d\mathrm{vol}(g(t)) = \lim_{\epsilon \to 0} \int_{B'_{(q,t)} \setminus \nu_\epsilon(S_{(q,t)}) \cap B'_{(q,t)}} \psi d\mathrm{vol}(g(t)).$$

We now have:

CLAIM 7.22. Let $\phi \colon B'_{(q,t)} \to \mathbb{R}$ be a non-negative, smooth function with compact support. Then
$$\int_{B'_{(q,t)}} \beta_{(q,t)} \triangle \phi d\mathrm{vol}(g(t)) \leq \int_{(B'_{(q,t)})^*} \phi \triangle \beta_{(q,t)} d\mathrm{vol}(g(t)).$$

REMARK 7.23. Here $\triangle$ denotes the Laplacian with respect to the metric $g(t)$.

PROOF. Since $f_k \to \beta_{(q,t)}$ uniformly on $B'_{(q,t)}$ we have

$$\int_{B'_{(q,t)}} \beta_{(q,t)} \triangle \phi d\mathrm{vol}(g(t)) = \lim_{k\to\infty} \int_{B'_{(q,t)}} f_k \triangle \phi d\mathrm{vol}(g(t)).$$

Since f_k is strictly convex with respect to the metric $g(t)$, $\triangle f_k \leq 0$ on all of $B'_{(q,t)}$. Since $\phi \geq 0$, for every ϵ and k we have

$$\int_{\nu_\epsilon(S_{(q,t)}) \cap B'_{(q,t)}} \phi \triangle f_k d\mathrm{vol}(g(t)) \leq 0.$$

Hence, for every k and for every ϵ we have

$$\int_{B'_{(q,t)}} f_k \triangle \phi d\mathrm{vol}(g(t)) = \int_{B'_{(q,t)}} \phi \triangle f_k d\mathrm{vol}(g(t))$$
$$\leq \int_{B'_{(q,t)} \setminus \left(B'_{(q,t)} \cap \nu_\epsilon(S_{(q,t)})\right)} \phi \triangle f_k d\mathrm{vol}(g(t)).$$

Taking the limit as $k \to \infty$, using the fact that $f_k \to \beta_{(q,t)}$ uniformly on $B'_{(q,t)}$ and that restricted to $B'_{(q,t)} \setminus (B'_{(q,t)} \cap \nu_\epsilon(S_{(q,t)}))$ the f_k converge uniformly in the C^∞-topology to $\beta_{(q,t)}$, yields

$$\int_{B'_{(q,t)}} \beta_{(q,t)} \triangle \phi d\mathrm{vol}(g(t)) \leq \int_{B'_{(q,t)} \setminus \left(B'_{(q,t)} \cap \nu_\epsilon(S_{(q,t)})\right)} \phi \triangle \beta_q d\mathrm{vol}(g(t)).$$

Now taking the limit as $\epsilon \to 0$ establishes the claim. □

COROLLARY 7.24. *Let $\phi \colon B'_{(q,t)} \to \mathbb{R}$ be a non-negative, smooth function with compact support. Then*

$$\int_{B'_{(q,t)}} l^\tau \triangle \phi d\mathrm{vol}(g(t)) \leq \int_{(B'_{(q,t)})^*} \phi \triangle l^\tau d\mathrm{vol}(g(t)).$$

PROOF. Recall that $\beta_{(q,t)} = L^\tau + \psi_{(q,t)}$ and that $\psi_{(q,t)}$ is a C^∞-function. Hence,

$$\int_{B'_{(q,t)}} \psi_{(q,t)} \triangle \phi d\mathrm{vol}(g(t)) = \int_{(B'_{(q,t)})^*} \phi \triangle \psi_{(q,t)} d\mathrm{vol}(g(t)).$$

Subtracting this equality from the inequality in the previous claim and dividing by $2\sqrt{\tau}$ gives the result. □

Now we turn to the proof proper of Theorem 7.13.

PROOF. Let $\phi \colon M \to \mathbb{R}$ be a non-negative, smooth function of compact support. Cover M by open subsets of the form $B'_{(q,t)}$ as above. Using a partition of unity we can write $\phi = \sum_i \phi_i$ where each ϕ_i is a non-negative smooth function supported in some $B'_{(q_i,t)}$. Since the inequalities we are trying to establish are linear in ϕ, it suffices to prove the result for each ϕ_i.

2. A BOUND FOR $\min l_x^\tau$

This allows us to assume (and we shall assume) that ϕ is supported in $B'_{(q,t)}$ for some $q \in M$.

Since l_x^τ is a locally Lipschitz function, the restriction of $|\nabla l_x^\tau|^2$ to $B'_{(q,t)}$ is an L^∞_{loc}-function. Similarly, $\partial l_x/\partial \tau$ is an L^∞_{loc}-function. Hence

$$\int_{B'_{(q,t)}} \phi \cdot \left(\frac{\partial l_x}{\partial \tau} + |\nabla l_x^\tau|^2 - R + \frac{n}{2\tau} \right) d\text{vol}(g(t))$$
$$= \int_{(B'_{(q,t)})^*} \phi \left(\frac{\partial l_x}{\partial \tau} + |\nabla l_x^\tau|^2 - R + \frac{n}{2\tau} \right) d\text{vol}(g(t)).$$

On the other hand, by Corollary 7.24 we have

$$\int_{B'_{(q,t)}} l_x^\tau \triangle \phi \, d\text{vol}(g(t)) \leq \int_{(B'_{(q,t)})^*} \phi \triangle l_x^\tau d\text{vol}(g(t)).$$

Putting these together we see

$$\int_{B'_{(q,t)}} \phi \left(\frac{\partial l_x}{\partial \tau} + |\nabla l_x^\tau|^2 - R + \frac{n}{2\tau} \right) - l_x^\tau \triangle \phi \, d\text{vol}(g(t))$$
$$\geq \int_{(B'_{(q,t)})^*} \phi \left(\frac{\partial l_x}{\partial \tau} + |\nabla l_x^\tau|^2 - R + \frac{n}{2\tau} - \triangle l_x^\tau \right) d\text{vol}(g(t)).$$

It follows immediately from the second inequality in Corollary 6.51 that, since $\phi \geq 0$ and $\left(B'_{(q,t)}\right)^* \subset \mathcal{U}_x(\tau)$, we have

$$\int_{(B'_{(q,t)})^*} \phi \left(\frac{\partial l_x}{\partial \tau} + |\nabla l_x^\tau|^2 - R + \frac{n}{2\tau} - \triangle l_x^\tau \right) d\text{vol}(g(t)) \geq 0.$$

This proves the first inequality in the statement of the theorem.

The second inequality in the statement of the theorem is proved in the same way using the third inequality in Corollary 6.51.

Now let us consider the distributions

$$D_1 = \frac{\partial l_x}{\partial \tau} + |\nabla l_x^\tau|^2 - R + \frac{n}{2\tau} - \triangle l_x^\tau$$

and

$$D_2 = -|\nabla l_x^\tau|^2 + R + \frac{l_x^\tau - n}{\tau} + 2\triangle l_x^\tau$$

on $M \times \{\tau\}$. According to Corollary 6.51 the following equality holds on $\mathcal{U}_x(\tau)$:

$$2\frac{\partial l_x}{\partial \tau} + |\nabla l_x^\tau|^2 - R + \frac{l_x^\tau}{\tau} = 0.$$

By Proposition 7.5 the open set $\mathcal{U}_x(\tau)$ has full measure in M and $|\nabla l_x^\tau|^2$ and $\partial l_x/\partial \tau$ are locally essentially bounded. Thus, this equality is an equality of locally essentially bounded, measurable functions, i.e., elements of $L^\infty_{\text{loc}}(M)$,

and hence is an equality of distributions on M. Subtracting $2D_1$ from this equality yields D_2. Thus,
$$D_2 = -2D_1,$$
as distributions on M. This shows that D_2 vanishes as a distribution if and only if D_1 does. But if $D_2 = 0$ as a distribution for some τ, then by elliptic regularity l_x^τ is smooth on $M \times \{\tau\}$ and the equality is the naïve one for smooth functions. Thus, if $D_2 = 0$ for all τ, then l_x^τ and $\partial l/\partial \tau$ are C^∞ functions on each slice $M \times \{\tau\}$ and both D_1 and D_2 hold in the naïve sense on each slice $M \times \{\tau\}$. It follows from a standard bootstrap argument that in this case l_x^τ is smooth on all of space-time. $\square$

3. Reduced volume

We have established that for a Ricci flow $(M, g(t))$, $0 \leq t \leq T$, and a point $x = (p, T) \in M \times [0, T]$ the reduced length function l_x is defined on all of $M \times (0, T)$. This allows us to defined the reduced volume of $M \times \{\tau\}$ for any $\tau \in (0, T)$ Recall that the *reduced volume* of M is defined to be

$$\widetilde{V}_x(M, \tau) = \int_M \tau^{-\frac{n}{2}} exp(-l_x(q, \tau)) dq.$$

This function is defined for $0 < \tau < T$.

There is one simple case where we can make an explicit computation.

LEMMA 7.25. *If $(M, g(t))$ is flat Euclidean n-space (independent of t), then for any $x \in \mathbb{R}^n \times (-\infty, \infty)$ we have*

$$\widetilde{V}_x(M, \tau) = (4\pi)^{n/2}$$

for all $\tau > 0$.

PROOF. By symmetry we can assume that $x = (0, T) \in \mathbb{R}^n \times [0, T]$, where $0 \in \mathbb{R}^n$ is the origin. We have already seen that the $\mathcal{L}$-geodesics in flat space are the usual geodesics when parameterized by $s = \sqrt{\tau}$. Thus, identifying $\mathbb{R}^n$ with $T_0\mathbb{R}^n$, for any $X \in \mathbb{R}^n$ we have $\gamma_X(\tau) = 2\sqrt{\tau}X$, and hence $\mathcal{L}exp(X, \overline{\tau}) = 2\sqrt{\tau}X$. Thus, for any $\tau > 0$ we have $\mathcal{U}(\tau) = T_pM$, and for any $X \in T_0\mathbb{R}^n$ we have $\mathcal{J}(X, \tau) = 2^n \tau^{n/2}$. Also, $L_x(X, \tau) = 2\sqrt{\tau}|X|^2$, so that $l_x(X, \tau) = |X|^2$. Thus, for any $\tau > 0$

$$\widetilde{V}_x(\mathbb{R}^n, \tau) = \int_{\mathbb{R}^n} \tau^{-n/2} e^{-|X|^2} 2^n \tau^{n/2} dX = (4\pi)^{n/2}.$$

$\square$

In the case when M is non-compact, it is not clear *a priori* that the integral defining the reduced volume is finite in general. In fact, as the next proposition shows, it is always finite and indeed, it is bounded above by the integral for $\mathbb{R}^n$.

3. REDUCED VOLUME

THEOREM 7.26. *Let $(M, g(t))$, $0 \leq t \leq T$, be a Ricci flow of bounded curvature with the property that for each $t \in [0, T]$ the Riemannian manifold $(M, g(t))$ is complete. Fix a point $x = (p, T) \in M \times [0, T]$. For every $0 < \tau < T$ the reduced volume*

$$\widetilde{V}_x(M, \tau) = \int_M \tau^{-\frac{n}{2}} \exp(-l_x(q, \tau)) dq$$

is absolutely convergent and $\widetilde{V}_x(M, \tau) \leq (4\pi)^{\frac{n}{2}}$. The function $\widetilde{V}_x(M, \tau)$ is a non-increasing function of τ with

$$\lim_{\tau \to 0} \widetilde{V}_x(M, \tau) = (4\pi)^{\frac{n}{2}}.$$

PROOF. By Proposition 7.5, $\mathcal{U}_x(\tau)$ is an open subset of full measure in M. Hence,

$$\widetilde{V}_x(M, \tau) = \int_{\mathcal{U}_x(\tau)} \tau^{-\frac{n}{2}} \exp(-l_x(q, \tau)) dq.$$

Take linear orthonormal coordinates $(z^1, \ldots, z^n)$ on $T_p M$. It follows from the previous equality and Lemma 6.71 that

$$\widetilde{V}_x(M, \tau) = \int_{\widetilde{\mathcal{U}}_x(\tau)} f(Z, \tau) dz^1 \cdots dz^n,$$

where $f(Z, \tau) = \tau^{-\frac{n}{2}} e^{-\widetilde{l}(Z, \tau)} \mathcal{J}(Z, \tau)$. By Proposition 6.81 for each Z the integrand, $f(Z, \tau)$, is a non-increasing function of τ and the function converges uniformly on compact sets as $\tau \to 0$ to $2^n e^{-|Z|^2}$. This implies that $f(Z, \tau) \leq 2^n e^{-|Z|^2}$ for all $\tau > 0$, and hence that

$$\int_{\widetilde{\mathcal{U}}_x(\tau)} f(Z, \tau) dz^1 \cdots dz^n$$

converges absolutely for each $\tau > 0$, and the integral has value at most $(4\pi)^{n/2}$.

Fix $0 < \tau_0 < T$. According to Theorem 6.80 (with $A = M \times (T - \tau_0, T)$), the reduced volume $\widetilde{V}_x(M, \tau)$ is a non-increasing function of τ on $(0, \tau_0]$. Since this is true for any $0 < \tau_0 < T$, it follows that $\widetilde{V}_x(M, \tau)$ is a non-increasing function of τ for all $\tau \in (0, T)$. (This of course is a consequence of the monotonicity of $f(Z, \tau)$ in τ and the fact that $\widetilde{\mathcal{U}}_x(\tau) \subset \widetilde{\mathcal{U}}_x(\tau')$ for $\tau' < \tau$.)

To show that $\lim_{\tau \to 0} \widetilde{V}_x(M, \tau) = (4\pi)^{n/2}$ we need only see that for each $A < \infty$ for all $\tau > 0$ sufficiently small, $\widetilde{\mathcal{U}}_x(\tau)$ contains the ball of radius A centered at the origin in $T_p M$. Since the curvature is bounded, this is exactly the content of Corollary 6.79. □

3.1. Converse to Lemma 7.25. In Lemma 7.25 we showed that for the trivial flow on flat Euclidean n-space and for any point $x \in \mathbb{R}^n \times \{T\}$ the reduced volume $\widetilde{V}_x(\mathbb{R}^n, \tau)$ is independent of $\tau > 0$ and is equal to $(4\pi)^{n/2}$. In this subsection we use the monotonicity results of the last subsection to establish the converse to Lemma 7.25, namely to show that if $(M, g(t)), 0 \leq t \leq T$, is a Ricci flow complete with bounded curvature and if $\widetilde{V}_x(M, \overline{\tau}) = (4\pi)^{n/2}$ for some $\overline{\tau} > 0$ and some $x \in M \times \{T\}$, then the flow on the interval $[T - \overline{\tau}, T]$ is the trivial flow on flat Euclidean n-space.

PROPOSITION 7.27. *Suppose that $(M, g(\tau))$, $0 \leq \tau \leq T$, is a solution to the backward Ricci flow equation, complete and of bounded curvature. Let $x = (p, T) \in M \times \{T\}$, and suppose that $0 < \overline{\tau} < T$. If $\widetilde{V}_x(M, \overline{\tau}) = (4\pi)^{n/2}$, then the backward Ricci flow on the interval $[0, \overline{\tau}]$ is the trivial flow on flat Euclidean space.*

PROOF. If $\widetilde{V}_x(M, \overline{\tau}) = (4\pi)^{n/2}$, then by Lemma 7.25, $\widetilde{V}_x(M, \tau)$ is constant on the interval $(0, \overline{\tau}]$. Hence, it follows from the proof of Theorem 7.26 that the closure of $\widetilde{\mathcal{U}}(\tau)$ is all of T_pM for all $\tau \in (0, \overline{\tau}]$ and that $f(Z, \tau) = e^{-|Z|^2} 2^n$ for all $Z \in T_pM$ and all $\tau \leq \overline{\tau}$. In particular,
$$\frac{\partial \ln(f(Z, \tau))}{\partial \tau} = 0.$$
From the proof of Proposition 6.81 this means that Inequality (6.20) is an equality and consequently, so is Inequality (6.19). Thus, by Proposition 6.37 (with $\tau_1 = 0$) each of the vector fields $Y_\alpha(\tau) = \widetilde{Y}_\alpha(\tau)$ is both a Jacobi field and adapted. By Proposition 6.43 we then have
$$\mathrm{Ric} + \mathrm{Hess}(l_x^\tau) = \frac{g}{2\tau}.$$
In particular, l_x is smooth. Let $\varphi_\tau \colon M \to M$, $0 < \tau \leq \overline{\tau}$, be the one-parameter family of diffeomorphisms obtained by solving
$$\frac{d\varphi_\tau}{d\tau} = \nabla l_x(\cdot, \tau) \quad \text{and} \quad \varphi_{\overline{\tau}} = \mathrm{Id}.$$
We now consider
$$h(\tau) = \frac{\overline{\tau}}{\tau} \varphi_\tau^* g(\tau).$$
We compute
$$\frac{\partial h}{\partial \tau} = -\frac{\overline{\tau}}{\tau^2} \varphi_\tau^* g(\tau) + \frac{\overline{\tau}}{\tau} \varphi_\tau^* \mathcal{L}_{\frac{d\varphi_\tau}{d\tau}}(g(\tau)) + \frac{\overline{\tau}}{\tau} \varphi_t^* 2\mathrm{Ric}(g(\tau))$$
$$= -\frac{\overline{\tau}}{\tau^2} \varphi_\tau^* g(\tau) + \frac{\overline{\tau}}{\tau} \varphi_\tau^* 2\mathrm{Hess}(l_x^\tau) + \frac{\overline{\tau}}{\tau} \varphi_\tau^* \left(\frac{1}{\tau} g(\tau) - 2\mathrm{Hess}(l_x^\tau) \right)$$
$$= 0.$$
That is to say the family of metrics $h(\tau)$ is constant in τ: for all $\tau \in (0, \overline{\tau}]$ we have $h(\tau) = h(\overline{\tau}) = g(\overline{\tau})$. It then follows that
$$g(\tau) = \frac{\tau}{\overline{\tau}} (\varphi_\tau^{-1})^* g(\overline{\tau}),$$
which means that the entire flow in the interval $(0, \overline{\tau}]$ differs by diffeomorphism and scaling from $g(\overline{\tau})$. Suppose that $g(\overline{\tau})$ is not flat, i.e., suppose that

there is some $(x,\overline{\tau})$ with $|\mathrm{Rm}(x,\overline{\tau})| = K > 0$. Then from the flow equation we see that $|\mathrm{Rm}(\varphi_\tau^{-1}(x),\tau)| = K\overline{\tau}^2/\tau^2$, and these curvatures are not bounded as $\tau \to 0$. This is a contradiction. We conclude that $g(\overline{\tau})$ is flat, and hence, again by the flow equation so are all the $g(\tau)$ for $0 < \tau \le \overline{\tau}$, and by continuity, so is $g(0)$. Thus, $(M, g(\tau))$ is isometric to a quotient of $\mathbb{R}^n$ by a free, properly discontinuous group action. Lastly, since $\widetilde{V}_x(M,\tau) = (4\pi)^{n/2}$, it follows that $(M, g(\tau))$ is isometric to $\mathbb{R}^n$ for every $\tau \in [0,\overline{\tau}]$. Of course, it then follows that the flow is the constant flow. □

CHAPTER 8

Non-collapsed results

In this chapter we apply the results for the reduced length function and reduced volume established in the last two sections to prove non-collapsing results. In the first section we give a general result that applies to generalized Ricci flows and will eventually be applied to Ricci flows with surgery to prove the requisite non-collapsing. In the second section we give a non-collapsing result for Ricci flows on compact 3-manifolds with normalized initial metrics.

1. A non-collapsing result for generalized Ricci flows

The main result of this chapter is a κ-non-collapsed result.

THEOREM 8.1. *Fix positive constants $\overline{\tau}_0 < \infty$, $l_0 < \infty$, and $V > 0$. Then there is $\kappa > 0$ depending on $\overline{\tau}_0$, V, and l_0 and the dimension n such that the following holds. Let $(\mathcal{M}, G)$ be a generalized n-dimensional Ricci flow, and let $0 < \tau_0 \leq \overline{\tau}_0$. Let $x \in \mathcal{M}$ be fixed. Set $T = \mathbf{t}(x)$. Suppose that $0 < r \leq \sqrt{\tau_0}$ is given. These data are required to satisfy:*
  (1) *The ball $B(x, T, r) \subset M_T$ has compact closure.*
  (2) *There is an embedding $B(x, T, r) \times [T - r^2, T] \subset \mathcal{M}$ compatible with $\mathbf{t}$ and with the vector field.*
  (3) *$|\mathrm{Rm}| \leq r^{-2}$ on the image of the embedding in (2).*
  (4) *There is an open subset $\widetilde{W} \subset \widetilde{\mathcal{U}}_x(\tau_0) \subset T_x M_T$ with the property that for every $\mathcal{L}$-geodesic $\gamma \colon [0, \tau_0] \to \mathcal{M}$ with initial condition contained in $\widetilde{W}$, the l-length of γ is at most l_0.*
  (5) *For each $\tau \in [0, \tau_0]$, let $W(\tau) = \mathcal{L}\exp_x^\tau(\widetilde{W})$. The volume of the image $W(\tau_0) \subset M_{T-\tau_0}$ is at least V.*

Then
$$\mathrm{Vol}(B(x,T,r)) \geq \kappa r^n.$$

See FIG. 1.

In this section we denote by $g(\tau)$, $0 \leq \tau \leq r^2$, the family of metrics on $B(x, T, r)$ induced from pulling back G under the embedding $B(x, T, r) \times [T - r^2, T] \to \mathcal{M}$. Of course, this family of metrics satisfies the backward Ricci flow equation.

PROOF. Clearly from the definition of the reduced volume, we have
$$\widetilde{V}_x(W(\tau_0)) \geq \tau_0^{-n/2} V e^{-l_0} \geq \overline{\tau}_0^{-n/2} V e^{-l_0}.$$

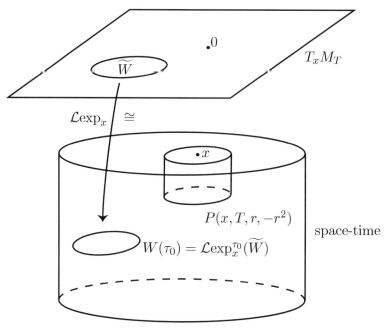

FIGURE 1. Non-collapsing.

By the monotonicity result (Theorem 6.80) it follows that for any $\tau \leq \tau_0$, and in particular for any $\tau \leq r^2$, we have

(8.1) $$\widetilde{V}_x(W(\tau)) \geq \overline{\tau}_0^{-n/2} V e^{-l_0}.$$

Let $\varepsilon = \sqrt[n]{\operatorname{Vol}(B(x,T,r))}/r$, so that $\operatorname{Vol} B(x,T,r) = \varepsilon^n r^n$. The basic result we need to establish in order to prove this theorem is the following:

PROPOSITION 8.2. *There is a positive constant $\varepsilon_0 \leq 1/4n(n-1)$ depending on $\overline{\tau}_0$ and l_0 such that if $\varepsilon \leq \varepsilon_0$ then, setting $\tau_1 = \varepsilon r^2$, we have $\widetilde{V}_x(W(\tau_1)) < 3\varepsilon^{\frac{n}{2}}$.*

Given this proposition, it follows immediately that either $\varepsilon > \varepsilon_0$ or

$$\varepsilon \geq \left(\frac{\widetilde{V}_x(W(\tau_1))}{3}\right)^{2/n} \geq \frac{1}{3^{2/n}\overline{\tau}_0} V^{2/n} e^{-2l_0/n}.$$

Since $\kappa = \varepsilon^n$, this proves the theorem.

PROOF. We divide $\widetilde{W}$ into

$$\widetilde{W}_{\operatorname{sm}} = \widetilde{W} \cap \left\{Z \in T_x M_T \big| |Z| \leq \frac{1}{8}\varepsilon^{-1/2}\right\}$$

1. A NON-COLLAPSING RESULT FOR GENERALIZED RICCI FLOWS

and
$$\widetilde{W}_{\mathrm{lg}} = \widetilde{W} \setminus \widetilde{W}_{\mathrm{sm}},$$
(see FIG. 2).

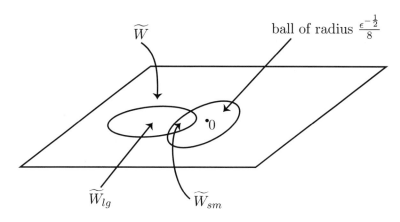

FIGURE 2. $\widetilde{W}_{lg}$ and $\widetilde{W}_{sm}$.

We set $W_{\mathrm{sm}}(\tau_1) = \mathcal{L}\exp_x^{\tau_1}(\widetilde{W}_{\mathrm{sm}})$ and $W_{\mathrm{lg}}(\tau_1) = \mathcal{L}\exp_x^{\tau_1}(\widetilde{W}_{\mathrm{lg}})$. Clearly, since $W(\tau_1)$ is the union of $W_{\mathrm{sm}}(\tau_1)$ and $W_{\mathrm{lg}}(\tau_1)$ and since these subsets are disjoint measurable subsets, we have

$$\widetilde{V}_x(W(\tau_1)) = \widetilde{V}_x(W_{\mathrm{sm}}(\tau_1)) + \widetilde{V}_x(W_{\mathrm{lg}}(\tau_1)).$$

We shall show that there is ε_0 such that either $\varepsilon > \varepsilon_0$ or $\widetilde{V}_x(W_{\mathrm{sm}}(\tau_1)) \leq 2\varepsilon^{n/2}$ and $\widetilde{V}_x(W_{\mathrm{lg}}(\tau_1)) \leq \varepsilon^{n/2}$. This will establish Proposition 8.2 and hence Theorem 8.1.

1.1. Upper bound for $\widetilde{V}_x(W_{\mathrm{sm}}(\tau_1))$. The idea here is that $\mathcal{L}$-geodesics with initial vector in $\widetilde{W}_{\mathrm{sm}}$ remain in the parabolic neighborhood

$$P = B(x, T, r) \times [T - r^2, T]$$

for all parameter values $\tau \in [0, r^2]$. Once we know this it is easy to see that their $\mathcal{L}$-lengths are bounded from below. Then if the volume of $B(x, T, r)$ was arbitrarily small, the reduced volume of $W_{\mathrm{sm}}(\tau_1)$ would be arbitrarily small.

LEMMA 8.3. *Setting $\tau_1 = \varepsilon r^2$, there is a constant $\varepsilon_0 > 0$ depending on $\overline{\tau}_0$ such that, if $\varepsilon \leq \varepsilon_0$, we have*

$$\int_{\widetilde{W}_{\mathrm{sm}}} \tau_1^{-n/2} e^{-\widetilde{l}(Z,\tau_1)} \mathcal{J}(Z, \tau_1) dZ \leq 2\varepsilon^{\frac{n}{2}}.$$

Of course, we have

$$\widetilde{V}_x(W_{\mathrm{sm}}(\tau_1)) = \int_{W_{\mathrm{sm}}(\tau_1)} \tau_1^{-n/2} e^{-l(q,\tau_1)} d\mathrm{vol}_{g(\tau_1)}$$

$$= \int_{\widetilde{W}_{\mathrm{sm}}} \tau_1^{-n/2} e^{-\widetilde{l}(Z,\tau_1)} \mathcal{J}(Z,\tau_1) dZ,$$

so that it will follow immediately from the lemma that:

COROLLARY 8.4. *There is a constant $\varepsilon_0 > 0$ depending on $\overline{\tau}_0$ such that, if $\varepsilon \leq \varepsilon_0$, we have*

$$\widetilde{V}_x(W_{\mathrm{sm}}(\tau_1)) \leq 2\varepsilon^{\frac{n}{2}}.$$

PROOF. *(Of Lemma 8.3)* In order to establish Lemma 8.3 we need two preliminary estimates:

CLAIM 8.5. *There is a universal positive constant ε_0' such that, if $\varepsilon \leq \varepsilon_0'$, then there is a constant $C_1 < \infty$ depending only on the dimension n such that the following hold for all $y \in B(x,T,r/2)$, and for all $t \in [T-\tau_1,T]$:*

(1)
$$|\nabla R(y,t)| \leq \frac{C_1}{r^3},$$

(2)
$$(1 - C_1\varepsilon) \leq \frac{g(y,t)}{g(y,T)} \leq (1 + C_1\varepsilon).$$

PROOF. Recall that by hypothesis $|\mathrm{Rm}(y,t)| \leq 1/r^2$ on the parabolic neighborhood $B(x,T,r) \times [T-r^2,T]$. Rescale the flow by multiplying the metric and time by r^{-2} resulting in a ball $\widetilde{B}$ of radius 1 and a flow defined for a time interval of length 1 with $|\mathrm{Rm}| \leq 1$ on the entire parabolic neighborhood $B(x,T,1) \times [T-1,T]$. Then by Theorem 3.28 there is a universal constant C_1 such that $|\nabla R(y,t)| \leq C_1$ for all $(y,t) \in B(x,T,1/2) \times [T-1/2,T]$. Rescaling back by r^2, we see that on the original flow $|\nabla R(y,t)| \leq C_1/r^3$ for all $(y,t) \in B(x,T,r/2) \times [T-r^2/2,T]$. Taking $\varepsilon_0' \leq 1/2$ gives the first item in the claim.

Since $|\mathrm{Ric}| \leq (n-1)/r^2$ for all $(y,t) \in B \times [T-r^2,T]$ it follows by integrating that

$$e^{-2(n-1)(T-t)/r^2} \leq \frac{g(x,t)}{g(x,T)} \leq e^{2(n-1)(T-t)/r^2}.$$

Thus, for $t \in [T-\tau_1,T]$ we have

$$e^{-2(n-1)\varepsilon} \leq \frac{g(x,t)}{g(x,T)} \leq e^{2(n-1)\varepsilon}.$$

From this the second item in the claim is immediate. □

1. A NON-COLLAPSING RESULT FOR GENERALIZED RICCI FLOWS

At this point we view the $\mathcal{L}$-geodesics as paths $\gamma\colon [0,\tau_1] \to B(x,T,r)$ (with the understanding that the path in space-time is given by the composition of the path $(\gamma(\tau), T-\tau)$ in $B(x,T,r) \times [T-r^2, T]$ followed by the given inclusion of this product into $\mathcal{M}$.

The next step in the proof is to show that for any $Z \in \widetilde{W}_{\mathrm{sm}}$ the $\mathcal{L}$-geodesic γ_Z (the one having $\lim_{\tau \to 0} \sqrt{\tau} X_{\gamma_Z}(\tau) = Z$) remains in $B(x,T,r/2)$ up to time τ_1. Because of this, as we shall see, these paths contribute a small amount to the reduced volume since $B(x,T,r/2)$ has small volume. We set $X(\tau) = X_{\gamma_Z}(\tau)$.

CLAIM 8.6. *There is a positive constant $\varepsilon_0 \leq 1/4n(n-1)$ depending on $\overline{\tau}_0$, such that the following holds. Suppose that $\varepsilon \leq \varepsilon_0$ and $\tau_1' \leq \tau_1 = \varepsilon r^2$. Let $Z \in T_x M_T$ and let γ_Z be the associated $\mathcal{L}$-geodesic from x. Suppose that $\gamma_Z(\tau) \in B(x,T,r/2)$ for all $\tau < \tau_1'$. Then for all $\tau < \tau_1'$ we have*

$$\big|\|\sqrt{\tau} X(\tau)\|_{g(T)} - |Z|\big| \leq 2\varepsilon(1+|Z|).$$

PROOF. First we make sure that ε_0 is less than or equal to the universal constant ε_0' of the last claim. For all $(y,t) \in B(x,T,r) \times [T-r^2, T]$ we have $|\mathrm{Rm}(y,t)| \leq r^{-2}$ and $|\nabla R(y,t)| \leq C_1/r^3$ for some universal constant C_1. Of course, $r^2 \leq \overline{\tau}$. Thus, at the expense of replacing C_1 by a larger constant, we can (and shall) assume that $C_1/r^3 > (n-1)r^{-2} \geq |\mathrm{Ric}(y,t)|$ for all $(y,t) \in B(x,T,r) \times [T-r^2, T]$. Thus, we can take the constant C_0 in the hypothesis of Lemma 6.60 to be C_1/r^3. We take the constant $\overline{\tau}$ in the hypothesis of that lemma to be εr^2. Then, we have that

$$\max_{0 \leq \tau \leq \tau_1'} \sqrt{\tau} |X(\tau)| \leq e^{2C_1\varepsilon^2}|Z| + \frac{e^{2C_1\varepsilon^2}-1}{2}\sqrt{\varepsilon}r,$$

and

$$|Z| \leq e^{2C_1\varepsilon^2} \min_{0 \leq \tau \leq \tau_1'} \sqrt{\tau}|X(\tau)| + \frac{e^{2C_1\varepsilon^2}-1}{2}\sqrt{\varepsilon}r.$$

By choosing $\varepsilon_0 > 0$ sufficiently small (as determined by the universal constant C_1 and by $\overline{\tau}_0$), we have

$$\max_{0 \leq \tau \leq \tau_1'} \sqrt{\tau}|X(\tau)|_{g(T-\tau)} \leq (1+\frac{\varepsilon}{2})|Z| + \frac{\varepsilon}{2},$$

and

$$|Z| \leq (1+\frac{\varepsilon}{2}) \min_{0 \leq \tau \leq \tau_1'} \sqrt{\tau}|X(\tau)|_{g(T-\tau)} + \frac{\varepsilon}{2}.$$

It is now immediate that

$$\big|\|\sqrt{\tau}X(\tau)\|_{g(T-\tau)} - |Z|\big| \leq \varepsilon(1+|Z|).$$

Again choosing ε_0 sufficiently small the result now follows from the second inequality in Claim 8.5 □

Now we are ready to establish that the $\mathcal{L}$-geodesics whose initial conditions are elements of $\widetilde{W}_{\mathrm{sm}}$ do not leave $B(x,T,r/2) \times [T-r^2, T]$ for any $\tau \leq \tau_1$.

CLAIM 8.7. *Suppose $\varepsilon_0 \leq 1/4n(n-1)$ is the constant from the last claim. Set $\tau_1 = \varepsilon r^2$, and suppose that $\varepsilon \leq \varepsilon_0$. Lastly, assume that $|Z| \leq \frac{1}{8\sqrt{\varepsilon}}$. Then $\gamma_Z(\tau) \in B(x,T,r/2)$ for all $\tau \leq \tau_1$.*

PROOF. Since $\varepsilon \leq \varepsilon_0 \leq 1/4n(n-1) \leq 1/8$, by the last claim we have

$$|\sqrt{\tau}X(\tau)|_{g(T)} \leq (1+2\varepsilon)|Z| + 2\varepsilon \leq \frac{5}{4}|Z| + \frac{3}{32\sqrt{\varepsilon}},$$

provided that $\gamma|_{[0,\tau)}$ is contained in $B(x,T,r/2) \times [T-\tau,T]$. Since $|Z| \leq (8\sqrt{\varepsilon})^{-1}$ we conclude that

$$|\sqrt{\tau}X(\tau)|_{g(T)} \leq \frac{1}{4\sqrt{\varepsilon}},$$

as long as $\gamma([0,\tau))$ is contained in $B(x,T,r/2) \times [T-\tau,T]$.

Suppose that there is $\tau' < \tau_1 = \varepsilon r^2$ for which γ_Z exits $B(x,T,r/2) \times [T-r^2,T]$. We take τ' to be the first such time. Then we have

$$|\gamma_Z(\tau') - x|_{g(T)} \leq \int_0^{\tau'} |X(\tau)|_{g(T)} d\tau \leq \frac{1}{4\varepsilon^{\frac{1}{2}}} \int_0^{\tau'} \frac{d\tau}{\sqrt{\tau}} = \frac{1}{2\varepsilon^{\frac{1}{2}}} \sqrt{\tau'} < r/2.$$

This contradiction implies that $\gamma_Z(\tau) \in B(x,T,r/2)$ for all $\tau < \tau_1 = \varepsilon r^2$. □

Now we assume that $\varepsilon_0 > 0$ depending on $\overline{\tau}_0$ is as above and that $\varepsilon \leq \varepsilon_0$, and we shall estimate

$$\widetilde{V}_x(W_{\mathrm{sm}}(\tau_1)) = \int_{W_{\mathrm{sm}}(\tau_1)} (\tau_1)^{-\frac{n}{2}} e^{-l(q,\tau_1)} d\mathrm{vol}_{g(\tau_1)}.$$

In order to do this we estimate $l_x(q,\tau_1)$ on $W_{\mathrm{sm}}(\tau_1)$. By hypothesis $|\mathrm{Rm}| \leq 1/r^2$ on $B(x,T,r/2) \times [0,\tau_1]$ and by Lemma 8.7 every $\mathcal{L}$-geodesic γ_Z, defined on $[0,\tau_1]$, with initial conditions Z satisfying $|Z| \leq \frac{1}{8}\varepsilon^{-\frac{1}{2}}$ remains in $B(x,T,r/2)$. Thus, for such γ_Z we have $R(\gamma_Z(\tau)) \geq -n(n-1)/r^2$. Thus, for any $q \in W_{\mathrm{sm}}(\tau_1)$ we have

$$L_x(q,\tau_1) = \int_0^{\tau_1} \sqrt{\tau}(R + |X(\tau)|^2) d\tau \geq -\frac{2n(n-1)}{3r^2}(\tau_1)^{\frac{3}{2}} = -\frac{2n(n-1)}{3}\varepsilon^{\frac{3}{2}}r,$$

and hence

$$l + x(q,\tau_1) = \frac{L_x(q,\tau_1)}{2\sqrt{\tau_1}} \geq -\frac{n(n-1)}{3}\varepsilon.$$

Since $W_{\mathrm{sm}}(\tau) \subset B(x,T,r/2) \subset B(x,T,r)$, we have:

(8.2) $\quad \widetilde{V}_x(W_{\mathrm{sm}}(\tau_1)) \leq \varepsilon^{-\frac{n}{2}} r^{-n} e^{n(n-1)\varepsilon/3} \mathrm{Vol}_{g(T-\tau_1)} W_{\mathrm{sm}}(\tau)$

$$\leq \varepsilon^{-\frac{n}{2}} r^{-n} e^{n(n-1)\varepsilon/3} \mathrm{Vol}_{g(T-\tau_1)} B(x,T,r).$$

CLAIM 8.8. *There is a universal constant $\varepsilon_0 > 0$ such that if $\varepsilon \leq \varepsilon_0$, for any open subset U of $B(x,T,r)$, and for any $0 \leq \tau_1 \leq \overline{\tau}_0$, we have*

$$0.9 \leq \mathrm{Vol}_{g(T)} U / \mathrm{Vol}_{g(T-\tau_1)} U \leq 1.1.$$

PROOF. This is immediate from the second item in Claim 8.5. □

1. A NON-COLLAPSING RESULT FOR GENERALIZED RICCI FLOWS

Now assume that ε_0 also satisfies this claim. Plugging this into Equation (8.2), and using the fact that $\varepsilon \leq \varepsilon_0 \leq 1/4n(n-1)$, which implies $n(n-1)\varepsilon/3 \leq 1/12$, and the fact that from the definition we have

$$\text{Vol}_{g(T)} B(x,T,r) = \varepsilon^n r^n,$$

gives

$$\widetilde{V}_x(W_{\text{sm}}(\tau_1)) \leq \varepsilon^{-\frac{n}{2}} r^{-n} e^{n(n-1)\varepsilon/3}(1.1)\text{Vol}_{g(T)}B(x,T,r) \leq (1.1)\varepsilon^{\frac{n}{2}} e^{\frac{1}{12}}.$$

Thus,

$$\widetilde{V}_x(W_{\text{sm}}(\tau_1)) \leq 2\varepsilon^{\frac{n}{2}}.$$

This completes the proof of Lemma 8.3. □

1.2. Upper bound for $\widetilde{V}_x(W_{\text{lg}}(\tau_1))$. Here the basic point is to approximate the reduced volume integrand by the heat kernel, which drops off exponentially fast as we go away from the origin.

Recall that $\text{Vol}\, B(x,T,r) = \varepsilon^n r^n$ and $\tau_1 = \varepsilon r^2$.

LEMMA 8.9. *There is a universal positive constant $\varepsilon_0 > 0$ such that if $\varepsilon \leq \varepsilon_0$, we have*

$$\widetilde{V}_x(W_{\text{lg}}(\tau_1)) \leq \int_{\widetilde{U}(\tau_1)\cap\{Z\,\big|\,|Z|\geq \frac{1}{8}\varepsilon^{-\frac{1}{2}}\}} (\tau_1)^{-\frac{n}{2}} e^{-\widetilde{l}(q,\tau_1)} \mathcal{J}(Z,\tau_1) dZ \leq \varepsilon^{\frac{n}{2}}.$$

PROOF. By the monotonicity result (Proposition 6.81), we see that the restriction of the function $\tau_1^{-\frac{n}{2}} e^{-\widetilde{l}(Z,\tau_1)} \mathcal{J}(Z,\tau_1)$ to $\widetilde{U}(\tau_1)$ is less than or equal to the restriction of the function $2^n e^{-|Z|^2}$ to the same subset. This means that

$$\widetilde{V}_x(W_{\text{lg}}(\tau_1)) \leq \int_{\widetilde{U}(\tau_1)\setminus \widetilde{U}(\tau_1)\cap B(0,\frac{1}{8}\varepsilon^{-1/2})} 2^n e^{-|Z|^2} dZ$$

$$\leq \int_{T_p\mathcal{M}_T \setminus B(0,\frac{1}{8}\varepsilon^{-1/2})} 2^n e^{-|Z|^2} dZ.$$

So it suffices to estimate this latter integral.

Fix some $a > 0$ and let $I(a) = \int_{B(0,a)} 2^n e^{-|Z|^2} dZ$. Let $R(a/\sqrt{n})$ be the n-cube centered at the origin with side lengths $2a/\sqrt{n}$. Then $R(a/\sqrt{n}) \subset B(0,a)$, so that

$$I(a) \geq \int_{R(a/\sqrt{n})} 2^n e^{-|Z|^2} dZ$$

$$= \prod_{i=1}^{n} \left(\int_{-a/\sqrt{n}}^{a/\sqrt{n}} 2e^{-z_i^2} dz_i\right)$$

$$= \left(\int_0^{2\pi}\int_0^{a/\sqrt{n}} 4e^{-r^2} r\, dr\, d\theta\right)^{n/2}.$$

Now
$$\int_0^{2\pi}\int_0^{a/\sqrt{n}} 4e^{-r^2}rdrd\theta = 4\pi(1-e^{-\frac{a^2}{n}}).$$
Applying this with $a = (8\sqrt{\varepsilon})^{-1}$ we have
$$\widetilde{V}_x(W_{\lg}(\tau_1)) \leq \int_{\mathbb{R}^n} 2^n e^{-|Z|^2}dZ - I(1/8\sqrt{\varepsilon})$$
$$\leq (4\pi)^{n/2}\left(1-\left(1-e^{-1/(64n\varepsilon)}\right)^{n/2}\right).$$

Thus,
$$\widetilde{V}_x(W_{\lg}(\tau_1)) \leq (4\pi)^{n/2}\frac{n}{2}e^{-1/(64n\varepsilon)}.$$

There is $\varepsilon_0 > 0$ such that the expression on the right-hand side is less than $\varepsilon^{n/2}$ if $\varepsilon \leq \varepsilon_0$. This completes the proof of Lemma 8.9. □

Putting Lemmas 8.3 and 8.9 together establishes Proposition 8.2. □

As we have already remarked, Proposition 8.2 immediately implies Theorem 8.1. This completes the proof of Theorem 8.1. □

2. Application to compact Ricci flows

Now let us apply this result to Ricci flows with normalized initial metrics to show that they are universally κ-non-collapsed on any fixed, finite time interval. In this section we specialize to 3-dimensional Ricci flows. We do not need this result in what follows for we shall prove a more delicate result in the context of Ricci flows with surgery. Still, this result is much simpler and serves as a paradigm of what will come.

THEOREM 8.10. *Fix positive constants $\omega > 0$ and $T_0 < \infty$. Then there is $\kappa > 0$ depending only on these constants such that the following holds. Let $(M,g(t))$, $0 \leq t < T \leq T_0$, be a 3-dimensional Ricci flow with M compact and with $|\mathrm{Rm}(p,0)| \leq 1$ and $\mathrm{Vol}\,B(p,0,1) \geq \omega$ for all $p \in M$. Then for any $t_0 \leq T$, any $r > 0$ with $r^2 \leq t_0$ and any $(p,t_0) \in M \times \{t_0\}$, if $|\mathrm{Rm}(q,t)| \leq r^{-2}$ on $B(p,t_0,r) \times [t_0 - r^2, t_0]$ then $\mathrm{Vol}\,B(p,t_0,r) \geq \kappa r^3$.*

PROOF. Fix any $x = (p,t_0) \in M \times [0,T]$. First, we claim that we can suppose that $t_0 \geq 1$. For if not, then rescale the flow by $Q = 1/t_0$. This does not affect the curvature inequality at time 0. Furthermore, there is $\omega' > 0$ depending only on ω such that for any ball B at time 0 and of radius 1 in the rescaled flow we have $\mathrm{Vol}\,B \geq \omega'$. The reason for the latter fact is the following: By the Bishop-Gromov inequality (Theorem 1.34) there is $\omega' > 0$ depending only on ω such that for any $q \in M$ and any $r \leq 1$ we have $\mathrm{Vol}\,B(q,0,r) \geq \omega' r^3$. Of course, the rescaling increases T, but simply restrict to the rescaled flow on $[0,1]$.

Next, we claim that we can assume that $r \leq \sqrt{t_0}/2$. If r does not satisfy this inequality, then we replace r with $r' = \sqrt{t_0}/2$. Of course, the curvature

2. APPLICATION TO COMPACT RICCI FLOWS

inequalities hold for r' if they hold for r. Suppose that we have established the result for r'. Then

$$\text{Vol } B(p, T, r) \geq \text{Vol } B(p, T, r') \geq \kappa (r')^3 \geq \kappa \left(\frac{r}{2}\right)^3 = \frac{\kappa}{8} r^3.$$

From now on we assume that $t_0 \geq 1$ and $r \leq \sqrt{t_0}/2$. According to Proposition 4.11, for any $(p, t) \in M \times [0, 2^{-4}]$ we have $|\text{Rm}(p, t)| \leq 2$ and $\text{Vol } B(p, t, r) \geq \kappa_0 r^3$ for all $r \leq 1$.

Once we know that $|\text{Rm}|$ is universally bounded on $M \times [0, 2^{-4}]$ it follows that there is a universal constant C_1 such that $C_1^{-1} g(q, 0) \leq g(q, t) \leq C_1 g(q, 0)$ for all $q \in M$ and all $t \in [0, 2^{-4}]$. This means that there is a universal constant $C < \infty$ such that the following holds. For any points $q_0, q \in M$ with $d_0(q_0, q) \leq 1$ let $\gamma_{q_0, q}$ be the path in $M \times [2^{-5}, 2^{-4}]$ given by

$$\gamma_{q_0, q}(\tau) = (A_{q_0, q}(\tau), 2^{-4} - \tau), \ 0 \leq \tau \leq 2^{-5},$$

where $A_{q_0, q}$ is a shortest $g(0)$-geodesic from q_0 to q. Then $\mathcal{L}(\gamma_{q_0, q}) \leq C$.

By Theorem 7.10 there is a point $q_0 \in M$ and an $\mathcal{L}$-geodesic γ_0 from $x = (p, t_0)$ to $(q_0, 2^{-4})$ with $l(\gamma_0) \leq 3/2$. Since $t_0 \geq 1$, this means that there is a universal constant $C' < \infty$ such that for each point $q \in B(q_0, 0, 1)$ the path which is the composite of γ_0 followed by $\gamma_{q_0, q}$ has $\widetilde{l}$-length at most C'. Setting $\tau_0 = t_0 - 2^{-5}$, this implies that $l_x(q, \tau_0) \leq C'$ for every $q \in B(q_0, 0, 1)$. This ball has volume at least κ_0. By Proposition 7.5, the open subset $\mathcal{U}_x(\tau_0)$ is of full measure in $M \times \{2^{-5}\}$. Hence, $W(\tau_0) = (B(q_0, 0, 1) \times \{2^{-5}\}) \cap \mathcal{U}_x(\tau_0)$ also has volume at least κ_0. Since $r^2 \leq t_0/4 < \tau_0$, Theorem 8.1 now gives the result. (See FIG. 3.) □

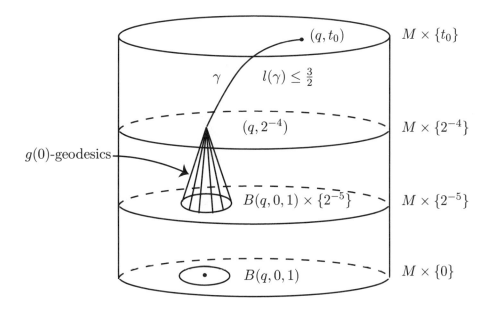

FIGURE 3. Non-collapsing of Ricci flows.

CHAPTER 9

κ-non-collapsed ancient solutions

In this chapter we discuss the qualitative properties of κ-non-collapsed, ancient solutions. One of the most important is the existence of a gradient shrinking soliton that is asymptotic at $-\infty$ to the solution. The other main qualitative result is the compactness result (up to scaling) for these solutions. Also extremely important for us is classification of 3-dimensional gradient shrinking solitons – up to finite covers there are only two: a shrinking family of round S^3's and a shrinking family of products of round S^2's with $\mathbb{R}$. This leads to a rough classification of all 3-dimensional κ-non-collapsed, ancient solutions. The κ-solutions are in turn the models for singularity development in 3-dimensional Ricci flows on compact manifolds, and eventually for singularity development in 3-dimensional Ricci flows with surgery.

1. Preliminaries

Our objects of study are Ricci flows $(M, g(t))$, $-\infty < t \leq 0$, with each $(M, g(t))$ being a complete manifold of bounded non-negative curvature. The first remark to make is that the appropriate notion of non-negative curvature is that the Riemann curvature operator

$$\mathrm{Rm}\colon \wedge^2 TM \to \wedge^2 TM,$$

which is a symmetric operator, is non-negative. In general, this implies, but is stronger than, the condition that the sectional curvatures are all non-negative. In case the dimension of M is at most 3, every element of $\wedge^2 TM$ is represented by a 2-plane (with area form) and hence the Riemann curvature operator is non-negative if and only if all the sectional curvatures are non-negative. When the curvature operator is non-negative, it is if and only if scalar curvature is bounded.

Since the $(M, g(t))$ have non-negative Ricci curvature, it follows immediately from the Ricci flow equation that the metric is non-increasing in time in the sense that for any point $p \in M$ and any $v \in T_pM$ the function $|v|^2_{g(t)}$ is a non-increasing function of t.

There are stronger results assuming the curvature operator is bounded and non-negative. These are consequences of the Harnack inequality (see [**32**]). As was established in Corollary 4.39, since the flow exists for $t \in (-\infty, 0]$ and since the curvature operator is non-negative and bounded for each $(q,t) \in M \times (-\infty, 0]$, it follows that $\partial R(q,t)/\partial t \geq 0$ for all q and t.

That is to say, for each $q \in M$ the scalar curvature $R(q,t)$ is a non-decreasing function of t.

1.1. Definition. Now we turn to the definition of what it means for a Ricci flow to be κ-non-collapsed.

DEFINITION 9.1. Fix $\kappa > 0$. Let $(M, g(t))$, $a < t \leq b$, be a Ricci flow of complete n-manifolds. Fix $r_0 > 0$. We say that $(M, g(t))$ is κ-*non-collapsed on scales at most* r_0 if the following holds for any $(p,t) \in M \times (a,b]$ and any $0 < r \leq r_0$ with the property that $a \leq t - r^2$. Whenever $|\text{Rm}(q,t')| \leq r^{-2}$ for all $q \in B(p,t,r)$ and all $t' \in (t-r^2, t]$, then $\text{Vol}\, B(p,t,r) \geq \kappa r^n$. We say that $(M, g(t))$ is κ-*non-collapsed*, or equivalently κ-*non-collapsed on all scales* if it is κ-non-collapsed on scales at most r_0 for every $r_0 < \infty$.

DEFINITION 9.2. An *ancient solution* is a Ricci flow $(M, g(t))$ defined for $-\infty < t \leq 0$ such that for each t, $(M, g(t))$ is a connected, complete, non-flat Riemannian manifold whose curvature operator is bounded and non-negative. For any $\kappa > 0$, an ancient solution is κ-*non-collapsed* if it is κ-non-collapsed on all scales. We also use the terminology κ-*solution* for a κ-non-collapsed, ancient solution.

Notice that a κ-solution is a κ'-solution for any $0 < \kappa' \leq \kappa$.

1.2. Examples. Here are some examples of κ-solutions:

EXAMPLE 9.3. Let (S^2, g_0) be the standard round 2-sphere of scalar curvature 1 (and hence Ricci tensor $g_0/2$). Set $g(t) = (1-t)g_0$. Then $\partial g(t)/\partial t = -2\text{Ric}(g(t))$, $-\infty < t \leq 0$. This Ricci flow is an ancient solution which is κ-non-collapsed on all scales for any κ at most the volume of the ball of radius 1 in the unit 2-sphere.

According to a result of Hamilton which we shall prove below (Corollary 9.50):

THEOREM 9.4. *Every orientable, 2-dimensional κ-solution is a rescaling of the previous example, i.e., is a family of shrinking round 2-spheres.*

EXAMPLE 9.5. Let (S^n, g_0) be the standard round n-sphere of scalar curvature $n/2$. Set $g(t) = (1-t)g_0$. This is a κ-solution for any κ which is at most the volume of the ball of radius 1 in the unit n-sphere. If Γ is a finite subgroup of the isometries of S^n acting freely on S^n, then the quotient S^n/Γ inherits an induced family of metrics $\overline{g}(t)$ satisfying the Ricci flow equation. The result is a κ-solution for any κ at most $1/|\Gamma|$ times the volume of the ball of radius 1 in the unit sphere.

EXAMPLE 9.6. Consider the product $S^2 \times \mathbb{R}$, with the metric $g(t) = (1-t)g_0 + ds^2$. This is a κ-solution for any κ at most the volume of a ball of radius 1 in the product of the unit 2-sphere with $\mathbb{R}$.

EXAMPLE 9.7. The quotient $S^2 \times \mathbb{R}/\langle \iota \rangle$, where the involution ι is the product of the antipodal map on S^2 with $s \mapsto -s$ on the $\mathbb{R}$ factor, is an orientable κ-solution for some $\kappa > 0$.

EXAMPLE 9.8. Consider the metric product $(S^2, g_0) \times (S_R^1, ds^2)$ where (S_R^1, ds^2) is the circle of radius R. We define $g(t) = (1-t)g_0 + ds^2$. This is an ancient solution to the Ricci flow. But it is not κ-non-collapsed for any $\kappa > 0$. The reason is that

$$|\text{Rm}(p,t)| = \frac{1}{1-t},$$

and

$$\frac{\text{Vol}_{g(t)}\, B(p, \sqrt{1-t})}{(1-t)^{3/2}} \leq \frac{\text{Vol}_{g(t)}(S^2 \times S_R^1)}{(1-t)^{3/2}} = \frac{2\pi R(1-t)4\pi}{(1-t)^{3/2}} = \frac{8\pi^2 R}{\sqrt{1-t}}.$$

Thus, as $t \to -\infty$ this ratio goes to zero.

1.3. A consequence of Hamilton's Harnack inequality. In order to prove the existence of an asymptotic gradient shrinking soliton associated to every κ-solution, we need the following inequality which is a consequence of Hamilton's Harnack inequality for Ricci flows with non-negative curvature operator.

PROPOSITION 9.9. *Let $(M, g(t))$, $-\tau_0 \leq t \leq 0$, be an n-dimensional Ricci flow such that for each $t \in [-\tau_0, 0]$ the Riemannian manifold $(M, g(t))$ is complete with non-negative, bounded curvature operator. Let $\tau = -t$. Fix a point $p \in M$ and let $x = (p, 0) \in M \times [-\tau_0, 0]$. Then for any $0 < c < 1$ and any $\tau \leq (1-c)\tau_0$ we have*

$$|\nabla l_x(q, \tau))|^2 + R(q, \tau) \leq \frac{(1+2c^{-1})l_x(q,\tau)}{\tau}, \quad \text{and}$$

$$R(q, \tau) - \frac{(1+c^{-1})l_x(q,\tau)}{\tau} \leq \frac{\partial l_x}{\partial \tau}$$

where these inequalities hold on the open subset of full measure of $M \times [-(1-c)\tau_0, 0)$ on which l_x is a smooth function.

PROOF. Recall that from Equation 6.15 we have

$$H(X) = -\frac{\partial R}{\partial \tau} - \frac{R}{\tau} - 2\langle \nabla R, X \rangle + 2\text{Ric}(X, X).$$

Using Hamilton's Harnack's inequality (Theorem 4.37) with $\chi = -X$, we have

$$-\frac{\partial R}{\partial \tau} - \frac{R}{\tau_0 - \tau} - 2\langle \nabla R, X \rangle + 2\text{Ric}(X, X) \geq 0.$$

Together these imply

$$H(X) \geq \left(\frac{1}{\tau - \tau_0} - \frac{1}{\tau}\right) R = \frac{\tau_0}{\tau(\tau - \tau_0)} R.$$

Restricting to $\tau \leq (1-c)\tau_0$ gives

$$H(X) \geq -\frac{c^{-1}}{\tau}R.$$

Take a minimal $\mathcal{L}$-geodesic from x to $(q, \overline{\tau})$; we have

$$(9.1) \quad K^{\overline{\tau}}(\gamma) = \int_0^{\overline{\tau}} \tau^{3/2} H(X) d\tau \geq -c^{-1} \int_0^{\overline{\tau}} \sqrt{\tau} R d\tau \geq -2c^{-1}\sqrt{\overline{\tau}} l_x(q, \overline{\tau}).$$

Together with the second equality in Theorem 6.50, this gives

$$4\overline{\tau}|\nabla l_x(q,\overline{\tau})|^2 = -4\overline{\tau} R(q,\overline{\tau}) + 4l_x(q,\overline{\tau}) - \frac{4}{\sqrt{\overline{\tau}}} K^{\overline{\tau}}(\gamma)$$
$$\leq -4\overline{\tau} R(q,\overline{\tau}) + 4l_x(q,\overline{\tau}) + 8c^{-1} l_x(q,\overline{\tau}).$$

Dividing through by $4\overline{\tau}$, and replacing $\overline{\tau}$ with τ yields the first inequality in the statement of the proposition:

$$|\nabla l_x(q,\tau)|^2 + R(q,\tau) \leq \frac{(1 + 2c^{-1})l_x(q,\tau)}{\tau}$$

for all $0 < \tau \leq (1-c)\tau_0$. This is an equation of smooth functions on the open dense subset $\mathcal{U}(\overline{\tau})$ but it extends as an equation of L^∞_{loc}-functions on all of M.

As to the second inequality in the statement, by the first equation in Theorem 6.50 we have

$$\frac{\partial l_x(q,\tau)}{\partial \tau} = R(q,\tau) - \frac{l_x(q,\tau)}{\tau} + \frac{1}{2\tau^{3/2}} K^\tau(\gamma).$$

The estimate on K^τ in Equation (9.1) then gives

$$R(q,\tau) - \frac{(1+c^{-1})l_x(q,\tau)}{\tau} \leq \frac{\partial l_x(q,\tau)}{\partial \tau}.$$

This establishes the second inequality. $\square$

COROLLARY 9.10. *Let $(M, g(t))$, $-\infty < t \leq 0$, be a Ricci flow on a complete, n-dimensional manifold with bounded, non-negative curvature operator. Fix a point $p \in M$ and let $x = (p, 0) \in M \times (-\infty, 0]$. Then for any $\tau > 0$ we have*

$$|\nabla l_x(q,\tau))|^2 + R(q,\tau) \leq \frac{3l_x(q,\tau)}{\tau},$$

$$-\frac{2l_x(q,\tau)}{\tau} \leq \frac{\partial l_x(q,\tau)}{\partial \tau} \leq \frac{l_x(q,\tau)}{\tau},$$

where these inequalities are valid in the sense of smooth functions on the open subset of full measure of $M \times \{\tau\}$ on which l_x is a smooth function, and are valid as inequalities of L^∞_{loc}-functions on all of $M \times \{\tau\}$.

PROOF. Fix τ and take a sequence of $\tau_0 \to \infty$, allowing us to take $c \to 1$, and apply the previous proposition. This gives the first inequality and gives the lower bound for $\partial l_x/\partial \tau$ in the second inequality.

To establish the upper bound in the second inequality we consider the path that is the concatenation of a minimal $\mathcal{L}$-geodesic γ from x to (q, τ) followed by the path $\mu(\tau') = (q, \tau')$ for $\tau' \geq \tau$. Then

$$l_x(\gamma * \mu|_{[\tau,\tau_1]}) = \frac{1}{2\sqrt{\tau_1}}\left(\mathcal{L}(\gamma) + \int_\tau^{\tau_1} \sqrt{\tau'} R(q, \tau') d\tau'\right).$$

Differentiating at $\tau_1 = \tau$ gives

$$\left.\frac{\partial l_x(\gamma * \mu)}{\partial \tau}\right|_{\tau_1 = \tau} = -\frac{1}{4\tau^{3/2}}\mathcal{L}(\gamma) + \frac{1}{2\sqrt{\tau}}\sqrt{\tau} R(q, \tau)$$

$$= -\frac{l_x(q, \tau)}{2\tau} + \frac{R(q, \tau)}{2}.$$

By the first inequality in this statement, we have

$$-\frac{l_x(q, \tau)}{2\tau} + \frac{R(q, \tau)}{2} \leq \frac{l_x(q, \tau)}{\tau}.$$

Since $l_x(q, \tau') \leq \tilde{l}(\gamma * \mu|_{[\tau,\tau']})$ for all $\tau' \geq \tau$, this establishes the claimed upper bound for $\partial l_x/\partial \tau$. $\square$

2. The asymptotic gradient shrinking soliton for κ-solutions

Fix $\kappa > 0$ and consider an n-dimensional κ-solution $(M, g(t))$, $-\infty < t \leq 0$. Our goal in this section is to establish the existence of an asymptotic gradient shrinking soliton associated to this κ-solution. Fix a reference point $p \in M$ and set $x = (p, 0) \in M \times (-\infty, 0]$. By Theorem 7.10 for every $\tau > 0$ there is a point $q(\tau) \in M$ at which the function $l_x(\cdot, \tau)$ achieves its minimum, and furthermore, we have

$$l_x(q(\tau), \tau) \leq \frac{n}{2}.$$

For $\bar{\tau} > 0$, define

$$g_{\bar{\tau}}(t) = \frac{1}{\bar{\tau}} g(\bar{\tau} t), \quad -\infty < t \leq 0.$$

Now we come to one of the main theorems about κ-solutions, a result that will eventually provide a qualitative description of all κ-solutions.

THEOREM 9.11. *Let $(M, g(t))$, $-\infty < t \leq 0$, be a κ-solution of dimension n. Fix $x = (p, 0) \in M \times (-\infty, 0]$. Suppose that $\{\bar{\tau}_k\}_{k=1}^\infty$ is a sequence tending to ∞ as $k \to \infty$. Then, after replacing $\{\bar{\tau}_k\}$ by a subsequence, the following holds. For each k denote by M_k the manifold M, by $g_k(t)$ the family of metrics $g_{\bar{\tau}_k}(t)$ on M_k, and by $q_k \in M_k$ the point $q(\bar{\tau}_k)$. The sequence of pointed flows $(M_k, g_k(t), (q_k, -1))$ defined for $t \in (-\infty, 0)$ converges smoothly to a non-flat based Ricci flow $(M_\infty, g_\infty(t), (q_\infty, -1))$ defined* for $t \in (-\infty, 0)$. *This limiting Ricci flow satisfies the gradient shrinking soliton equation in the sense that there is a smooth function* $f: M_\infty \times (-\infty, 0) \to \mathbb{R}$ *such that for every* $t \in (-\infty, 0)$ *we have*

$$\text{(9.2)} \qquad \text{Ric}_{g_\infty(t)} + \text{Hess}^{g_\infty(t)}(f(t)) + \frac{1}{2t} g_\infty(t) = 0.$$

Furthermore, $(M_\infty, g_\infty(t))$ *has non-negative curvature operator, is κ-non-collapsed, and satisfies* $\partial R_{g_\infty}(x, t)/\partial t \geq 0$ *for all* $x \in M_\infty$ *and all* $t < 0$.

See FIG. 1.

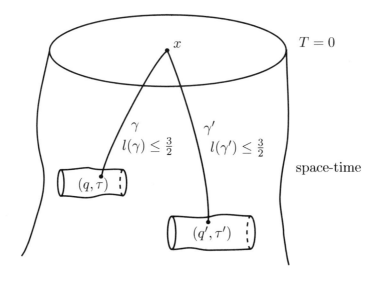

FIGURE 1. Gradient shrinking soliton.

REMARK 9.12. We are not claiming that the gradient shrinking soliton is a κ-solution (or more precisely an extension forward in time of a time-shifted version of a κ-solution) because we are not claiming that the time-slices have bounded curvature operator. Indeed, we do not know if this is true in general. We shall establish below (see Corollary 9.50 and Corollary 9.53) that in the case $n = 2, 3$, the gradient shrinking soliton does indeed have time-slices of bounded curvature, and hence is an extension of a κ-solution. We are

2. THE ASYMPTOTIC GRADIENT SHRINKING SOLITON FOR κ-SOLUTIONS

also not claiming at this point that the limiting flow is a gradient shrinking soliton in the sense that there is a one-parameter family of diffeomorphisms $\varphi_t \colon M_\infty \to M_\infty$, $t < 0$, with the property that $|t|\varphi_t^* g_\infty(-1) = g_\infty(t)$ and with the property that the φ_t are generated by the gradient vector field of a function. We shall also establish this result in dimensions 2 and 3 later in this chapter.

We will divide the proof of Theorem 9.11 into steps. First, we will show that the reduced length and norm of the curvature $|\mathrm{Rm}|$ are bounded throughout the sequence in some way. Then using the κ-non-collapsed assumption, by the compactness theorem (Theorem 5.15), we conclude that a subsequence of the sequence of flows converges geometrically to a limiting flow. Then, using the fact that the limit of the reduced volumes, denoted $\widetilde{V}_\infty(M_\infty \times \{t\})$, is constant, we show that the limit flow is a gradient shrinking soliton. Finally we argue that the limit is non-flat. The proof occupies the rest of this section.

2.1. Bounding the reduced length and the curvature.
Now let's carry this procedure out in detail. The first remark is that since rescaling does not affect the κ-non-collapsed hypothesis, all the Ricci flows $(M_k, g_k(t))$ are κ-non-collapsed on all scales. Next, we have the effect on reduced volume.

CLAIM 9.13. *For each $k \geq 1$ denote by $x_k \in M_k$ the point $(p, 0) \in M_k$. Let $\widetilde{V}_{x_k}(\tau) = \widetilde{V}_{x_k}(M_k \times \{\tau\})$ denote the reduced volume function for the Ricci flow $(M_k, g_k(t))$ from the point x_k, and let $\widetilde{V}_x(\tau)$ denote the reduced volume of $M \times \{\tau\}$ for the Ricci flow $(M, g(t))$ from the point x. Then*

$$\widetilde{V}_{x_k}(\tau) = \widetilde{V}_x(\overline{\tau}_k \tau).$$

PROOF. This is a special case of the reparameterization equation for reduced volume (Lemma 6.75). □

By Theorem 7.26 the reduced volume function $\widetilde{V}_x(\tau)$ is a non-increasing function of τ with $\lim_{\tau \to 0} \widetilde{V}_x(\tau) = (4\pi)^{\frac{n}{2}}$. Since the integrand for $\widetilde{V}_x(\tau)$ is everywhere positive, it is clear that $\widetilde{V}_x(\tau) > 0$ for all τ. Hence, $\lim_{\tau \to \infty} \widetilde{V}_x(\tau)$ exists. By Proposition 7.27 either this limit as τ goes to infinity is less than $(4\pi)^{n/2}$ or the flow is the constant flow on flat Euclidean space. The latter is ruled out by our assumption that the manifolds are non-flat. It follows immediately from this and Claim 9.13 that:

COROLLARY 9.14. *There is a non-negative constant $V_\infty < (4\pi)^{n/2}$ such that for all $\tau \in (0, \infty)$, we have*

(9.3) $$\lim_{k \to \infty} \widetilde{V}_{x_k}(\tau) = V_\infty.$$

Now let us turn to the length functions l_{x_k}.

CLAIM 9.15. *For any $\tau > 0$ we have*

$$l_{x_k}(q_k, \tau) \leq \frac{n}{2\tau^2} + \frac{n\tau}{2}.$$

PROOF. By the choice of q_k we have $l_{x_k}(q_k, \tau_k) \leq \frac{n}{2}$. By the scale invariance of l (Corollary 6.74) we have $l_{x_k}(q_k, -1) \leq n/2$ for all k. Fix $0 < \tau < 1$. Integrating the inequality

$$\frac{-2l_x(q_k, \tau)}{\tau} \leq \frac{\partial l_{x_k}(q_k, \tau)}{\partial \tau}$$

from τ to 1 yields

$$l_{x_k}(q_k, \tau) \leq \frac{n}{2\tau^2}.$$

If $\tau > 1$, then integrating the second inequality in the second displayed line of Corollary 9.10 gives $l_{x_k}(q_k, \tau) \leq \frac{n\tau}{2}$. □

COROLLARY 9.16. *There is a positive continuous function $C_1(\tau)$ defined for $\tau > 0$ such that for any $q \in M_k$ we have:*

$$l_{x_k}(q, \tau) \leq \left(\sqrt{\frac{3}{\tau}} d_{g_k(-\tau)}(q_k, q) + C_1(\tau)\right)^2,$$

$$|\nabla l_{x_k}(q, \tau)| \leq \frac{3}{\tau} d_{g_k(-\tau)}(q_k, q) + \sqrt{\frac{3}{\tau}} C_1(\tau).$$

PROOF. By Corollary 9.10, for any $q \in M_k$ we have $|\nabla l_{x_k}(q, \tau)|^2 \leq 3 l_{x_k}(q, \tau)/\tau$. Since $l_{x_k}(q_k, \tau) \leq \frac{n}{2\tau_0^2} + \frac{n\tau}{2}$, integrating yields

$$l_{x_k}(q, \tau) \leq \left(\sqrt{\frac{3}{\tau}} d_{g_k(-\tau)}(q_k, q) + C_1(\tau)\right)^2,$$

with $C_1(\tau)$ being $\sqrt{(n/2\tau^2) + (n\tau/2)}$. The second statement follows from this and Proposition 9.9. □

It follows immediately from Corollary 9.16 that for each $A < \infty$ and $\tau_0 > 0$, the functions l_{x_k} are uniformly bounded (by a bound that is independent of k but depends on τ_0 and A) on the balls $B(q_k, -\tau_0, A)$. Once we know that the l_{x_k} are uniformly bounded on $B(q_k, -\tau_0, A)$, it follows from Corollary 9.10 that R_{g_k} are also uniformly bounded on the $B(q_k, -\tau_0, A)$. Invoking Corollary 4.39, we see that for any $A < \infty$ the scalar curvatures of the metrics g_k are uniformly bounded on $B_{g_k}(q_k, -\tau_0, A) \times (-\infty, -\tau_0]$. Since the metrics have non-negative curvature operator, this implies that the eigenvalues of this operator are uniformly bounded on these regions. Since we are assuming that the original Ricci flows are κ-non-collapsed on all scales, it follows from Theorem 5.15 that after passing to a subsequence there is a geometric limit $(M_\infty, g_\infty(t), (q_\infty, -1))$, $-\infty < t \leq -\tau_0$, which is a Ricci flow that is κ-non-collapsed on all scales.

Since this is true for every $\tau_0 > 0$, by a standard diagonalization argument passing to a further subsequence we get a geometric limit flow $(M_\infty, g_\infty(t), (q_\infty, -1))$, $-\infty < t < 0$.

Let us summarize our progress to this point.

COROLLARY 9.17. *After passing to a subsequence of the τ_k there is a smooth limiting flow of the $(M_k, g_k(t), (q_k, -1))$, $-\infty < t \leq 0$,*

$$(M_\infty, g_\infty(t), (q_\infty, -1)),$$

defined for $-\infty < t < 0$. For every $t < 0$ the Riemannian manifold $(M_\infty, g_\infty(t))$ is complete of non-negative curvature. The flow is κ-non-collapsed on all scales and satisfies $\partial R/\partial t \geq 0$.

PROOF. Since the flows in the sequence are all κ-non-collapsed on all scales and have non-negative curvature operator, the limiting flow is κ-non-collapsed on all scales and has non-negative curvature operator. By the consequence of Hamilton's Harnack inequality (Corollary 4.39), we have $\partial R/\partial t \geq 0$ for the original κ-solution. This condition also passes to the limit. □

2.2. The limit function. The next step in the proof is to construct the limiting function l_∞ of the l_{x_k} and show that it satisfies the gradient shrinking soliton equation.

By definition of the geometric limit, for any compact connected set $K \subset M_\infty$ containing q_∞ and any compact subinterval J of $(-\infty, 0)$ containing -1, for all k sufficiently large we have smooth embeddings $\psi_k \colon K \to M_k$ sending q_∞ to q_k so that the pullbacks of the restrictions of the family of metrics $g_k(t)$ for $t \in J$ to K converge uniformly in the C^∞-topology to the restriction of $g_\infty(t)$ on $K \times J$. Take an exhausting sequence $K_k \times J_k$ of such products of compact sets with closed intervals, and pass to a subsequence so that for all k the diffeomorphism ψ_k is defined on $K_k \times J_k$. We denote by l_k the pullback of l_{x_k} under these embeddings and by $h_k(t)$ the pullback of the family of metrics $g_k(t)$. We denote by ∇^{h_k} the gradient with respect to $h_k(t)$, and similarly $\triangle^{h_k}$ denotes the Laplacian for the metric $h_k(t)$. By construction, for any compact subset of $M_\infty \times (-\infty, 0)$ for all k sufficiently large the function l_k is defined on the compact set. We use ∇ and $\triangle$ to refer to the covariant derivative and the Laplacian in the limiting metric g_∞.

Now let us consider the functions l_{x_k}. According to Corollary 9.16, for any $A < \infty$ and any $0 < \tau_0 < T$, both l_{x_k} and $|\nabla l_{x_k}|$ are uniformly bounded on $B(q_k, -1, A) \times [-T, -\tau_0]$ independent of k. Hence, the l_{x_k} are uniformly Lipschitz on these subspaces. Doing this for each A, τ_0, and T and using a standard diagonalization argument then shows that, after transferring to the limit, the functions l_k are uniformly locally bounded and uniformly locally Lipschitz on $M_\infty \times (-\infty, 0)$ with respect to the limiting metric g_∞.

Fix $0 < \alpha < 1$. Passing to a further subsequence if necessary, we can arrange that the l_k converge strongly in $C^{0,\alpha}_{\mathrm{loc}}$ to a function l_∞ defined on $M_\infty \times (-\infty, 0)$. Furthermore, it follows that the restriction of l_∞ is locally Lipschitz, and hence the function l_∞ is an element of $W^{1,2}_{\mathrm{loc}}(M_\infty \times (-\infty, 0))$. Also, by passing to a further subsequence if necessary, we can assume that the l_k converge weakly in $W^{1,2}_{\mathrm{loc}}$ to l_∞.

COROLLARY 9.18. *For any $\tau > 0$ and any q we have*

$$|\nabla l_\infty(q,\tau)| \leq \frac{3}{\tau} d_{g_\infty}(-\tau)(q_\infty, q) + \sqrt{\frac{3}{\tau} C_1(\tau)},$$

where $C_1(\tau)$ is the continuous function from Corollary 9.16.

PROOF. This is immediate from Corollary 9.16 and Fatou's lemma. □

REMARK 9.19. **N.B.** We are not claiming that l_∞ is the reduced length function from a point of $M_\infty \times (-\infty, 0)$.

2.3. Differential inequalities for l_∞. The next step is to establish differential equalities for l_∞ related to, but stronger than, those that we established in Chapter 7 for l_x. Here is a crucial result.

PROPOSITION 9.20. *The function l_∞ is a smooth function on $M \times (-\infty, 0)$ and satisfies the following two differential equalities:*

(9.4) $$\frac{\partial l_\infty}{\partial \tau} + |\nabla l_\infty|^2 - R + \frac{n}{2\tau} - \triangle l_\infty = 0,$$

(9.5) $$2\triangle l_\infty - |\nabla l_\infty|^2 + R + \frac{l_\infty - n}{\tau} = 0.$$

The proof of this result is contained in Sections 2.4 through 2.6

2.4. Preliminary results toward the proof of Proposition 9.20. In this subsection we shall prove that the left-hand side of Equation (9.4) is a distribution and is ≥ 0 in the distributional sense. We shall also show that this distribution extends to a continuous linear functional on compactly supported functions in $W^{1,2}$.

The first step in the proof of this result is the following, somewhat delicate lemma.

LEMMA 9.21. *For any $t \in (-\infty, 0)$ we have*

$$\lim_{k \to \infty} |\nabla^{h_k} l_k|^2_{h_k} d\mathrm{vol}(h_k) = |\nabla l_\infty|^2_{g_\infty} d\mathrm{vol}(g_\infty)$$

in the sense of distributions on $M_\infty \times \{t\}$.

PROOF. It suffices to fix $0 < \tau_0 < |t|$. The inequality in one direction ($\geq$) is a general result. Here is the argument. Since the $|\nabla^{g_k} l_{x_k}|_{g_k}$ are uniformly essentially bounded on every $B(x_k, -\tau_0, A) \times [-T, -\tau_0]$, the $|\nabla^{h_k} l_k|_{h_k}$ are uniformly essentially bounded on $B(x_\infty, -\tau_0, A) \times [-T, -\tau_0]$. (Of course,

2. THE ASYMPTOTIC GRADIENT SHRINKING SOLITON FOR κ-SOLUTIONS

$\nabla^{h_k} l_k = dl_k = \nabla l_k$.) Since the h_k converge uniformly on compact sets to g_∞, it is clear that

$$(9.6) \qquad \lim_{k \to \infty} \left(|\nabla^{h_k} l_k|^2_{h_k} \mathrm{dvol}(h_k) - |\nabla l_k|^2_{g_\infty} \mathrm{dvol}(g_\infty) \right) = 0$$

in the sense of distributions on $M \times \{t\}$. Since the l_k converge uniformly on compact subsets to l_∞, it follows immediately from Fatou's lemma that

$$\lim_{k \to \infty} |\nabla l_k|^2_{g_\infty} \mathrm{dvol}(g_\infty) \geq |\nabla l_\infty|^2_{g_\infty} \mathrm{dvol}(g_\infty)$$

in the sense of distributions on $M_\infty \times \{t\}$. Thus, we have the following inequality of distributions:

$$\lim_{k \to \infty} |\nabla^{h_k} l_k|^2_{h_k} \mathrm{dvol}(h_k) \geq |\nabla l_\infty|^2_{g_\infty} \mathrm{dvol}(g_\infty).$$

We need to establish the opposite inequality which is not a general result, but rather relies on the bounds on $\triangle^{g_k} l_{x_k}$ (or equivalently on $\triangle^{h_k} l_k$) given in the second inequality in Theorem 7.13. We must show that for each $t \leq -\tau_0$ and for any φ, a non-negative, smooth function with compact support in $M_\infty \times \{t\}$, we have

$$\lim_{k \to \infty} \int_{M \times \{t\}} \varphi \left(|\nabla^{h_k} l_k|^2_{h_k} \mathrm{dvol}(h_k) - |\nabla l_\infty|^2_{g_\infty} \mathrm{dvol}(g_\infty) \right) \leq 0.$$

First, notice that since, on the support of φ, the metrics h_k converge uniformly in the C^∞-topology to g_∞ and since $|\nabla^{h_k} l_k|^2_{h_k}$ and $|\nabla l_\infty|^2_{g_\infty}$ are essentially bounded on the support of φ, we have

$$(9.7) \quad \lim_{k \to \infty} \int_{M \times \{t\}} \varphi \left(|\nabla^{h_k} l_k|^2_{h_k} \mathrm{dvol}(h_k) - |\nabla l_\infty|^2_{g_\infty} \mathrm{dvol}(g_\infty) \right)$$

$$= \lim_{k \to \infty} \int_{M \times \{t\}} \varphi (|\nabla^{h_k} l_k|^2_{h_k} - |\nabla l_\infty|^2_{h_k}) \mathrm{dvol}(h_k)$$

$$= \lim_{k \to \infty} \int_{M \times \{t\}} \langle \nabla^{h_k} l_k - \nabla l_\infty), \varphi \nabla^{h_k} l_k \rangle_{h_k} \mathrm{dvol}(h_k)$$

$$+ \int_{M \times \{t\}} \langle \nabla^{h_k} l_k - \nabla l_\infty), \varphi \nabla l_\infty \rangle_{h_k} \mathrm{dvol}(h_k).$$

We claim that, in the limit, the last term in this expression vanishes. Using the fact that the h_k converge uniformly in the C^∞-topology to g_∞ on the support of φ, and $|\nabla l_\infty|$ is bounded on this support we can rewrite the last term as

$$(9.8) \qquad \lim_{k \to \infty} \int_{M \times \{t\}} \langle \nabla (l_k - l_\infty), \varphi \nabla l_\infty \rangle_{g_\infty} \mathrm{dvol}(g_\infty).$$

Since $l_k - l_\infty$ goes to zero weakly in $W^{1,2}$ on the support of φ whereas l_∞ is an element of $W^{1,2}$ of this compact set, we see that the expression given in (9.8) vanishes and hence that

$$\lim_{k\to\infty} \int_{M\times\{t\}} \langle \nabla^{h_k}(l_k - l_\infty), \varphi \nabla l_\infty \rangle_{h_k} d\mathrm{vol}(h_k) = 0.$$

It remains to consider the first term in the last expression in Equation (9.7). (This is where we shall need the differential inequality for the $\triangle^{g_k} l_{x_k}$.) Since the l_k converge uniformly to l_∞ on the support of φ, we can choose positive constants ϵ_k tending to 0 as k tends to ∞ so that $l_\infty - l_k + \epsilon_k > 0$ on the support of φ. We can rewrite

$$\lim_{k\to\infty} \int_{M\times\{t\}} \langle \left(\nabla^{h_k} l_k - \nabla l_\infty \right), \varphi \nabla^{h_k} l_k \rangle_{h_k} d\mathrm{vol}(h_k)$$

$$= \lim_{k\to\infty} \int_{M\times\{t\}} \langle \nabla^{h_k}(l_k - l_\infty - \epsilon_k), \varphi \nabla^{h_k} l_k \rangle_{h_k} d\mathrm{vol}(h_k).$$

CLAIM 9.22.

$$\lim_{k\to\infty} \int_{M\times\{t\}} \langle \nabla^{h_k}(l_k - l_\infty - \epsilon_k), \varphi \nabla^{h_k} l_k \rangle_{h_k} d\mathrm{vol}(h_k) \leq 0.$$

PROOF. Since φ is a compactly supported, non-negative smooth function, it follows from Theorem 7.13 that we have the following inequality of distributions:

$$\varphi \triangle^{h_k} l_k \leq \frac{\varphi}{2} \left(|\nabla^{h_k} l_k|^2_{h_k} - R_{h_k} - \frac{l_k - n}{\tau} \right).$$

(Here R_{h_k} is the scalar curvature of h_k.) That is to say, for any non-negative C^∞-function f we have

$$\int_{M\times\{t\}} -\langle \nabla^{h_k} l_k, \nabla^{h_k}(\varphi \cdot f) \rangle_{h_k} d\mathrm{vol}(h_k)$$

$$\leq \int_{M\times\{t\}} \frac{\varphi f}{2} \left(|\nabla^{h_k} l_k|^2_{h_k} - R_{h_k} - \frac{l_k - n}{\tau} \right) d\mathrm{vol}(h_k).$$

We claim that the same inequality holds as long as f is a non-negative, locally Lipschitz function. The point is that given such a function f, we can find a sequence of non-negative C^∞-functions f_k on the support of φ (by say mollifying f) that converge to f strongly in the $W^{1,2}$-norm on the support of φ. The sought-after inequality holds for every f_k. Since both sides of the inequality are continuous in the $W^{1,2}$-norm of the function, the result holds for the limit function f as well.

Now we apply this with f being the non-negative locally Lipschitz function $l_\infty - l_k + \epsilon_k$. We conclude that

$$\int_{M\times\{t\}} \langle \nabla^{h_k}(\varphi(l_k - l_\infty - \epsilon_k)), \nabla^{h_k} l_k \rangle_{h_k} d\mathrm{vol}(h_k)$$

$$\leq \int_{M\times\{t\}} \frac{\varphi(l_\infty - l_k + \epsilon_k)}{2} \left(|\nabla^{h_k} l_k|^2_{h_k} - R_{h_k} - \frac{l_k - n}{\tau} \right) d\mathrm{vol}(h_k).$$

Now taking the limit as $k \to \infty$, we see that the right-hand side of this inequality tends to zero since $(l_\infty - l_k + \epsilon_k)$ tends uniformly to zero on the support of φ and $|\nabla^{h_k} l_k|^2_{h_k}$, R_k and l_k are all uniformly essentially bounded on the support of φ. Thus, the term

$$\int_{M\times\{t\}} \langle \nabla^{h_k}(\varphi(l_k - l_\infty - \epsilon_k)), \nabla^{h_k} l_k \rangle_{h_k} d\mathrm{vol}(h_k)$$

has a limsup ≤ 0 as k tends to ∞. Now we expand

$$\nabla^{h_k}(\varphi(l_k - l_\infty - \epsilon_k)) = \nabla^{h_k}(\varphi)(l_k - l_\infty - \epsilon_k) + \varphi \nabla^{h_k}(l_k - l_\infty - \epsilon_k).$$

The first term on the right-hand side converges to zero as $k \to \infty$ since $l_k - l_\infty - \epsilon_k$ tends uniformly to zero on the support of φ. This completes the proof of the claim. □

We have now established the inequalities in both directions and hence completed the proof of Lemma 9.21. □

LEMMA 9.23. *Consider the distribution*

$$\mathcal{D} = \frac{\partial l_\infty}{\partial \tau} + |\nabla l_\infty|^2 - R + \frac{n}{2\tau} - \Delta l_\infty$$

on $M_\infty \times (-\infty, 0)$. Then $\mathcal{D}$ extends to a continuous linear functional on the space of compactly supported $W^{1,2}$-functions on $M_\infty \times (-\infty, 0)$. Furthermore, if ψ is a non-negative Lipschitz function on $M_\infty \times (-\infty, 0)$ with compact support, then $\mathcal{D}(\psi) \leq 0$.

PROOF. Clearly, since the l_k converge uniformly on compact subsets of $M_\infty \times (-\infty, 0)$ to l_∞ and the metrics h_k converge smoothly to g_∞, uniformly on compact sets, it follows that the $\triangle^{h_k} l_k$ converge in the weak sense to $\triangle l_\infty$ and similarly, the $\partial l_k / \partial \tau$ converge in the weak sense to $\partial l_\infty / \partial \tau$. Hence, by taking limits from Theorem 7.13, using Lemma 9.21, we see that

(9.9) $$\mathcal{D} = \frac{\partial l_\infty}{\partial \tau} + |\nabla l_\infty|^2 - R + \frac{n}{2\tau} - \Delta l_\infty \geq 0$$

in the weak sense on $M \times (-\infty, 0)$.

Since R and $\frac{n}{2\tau}$ are C^∞-functions, it is clear that the distributions given by these terms extend to continuous linear functionals on the space of compactly supported $W^{1,2}$-functions. Similarly, since $|\nabla l_\infty|^2$ is an element of L^∞_{loc}, it also extends to a continuous linear functional on compactly supported $W^{1,2}$-functions. Since $|\partial l_\infty / \partial \tau|$ is an locally essentially bounded function,

$\partial l_\infty/\partial \tau$ extends to a continuous functional on the space of compactly supported $W^{1,2}$ functions. Lastly, we consider $\triangle l_\infty$. As we have seen, the value of the associated distribution on φ is given by

$$\int_{M\times(-\infty,,0)} -\langle \nabla\varphi, \nabla l_\infty\rangle_{g_\infty} d\mathrm{vol}(g_\infty)d\tau.$$

Since $|\nabla l_\infty|$ is a locally essentially bounded function, this expression also extends to a continuous linear functional on compactly supported $W^{1,2}$-functions.

Lastly, if ψ is an element of $W^{1,2}$ with compact support and hence can be approximated in the $W^{1,2}$-norm by non-negative smooth functions. The last statement is now immediate from Equation (9.9). □

This leads immediately to:

COROLLARY 9.24. *The functional*

$$\varphi \mapsto \mathcal{D}(e^{-l_\infty}\varphi)$$

is a distribution and its value on any non-negative, compactly supported C^∞-function φ is ≥ 0.

PROOF. If φ is a compactly supported non-negative C^∞-function, then $e^{-l_\infty}\varphi$ is a compactly supported non-negative Lipschitz function. Hence, this result is an immediate consequence of the previous corollary. □

2.5. Extension to non-compactly supported functions. The next step in this proof is to estimate the l_{x_k} uniformly from below in order to show that the integrals involved in the distributions in Proposition 9.20 are absolutely convergent so that they extend to continuous functionals on a certain space of functions that includes non-compactly supported functions.

LEMMA 9.25. *There is a constant $c_1 > 0$ depending only on the dimension n such that for any $p, q \in M_k$ we have*

$$l_{x_k}(p,\tau) \geq -l_{x_k}(q,\tau) - 1 + c_1 \frac{d^2_{g(-\tau)}(p,q)}{\tau}.$$

PROOF. Since both sides of this inequality and also Ricci flow are invariant if the metric and time are simultaneously rescaled, it suffices to consider the case when $\tau = 1$. Also, since $\mathcal{U}_x(1)$ is a dense subset, it suffices to assume that $p, q \in \mathcal{U}_x(1)$. Also, by symmetry, we can suppose that $l_{x_k}(q,1) \leq l_{x_k}(p,1)$.

Let γ_1 and γ_2 be the minimizing $\mathcal{L}$-geodesics from x to $(p,1)$ and $(q,1)$ respectively. We define a function $f\colon M_k \times M_k \times [0,\infty) \to \mathbb{R}$ by

$$f(a,b,\tau) = d_{g_k(-\tau)}(a,b).$$

2. THE ASYMPTOTIC GRADIENT SHRINKING SOLITON FOR κ-SOLUTIONS

Since $\gamma_1(0) = \gamma_2(0)$ we have

(9.10)
$$d_{g_k(-1)}(p, q) = f(p, q, 1)$$
$$= \int_0^1 \frac{d}{d\tau} f(\gamma_1(\tau), \gamma_2(\tau), \tau) d\tau$$
$$= \int_0^1 \left(\frac{\partial f}{\partial \tau}(\gamma_1(\tau), \gamma_2(\tau), \tau) + \langle \nabla f_a, \gamma_1'(\tau) \rangle\right.$$
$$\left. + \langle \nabla f_b, \gamma_2'(\tau) \rangle\right) d\tau,$$

where $\nabla_a f$ and $\nabla_b f$ refer respectively to the gradient of f with respect to the first copy of M_k in the domain and the second copy of M_k in the domain. Of course, $|\nabla f_a| = 1$ and $|\nabla f_b| = 1$.

By Corollary 6.29, $\gamma_1'(\tau) = \nabla l_{x_k}(\gamma_1(\tau), \tau)$ and $\gamma_2'(\tau) = \nabla l_{x_k}(\gamma_2(\tau), \tau)$. Since $R \geq 0$ we have

$$l_{x_k}(\gamma_1(\tau), \tau) = \frac{1}{2\sqrt{\tau}} \mathcal{L}_{x_k}(\gamma_1|_{[0,\tau]}) \leq \frac{1}{2\sqrt{\tau}} \mathcal{L}_{x_k}(\gamma_1) = \frac{1}{\sqrt{\tau}} l_{x_k}(p, 1).$$

Symmetrically, we have

$$l_{x_k}(\gamma_2(\tau), \tau) \leq \frac{1}{\sqrt{\tau}} l_{x_k}(q, 1).$$

From this inequality, Corollary 9.10, and the fact that $R \geq 0$, we have

(9.11)
$$\left|\langle \nabla f_a(\gamma_1(\tau), \gamma_2(\tau), \tau), \gamma_1'(\tau)\rangle\right| \leq |\gamma_1'(\tau)| = |\nabla l_{x_k}(\gamma_1(\tau), \tau)|$$
$$\leq \frac{\sqrt{3}}{\tau^{3/4}} \sqrt{l_{x_k}(p, 1)}$$
$$\leq \frac{\sqrt{3}}{\tau^{3/4}} \sqrt{l_{x_k}(p, 1) + 1}.$$

Symmetrically, we have
(9.12)
$$\left|\langle \nabla f_b(\gamma_1(\tau), \gamma_2(\tau), \tau), \gamma_2'(\tau)\rangle\right| \leq \frac{\sqrt{3}}{\tau^{3/4}} \sqrt{l_{x_k}(q, 1)} \leq \frac{\sqrt{3}}{\tau^{3/4}} \sqrt{l_{x_k}(q, 1) + 1}.$$

It follows from Corollary 9.10 that for any p, we have

$$|\nabla(\sqrt{l_{x_k}(p, \tau)})| \leq \frac{\sqrt{3}}{2\sqrt{\tau}}.$$

Set $r_0(\tau) = \tau^{3/4}(l_{x_k}(q, 1) + 1)^{-1/2}$. For any $p' \in B_{g_k}(\gamma_1(\tau), \tau, r_0(\tau))$ integrating gives

$$l_{x_k}^{1/2}(p', \tau) \leq l_{x_k}^{1/2}(\gamma_1(\tau), \tau) + \frac{\sqrt{3}}{2\sqrt{\tau}} r_0(\tau) \leq \left(\tau^{-1/4} + \frac{\sqrt{3}}{2}\tau^{1/4}\right) \sqrt{l_{x_k}(p, 1) + 1},$$

where in the last inequality we have used the fact that $1 \leq l_{x_k}(q,1) + 1 \leq l_{x_k}(p,1) + 1$. Again using Corollary 9.10 we have

$$R(p',\tau) \leq \frac{3}{\tau}\left(\tau^{-1/4} + \frac{\sqrt{3}}{2}\tau^{1/4}\right)^2 (l_{x_k}(p,1) + 1)$$

Now consider $q' \in B_{g_k(\tau)}(\gamma_2(\tau), \tau, r_0(\tau))$. Similarly to the above computations, we have

$$l_{x_k}^{1/2}(q',\tau) \leq l_{x_k}^{1/2}(q,1) + \frac{\sqrt{3}}{2\sqrt{\tau}}r_0(\tau),$$

so that

$$l_{x_k}^{1/2}(q',\tau) \leq \left(\tau^{-1/4} + \frac{\sqrt{3}}{2}\tau^{1/4}\right)\sqrt{l_{x_k}(q,1) + 1},$$

and

$$|\mathrm{Ric}(q',\tau)| \leq R(q',\tau) \leq \frac{3}{\tau}\left(\tau^{-1/4} + \frac{\sqrt{3}}{2}\tau^{1/4}\right)^2 (l_{x_k}(q,1) + 1).$$

We set

$$K = \frac{3}{\tau}\left(\tau^{-1/4} + \frac{\sqrt{3}}{2}\tau^{1/4}\right)^2 (l_{x_k}(q,1) + 1).$$

Now, noting that $\partial/\partial\tau$ here is $-\partial/\partial t$ of Proposition 3.21, we apply Proposition 3.21 to see that

$$\left|\frac{\partial}{\partial\tau}f(\gamma_1(\tau), \gamma_2(\tau), \tau)\right| \leq 2(n-1)\left(\frac{2}{3(n-1)}Kr_0(\tau) + r_0(\tau)^{-1}\right)$$
$$\leq \left(C_1\tau^{-3/4} + C_2\tau^{-1/4} + C_3\tau^{1/4}\right)\sqrt{l_{x_k}(q,1) + 1},$$

where C_1, C_2, C_3 are constants depending only on the dimension n.

Now plugging Equation (9.11) and (9.12) and the above inequality into Equation (9.10) we see that

$$d_{g(-1)}(p,q) \leq \int_0^1 \Big(\left(C_1\tau^{-3/4} + C_2\tau^{-1/4} + C_3\tau^{1/4}\right)\sqrt{l_{x_k}(q,1) + 1}$$
$$+ \sqrt{3}\tau^{-3/4}\sqrt{l_{x_k}(q,1) + 1} + \sqrt{3}\tau^{-3/4}\sqrt{l_{x_k}(p,1) + 1}\Big)d\tau.$$

This implies that

$$d_{g(-1)}(p,q) \leq C\left(\sqrt{l_{x_k}(q,1) + 1} + \sqrt{l_{x_k}(p,1) + 1}\right),$$

for some constant depending only on the dimension. Thus, since we are assuming that $l_{x_k}(p,1) \geq l_{x_k}(q,1)$ we have

$$d_{g(-1)}^2(p,q) \leq C^2\left(3(l_{x_k}(p,1) + 1) + (l_{x_k}(q,1) + 1)\right)$$
$$\leq 4C^2(l_{x_k}(p,1) + 1 + l_{x_k}(q,1)),$$

2. THE ASYMPTOTIC GRADIENT SHRINKING SOLITON FOR κ-SOLUTIONS

for some constant $C < \infty$ depending only on the dimension. The result now follows immediately. $\square$

COROLLARY 9.26. *For any $q' \in M$ and any $0 < \tau_0 \leq \tau'$ we have*

$$l_{x_k}(q', \tau') \geq -\frac{n}{2(\tau')^2} - \frac{\tau'}{2} - 1 + c_1 \frac{d^2_{g_k(-\tau_0)}(q_k, q')}{\tau'},$$

where c_1 is the constant from Lemma 9.25.

PROOF. By Claim 9.15

$$l_{x_k}(q_k, \tau') \leq \frac{n}{2(\tau')^2} + \frac{n\tau'}{2}.$$

Now applying Lemma 9.25 we see that for any $0 < \tau'$ and any $q' \in M_k$ we have

$$l_{x_k}(q', \tau') \geq -\frac{n}{2(\tau')^2} - \frac{n\tau'}{2} - 1 + c_1 \frac{d^2_{g_k(-\tau')}(q_k, q')}{\tau'}$$

$$\geq -\frac{n}{2(\tau')^2} - \frac{n\tau'}{2} - 1 + c_1 \frac{d^2_{g_k(-\tau_0)}(q_k, q')}{\tau'}.$$

In the last inequality, we use the fact that the Ricci curvature is positive so that the metric is decreasing under the Ricci flow. $\square$

Since the time-slices of all the flows in question have non-negative curvature, by Theorem 1.34 the volume of the ball of radius s is at most ωs^n where ω is the volume of the ball of radius 1 in $\mathbb{R}^n$. Since the l_k converge uniformly to l_∞ on compact sets and since the metrics h_k converge uniformly in the C^∞-topology on compact sets to g_∞, it follows that for any $\epsilon > 0$, for any $0 < \tau_0 \leq \tau' < \infty$ there is a radius r such that for every k and any $\tau \in [\tau_0, \tau']$ the integral

$$\int_{M_\infty \setminus B_{h_k(-\tau_0)}(q_k, r)} e^{-l_k(q, \tau)} dq < \epsilon.$$

It follows by Lebesgue dominated convergence that

$$\int_{M_\infty \setminus B_{g_\infty(-\tau_0)}(q_\infty, r)} e^{-l_\infty(q, \tau)} dq \leq \epsilon.$$

CLAIM 9.27. *Fix a compact interval $[-\tau, -\tau_0] \subset (-\infty, 0)$. Let f be a locally Lipschitz function that is defined on $M_\infty \times [-\tau, -\tau_0]$ and such that there is a constant C with the property that $f(q, \tau')$ by C times $\max(l_\infty(q, \tau'), 1)$. Then the distribution $\mathcal{D}_1 = fe^{-l_\infty}$ is absolutely convergent in the following sense. For any bounded smooth function φ defined on all of $M_\infty \times [-\tau, -\tau_0]$ and any sequence of compactly supported, non-negative smooth functions ψ_k, bounded above by 1 everywhere that are eventually 1 on every compact subset, the following limit exists and is finite:*

$$\lim_{k \to \infty} \mathcal{D}_1(\varphi \psi_k).$$

Furthermore, the limit is independent of the choice of the ψ_k with the given properties.

PROOF. It follows from the above discussion that there are constants $c > 0$ and a ball $B \subset M_\infty$ centered at q_∞ such that on $M_\infty \times [-\tau, -\tau_0] \setminus B \times [\tau, -\tau_0]$ the function l_∞ is greater than $cd^2_{g_\infty(-\tau_0)}(q_\infty, \cdot) - C'$. Thus, fe^{-l_∞} has fixed exponential decay at infinity. Since the Riemann curvature of $M_\infty \times \{\tau'\}$ is non-negative for every τ', the flow is distance decreasing, and there is a fixed polynomial upper bound to the growth rate of volume at infinity. This leads to the claimed convergence property. □

COROLLARY 9.28. *Each of the distributions $|\nabla l_\infty|^2 e^{-l_\infty}$, Re^{-l_∞}, and $|(\partial l_\infty / \partial \tau)|e^{-l_\infty}$ is absolutely convergent in the sense of the above claim.*

PROOF. By Corollary 9.10, the Lipschitz functions $|\nabla l_\infty|^2$, $|\partial l_\infty / \partial \tau|$ and R are at most a constant multiple of l_∞. Hence, the corollary follows from the previous claim. □

There is a slightly weaker statement that is true for $\triangle e^{-l_\infty}$.

CLAIM 9.29. *Suppose that φ and ψ_k are as in Claim 9.27, but in addition φ and all the ψ_k are uniformly Lipschitz. Then*

$$\lim_{k\to\infty} \int_{M_\infty} \varphi \psi_k \triangle e^{-l_\infty} d\mathrm{vol}_{g_\infty}$$

converges absolutely.

PROOF. This time the value of the distribution on a compactly supported smooth function ρ is given by the integral of

$$-\langle \nabla \rho, \nabla e^{-l_\infty} \rangle = \langle \nabla \rho, \nabla l_\infty \rangle e^{-l_\infty}.$$

Since $|\nabla l_\infty|$ is less than or equal to the maximum of 1 and $|\nabla l_\infty|^2$, it follows immediately, that if $|\nabla \rho|$ is bounded, then the integral is absolutely convergent. From this the claim follows easily. □

COROLLARY 9.30. *Fix $0 < \tau_0 < \tau_1 < \infty$. Let f be a non-negative, smooth bounded function on $M_\infty \times [\tau_0, \tau_1]$ with (spatial) gradient of bounded norm. Then*

$$\int_{M_\infty \times [\tau_0,\tau_1]} \left(\frac{\partial l_\infty}{\partial \tau} + |\nabla l_\infty|^2 - R + \frac{n}{2\tau} - \triangle l_\infty \right) f\tau^{-n/2} e^{-l_\infty} d\mathrm{vol}_{g_\infty} d\tau \geq 0.$$

PROOF. For the interval $[\tau_0, \tau']$ we construct a sequence of uniformly Lipschitz functions ψ_k on $M_\infty \times [\tau_0, \tau']$ that are non-negative, bounded above by 1 and eventually 1 on every compact set. Let $\rho(x)$ be a smooth bump function which is 1 for x less than $1/4$ and is 0 from $x \geq 3/4$ and is everywhere between 0 and 1. For any k sufficiently large let ψ_k be the composition of $\rho(d_{g_\infty(-\tau_0)}(q_\infty, \cdot) - k)$. Being compositions of ρ with Lipschitz functions with Lipschitz constant 1, the ψ_k are a uniformly Lipschitz family of functions on $M_\infty \times \{-\tau\}$. Clearly then they form a uniformly Lipschitz family on $M_\infty \times [\tau_0, \tau']$ as required. This allows us to define any of the above distributions on Lipschitz functions on $M_\infty \times [\tau_0, \tau']$.

Take a family ψ_k of uniformly Lipschitz functions, each bounded between 0 and 1 and eventually 1 on every compact subset of $M_\infty \times [\tau_0, \tau_1]$. Then the family $f\psi_k$ is a uniformly Lipschitz family of compactly supported functions. Hence, we can apply Claims 9.27 and 9.29 to establish that the integral in question is the limit of an absolutely convergent sequence. By Corollary 9.24 each term in the sequence is non-positive. □

2.6. Completion of the proof of Proposition 9.20.
Lebesgue dominated convergence implies that the following limit exists:

$$\lim_{k \to \infty} \widetilde{V}_k(\tau) \equiv \widetilde{V}_\infty(\tau) = \int_{M_\infty \times \{-\tau\}} \tau^{-n/2} e^{-l_\infty(q,\tau)} dvol_{g_\infty(\tau)}.$$

By Corollary 9.14, the function $\tau \to \widetilde{V}_\infty(\tau)$ is constant. On the other hand, note that for any $0 < \tau_0 < \tau_1 < \infty$, we have

$$\widetilde{V}_\infty(\tau_1) - \widetilde{V}_\infty(\tau_0) = \int_{\tau_0}^{\tau_1} \frac{d\widetilde{V}_\infty}{d\tau} d\tau$$

$$= \int_{\tau_0}^{\tau_1} \int_{M_\infty} \left(\frac{\partial l_\infty}{\partial \tau} - R + \frac{n}{2\tau}\right) \left(\tau^{-n/2} e^{-l_\infty(q,\tau)} dvol_{g_\infty(\tau)}\right).$$

According to Corollary 9.28 this is an absolutely convergent integral, and so this integral is zero.

CLAIM 9.31.

$$\int_{M_\infty \times [\tau_0, \tau_1]} \triangle e^{-l_\infty} dvol_{g_\infty} d\tau = \int_{M_\infty \times [\tau_0, \tau_1]} \left(|\nabla l_\infty|^2 - \triangle l_\infty\right) e^{-l_\infty} dvol_{g_\infty} d\tau$$
$$= 0.$$

PROOF. Since we are integrating against the constant function 1, this result is clear, given the convergence result, Corollary 9.28, necessary to show that this integral is well defined. □

Adding these two results together gives us
(9.13)
$$\int_{\tau_0}^{\tau_1} \int_{M_\infty \times \{-\tau\}} \left(\frac{\partial l_\infty}{\partial \tau} + |\nabla l_\infty|^2 - R + \frac{n}{2\tau} - \triangle l_\infty\right) \tau^{-n/2} e^{-l_\infty} dvol_{g_\infty} = 0.$$

Now let φ be any compactly supported, non-negative smooth function. By scaling by a positive constant, we can assume that $\varphi \leq 1$ everywhere.

Let $\widetilde{\mathcal{D}}$ denote the distribution given by

$$\widetilde{\mathcal{D}}(\varphi) = \int_{M_\infty \times [\tau_0, \tau_1]} \varphi \left(\frac{\partial l_\infty}{\partial \tau} + |\nabla l_\infty|^2 - R + \frac{n}{2\tau} - \triangle l_\infty \right) \tau^{-n/2} e^{-l_\infty} \, d\mathrm{vol}_{y_\infty} d\tau$$

Then we have seen that $\widetilde{\mathcal{D}}$ extends to a functional on bounded smooth functions of bounded gradient. Furthermore, according to Equation (9.13), we have $\widetilde{\mathcal{D}}(1) = 0$. Thus,

$$0 = \mathcal{D}(1) = \mathcal{D}(\varphi) + \mathcal{D}(1 - \varphi).$$

Since both φ and $1 - \varphi$ are non-negative, it follows from Corollary 9.30, that $\mathcal{D}(\varphi)$ and $\mathcal{D}(1 - \varphi)$ are each ≥ 0. Since their sum is zero, it must be the case that each is individually zero.

This proves that the Inequality (9.9) is actually an equality in the weak sense, i.e., an equality of distributions on $M_\infty \times [\tau_0, \tau')$. Taking limits we see:

$$(9.14) \qquad \widetilde{\mathcal{D}} = \left(\frac{\partial l_\infty}{\partial \tau} + |\nabla l_\infty|^2 - R + \frac{n}{2\tau} - \triangle l_\infty \right) \tau^{-n/2} e^{-l_\infty} = 0,$$

in the weak sense on all of $M \times (-\infty, 0)$. Of course, this implies that

$$\frac{\partial l_\infty}{\partial \tau} + |\nabla l_\infty|^2 - R + \frac{n}{2\tau} - \triangle l_\infty = 0$$

in the weak sense.

It now follows by parabolic regularity that l_∞ is a smooth function on $M_\infty \times (-\infty, 0)$ and that Equation (9.14) holds in the usual sense.

Now from the last two equations in Corollary 6.51 and the convergence of the l_{x_k} to l_∞, we conclude that the following equation also holds:

$$(9.15) \qquad 2\triangle l_\infty - |\nabla l_\infty|^2 + R + \frac{l_\infty - n}{\tau} = 0.$$

This completes the proof of Proposition 9.20.

2.7. The gradient shrinking soliton equation. Now we return to the proof of Theorem 9.11. We have shown that the limiting Ricci flow referred to in that result exists, and we have established that the limit l_∞ of the length functions l_{x_k} is a smooth function and satisfies the differential equalities given in Proposition 9.20. We shall use these to establish the gradient shrinking soliton equation, Equation (9.2), for the limit for $f = l_\infty$.

PROPOSITION 9.32. *The following equation holds on $M_\infty \times (-\infty, 0)$:*

$$\mathrm{Ric}_{g_\infty(t)} + \mathrm{Hess}^{g_\infty(t)}(l_\infty(\cdot, \tau)) - \frac{1}{2\tau} g_\infty(t) = 0,$$

where $\tau = -t$,

PROOF. This result will follow immediately from:

2. THE ASYMPTOTIC GRADIENT SHRINKING SOLITON FOR κ-SOLUTIONS

LEMMA 9.33. *Let $(M, g(t))$, $0 \le t \le T$, be an n-dimensional Ricci flow, and let $f\colon M \times [0,T] \to \mathbb{R}$ be a smooth function. As usual set $\tau = T - t$. Then the function*
$$u = (4\pi\tau)^{-\frac{n}{2}} e^{-f}$$
satisfies the conjugate heat equation
$$-\frac{\partial u}{\partial t} - \triangle u + Ru = 0,$$
if and only if we have
$$\frac{\partial f}{\partial t} + \triangle f - |\nabla f|^2 + R - \frac{n}{2\tau} = 0.$$
Assuming that u satisfies the conjugate heat equation, then setting
$$v = \left[\tau\left(2\triangle f - |\nabla f|^2 + R\right) + f - n\right] u,$$
we have
$$-\frac{\partial v}{\partial t} - \triangle v + Rv = -2\tau \left|\mathrm{Ric}_g + \mathrm{Hess}^g(f) - \frac{1}{2\tau} g\right|^2 u.$$

Let us assume the lemma for a moment and use it to complete the proof of the proposition.

We apply the lemma to the limiting Ricci flow $(M_\infty, g_\infty(t))$ with the function $f = l_\infty$. According to Proposition 9.20 and the first statement in Lemma 9.33, the function u satisfies the conjugate heat equation. Thus, according to the second statement in Lemma 9.33, setting
$$v = \left[\tau\left(2\triangle f - |\nabla f|^2 + R\right) + f - n\right] u,$$
we have
$$\frac{\partial v}{\partial \tau} - \triangle v + Rv = -2\tau \left|\mathrm{Ric}_g + \mathrm{Hess}(f) - \frac{1}{2\tau} g\right|^2 u.$$
On the other hand, the second equality in Proposition 9.20 shows that $v = 0$. Since u is nowhere zero, this implies that
$$\mathrm{Ric}_{g_\infty} + \mathrm{Hess}^{g_\infty}(f) - \frac{1}{2\tau} g_\infty = 0.$$
This completes the proof of the proposition assuming the lemma. $\square$

Now we turn to the proof of the lemma.

PROOF. *(of Lemma 9.33)* Direct computation shows that
$$-\frac{\partial u}{\partial t} - \triangle u + Ru = \left(-\frac{n}{2\tau} + \frac{\partial f}{\partial t} + \triangle f - |\nabla f|^2 + R\right) u.$$
From this, the first statement of the lemma is clear. Let
$$H = \left[\tau(2\triangle f - |\nabla f|^2 + R) + f - n\right]$$
so that $v = Hu$. Then, of course,
$$\frac{\partial v}{\partial t} = \frac{\partial H}{\partial t} u + H \frac{\partial u}{\partial t} \quad \text{and}$$

$$\triangle v = \triangle H \cdot u + 2\langle \nabla H, \nabla u\rangle + H\triangle u.$$

Since u satisfies the conjugate heat equation, we have
$$\frac{\partial v}{\partial t} - \triangle v + Rv = \left(-\frac{\partial H}{\partial t} - \triangle H\right) u - 2\langle \nabla H, \nabla u\rangle.$$

Differentiating the definition of H yields
$$(9.16) \quad \frac{\partial H}{\partial t} = -(2\triangle f - |\nabla f|^2 + R) + \frac{\partial f}{\partial t} + \tau\left(2\frac{\partial}{\partial t}\triangle f - \frac{\partial}{\partial t}(|\nabla f|^2) + \frac{\partial R}{\partial t}\right).$$

CLAIM 9.34.
$$\frac{\partial}{\partial t}\triangle f = \triangle(\frac{\partial f}{\partial t}) + 2\langle \mathrm{Ric}, \mathrm{Hess}(f)\rangle.$$

PROOF. We work in local coordinates. We have
$$\triangle f = g^{ij}\nabla_i\nabla_j f = g^{ij}(\partial_i\partial_j f - \Gamma^k_{ij}\partial_k f),$$
so that from the Ricci flow equation we have
$$\frac{\partial}{\partial t}\triangle f = 2\mathrm{Ric}^{ij}\mathrm{Hess}(f)_{ij} + g^{ij}\frac{\partial}{\partial t}(\mathrm{Hess}(f)_{ij})$$
$$= 2\mathrm{Ric}^{ij}\mathrm{Hess}(f)_{ij} + g^{ij}\mathrm{Hess}\left(\frac{\partial f}{\partial t}\right)_{ij} - g^{ij}\frac{\partial \Gamma^k_{ij}}{\partial t}\partial_k f.$$

Since the first term is $2\langle \mathrm{Ric}, \mathrm{Hess}(f)\rangle$ and the second is $\triangle(\frac{\partial f}{\partial t})$, to complete the proof of the claim, we must show that the last term of this equation vanishes. In order to simplify the computations, we assume that the metric is standard to second order at the point and time under consideration. Then, using the Ricci flow equation, the definition of the Christoffel symbols in terms of the metric, and the fact that g_{ij} is the identity matrix at the given point and time and that its covariant derivatives in all spatial directions vanish at this point and time, we get
$$g^{ij}\frac{\partial \Gamma^k_{ij}}{\partial t} = g^{kl}g^{ij}\left(-(\nabla_j\mathrm{Ric})_{li} - (\nabla_i\mathrm{Ric})_{lj} + (\nabla_l\mathrm{Ric})_{ij}\right).$$

This expression vanishes by the second Bianchi identity (Claim 1.5). This completes the proof of the claim. $\square$

We also have
$$\frac{\partial}{\partial t}(|\nabla f|^2) = 2\mathrm{Ric}(\nabla f, \nabla f) + 2\langle \nabla\frac{\partial f}{\partial t}, \nabla f\rangle.$$

(Here ∇f is a one-form, which explains the positive sign in the Ricci term.)

Plugging this and Claim 9.34 into Equation (9.16) yields
$$\frac{\partial H}{\partial t} = -2\triangle f + |\nabla f|^2 - R + \frac{\partial f}{\partial t}$$
$$+ \tau\left(4\langle \mathrm{Ric}, \mathrm{Hess}(f)\rangle + 2\triangle\frac{\partial f}{\partial t} - 2\mathrm{Ric}(\nabla f, \nabla f) - 2\langle \nabla\frac{\partial f}{\partial t}, \nabla f\rangle + \frac{\partial R}{\partial t}\right).$$

2. THE ASYMPTOTIC GRADIENT SHRINKING SOLITON FOR κ-SOLUTIONS

Also,
$$\triangle H = \triangle f + \tau\left(2\triangle^2 f - \triangle(|\nabla f|^2) + \triangle R\right).$$

Since u satisfies the conjugate heat equation, from the first part of the lemma we have

(9.17) $$\frac{\partial f}{\partial t} = -\triangle f + |\nabla f|^2 - R + \frac{n}{2\tau}.$$

Putting all this together and using the Equation (3.7) for $\partial R/\partial t$ yields

$$\frac{\partial H}{\partial t} + \triangle H = -\triangle f + |\nabla f|^2 + \frac{\partial f}{\partial t} - R$$
$$+ \tau\Big(4\langle\text{Ric},\text{Hess}(f)\rangle + 2\triangle\frac{\partial f}{\partial t} + 2\triangle^2 f - 2\text{Ric}(\nabla f, \nabla f)$$
$$- \triangle(|\nabla f|^2) - 2\langle\nabla\frac{\partial f}{\partial t},\nabla f\rangle + 2\triangle R + 2|\text{Ric}|^2\Big)$$
$$= -\triangle f + |\nabla f|^2 + \frac{\partial f}{\partial t} - R$$
$$+ \tau\Big(4\langle\text{Ric},\text{Hess}(f)\rangle + 2\triangle(|\nabla f|^2 - R - 2\text{Ric}(\nabla f, \nabla f)$$
$$- \triangle(|\nabla f|^2) - 2\langle\nabla\frac{\partial f}{\partial t},\nabla f\rangle + 2\triangle R + 2|\text{Ric}|^2\Big)$$
$$= -\triangle f + |\nabla f|^2 + \frac{\partial f}{\partial t} - R + \tau\Big[4\langle\text{Ric},\text{Hess}(f)\rangle + \triangle(|\nabla f|^2)$$
$$- 2\text{Ric}(\nabla f, \nabla f) + 2\langle\nabla(\triangle f), \nabla f\rangle - 2\langle\nabla(|\nabla f|^2), \nabla f\rangle$$
$$+ 2\langle\nabla R, \nabla f\rangle + 2|\text{Ric}|^2\Big].$$

Similarly, we have
$$\frac{2\langle\nabla u, \nabla H\rangle}{u} = -2\langle\nabla f, \nabla H\rangle$$
$$= -2|\nabla f|^2 - 2\tau\langle\nabla f, (\nabla(2\triangle f) - |\nabla f|^2 + R)\rangle$$
$$= -2|\nabla f|^2 - \tau\left(4\langle\nabla f, \nabla(\triangle f)\rangle - 2\langle\nabla f, \nabla(|\nabla f|^2)\rangle\right.$$
$$\left.+ 2\langle\nabla f, \nabla R\rangle\right).$$

Thus,
$$\frac{\partial H}{\partial t} + \triangle H + \frac{2\langle\nabla u, \nabla H\rangle}{u} = -\triangle f - |\nabla f|^2 + \frac{\partial f}{\partial t} - R + \tau\Big[4\langle\text{Ric},\text{Hess}(f)\rangle$$
$$+ \triangle(|\nabla f|^2) - 2\text{Ric}(\nabla f, \nabla f) + 2|\text{Ric}|^2$$
$$- 2\langle\nabla f, \nabla(\triangle f)\rangle\Big].$$

CLAIM 9.35. *The following equality holds:*
$$\triangle(|\nabla f|^2) = 2\langle\nabla(\triangle f), \nabla f\rangle + 2\text{Ric}(\nabla f, \nabla f) + 2|\text{Hess}(f)|^2.$$

PROOF. We have
$$\triangle(|\nabla f|^2) = \triangle\langle\nabla f, \nabla f\rangle = \triangle\langle df, df\rangle = 2\langle\triangle df, df\rangle + 2\langle\nabla df, \nabla df\rangle.$$

The last term is $|\text{Hess}(f)|^2$. According to Lemma 1.10 we have $\triangle df = d(\triangle f) + \text{Ric}(\nabla f, \cdot)$. Plugging this in gives

$$\triangle(|\nabla f|^2) = 2\langle d(\triangle f), df\rangle + 2\langle \text{Ric}(\nabla f, \cdot), df\rangle + 2|\text{Hess}(f)|^2,$$

which is clearly another way of writing the claimed result. □

Using this we can simplify the above to

$$\frac{\partial H}{\partial t} + \triangle H + \frac{2\langle\nabla u, \nabla H\rangle}{u} = -\triangle f - |\nabla f|^2 + \frac{\partial f}{\partial t} - R$$
$$+ \tau\left(4\langle\text{Ric}, \text{Hess}(f)\rangle + 2|\text{Hess}(f)|^2 + 2|\text{Ric}|^2\right).$$

Now using Equation (9.17) we have

$$\frac{\partial H}{\partial t} + \triangle H + \frac{2\langle\nabla u, \nabla H\rangle}{u} = -2\triangle f - 2R + \frac{n}{2\tau}$$
$$+ \tau\left(4\langle\text{Ric}, \text{Hess}(f)\rangle + 2|\text{Ric}|^2 + 2|\text{Hess}(f)|^2\right)$$
$$= 2\tau\Big(2\langle\text{Ric}, \text{Hess}(f)\rangle + |\text{Ric}|^2 + |\text{Hess}(f)|^2$$
$$- \frac{\triangle f}{\tau} - \frac{R}{\tau} + \frac{n}{4\tau^2}\Big)$$
$$= 2\tau\Big|\text{Ric} + \text{Hess}(f) - \frac{1}{2\tau}g_\infty\Big|^2.$$

Since

$$-\frac{\partial v}{\partial t} - \triangle v + Rv = -u\left(\frac{\partial H}{\partial t} + \triangle H + \frac{2\langle\nabla u, \nabla H\rangle}{u}\right),$$

this proves the lemma. □

At this point, setting $f = l_\infty$, we have established all the results claimed in Theorem 9.11 except for the fact that the limit is not flat. This we establish in the next section.

2.8. Completion of the proof of Theorem 9.11. To complete the proof of Theorem 9.11 we need only show that for every $t \in (-\infty, 0)$ the manifold $(M_\infty, g_\infty(t))$ is non-flat.

CLAIM 9.36. *If, for some $t < 0$, the Riemannian manifold $(M_\infty, g_\infty(t))$ is flat, then there is an isometry from $\mathbb{R}^n$ to $(M_\infty, g_\infty(t))$ and the pullback under this isometry of the function $l_\infty(x, \tau)$ is the function $|x|^2/4\tau + \langle x, a\rangle + b\cdot\tau$ for some $a \in \mathbb{R}^n$ and $b \in \mathbb{R}$.*

PROOF. We know $f = l_\infty(\cdot, \tau)$ solves the equation given in Lemma 9.33 and hence by the above argument, f also satisfies the equation given in Proposition 9.32. If the limit is flat, then the equation becomes

$$\text{Hess}(f) = \frac{1}{2\tau}g.$$

The universal covering of $(M_\infty, g_\infty(t))$ is isometric to $\mathbb{R}^n$. Choose an identification with $\mathbb{R}^n$, and lift f to the universal cover. Call the result $\widetilde{f}$. Then $\widetilde{f}$

satisfies $\text{Hess}(\widetilde{f}) = \frac{1}{2\tau}\widetilde{g}$, where $\widetilde{g}$ is the usual Euclidean metric on $\mathbb{R}^n$. This means that $\widetilde{f} - |x|^2/4\tau$ is an affine linear function. Clearly, then $\widetilde{f}$ is not invariant under any free action of a non-trivial group, so that the universal covering in question is the trivial cover. This completes the proof of the claim. □

If $(M_\infty, g_\infty(t))$ is flat for some $t < 0$, then by the above $(M_\infty, g_\infty(t))$ is isometric to $\mathbb{R}^n$. According to Proposition 7.27 this implies that $\widetilde{V}_\infty(\tau) = (4\pi)^{n/2}$. This contradicts Corollary 9.14, and the contradiction establishes that $(M_\infty, g_\infty(t))$ is not flat for any $t < 0$. Together with Proposition 9.32, this completes the proof of Theorem 9.11. The flow $(M_\infty, g_\infty(t))$, $-\infty < t < 0$, is a non-flat, κ-non-collapsed Ricci flow with non-negative curvature operator that satisfies the gradient shrinking soliton equation, Equation (9.2).

To emphasize once again, we do not claim that $(M_\infty, g_\infty(t))$ is a κ-solution, since we do not claim that each time-slice has bounded curvature operator.

3. Splitting results at infinity

3.1. Point-picking. There is a very simple, general result about Riemannian manifolds that we shall use in various contexts to prove that certain types of Ricci flows split at infinity as a product with $\mathbb{R}$.

LEMMA 9.37. *Let (M, g) be a Riemannian manifold and let $p \in M$ and $r > 0$ be given. Suppose that $B(p, 2r)$ has compact closure in M and suppose that $f\colon B(p, 2r) \times (-2r, 0] \to \mathbb{R}$ is a continuous, bounded function with $f(p, 0) > 0$. Then there is a point $(q, t) \in B(p, 2r) \times (-2r, 0]$ with the following properties:*
  (1) $f(q, t) \geq f(p, 0)$.
  (2) *Setting* $\alpha = f(p, 0)/f(q, t)$ *we have* $d(p, q) \leq 2r(1 - \alpha)$ *and* $t \geq -2r(1 - \alpha)$.
  (3) $f(q', t') < 2f(q, t)$ *for all* $(q', t') \in B(q, \alpha r) \times (t - \alpha r, t]$.

PROOF. Consider sequences of points $x_0 = (p, 0), x_1 = (p_1, t_1), \ldots, x_j = (p_j, t_j)$ in $B(p, 2r) \times (-2r, 0]$ with the following properties:
  (1) $f(x_i) \geq 2f(x_{i-1})$.
  (2) Setting $r_i = rf(x_0)/f(x_{i-1})$, then $r_i \leq 2^{i-1}r$, and we have that
  $$x_i \in B(p_{i-1}, r_i) \times (t_{i-1} - r_i, t_{i-1}].$$

Of course, there is exactly one such sequence with $j = 0$: it has $x_0 = (p, 0)$. Suppose we have such a sequence defined for some $j \geq 0$. If follows immediately from the properties of the sequence that $f(p_j, t_j) \geq 2^j f(p, 0)$, that
$$t_j \geq -r(1 + 2^{-1} + \cdots + 2^{1-j}),$$

and that $r_{j+1} \leq 2^{-j}r$. It also follows immediately from the triangle inequality that $d(p, p_j) \leq r(1 + 2^{-1} + \cdots + 2^{1-j})$. This means that
$$B(p_j, r_{j+1}) \times (t_j - r_{j+1}, t_j] \subset B(p, 2r) \times (-2r, 0].$$
Either the point x_j satisfies the conclusion of the lemma, or we can find $x_{j+1} \in B(p_j, r_{j+1}) \times (t_j - r_{j+1}, t_j]$ with $f(x_{j+1}) \geq 2f(x_j)$. In the latter case we extend our sequence by one term. This shows that either the process terminates at some j, in which case x_j satisfies the conclusion of the lemma, or it continues indefinitely. But it cannot continue indefinitely since f is bounded on $B(p, 2r) \times (-2r, 0]$. □

One special case worth stating separately is when f is independent of t.

COROLLARY 9.38. *Let (M, g) be a Riemannian manifold and let $p \in M$ and $r > 0$ be given. Suppose that $B(p, 2r)$ has compact closure in M and suppose that $f \colon B(p, 2r) \to \mathbb{R}$ is a continuous, bounded function with $f(p) > 0$. Then there is a point $q \in B(x, 2r)$ with the following properties:*

(1) *$f(q) \geq f(p)$.*
(2) *Setting $\alpha = f(p)/f(q)$ we have $d(p, q) \leq 2r(1-\alpha)$ and $f(q') < 2f(q)$ for all $q' \in B(q, \alpha r)$.*

PROOF. Apply the previous lemma to $\widehat{f} \colon B(p, 2r) \times (-2r, 0] \to \mathbb{R}$ defined by $\widehat{f}(p, t) = f(p)$. □

3.2. Splitting results. Here we prove a splitting result for ancient solutions of non-negative curvature. They are both based on Theorem 5.35.

PROPOSITION 9.39. *Suppose that $(M, g(t))$, $-\infty < t < 0$, is a κ-non-collapsed Ricci flow of dimension[1] $n \leq 3$. Suppose that $(M, g(t))$ is a complete, non-compact, non-flat Riemannian manifold with non-negative curvature operator for each t. Suppose that $\partial R(q, t)/\partial t \geq 0$ for all $q \in M$ and all $t < 0$. Fix $p \in M$. Suppose that there is a sequence of points $p_i \in M$ going to infinity with the property that*
$$\lim_{i \to \infty} R(p_i, -1) d^2_{g(-1)}(p, p_i) = \infty.$$
Then there is a sequence of points $q_i \in M$ tending to infinity such that, setting $Q_i = R(q_i, -1)$, we have $\lim_{i \to \infty} d^2(p, q_i) Q_i = \infty$. Furthermore, setting $g_i(t) = Q_i g(Q_i^{-1}(t+1) - 1)$, the sequence of based flows
$$(M, g_i(t), (q_i, -1)), \quad -\infty < t \leq -1,$$
converges smoothly to $(N^{n-1}, h(t)) \times (\mathbb{R}, ds^2)$, a product Ricci flow defined for $-\infty < t \leq -1$ with $(N^{n-1}, h(-1))$ being non-flat and of bounded, non-negative curvature.

COROLLARY 9.40. *There is no 2-dimensional flow satisfying the hypotheses of Proposition 9.39.*

---

[1] This result in fact holds in all dimensions.

PROOF. (of Proposition 9.39) Take a sequence $p_i \in M$ such that
$$d^2_{g(-1)}(p, p_i) R(p_i, -1) \to \infty$$
as $i \to \infty$. We set $d_i = d_{g(-1)}(p, p_i)$ and we set $B_i = B(p_i, -1, d_i/2)$, and we let $f \colon B_i \to \mathbb{R}$ be the square root of the scalar curvature. Since $(M, g(-1))$ is complete, B_i has compact closure in M, and consequently f is a bounded continuous function on B_i. Applying Corollary 9.38 to $(B_i, g(-1))$ and f, we conclude that there is a point $q_i \in B_i$ with the following properties:

(1) $R(q_i, -1) \geq R(p_i, -1)$.
(2) $B'_i = B(q_i, -1, (d_i R(p_i, -1)^{1/2})/(4R(q_i, t_i)^{1/2}) \subset B(p_i, -1, d_i/2)$.
(3) $R(q', -1) \leq 4R(q_i, -1)$ for all $(q', -1) \in B'_i$.

Since $d_{g(-1)}(p, q_i) \geq d_i/2$, it is also the case that $d^2_{g(-1)}(p, q_i) R(q_i, -1)$ tends to infinity as i tends to infinity. Because of our assumption on the time derivative of R, it follows that $R(q', t) \leq 4R(q_i, -1)$ for all $q' \in B'_i$ and for all $t \leq -1$.

Set $Q_i = R(q_i, -1)$. Let $M_i = M$, and set $x_i = (q_i, -1)$. Lastly, set $g_i(t) = Q_i g(Q_i^{-1}(t+1) - 1)$. We consider the based Ricci flows
$$(M_i, g_i(t), x_i), \quad -\infty < t \leq -1.$$
We see that $R_{g_i}(q', t) \leq 4$ for all
$$(q', t) \in B_{g_i}(q_i, -1, d_i R(p_i, -1)^{1/2}/4) \times (-\infty, -1].$$
Since the original Ricci flows are κ-non-collapsed, the same is true for the rescaled flows. Since $d_i R(p_i, -1)^{1/2}/4 \to \infty$, by Theorem 5.15 there is a geometric limit flow $(M_\infty, g_\infty(t), (q_\infty, -1))$ defined for $t \in (-\infty, -1]$. Of course, by taking limits we see that $(M_\infty, g_\infty(t))$ is κ-non-collapsed, its scalar curvature is bounded above by 4, and its curvature operator is non-negative. It follows that $(M_\infty, g_\infty(t))$ has bounded curvature.

To complete the proof we show that the Ricci flow $(M_\infty, g_\infty(t))$ splits as a product of a line with a Ricci flow of one lower dimension. By construction $(M_\infty, g_\infty(-1))$ is the geometric limit constructed from $(M, g(-1))$ in the following manner. We have a sequence of points q_i tending to infinity in M and constants $\lambda_i = R(q_i, -1)$ with the property that $\lambda_i d^2_{g(-1)}(p, q_i)$ tending to infinity such that $(M_\infty, g_\infty(-1))$ is the geometric limit of $(M, \lambda_i g(-1), q_i)$. Thus, according to Theorem 5.35, the limit $(M_\infty, g_\infty(-1))$ splits as a Riemannian product with a line. If the dimension of M_∞ is 2, then this is a contradiction: We have that $(M_\infty, g_\infty(-1))$ splits as the Riemannian product of a line and a one-manifold and hence is flat, but $R(q_\infty, -1) = 1$. Suppose that the dimension of M_∞ is 3. Since $(M_\infty, g_\infty(-1))$ splits as a product with a line, it follows from the maximum principle (Corollary 4.19) that the entire flow splits as a product with a line, and the Ricci flow on the surface has strictly positive curvature. □

4. Classification of gradient shrinking solitons

In this section we fix $\kappa > 0$ and classify κ-solutions $(M, g_\infty(t))$, $-\infty < t < 0$, that satisfy the gradient shrinking soliton equation at the time-slice $t = -1$ in the sense that there is a function $f\colon M \to \mathbb{R}$ such that

$$(9.18) \qquad \mathrm{Ric}_{g_\infty(-1)} + \mathrm{Hess}^{g_\infty(-1)}(f) - \frac{1}{2}g_\infty(-1) = 0.$$

This will give a classification of the 2- and 3-dimensional asymptotic gradient shrinking solitons constructed in Theorem 9.11.

Let us give some examples in dimensions 2 and 3 of ancient solutions that have such functions. It turns out, as we shall see below, that in dimensions 2 and 3 the only such are compact manifolds of constant positive curvature – i.e., Riemannian manifolds finitely covered by the round sphere. We can create another, non-flat gradient shrinking soliton in dimension 3 by taking (M, g_{-1}) equal to the product of (S^2, h_{-1}), the round sphere of Gaussian curvature $1/2$, with the real line (with the metric on the real line denoted ds^2) and setting $g(t) = |t|h_{-1} + ds^2$ for all $t < 0$. We define $f\colon M \times (-\infty, 0) \to \mathbb{R}$ by $f(p, t) = s^2/4|t|$ where $s\colon M \to \mathbb{R}$ is the projection onto the second factor. Then it is easy to see that

$$\mathrm{Ric}_{g(t)} + \mathrm{Hess}^{g(t)}(f) - \frac{1}{2|t|}g(t) = 0,$$

so that this is a gradient shrinking soliton. There is a free, orientation-preserving involution on this Ricci flow: the product of the sign change on $\mathbb{R}$ with the antipodal map on S^2. This preserves the family of metrics and hence there is an induced Ricci flow on the quotient. Since this involution also preserves the function f, the quotient is also a gradient shrinking soliton. These are the basic 3-dimensional examples. As the following theorem shows, they are all the κ-non-collapsed gradient shrinking solitons in dimension 3.

First we need a definition for a single Riemannian manifold analogous to a definition we have already made for Ricci flows.

DEFINITION 9.41. Let (M, g) be an n-dimensional complete Riemannian manifold and fix $\kappa > 0$. We say that (M, g) is κ-*non-collapsed* if for every $p \in M$ and any $r > 0$, if $|\mathrm{Rm}_g| \leq r^{-2}$ on $B(p, r)$ then $\mathrm{Vol}\, B(p, r) \geq \kappa r^n$.

Here is the theorem that we shall prove:

THEOREM 9.42. *Let (M, g) be a complete, non-flat Riemannian manifold of bounded non-negative curvature of dimension 2 or 3. Suppose that the Riemannian manifold (M, g) is κ-non-collapsed. Lastly, suppose that there is a C^2-function $f\colon M \to \mathbb{R}$ such that*

$$\mathrm{Ric}_g + \mathrm{Hess}^g(f) = \frac{1}{2}g.$$

Then there is a Ricci flow $(M, G(t))$, $-\infty < t < 0$, with $G(-1) = g$ and with $(M, G(t))$ isometric to $(M, |t|g)$ for every $t < 0$. In addition, $(M, G(t))$ is of one of the following three types:

(1) The flow $(M, G(t))$, $-\infty < t < 0$, is a shrinking family of compact, round (constant positive curvature) manifolds.
(2) The flow $(M, G(t))$, $-\infty < t < 0$, is a product of a shrinking family of round 2-spheres with the real line.
(3) $(M, G(t))$ is isomorphic to the quotient family of metrics of the product of a shrinking family of round 2-spheres and the real line under the action of an isometric involution.

Now let us begin the proof of Theorem 9.42

4.1. Integrating ∇f. Since the curvature of (M, g) is bounded, it follows immediately from the gradient shrinking soliton equation that $\text{Hess}^g(f)$ is bounded. Fix a point $p \in M$. For any $q \in M$ let $\gamma(s)$ be a minimal geodesic from p to q parameterized at unit length. Since

$$\frac{d}{ds}(|\nabla f(\gamma(s))|)^2 = 2\langle \text{Hess}(f)(\gamma'(s)), \nabla f(\gamma(s)) \rangle,$$

it follows that

$$\frac{d}{ds}(|\nabla f(\gamma(s))|) \leq C,$$

where C is an upper bound for $|\text{Hess}(f)|$. By integrating, it follows that

$$|\nabla f(q)| \leq C d_g(p, q) + |\nabla f(p)|.$$

This means that any flow line $\lambda(t)$ for ∇f satisfies

$$\frac{d}{dt} d_g(p, \lambda(t)) \leq C d_g(p, \lambda(t)) + |\nabla f(p)|,$$

and hence these flow lines do not escape to infinity in finite time. It follows that there is a flow $\Phi_t \colon M \to M$ defined for all time with $\Phi_0 = \text{Id}$ and $\partial \Phi_t / \partial t = \nabla f$. We consider the one-parameter family of diffeomorphisms $\Phi_{-\log(|t|)} \colon M \to M$ and define

(9.19) $$G(t) = |t|\Phi^*_{-\log(|t|)} g. \quad -\infty < t < 0.$$

We compute

$$\frac{\partial G}{\partial t} = -\Phi^*_{h(t)} g + 2\Phi^*_{h(t)} \text{Hess}^g(f) = -2\Phi^*_{h(t)} \text{Ric}(g) = -2\text{Ric}(G(t)),$$

so that $G(t)$ is a Ricci flow. Clearly, every time-slice is a complete, non-flat manifold of non-negative bounded curvature. It is clear from the construction that $G(-1) = g$ and that $(M, G(t))$ is isometric to $(M, |t|g)$. This shows that (M, g) is the -1 time-slice of a Ricci flow $(M, G(t))$ defined for all $t < 0$, and that, furthermore, all the manifolds $(M, G(t))$ are equivalent up to diffeomorphism and scaling by $|t|$.

4.2. Case 1: M is compact and the curvature is strictly positive.

CLAIM 9.43. *Suppose that (M,g) and $f\colon M \to \mathbb{R}$ satisfies the hypotheses of Theorem 9.42 and that M is compact and of positive curvature. Then the Ricci flow $(M, G(t))$ with $G(-1) = g$ given in Equation (9.19) is a shrinking family of compact round manifolds.*

PROOF. The manifold $(M, G(t))$ given in Equation (9.19) is equivalent up to diffeomorphism and scaling by $|t|$ to (M, g). If the dimension of M is 3, then according to Hamilton's pinching toward positive curvature result (Theorem 4.23), the Ricci flow becomes singular in finite time and as it becomes singular the metric approaches constant curvature in the sense that the ratio of the largest sectional curvature to the smallest goes to 1. But this ratio is invariant under scaling and diffeomorphism, so that it must be the case that for each t, all the sectional curvatures of the metric $G(t)$ are equal; i.e., for each t the metric $G(t)$ is round. If the dimension of M is 2, then the results go back to Hamilton in [**31**]. According to Proposition 5.21 on p. 118 of [**13**], M is a shrinking family of constant positive curvature surfaces, which must be either S^2 or $\mathbb{R}P^2$. This completes the analysis in the compact case. □

From this result, we can easily deduce a complete classification of κ-solutions with compact asymptotic gradient shrinking soliton.

COROLLARY 9.44. *Suppose that $(M, g(t))$ is a κ-solution of dimension 3 with a compact asymptotic gradient shrinking soliton. Then the Ricci flow $(M, g(t))$ is isomorphic to a time-shifted version of its asymptotic gradient shrinking soliton.*

PROOF. We suppose that the compact asymptotic gradient shrinking soliton is the limit of the $(M, g_{\tau_n}(t), (q_n, -1))$ for some sequence of $\tau_n \to \infty$. Since by the discussion in the compact case, this limit is of constant positive curvature, it follows that for all n sufficiently large, M is diffeomorphic to the limit manifold and the metric $g_{\tau_n}(-1)$ is close to a metric of constant positive curvature. In particular, for all n sufficiently large, $(M, g_{\tau_n}(-1))$ is compact and of strictly positive curvature. Furthermore, as $n \to \infty$, $\tau_n \to \infty$ and Riemannian manifolds $(M, g_{\tau_n}(-1))$ become closer and closer to round in the sense that the ratio of its largest sectional curvature to its smallest sectional curvature goes to 1. Since this is a scale invariant ratio, the same is true for the sequence of Riemannian manifolds $(M, g(-\tau_n))$. In the case when the dimension of M is 3, by Hamilton's pinching toward round result or Ivey's theorem (see Theorem 4.23), this implies that the $(M, g(t))$ are all exactly round.

This proves that $(M, g(t))$ is a shrinking family of round metrics. The only invariants of such a family are the diffeomorphism type of M and the time Ω at which the flow becomes singular. Of course, M is diffeomorphic to its asymptotic soliton. Hence, the only remaining invariant is the singular time, and hence $(M, g(t))$ is equivalent to a time-shifted version of its asymptotic soliton. $\square$

4.3. Case 2: Non-strictly positively curved.

CLAIM 9.45. *Suppose that (M,g) and $f\colon M \to \mathbb{R}$ are as in the statement of Theorem 9.42 and that (M,g) does not have strictly positive curvature. Then $n = 3$ and the Ricci flow $(M, G(t))$ with $G(-1) = g$ given in Equation (9.19) has a one- or two-sheeted covering that is a product of a 2-dimensional κ-non-collapsed Ricci flow of positive curvature and a constant flat copy of $\mathbb{R}$. The curvature is bounded on each time-slice.*

PROOF. According to Corollary 4.20, Hamilton's strong maximum principle, the Ricci flow $(M, G(t))$ has a one- or two-sheeted covering that splits as a product of an evolving family of manifolds of one dimension less of positive curvature and a constant one-manifold. It follows immediately that $n = 3$. Let $\widetilde{f}$ be the lifting of f to this one- or two-sheeted covering. Let Y be a unit tangent vector in the direction of the one-manifold. Then it follows from Equation (9.18) that the value of the Hessian of $\widetilde{f}$ of (Y,Y) is 1. If the flat one-manifold factor is a circle then there can be no such function $\widetilde{f}$. Hence, it follows that the one- or two-sheeted covering is a product of an evolving surface with a constant copy of $\mathbb{R}$. Since (M, g) is κ-non-collapsed and of bounded curvature, $(M, G(t))$ is κ-non-collapsed and each time-slice has positive bounded curvature. These statements are also true for the flow of surfaces. $\square$

4.4. Case 3: M is non-compact and strictly positively curved.

Here the main result is that this case does not occur.

PROPOSITION 9.46. *There is no 2- or 3-dimensional Ricci flow satisfying the hypotheses of Theorem 9.42 with (M,g) non-compact and of positive curvature.*

We suppose that we have (M,g) as in Theorem 9.42 with (M,g) being non-compact and of positive curvature. Let n be the dimension of M, so that n is either 2 or 3. Taking the trace of the gradient shrinking soliton equation yields
$$R + \triangle f - \frac{n}{2} = 0,$$
and consequently that
$$dR + d(\triangle f) = 0.$$
Using Lemma 1.10 we rewrite this equation as
(9.20) $$dR + \triangle(df) - \mathrm{Ric}(\nabla f, \cdot) = 0.$$

On the other hand, taking the divergence of the gradient shrinking soliton equation and using the fact that $\nabla^* g = 0$ gives

$$\nabla^* \text{Ric} + \nabla^* \text{Hess}(f) = 0.$$

Of course,

$$\nabla^* \text{Hess}(f) = \nabla^*(\nabla \nabla f) = (\nabla^* \nabla)\nabla f = \triangle(df),$$

so that

$$\triangle(df) = -\nabla^* \text{Ric}.$$

Plugging this into Equation 9.20 gives

$$dR - \nabla^* \text{Ric} - \text{Ric}(\nabla f, \cdot) = 0.$$

Now invoking Lemma 1.9 we have

(9.21) $$dR = 2\text{Ric}(\nabla f, \cdot).$$

Fix a point $p \in M$. Let $\gamma(s)$; $0 \leq s \leq \overline{s}$, be a shortest geodesic (with respect to the metric g), parameterized at unit speed, emanating from p, and set $X(s) = \gamma'(s)$.

CLAIM 9.47. *There is a constant C independent of the choice of γ and of $\overline{s}$ such that*

$$\int_0^{\overline{s}} \text{Ric}(X, X) ds \leq C.$$

PROOF. Since the curvature is bounded, clearly it suffices to assume that $\overline{s} \gg 1$. Since γ is length-minimizing and parameterized at unit speed, it follows that it is a local minimum for the energy functional $E(\gamma) = \frac{1}{2}\int_0^{\overline{s}} |\gamma'(s)|^2 ds$ among all paths with the same end points. Thus, letting $\gamma_u(s) = \gamma(s, u)$ be a one-parameter family of variations (fixed at the endpoints) with $\gamma_0 = \gamma$ and with $d\gamma/du|_{u=0} = Y$, we see

$$0 \leq \delta_Y^2 E(\gamma_u) = \int_0^{\overline{s}} |\nabla_X Y|^2 + \langle \mathcal{R}(Y, X)Y, X \rangle ds.$$

We conclude that

(9.22) $$\int_0^{\overline{s}} \langle -\mathcal{R}(Y, X)Y, X \rangle ds \leq \int_0^{\overline{s}} |\nabla_X Y|^2 ds.$$

Fix an orthonormal basis $\{E_i\}_{i=1}^n$ at p with $E_n = X$, and let $\widetilde{E}_i$ denote the parallel translation of E_i along γ. (Of course, $\widetilde{E}_n = X$.) Then, for $i \leq n-1$, we define

$$Y_i = \begin{cases} s\widetilde{E}_i & \text{if } 0 \leq s \leq 1, \\ \widetilde{E}_i & \text{if } 1 \leq s \leq \overline{s} - 1, \\ (\overline{s} - s)\widetilde{E}_i & \text{if } \overline{s} - 1 \leq s \leq \overline{s}. \end{cases}$$

4. CLASSIFICATION OF GRADIENT SHRINKING SOLITONS

Adding up Equation (9.22) for each i gives

$$-\sum_{i=1}^{n-1}\int_0^{\bar{s}}\langle \mathcal{R}(Y_i,X)Y_i,X\rangle ds \le \sum_{i=1}^{n-1}\int_0^{\bar{s}}|\nabla_X Y_i|^2 ds.$$

Of course, since the $\widetilde{E}_i$ are parallel along γ, we have

$$|\nabla_X Y_i|^2 = \begin{cases} 1 & \text{if } 0\le s\le 1, \\ 0 & \text{if } 1\le s\le \bar{s}-1, \\ 1 & \text{if } \bar{s}-1\le s\le \bar{s}, \end{cases}$$

so that

$$\sum_{i=1}^{n-1}\int_0^{\bar{s}}|\nabla_X Y_i|^2 = 2(n-1).$$

On the other hand,

$$-\sum_{i=1}^{n-1}\langle \mathcal{R}(Y_i,X)(Y_i),X\rangle = \begin{cases} s^2\operatorname{Ric}(X,X) & \text{if } 0\le s\le 1, \\ \operatorname{Ric}(X,X) & \text{if } 1\le s\le \bar{s}-1, \\ (\bar{s}-s)^2\operatorname{Ric}(X,X) & \text{if } \bar{s}-1\le s\le \bar{s}. \end{cases}$$

Since the curvature is bounded and $|X|=1$, we see that

$$\int_0^{\bar{s}}(1-s^2)\operatorname{Ric}(X,X)ds + \int_{\bar{s}-1}^{\bar{s}}(\bar{s}-s)^2\operatorname{Ric}(X,X)$$

is bounded independent of γ and of $\bar{s}$. This concludes the proof of the claim. □

CLAIM 9.48. $|\operatorname{Ric}(X,\cdot)|^2 \le R\cdot \operatorname{Ric}(X,X)$.

PROOF. This is obvious if $n=2$, so we may as well assume that $n=3$. We diagonalize Ric in an orthonormal basis $\{e_i\}$. Let $\lambda_i \ge 0$ be the eigenvalues. Write $X = X^i e_i$ with $\sum_i (X^i)^2 = 1$. Then

$$\operatorname{Ric}(X,\cdot) = X^i \lambda_i (e_i)^*,$$

so that $|\operatorname{Ric}(X,\cdot)|^2 = \sum_i (X^i)^2 \lambda_i^2$. Of course, since the $\lambda_i \ge 0$, this gives

$$R\cdot \operatorname{Ric}(X,X) = \left(\sum_i \lambda_i\right)\sum_i \lambda_i (X^i)^2 \ge \sum_i \lambda_i^2 (X^i)^2,$$

establishing the claim. □

Now we compute, using Cauchy-Schwarz,

$$\left(\int_0^{\bar{s}}|\operatorname{Ric}(X,\widetilde{E}_i)|ds\right)^2 \le \bar{s}\int_0^{\bar{s}}|\operatorname{Ric}(X,\widetilde{E}_i)|^2 ds \le \bar{s}\int_0^{\bar{s}}|\operatorname{Ric}(x,\cdot)|^2 ds$$

$$\le \bar{s}\int_0^{\bar{s}} R\cdot \operatorname{Ric}(X,X)ds.$$

Since R is bounded, it follows from the first claim that there is a constant C' independent of γ and $\bar{s}$ with

$$(9.23) \qquad \int_0^{\bar{s}} |\text{Ric}(X, \widetilde{E}_i)| ds \leq C'\sqrt{\bar{s}}.$$

Since γ is a geodesic in the metric g, we have $\nabla_X X = 0$. Hence,

$$\frac{d^2 f(\gamma(s))}{ds^2} = X(X(f)) = \text{Hess}(f)(X, X).$$

Applying the gradient shrinking soliton equation to the pair (X, X) gives

$$\frac{d^2 f(\gamma(s))}{ds^2} = \frac{1}{2} - \text{Ric}_g(X, X).$$

Integrating we see

$$\frac{df(\gamma(s))}{ds}\Big|_{s=\bar{s}} = \frac{df(\gamma(s))}{ds}\Big|_{s=0} + \frac{\bar{s}}{2} - \int_0^{\bar{s}} \text{Ric}(X, X) ds.$$

It follows that

$$(9.24) \qquad X(f)(\gamma(\bar{s})) \geq \frac{\bar{s}}{2} - C'',$$

for some constant C'' depending only on (M, g) and f. Similarly, applying the gradient shrinking soliton equation to the pair $(X, \widetilde{E}_i)$, using Equation (9.23) and the fact that $\nabla_X \widetilde{E}_i = 0$ gives

$$(9.25) \qquad |\widetilde{E}_i(f)(\gamma(\bar{s}))| \leq C''(\sqrt{\bar{s}} + 1).$$

These two inequalities imply that for $\bar{s}$ sufficiently large, f has no critical points and that ∇f makes a small angle with the gradient of the distance function from p, and $|\nabla f|$ goes to infinity as the distance from p increases. In particular, f is a proper function going off to $+\infty$ as we approach infinity in M.

Now apply Equation (9.21) to see that R is increasing along the gradient curves of f. Hence, there is a sequence p_k tending to infinity in M with $\lim_k R(p_k) = \text{limsup}_{q \in M} R_g(q) > 0$.

The Ricci flow $(M, G(t))$, $-\infty < t < 0$, given in Equation (9.19) has the property that $G(-1) = g$ and that $(M, G(t))$ is isometric to $(M, |t|g)$. Since the original Riemannian manifold (M, g) given in the statement of Theorem 9.42 is κ-non-collapsed, it follows that, for every $t < 0$, the Riemannian manifold $(M, G(t))$ is κ-non-collapsed. Consequently, the Ricci flow $(M, G(t))$ is κ-non-collapsed. It clearly has bounded non-negative curvature on each time-slice and is non-flat. Fix a point $p \in M$. There is a sequence of points p_i tending to infinity with $R(p_i, -1)$ bounded away from zero. It follows that $\lim_{i \to \infty} R(p_i, -1) d^2_{g(-1)}(p, p_i) = \infty$. Thus, this flow satisfies all the hypotheses of Proposition 9.39. Hence, by Corollary 9.40 we see that n cannot be equal to 2. Furthermore, by Proposition 9.39, when $n = 3$ there is another subsequence q_i tending to infinity in M such that there is a geometric limit $(M_\infty, g_\infty(t), (q_\infty, -1))$, $-\infty < t \le -1$, of the flows $(M, G(t), (q_i, -1))$ defined for all $t < 0$ and this limit splits as a product of a surface flow $(\Sigma^2, h(t))$ times the real line where the surfaces $(\Sigma^2, h(t))$ are all of positive, bounded curvature and the surface flow is κ-non-collapsed. Since there is a constant $C < \infty$ such that the curvature of $(M, G(t))$, $-\infty < t \le t_0 < 0$, is bounded by $C/|t_0|$, this limit actually exists for $-\infty < t < 0$ with the same properties.

Let us summarize our progress to date.

COROLLARY 9.49. *There is no non-compact, 2-dimensional Riemannian manifold (M, g) satisfying the hypotheses of Theorem 9.42. For any non-compact 3-manifold (M, g) of positive curvature satisfying the hypotheses of Theorem 9.42, there is a sequence of points $q_i \in M$ tending to infinity such that $\lim_{i \to \infty} R_g(q_i) = \sup_{p \in M}$ such that the based Ricci flows $(M, G(t), (q_i, -1))$ converge to a Ricci flow $(M_\infty, G_\infty(t), (q_\infty, -1))$ defined for $-\infty < t < 0$ that splits as a product of a line and a family of surfaces, each of positive, bounded curvature $(\Sigma^2, h(t))$. Furthermore, the flow of surfaces is κ-non-collapsed.*

PROOF. In Claim 9.45 we saw that every 2-dimensional (M, g) satisfying the hypotheses of Theorem 9.42 has strictly positive curvature. The argument that we just completed shows that there is no non-compact 2-dimensional example of strictly positive curvature.

The final statement is exactly what we just established. □

COROLLARY 9.50. (1) *Let $(M, g(t))$ be a 2-dimensional Ricci flow satisfying all the hypotheses of Proposition 9.50 except possibly the non-compactness hypothesis. Then M is compact and for any $a > 0$ the restriction of the flow to any interval of the form $(-\infty, -a]$ followed by a shift of time by $+a$ is a κ-solution.*
(2) *Any asymptotic gradient shrinking soliton for a 2-dimensional κ-solution is a shrinking family of round surfaces.*
(3) *Let $(M, g(t))$, $-\infty < t \le 0$, be a 2-dimensional κ-solution. Then $(M, g(t))$ is a shrinking family of compact, round surfaces.*

PROOF. Let $(M, g(t))$ be a 2-dimensional Ricci flow satisfying all the hypotheses of Proposition 9.39 except possibly non-compactness. It then follows from Corollary 9.40 that M is compact. This proves the first item.

Now suppose that $(M, g(t))$ is an asymptotic soliton for a κ-solution of dimension 2. If $(M, g(-1))$ does not have bounded curvature, then there is a sequence $p_i \to \infty$ so that $\lim_{i \to \infty} R(p_i, -1) = \infty$. By this and Theorem 9.11 the Ricci flow $(M, g(t))$ satisfies all the hypotheses of Proposition 9.39. But this contradicts Corollary 9.40. We conclude that $(M, g(-1))$ has bounded curvature. According to Corollary 9.49 this means that $(M, g(t))$ is compact. Results going back to Hamilton in [**31**] imply that this compact asymptotic shrinking soliton is a shrinking family of compact, round surfaces. For example, this result is contained in Proposition 5.21 on p. 118 of [**13**]. This proves the second item.

Now suppose that $(M, g(t))$ is a 2-dimensional κ-solution. By the second item any asymptotic gradient shrinking soliton for this κ-solution is compact. It follows that M is compact. We know that as t goes to $-\infty$ the Riemannian surfaces $(M, g(t))$ are converging to compact, round surfaces. Extend the flow forward from 0 to a maximal time $\Omega < \infty$. By Theorem 5.64 on p. 149 of [**13**] the surfaces $(M, g(t))$ are also becoming round as t approaches Ω from below. Also, according to Proposition 5.39 on p. 134 of [**13**] the entropy of the flow is weakly monotone decreasing and is strictly decreasing unless the flow is a gradient shrinking soliton. But we have seen that the limits at both $-\infty$ and Ω are round manifolds, and hence of the same entropy. It follows that the κ-solution is a shrinking family of compact, round surfaces. □

Now that we have shown that every 2-dimensional κ-solution is a shrinking family of round surfaces, we can complete the proof of Proposition 9.46. Let (M, g) be a non-compact manifold of positive curvature satisfying the hypotheses of Theorem 9.42. According to Corollary 9.50 the limiting Ricci flow $(M_\infty, G_\infty(t))$ referred to in Corollary 9.49 is the product of a line and a shrinking family of round surfaces. Since (M, g) is non-compact and has positive curvature, it is diffeomorphic to $\mathbb{R}^3$ and hence does not contain an embedded copy of a projective plane. It follows that the round surfaces are in fact round 2-spheres. Thus, $(M_\infty, G_\infty(t))$, $-\infty < t < 0$, splits as the product of a shrinking family $(S^2, h(t))$, $-\infty < t < 0$, of round 2-spheres and the real line.

CLAIM 9.51. *The scalar curvature of $(S^2, h(-1))$ is equal to 1.*

PROOF. Since the shrinking family of round 2-spheres $(S^2, h(t))$ exists for all $-\infty < t < 0$, it follows that the scalar curvature of $(S^2, h(-1))$ is at most 1. On the other hand, since the scalar curvature is increasing along the gradient flow lines of f, the infimum of the scalar curvature of (M, g), $R_{\inf}$, is positive. Thus, the infimum of the scalar curvature of $(M, G(t))$ is $R_{\inf}/|t|$ and goes to infinity as $|t|$ approaches 0. Thus, the infimum of the scalar curvature of $(S^2, h(t))$ goes to infinity as t approaches zero. This means that the shrinking family of 2-spheres becomes singular as t approaches zero, and consequently the scalar curvature of $(S^2, h(-1))$ is equal to 1. □

It follows that for any p in a neighborhood of infinity of (M, g), we have
$$R_g(p) < 1.$$
For any unit vector Y at any point of $M \setminus K$ we have
$$\text{Hess}(f)(Y, Y) = \frac{1}{2} - \text{Ric}(Y, Y) \geq \frac{1}{2} - \frac{R}{2} > 0.$$

(On a manifold with non-negative curvature $\mathrm{Ric}(Y,Y) \le R/2$ for any unit tangent vector Y.) This means that for u sufficiently large the level surfaces of $N_u = f^{-1}(u)$ are convex and hence have increasing area as u increases.

According to Equations (9.24) and (9.25) the angle between ∇f and the gradient of the distance function from p goes to zero as we go to infinity. According to Theorem 5.35 the gradient of the distance function from p converges to the unit vector field in the $\mathbb{R}$-direction of the product structure. It follows that the unit vector in the ∇f-direction converges to the unit vector in the $\mathbb{R}$-direction. Hence, as u tends to ∞ the level surfaces $f^{-1}(u)$ converge in the C^1-sense to $\Sigma \times \{0\}$. Thus, the areas of these level surfaces converge to the area of $(\Sigma, h(-1))$ which is 8π since the scalar curvature of this limiting surface is $\mathrm{limsup}_{p \in M} R(p, -1) = 1$. It follows that the area of $f^{-1}(u)$ is less than 8π for all u sufficiently large.

Now let us estimate the intrinsic curvature of $N = N_u = f^{-1}(u)$. Let K_N denote the sectional curvature of the induced metric on N, whereas K_M is the sectional curvature of M. We also denote by R_N the scalar curvature of the induced metric on N. Fix an orthonormal basis $\{e_1, e_2, e_3\}$ at a point of N, where $e_3 = \nabla f/|\nabla f|$. Then by the Gauss-Codazzi formula we have

$$R_N = 2K_N(e_1, e_2) = 2(K_M(e_1, e_2) + \det S)$$

where S is the shape operator

$$S = \frac{\mathrm{Hess}(f|TN)}{|\nabla f|}.$$

Clearly, we have $R - 2\mathrm{Ric}(e_3, e_3) = 2K_M(e_1, e_2)$, so that

$$R_N = R - 2\mathrm{Ric}(e_3, e_3) + 2\det S.$$

We can assume that the basis is chosen so that $\mathrm{Ric}|_{TN}$ is diagonal; i.e., in the given basis we have

$$\mathrm{Ric} = \begin{pmatrix} r_1 & 0 & c_1 \\ 0 & r_2 & c_2 \\ c_1 & c_2 & r_3 \end{pmatrix}.$$

From the gradient shrinking soliton equation we have $\mathrm{Hess}(f) = (1/2)g - \mathrm{Ric}$ so that

$$\det(\mathrm{Hess}(f|TN)) = \left(\frac{1}{2} - r_1\right)\left(\frac{1}{2} - r_2\right)$$
$$\le \frac{1}{4}(1 - r_1 - r_2)^2$$
$$= \frac{1}{4}(1 - R + \mathrm{Ric}(e_3, e_3))^2.$$

Thus, it follows that

(9.26) $$R_N \le R - 2\mathrm{Ric}(e_3, e_3) + \frac{(1 - R + \mathrm{Ric}(e_3, e_3))^2}{2|\nabla f|^2}.$$

It follows from Equation (9.24) that $|\nabla f(x)| \to \infty$ as x goes to infinity in M. Thus, since the curvature of $(M, g(-1))$ is bounded, provided that u is sufficiently large, we have $1 - R + \operatorname{Ric}(e_3, e_3) < 2|\nabla f|^2$. Since the left-hand side of this inequality is positive (since $R < 1$), it follows that

$$(1 - R + \operatorname{Ric}(e_3, e_3))^2 < 2(1 - R + \operatorname{Ric}(e_3, e_3))|\nabla f|^2.$$

Plugging this into Equation (9.26) gives that

$$R_N < 1 - \operatorname{Ric}(e_3, e_3) \leq 1,$$

assuming that u is sufficiently large.

This contradicts the Gauss-Bonnet theorem for the surface N: Its area is less than 8π, and the scalar curvature of the induced metric is less than 1, meaning that its Gaussian curvature is less than $1/2$; yet N is diffeomorphic to a 2-sphere. This completes the proof of Proposition 9.46, that is to say this shows that there are no non-compact positive curved examples satisfying the hypotheses of Theorem 9.42.

4.5. Case of non-positive curvature revisited. We return now to the second case of Theorem 9.42. We extend (M, g) to a Ricci flow $(M, G(t))$ defined for $-\infty < t < 0$ as given in Equation (9.19). By Claim 9.45, M has either a one- or 2-sheeted covering $\widetilde{M}$ such that $(\widetilde{M}, \widetilde{G}(t))$ is a metric product of a surface and a one-manifold for all $t < 0$. The evolving metric on the surface is itself a κ-solution and hence by Corollary 9.50 the surfaces are compact and the metrics are all round. Thus, in this case, for any $t < 0$, the manifold $(\widetilde{M}, \widetilde{G}(t))$ is a metric product of a round S^2 or $\mathbb{R}P^2$ and a flat copy of $\mathbb{R}$. The conclusion in this case is that the one- or two-sheeted covering $(\widetilde{M}, \widetilde{G}(t))$ is a product of a round S^2 or $\mathbb{R}P^2$ and the line for all $t < 0$.

4.6. Completion of the proof of Theorem 9.42.

COROLLARY 9.52. *Let $(M, g(t))$ be a 3-dimensional Ricci flow satisfying the hypotheses of Proposition 9.39. Then the limit constructed in that proposition splits as a product of a shrinking family of compact round surfaces with a line. In particular, for any non-compact gradient shrinking soliton of a 3-dimensional κ-solution the limit constructed in Proposition 9.39 is the product of a shrinking family of round surfaces and the real line.*

PROOF. Let $(M, g(t))$ be a 3-dimensional Ricci flow satisfying the hypotheses of Proposition 9.39 and let $(N^2, h(t)) \times (\mathbb{R}, ds^2)$ be the limit constructed in that proposition. Since this limit is κ-non-collapsed, $(N, h(t))$ is κ'-non-collapsed for some $\kappa' > 0$ depending only on κ. Since the limit is not flat and has non-negative curvature, the same is true for $(N, h(t))$. Since $\partial R/\partial t \geq 0$ for the limit, the same is true for $(N, h(t))$. That is to say $(N, h(t))$ satisfies all the hypotheses of Proposition 9.39 except possibly non-compactness. It now follows from Corollary 9.50 that $(N, h(t))$ is a shrinking family of compact, round surfaces. $\square$

4. CLASSIFICATION OF GRADIENT SHRINKING SOLITONS

COROLLARY 9.53. *Let $(M, g(t))$, $-\infty < t < 0$, be an asymptotic gradient shrinking soliton for a 3-dimensional κ-solution. Then for each $t < 0$, the Riemannian manifold $(M, g(t))$ has bounded curvature. In particular, for any $a > 0$ the flow $(M, g(t))$, $-\infty < t \leq -a$, followed by a shift of time by $+a$ is a κ-solution.*

PROOF. If an asymptotic gradient shrinking soliton $(M, g(t))$ of a 3-dimensional κ-solution does not have strictly positive curvature, then according to Corollary 4.20, $(M, g(t))$ has a covering that splits as a product of a 2-dimensional Ricci flow and a line. The 2-dimensional Ricci flow satisfies all the hypotheses of Proposition 9.39 except possibly compactness, and hence by Corollary 9.50 it is a shrinking family of round surfaces. In this case, it is clear that each time-slice of $(M, g(t))$ has bounded curvature.

Now we consider the remaining case when $(M, g(t))$ has strictly positive curvature. Assume that $(M, g(t))$ has unbounded curvature. Then there is a sequence of points p_i tending to infinity in M such that $R(p_i, t)$ tends to infinity. By Corollary 9.52 we can replace the points p_i by points q_i with $Q_i = R(q_i, t) \geq R(p_i, t)$ so that the based Riemannian manifolds $(M, Q_i g(t), q_i)$ converge to a product of a round surface $(N, h(t))$ with $\mathbb{R}$. The surface N is either diffeomorphic to S^2 or $\mathbb{R}P^2$. Since $(M, g(t))$ has positive curvature, by Theorem 2.7, it is diffeomorphic to $\mathbb{R}^3$, and hence it contains no embedded $\mathbb{R}P^2$. It follows that $(N, h(t))$ is a round 2-sphere.

Fix $\epsilon > 0$ sufficiently small as in Proposition 2.19. Then the limiting statement means that, for every i sufficiently large, there is an ϵ-neck in $(M, g(t))$ centered at q_i with scale $Q_i^{-1/2}$. This contradicts Proposition 2.19, establishing that for each $t < 0$ the curvature of $(M, g(t))$ is bounded. □

COROLLARY 9.54. *Let $(M, g(t))$, $-\infty < t \leq 0$, be a κ-solution of dimension 3. Then any asymptotic gradient shrinking soliton $(M_\infty, g_\infty(t))$ for this κ-solution, as constructed in Theorem 9.11, is of one of the three types listed in Theorem 9.42.*

PROOF. Let $(M_\infty, g_\infty(t))$, $-\infty < t < 0$, be an asymptotic gradient shrinking soliton for $(M, g(t))$. According to Corollary 9.53, this soliton is a κ-solution, implying that $(M_\infty, g_\infty(-1))$ is a complete Riemannian manifold of bounded, non-negative curvature. Suppose that $B(p, -1, r) \subset M_\infty$ is a metric ball and $|\mathrm{Rm}_{g_\infty}|(x, -1) \leq r^{-2}$ for all $x \in B(p, -1, r)$. Since $\partial R_{g_\infty}(x, t)/\partial t \geq 0$, it follows that $R(x, t) \leq 3r^{-2}$ on $B(p, -1, r) \times (-1 - r^2, -1]$, and hence that $|\mathrm{Rm}_{g_\infty}| \leq 3r^{-2}$ on this same region. Since the Ricci flow $(M_\infty, g_\infty(t))$ is κ-non-collapsed, it follows that $\mathrm{Vol}\, B(p, -1, r/\sqrt{3}) \geq \kappa(r/\sqrt{3})^3$. Hence, $\mathrm{Vol}\, B(p, -1, r) \geq (\kappa/3\sqrt{3})r^3$. This proves that the manifold $(M_\infty, g_\infty(-1))$ is κ'-non-collapsed for some $\kappa' > 0$ depending only on κ. On the other hand, according to Theorem 9.11 there is a function $f(\cdot, -1)$ from M_∞ to $\mathbb{R}$ satisfying the gradient shrinking soliton equation at the time-slice -1. Thus, Theorem 9.42 applies to $(M_\infty, g_\infty(-1))$ to produce a Ricci flow $G(t)$, $-\infty < t < 0$, of one of the three types listed in that theorem and with $G(-1) = g_\infty(-1)$.

Now we must show that $G(t) = g_\infty(t)$ for all $t < 0$. In the first case when M_∞ is compact, this is clear by uniqueness of the Ricci flow in the compact case. Suppose that $(M_\infty, G(t))$ is of the second type listed in Theorem 9.42. Then $(M_\infty, g_\infty(-1))$ is a product of a round 2-sphere and the real line. By Corollary 4.20 this implies that the entire flow $(M_\infty, g_\infty(t))$ splits as the product of a flow of compact 2-spheres and the real line. Again by uniqueness in the compact case, this family of 2-spheres must be a shrinking family of round 2-spheres. In the third case, one passes to a finite sheeted covering of the second type, and applies the second case. $\square$

4.7. Asymptotic curvature. There is an elementary result that will be needed in what follows.

DEFINITION 9.55. Let (M, g) be a complete, connected, non-compact Riemannian manifold of non-negative curvature. Fix a point $p \in M$. We define the *asymptotic scalar curvature*

$$\mathcal{R}(M, g) = \limsup_{x \to \infty} R(x) d^2(x, p).$$

Clearly, this limit is independent of p.

PROPOSITION 9.56. *Suppose that $(M, g(t))$, $-\infty < t < 0$, is a connected, non-compact κ-solution of dimension[2] at most 3. Then $\mathcal{R}(M, g(t)) = +\infty$ for every $t < 0$.*

PROOF. By Corollary 9.50 the only 2-dimensional κ-solutions are compact, so that the result is vacuously true in this case. Suppose that $(M, g(t))$ is 3-dimensional. If $(M, g(t))$ does not have strictly positive curvature, then, since it is not flat, by Corollary 4.20 it must be 3-dimensional and it has a finite-sheeted covering space that splits as a product $(Q, h(0)) \times (\mathbb{R}, ds^2)$ with $(Q, h(0))$ being a surface of strictly positive curvature and T being a flat one-manifold. Clearly, in this case the asymptotic curvature is infinite.

Thus, without loss of generality we can assume that $(M, g(t))$ has strictly positive curvature. Let us first consider the case when $\mathcal{R}(M, g(t))$ has a finite, non-zero value. Fix a point $p \in M$. Take a sequence of points x_n tending to infinity and set $\lambda_n = d_0^2(x_n, p)$ and $Q_n = R(x_n, t)$. We choose this sequence such that

$$\lim_{n \to \infty} Q_n \lambda_n = \mathcal{R}(M, g(t)).$$

We consider the sequence of Ricci flows $(M, h_n(t), (x_n, 0))$, where

$$h_n(t) = Q_n g(Q_n^{-1} t).$$

Fix $0 < a < \sqrt{\mathcal{R}(M, g(t))} < b < \infty$. Consider the annuli

$$A_n = \{y \in M \mid a < d_{h_n(0)}(y, p) < b\}.$$

---

[2] This result, in fact, holds in all dimensions.

4. CLASSIFICATION OF GRADIENT SHRINKING SOLITONS

Because of the choice of sequence, for all n sufficiently large, the scalar curvature of the restriction of $h_n(0)$ to A_n is bounded independent of n. Furthermore, since $d_{h_n}(p, x_n)$ converges to $\sqrt{\mathcal{R}(M, g(t))}$, there is $\alpha > 0$ such that for all n sufficiently large, the annulus A_n contains $B_{h_n}(x_n, 0, \alpha)$. Consequently, we have a bound, independent of n, for the scalar curvature of $h_n(0)$ on these balls. By the hypothesis that $\partial R/\partial t \geq 0$, there is a bound, independent of n, for the scalar curvature of h_n on $B_{h_n}(x_n, 0, \alpha) \times (-\infty, 0]$. Using the fact that the flows have non-negative curvature, this means that there is a bound, independent of n, for $|\mathrm{Rm}_{h_n}(y, 0)|$ on $B_{h_n}(x_n, 0, \alpha) \times (-\infty, 0]$. This means that by Shi's theorem (Theorem 3.28), there are bounds, independent of n, for every covariant derivative of the curvature on $B_{h_n}(x_n, 0, \alpha/2) \times (-\infty, 0]$.

Since the original flow is κ-non-collapsed on all scales, it follows that the rescaled flows are also κ non-collapsed on all scales. Since the curvature is bounded, independent of n, on $B_{h_n}(x_n, 0, \alpha)$, this implies that there is $\delta > 0$, independent of n, such that for all n sufficiently large, every ball of radius δ centered at any point of $B_{h_n}(x_n, 0, \alpha/2)$ has volume at least $\kappa \delta^3$. Now applying Theorem 5.6 we see that a subsequence converges geometrically to a limit which will automatically be a metric ball $B_{g_\infty}(x_\infty, 0, \alpha/2)$. In fact, by Hamilton's result (Proposition 5.14) there is a limiting flow on $B_{g_\infty}(x_\infty, 0, \alpha/4) \times (-\infty, 0]$. Notice that the limiting flow is not flat since $R(x_\infty, 0) = 1$.

On the other hand, according to Lemma 5.31 the Gromov-Hausdorff limit of a subsequence $(M, \lambda_n^{-1} g_n(0), x_n)$ is the Tits cone, i.e., the cone over $S_\infty(M, p)$. Since $Q_n = \mathcal{R}(M, g(t)) \lambda_n^{-1}$, the rescalings $(M, Q_n g_n(0), x_n)$ also converge to a cone, say (C, h, y_∞), which is in fact simply a rescaling of the Tits cone by a factor $\mathcal{R}(M, g(t))$. Pass to a subsequence so that both the geometric limit on the ball of radius $\alpha/2$ and the Gromov-Hausdorff limit exist. Then the geometric limit $B_{g_\infty}(x_\infty, 0, \alpha/2)$ is isometric to an open ball in the cone. Since we have a limiting Ricci flow

$$(B_{g_\infty}(x_\infty, 0, \alpha/2), g_\infty(t)), \quad -\infty < t \leq 0,$$

this contradicts Proposition 4.22. This completes the proof that it is not possible for the asymptotic curvature to be finite and non-zero.

Lastly, we consider the possibility that the asymptotic curvature is zero. Again we fix $p \in M$. Take any sequence of points x_n tending to infinity and let $\lambda_n = d_0^2(p, x_n)$. Form the sequence of based Ricci flows $(M, h_n(t), (x_n, 0))$ where $h_n(t) = \lambda_n^{-1} g(\lambda_n t)$. On the one hand, the Gromov-Hausdorff limit (of a subsequence) is the Tits cone. On the other hand, the curvature condition tells us the following: For any $0 < a < 1 < b$ on the regions

$$\{y \in M \mid a < d_{h_n(0)}(y, p) < b\},$$

the curvature tends uniformly to zero as n tends to infinity. Arguing as in the previous case, Shi's theorem, Hamilton's result, Theorem 5.14, and the fact that the original flow is κ non-collapsed on all scales tells us that we can pass to a subsequence so that these annuli centered at x_n converge geometrically to a limit. Of course, the limit is flat. Since this holds for all $0 < a < 1 < b$, this implies that the Tits cone is smooth and flat except possibly at its cone point. In particular, the sphere at infinity, $S_\infty(M, p)$, is a smooth surface of constant curvature $+1$.

CLAIM 9.57. *$S_\infty(M, p)$ is isometric to a round 2-sphere.*

PROOF. Since M is orientable the complement of the cone point in the Tits cone is an orientable manifold and hence $S_\infty(M, p)$ is an orientable surface. Since we have already established that it has a metric of constant positive curvature, it must be diffeomorphic to S^2, and hence isometric to a round sphere. (In higher dimensions one can prove that $S_\infty(M, p)$ is simply connected, and hence isometric to a round sphere.) □

It follows that the Tits cone is a smooth flat manifold even at the origin, and hence is isometric to Euclidean 3-space. This means that in the limit, for any $r > 0$ the volume of the ball of radius r centered at the cone point is exactly $\omega_3 r^3$, where ω_3 is the volume of the unit ball in $\mathbb{R}^3$. Consequently,

$$\lim_{n\to\infty} \text{Vol}\left(B_g(p, 0, \sqrt{\lambda_n}r) \setminus B_g(p, 0, 1)\right) \to \omega_3 \lambda_n^{3/2} r^3.$$

By Theorem 1.34 and the fact that the Ricci curvature is non-negative, this implies that

$$\text{Vol}\, B_g(p, 0, R) = \omega_3 R^3$$

for all $R < \infty$. Since the Ricci curvature is non-negative, this means that $(M, g(t))$ is Ricci-flat, and hence flat. But this contradicts the fact that $(M, g(t))$ is a κ-solution and hence is not flat.

Having ruled out the other two cases, we are left with only one possibility: $\mathcal{R}(M, g(t)) = \infty$. □

5. Universal κ

The first consequence of the existence of an asymptotic gradient shrinking soliton is that there is a universal κ for all 3-dimensional κ-solutions, except those of constant positive curvature.

PROPOSITION 9.58. *There is a $\kappa_0 > 0$ such that any non-round 3-dimensional κ-solution is a κ_0-solution.*

PROOF. Let $(M, g(t))$ be a non-round 3-dimensional κ-solution. By Corollary 9.44 since $(M, g(t))$ is not a family of round manifolds, the asymptotic soliton for the κ-solution cannot be compact. Thus, according to Corollary 9.42 there are only two possibilities for the asymptotic soliton $(M_\infty, g_\infty(t))$ – either $(M_\infty, g_\infty(t))$ is the product of a round 2-sphere of Gaussian curvature $1/2|t|$ with a line or has a two-sheeted covering by such a product. In fact, there are three possibilities: $S^2 \times \mathbb{R}$, $\mathbb{R}P^2 \times \mathbb{R}$ or the twisted $\mathbb{R}$-bundle over $\mathbb{R}P^2$ whose total space is diffeomorphic to the complement of a point in $\mathbb{R}P^3$.

Fix a point $x = (p, 0) \in M \times \{0\}$. Let $\overline{\tau}_k$ be a sequence converging to ∞, and $q_k \in M$ a point with $l_x(q_k, \overline{\tau}_k) \leq 3/2$. The existence of an asymptotic soliton means that, possibly after passing to a subsequence, there is a gradient shrinking soliton $(M_\infty, g_\infty(t))$ and a ball B of radius 1 in $(M_\infty, g_\infty(-1))$ centered at a point $q_\infty \in M_\infty$ and a sequence of embeddings $\psi_k \colon B \to M$ such that $\psi_k(q_\infty) = q_k$ and such that the map

$$B \times [-2, -1] \to M \times [-2\overline{\tau}_k, -\overline{\tau}_k]$$

given by $(b, t) \mapsto (\psi_k(b), \overline{\tau}_k t)$ has the property that the pullback of $\overline{\tau}_k^{-1} g(\overline{\tau}_k t)$ converges smoothly and uniformly as $k \to \infty$ to the restriction of $g_\infty(t)$ to $B \times [-2, -1]$. Let $(M_k, g_k(t))$ be this rescaling of the κ-solution by $\overline{\tau}_k$. Then the embeddings $\psi_k \times \mathrm{id} \colon B \times (-2, -1] \to (M_k \times [-2, -1]$ converge as $k \to \infty$ to a one-parameter family of isometries. That is to say, the image $\psi_k(B \times [-2, -1]) \subset M_k \times [-2, -1]$ is an almost isometric embedding. Since the reduced length from x to $(\psi_k(a), -1)$ is at most $3/2$, from the invariance of reduced length under rescalings (see Corollary 6.74) it follows that the reduced length function on $\psi_k(B \times \{-2\})$ is bounded independent of k. Similarly, the volume of $\psi_k(B \times \{-2\})$ is bounded independent of k. This means the reduced volume of $\psi_k(B \times \{-2\})$ in $(M_k, g_k(t))$ is bounded independent of k. Now according to Theorem 8.1 this implies that $(M_k, g_k(t))$ is κ_0-non-collapsed at $(p, 0)$ on scales $\leq \sqrt{2}$ for some κ_0 depending only on the geometry of the three possibilities for $(M_\infty, g_\infty(t))$, $-2 \leq t \leq -1$. Being κ_0-non-collapsed is invariant under rescalings, so that it follows immediately that $(M, g(t))$ is κ_0-non-collapsed on scales $\leq \sqrt{2\overline{\tau}_k}$. Since this is true for all k, it follows that $(M, g(t))$ is κ_0-non-collapsed on all scales at $(p, 0)$.

This result holds of course for every $p \in M$, showing that at $t = 0$ the flow is κ_0-non-collapsed. To prove this result at points of the form $(p, t) \in M \times (-\infty, 0]$ we simply shift the original κ-solution upward by $|t|$ and remove the part of the flow at positive time. This produces a new κ-solution and the point in question has been shifted to the time-zero slice, so that we can apply the previous results. $\square$

6. Asymptotic volume

Let $(M, g(t))$ be an n-dimensional κ-solution. For any $t \leq 0$ and any point $p \in M$ we consider $(\mathrm{Vol}\, B_{g(t)}(p, r))/r^n$. According to the Bishop-Gromov Theorem (Theorem 1.34), this is a non-increasing function of r. We define the *asymptotic volume* $\mathcal{V}_\infty(M, g(t))$, or $\mathcal{V}_\infty(t)$ if the flow is clear from the context, to be the limit as $r \to \infty$ of this function. Clearly, this limit is independent of $p \in M$.

THEOREM 9.59. *For[3] any $\kappa > 0$ and any κ-solution $(M, g(t))$ the asymptotic volume $\mathcal{V}_\infty(M, g(t))$ is identically zero.*

PROOF. The proof is by induction on the dimension n of the solution. For $n = 2$ by Corollary 9.50 there are only compact κ-solutions, which clearly have zero asymptotic volume. Suppose that we have established the result for $n - 1 \geq 2$ and let us prove it for n.

According to Proposition 9.39 there is a sequence of points $p_n \in M$ tending to infinity such that setting $Q_n = R(p_n, 0)$ the sequence of Ricci flows
$$(M, Q_n g(Q_n^{-1} t), (q_n, 0))$$
converges geometrically to a limit $(M_\infty, g_\infty(t), (q_\infty, 0))$, and this limit splits off a line: $(M_\infty, g_\infty(t)) = (N, h(t)) \times \mathbb{R}$. Since the ball of radius R about a point $(x, t) \in N \times \mathbb{R}$ is contained in the product of the ball of radius R in N centered at x and an interval of length $2R$, it follows that $(N, h(t))$ is a $\kappa/2$-ancient solution. Hence, by induction, for every t, the asymptotic volume of $(N, h(t))$ is zero, and hence so is that of $(M, g(t))$. □

6.1. Volume comparison. One important consequence of the asymptotic volume result is a volume comparison result.

PROPOSITION 9.60. *Fix the dimension n. For every $\nu > 0$ there is $A < \infty$ such that the following holds. Suppose that $(M_k, g_k(t))$, $-t_k \leq t \leq 0$, is a sequence of (not necessarily complete) n-dimensional Ricci flows of nonnegative curvature operator. Suppose in addition we have points $p_k \in M_k$ and radii $r_k > 0$ with the property that for each k the ball $B(p_k, 0, r_k)$ has compact closure in M_k. Let $Q_k = R(p_k, 0)$ and suppose that $R(q, t) \leq 4Q_k$ for all $q \in B(p_k, 0, r_k)$ and for all $t \in [-t_k, 0]$, and suppose that $t_k Q_k \to \infty$ and $r_k^2 Q_k \to \infty$ as $k \to \infty$. Then $\operatorname{Vol} B(p_k, 0, A/\sqrt{Q_k}) < \nu (A/\sqrt{Q_k})^n$ for all k sufficiently large.*

PROOF. Suppose that the result fails for some $\nu > 0$. Then there is a sequence $(M_k, g_k(t))$, $-t_k \leq t \leq 0$, of n-dimensional Ricci flows, points $p_k \in M_k$, and radii r_k as in the statement of the lemma such that for every $A < \infty$ there is an arbitrarily large k with $\operatorname{Vol} B(p_k, 0, A/\sqrt{Q_k}) \geq \nu (A/\sqrt{Q_k})^n$. Pass to a subsequence so that for each $A < \infty$ we have
$$\operatorname{Vol} B(p_k, 0, A/\sqrt{Q_k}) \geq \nu (A/\sqrt{Q_k})^n$$
for all k sufficiently large. Consider now the flows $h_k(t) = Q_k g_k(Q_k^{-1} t)$, defined for $-Q_k t_k \leq t \leq 0$. Then for every $A < \infty$ for all k sufficiently large we have $R_{h_k}(q, t) \leq 4$ for all $q \in B_{h_k}(p_k, 0, A)$ and all $t \in (-t_k Q_k, 0]$. Also,

---

[3]This theorem and all the other results of this section are valid in all dimensions. Our proofs use Theorem 9.56 and Proposition 9.39 which are also valid in all dimensions but which we proved only in dimensions 2 and 3. Thus, while we state the results of this section for all dimensions, strictly speaking we give proofs only for dimensions 2 and 3. These are the only cases we need in what follows.

for every $A < \infty$ for all k sufficiently large we have $\operatorname{Vol} B(p_k, 0, A) \geq \nu A^n$. According to Theorem 5.15 we can then pass to a subsequence that has a geometric limit which is an ancient flow of complete Riemannian manifolds. Clearly, the time-slices of the limit have non-negative curvature operator, and the scalar curvature is bounded (by 4) and is equal to 1 at the base point of the limit. Also, the asymptotic volume $\mathcal{V}(0) \geq \nu$.

CLAIM 9.61. *Suppose that $(M, g(t))$ is an ancient Ricci flow such that for each $t \leq 0$ the Riemannian manifold $(M, g(t))$ is complete and has bounded, non-negative curvature operator. Let $\mathcal{V}(t)$ be the asymptotic volume of the manifold $(M, g(t))$.*

(1) *The asymptotic volume $\mathcal{V}(t)$ is a non-increasing function of t.*
(2) *If $\mathcal{V}(t) = V > 0$ then every metric ball $B(x, t, r)$ has volume at least $V r^n$.*

PROOF. We begin with the proof of the first item. Fix $a < b \leq 0$. By hypothesis there is a constant $K < \infty$ such that the scalar curvature of $(M, g(0))$ is bounded by $(n-1)K$. By the Harnack inequality (Corollary 4.39) the scalar curvature of $(M, g(t))$ is bounded by $(n-1)K$ for all $t \leq 0$. Hence, since the $(M, g(t))$ have non-negative curvature, we have $\operatorname{Ric}(p, t) \leq (n-1)K$ for all p and t. Set $A = 4(n-1)\sqrt{\frac{2K}{3}}$. Then by Corollary 3.26 we have

$$d_a(p_0, p_1) \leq d_b(p_0, p_1) + A(b - a).$$

This means that for any $r > 0$ we have

$$B(p_0, b, r) \subset B(p_0, a, r + A(b - a)).$$

On the other hand, since $d\operatorname{Vol}/dt = -R d\operatorname{Vol}$, it follows that, in the case of non-negative curvature, the volume of any open set is non-increasing in time. Consequently,

$$\operatorname{Vol}_{g(b)} B(p_0, b, r) \leq \operatorname{Vol}_{g(a)} B(p_0, a, r + A(b - a)),$$

and hence

$$\frac{\operatorname{Vol}_{g(b)} B(p_0, b, r)}{r^n} \leq \frac{\operatorname{Vol}_{g(a)} B(p_0, a, r + A(b - a))}{(r + A(b - a))^n} \frac{(r + A(b - a))^n}{r^n}.$$

Taking the limit as $r \to \infty$ gives

$$\mathcal{V}(b) \leq \mathcal{V}(a).$$

The second item of the claim is immediate from the Bishop-Gromov inequality (Theorem 1.34). □

Now we return to the proof of the proposition. Under the assumption that there is a counterexample to the proposition for some $\nu > 0$, we have constructed a limit that is an ancient Ricci flow with bounded, non-negative curvature with $\mathcal{V}(0) \geq \nu$. Since $\mathcal{V}(0) \geq \nu$, it follows from the claim that

$\mathcal{V}(t) \geq \nu$ for all $t \leq 0$ and hence, also by the claim, we see that $(M, g(t))$ is ν-non-collapsed for all t. This completes the proof that the limit is a ν-solution. This contradicts Theorem 9.59 applied with $\kappa = \nu$, and proves the proposition. □

This proposition has two useful corollaries about balls in κ-solutions with volumes bounded away from zero. The first says that the normalized curvature is bounded on such balls.

COROLLARY 9.62. *For any $\nu > 0$ there is a $C = C(\nu) < \infty$ depending only on the dimension n such that the following holds. Suppose that $(M, g(t))$, $-\infty < t \leq 0$, is an n-dimensional Ricci flow with each $(M, g(t))$ being complete and with bounded, non-negative curvature operator. Suppose $p \in M$, and $r > 0$ are such that $\operatorname{Vol} B(p, 0, r) \geq \nu r^n$. Then $r^2 R(q, 0) \leq C$ for all $q \in B(p, 0, r)$.*

PROOF. Suppose that the result fails for some $\nu > 0$. Then there is a sequence $(M_k, g_k(t))$ of n-dimensional Ricci flows, complete, with bounded non-negative curvature operator and points $p_k \in M_k$, constants $r_k > 0$, and points $q_k \in B(p_k, 0, r_k)$ such that:

(1) $\operatorname{Vol} B(p_k, 0, r_k) \geq \nu r_k^n$, and
(2) setting $Q_k = R(q_k, 0)$ we have $r_k^2 Q_k \to \infty$ as $k \to \infty$.

Using Lemma 9.37 we can find points $q_k' \in B(p_k, 0, 2r_k)$ and constants $s_k \leq r_k$, such that setting $Q_k' = R(q_k', 0)$ we have $Q_k' s_k^2 = Q_k r_k^2$ and $R(q, 0) < 4 Q_k'$ for all $q \in B(q_k', 0, s_k)$. Of course, $Q_k' s_k^2 \to \infty$ as $k \to \infty$. Since $d_0(p_k, q_k') < 2r_k$, we have $B(p_k, 0, r_k) \subset B(q_k', 0, 3r_k)$ so that

$$\operatorname{Vol} B(q_k', 0, 3r_k) \geq \operatorname{Vol} B(p_k, 0, r_k) \geq \nu r_k^n = (\nu/3^n)(3r_k)^n.$$

Since the sectional curvatures of $(M, g_k(0))$ are non-negative, it follows from the Bishop-Gromov inequality (Theorem 1.34) that $\operatorname{Vol} B(q_k', 0, s) \geq (\nu/3^n) s^n$ for any $s \leq s_k$.

Of course, by Corollary 4.39, we have $R(q, t) < 4 Q_k'$ for all $t \leq 0$ and all $q \in B(q_k', 0, s_k)$. Now consider the sequence of based, rescaled flows

$$(M_k, Q_k' g(Q_k'^{-1} t), (q_k', 0)).$$

In these manifolds all balls centered at $(q_k', 0)$ of radii at most $\sqrt{Q_k'} s_k$ are $(\nu/3^n)$ non-collapsed. Also, the curvatures of these manifolds are non-negative and the scalar curvature is bounded by 4. It follows that by passing to a subsequence we can extract a geometric limit. Since $Q_k' s_k^2 \to \infty$ as $k \to \infty$ the asymptotic volume of this limit is at least $\nu/3^n$. But this geometric limit is a $\nu/3^n$-non-collapsed ancient solution with non-negative curvature operator with scalar curvature bounded by 4. This contradicts Theorem 9.59. □

The second corollary gives curvature bounds at all points in terms of the distance to the center of the ball.

COROLLARY 9.63. *Fix the dimension n. Given $\nu > 0$, there is a function $K(A) < \infty$, defined for $A \in (0, \infty)$, such that if $(M, g(t))$, $-\infty < t \leq 0$, is an n-dimensional Ricci flow, complete of bounded, non-negative curvature operator, $p \in M$ is a point and $0 < r < \infty$ is such that $\operatorname{Vol} B(p, 0, r) \geq \nu r^n$, then for all $q \in M$ we have*

$$(r + d_0(p, q))^2 R(q, 0) \leq K(d_0(p, q)/r).$$

PROOF. Fix $q \in M$ and let $d = d_0(p, q)$. We have

$$\operatorname{Vol} B(q, 0, r + d) \geq \operatorname{Vol} B(p, 0, r) \geq \nu r^n = \frac{\nu}{(1 + (d/r))^n}(r + d)^n.$$

Let $K(A) = C(\nu/^n)$, where C is the constant provided by the previous corollary. The result is immediate from the previous corollary. □

7. Compactness of the space of 3-dimensional κ-solutions

This section is devoted to proving the following result.

THEOREM 9.64. *Let $(M_k, g_k(t), (p_k, 0))$ be a sequence of 3-dimensional based κ-solutions satisfying $R(p_k, 0) = 1$. Then there is a subsequence converging smoothly to a based κ-solution.*

The main point in proving this theorem is to establish the uniform curvature bounds given in the next lemma.

LEMMA 9.65. *For each $r < \infty$ there is a constant $C(r) < \infty$, such that the following holds. Let $(M, g(t), (p, 0))$ be a based 3-dimensional κ-solution satisfying $R(p, 0) = 1$. Then $R(q, 0) \leq C(r)$ for all $q \in B(p, 0, r)$.*

PROOF. Fix a based 3-dimensional κ-solution $(M, g(t), (p, 0))$. By Theorem 9.56 we have

$$\sup_{q \in M} d_0(p, q)^2 R(q, 0) = \infty.$$

Let q be a closest point to p satisfying

$$d_0(p, q)^2 R(q, 0) = 1.$$

We set $d = d_0(p, q)$, and we set $Q = R(q, 0)$. Of course, $d^2 Q = 1$. We carry this notation and these assumptions through the next five claims. The goal of these claims is to show that $R(q', 0)$ is uniformly bounded for q' near $(p, 0)$ so that in fact the distance d from the point q to p is uniformly bounded from below by a positive constant (see Claim 9.69 for a more precise statement). Once we have this the lemma will follow easily. To establish this uniform bound requires a sequence of claims.

CLAIM 9.66. *There is a universal (i.e., independent of the 3-dimensional κ-solution) upper bound C for $R(q', 0)/R(q, 0)$ for all $q' \in B(q, 0, 2d)$.*

PROOF. Suppose not. Then there is a sequence $(M_k, g_k(t), (p_k, 0))$ of based 3-dimensional κ-solutions with $R(p_k, 0) = 1$, points q_k in $(M_k, g_k(0))$ closest to p_k satisfying $d_k^2 R(q_k, 0) = 1$, where $d_k = d_0(p_k, q_k)$, and points $q_k' \in B(q_k, 0, 2d_k)$ with

$$\lim_{k \to \infty} (2d_k)^2 R(q_k', 0) = \infty.$$

Then according to Corollary 9.62 for every $\nu > 0$ for all k sufficiently large, we have

(9.27) $$\operatorname{Vol} B(q_k, 0, 2d_k) < \nu (2d_k)^3.$$

Therefore, by passing to a subsequence, we can assume that for each $\nu > 0$

(9.28) $$\operatorname{Vol} B(q_k, 0, 2d_k) < \nu (2d_k)^3$$

for all k sufficiently large. Let ω_3 be the volume of the unit ball in $\mathbb{R}^3$. Then for all k sufficiently large, $\operatorname{Vol} B(q_k, 0, 2d_k) < [\omega_3/2](2d_k)^3$. Since the sectional curvatures of $(M_k, g_k(0))$ are non-negative, by the Bishop-Gromov inequality (Theorem 1.34), it follows that for every k sufficiently large there is $r_k < 2d_k$ such that

(9.29) $$\operatorname{Vol} B(q_k, 0, r_k) = [\omega_3/2] r_k^3.$$

Of course, because of Equation (9.28) we see that $\lim_{k \to \infty} r_k/d_k = 0$. Then, according to Corollary 9.63, for all $q \in M_k$, we have

$$(r_k + d_{g_k(0)}(q_k, q))^2 R(q, 0) \leq K(d_{g_k(0)}(q_k, q)/r_k),$$

where K is as given in Corollary 9.63. Form the sequence $(M_k, g_k'(t), (q_k, 0))$, where $g_k'(t) = r_k^{-2} g_k(r_k^2 t)$. This is a sequence of based Ricci flows. For each $A < \infty$ we have

$$(1 + A)^2 R_{g_k'}(q, 0) \leq K(A)$$

for all $q \in B_{g_k'(0)}(q_k, 0, A)$. Hence, by the consequence of Hamilton's Harnack inequality (Corollary 4.39)

$$R_{g_k'}(q, t) \leq K(A),$$

for all $(q, t) \in B_{g_k'(0)}(q_k, 0, A) \times (-\infty, 0]$. Using this and the fact that all the flows are κ-non-collapsed, Theorem 5.15 implies that, after passing to a subsequence, the sequence $(M_k, g_k'(t), (q_k, 0))$ converges geometrically to a limiting Ricci flow $(M_\infty, g_\infty(t), (q_\infty, 0))$ consisting of non-negatively curved, complete manifolds κ-non-collapsed on all scales (though possibly with unbounded curvature).

Furthermore, Equation (9.29) passes to the limit to give

(9.30) $$\operatorname{Vol} B_{g_\infty}(q_\infty, 0, 1) = \omega_3/2.$$

Since $r_k/d_k \to 0$ as $k \to \infty$ and since $R_{g_k}(q_k, 0) = d_k^{-2}$, we see that $R_{g_\infty}(q_\infty, 0) = 0$. By the strong maximum principle for scalar curvature (Theorem 4.18), this implies that the limit $(M_\infty, g_\infty(0))$ is flat. But Equation (9.30) tells us that this limit is not $\mathbb{R}^3$. Since it is complete and flat, it must be a quotient of $\mathbb{R}^3$ by an action of a non-trivial group of isometries acting freely and properly discontinuously. But the quotient of $\mathbb{R}^3$ by any non-trivial group of isometries acting freely and properly discontinuously has zero asymptotic volume. [Proof: It suffices to prove the claim in the special case when the group is infinite cyclic. The generator of this group has an axis α on which it acts by translation and on the orthogonal subspace its acts by an isometry. Consider the circle in the quotient that is the image of α, and let L_α be its length. The volume of the ball of radius r about L_α is $\pi r^2 L_\alpha$. Clearly then, for any $p \in \alpha$, the volume of the ball of radius r about p is at most $\pi L_\alpha r^2$. This proves that the asymptotic volume of the quotient is zero.]

We have now shown that $(M_\infty, g_\infty(0))$ has zero curvature and zero asymptotic volume. But this implies that it is not κ-non-collapsed on all scales, which is a contradiction. This contradiction completes the proof of Claim 9.66. □

This claim establishes the existence of a universal constant $C < \infty$ (universal in the sense that it is independent of the 3-dimensional κ-solution) such that $R(q', 0) \leq CQ$ for all $q' \in B(q, 0, 2d)$. Since the curvature of $(M, g(t))$ is non-negative and bounded, we know from the Harnack inequality (Corollary 4.39) that $R(q', t) \leq CQ$ for all $q' \in B(q, 0, 2d)$ and all $t \leq 0$. Hence, the Ricci curvature $\mathrm{Ric}(q', t) \leq CQ$ for all $q' \in B(q, 0, 2d)$ and all $t \leq 0$.

CLAIM 9.67. *Given any constant $c > 0$ there is a constant $\widetilde{C} = \widetilde{C}(c)$, depending only on c and not on the 3-dimensional κ-solution, so that*

$$d_{g(-cQ^{-1})}(p, q) \leq \widetilde{C} Q^{-1/2}.$$

PROOF. Let $\gamma \colon [0, d] \to M$ be a $g(0)$-geodesic from p to q, parameterized at unit speed. Denote by $\ell_t(\gamma)$ the length of γ under the metric $g(t)$. We have $d_t(p, q) \leq \ell_t(\gamma)$. We estimate $\ell_t(\gamma)$ using the fact that $|\mathrm{Ric}| \leq CQ$ on the image of γ at all times.

$$\frac{d}{dt}\ell_{t_0}(\gamma) = \frac{d}{dt}\left(\int_0^d \sqrt{\langle \gamma'(s), \gamma'(s)\rangle_{g(t)}}\,ds\right)\bigg|_{t=t_0}$$

$$= \int_0^d \frac{-\mathrm{Ric}_{g(t_0)}(\gamma'(s), \gamma'(s))}{\sqrt{\langle \gamma'(s), \gamma'(s)\rangle_{g(t_0)}}}\,ds$$

$$\geq -CQ \int_0^d |\gamma'(s)|_{g(t_0)}\,ds = -CQ\ell_{t_0}(\gamma).$$

Integrating yields

$$\ell_{-t}(\gamma) \leq e^{CQt}\ell_0(\gamma) = e^{CQt}Q^{-1/2}.$$

(Recall $d^2 Q = 1$.) Plugging in $t = cQ^{-1}$ gives us

$$d_{-cQ^{-1}}(p, q) \leq \ell_{-cQ^{-1}}(\gamma) \leq e^{cC}Q^{-1/2}.$$

Setting $\widetilde{C} = e^{cC}$ completes the proof of the claim. □

The integrated form of Hamilton's Harnack inequality, Theorem 4.40, tells us that

$$\log\left(\frac{R(p,0)}{R(q,-cQ^{-1})}\right) \geq -\frac{d^2_{-cQ^{-1}}(p,q)}{2cQ^{-1}}.$$

According to the above claim, this in turn tells us

$$\log\left(\frac{R(p,0)}{R(q,-cQ^{-1})}\right) \geq -\widetilde{C}^2/2c.$$

Since $R(p,0) = 1$, it immediately follows that $R(q, -cQ^{-1}) \leq \exp(\widetilde{C}^2/(2c))$.

CLAIM 9.68. *There is a universal (i.e., independent of the 3-dimensional κ-solution) upper bound for $Q = R(q,0)$.*

PROOF. Let $G' = QG$ and $t' = Qt$. Then $R_{G'}(q',0) \leq C$ for all $q' \in B_{G'}(q,0,2)$. Consequently, $R_{G'}(q',t') \leq C$ for all $q' \in B_{G'}(q,0,2)$ and all $t' \leq 0$. Thus, by Shi's derivative estimates (Theorem 3.28) applied with $T = 2$ and $r = 2$, there is a universal constant C_1 such that for all $-1 \leq t' \leq 0$,

$$|\triangle\mathrm{Rm}_{G'}(q,t')|_{G'} \leq C_1,$$

where the Laplacian is taken with respect to the metric G'. Rescaling by Q^{-1} we see that for all $-Q^{-1} \leq t \leq 0$ we have

$$|\triangle\mathrm{Rm}_G(q,t)| \leq C_1 Q^2,$$

where the Laplacian is taken with respect to the metric G. Since the metric is non-negatively curved, by Corollary 4.39 we have $2|\mathrm{Ric}(q,t)|^2 \leq 2Q^2$ for all $t \leq 0$. From these two facts we conclude from the flow equation (3.7) that there is a constant $1 < C'' < \infty$ with the property that $\partial R/\partial t(q,t) \leq C''Q^2$ for all $-Q^{-1} < t \leq 0$. Thus for any $0 < c < 1$, we have $Q = R(q,0) \leq cC''Q + R(q,-cQ^{-1}) \leq cC''Q + e^{(\widetilde{C}^2(c)/2c)}$. Now we take $c = (2C'')^{-1}$ and $\widetilde{C} = \widetilde{C}(c)$. Plugging these values into the previous inequality yields

$$Q \leq 2e^{(\widetilde{C}^2 C'')}.$$

□

This leads immediately to:

CLAIM 9.69. *There are universal constants $\delta > 0$ and $C_1 < \infty$ (independent of the based 3-dimensional κ-solution $(M, g(t), (p,0))$ with $R(p,0) = 1$) such that $d(p,q) \geq \delta$. In addition, $R(q',t) \leq C_1$ for all $q' \in B(p,0,d)$ and all $t \leq 0$.*

PROOF. Since, according to the previous claim, Q is universally bounded above and $d^2Q = 1$, the existence of $\delta > 0$ as required is clear. Since

$B(p, 0, d) \subset B(q, 0, 2d)$, since, by Claim 9.66, $R(q', 0)/R(q, 0)$ is universally bounded on $B(q, 0, 2d)$, and since $R(q, 0)$ is universally bounded by Lemma 9.68, the second statement is clear for all

$$(q', 0) \in B(p, 0, d) \subset B(q, 0, 2d).$$

Given this, the fact that the second statement holds for all

$$(q', t) \in B(p, 0, d) \times (-\infty, 0]$$

then follows immediately from Corollary 4.39, the derivative inequality for $\partial R(q, t)/\partial t$. □

This, in turn, leads immediately to:

COROLLARY 9.70. *Fix $\delta > 0$ the universal constant of the last claim. Then $R(q', t) \leq \delta^{-2}$ for all $q' \in B(p, 0, \delta)$ and all $t \leq 0$.*

Now we return to the proof of Lemma 9.65. Since $(M, g(t))$ is κ-non-collapsed, it follows from the previous corollary that $\text{Vol } B(p, 0, \delta) \geq \kappa \delta^3$. Hence, according to Corollary 9.63 for each $A < \infty$ there is a constant $K(A)$ such that $R(q', 0) \leq K(A/\delta)/(\delta + A)^2$ for all $q' \in B(p, 0, A)$. Since δ is a universal positive constant, this completes the proof of Lemma 9.65. □

Now let us turn to the proof of Theorem 9.64, the compactness result for κ-solutions.

PROOF. Let $(M_k, g_k(t), (p_k, 0))$ be a sequence of based 3-dimensional κ-solutions with $R(p_k, 0) = 1$ for all k. The immediate consequence of Lemma 9.65 and Corollary 4.39 is the following. For every $r < \infty$ there is a constant $C(r) < \infty$ such that $R(q, t) \leq C(r)$ for all $q \in B(p_k, 0, r)$ and for all $t \leq 0$. Of course, since, in addition, the elements in the sequence are κ-non-collapsed, by Theorem 5.15 this implies that there is a subsequence of the $(M_k, g_k(t), (p_k, 0))$ that converges geometrically to an ancient flow $(M_\infty, g_\infty(t), (p_\infty, 0))$. Being a geometric limit of κ-solutions, this limit is complete and κ-non-collapsed, and each time-slice is of non-negative curvature. Also, it is not flat since, by construction, $R(p_\infty, 0) = 1$. Of course, it also follows from the limiting procedure that $\partial R(q, t)/\partial t \geq 0$ for every $(q, t) \in M_\infty \times (-\infty, 0]$. Thus, according to Corollary 9.53 the limit $(M_\infty, g_\infty(t))$ has bounded curvature for each $t \leq 0$. Hence, the limit is a κ-solution. This completes the proof of Theorem 9.64. □

COROLLARY 9.71. *Given $\kappa > 0$, there is $C < \infty$ such that for any 3-dimensional κ-solution $(M, g(t))$, $-\infty < t \leq 0$, we have*

$$\text{(9.31)} \qquad \sup_{(x,t)} \frac{|\nabla R(x, t)|}{R(x, t)^{3/2}} < C,$$

$$\text{(9.32)} \qquad \sup_{(x,t)} \frac{|\frac{d}{dt} R(x, t)|}{R(x, t)^2} < C.$$

PROOF. Notice that the two inequalities are scale invariant. Thus, this result is immediate from the compactness theorem, Theorem 9.64. □

Because of Proposition 9.58, and the fact that the previous corollary obviously holds for any shrinking family of round metrics, we can take the constant C in the above corollary to be independent of $\kappa > 0$.

Notice that, using Equation (3.7), we can rewrite the second inequality in the above corollary as

$$\sup_{(x,t)} \frac{|\triangle R + 2|\text{Ric}|^2|}{R(x,t)^2} < C.$$

8. Qualitative description of κ-solutions

In Chapter 2 we defined the notion of an ϵ-neck. In this section we define a stronger version of these, called strong ϵ-necks. We also introduce other types of canonical neighborhoods – ϵ-caps, ϵ-round components and C-components. These definitions pave the way for a qualitative description of κ-solutions.

8.1. Strong canonical neighborhoods. The next manifold we introduce is one with controlled topology (diffeomorphic either to the 3-disk or a punctured $\mathbb{R}P^3$) with the property that the complement of a compact submanifold is an ϵ-neck.

DEFINITION 9.72. Fix constants $0 < \epsilon < 1/2$ and $C < \infty$. Let (M, g) be a Riemannian 3-manifold. A (C, ϵ)-cap in (M, g) is an open submanifold $(\mathcal{C}, g|_{\mathcal{C}})$ together with an open submanifold $N \subset \mathcal{C}$ with the following properties:

(1) $\mathcal{C}$ is diffeomorphic either to an open 3-ball or to a punctured $\mathbb{R}P^3$.
(2) N is an ϵ-neck with compact complement in $\mathcal{C}$.
(3) $\overline{Y} = \mathcal{C} \setminus N$ is a compact submanifold with boundary. Its interior, Y, is called the *core* of $\mathcal{C}$. The frontier of Y, which is $\partial \overline{Y}$, is a central 2-sphere of an ϵ-neck contained in $\mathcal{C}$.
(4) The scalar curvature $R(y) > 0$ for every $y \in \mathcal{C}$ and

$$\text{diam}(\mathcal{C}, g|_{\mathcal{C}}) < C \left(\sup_{y \in \mathcal{C}} R(y)\right)^{-1/2}.$$

(5) $\sup_{x,y \in \mathcal{C}} [R(y)/R(x)] < C$.
(6) $\text{Vol}\,\mathcal{C} < C(\sup_{y \in \mathcal{C}} R(y))^{-3/2}$.
(7) For any $y \in Y$ let r_y be defined so that $\sup_{y' \in B(y, r_y)} R(y') = r_y^{-2}$. Then for each $y \in Y$, the ball $B(y, r_y)$ lies in $\mathcal{C}$ and indeed has compact closure in $\mathcal{C}$. Furthermore,

$$C^{-1} < \inf_{y \in Y} \frac{\text{Vol}\,B(y, r_y)}{r_y^3}.$$

(8) Lastly,
$$\sup_{y \in \mathcal{C}} \frac{|\nabla R(y)|}{R(y)^{3/2}} < C$$
and
$$\sup_{y \in \mathcal{C}} \frac{|\triangle R(y) + 2|\mathrm{Ric}|^2|}{R(y)^2} < C.$$

REMARK 9.73. If the ball $B(y, r_y)$ meets the complement of the core of $\mathcal{C}$, then it contains a point whose scalar curvature is close to $R(x)$, and hence r_y is bounded above by, say $2R(x)^{-1}$. Since $\epsilon < 1/2$, using the fact that y is contained in the core of $\mathcal{C}$ it follows that $B(y, r_y)$ is contained in $\mathcal{C}$ and has compact closure in $\mathcal{C}$.

Implicitly, we always orient the ϵ-neck structure on N so that the closure of its negative end meets the core of $\mathcal{C}$. See FIG. 1 in the Introduction.

Condition (8) in the above definition may seem unnatural, but here is the reason for it.

CLAIM 9.74. *Suppose that $(M, g(t))$ is a Ricci flow and that $(\mathcal{C}, g(t)|_\mathcal{C})$ is a subset of a t time-slice. Then Condition (8) above is equivalent to*
$$\sup_{(x,t) \in \mathcal{C}} \frac{\left|\frac{\partial R(x,t)}{\partial t}\right|}{R^2(x,t)} < C.$$

PROOF. This is immediate from Equation (3.7). □

Notice that the definition of a (C, ϵ)-cap is a scale invariant notion.

DEFINITION 9.75. Fix a positive constant C. A compact connected Riemannian manifold (M, g) is called *a C-component* if:
(1) M is diffeomorphic to either S^3 or $\mathbb{R}P^3$.
(2) (M, g) has positive sectional curvature.
(3)
$$C^{-1} < \frac{\inf_P K(P)}{\sup_{y \in M} R(y)}$$
where P varies over all 2-planes in TX (and $K(P)$ denotes the sectional curvature in the P-direction).
(4)
$$C^{-1}\sup_{y \in M}\left(R(y)^{-1/2}\right) < \mathrm{diam}(M) < C\inf_{y \in M}\left(R(y)^{-1/2}\right).$$

DEFINITION 9.76. Fix $\epsilon > 0$. Let (M, g) be a compact, connected 3-manifold. Then (M, g) is *within ϵ of round in the $C^{[1/\epsilon]}$-topology* if there exist a constant $R > 0$, a compact manifold (Z, g_0) of constant curvature $+1$, and a diffeomorphism $\varphi \colon Z \to M$ with the property that the pull-back under φ of Rg is within ϵ in the $C^{[1/\epsilon]}$-topology of g_0.

Notice that both of these notions are scale invariant notions.

DEFINITION 9.77. Fix $C < \infty$ and $\epsilon > 0$. For any Riemannian manifold (M, g), an open neighborhood U of a point $x \in M$ is a (C, ϵ)-*canonical neighborhood* if one of the following holds:
  (1) U is an ϵ-neck in (M, g) centered at x.
  (2) U is a (C, ϵ)-cap in (M, g) whose core contains x.
  (3) U is a C-component of (M, g).
  (4) U is an ϵ-round component of (M, g).

Whether or not a point $x \in M$ has a (C, ϵ)-canonical neighborhood in M is a scale invariant notion.

The notion of (C, ϵ)-canonical neighborhoods is sufficient for some purposes, but often we need a stronger notion.

DEFINITION 9.78. Fix constants $C < \infty$ and $\epsilon > 0$. Let $(\mathcal{M}, G)$ be a generalized Ricci flow. An *evolving ϵ-neck defined for an interval of normalized time of length $t' > 0$ centered at a point $x \in \mathcal{M}$ with $\mathbf{t}(x) = t$* is an embedding $\psi \colon S^2 \times (-\epsilon^{-1}, \epsilon^{-1}) \xrightarrow{\cong} N \subset M_t$ with $x \in \psi(S^2 \times \{0\})$ satisfying the following properties:
  (1) There is an embedding $N \times (t - R(x)^{-1}t', t] \to \mathcal{M}$ compatible with time and the vector field.
  (2) The pullback under ψ of the one-parameter family of metrics on N determined by restricting $R(x)G$ to the image of this embedding is within ϵ in the $C^{[1/\epsilon]}$-topology of the standard family $(h(t), ds^2)$ on the interval $-t' < t \leq 0$, where $h(t)$ is the round metric of scalar curvature $1/(1-t)$ on S^2 and ds^2 is the usual Euclidean metric on the interval (see Definition 2.16 for the notion of two families of metrics being close).

A *strong ϵ-neck* is the image of an evolving ϵ-neck which is defined for an interval of normalized time of length 1.

Both of these notions are scale invariant notions.

Let $(\mathcal{M}, G)$ be a generalized Ricci flow. Let $x \in \mathcal{M}$ be a point with $\mathbf{t}(x) = t$. We say that an open neighborhood U of x in M_t is a *strong (C, ϵ)-canonical neighborhood* of x if one of the following holds:
  (1) U is a strong ϵ-neck in $(\mathcal{M}, G)$ centered at x.
  (2) U is a (C, ϵ)-cap in M_t whose core contains x.
  (3) U is a C-component of M_t.
  (4) U is an ϵ-round component of M_t.

Whether or not a point x in a generalized Ricci flow has a strong (C, ϵ)-canonical neighborhood is a scale invariant notion.

PROPOSITION 9.79. *The following holds for any $\epsilon < 1/4$ and any $C < \infty$. Let $(\mathcal{M}, G)$ be a generalized Ricci flow and let $x \in \mathcal{M}$ be a point with $\mathbf{t}(x) = t$.*

8. QUALITATIVE DESCRIPTION OF κ-SOLUTIONS

(1) Suppose that $U \subset M_t$ is a (C,ϵ)-canonical neighborhood for x. Then for any horizontal metric G' sufficiently close to $G|_U$ in the $C^{[1/\epsilon]}$-topology, $(U, G'|_U)$ is a (C,ϵ)-canonical neighborhood for any $x' \in U$ sufficiently close to x.

(2) Suppose that in $(\mathcal{M}, G)$ there is an evolving ϵ-neck U centered at (x,t) defined for an interval of normalized time of length $a > 1$. Then any Ricci flow on
$$U \times (t - aR(x,t)^{-1}, t]$$
sufficiently close in the $C^{[1/\epsilon]}$-topology to the pullback of G contains a strong ϵ-neck centered at (x,t).

(3) Given (C,ϵ) and (C',ϵ') with $C' > C$ and $\epsilon' > \epsilon$ there is $\delta > 0$ such that the following holds. Suppose that $R(x) \le 2$. Let (U,g) be a (C,ϵ)-canonical neighborhood of x then for any metric g' within δ of g in the $C^{[1/\epsilon]}$-topology (U, g') contains a (C', ϵ')-neighborhood of x.

(4) Suppose that $g(t)$, $-1 < t \le 0$, is a one-parameter family of metrics on (U,g) that is a strong ϵ-neck centered at $(x,0)$ and $R_g(x,0) = 1$. Then any one-parameter family $g'(t)$ within δ in the $C^{[1/\epsilon]}$-topology of g with $R_{g'}(x,0) = 1$ is a strong ϵ'-neck.

PROOF. Since $\epsilon < 1/4$, the diameter of (U,g), the volume of (U,g), the supremum over $x \in U$ of $R(x)$, the supremum over x and y in U of $R(y)/R(x)$, and the infimum over all 2-planes P in $\mathcal{H}TU$ of $K(P)$ are all continuous functions of the horizontal metric G in the $C^{[1/\epsilon]}$-topology.

Let us consider the first statement. Suppose $(U, G|_U)$ is a C-component or an ϵ-round component. Since the defining inequalities are strict, and, as we just remarked, the quantities in these inequalities vary continuously with the metric in the $C^{[1/\epsilon]}$-topology, the result is clear in this case.

Let us consider the case when $U \subset M_t$ is an ϵ-neck centered at x. Let $\psi \colon S^2 \times (-\epsilon^{-1}, \epsilon^{-1}) \to U$ be the map giving the ϵ-neck structure. Then for all horizontal metrics G' sufficiently close to G in the $C^{[1/\epsilon]}$-topology, the same map ψ determines an ϵ-neck centered at x for the structure $(U, G'|_U)$. Now let us consider moving x to a nearby point x', say x' is the image of $(a, s) \in S^2 \times (-\epsilon^{-1}, \epsilon^{-1})$. We pre-compose ψ by a map which is the product of the identity in the S^2-factor with a diffeomorphism α on $(-\epsilon^{-1}, \epsilon^{-1})$ that is the identity near the ends and moves 0 to s. As x' approaches x, s tends to zero, and hence we can choose α so that it tends to the identity in the C^∞-topology. Thus, for x' sufficiently close to x, this composition will determine an ϵ-neck structure centered at x'. Lastly, let us consider the case when $(U, G|_U)$ is a (C, ϵ)-cap whose core Y contains x. Let G' be a horizontal metric sufficiently close to $G|_U$ in the $C^{[1/\epsilon]}$-topology. Let $N \subset U$ be the ϵ-neck $U \setminus \overline{Y}$. We have just seen that $(N, G'|_N)$ is an ϵ-neck. Similarly, if $N' \subset U$ is an ϵ-neck with central 2-sphere $\partial \overline{Y}$, then $(N', G'|_{N'})$ is an ϵ-neck if G' is sufficiently close to G in the $C^{[1/\epsilon]}$-topology.

Thus, Conditions (1), (2), and (3) in the definition of a (C, ϵ)-cap hold for (U, G'_U). Since the curvature, volume and diameter inequalities in Conditions (4), (5), and (6) are strict, they also hold for g'. To verify that Condition (7) holds for C', we need only remark that r_y is a continuous function of the metric. Lastly, since the derivative inequalities for the curvature in Condition (8) are strict inequalities and $\epsilon^{-1} > 4$, if these inequalities hold for all horizontal metrics G' sufficiently close to G in the $C^{[1/\epsilon]}$-topology. This completes the examination of all cases and proves the first statement.

The second statement is proved in the same way using the fact that if $g'(t)$ is sufficiently close to g in the $C^{[1/\epsilon]}$-topology and if x' is sufficiently close to x, then $R_{g'}(x')^{-1} < aR_g(x)^{-1}$.

Now let us turn to the third statement. The result is clear for ϵ-necks. Also, since $R(x) \leq 2$ the result is clear for ϵ-round components and C-components as well. Lastly, we consider a (C, ϵ)-cap U whose core Y contains x. Clearly, since $R(x)$ is bounded above by 2, for $\delta > 0$ sufficiently small, any metric g' within δ of g will satisfy the diameter, volume and curvature and the derivative of the curvature inequalities with C' replacing C. Let N be the ϵ-neck in (U, g) containing the end of U. Assuming that δ is sufficiently small, let N' be the image of $S^2 \times \left(-\epsilon^{-1}, 2(\epsilon')^{-1} - \epsilon^{-1}\right)$. Then (N', g') becomes an ϵ'-neck structure once we shift the parameter in the s-direction by $\epsilon^{-1} - (\epsilon')^{-1}$. We let $U' = \overline{Y} \cup N'$. Clearly, the ϵ-neck with central 2-sphere $\partial \overline{Y}$ will also determine an ϵ'-neck with the same central 2-sphere provided that $\delta > 0$ is sufficiently small. Thus, for $\delta > 0$ sufficiently small, for any (C, ϵ) the result of this operation is a (C', ϵ')-cap with the same core.

The fourth statement is immediate. □

COROLLARY 9.80. *In an ancient solution $(M, g(t))$ the set of points that are centers of strong ϵ-necks is an open subset.*

PROOF. Let T be the final time of the flow. Suppose that (x, t) is the center of a strong ϵ-neck $U \times (t - R(x,t)^{-1}, t] \subset M \times (-\infty, 0]$. This neck extends backwards for all time and forwards until the final time T giving an embedding $U \times (-\infty, T] \to M \times (-\infty, T]$. There is $a > 1$ such that for all t' sufficiently close to t the restriction of this embedding determines an evolving ϵ-neck centered at (x, t') defined for an interval of normalized time of length a. Composing this neck structure with a self-diffeomorphism of U moving x' to x, as described above, shows that all (x', t') sufficiently close to (x, t) are centers of strong ϵ-necks. □

DEFINITION 9.81. *An ϵ-tube $\mathcal{T}$ in a Riemannian 3-manifold M is a submanifold diffeomorphic to the product of S^2 with a non-degenerate interval with the following properties:*

(1) *Each boundary component S of $\mathcal{T}$ is the central 2-sphere of an ϵ-neck $N(S)$ in M.*

8. QUALITATIVE DESCRIPTION OF κ-SOLUTIONS

(2) $\mathcal{T}$ is a union of ϵ-necks and the closed half ϵ-necks whose boundary sphere is a component of $\partial\mathcal{T}$. Furthermore, the central 2-sphere of each of the ϵ-necks is isotopic in $\mathcal{T}$ to the S^2-factors of the product structure.

An *open ϵ-tube* is one without boundary. It is a union of ϵ-necks with the central spheres that are isotopic to the 2-spheres of the product structure.

A *C-capped ϵ-tube* in M is a connected submanifold that is the union of a (C,ϵ)-cap $\mathcal{C}$ and an open ϵ-tube where the intersection of $\mathcal{C}$ with the ϵ-tube is diffeomorphic to $S^2 \times (0,1)$ and contains an end of the ϵ-tube and an end of the cap. A *doubly C-capped ϵ-tube* in M is a closed, connected submanifold of M that is the union of two (C,ϵ)-caps $\mathcal{C}_1$ and $\mathcal{C}_2$ and an open ϵ-tube. Furthermore, we require (i) that the cores Y_1 and Y_2 of $\mathcal{C}_1$ and $\mathcal{C}_2$ have disjoint closures, (ii) that the union of either $\mathcal{C}_i$ with the ϵ-tube is a capped ϵ-tube and $\mathcal{C}_1$ and $\mathcal{C}_2$ contain the opposite ends of the ϵ-tube. There is one further closely related notion, that of an *ϵ-fibration*. By definition an ϵ-fibration is a closed, connected manifold that fibers over the circle with fibers S^2 that is also a union of ϵ-necks with the property that the central 2-sphere of each neck is isotopic to a fiber of the fibration structure. We shall not see this notion again until the appendix, but because it is clearly closely related to the notion of an ϵ-tube, we introduce it here.

See FIG. 2.

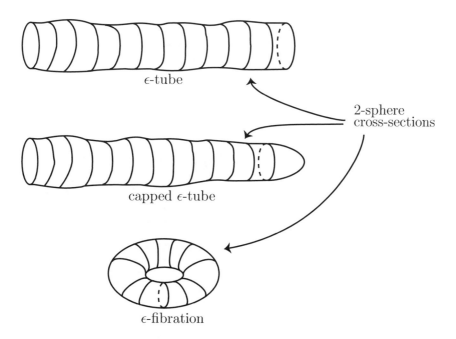

FIGURE 2. ϵ-canonical neighborhoods.

DEFINITION 9.82. A *strong ϵ-tube* in a generalized Ricci flow is an ϵ-tube with the property that each point of the tube is the center of a strong ϵ-neck in the generalized flow.

8.2. Canonical neighborhoods for κ solutions.

PROPOSITION 9.83. *Let $(M, g(t))$ be a 3-dimensional κ-solution. Then one of the following holds:*
  (1) *For every $t \leq 0$ the manifold $(M, g(t))$ has positive curvature.*
  (2) *$(M, g(t))$ is the product of an evolving family of round S^2's with a line.*
  (3) *M is diffeomorphic to a line bundle over $\mathbb{R}P^2$, and there is a finite covering of $(M, g(t))$ that is a flow as in (2).*

PROOF. Suppose that $(M, g(t))$ does not have positive curvature for some t. Then, by the application of the strong maximum principle given in Corollary 4.20, there is a covering $\widetilde{M}$ of M, with either one or two sheets, such that $(\widetilde{M}, \widetilde{g}(t))$ is the product of an evolving family of round surfaces with a flat 1-manifold (either a circle or the real line). Of course, the covering must be a κ-solution. In the case in which $(\widetilde{M}, \widetilde{g}(t))$ is isometric to the product of an evolving family of round surfaces and a circle, that circle has a fixed length, say $L < \infty$. Since the curvature of the surface in the t time-slice goes to zero as $t \to -\infty$, we see that the flow is not κ-non-collapsed on all scales for any $\kappa > 0$. Thus, $(M, g(t))$ has either a trivial cover or a double cover isometric to the product of a shrinking family of round surfaces with $\mathbb{R}$. If the round surface is S^2, then we have established the result. If the round surface is $\mathbb{R}P^2$ a further double covering is a product of round 2-spheres with $\mathbb{R}$. This proves the proposition. □

LEMMA 9.84. *Let $(M, g(t))$ be a non-compact 3-dimensional κ-solution of positive curvature and let $p \in M$. Then there is $D' < \infty$, possibly depending on $(M, g(0))$ and p, such that every point of*
$$M \times \{0\} \setminus B(p, 0, D'R(p, 0)^{-1/2})$$
is the center of an evolving ϵ-neck in $(M, g(t))$ defined for an interval of normalized time of length 2. Furthermore, there is $D_1' < \infty$ such that for any point $x \in B(p, 0, D'R(p, 0)^{-1/2})$ and any 2-plane P_x in $T_x M$ we have $(D_1')^{-1} < K(P_x)/R(p, 0) < D_1'$ where $K(P_x)$ denotes the sectional curvature in the direction of the 2-plane P_x.

PROOF. Given $(M, g(t))$ and p, suppose that no such $D' < \infty$ exists. Because the statement is scale invariant, we can arrange that $R(p, 0) = 1$. Then we can find a sequence of points $p_k \in M$ with $d_0(p, p_k) \to \infty$ as $k \to \infty$ such that no p_k is the center of an evolving ϵ-neck in $(M, g(0))$ defined for an interval of normalized time of length 2. By passing to a subsequence we can assume that one of two possibilities holds: either $d_0^2(p, p_k) R(p_k, 0) \to \infty$

as $k \to \infty$ or $\lim_{k\to\infty} d_0^2(p,p_k)R(p_k,0) = \ell < \infty$. In the first case, set $\lambda_k = R(p_k,0)$ and consider the based flows $(M, \lambda_k g(\lambda_k^{-1}t), (p_k, 0))$. According to Theorem 9.64, after passing to a subsequence there is a geometric limit. Thus, by Theorem 5.35 and Corollary 4.19 the limit splits as a product of a 2-dimensional κ-solution and $\mathbb{R}$. By Corollary 9.50 it follows that the limit is the standard round evolving cylinder. This implies that for all k sufficiently large $(p_k, 0)$ is the center of an evolving ϵ-neck in $(M, g(t))$ defined for an interval of normalized time of length 2. This contradiction establishes the existence of D' as required in this case.

Now suppose that $\lim_{k\to\infty} d_0^2(p,p_k)R(p_k,0) = \ell < \infty$. Of course, since $d_0(p, p_k) \to \infty$, it must be the case that $R(p_k, 0) \to 0$ as $k \to \infty$. Set $Q_k = R(p_k, 0)$. By passing to a subsequence we can arrange that $d_0^2(p, p_k)Q_k < \ell + 1$ for all k. Consider the κ-solutions $(M_k, g_k(t)) = (M, Q_k g(Q_k^{-1}t))$. For each k we have $p \in B_{g_k}(p_k, 0, \ell+1)$, and $R_{g_k}(p, 0) = Q_k^{-1} \to \infty$ as $k \to \infty$. This contradicts Lemma 9.65, and completes the proof of the existence of D as required in this case as well.

The existence of D_1' is immediate since the closure of the ball is compact and the manifold has positive curvature. □

In fact a much stronger result is true. The constants D' and D_1' in the above lemma can be chosen independent of the non-compact κ-solutions.

PROPOSITION 9.85. *For any $0 < \epsilon$ sufficiently small there are constants $D = D(\epsilon) < \infty$ and $D_1 = D_1(\epsilon) < \infty$ such that the following holds for any non-compact 3-dimensional κ-solution $(M, g(t))$ of positive curvature. Let $p \in M$ be a soul of $(M, g(0))$. Then:*

(1) *Every point in $M \setminus B(p, 0, DR(p,0)^{-1/2})$ is the center of a strong ϵ-neck in $(M, g(t))$. Furthermore, for any $x \in B(p, 0, DR(p,0)^{-1/2})$ and any 2-plane P_x in T_xM we have*

$$D_1^{-1} < K(P_x)/R(p,0) < D_1.$$

Also,

$$D_1^{-3/2}R(p,0)^{-3/2} < \mathrm{Vol}(B(p,0,DR(p,0)^{-1/2})) < D_1^{3/2}R(p,0)^{-3/2}.$$

(2) *Let f denote the distance function from p. For any ϵ-neck $N \subset (M, g(0))$, the middle two-thirds of N is disjoint from p, and the central 2-sphere S_N of N is (topologically) isotopic in $M \setminus \{p\}$ to $f^{-1}(a)$ for any $a > 0$. In particular, given two disjoint central 2-spheres of ϵ-necks in $(M, g(0))$ the region of M bounded by these 2-spheres is diffeomorphic to $S^2 \times [0,1]$.*

REMARK 9.86. In part 1 of this theorem one can replace p by any point $p' \in M$ that is not the center of a strong ϵ-neck.

PROOF. First suppose that no D exists so that the first statement holds. Then there is a sequence of such solutions $(M_k, g_k(t))$, with $p_k \in M_k$ being a soul of $(M_k, g_k(0))$ and points $q_k \in M_k$ with $d_0^2(p_k, q_k)R(p_k, 0) \to \infty$ as $k \to \infty$ such that q_k is not the center of a strong ϵ-neck in $(M_k, g_k(0))$. By rescaling we can assume that $R(p_k, 0) = 1$ for all k, and hence that $d_0(p_k, q_k) \to \infty$. Then, according to Theorem 9.64, by passing to a subsequence we can assume that there is a geometric limit $(M_\infty, g_\infty(t), (p_\infty, 0))$ with $R(p_\infty, 0) = 1$. By Lemma A.10, provided that ϵ is sufficiently small for all k the soul $(p_k, 0)$ is not the center of a strong 2ϵ-neck in $(M_k, g_k(t))$. Hence, invoking part 4 of Proposition 9.79 and using the fact that $R(p_k, 0) = 1$ for all k and hence $R(p_\infty, 0) = 1$, we see that $(p_\infty, 0)$ is not the center of a strong ϵ-neck in $(M_\infty, g_\infty(t))$. Since the manifolds M_k are non-compact and have metrics of positive curvature, they are diffeomorphic to $\mathbb{R}^3$ and in particular, do not contain embedded copies of $\mathbb{R}P^2$. Thus, the limit $(M_\infty, g_\infty(t))$ is a non-compact κ-solution containing no embedded copy of $\mathbb{R}P^2$. Thus, by Proposition 9.83 either it is positively curved or it is a Riemannian product S^2 times $\mathbb{R}$. In the second case every point is the center of a strong ϵ-neck. Since we have seen that the point $(p_\infty, 0)$ is not the center of a strong ϵ-neck, it follows that the limit is a positively curved κ-solution.

Then according to the previous lemma there is D', which depends only on $(M_\infty, g_\infty(0))$ and p_∞, such that every point outside $B(p_\infty, 0, D')$ is the center of an evolving $\epsilon/2$-neck defined for an interval of normalized time of length 2.

Since $(M_k, g_k(t), (p_k, 0))$ converge geometrically to $(M_\infty, g_\infty(t), (p_\infty, 0))$, by part 2 of Proposition 9.79 for any $L < \infty$, for all k sufficiently large, all points of

$$B(p_k, 0, L) \setminus B(p_k, 0, 2D')$$

are centers of strong ϵ-necks in $(M_k, g_k(t))$. In particular, for all k sufficiently large, $d_0(p_k, q_k) > L$. Let L_k be a sequence tending to infinity as $k \to \infty$. Passing to a subsequence, we can suppose that every point of $(B(p_k, 0, L_k) \setminus B(p_k, 0, 2D')) \subset M_k$ is the center of a strong ϵ-neck in $(M_k, g_k(0))$. Of course, for all k sufficiently large, $q_k \in M_k \setminus B(p_k, 0, 2D'))$. By Corollary 9.80 the subset of points in $M_k \times \{0\}$ that are centers of strong ϵ-necks is an open set. Thus, replacing q_k with another point if necessary, we can suppose that q_k is a closest point to p_k contained in $M_k \setminus B(p_k, 0, 2D')$ with the property that q_k is not the center of a strong ϵ-neck. Then $q_k \in M_k \setminus B(p_k, 0, L_k)$ and $(q_k, 0)$ is in the closure of the set of points in M_k that are centers of strong ϵ-necks in $(M_k, g_k(t))$, and hence by part 3 of Proposition 9.79 each $(q_k, 0)$ is the center of a 2ϵ-neck in $(M_k, g_k(t))$.

Let γ_k be a minimizing geodesic connecting $(p_k, 0)$ to $(q_k, 0)$, and let μ_k be a minimizing geodesic ray from $(q_k, 0)$ to infinity. Set $Q_k = R(q_k, 0)$. Since $(q_k, 0)$ is the center of a 2ϵ-neck, from Lemma 2.20 we see that, provided that ϵ is sufficiently small, the 2ϵ-neck centered at q_k separates p from ∞, so that γ_k and μ_k exit this 2ϵ-neck at opposite ends. According to

Theorem 9.64, after passing to a subsequence, the based, rescaled flows

$$(M_k, Q_k g(Q_k^{-1} t), (q_k, 0))$$

converge geometrically to a limit. Let $(q_\infty, 0)$ be the base point of the resulting limit. By part 3 of Proposition 9.79, it is the center of a 4ϵ-neck in the limit.

CLAIM 9.87. $d_0^2(p_k, q_k) Q_k \to \infty$ as $k \to \infty$.

PROOF. Suppose not. Then by passing to a subsequence we can suppose that these products are bounded independent of k. Then since $d_0(p_k, q_k) \to \infty$, we see that $Q_k \to 0$. Thus, in the rescaled flows $(M_k, Q_k g_k(Q_k^{-1} t))$ the curvature at $(p_k, 0)$ goes to infinity. But this is impossible since the $Q_k g_k$-distance from $(p_k, 0)$ to $(q_k, 0)$ is $\sqrt{Q_k} d_0(p_k, q_k)$, which is bounded independent of k, and the scalar curvature of $(p, 0)$ in the metric $Q_k g_k(0)$ is $R(p_k, 0) Q_k^{-1} = Q_k^{-1}$, which tends to ∞. Unbounded curvature at bounded distance contradicts Lemma 9.65, and this establishes the claim. □

A subsequence of the based flows $(M_k, Q_k g_k(Q_k^{-1} t), (q_k, 0))$ converges geometrically to a κ-solution. According to Theorem 5.35 and Corollary 4.19, this limiting flow is the product of a 2-dimensional κ-solution with a line. Since M is orientable, this 2-dimensional κ-solution is an evolving family of round 2-spheres. This implies that for all k sufficiently large, $(q_k, 0)$ is the center of a strong ϵ-neck in $(M_k, g_k(t))$. This is a contradiction and proves the existence of $D < \infty$ as stated in the proposition.

Let $(M_k, g_k(t), (p_k, 0))$ be a sequence of non-compact Ricci flows based at a soul p_k of $(M_k, g_k(0))$. We rescale so that $R(p_k, 0) = 1$. By Lemma A.10, if ϵ is sufficiently small, then p_k cannot be the center of an ϵ-neck. It follows from Proposition 9.79 that for any limit of a subsequence the point p_∞, which is the limit of the p_k, is not the center of an 2ϵ-neck in the limit. Since the limit manifold is orientable, it is either contractible with strictly positive curvature or is a metric product of a round 2-sphere and the line. It follows that the limit manifold has strictly positive curvature at $(p_\infty, 0)$, and hence positive curvature everywhere. The existence of $D_1 < \infty$ as required is now immediate from Theorem 9.64.

The fact that any soul is disjoint from the middle two-thirds of any ϵ-neck and the fact that the central 2-spheres of all ϵ-necks are isotopic in $M \setminus \{p\}$ are contained in Lemma A.10 and Corollary 2.20. □

COROLLARY 9.88. *There is $\bar{\epsilon}_2 > 0$ such that for any $0 < \epsilon \leq \bar{\epsilon}_2$ the following holds. There is $C_0 = C_0(\epsilon)$ such that for any $\kappa > 0$ and any non-compact 3-dimensional κ-solution not containing an embedded $\mathbb{R}P^2$ with trivial normal bundle, the zero time-slice is either a strong ϵ-tube or a C_0-capped strong ϵ-tube.*

PROOF. For $\epsilon > 0$ sufficiently small let $D(\epsilon)$ and $D_1(\epsilon)$ be as in the previous corollary. At the expense of increasing these, we can assume that they are at least the constant C in Corollary 9.71. We set

$$C_0(\epsilon) = \max(D(\epsilon), D_1(\epsilon)).$$

If the non-compact κ-solution has positive curvature, then the corollary follows immediately from Proposition 9.85 and Corollary 9.71. If the κ-solution is the product of an evolving round S^2 with the line, then every point of the zero time-slice is the center of a strong ϵ-neck for every $\epsilon > 0$ so that the zero time-slice of the solution is a strong ϵ-tube. Suppose the solution is double covered by the product of an evolving round 2-sphere and the line. Let ι be the involution and take the product coordinates so that $S^2 \times \{0\}$ is the invariant 2-sphere of ι in the zero time-slice. Then any point in the zero time-slice at distance at least $3\epsilon^{-1}$ from $P = (S^2 \times \{0\})/\iota$ is the center of a strong ϵ-neck. Furthermore, an appropriate neighborhood of P in the time zero slice is a (C, ϵ)-cap whose core contains the $3\epsilon^{-1}$ neighborhood of P. The derivative bounds in this case come from the fact that the metric is close in the $C^{[1/\epsilon]}$-topology to the standard evolving flow. This proves the corollary in this case and hence completes the proof. □

Now let us consider compact κ-solutions.

THEOREM 9.89. *There is $\overline{\epsilon}_3 > 0$ such that for every $0 < \epsilon \leq \overline{\epsilon}_3$ there is $C_1 = C_1(\epsilon) < \infty$ such that one of the following holds for any $\kappa > 0$ and any compact 3-dimensional κ-solution $(M, g(t))$.*

(1) *The manifold M is compact and of constant positive sectional curvature.*
(2) *The diameter of $(M, g(0))$ is less than $C_1 \cdot (\max_{x \in M} R(x, 0))^{-1/2}$, and M is diffeomorphic to either S^3 or $\mathbb{R}P^3$.*
(3) *$(M, g(0))$ is a double C_1-capped strong ϵ-tube.*

PROOF. First notice that if $(M, g(t))$ is not of strictly positive curvature, then the universal covering of $(M, g(0))$ is a Riemannian product $S^2 \times \mathbb{R}$, and hence $(M, g(0))$ is either non-compact or finitely covered by the product flow on $S^2 \times S^1$. The former case is ruled out since we are assuming that M is compact and the latter case is ruled out because such flows are not κ-non-collapsed for any $\kappa > 0$. We conclude that $(M, g(t))$ is of positive curvature. This implies that the fundamental group of M is finite. If there were an embedded $\mathbb{R}P^2$ in M with trivial normal bundle, that $\mathbb{R}P^2$ could not separate (since the Euler characteristic of $\mathbb{R}P^2$ is 1, it is not the boundary of a compact 3-manifold). But a non-separating surface in M induces a surjective homomorphism of $H_1(M)$ onto $\mathbb{Z}$. We conclude from this that M does not contain an embedded $\mathbb{R}P^2$ with trivial normal bundle.

We assume that $(M, g(0))$ is not round so that by Proposition 9.58 there is a universal $\kappa_0 > 0$ such that $(M, g(0))$ is a κ_0-solution. Let $C_0(\epsilon)$ be the constant from Corollary 9.88.

CLAIM 9.90. *Assuming that $(M, g(0))$ is compact but not of constant positive sectional curvature, for each $\epsilon > 0$ there is C_1 such that if the diameter of $(M, g(0))$ is greater than $C_1(\max_{x \in M} R(x, 0))^{-1/2}$, then every point of $(M, g(0))$ is either contained in the core of $(C_0(\epsilon), \epsilon)$-cap or is the center of a strong ϵ-neck in $(M, g(t))$.*

PROOF. Suppose that for some $\epsilon > 0$ there is no such C_1. We take a sequence of constants C'_k that diverges to $+\infty$ as $k \to \infty$ and a sequence $(M_k, g_k(t), (p_k, 0))$ of based κ_0-solutions such that the diameter of $(M_k, 0)$ is greater than $C'_k R^{-1/2}(p_k, 0)$ and yet $(p_k, 0)$ is not contained in the core of a $(C_0(\epsilon), \epsilon)$-cap nor is the center of a strong ϵ-neck. We scale $(M_k, g_k(t))$ by $R(p_k, 0)$. This allows us to assume that $R(p_k, 0) = 1$ for all k. According to Theorem 9.64, after passing to a subsequence we can assume these based κ-solutions converge to a based κ-solution $(M_\infty, g_\infty(t), (p_\infty, 0))$. Since the diameters of the $(M_k, g_k(0))$ go to infinity, M_∞ is non-compact. According to Corollary 9.88 the point p_∞ is either the center of a strong ϵ-neck, or is contained in the core of a $(C_0(\epsilon), \epsilon)$-cap. Since $R(p_k, 0) = 1$ for all k, it follows from parts 1 and 4 of Proposition 9.79 that for all k sufficiently large, $(p_k, 0)$ is either the center of a strong ϵ-neck in $(M_k, g_k(t))$ or is contained in the core of a $(C_0(\epsilon), \epsilon)$-cap. This is a contradiction, proving the claim. $\square$

Now it follows from Proposition A.25 that if the diameter of $(M, g(0))$ is greater than $C_1(\max_{x \in M} R(x, 0))^{-1/2}$ and if it is not of constant positive curvature, then M is diffeomorphic to either S^3, $\mathbb{R}P^3$, $\mathbb{R}P^3 \# \mathbb{R}P^3$ or is an S^2-fibration over S^1. On the other hand, since M is compact of positive curvature its fundamental group is finite, see Theorem 4.1 on p. 154 of [**57**]. This rules out the last two cases. This implies that when $(M, g(0))$ has diameter greater than $C_1(\max_{x \in M} R(x, 0))^{-1/2}$ and is not of constant positive curvature, it is a double C_0-capped ϵ-tube.

We must consider the case when $(M, g(0))$ is not of constant positive curvature and its diameter is less than or equal to $C_1(\max_{x \in M} R(x, 0))^{-1/2}$. Since $(M, g(0))$ is not round, by Corollary 9.44 its asymptotic soliton is not compact. Thus, by Theorem 9.42 its asymptotic soliton is either $S^2 \times \mathbb{R}$ or is double covered by this product. This means that for t sufficiently negative the diameter of $(M, g(t))$ is greater than $C_1(\max_{x \in M} R(x, 0))^{-1/2}$. Invoking the previous result for this negative time tells us that M is diffeomorphic to S^3 or $\mathbb{R}P^3$. $\square$

PROPOSITION 9.91. *Let $\bar{\epsilon}_2$ and $\bar{\epsilon}_3$ be as in Corollary 9.88 and Theorem 9.89, respectively. For each $0 < \epsilon \leq \min(\bar{\epsilon}_2, \bar{\epsilon}_3)$ let $C_1 = C_1(\epsilon)$ be as in Theorem 9.89. There is $C_2 = C_2(\epsilon) < \infty$ such that for any $\kappa > 0$*

and any compact κ-solution $(M,g(t))$ the following holds. If $(M,g(0))$ is not of constant positive curvature and if $(M,g(0))$ is of diameter less than $C_1(\max_{x\in M} R(x,0))^{-1/2}$, then for any $x \in M$ we have

$$C_2^{-1} R(x,0)^{-3/2} < \text{Vol}(M,g(0)) < C_2 R(x,0)^{-3/2}.$$

In addition, for any $y \in M$ and any 2-plane P_y in $T_y M$ we have

$$C_2^{-1} < \frac{K(P_y)}{R(x,0)} < C_2,$$

where $K(P_y)$ is the sectional curvature in the P_y-direction.

PROOF. The result is immediate from Corollary 9.58 and Theorem 9.64.
□

REMARK 9.92. For a round κ-solution $(M,g(t))$ we have $R(x,0) = R(y,0)$ for all $x,y \in M$, and the volume of $(M,g(0))$ is bounded above by a constant times $R(x,0)^{-3/2}$. There is no universal lower bound to the volume in terms of the curvature. The lower bound takes the form $C_2|\pi_1(M)|^{-1}R(x,0)^{-3/2}$, where $|\pi_1(M)|$ is the order of the fundamental group $\pi_1(M)$.

Let us summarize our results.

THEOREM 9.93. There is $\bar{\epsilon} > 0$ such that the following is true for any $0 < \epsilon < \bar{\epsilon}$. There is $C = C(\epsilon)$ such that for any $\kappa > 0$ and any κ-solution $(M,g(t))$ one of the following holds.
  (1) $(M,g(t))$ is round for all $t \leq 0$. In this case M is diffeomorphic to the quotient of S^3 by a finite subgroup of $SO(4)$ acting freely.
  (2) $(M,g(0))$ is compact and of positive curvature. For any $x,y \in M$ and any 2-plane P_y in $T_y M$ we have

$$\begin{aligned} C^{-1/2} R(x,0)^{-1} &< \text{diam}(M,g(0)) < CR(x,0)^{-1/2}, \\ C^{-1} R(x,0)^{-3/2} &< \text{Vol}(M,g(0)) < CR(x,0)^{-3/2}, \\ C^{-1} R(x,0) &< K(P_y) < CR(x,0). \end{aligned}$$

  In this case M is diffeomorphic either to S^3 or to $\mathbb{R}P^3$.
  (3) $(M,g(0))$ is of positive curvature and is a double C-capped strong ϵ-tube, and in particular M is diffeomorphic to S^3 or to $\mathbb{R}P^3$.
  (4) $(M,g(0))$ is of positive curvature and is a C-capped strong ϵ-tube and M is diffeomorphic to $\mathbb{R}^3$.
  (5) $(M,g(0))$ is isometric to the quotient of the product of a round S^2 and $\mathbb{R}$ by a free, orientation-preserving involution. It is a C-capped strong ϵ-tube and is diffeomorphic to a punctured $\mathbb{R}P^3$.
  (6) $(M,g(0))$ is isometric to the product of a round S^2 and $\mathbb{R}$ and is a strong ϵ-tube.
  (7) $(M,g(0))$ is isometric to a product $\mathbb{R}P^2 \times \mathbb{R}$, where the metric on $\mathbb{R}P^2$ is of constant Gaussian curvature.

In particular, in all cases except the first two and the last one, all points of $(M, g(0))$ are either contained in the core of a (C, ϵ)-cap or are the centers of a strong ϵ-neck in $(M, g(0))$.

Lastly, in all cases we have

$$\sup_{p \in M, t \leq 0} \frac{|\nabla R(p,t)|}{R(p,t)^{3/2}} < C, \tag{9.33}$$

$$\sup_{p \in M, t \leq 0} \frac{|\partial R(p,t)/\partial t|}{R(p,t)^2} < C. \tag{9.34}$$

An immediate consequence of this result is:

COROLLARY 9.94. *For every $0 < \epsilon \leq \bar{\epsilon}'$ there is $C = C(\epsilon) < \infty$ such that every point in a κ-solution has a strong (C, ϵ)-canonical neighborhood unless the κ-solution is a product $\mathbb{R}P^2 \times \mathbb{R}$.*

COROLLARY 9.95. *Fix $0 < \epsilon \leq \bar{\epsilon}'$, and let $C(\epsilon)$ be as in the last corollary. Suppose that $(\mathcal{M}_n, G_n, x_n)$ is a sequence of based, generalized Ricci flows with $\mathbf{t}(x_n) = 0$ for all n. Suppose that none of the time-slices of the $\mathcal{M}_n$ contain embedded $\mathbb{R}P^2$'s with trivial normal bundle. Suppose also that there is a smooth limiting flow $(M_\infty, g_\infty(t), (x_\infty, 0))$ defined for $-\infty < t \leq 0$ that is a κ-solution. Then for all n sufficiently large, the point x_n has a strong (C, ϵ)-canonical neighborhood in $(\mathcal{M}_n, G_n, x_n)$.*

PROOF. The limiting manifold M_∞ cannot contain an embedded $\mathbb{R}P^2$ with trivial normal bundle. Hence, by the previous corollary, the point $(x_\infty, 0)$ has a strong (C, ϵ)-canonical neighborhood in the limiting flow. If the limiting κ-solution is round, then for all n sufficiently large, x_n is contained in a component of the zero time-slice that is ϵ-round. If $(x_\infty, 0)$ is contained in a C-component of the zero time-slice of the limiting κ-solution, then for all n sufficiently large x_n is contained in a C-component of the zero time-slice of $\mathcal{M}_n$. Suppose that $(x_\infty, 0)$ is the center of a strong ϵ-neck in the limiting flow. This neck extends backwards in the limiting solution some amount past an interval of normalized time of length 1, where by continuity it is an evolving ϵ-neck defined backwards for an interval of normalized time of length greater than 1. Then by part 2 of Proposition 9.79, any family of metrics on this neck sufficiently close to the limiting metric will determine a strong ϵ-neck. This implies that for all n sufficiently large, x_n is the center of a strong ϵ-neck in $(\mathcal{M}_n, G_n)$. Lastly, if $(x_\infty, 0)$ is contained in the core of a (C, ϵ)-cap in the limiting flow, then by part 1 of Proposition 9.79 for all n sufficiently large, x_n is contained in the core of a (C, ϵ)-cap in $(\mathcal{M}_n, G_n)$. □

CHAPTER 10

Bounded curvature at bounded distance

This chapter is devoted to Perelman's result about bounded curvature at bounded distance for blow-up limits. Crucial to the argument is that each member of the sequence of generalized Ricci flows has curvature pinched toward positive and also has strong canonical neighborhoods.

1. Pinching toward positive: the definitions

In this section we give the definition of what it means for a generalized Ricci flow to have curvature pinched toward positive. This is the obvious generalization of the corresponding notion for Ricci flows.

DEFINITION 10.1. Let $(\mathcal{M}, G)$ be a generalized 3-dimensional Ricci flow whose domain of definition is contained in $[0, \infty)$. For each $x \in \mathcal{M}$, let $\nu(x)$ be the smallest eigenvalue of $\mathrm{Rm}(x)$ on $\wedge^2 T_x M_{\mathbf{t}(x)}$, as measured with respect to a $G(x)$-orthonormal basis for the horizontal space at x, and set $X(x) = \max(0, -\nu(x))$. We say that $(\mathcal{M}, G)$ has curvature *pinched toward positive* if, for all $x \in \mathcal{M}$, if the following two inequalities hold:

$$R(x) \geq \frac{-6}{1 + 4\mathbf{t}(x)},$$
$$R(x) \geq 2X(x)\left(\log X(x) + \log(1 + \mathbf{t}(x)) - 3\right), \quad \text{if } 0 < X(x).$$

According to Theorem 4.32, if $(M, g(t))$, $0 \leq a \leq t < T$, is Ricci flow with M a compact 3-manifold, and if the two conditions given in the definition hold at the initial time a, then they hold for all $t \in [a, T)$. In particular, if $a = 0$ and if $|\mathrm{Rm}(p, 0)| \leq 1$ for all $p \in M$, then the curvature of the flow is pinched toward positive.

Next we fix $\epsilon_0 > 0$ sufficiently small such that for any $0 < \epsilon \leq \epsilon_0$ all the results of the Appendix hold for 2ϵ and $\alpha = 10^{-2}$, and Proposition 2.19 holds for 2ϵ.

2. The statement of the theorem

Here is the statement of the main theorem of this chapter, the theorem that establishes bounded curvature at bounded distance for blow-up limits.

THEOREM 10.2. *Fix $0 < \epsilon \leq \epsilon_0$ and $C < \infty$. Then for each $A < \infty$ there are $D_0 < \infty$ and $D < \infty$ depending on A, ϵ and C such that the following holds. Suppose that $(\mathcal{M}, G)$ is a generalized 3-dimensional Ricci flow whose* *interval of definition is contained in $[0,\infty)$, and suppose that $x \in \mathcal{M}$. Set $t = \mathbf{t}(x)$. We suppose that these data satisfy the following:*

(1) *$(\mathcal{M}, G)$ has curvature pinched toward positive.*
(2) *Every point $y \subset \mathcal{M}$ with $R(y) \geq 4R(x)$ and $\mathbf{t}(y) \leq t$ has a strong (C, ϵ)-canonical neighborhood.*

If $R(x) \geq D_0$, then $R(y) \leq DR(x)$ for all $y \in B(x, t, AR(x)^{-1/2})$.

This chapter is devoted to the proof of this theorem. The proof is by contradiction. Suppose that there is some $A_0 < \infty$ for which the result fails. Then there are a sequence of generalized 3-dimensional Ricci flows $(\mathcal{M}_n, G_n)$ whose intervals of definition are contained in $[0, \infty)$ and whose curvatures are pinched toward positive. Also, there are points $x_n \in \mathcal{M}_n$ satisfying the second condition given in the theorem and points $y_n \in \mathcal{M}_n$ such that for all n we have:

(1) $\lim_{n\to\infty} R(x_n) = \infty$.
(2) $\mathbf{t}(y_n) = \mathbf{t}(x_n)$.
(3) $d(x_n, y_n) < A_0 R(x_n)^{-1/2}$.
(4)
$$\lim_{n\to\infty} \frac{R(y_n)}{R(x_n)} = \infty.$$

For the rest of this chapter we assume that such a sequence of generalized Ricci flows exists. We shall eventually derive a contradiction.

Let us sketch how the argument goes. We show that there is a (partial) geometric blow-up limit of the sequence $(\mathcal{M}_n, G_n)$ based at the x_n. We shall see that the following hold for this limit. It is an incomplete manifold U_∞ diffeomorphic to $S^2 \times (0, 1)$ with the property that the diameter of U_∞ is finite and the curvature goes to infinity at one end of U_∞, an end denoted $\mathcal{E}$, while remaining bounded at the other end. (The non-compact manifold in question is diffeomorphic to $S^2 \times (0, 1)$ and, consequently, it has two ends.) Every point of U_∞ sufficiently close to $\mathcal{E}$ is the center of a 2ϵ-neck in U_∞. In fact, there is a partial geometric limiting flow on U_∞ so that these points are centers of evolving 2ϵ-necks. Having constructed this incomplete blow-up limit of the original sequence we then consider further blow-up limits about the end $\mathcal{E}$, the end where the scalar curvature goes to infinity. On the one hand, a direct argument shows that a sequence of rescalings of U_∞ around points converging to the end $\mathcal{E}$ converge in the Gromov-Hausdorff sense to a cone. On the other hand, a slightly different sequence of rescalings at the same points converges geometrically to a limiting non-flat Ricci flow. Since both limits are non-degenerate 3-dimensional spaces, we show that the ratio of the rescaling factors used to construct them converges to a finite, non-zero limit. This means that the two limits differ only by an overall constant factor. That is to say the geometric blow-up limit is isometric to an open subset of a non-flat cone. This contradicts Hamilton's result

(Proposition 4.22) which says that it is not possible to flow under the Ricci flow to an open subset of a non-flat cone. Now we carry out all the steps in this argument.

3. The incomplete geometric limit

We fix a sequence $(\mathcal{M}_n, G_n, x_n)$ of generalized Ricci flows as above. The first step is to shift and rescale this sequence of generalized Ricci flows so that we can form an (incomplete) geometric limit which will be a tube of finite length with scalar curvature going to infinity at one end.

We shift the time parameter of $(\mathcal{M}_n, G_n)$ by $-\mathbf{t}(x_n)$. We change notation and denote these shifted flows by $(\mathcal{M}_n, G_n)$. This allows us to arrange that $\mathbf{t}(x_n) = 0$ for all n. Since shifting leaves the curvature unchanged, the shifted flows satisfy a weaker version of curvature pinched toward positive. Namely, for the shifted flows we have

$$R(x) \geq -6,$$
(10.1)
$$R(x) \geq 2X(x)\left(\log(X(x)) - 3\right).$$

We set $Q_n = R(x_n)$, and we denote by M_n the 0-time-slice of $\mathcal{M}_n$. We rescale $(\mathcal{M}_n, G_n)$ by Q_n. Denote by $(\mathcal{M}'_n, G'_n)$ the rescaled (and shifted) generalized flows. For the rest of this argument we implicity use the metrics G'_n. If we are referring to G_n we mention it explicitly.

3.1. The sequence of tubes. Let γ_n be a smooth path from x_n to y_n in $B_{G_n}(x_n, 0, A_0 Q_n^{-1/2})$. For all n sufficiently large we have $R_{G'_n}(y_n) \gg 1$. Thus, there is a point $z_n \in \gamma_n$ such that $R_{G'_n}(z_n) = 4$ and such that on the sub-path $\gamma_n|_{[z_n, y_n]}$ we have $R_{G'_n} \geq 4$. We replace γ_n by this sub-path. Now, with this replacement, according to the second condition in the statement of the theorem, every point of γ_n has a strong (C, ϵ) canonical neighborhood. As n tends to infinity the ratio of $R(y_n)/R(z_n)$ tends to infinity. This means that for all n sufficiently large, no point of γ_n can be contained in an ϵ-round component or a C-component, because if it were then all of γ_n would be contained in that component, contradicting the fact that the curvature ratio is arbitrarily large for large n. Hence, for n sufficiently large, every point of γ_n is either contained in the core of a (C, ϵ)-cap or is the center of a strong ϵ-neck. According to Proposition A.21, for all n sufficiently large γ_n is contained in an open submanifold X_n of the zero time-slice of $\mathcal{M}'_n$ that is one of the following:

(1) an ϵ-tube and both endpoints of γ_n are centers of ϵ-necks contained in X_n,
(2) a C-capped ϵ-tube with cap $\mathcal{C}$, and each endpoint of γ_n either is contained in the core Y of $\mathcal{C}$ or is the center of an ϵ-neck contained in X_n,
(3) a double C-capped ϵ-tube, or finally

(4) the union of two (C, ϵ)-caps.

The fourth possibility is incompatible with the fact that the ratio of the curvatures at the endpoints of γ_n grows arbitrarily large as n tends to infinity. Hence, this fourth possibility cannot occur for n sufficiently large. Thus, for all n sufficiently large, X_n is one of the first three types listed above.

CLAIM 10.3. *There is a geodesic $\hat{\gamma}_n$ in X_n with endpoints z_n and y_n. This geodesic is minimizing among all paths in X_n from z_n to y_n.*

PROOF. This is clear in the third case since X_n is a closed manifold.

Let us consider the first case. There are ϵ-necks $N(z_n)$ and $N(y_n)$ centered at z_n and y_n and contained in X_n. Suppose first that the central 2-spheres $S(z_n)$ and $S(y_n)$ of these necks are disjoint. Then they are the boundary of a compact submanifold X'_n of X_n. It follows easily from Lemma A.1 that any sequence of minimizing paths from z_n to y_n is contained in the union of X'_n with the middle halves of $N(z_n)$ and $N(y_n)$. Since this manifold has compact closure in X_n, the usual arguments show that one can extract a limit of a subsequence which is a minimizing geodesic in X_n from z_n to y_n. If $S(z_n) \cap S(y_n) \neq \emptyset$, then y_n is contained in the middle half of $N(z_n)$, and again it follows immediately from Lemma A.1 that there is a minimizing geodesic in $N(z_n)$ between these points.

Now let us consider the second case. If each of z_n and y_n is the center of an ϵ-neck in X_n, the argument as in the first case applies. If both points are contained in the core of $\mathcal{C}$ then, since that core has compact closure in X_n, the result is again immediate. Lastly, suppose that one of the points, we can assume by the symmetry of the roles of the points that it is z_n, is the center of an ϵ-neck $N(z_n)$ in X_n and the other is contained in the core of $\mathcal{C}$. Suppose that the central 2-sphere $S(z_n)$ of $N(z_n)$ meets the core Y of $\mathcal{C}$. Then z_n lies in the half of the neck $N = \mathcal{C} \setminus \overline{Y}$ whose closure contains the frontier of Y. Orient s_N so that this half is the positive half. Thus, by Lemma A.1 any minimizing sequence of paths from z_n to y_n is eventually contained in the union of the core of $\mathcal{C}$ and the positive three-quarters of this neck. Hence, as before we can pass to a limit and construct a minimizing geodesic in X_n connecting z_n to y_n. On the other hand, if $S(z_n)$ is disjoint from Y, then $S(z_n)$ separates X_n into a compact complementary component and a non-compact complementary component and the compact complementary component contains Y. Orient the s_N-direction so that the compact complementary component lies on the positive side of $S(z_n)$. Then any minimizing sequence of paths in X_n from z_n to y_n is eventually contained in the union of the compact complementary component of $N(z_n)$ and the positive 3/4's of $N(z_n)$. As before, this allows us to pass to a limit to obtain a minimizing geodesic in X_n. $\square$

3. THE INCOMPLETE GEOMETRIC LIMIT

This claim allows us to assume (as we now shall) that γ_n is a minimizing geodesic in X_n from z_n to y_n.

CLAIM 10.4. *For every n sufficiently large, there is a sub-geodesic γ'_n of γ_n with end points z'_n and y'_n such that the following hold:*
  (1) *The length of γ'_n is bounded independent of n.*
  (2) *$R(z'_n)$ is bounded independent of n.*
  (3) *$R(y'_n)$ tends to infinity as n tends to infinity.*
  (4) *γ'_n is contained in a strong ϵ-tube T_n that is the union of a balanced chain of strong ϵ-necks centered at points of γ'_n. The first element in this chain is a strong ϵ-neck $N(z'_n)$ centered at z'_n. The last element is a strong ϵ-neck containing y'_n.*
  (5) *For every $x \in T_n$, we have $R(x) > 3$ and x is the center of a strong ϵ-neck in the flow $(\mathcal{M}'_n, G'_n)$.*

PROOF. The first item is immediate since, for all n, the geodesic γ_n has G_n-length at most $A_0 Q_n^{-1/2}$ and hence G'_n-length at most A_0. Suppose that we have a (C, ϵ)-cap $\mathcal{C}$ whose core Y contains a point of γ_n. Let N be the ϵ-neck that is the complement of the closure of Y in $\mathcal{C}$, and let $\widehat{Y}$ be the union of Y and the closed negative half of N. We claim that $\widehat{Y}$ contains either z_n or y_n. By Corollary A.8, since Y contains a point of γ_n, the intersection of $\widehat{Y}$ with γ_n is a subinterval containing one of the end points of γ_n, i.e., either z_n or y_n. This means that any point w which is contained in a (C, ϵ)-cap whose core contains a point of γ_n must satisfy one of the following:
$$R(w) < CR(z'_n) \quad \text{or} \quad R(w) > C^{-1}R(y'_n).$$

We pass to a subsequence so that $R(y_n)/R(z_n) > 4C^2$ for all n, and we pass to a subinterval γ'_n of γ_n with endpoints z'_n and y'_n such that:
  (1) $R(z'_n) = 2CR(z_n)$.
  (2) $R(y'_n) = (2C)^{-1}R(y_n)$.
  (3) $R(z'_n) \leq R(w) \leq R(y'_n)$ for all $w \in \gamma'_n$.

Clearly, with these choices $R(z'_n)$ is bounded independent of n and $R(y'_n)$ tends to infinity as n tends to infinity. Also, no point of γ'_n is contained in the core of a (C, ϵ)-cap. Since every point of γ'_n has a strong (C, ϵ)-canonical neighborhood, it follows that every point of γ'_n is the center of a strong ϵ-neck. It now follows from Proposition A.19 that there is a balanced ϵ-chain consisting of strong ϵ-necks centered at points of γ'_n whose union contains γ'_n. (Even if the 2-spheres of these necks do not separate the zero time-slice of $\mathcal{M}'_n$, as we build the balanced ϵ-chain as described in Proposition A.19 the new necks we add can not meet the negative end of $N(z'_n)$ since the geodesic γ'_n is minimal.) We can take the first element in the balanced chain to be a strong ϵ-neck $N(z'_n)$ centered at z'_n, and the last element to be a strong ϵ-neck N_n^+ containing y'_n. The union of this chain is T_n. (See FIG. 1.)

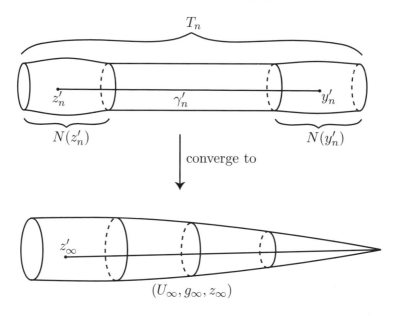

FIGURE 1. Limiting tube.

Next, we show that every point of T_n is the center of a strong ϵ-neck in $(\mathcal{M}_n, G_n)$. We must rule out the possibility that there is a point of T_n that is contained in the core of a (C, ϵ)-cap. Since T_n is a union of ϵ-necks centered at points of γ'_n we see that every point $w \in T_n$ has

$$(3C/2)R(z_n) < R(w) < (2/3C)R(y_n).$$

This tells us that no point of T_n is contained in a (C, ϵ)-cap whose core contains a point of γ_n. Thus, to complete the argument we need only see that if there is a point of T_n contained in the core of a (C, ϵ)-cap, then the core of that (C, ϵ)-cap also contains a point of γ_n. The scalar curvature inequality implies that both z_n and y_n are outside T_n. This means that γ_n traverses T_n from one end to the other. Let w_-, resp. w_+, be the point of γ_n that lies in the frontier of T_n contained in the closure of the $N(z'_n)$, resp. N_n^+. Since the scalar curvatures at these two points of γ satisfy the weak version of the above inequalities, we see that there are strong ϵ-necks $N(w_-)$ and $N(w_+)$ centered at them. Let $\widehat{T}_n$ be the union of T_n, $N(w_-)$ and $N(w_+)$. It is also a strong ϵ-tube, and every point $\hat{w}$ of $\widehat{T}_n$ satisfies

$$(1.1) CR(z_n) < R(\hat{w}) < (0.9)C^{-1}R(y_n).$$

Thus, z_n and y_n are disjoint from $\widehat{T}_n$ and hence γ crosses $\widehat{T}_n$ from one end to the other.

Now suppose that T_n meets the core Y of a (C, ϵ)-cap $\mathcal{C}$. Consider the boundary S of the closure of Y. If it is disjoint from T_n then T_n is contained in the core Y. For large n this is inconsistent with the fact that the ratio of the scalar curvature at the endpoints of γ'_n goes to infinity. Thus, we are left to consider the case when S contains a point of the tube T_n. In this case S is completely contained in $\widehat{T}_n$ and by Corollary A.3, S is isotopic to the 2-spheres of the product decomposition of $\widehat{T}_n$. Hence, S meets a point of γ_n and consequently the core Y contains a point of γ_n. But we have already seen that this is not possible.

Lastly, we must show that $R(x) > 3$ for every $x \in T_n$. We have just seen that every $x \in T_n$ is the center of an ϵ-neck. If x is contained in the ϵ-neck centered at z'_n or y'_n, then since $R(z'_n) \geq 4$ and $R(y'_n) \geq 4$, clearly $R(x) > 3$. We must consider the case when x is not contained in either of these ϵ-necks. In this case the central 2-sphere S_x of the ϵ-neck centered at x is contained in the compact submanifold of T'_n bounded by the central 2-spheres of the necks centered at z'_n and y'_n. These 2-spheres are disjoint and by Condition 4 in Proposition A.11 each is a homotopically non-trivial 2-sphere in T'_n. Hence, the compact manifold with their disjoint union as boundary is diffeomorphic to $S^2 \times [0,1]$ and, again according to Condition 4 of Proposition A.11, S_x is isotopic to the 2-sphere factor in this product decomposition. Since the intersection of γ'_n with this submanifold is an arc spanning from one boundary component to the other, S_x must meet γ'_n, in say w. By construction, since $w \in \gamma'_n$ we have $R(w) \geq 4$. This implies that $R(x) > 3$. This completes the proof of the claim. $\square$

3.2. Extracting a limit of a subsequence of the tubes. Passing to a subsequence we arrange that the $R(z'_n)$ converge. Now consider the subset $\mathcal{A} \subset \mathbb{R}$ consisting of all $A > 0$ such that there is a uniform bound, independent of n, for the curvature on $B(z'_n, A) \cap T_n$. The set $\mathcal{A}$ is non-empty since $R(z'_n)$ is bounded independent of n and for every n there is a strong ϵ-neck $N(z'_n)$ centered at z'_n contained in T_n. On the other hand, since $d_{G'_n}(z'_n, y'_n)$ is uniformly bounded and $R(y'_n) \to \infty$, there is a finite upper bound for $\mathcal{A}$. Let A_1 be the least upper bound of $\mathcal{A}$. We set $U_n = T_n \cap B(z'_n, A_1)$. This is an open subset of T_n containing z'_n. We let $g'_n = G'_n|U_n$.

CLAIM 10.5. *For all n sufficiently large, $3R(z'_n)^{-1/2}\epsilon^{-1}/2$ is less than A_1, and hence U_n contains the strong ϵ-neck $N(z'_n)$ centered at z'_n.*

PROOF. The curvature on $N(z'_n)$ is bounded independent of n. Consider a point w near the end of $N(z'_n)$ that separates y'_n from z'_n. It is also the center of a strong ϵ-neck $N(w)$. By Proposition A.11 and our assumption that $\epsilon \leq \overline{\epsilon}(10^{-2})$, the scalar curvature on $N(z'_n) \cup N(w)$ is between $(0.9)R(z'_n)$ and $(1.1)R(z'_n)$. Since, by construction, the negative end of $N(z'_n)$ contains an end of T_n, this implies that

$$N(z'_n) \cup N(w) \supset B(z'_n, 7R(z'_n)^{-1/2}\epsilon^{-1}/4) \cap T_n,$$

so that we see that $A_1 \geq 7\epsilon^{-1}\lim_{n \to \infty} R(z'_n)^{-1/2}/4 > 3R(z'_n)^{-1}\epsilon^{-1}/2$ for all n sufficiently large. Obviously then U_n contains $N(z'_n)$. $\square$

The next claim uses terminology from Definition 5.1.

CLAIM 10.6. *For any $\delta > 0$ there is a uniform bound, independent of n, for the curvature on $\mathrm{Reg}_\delta(U_n, g'_n)$.*

PROOF. To prove this it suffices to show that given $\delta > 0$ there is $A < A_1$ such that $\mathrm{Reg}_\delta(U_n, g'_n) \subset B(z'_n, A)$ for all n sufficiently large. Of course, if we establish this for every $\delta > 0$ sufficiently small, then it follows for all $\delta > 0$. First of all, by Corollary A.5 and Lemma A.2, the fact that $\epsilon \leq \bar{\epsilon}(10^{-2})$ implies that any point w with the property that the strong ϵ-neighborhood centered at w contains z'_n is contained in the ball of radius $(1.1)R(z'_n)^{-1/2}\epsilon^{-1} < A_1$ centered at z'_n. Thus, it suffices to consider points w_n in $\mathrm{Reg}_\delta(U_n, g'_n)$ with the property that the strong ϵ-neck centered at w_n does not contain z'_n. Fix such a w_n. Take a path $\mu_n(s)$ starting at w_n moving in the s-direction at unit speed measured in the s-coordinate of the ϵ-neck centered at w_n away from z'_n and ending at the frontier of this neck. Let u_1 be the final point of this path. The rescaled version of Lemma A.9 implies that the forward difference quotient for the distance from z'_n satisfies

$$(0.99)R(w_n)^{-1/2} \leq \frac{d}{ds}d(z'_n, \mu_n(s)) \leq (1.01)R(w_n)^{-1/2}.$$

Of course, since we are working in an ϵ-neck we also have

$$(1-\epsilon)R(w_n)^{-1/2} \leq \frac{d(d(w_n, \mu_n(s)))}{ds} \leq (1+\epsilon)R(w_n)^{-1/2}.$$

We continue the path μ_n moving in the s-direction of a neck centered at u_1. Applying Lemma A.9 again both to the distance from w_n and the distance from z'_n yields:

$$(0.99)R(u_1)^{-1/2} \leq \frac{d(d(z'_n, \mu_n(s)))}{ds} \leq (1.01)R(u_1)^{-1/2},$$

$$(0.99)R(u_1)^{-1/2} \leq \frac{d(d(w_n, \mu_n(s)))}{ds} \leq (1.01)R(u_1)^{-1/2}$$

on this part of the path μ_n. We repeat this process as many times as necessary until we reach a point $w'_n \in U_n$ at distance $\delta/2$ from w_n. This is possible since the ball of radius δ centered at w_n is contained in U_n. By the difference quotient inequalities, it follows that $d(z'_n, w'_n) - d(z'_n, w_n) > \delta/4$. Since $w'_n \in U_n$ and consequently that $d(z'_n, w'_n) < A_1$. It follows that $d(z'_n, w_n) \leq A_1 - \delta/4$. This proves that, for all n sufficiently large, $\mathrm{Reg}_\delta(U_n, g'_n) \subset B(z'_n, A_1 - \delta/4)$, and consequently that the curvature on $\mathrm{Reg}_\delta(U_n, g'_n)$ is bounded independent of n. □

By Shi's theorem (Theorem 3.28), the fact that each point of U_n is the center of a strong ϵ-neck means that there is a bound, independent of n, on all covariant derivatives of the curvature at any point of U_n in terms of the bound on the curvature at the center point. In particular, because of the previous result, we see that for any $\epsilon > 0$ and any $\ell \geq 0$ there is a uniform bound for $|\nabla^\ell \mathrm{Rm}|$ on $\mathrm{Reg}_\delta(U_n, g'_n)$. Clearly, since the base point z'_n has bounded curvature it lies in $\mathrm{Reg}_\delta(U_n, g'_n)$ for sufficiently small δ (how small being independent of n). Lastly, the fact that every point in U_n is the center of an ϵ-neighborhood implies that (U_n, g'_n) is κ-non-collapsed on scales $\leq r_0$ where both κ and r_0 are universal. Since the γ'_n have uniformly bounded lengths, the ϵ-tubes T'_n have uniformly bounded diameter. Also, we have seen that their curvatures are bounded from below by 3. It follows that their volumes are uniformly bounded. Now invoking Theorem 5.6 we see that after passing to a subsequence we have a geometric limit $(U_\infty, g_\infty, z_\infty)$ of a subsequence of (U_n, g'_n, z'_n).

3.3. Properties of the limiting tube. Now we come to a result establishing all the properties we need for the limiting manifold.

PROPOSITION 10.7. *The geometric limit $(U_\infty, g_\infty, z_\infty)$ is an incomplete Riemannian 3-manifold of finite diameter. There is a diffeomorphism ψ from U_∞ to $S^2 \times (0,1)$. There is a 2ϵ-neck centered at z_∞ whose central 2-sphere $S^2(z_\infty)$ maps under ψ to a 2-sphere isotopic to a 2-sphere factor in the product decomposition. The scalar curvature is bounded at one end of U_∞ but tends to infinity at the other end, the latter end which is denoted $\mathcal{E}$. Let $\mathcal{U}_\infty \subset U_\infty \times (-\infty, 0]$ be the open subset consisting of all (x,t) for which $-R(x)^{-1} < t \leq 0$. We have a generalized Ricci flow on $\mathcal{U}_\infty$ which is a partial geometric limit of a subsequence of the generalized Ricci flows $(\mathcal{M}'_n, G'_n, z'_n)$. In particular, the zero time-slice of the limit flow is (U_∞, g_∞). The Riemannian curvature is non-negative at all points of the limiting smooth flow on $\mathcal{U}_\infty$. Every point $x \in U_\infty \times \{0\}$ which is not separated from $\mathcal{E}$ by $S^2(z_\infty)$ is the center of an evolving 2ϵ-neck $N(x)$ defined for an interval of normalized time of length $1/2$. Furthermore, the central 2-sphere of $N(x)$ is isotopic to the 2-sphere factor of U_∞ under the diffeomorphism ψ (see* FIG. 1*).*

The proof of this proposition occupies the rest of Section 3.

PROOF. Let $V_1 \subset V_2 \subset \cdots \subset U_\infty$ be the open subsets and $\varphi_n \colon V_n \to U_n$ be the maps having all the properties stated in Definition 5.3 so as to exhibit $(U_\infty, g_\infty, z_\infty)$ as the geometric limit of the (U_n, g'_n, z'_n).

Since the U_n are all contained in $B(z'_n, A_1)$, it follows that any point of U_∞ is within A_1 of the limiting base point z_∞. This proves that the diameter of U_∞ is bounded.

For each n there is the ϵ-neck $N(z'_n)$ centered at z'_n contained in U_n. The middle two-thirds, N'_n, of this neck has closure contained in $\mathrm{Reg}_\delta(U_n, g_n)$ for some $\delta > 0$ independent of n (in fact, restricting to n sufficiently large, δ can be taken to be approximately equal to $R(z_\infty)^{-1/2}\epsilon^{-1}/3$). This means that for some n sufficiently large and for all $m \geq n$, the image $\varphi_m(V_n) \subset U_m$ contains N'_m. For any fixed n as m tends to infinity the metrics $\varphi_m^* g_m|_{V_n}$ converge uniformly in the C^∞-topology to $g_\infty|_{V_n}$. Thus, it follows from Proposition 9.79 that for all m sufficiently large, $\varphi_m^{-1}(N'_m)$ is a $3\epsilon/2$-neck centered at z_∞. We fix such a neck $N'(z_\infty) \subset U_\infty$. Let $S(z_\infty)$ be the central 2-sphere of $N'(z_\infty)$. For each n sufficiently large, $\varphi_n(S(z_\infty))$ separates U_n into two components, one, say W_n^- contained in $N(z_n')$ and the other, W_n^+ containing all of $U_n \setminus N(z_n')$. It follows that $S(z_\infty)$ separates U_∞ into two components, one, denoted W_∞^-, where the curvature is bounded (and where, in fact, the curvature is close to $R(z_\infty)$) and the other, denoted W_∞^+, where it is unbounded.

CLAIM 10.8. *Any point $q \in W_\infty^+$ is the center of a 2ϵ-neck in U_∞.*

PROOF. Fix a point $q \in W_\infty^+$. For all n sufficiently large let $q_n = \varphi_n(q)$. Then for all n sufficiently large, $q_n \in W_n^+$ and $\lim_{n\to\infty} R(q_n) = R(q)$. This means that for all n sufficiently large $R(y_n') >> R(q_n)$), and hence the $3\epsilon/2$-neck centered at $q_n \in U_n$ is disjoint from $N(y_n')$. Thus, by the rescaled version of Corollary A.5, the distance from the $3\epsilon/2$-neck centered at q_n to $N(y_n')$ is bounded below by $(0.99)\epsilon^{-1} R(q_n)^{-1/2}/4 \geq \epsilon^{-1} R(q_\infty)^{-1/2}/12$. Also, since $q_n \in W_n$, this $3\epsilon/2$-neck $N'(q_n)$ centered at q_n does not extend past the 2-sphere at $s^{-1}(-3\epsilon^{-1}/4)$ in the ϵ-neck $N(z_n')$. It follows that for all n sufficiently large, this $3\epsilon/2$-neck has compact closure contained in $\text{Reg}_\delta(U_n, g_n)$ for some δ independent of n, and hence there is m such that for all n sufficiently large $N'(q_n)$ is contained in the image $\varphi_n(V_m)$. Again using the fact that $\varphi_n^*(g_n|_{V_m})$ converges in the C^∞-topology to $g_\infty|_{V_m}$ as n tends to infinity, we see, by Proposition 9.79 that for all n sufficiently large $\varphi_n^{-1}(N_m)$ contains a 2ϵ-neck in U_∞ centered at q. □

It now follows from Proposition A.21 that W_∞^+ is contained in a 2ϵ-tube T_∞ that is contained in U_∞. Furthermore, the frontier of W_∞^+ in T_∞ is the 2-sphere $S(z_\infty)$ which is isotopic to the central 2-spheres of the 2ϵ-necks making up T_∞. Hence, the closure $\overline{W}_\infty^+$ of W_∞^+ is a 2ϵ-tube with boundary $S(z_\infty)$. In particular, $\overline{W}_\infty^+$ is diffeomorphic to $S^2 \times [0, 1)$.

Now we consider the closure $\overline{W}_\infty^-$ of W_∞^-. Since the closure of each W_n^- is the closed negative half of the ϵ-neck $N(z_n')$ and the curvatures of the z_n' have a finite, positive limit, the limit $\overline{W}_\infty^-$ is diffeomorphic to a product $S^2 \times (-1, 0]$. Hence, U_∞ is the union of $\overline{W}_\infty^+$ and $\overline{W}_\infty^-$ along their common boundary. It follows immediately that U_∞ is diffeomorphic to $S^2 \times (0, 1)$.

CLAIM 10.9. *The curvature is bounded in a neighborhood of one end of U_∞ and goes to infinity at the other end.*

PROOF. A neighborhood of one end of U_∞, the end $\overline{W}_\infty^-$, is the limit of the negative halves of ϵ-necks centered at z_n'. Thus, the curvature is bounded on this neighborhood, and in fact is approximately equal to $R(z_\infty)$. Let x_k be any sequence of points in U_∞ tending to the other end. We show that $R(x_k)$ tends to ∞ as k does. The point is that since the sequence is tending to the end, the distance from x_k to the end of U_∞ is going to zero. Yet, each

x_k is the center of an ϵ-neck in U_∞. The only way this is possible is if the scales of these ϵ-necks are converging to zero as k goes to infinity. This is equivalent to the statement that $R(x_k)$ tends to ∞ as k goes to infinity. $\square$

The next step in the proof of Proposition 10.7 is to extend the flow backwards a certain amount. As stated in the proposition, the amount of backward time that we can extend the flow is not uniform over all of U_∞, but rather depends on the curvature of the point at time zero.

CLAIM 10.10. *For each $x \in U_n \subset M_n$ there is a flow line in $\mathcal{M}_n$ through x defined on the time-interval $(-R(x)^{-1}, 0]$. Furthermore, the scalar curvature at any point of this flow line is less than or equal to the scalar curvature at x.*

PROOF. Since $x \in U_n \subset T_n$, there is a strong ϵ-neck in $\mathcal{M}_n$ centered at x. Both statements follow immediately from that. $\square$

Let $X \subset U_\infty$ be an open submanifold with compact closure and set

$$t_0(X) = \sup_{x \in X}(-R_{g_\infty}(x)^{-1}).$$

Then for all n sufficiently large φ_n is defined on X and the scalar curvature of the flow $g_n(t)$ on $\varphi_n(X) \times (t_0, 0]$ is uniformly bounded independent of n. Thus, according to Proposition 5.14, by passing to a subsequence we can arrange that there is a limiting flow defined on $X \times (t_0, 0]$. Let $\mathcal{U}_\infty \subset U_\infty \times (-\infty, 0]$ consist of all pairs (x, t) with the property that $-R_{g_\infty}(x, 0)^{-1} < t \le 0$. Cover $\mathcal{U}_\infty$ by countably many such boxes of the type $X \times (-t_0(X), 0]$ as described above, and take a diagonal subsequence. This allows us to pass to a subsequence so that the limiting flow exists (as a generalized Ricci flow) on $\mathcal{U}_\infty$.

CLAIM 10.11. *The curvature of the generalized Ricci flow on $\mathcal{U}_\infty$ is non-negative.*

PROOF. This claim follows from the fact that the original sequence $(\mathcal{M}_n, G_n)$ consists of generalized flows whose curvatures are pinched toward positive in the weak sense given in Equation (10.1) and the fact that $Q_n \to \infty$ as $n \to \infty$. (See Theorem 5.33.) $\square$

This completes the proof that all the properties claimed in Proposition 10.7 hold for the geometric limit $(U_\infty, g_\infty, z_\infty)$. This completes the proof of that proposition. $\square$

4. Cone limits near the end $\mathcal{E}$ for rescalings of U_∞

The next step is to study the nature of the limit U_∞ given in Proposition 10.7. We shall show that an appropriate blow-up limit (limit in the Gromov-Hausdorff sense) around the end is a cone.

Let (X, d_X) be a metric space. Recall that the cone on X, denoted $C(X)$, is the quotient space $X \times [0, \infty)$ under the identification $(x, 0) \cong (y, 0)$ for all $x, y \in X$. The image of $X \times \{0\}$ is the *cone point* of the cone. The metric on $C(X)$ is given by

$$(10.2) \quad d((x, s_1), (y, s_2)) = s_1^2 + s_2^2 - 2s_1 s_2 \cos(\min(d_X(x,y), \pi)).$$

The open cone $C'(X)$ is the complement of the cone point in $C(X)$ with the induced metric.

The purpose of this section is to prove the following result.

PROPOSITION 10.12. *Let $(U_\infty, g_\infty, z_\infty)$ be as in the conclusion of Proposition 10.7. Let $Q_\infty = R_{g_\infty}(z_\infty)$ and let $\mathcal{E}$ be the end of U_∞ where the scalar curvature is unbounded. Let λ_n be any sequence of positive numbers with $\lim_{n \to \infty} \lambda_n = +\infty$. Then there is a sequence x_n in U_∞ such that for each n the distance from x_n to $\mathcal{E}$ is $\lambda_n^{-1/2}$, and such that the pointed Riemannian manifolds $(U_\infty, \lambda_n g_\infty, x_n)$ converge in the Gromov-Hausdorff sense to an open cone, an open cone not homeomorphic to an open ray (i.e., not homeomorphic to the open cone on a point). (see FIG. 2).*

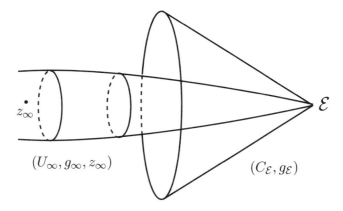

FIGURE 2. Limiting cone.

The rest of this section is devoted to the proof of this result.

4.1. Directions at $\mathcal{E}$. We orient the direction down the tube U_∞ so that $\mathcal{E}$ is at the positive end. This gives an s_N-direction for each 2ϵ-neck N contained in U_∞.

Fix a point $x \in U_\infty$. We say a ray γ with endpoint x limiting to $\mathcal{E}$ is a *minimizing geodesic ray* if for every $y \in \gamma$ the segment on γ from x to y is a minimizing geodesic segment; i.e., the length of this geodesic segment is equal to $d(x, y)$.

CLAIM 10.13. *There is a minimizing geodesic ray to $\mathcal{E}$ from each $x \in U_\infty$ with $R(x) \geq 2Q_\infty$.*

4. CONE LIMITS NEAR THE END $\mathcal{E}$ FOR RESCALINGS OF U_∞ 257

PROOF. Fix x with $R(x) \geq 2Q_\infty$ and fix a 2ϵ-neck N_x centered at x. Let S_x^2 be the central 2-sphere of this neck. Take a sequence of points q_n tending to the end $\mathcal{E}$, each being closer to the end than x in the sense that S_x^2 does not separate any q_n from the end $\mathcal{E}$. We claim that there is a minimizing geodesic from x to each q_n. The reason is that by Lemma A.7 any minimizing sequence of arcs from x to q_n cannot exit from the minus end of N_x nor the plus end of a 2ϵ-neck centered at q_n. Consider a sequence of paths from x to q_n minimizing the distance. Hence these paths all lie in a fixed compact subset of U_n. After replacing the sequence by a subsequence, we can pass to a limit, which is clearly a minimizing geodesic from x to q_n. Consider minimizing geodesics μ_n from x to q_n. The same argument shows that, after passing to a subsequence, the μ_n converge to a minimizing geodesic ray from x to $\mathcal{E}$. □

CLAIM 10.14. *(1) Any minimizing geodesic ray from x to the end $\mathcal{E}$ is a shortest ray from x to the end $\mathcal{E}$, and conversely any shortest ray from x to the end $\mathcal{E}$ is a minimizing geodesic ray.*

(2) The length of a shortest ray from x to $\mathcal{E}$ is the distance (see Section 4 of Chapter 2) from x to $\mathcal{E}$.

PROOF. The implication in (1) in one direction is clear: If γ is a ray from x to the end $\mathcal{E}$, and for some $y \in \gamma$ the segment on γ from x to y is not minimizing, then there is a shorter geodesic segment μ from x to y. The union of this together with the ray on γ from y to the end is a shorter ray from x to the end.

Let us establish the opposite implication. Suppose that γ is a minimizing geodesic ray from x to the end $\mathcal{E}$ and that there is a $\delta > 0$ and a shortest geodesic ray γ' from x to the end $\mathcal{E}$ with $|\gamma'| = |\gamma| - \delta$. As we have just seen, γ' is a minimizing geodesic ray. Take a sequence of points q_i tending to the end $\mathcal{E}$ and let S_i^2 be the central 2-sphere in the 2ϵ-neck centered at q_i. Of course, for all i sufficiently large, both γ' and γ must cross S_i^2. Since the scalar curvature tends to infinity at the end $\mathcal{E}$, it follows from Lemma A.4 that for all i sufficiently large, the extrinsic diameter of S_i^2 is less than $\delta/3$. Let p_i be a point of intersection of γ with S_i^2. For all i sufficiently large the length d_i of the sub-ray in γ from p_i to the end $\mathcal{E}$ is at most $\delta/3$. Let p_i' be a point of intersection of γ' with S_i^2 and let d_i' be the length of the ray in γ' from p_i' to the end $\mathcal{E}$. Let λ be the sub-geodesic of γ from x to p_i and λ' the sub-geodesic of γ' from x to p_i'. Let β be a minimizing geodesic from p_i' to p_i. Of course, $|\beta| < \delta/3$ so that by the minimality of λ and λ' we have

$$-\delta/3 < |\lambda| - |\lambda'| < \delta/3.$$

Since $|\lambda'| + d_i' = |\lambda| + d_i - \delta$, we have

$$2\delta/3 \leq d_i - d_i'.$$

This is absurd since $d_i' > 0$ and $d_i < \delta/3$.

(2) follows immediately from (1) and the definition. $\square$

Given this result, the usual arguments show:

COROLLARY 10.15. *If γ is a minimizing geodesic ray from x to the end $\mathcal{E}$, then for any $y \in \gamma \setminus \{x\}$ the sub-ray of γ from y to the end, is the unique shortest geodesic from y to the end.*

Also, we have a version of the triangle inequality for distances to $\mathcal{E}$.

LEMMA 10.16. *Let x and y be points of M. Then the three distances $d(x,y)$, $d(x,\mathcal{E})$ and $d(y,\mathcal{E})$ satisfy the triangle inequality.*

PROOF. From the definitions it is clear that $d(x,y) + d(y,\mathcal{E}) \geq d(x,\mathcal{E})$, and symmetrically, reversing the roles of x and y. The remaining inequality that we must establish is the following: $d(x,\mathcal{E}) + d(y,\mathcal{E}) \geq d(x,y)$. Let q_n be any sequence of points converging to $\mathcal{E}$. Since the end is at finite distance, it is clear that $d(x,\mathcal{E}) = \lim_{n\to\infty} d(x,q_n)$. The remaining inequality follows from this and the usual triangle inequality applied to $d(x,q_n)$, $d(y,q_n)$ and $d(x,y)$. $\square$

DEFINITION 10.17. We say that two minimizing geodesic rays limiting to $\mathcal{E}$ are *equivalent* if one is contained in the other. From the unique continuation of geodesics it is easy to see that this generates an equivalence relation. An equivalence class is a *direction at* $\mathcal{E}$, and the set of equivalence classes is the *set of directions at* $\mathcal{E}$.

LEMMA 10.18. *There is more than one direction at $\mathcal{E}$.*

PROOF. Take a minimal geodesic ray γ from a point x limiting to the end and let y be a point closer to $\mathcal{E}$ than x and not lying on γ. Then a minimal geodesic ray from y to $\mathcal{E}$ gives a direction at $\mathcal{E}$ distinct from the direction determined by γ. $\square$

REMARK 10.19. In fact, the general theory of positively curved spaces implies that the space of directions is homeomorphic to S^2. Since we do not need this stronger result we do not prove it.

4.2. The metric on the space of directions at $\mathcal{E}$.

DEFINITION 10.20. Let γ and μ be minimizing geodesic rays limiting to $\mathcal{E}$, of lengths a and b, parameterized by the distance from the end. For $0 < s \leq a$ and $0 < s' \leq b$ construct a triangle $\alpha_s e \beta_{s'}$ in the Euclidean plane with $|\alpha_s e| = s$, $|e\beta_{s'}| = s'$ and $|\alpha_s \beta_{s'}| = d(\gamma(s), \mu(s'))$. We define $\theta(\gamma, s, \mu, s')$ to be the angle at e of the triangle $\alpha_s e \beta_{s'}$.

LEMMA 10.21. *For all γ, s, μ, s' as in the previous definition we have*
$$0 \leq \theta(\gamma, s, \mu, s') \leq \pi.$$

Furthermore, $\theta(\gamma, s, \mu, s')$ is a non-increasing function of s when γ, μ, s' are held fixed, and symmetrically it is a non-increasing function of s' when γ, s, μ are held fixed. In particular, fixing γ and μ, the function $\theta(\gamma, s, \mu, s')$ is non-decreasing as s and s' tend to zero. Thus, there is a well-defined limit as s and s' go to zero, denoted $\theta(\gamma, \mu)$. This limit is greater than or equal to $\theta(\gamma, s, \mu, s')$ for all s and s' for which the latter is defined. We have $0 \leq \theta(\gamma, \mu) \leq \pi$. The angle $\theta(\gamma, \mu) = 0$ if and only if γ and μ are equivalent. Furthermore, if γ is equivalent to γ' and μ is equivalent to μ', then $\theta(\gamma, \mu) = \theta(\gamma', \mu')$.

PROOF. By restricting γ and μ to slightly smaller rays, we can assume that each is the unique shortest ray from its endpoint to the end $\mathcal{E}$. Let x, resp., y be the endpoint of γ, resp., μ. Now let q_n be any sequence of points in U_∞ limiting to the end $\mathcal{E}$, and consider minimizing geodesic rays γ_n from q_n to x and μ_n from q_n to y, each parameterized by the distance from q_n. By passing to a subsequence we can assume that each of the sequences $\{\gamma_n\}$ and $\{\mu_n\}$ converge to a minimizing geodesic ray, which by uniqueness, implies that the first sequence limits to γ and the second to μ. For s, s' sufficiently small, let $\theta_n(s, s')$ be the angle at $\widetilde{q}_n$ of the Euclidean triangle $\alpha_n \widetilde{q}_n \beta_n$, where $|\alpha_n \widetilde{q}_n| = d(\gamma_n(s), q_n)$, $|\beta_n \widetilde{q}_n| = d(\mu_n(s'), q_n)$ and $|\alpha_n \beta_n| = d(\gamma_n(s), \mu_n(s'))$. Clearly, for fixed s and s' sufficiently small, $\theta_n(s, s')$ converges as $n \to \infty$ to $\theta(\gamma, s, \mu, s')$. By the Toponogov property (Theorem 2.4) for manifolds with non-negative curvature, for each n the function $\theta_n(s, s')$ is a non-increasing function of each variable, when the other is held fixed. This property then passes to the limit, giving the first statement in the lemma.

By the monotonicity, $\theta(\gamma, \mu) = 0$ if and only if for all s, s' sufficiently small we have $\theta(\gamma, s, \mu, s') = 0$, which means one of γ and μ is contained in the other.

It is obvious that the last statement holds. □

It follows that $\theta(\gamma, \mu)$ yields a well-defined function on the set of pairs of directions at $\mathcal{E}$. It is clearly a symmetric, non-negative function which is positive off of the diagonal. The next lemma shows that it is a metric by establishing the triangle inequality for θ.

LEMMA 10.22. *If γ, μ, ν are minimizing geodesic rays limiting to $\mathcal{E}$, then*

$$\theta(\gamma, \mu) + \theta(\mu, \nu) \geq \theta(\gamma, \nu).$$

PROOF. By Corollary 10.15, after replacing γ, μ, ν by equivalent, shorter geodesic arcs, we can assume that they are the unique minimizing geodesics from their end points, say x, y, z respectively, to $\mathcal{E}$. Let q_n be a sequence of points limiting to $\mathcal{E}$, and let γ_n, μ_n, ν_n be minimizing geodesics from x, y, z to q_n. Denote by $\theta_n(x, y), \theta_n(y, z)$, and $\theta_n(x, z)$, respectively, the angles at $\widetilde{q}_n$ of the triangles in $\mathbb{R}^2$ with the edge lengths: $\{d(x, y), d(x, q_n), d(y, q_n)\}$,

$\{d(y,z), d(y,q_n), d(z,q_n)\}$, and $\{d(z,x), d(z,q_n), d(x,q_n)\}$. Then by Corollary 2.6 we have $\theta_n(x,y) + \theta_n(y,z) \geq \theta_n(x,z)$. Passing to the limit as n goes to ∞ and then the limit as x, y and z tend to $\mathcal{E}$, gives the result. $\square$

DEFINITION 10.23. Let $X(\mathcal{E})$ denote the set of directions at $\mathcal{E}$. We define the metric on $X(\mathcal{E})$ by setting $d([\gamma], [\mu]) = \theta(\gamma, \mu)$. We call this the *(metric) space of realized directions at $\mathcal{E}$*. The *metric space of directions at $\mathcal{E}$* is the completion $\overline{X}(\mathcal{E})$ of $X(\mathcal{E})$ with respect to the given metric. We denote by $(C_\mathcal{E}, g_\mathcal{E})$ the cone on $\overline{X}(\mathcal{E})$ with the cone metric as given in Equation (10.2). (See FIG. 2.)

PROPOSITION 10.24. $(C_\mathcal{E}, g_\mathcal{E})$ *is a metric cone that is not homeomorphic to a ray.*

PROOF. By construction $(C_\mathcal{E}, g_\mathcal{E})$ is a metric cone. That it is not homeomorphic to a ray follows immediately from Lemma 10.18. $\square$

4.3. Comparison results for distances.

LEMMA 10.25. *Suppose that γ and μ are unique shortest geodesic rays from points x and y to the end $\mathcal{E}$. Let $[\gamma]$ and $[\mu]$ be the points of $X(\mathcal{E})$ represented by these two geodesics rays. Let a, resp. b, be the distance from x, resp. y, to $\mathcal{E}$. Denote by x', resp. y', the image in $C_\mathcal{E}$ of the point $([\gamma], a)$, resp. $([\mu], b)$, of $X(\mathcal{E}) \times [0, \infty)$. Then*

$$d_{g_\infty}(x,y) \leq d_{g_\mathcal{E}}(x',y').$$

PROOF. By the definition of the cone metric we have

$$d_{g_\mathcal{E}}(x',y') = a^2 + b^2 - 2ab\cos(\theta(\gamma,\mu)).$$

On the other hand by Definition 10.20 and the law of cosines for Euclidean triangles, we have

$$d_{g_\infty}(x,y) = a^2 + b^2 - 2ab\cos(\theta(\gamma,a,\mu,b)).$$

The result is now immediate from the fact, proved in Lemma 10.20 that

$$0 \leq \theta(\gamma,a,\mu,b) \leq \theta(\gamma,\mu) \leq \pi,$$

and the fact that the cosine is a monotone decreasing function on the interval $[0, \pi]$. $\square$

COROLLARY 10.26. *Let γ, μ, x, y be as in the previous lemma. Fix $\lambda > 0$. Let $a = d_{\lambda g_\infty}(x, \mathcal{E})$ and $b = d_{\lambda g_\infty}(y, \mathcal{E})$. Set x'_λ and y'_λ equal to the points in the cone $([\gamma], a)$ and $([\mu], b)$. Then we have*

$$d_{\lambda g_\infty}(x,y) \leq d_{g_\mathcal{E}}(x'_\lambda, y'_\lambda).$$

PROOF. This follows by applying the previous lemma to the rescaled manifold $(U_\infty, \lambda g_\infty)$, and noticing that rescaling does not affect the cone $C_\mathcal{E}$ nor its metric. $\square$

LEMMA 10.27. *For any $\delta > 0$ there is $K = K(\delta) < \infty$ so that for any set of realized directions at $\mathcal{E}$ of cardinality K, $\ell_1, \ldots, \ell_K$, it must be the case that there are j and j' with $j \neq j'$ such that $\theta(\ell_j, \ell_{j'}) < \delta$.*

PROOF. Let K be such that, given K points in the central 2-sphere of any 2ϵ-tube of scale 1, at least two are within distance $\delta/2$ of each other. Now suppose that we have K directions $\ell_1, \ldots, \ell_K$ at $\mathcal{E}$. Let $\gamma_1, \ldots, \gamma_K$ be minimizing geodesic rays limiting to $\mathcal{E}$ that represent these directions. Choose a point x sufficiently close to the end $\mathcal{E}$ so that all the γ_j cross the central 2-sphere S^2 of the 2ϵ-neck centered at x. By replacing the γ_j with sub-rays we can assume that for each j the endpoint x_j of γ_j lies in S^2. Let d_j be the length of γ_j. By taking x sufficiently close to $\mathcal{E}$ we can also assume the following. For each j and j', the angle at e of the Euclidean triangle $\alpha_j e \alpha_{j'}$, where $|\alpha_j e| = d_j; |\alpha_{j'} e| = d_{j'}$ and $|\alpha_j \alpha_{j'}| = d(x_j, x_{j'})$ is within $\delta/2$ of $\theta(\ell_j, \ell_{j'})$. Now there must be $j \neq j'$ with $d(x_i, x_j) < (\delta/2) r_i$ where r_i is the scale of N_i. Since $d_j, d_{j'} > \epsilon^{-1} r_i / 2$, it follows that the angle at e of $\alpha_j e \alpha_{j'}$ is less than $\delta/2$. Consequently, $\theta(\ell_j, \ell_{j'}) < \delta$. □

Recall that a δ-net in a metric space X is a finite set of points such that X is contained in the union of the δ-neighborhoods of these points. The above lemma immediately yields:

COROLLARY 10.28. *The metric completion $\overline{X}(\mathcal{E})$ of the space of directions at $\mathcal{E}$ is a compact space. For every $\delta > 0$ this space has a δ-net consisting of realized directions. For every $0 < r < R < \infty$ the annular region $A_{\mathcal{E}}(r, R) = \overline{X}(\mathcal{E}) \times [r, R]$ in $C_{\mathcal{E}}$ has a δ-net consisting of points (ℓ_i, s_i) where for each i we have ℓ_i is a realizable direction and $r < s_i < R$.*

4.4. Completion of the proof of a cone limit at $\mathcal{E}$. Now we are ready to prove Proposition 10.12. In fact, we prove a version of the proposition that identifies the sequence of points x_n and also identifies the cone to which the rescaled manifolds converge.

PROPOSITION 10.29. *Let (U_∞, g_∞) be an incomplete Riemannian 3-manifold of non-negative curvature with an end $\mathcal{E}$ as in the hypothesis of Proposition 10.12. Fix a minimizing geodesic ray γ limiting to $\mathcal{E}$. Let λ_n be any sequence of positive numbers tending to infinity. For each n sufficiently large let $x_n \in \gamma$ be the point at distance $\lambda_n^{-1/2}$ from the end $\mathcal{E}$. Then the based metric spaces $(U_\infty, \lambda_n g_\infty, x_n)$ converge in the Gromov-Hausdorff sense to $(C'_{\mathcal{E}}, g_{\mathcal{E}}, ([\gamma], 1))$. Under this convergence the distance function from the end $\mathcal{E}$ in $(U_\infty, \lambda_n g_\infty)$ converges to the distance function from the cone point in the open cone.*

PROOF. It suffices to prove that given any subsequence of the original sequence, the result holds for a further subsequence. So let us replace the given sequence by a subsequence. Recall that for each $0 < r < R < \infty$ we have

$A_\mathcal{E}(r, R) \subset C'_\mathcal{E}$, the compact annulus which is the image of $\overline{X}(\mathcal{E}) \times [r, R]$. The statement about the non-compact spaces converging in the Gromov-Hausdorff topology, means that for each compact subspace K of $C'_\mathcal{E}$ containing the base point, for all n sufficiently large, there are compact subspaces $K_n \subset (U_\infty, \lambda_n g_\infty)$ containing x_n with the property that the (K_n, x_n) converge in the Gromov-Hausdorff topology to (K, x) (see Section D of Chapter 3, p. 39, of [**25**]).

Because of this, it suffices to fix $0 < r < 1 < R < \infty$ arbitrarily and prove the convergence result for $A_\mathcal{E}(r, R)$. Since the Gromov-Hausdorff distance from a compact pointed metric space to a δ-net in it containing the base point is at most δ, it suffices to prove that for $\delta > 0$ there is a δ-net $(\mathcal{N}, h)$ in $A_\mathcal{E}(r, R)$, with $([\gamma], 1) \in \mathcal{N}$ such that for all n sufficiently large there are embeddings φ_n of $\mathcal{N}$ into $A_n(r, R) = \overline{B}_{\lambda_n g_\infty}(\mathcal{E}, R) \setminus B_{\lambda_n g_\infty}(\mathcal{E}, r)$ with the following four properties:

(1) $\varphi_n^*(\lambda_n g_\infty)$ converge to h as $n \to \infty$,
(2) $\varphi_n([\gamma], 1) = x_n$,
(3) $\varphi_n(\mathcal{N})$ is a δ-net in $A_n(r, R)$, and
(4) denoting the cone point by $c \in C_\mathcal{E}$, if $d(p, c) = r$ then $d(\varphi_n(p), \mathcal{E}) = r$.

According to Corollary 10.28 there is a δ-net $\mathcal{N} \subset A_\mathcal{E}(r, R)$ consisting of points (ℓ_i, s_i) where the ℓ_i are realizable directions and $r < s_i < R$. Add $([\gamma], 1)$ to $\mathcal{N}$ if necessary so that we can assume that $([\gamma], 1) \in \mathcal{N}$. Let γ_i be a minimizing geodesic realizing ℓ_i and let d_i be its length.

Fix n sufficiently large so that $\lambda_n^{-1/2} R \leq d_i$ for all i. We define $\varphi_n \colon \mathcal{N} \to A_n(r, R)$ as follows. For any $a_i = ([\gamma_i], s_i) \in \mathcal{N}$ we let $\varphi_n(a_i) = \gamma_i(\lambda_n^{-1/2} s_i)$. (Since $\lambda_n^{-1/2} s \leq \lambda_n^{-1/2} R \leq d_i$, the geodesic γ_i is defined at $\lambda_n^{-1/2} s_i$.) This defines the embeddings φ_n for all n sufficiently large. Notice that

$$d_{g_\infty}(\varphi_n(\ell_i, s_i), \varphi_n(\ell_j, s_j)) = \lambda_n^{-1} s_i^2 + \lambda_n^{-1} s_j^2 - 2\lambda_n^{-1} s_i s_j \theta_{ij}(\lambda_n^{-1/2} s_i, \lambda_n^{-1/2} s_j),$$

or equivalently

$$d_{\lambda_n g_\infty}(\varphi_n(\ell_i, s_i), \varphi_n(\ell_j, s_j)) = s_i^2 + s_j^2 - 2 s_i s_j \theta_{ij}(\lambda_n^{-1/2} s_i, \lambda_n^{-1/2} s_j).$$

Here, $\theta_{ij}(\lambda_n^{-1/2} s_i, \lambda_n^{-1/2} s_j)$ is the comparison angle between the point at distance $\lambda_n^{-1/2} s_i$ along γ_i, the point p, and the point at distance $\lambda_n^{-1/2} s_j$ along γ_j. (Because of the convergence result on angles (Lemma 10.21), for all i and j we have

$$\lim_{n \to \infty} d_{\lambda_n g_\infty}(\varphi_n(\ell_i, s_i), \varphi_n(\ell_j, s_j)) = s_i^2 + s_j^2 - 2 s_i s_j \cos(\theta(\gamma_i, \gamma_j))$$
$$= d_{g_\mathcal{E}}((\ell_i, s_i), (\ell_j, s_j)).$$

This establishes the existence of the φ_n for all n sufficiently large satisfying the first condition. Clearly, from the definition $\varphi_n([\gamma], 1) = x_n$, and for all $p \in \mathcal{N}$ we have $d(\varphi_n(p), \mathcal{E}) = d(p, c)$.

It remains to check that for all n sufficiently large $\varphi_n(\mathcal{N})$ is a δ-net in $A_n(r, R)$. For n sufficiently large let $z \in A_n(r, R)$ and let γ_z be a minimizing geodesic ray from z to $\mathcal{E}$ parameterized by the distance from the end. Set $d_n = d_{\lambda_n g_\infty}(z, \mathcal{E})$, so that $r \leq d_n \leq R$. Fix n sufficiently large so that $\lambda_n^{-1/2} R < d_i$ for all i. The point $([\gamma_z], d_n) \in C_{\mathcal{E}}$ is contained in $A_{\mathcal{E}}(r, R)$ and hence there is an element $a = ([\gamma_i], s_i) \in \mathcal{N}$ within distance δ of $([\gamma_z], d_n)$ in $C_{\mathcal{E}}$. Since $s_i \leq R$, $\lambda_n^{-1/2} s_i \leq d_i$ and hence $x = \gamma_i(\lambda_n^{-1/2} s_i)$ is defined. By Corollary 10.26 we have

$$d_{\lambda_n g_\infty}(x, z) \leq d_{g_{\mathcal{E}}}\left(([\gamma], d_n), ([\gamma_i], s_i)\right) \leq \delta.$$

This completes the proof that for n sufficiently large the image $\varphi_n(\mathcal{N})$ is a δ-net in $A_n(r, R)$.

This shows that the $(U_\infty, \lambda_n g_\infty, x_n)$ converge in the Gromov-Hausdorff topology to $(C'_{\mathcal{E}}, g_{\mathcal{E}}, ([\gamma], 1))$. $\square$

REMARK 10.30. Notice that since the manifolds $(U_\infty, \lambda_n g_\infty, x_n)$ are not complete, there can be more than one Gromov-Hausdorff limit. For example we could take the full cone as a limit. Indeed, the cone is the only Gromov-Hausdorff limit that is complete as a metric space.

5. Comparison of the two types of limits

Let us recap the progress to date. We constructed an incomplete geometric blow-up limit $(\mathcal{U}_\infty, G_\infty, z_\infty)$ for our original sequence. It has non-negative Riemann curvature. We showed that the zero time-slice U_∞ of the limit is diffeomorphic to a tube $S^2 \times (0, 1)$ and that at one end of the tube the scalar curvature goes to infinity. Also, any point sufficiently near this end is the center of an evolving 2ϵ-neck defined for an interval of normalized time of length $1/2$ in the limiting flow. Then we took a further blow-up limit. We chose a sequence of points $x_n \in U_\infty$ tending to the end $\mathcal{E}$ where the scalar curvature goes to infinity. Then we formed $(U_\infty, \lambda_n g_\infty, x_n)$ where the distance from x_n to the end $\mathcal{E}$ is $\lambda_n^{-1/2}$. By fairly general principles (in fact it is a general theorem about manifolds of non-negative curvature) we showed that this sequence converges in the Gromov-Hausdorff sense to a cone.

The next step is to show that this second blow-up limit also exists as a geometric limit away from the cone point. Take a sequence of points $x_n \in U_\infty$ tending to $\mathcal{E}$. We let $\lambda'_n = R(x_n)$, and we consider the based Riemannian manifolds $(U_\infty, \lambda'_n g_\infty(0), x_n)$. Let $B_n \subset U_n$ be the metric ball of radius $\epsilon^{-1}/3$ centered at x_n in $(U_\infty, \lambda'_n g_\infty(0))$. Since this ball is contained in a 2ϵ-neck centered at x_n, the curvature on this ball is bounded, and this ball has compact closure in U_∞. Also, for each $y \in B_n$, there is a rescaled flow $\lambda' g(t)$ defined on $\{y\} \times (-1/2, 0]$ whose curvature on $B_n \times (-1/2, 0]$ is bounded. Hence, by Theorem 5.11 we can pass to a subsequence and extract a geometric limit. In fact, by Proposition 5.14 there is even a geometric limiting flow defined on the time interval $(-1/2, 0]$.

We must compare the zero time-slice of this geometric limiting flow with the corresponding open subset of the Gromov-Hausdorff limit constructed in the previous section. Of course, one obvious difference is that we have used different blow-up factors: $d(x_n, \mathcal{E})^{-2}$ in the first case and $R(x_n)$ in the second case. So one important ingredient in comparing the limits will be to compare these factors, at least in the limit.

5.1. Comparison of the blow-up factors. Now let us compare the two limits: (i) the Gromov-Hausdorff limit of the sequence $(U_\infty, \lambda_n g_\infty, x_n)$ and (ii) the geometric limit of the sequence $(U_\infty, \lambda'_n g_\infty, x_n)$ constructed above.

CLAIM 10.31. *The ratio $\rho_n = \lambda'_n/\lambda_n$ is bounded above and below by positive constants.*

PROOF. Since there is a 2ϵ-neck centered at x_n, according to Proposition A.11 the distance $\lambda_n^{-1/2}$ from x_n to $\mathcal{E}$ is at least $R(x_n)^{-1/2}\epsilon^{-1}/2 = (\lambda'_n)^{-1/2}\epsilon^{-1}/2$. Thus,
$$\rho_n^{-1} = \lambda_n/\lambda'_n \leq 4\epsilon^2.$$

On the other hand, suppose that $\rho_n = \lambda'_n/\lambda_n \to \infty$ as $n \to \infty$. Rescale by λ'_n so that $R(x_n) = 1$. The distance from x_n to $\mathcal{E}$ is $\sqrt{\rho_n}$. Then by Lemma A.4 with respect to this metric there is a sphere of diameter at most 2π through x_n that separates all points at distance at most $\sqrt{\rho_n} - \epsilon^{-1}$ from $\mathcal{E}$ from all points at distance at least $\sqrt{\rho_n} + \epsilon^{-1}$ from $\mathcal{E}$. Now rescale the metric by ρ_n. In the rescaled metric there is a 2-sphere of diameter at most $2\pi/\sqrt{\rho_n}$ through x_n that separates all points at distance at most $1 - \epsilon^{-1}/\sqrt{\rho_n}$ from $\mathcal{E}$ from all points at distance at least $1 + \epsilon^{-1}/\sqrt{\rho_n}$ from $\mathcal{E}$. Taking the Gromov-Hausdorff limit of these spaces, we see that the base point x_∞ separates all points of distance less than 1 from $\mathcal{E}$ from all points of distance greater than 1 from $\mathcal{E}$. This is impossible since the Gromov-Hausdorff limit is a cone that is not the cone on a single point. □

5.2. Completion of the comparison of the blow-up limits. Once we know that the λ_n/λ'_n are bounded above and below by positive constants, we can pass to a subsequence so that these ratios converge to a finite positive limit. This means that the Gromov-Hausdorff limit of the sequence of based metric spaces $(U_\infty, \lambda'_n g_\infty, x_n)$ is a cone, namely the Gromov-Hausdorff limiting cone constructed in Section 4, rescaled by $\lim_{n\to\infty} \rho_n$. In particular, the balls of radius $\epsilon^{-1}/2$ around the base points in this sequence converge in the Gromov-Hausdorff sense to the ball of radius $\epsilon^{-1}/2$ about the base point of a cone.

But we have already seen that the balls of radius $\epsilon^{-1}/2$ centered at the base points converge geometrically to a limiting manifold. That is to say, on every ball of radius less than $\epsilon^{-1}/2$ centered at the base point the metrics converge uniformly in the C^∞-topology to a limiting smooth metric. Thus, on every ball of radius less than $\epsilon^{-1}/2$ centered at the base point the limiting smooth metric is isometric to the metric of the Gromov-Hausdorff limit. This means that the limiting smooth metric on the ball B_∞ of radius $\epsilon^{-1}/2$ centered at the base point is isometric to an open subset of a cone. Notice that the scalar curvature of the limiting smooth metric at the base point is 1, so that this cone is a non-flat cone.

6. The final contradiction

We have now shown that the smooth limit of the balls of radius $\epsilon^{-1}/2$ centered at the base points of $(U_\infty, \lambda'_n g_\infty, x_n)$ is isometric to an open subset of a non-flat cone, and is also the zero time-slice of a Ricci flow defined for the time interval $(-1/2, 0]$. This contradicts Proposition 4.22, one of the consequences of the maximum principle established by Hamilton. The contradiction shows that the limit $(U_\infty, g_\infty, x_\infty)$ cannot exist. The only assumption that we made in order to construct this limit was that Theorem 10.2 did not hold for some $A_0 < \infty$. Thus, we have established Theorem 10.2 by contradiction.

CHAPTER 11

Geometric limits of generalized Ricci flows

In this chapter we apply the main result of the last section, bounded curvature at bounded distance, to blow-up limits in order to establish the existence of a smooth limit for sequences of generalized Ricci flows. In the first section we establish a blow-up limit that is defined for some interval of time of positive length, where the length of the interval of time is allowed to depend on the limit. In the second section we give conditions under which this blow-up limit can be extended backwards to make an ancient Ricci flow. In the third section we construct limits at the singular time of a generalized Ricci flow satisfying appropriate conditions. We characterize the ends of the components of these limits. We show that they are ϵ-horns – the ends are diffeomorphic to $S^2 \times [0,1)$ and the scalar curvature goes to infinity at the end. In the fourth section we prove for any $\delta > 0$ that there are δ-necks sufficiently deep in any ϵ-horn, provided that the curvature at the other end of the horn is not too large. Throughout this chapter we fix $\epsilon > 0$ sufficiently small such that all the results of the Appendix hold for 2ϵ and $\alpha = 10^{-2}$, and Proposition 2.19 holds for 2ϵ.

1. A smooth blow-up limit defined for a small time

We begin with a theorem that produces a blow-up limit flow defined on some small time interval.

THEOREM 11.1. *Fix canonical neighborhood constants* (C, ϵ), *and non-collapsing constants* $r > 0, \kappa > 0$. *Let* $(\mathcal{M}_n, G_n, x_n)$ *be a sequence of based generalized 3-dimensional Ricci flows. We set* $t_n = \mathbf{t}(x_n)$ *and* $Q_n = R(x_n)$. *We denote by* M_n *the* t_n *time-slice of* $\mathcal{M}_n$. *We suppose that:*

(1) *Each* $(\mathcal{M}_n, G_n)$ *either has a time interval of definition contained in* $[0, \infty)$ *and has curvature pinched toward positive, or has non-negative curvature.*
(2) *Every point* $y_n \in (\mathcal{M}_n, G_n)$ *with* $\mathbf{t}(y_n) \le t_n$ *and with* $R(y_n) \ge 4R(x_n)$ *has a strong* (C, ϵ)-*canonical neighborhood.*
(3) $\lim_{n \to \infty} Q_n = \infty$.
(4) *For each* $A < \infty$ *the following holds for all* n *sufficiently large. The ball* $B(x_n, t_n, AQ_n^{-1/2})$ *has compact closure in* M_n *and the flow is* κ-*non-collapsed on scales* $\le r$ *at each point of* $B(x_n, t_n, AQ_n^{-1/2})$.

(5) *There is $\mu > 0$ such that for every $A < \infty$ the following holds for all n sufficiently large. For every $y_n \in B(x_n, t_n, AQ_n^{-1/2})$ the maximal flow line through y_n extends backwards for a time at least $\mu \left(\max(Q_n, R(y_n))\right)^{-1}$.*

Then, after passing to a subsequence and shifting the times of each of the generalized flows so that $t_n = 0$ for every n, there is a geometric limit $(M_\infty, g_\infty, x_\infty)$ of the sequence $(M_n, Q_n G_n(0), x_n)$ of based Riemannian manifolds. This limit is a complete 3-dimensional Riemannian manifold of bounded, non-negative curvature. Furthermore, for some $t_0 > 0$ which depends on the curvature bound for (M_∞, g_∞) and on μ, there is a geometric limit Ricci flow defined on $(M_\infty, g_\infty(t))$, $-t_0 \leq t \leq 0$, with $g_\infty(0) = g_\infty$.

Before beginning the proof of this theorem we establish a lemma that we shall need both in its proof and also for later applications.

LEMMA 11.2. *Let $(\mathcal{M}, G)$ be a generalized 3-dimensional Ricci flow. Suppose that $r_0 > 0$ and that any $z \in \mathcal{M}$ with $R(z) \geq r_0^{-2}$ has a strong (C, ϵ)-canonical neighborhood. Suppose $z \in \mathcal{M}$ and $\mathbf{t}(z) = t_0$. Set*

$$r = \frac{1}{2C\sqrt{\max(R(z), r_0^{-2})}}$$

and

$$\Delta t = \frac{1}{16C\left(R(z) + r_0^{-2}\right)}.$$

Suppose that $r' \leq r$ and that $|t' - t_0| \leq \Delta t$ and let I be the interval with endpoints t_0 and t'. Suppose that there is an embedding of $j \colon B(z, t_0, r') \times I$ into $\mathcal{M}$ compatible with time and with the vector field. Then $R(y) \leq 2\left(R(z) + r_0^{-2}\right)$ for all y in the image of j.

PROOF. We first prove that for any $y \in B(z, t_0, r)$ we have

(11.1) $$R(y) \leq \frac{16}{9}(R(z) + r_0^{-2}).$$

Let $\gamma \colon [0, s_0] \to B(z, t_0, r)$ be a path of length $s_0 < r$ connecting $z = \gamma(0)$ to $y = \gamma(s_0)$. We take γ parameterized by arc length. For any $s \in [0, s_0]$ let $R(s) = R(\gamma(s))$. According to the strong (C, ϵ)-canonical neighborhood assumption at any point where $R(s) \geq r_0^{-2}$ we have $|R'(s)| \leq CR^{3/2}(s)$. Let $J \subset [0, s_0]$ be the closed subset consisting of $s \in [0, s_0]$ for which $R(s) \geq r_0^{-2}$. There are three possibilities. If $s_0 \notin J$ then $R(y) \leq r_0^{-2}$ and we have established Inequality (11.1). If $J = [0, s_0]$, then we have $|R'(s)| \leq CR^{3/2}(s)$ for all s in J. Using this differential inequality and the fact that the interval has length at most $\frac{1}{2C\sqrt{R(z)}}$, we see that $R(y) \leq 16R(z)/9$, again establishing Inequality (11.1). The last possibility is that $J \neq [0, s_0]$ but $s_0 \in J$. We restrict attention to the maximal interval of J containing s_0. This interval has length at most $\frac{r_0}{2C}$ and at its initial point R takes the value r_0^{-2}. For every s in this interval by our assumptions we again have the inequality $|R'(s)| \leq CR^{3/2}(s)$; it follows immediately that $R(y) \leq 16r_0^{-2}/9$. This establishes Inequality (11.1) in all cases.

Now consider the vertical path $j(\{y\} \times I)$. Let $R(t) = R(j(y,t))$. Again by the strong canonical neighborhood assumption $|R'(t)| \leq CR^2(t)$ at all points where $R(t) \geq r_0^{-2}$. Consider the closed subset K of I where $R(t) \geq r_0^{-2}$. There are three cases to consider: $t' \notin K$, $t' \in K \neq I$, or $K = I$. In the first case, $R(y,t') \leq r_0^{-2}$ and we have established the result. In the second case, let K' be the maximal subinterval of K containing t'. On the interval K' we have $|R'(t)| \leq CR^2(t)$ and at one endpoint $R(t) = r_0^{-2}$. Since this interval has length at most $r_0^2/16C$, it follows easily that $R(t') \leq 16r_0^{-2}/15$, establishing the result. In the last case where $K = I$, then, by what we established above, the initial condition is $R(t_0) = R(y) \leq 16(R(z) + r_0^{-2})/9$, and the differential inequality $|R'(t)| \leq CR^2(t)$ holds for all $t \in I$. Since the length of I is at most $\frac{1}{16C(R(y)+r_0^{-2})}$ we see directly that $R(t') \leq 2(R(z)+r_0^{-2})$, completing the proof in this case as well. □

Now we begin the proof of Theorem 11.1.

PROOF. (of Theorem 11.1) We shift the times for the flows so that $t_n = 0$ for all n. Since Q_n tends to ∞ as n tends to ∞, according to Theorem 10.2 for any $A < \infty$, there is a bound $Q(A) < \infty$ on the scalar curvature of $Q_nG_n(0)$ on $B_{Q_nG_n}(x_n, 0, A)$ for all n sufficiently large. According to the hypothesis of Theorem 11.1, this means that there is $t_0(A) > 0$ and, for each n sufficiently large, an embedding of $B_{Q_nG_n}(x_n, 0, A) \times [-t_0(A), 0]$ into $\mathcal{M}_n$ compatible with time and with the vector field. In fact, we can choose $t_0(A)$ so that more is true.

COROLLARY 11.3. *For each $A < \infty$, let $Q(A)$ be a bound on the scalar curvature of the restriction of Q_nG_n to $B_{Q_nG_n}(x_n, 0, A)$ for all n sufficiently large. Then there exist a constant $t'_0(A) > 0$ depending on $t_0(A)$ and $Q(A)$, and a constant $Q'(A) < \infty$ depending only on $Q(A)$, and, for all n sufficiently large, an embedding*

$$B_{Q_nG_n}(x_n, 0, A) \times (-t'_0(A), 0] \to \mathcal{M}_n$$

compatible with time and with the vector field with the property that the scalar curvature of the restriction of Q_nG_n to the image of this subset is bounded by $Q'(A)$.

PROOF. This is immediate from Lemma 11.2 and Assumption (5) in the hypothesis of the theorem. □

Now since the curvatures of the Q_nG_n are pinched toward positive or are non-negative, bounding the scalar curvature above gives a bound on $|\text{Rm}_{Q_nG_n}|$ on the product $B_{Q_nG_n}(x_n, 0, A) \times (-t'_0(A), 0]$. Now we invoke Shi's theorem (Theorem 3.28):

COROLLARY 11.4. *For each $A < \infty$ and for each integer $\ell \geq 0$, there is a constant C_2 such that for all n sufficiently large we have*
$$|\nabla^\ell \mathrm{Rm}_{Q_n G_n}(x)| \leq C_2$$
for all $x \in B_{Q_n G_n}(x_n, 0, A)$.

Also, by the curvature bound and the κ-non-collapsed hypothesis we have the following:

CLAIM 11.5. *There is $\eta > 0$ such that for all n sufficiently large*
$$\mathrm{Vol}(B_{Q_n G_n}(x_n, 0, \eta)) \geq \kappa \eta^3.$$

Now we are in a position to apply Corollary 5.10. This implies that, after passing to a subsequence, there is a geometric limit $(M_\infty, g_\infty, x_\infty)$ of the sequence of based Riemannian manifolds $(M_n, Q_n G_n(0), x_n)$. The geometric limit is a complete Riemannian manifold. If the $(\mathcal{M}_n, G_n)$ satisfy the curvature pinched toward positive hypothesis, by Theorem 5.33, the limit Riemannian manifold (M_∞, g_∞) has non-negative curvature. If the $(\mathcal{M}_n, G_n)$ have non-negative curvature, then it is obvious that the limit has non-negative curvature. By construction $R(x_\infty) = 1$.

In fact, by Proposition 5.14 for each $A < \infty$, there is $t(A) > 0$ and, after passing to a subsequence, geometrically limit flow defined on $B(x_\infty, 0, A) \times (-t(A), 0]$.

CLAIM 11.6. *Any point in (M_∞, g_∞) of curvature greater than 4 has a $(2C, 2\epsilon)$-canonical neighborhood.*

PROOF. Since the sequence $(M_n, Q_n G_n(0), x_n)$ converges geometrically to $(M_\infty, g_\infty, x_\infty)$, there is an exhaustive sequence $V_1 \subset V_2 \subset \cdots \subset M_\infty$ of open subsets of M_∞, with compact closure, each containing x_∞, and for each n an embedding φ_n of V_n into the zero time-slice of $\mathcal{M}_n$ such that $\varphi_n(x_\infty) = x_n$ and such that the Riemannian metrics $\varphi_n^* G_n$ converge uniformly on compact sets to g_∞. Let $q \in M_\infty$ be a point with $R_{g_\infty}(q) > 4$. Then for all n sufficiently large, $q \in V_n$, so that $q_n = \varphi_n(q)$ is defined, and $R_{Q_n G_n}(q_n) > 4$. Thus, q_n has an (C, ϵ)-canonical neighborhood, U_n, in $\mathcal{M}_n$; and, since $R(q_n) > 4$ for all n, there is a uniform bound to the distance from any point of U_n to q_n. Thus, there exists m such that for all n sufficiently large $\varphi_n(V_m)$ contains U_n. Clearly as n goes to infinity the Riemannian metrics $\varphi_n^*(G_n)|_{\varphi_n^{-1}(U_m)}$ converge smoothly to $g_\infty|_{\varphi_n^{-1}(U_n)}$. Thus, by Proposition 9.79 for all n sufficiently large the restriction of g_∞ to $\varphi_n^{-1}(U_n)$ contains a $(2C, 2\epsilon)$-canonical neighborhood of q. □

CLAIM 11.7. *The limit Riemannian manifold (M_∞, g_∞) has bounded curvature.*

PROOF. First, suppose that (M_∞, g_∞) does not have strictly positive curvature. Suppose that $y \in M_\infty$ has the property that $\mathrm{Rm}(y)$ has a zero eigenvalue. Fix $A < \infty$ greater than $d_{g_\infty}(x_\infty, y)$. Then applying Corollary 4.19 to the limit flow on $B(x_\infty, 0, A) \times (-t(A), 0]$, we see that the Riemannian manifold $(B(x_\infty, 0, A), g_\infty)$ is locally a Riemannian product of a compact surface of positive curvature with a one-manifold. Since this is true for every $A < \infty$ sufficiently large, the same is true for (M_∞, g_∞). Hence (M_∞, g_∞) has a one- or two-sheeted covering that is a global Riemannian product of a compact surface and one-manifold. Clearly, in this case the curvature of (M_∞, g_∞) is bounded.

If M_∞ is compact, then it is clear that the curvature is bounded.

It remains to consider the case where (M_∞, g_∞) is non-compact and of strictly positive curvature. Since any point of curvature greater than 4 has a $(2C, 2\epsilon)$-canonical neighborhood, and since M_∞ is non-compact, it follows that the only possible canonical neighborhoods for $x \in M_\infty$ are a 2ϵ-neck centered at x or $(2C, 2\epsilon)$-cap whose core contains x. Each of these canonical neighborhoods contains a 2ϵ-neck. Thus, if (M_∞, g_∞) has unbounded and positive Riemann curvature or equivalently, it has unbounded scalar curvature, then it has $(2C, 2\epsilon)$-canonical neighborhoods of arbitrarily small scale, and hence 2ϵ-necks of arbitrarily small scale. But this contradicts Proposition 2.19. It follows from this contradiction that the curvature of (M_∞, g_∞) is bounded. $\square$

To complete the proof of Theorem 11.1 it remains to extend the limit for the 0 time-slices of the $(\mathcal{M}_n, G_n)$ that we have just constructed to a limit flow defined for some positive amount of time backward. Since the curvature of (M_∞, g_∞) is bounded, this implies that there is a bound, Q, such that for any $A < \infty$ the curvature of the restriction of $Q_n G_n$ to $B_{Q_n G_n}(x_n, 0, A)$ is bounded by Q for all n sufficiently large. Thus, we can take the constant $Q(A)$ in Corollary 11.3 to be independent of A. According to that corollary this implies that there is a $t'_0 > 0$ and $Q' < \infty$ such that for every A there is an embedding $B_{Q_n G_n}(x_N, 0, A) \times (-t'_0, 0] \to \mathcal{M}_n$ compatible with time and with the vector field so that the scalar curvature of the restriction of $Q_n G_n$ to the image is bounded by Q' for all n sufficiently large. This uniform bound on the scalar curvature yields a uniform bound, uniform in the sense of being independent of n, on $|\mathrm{Rm}_{Q_n G_n}|$ on the image of the embedding $B_{Q_n G_n}(x_N, 0, A) \times (-t'_0, 0]$.

Then by Hamilton's result, Proposition 5.14, we see that, after passing to a further subsequence, there is a limit flow defined on $(-t'_0, 0]$. Of course, the zero time-slice of this limit flow is the limit (M_∞, g_∞). This completes the proof of Theorem 11.1. $\square$

2. Long-time blow-up limits

Now we wish to establish conditions under which we can, after passing to a further subsequence, establish the existence of a geometric limit flow defined on $-\infty < t \leq 0$. Here is the main result.

THEOREM 11.8. *Suppose that $\{(\mathcal{M}_n, G_n, x_n)\}_{n=1}^{\infty}$ is a sequence of generalized 3-dimensional Ricci flows satisfying the hypotheses of Theorem 11.1. Suppose in addition that there is T_0 with $0 < T_0 \le \infty$ such that the following holds. For any $T < T_0$, for each $A < \infty$, and all n sufficiently large, there is an embedding $B(x_n, t_n, AQ_n^{-1/2}) \times (t_n - TQ_n^{-1}, t_n]$ into $\mathcal{M}_n$ compatible with time and with the vector field and at every point of the image the generalized flow is κ-non-collapsed on scales $\le r$. Then, after shifting the times of the generalized flows so that $t_n = 0$ for all n and passing to a subsequence there is a geometric limit Ricci flow*

$$(M_\infty, g_\infty(t), x_\infty), \quad -T_0 < t \le 0,$$

for the rescaled generalized flows $(Q_n\mathcal{M}_n, Q_nG_n, x_n)$. This limit flow is complete and of non-negative curvature. Furthermore, the curvature is locally bounded in time. If in addition $T_0 = \infty$, then it is a κ-solution.

REMARK 11.9. Let us point out the differences between this result and Theorem 11.1. The hypotheses of this theorem include all the hypotheses of Theorem 11.1. The main difference between the conclusions is that in Theorem 11.1 the amount of backward time for which the limit flow is defined depends on the curvature bound for the final time-slice of the limit (as well as how far back the flows in the sequence are defined). This amount of backward time tends to zero as the curvature of the final time-slice limit tends to infinity. Here, the amount of backward time for which the limit flow is defined depends only on how far backwards the flows in the sequence are defined.

PROOF. In Theorem 11.1 we proved that, after passing to a subsequence, there is a geometric limit Ricci flow, complete of bounded non-negative curvature,

$$(M_\infty, g_\infty(t), x_\infty), \quad -t_0 \le t \le 0,$$

defined for some $t_0 > 0$. Our next step is to extend the limit flow all the way back to time $-T_0$.

PROPOSITION 11.10. *With the notation of, and under the hypotheses of Theorem 11.8, suppose that there is a geometric limit flow $(M_\infty, g_\infty(t))$ defined for $-T < t \le 0$ which has non-negative curvature locally bounded in time. Suppose that $T < T_0$. Then the curvature of the limit flow is bounded and the geometric limit flow can be extended to a flow with bounded curvature defined on $(-(T+\delta), 0]$ for some $\delta > 0$.*

PROOF. The argument is by contradiction, so we suppose that there is a $T < T_0$ as in the statement of the proposition. Then the geometric limit flow on $(-T, 0]$ is complete of non-negative curvature and with the curvature locally bounded in time. First suppose that the scalar curvature is bounded by, say $Q < \infty$. Fix $T' < T$. The Riemannian manifold $(M_\infty, g_\infty(T'))$ is complete of non-negative curvature with the scalar curvature, and hence the norm of the Riemann curvature, bounded by Q. Thus, for any $A < \infty$ for all n sufficiently large, the norm of the Riemann curvature of $Q_n G_n(-T')$ on $B_{Q_n G_n}(x_n, -T, A)$ is bounded above by $2Q$. Also, arguing as in the proof of Theorem 11.1 we see that any point $y \in M_\infty$ with $R(y, -T') > 4$ has a $(2C, 2\epsilon)$-canonical neighborhood. Hence, applying Lemma 11.2 as in the argument in the proof of Corollary 11.3 shows that for all n sufficiently large, every point in $B_{Q_n G_n}(x_n, -T', A)$ has a uniform size parabolic neighborhood on which the Riemann curvature is uniformly bounded, where both the time interval in the parabolic neighborhood and the curvature bound on this neighborhood depend only on C and the curvature bound on Q for the limit flow. According to Hamilton's result (Proposition 5.14) this implies that, by passing to a further subsequence, we can extend the limit flow backward beyond $-T'$ a uniform amount of time, say 2δ. Taking $T' > T - \delta$ then gives the desired extension under the condition that the scalar curvature is bounded on $(-T, 0]$.

It remains to show that, provided that $T < T_0$, the scalar curvature of the limit flow $(M_\infty, g_\infty(t))$, $-T < t \leq 0$, is bounded. To establish this we need a couple of preliminary results.

LEMMA 11.11. *Suppose that there is a geometric limit flow defined on $(-T, 0]$ for some $0 < T \leq T_0$ with $T < \infty$. We suppose that this limit is complete with non-negative curvature, and with curvature locally bounded in time. Let $X \subset M_\infty$ be a compact, connected subset. If $\min_{x \in X}(R_{g_\infty}(x, t))$ is bounded, independent of t, for all $t \in (-T, 0]$, then there is a finite upper bound on $R_{g_\infty}(x, t)$ for all $x \in X$ and all $t \in (-T, 0]$.*

PROOF. Let us begin with:

CLAIM 11.12. *Let Q be an upper bound on $R(x, 0)$ for all $x \in M_\infty$. Then for any points $x, y \in M_\infty$ and any $t \in (-T, 0]$ we have*

$$d_t(x, y) \leq d_0(x, y) + 16\sqrt{\frac{Q}{3}} T.$$

PROOF. Fix $-t_0 \in (-T, 0]$. Then for any $\epsilon > 0$ sufficiently small, by the Harnack inequality (the second result in Theorem 4.37) we have

$$\frac{\partial R}{\partial t}(x, t) \geq -\frac{R(x, t)}{t + T - \epsilon}.$$

Taking the limit as $\epsilon \to 0$ gives

$$\frac{\partial R}{\partial t}(x, t) \geq -\frac{R(x, t)}{t + T},$$

and hence, fixing x,

$$\frac{dR(x, t)}{R(x, t)} \geq \frac{-dt}{(t + T)}.$$

Integrating from $-t_0$ to 0 shows that
$$\log(R(x,0)) - \log(R(x,-t_0)) \geq \log(T-t_0) - \log(T),$$
and since $R(x,0) \leq Q$, this implies
$$R(x,-t_0) \leq Q \frac{T}{T-t_0}.$$
Recalling that $n = 3$ and that the curvature is non-negative we see that
$$\mathrm{Ric}(x,-t_0) \leq (n-1)\frac{QT}{2}\frac{1}{T-t_0}.$$
Hence by Corollary 3.26, for all $-t_0 \in (-T, 0]$ we have that
$$\mathrm{dist}_{-t_0}(x,y) \leq \mathrm{dist}_0(x,y) + 8\int_{-t_0}^0 \sqrt{\frac{QT}{3(T+t)}} \leq \mathrm{dist}_0(x,y) + 16\sqrt{\frac{Q}{3}}T.$$
$\square$

It follows immediately from this claim that any compact subset $X \subset M_\infty$ has uniformly bounded diameter under all the metrics $g_\infty(t)$, $-T < t \leq 0$.

By the hypothesis of the lemma there is a constant $C' < \infty$ such that for each $t \in (-T, 0]$ there is $y_t \in X$ with $R_{g_\infty}(y_t, t) \leq C'$. Suppose that the conclusion of the lemma does not hold. Then there is a sequence $t_m \to -T$ as $m \to \infty$ and points $z_m \in X$ such that $R_{g_\infty}(z_m, t_m) \to \infty$ as $m \to \infty$. In this case, possibly after redefining the constant C', we can also assume that there is a point y_m such that $2 \leq R(y_m, t_m) \leq C'$. Since the sequence $(\mathcal{M}_n, Q_n G_n, x_n)$ converges smoothly to $(M_\infty, g_\infty(t), x_\infty)$ for $t \in (-T, 0]$, it follows that for each m there are sequences $\{y_{m,n} \in \mathcal{M}_n\}_{n=1}^\infty$ and $\{z_{m,n} \in \mathcal{M}_n\}_{n=1}^\infty$ with $\mathbf{t}(y_{m,n}) = \mathbf{t}(z_{m,n}) = t_m$ converging to (y_m, t_m) and (z_m, t_m) respectively. Thus, for all m there is $n_0 = n_0(m)$ such that for all $n \geq n_0$ we have:

(1) $1 \leq R_{Q_n G_n}(y_{m,n}) \leq 2C'$,
(2) $R_{Q_n G_n}(z_{m,n}) \geq R_{g_\infty}(z_m, t_m)/2$,
(3) $d_{Q_n G_n}((y_{m,n}), (z_{m,n})) \leq 2\mathrm{diam}_{g_\infty(t_m)}(X)$.

It follows from the third condition and the fact that X has uniformly bounded diameter under all the metrics $g_\infty(t)$ for $t \in (-T, 0]$, that the distance $d_{Q_{m,n} G_{m,n}}(z_{m,n}, y_{m,n})$ is bounded independent of m and n as long as $n \geq n_0$. Because of the fact that $R_{Q_n G_n}(y_{m,n}) \geq 1 = R(x_n)$, it follows that any point $z \in \mathcal{M}_n$ with $\mathbf{t}(z_m) \leq t_m$ and with $R(z) \geq 4R(y_{m,n})$ has a strong (C, ϵ)-canonical neighborhood. This then contradicts Theorem 10.2 and completes the proof of the lemma. $\square$

Clearly, this argument will be enough to handle the case when M_∞ is compact. The case when M_∞ is non-compact uses additional results.

LEMMA 11.13. *Let (M,g) be a complete, connected, non-compact manifold of non-negative sectional curvature and let $x_0 \in M$ be a point. Then there is $D > 0$, such that for any $y \in M$ with $d(x_0, y) = d \geq D$, there is $x \in M$ with $d(y, x) = d$ and with $d(x_0, x) > 3d/2$.*

PROOF. Suppose that the result is false for (M, g) and $x \in M$. Then there is a sequence $y_n \in M$ with the following property. Let $d_n = d(x, y_n)$. Then $\lim_{n\to\infty} d_n = \infty$ and yet $B(y_n, d_n) \subset B(x, 3d_n/2)$ for every n. Let γ_n be a minimal geodesic from x to y_n. By passing to a subsequence we arrange that the γ_n converge to a minimal geodesic ray γ from x to infinity in M. In particular, the angle at x between γ_n and γ tends to zero as $n \to \infty$. Let w_n be the point on γ at distance d_n from x, and let $\alpha_n = d(y_n, w_n)$. Because (M, g) has non-negative curvature, by Corollary 2.5, $\lim_{n\to\infty} \alpha_n/d_n = 0$. In particular, for all n sufficiently large, $\alpha_n < d_n$. This implies that there is a point z_n on the sub-ray of γ with endpoint w_n at distance d_n from y_n. By the triangle inequality, $d(w_n, z_n) \geq d_n - \alpha_n$. Since γ is a minimal geodesic ray, $d(z, z_n) = d(z, w_n) + d(w_n, z_n) \geq 2d_n - \alpha_n$. Since $\alpha_n/d_n \to 0$ as $n \to \infty$, it follows that for all n sufficiently large $d(z, z_n) > 3d_n/2$. This contradiction proves the lemma. □

CLAIM 11.14. *Fix $D < \infty$ greater than or equal to the constant given in the previous lemma for the Riemannian manifold $(M_\infty, g_\infty(0))$ and the point x_∞. We also choose $D \geq 32\sqrt{\frac{Q}{3}}T$. Then for any $y \in M_\infty \setminus B(x_\infty, 0, D)$ the scalar curvature $R_{g_\infty}(y, t)$ is uniformly bounded for all $t \in (-T, 0]$.*

PROOF. Suppose this does not hold for some $y \in M_\infty \setminus B(x_\infty, 0, D)$. Let $d = d_0(x_\infty, y)$. Of course, $d \geq D$. Thus, by the lemma there is $z \in M_\infty$ with $d_0(y, z) = d$ and $d_0(x_\infty, z) > 3d/2$. Since the scalar curvature $R(y, t)$ is not uniformly bounded for all $t \in (-T, 0]$, there is t for which $R(y, t)$ is arbitrarily large and hence (y, t) has an $(2C, 2\epsilon)$-canonical neighborhood of arbitrarily small scale. By Claim 11.12 we have $d_t(x_\infty, y) \leq d + 8\sqrt{\frac{Q}{3}}T$ and $d_t(y, z) \leq d + 8\sqrt{\frac{Q}{3}}T$. Of course, since Ric ≥ 0 the metric is non-increasing in time and hence $d \leq \min(d_t(y, z), d_t(x_\infty, y))$ and $3d/2 \leq d_t(x_0, z)$. Since y has a $(2C, 2\epsilon)$-canonical neighborhood in $(M_\infty, g_\infty(t))$, either y is the center of a 2ϵ-neck in $(M_\infty, g_\infty(t))$ or y is contained in the core of a $(2C, 2\epsilon)$-cap in $(M_\infty, g_\infty(t))$. (The other two possibilities for canonical neighborhoods require that M_∞ be compact.)

CLAIM 11.15. *y cannot lie in the core of a $(2C, 2\epsilon)$-cap in $(M_\infty, g_\infty(t))$, and hence it is the center of a 2ϵ-neck N in $(M_\infty, g_\infty(t))$. Furthermore, minimal $g(t)$-geodesics from y to x_∞ and z exit out of opposite ends of N (see FIG. 1).*

PROOF. Let $\mathcal{C}$ be a $(2C, 2\epsilon)$-canonical neighborhood of y in $(M_\infty, g_\infty(t))$. Since $R(y, t)$ can be arbitrarily large, we can assume that $d \gg 2CR(y)^{-1/2}$,

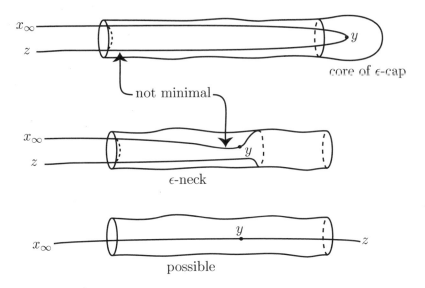

FIGURE 1. Minimal geodesics in necks and caps.

which is a bound on the diameter of $\mathcal{C}$. This implies that minimal $g(t)$-geodesics γ_{x_∞} and γ_z connecting y to x_∞ and to z, respectively, must exit from $\mathcal{C}$. Let a be a point on $\gamma_{x_\infty} \cap \mathcal{C}$ close to the complement of $\mathcal{C}$. Let b be a point at the same $g(t)$-distance from y on γ_z. In the case that $\mathcal{C}$ is a cap or that it is a 2ϵ-neck and γ_{x_∞} and γ_z exit from the same end, then $d_t(b,y)/d_t(a,y) < 4\pi\epsilon$. This means that the angle θ of the Euclidean triangle with these side lengths at the point corresponding to y satisfies

$$\cos(\theta) \geq 1 - \frac{(4\pi\epsilon)^2}{2}.$$

Recall that Q is the maximum value of $R(x,0)$, and that by Claim 11.12 we have

$$d \leq d_t(x_\infty, y) \leq d + 16\sqrt{\frac{Q}{3}}T,$$

with the same inequalities holding with $d_t(z,y)$ replacing $d_t(x_\infty, y)$. Also, by construction $d \geq 32\sqrt{\frac{Q}{3}}T$. We set $a_0 = d_t(x_\infty, y)$ and $a_1 = d_t(z,y)$. Then by the Toponogov property we have

$$d_t(x,z)^2 \leq a_0^2 + a_1^2 - 2a_0 a_1 \left(1 - \frac{(4\pi\epsilon)^2}{2}\right) = (a_0 - a_1)^2 + (4\pi\epsilon)^2 a_0 a_1.$$

Since $|a_0 - a_1| \leq d/2$ and $a_0, a_1 \leq 3d/2$ and $\epsilon < 1/8\pi$, it follows that $d_t(x,z) < d$. Since distances do not increase under the flow, it follows that $d_0(x,z) < d$. This contradicts the fact that $d_0(x,z) = d$. □

Since y is the center of a $(2C, 2\epsilon)$-neck N in $(M_\infty, g_\infty(t))$ and minimal $g(t)$-geodesics from y to z and to x_∞ exit out of opposite ends of N, it follows that $B(y, t, 4\pi R(y,t)^{-1/2})$ separates x_∞ and z. Since the curvature of the time-slices is non-negative, the Ricci flow does not increase distances. Hence, $B(y, 0, 4\pi R(y,t)^{-1/2})$ separates z from x_∞. (Notice that since $d > 4\pi R(y,t)^{-1/2}$, neither z nor x_∞ lies in this ball.) Thus, if $R(y,t)$ is unbounded as $t \to -T$ then arbitrarily small $g(0)$-balls centered at y separate z and x_∞. Since y is distinct from x_∞ and z, this is clearly impossible. □

Next we establish that the curvature near the base point x_∞ is bounded for all $t \in (-T, 0]$.

COROLLARY 11.16. *Suppose there is a geometric limit flow* $(M_\infty, g_\infty(t))$ *of a subsequence defined on* $(-T, 0]$ *for some* $T < \infty$. *Suppose that the limit flow is complete of non-negative curvature with the curvature locally bounded in time. Then for every* $A < \infty$ *the scalar curvature* $R_{g_\infty}(y, t)$ *is uniformly bounded for all* $(y, t) \in B(x_\infty, 0, A) \times (-T, 0]$.

PROOF. First we pass to a subsequence so that a geometric limit flow

$$(M_\infty, g_\infty(t), (x_\infty, 0))$$

exists on $(-T, 0]$. We let Q be the upper bound for $R(x, 0)$ for all $x \in M_\infty$. We now divide the argument into two cases: (i) M_∞ is compact, and (ii) M_∞ is non-compact.

Suppose that M_∞ is compact. By Proposition 4.1 we know that

$$\min_{x \in M_\infty}(R_{g_\infty}(x, t))$$

is a non-decreasing function of t. Since $R_{g_\infty}(x_\infty, 0) = 1$, it follows that for each $t \in (-T, 0]$, we have $\min_{x \in M_\infty} R(x, t) \leq 1$, and hence there is a point $x_t \in M_\infty$ with $R(x_t, t) \leq 1$. Now we can apply Lemma 11.11 to see that the scalar curvature of g_∞ is bounded on all of $M_\infty \times (-T, 0]$.

If M_∞ is non-compact, choose D as in Claim 11.14. According to that claim every point in the boundary of $B(x_\infty, 0, D)$ has bounded curvature under $g_\infty(t)$ for all $t \in (-T, 0]$. In particular, for each $t \in (-T, 0]$ the minimum of $R(x, t)$ over $\overline{B}(x_\infty, 0, D)$ is bounded independent of t. Now apply Lemma 11.11 to the closure of $B(x_\infty, 0, D)$. We conclude that the curvature of $B(x_\infty, 0, D)$ is uniformly bounded for all $g_\infty(t)$ for all $t \in (-T, 0]$. In particular, $R(x_\infty, t)$ is uniformly bounded for all $t \in (-T, 0]$.

Now for any $A < \infty$ we apply Lemma 11.11 to the compact subset $\overline{B}(x_\infty, 0, A)$ to conclude that the curvature is uniformly bounded on $B(x_\infty, 0, A) \times (-T, 0]$. This completes the proof of the corollary. □

Now let us return to the proof of Proposition 11.10.

CLAIM 11.17. *For each* $A < \infty$ *and for all n sufficiently large, there are* $\delta > 0$ *with* $\delta \leq T_0 - T$ *and a bound, independent of n, on the scalar curvature of the restriction of* $Q_n G_n$ *to* $B_{Q_n G_n}(x_n, 0, A) \times [-(T+\delta), 0]$.

PROOF. Fix $A < \infty$ and let K be the bound for the scalar curvature of g_∞ on $B(x_\infty, 0, 2A) \times (-T, 0]$ from Corollary 11.16. Lemma 11.2 shows that there are $\delta > 0$ and a bound in terms of K and C on the scalar curvature of the restriction of $Q_n G_n$ to $B_{Q_n G_n}(x_n, 0, A) \times [-(T+\delta), 0]$. □

Since the scalar curvature is bounded, by the assumption that either the curvature is pinched toward positive or the Riemann curvature is non-negative, this implies that the sectional curvatures of $Q_n G_n$ are also uniformly bounded on the products $B_{Q_n G_n}(x_n, 0, A) \times [-(T+\delta), 0]$ for all n sufficiently large. Consequently, it follows that by passing to a further subsequence we can arrange that the $-T$ time-slices of the $(\mathcal{M}_n, G_n, x_n)$ converge to a limit $(M_\infty, g_\infty(-T))$. This limit manifold satisfies the hypothesis of Proposition 2.19 and hence, by that proposition, it has bounded sectional curvature. This means that there is a $\delta > 0$ such that for all n sufficiently large and for any $A < \infty$ the scalar curvatures (and hence the Riemann curvatures) of the restriction of $Q_n G_n$ to $B_{Q_n G_n}(x_n, 0, A) \times [-(T+\delta), 0]$ are bounded independent of n. This allows us to pass to a further subsequence for which there is a geometric limit defined on $(-(T+\delta/2), -T]$. This geometric limit is complete of bounded, non-negative curvature. Hence, we have now constructed a limit flow on $(-(T+\delta/2), 0]$ with the property that for each $t \in (-(T+\delta/2), 0]$ the Riemannian manifold $(M, g(t))$ is complete and of bounded non-negative curvature. (We still don't know whether the entire flow is of bounded curvature.) But now invoking Hamilton's Harnack inequality (Theorem 4.37), we see that the curvature is bounded on $[-T, 0]$. Since we already know it is bounded in $(-T+\delta/2, -T]$, this completes the proof of the proposition. □

It follows immediately from Proposition 11.10 that there is a geometric limit flow defined on $(-T_0, 0]$. The geometric limit flow on $(-T_0, 0]$ is complete of non-negative curvature, locally bounded in time.

It remains to prove the last statement in the theorem. So let us suppose that $T_0 = \infty$. We have just established the existence of a geometric limit flow defined for $t \in (-\infty, 0]$. Since the $(\mathcal{M}_n, G_n)$ either have curvature pinched toward positive or are of non-negative curvature, it follows from Theorem 5.33 that all time-slices of the limit flow are complete manifolds of non-negative curvature. Since points of scalar curvature greater than 4 have $(2C, 2\epsilon)$-canonical neighborhoods, it follows from Proposition 2.19 that the curvature is bounded on each time-slice, and hence universally bounded by the Harnack inequality (Theorem 4.37). Since for any $A < \infty$ and every $T < \infty$ the parabolic neighborhoods $B_{Q_n G_n}(x_n, 0, A) \times [-T, 0]$ are κ-non-collapsed on scales $Q_n r$ for every n sufficiently large, the limit is κ-non-collapsed on scales $\leq \lim_{n \to \infty} Q_n r$. Since $r > 0$ and $\lim_{n \to \infty} Q_n = \infty$, it follows that the limit flow is κ-non-collapsed on all scales. Since

$R_{Q_nG_n}(x_n) = 1$, $R_{g_\infty}(x_\infty, 0) = 1$ and the limit flow is non-flat. This establishes all the properties needed to show that the limit is a κ-solution. This completes the proof of Theorem 11.8. □

3. Incomplete smooth limits at singular times

Now we wish to consider smooth limits where we do not blow up, i.e., do not rescale the metric. In this case the limits that occur can be incomplete, but we have strong control over their ends.

3.1. Assumptions. We shall assume the following about the generalized Ricci flow $(\mathcal{M}, G)$:

ASSUMPTIONS 11.18. (a) The singular times form a discrete subset of $\mathbb{R}$, and each time slice of the flow at a non-singular time is a compact 3-manifold.
  (b) The time interval of definition of the generalized Ricci flow $(\mathcal{M}, G)$ is contained in $[0, \infty)$ and its curvature is pinched toward positive.
  (c) There are $r_0 > 0$ and $C < \infty$, such that any point $x \in \mathcal{M}$ with $R(x) \geq r_0^{-2}$ has a strong (C, ϵ)-canonical neighborhood. In particular, for every $x \in \mathcal{M}$ with $R(x) \geq r_0^{-2}$ the following two inequalities hold:
$$\left|\frac{\partial R(x)}{\partial t}\right| < CR^2(x),$$
$$|\nabla R(x)| < CR^{3/2}(x).$$

With these assumptions we can say quite a bit about the limit metric at time T.

THEOREM 11.19. *Suppose that $(\mathcal{M}, G)$ is a generalized Ricci flow defined for $0 \leq t < T < \infty$ satisfying Assumptions 11.18. Let $T^- < T$ be such that there is a diffeomorphism $\rho\colon M_{T^-} \times [T^-, T) \to \mathbf{t}^{-1}([T^-, T))$ compatible with time and with the vector field. Set $M = M_{T^-}$ and let $g(t)$, $T^- \leq t < T$, be the family of metrics $\rho^* G(t)$ on M. Let $\Omega \subset M$ be the subset defined by*
$$\Omega = \{x \in M | \liminf_{t \to T} R_g(x, t) < \infty\}.$$
Then $\Omega \subset M$ is an open subset and there is a Riemannian metric $g(T)$ with the following properties:
  (1) *As $t \to T$ the metrics $g(t)|_\Omega$ limit to $g(T)$ uniformly in the C^∞-topology on every compact subset of Ω.*
  (2) *The scalar curvature $R(g(T))$ is a proper function from $\Omega \to \mathbb{R}$ and is bounded below.*
  (3) *Let*
$$\widehat{\mathcal{M}} = \mathcal{M} \cup_{\Omega \times [T^-, T)} (\Omega \times [T^-, T]).$$
  Then the generalized Ricci flow $(\mathcal{M}, G)$ extends to a generalized Ricci flow $(\widehat{\mathcal{M}}, \widehat{G})$.

(4) Every end of a connected component of Ω is contained in a strong 2ϵ-tube.
(5) Any $x \in \Omega \times \{T\}$ with $R(x) > r_0^{-2}$ has a strong $(2C, 2\epsilon)$-canonical neighborhood in $\widehat{\mathcal{M}}$.

REMARK 11.20. Recall that by definition a function f is proper if the pre-image under f of every compact set is compact.

In order to prove this result we establish a sequence of lemmas. The first in the series establishes that Ω is an open subset and also establishes the first two of the above five conclusions.

LEMMA 11.21. *Suppose that $(\mathcal{M}, G)$ is a generalized Ricci flow defined for $0 \leq t < T < \infty$ satisfying the three assumptions given in 11.18. Let $T' < T$ be as in the previous theorem, set $M = M_{T^-}$, and let $g(t)$ be the family of metrics on M and let $\Omega \subset M$, each being as defined in the previous theorem. Then $\Omega \subset M$ is an open subset of M. Furthermore, the restriction of the family $g(t)$ to Ω converges in the C^∞-topology, uniformly on compact sets of Ω, to a Riemannian metric $g(T)$. Lastly, $R(g(T))$ is a proper function, bounded below, from Ω to $\mathbb{R}$.*

PROOF. We pull back G to $M \times [T^-, T)$ in order to define a Ricci flow $(M, g(t))$, $T^- \leq t < T$. Suppose that $x \in \Omega$. Then there is a sequence $t_n \to T$ as $n \to \infty$ such that $R(x, t_n)$ is bounded above, independent of n, by say Q. For all n sufficiently large we have $T - t_n \leq \frac{1}{16C(Q^2 + r_0^{-2})}$. Fix such an n. Then, according to Lemma 11.2, there is $r > 0$ such that $R(y, t)$ is uniformly bounded for $y \in B(x, t_n, r) \times [t_n, T)$. This means that $B(x, t_n, r) \subset \Omega$, proving that Ω is open in M.

Furthermore, since $R(y, t)$ is bounded on $B(x, t_n, r) \times [t_n, T)$, it follows from the curvature pinching toward positive hypothesis that $|\text{Rm}(y, t)|$ is bounded on $B(x, t_n, r) \times [t_n, T)$. Now applying Theorem 3.28 we see that in fact Rm is bounded in the C^∞-topology on $B(x, t_n, r) \times [(t_n + T)/2, T)$. The same is of course also true for Ric and hence for $\frac{\partial g}{\partial t}$ in the C^∞-topology. It then follows that there is a continuous extension of g to $B(x, t_n, r) \times [t_n, T]$. Since this is true for every $x \in \Omega$ we see that $g(t)$ converges in the C^∞-topology, uniformly on compact subsets of Ω, to $g(T)$.

Lastly, let us consider the function $R(g(T))$ on Ω. Since the metric $g(T)$ is a smooth metric on $\Omega(T)$, this is a smooth function. Clearly, by the curvature pinching toward positive hypothesis, this function is bounded below. We must show that it is proper. Since M is compact, it suffices to show that if x_n is a sequence in $\Omega \subset M$ converging to a point $x \in M \setminus \Omega$ then $R(x_n, T)$ is unbounded. Suppose that $R(x_n, T)$ is bounded independent of n. It follows from Lemma 11.2 that there is a positive constant Δt such that $R(x_n, t)$ is uniformly bounded for all n and all $t \in [T - \Delta t, T)$, and hence, by the same result, there is $r > 0$ such that $R(y_n, t)$ is bounded for all n, all $y_n \in B(x_n, T - \Delta t, r)$, and all $t \in [T - \Delta t, T)$. Since the $x_n \to x \in M$, it follows that, for all n sufficiently large, $x \in B(x_n, T - \Delta t, r)$, and hence $R(x,t)$ is uniformly bounded as $t \to T$. This contradicts the fact that $x \notin \Omega$. □

DEFINITION 11.22. Let
$$\widehat{\mathcal{M}} = \mathcal{M} \cup_{\Omega \times [T^-,T)} \left(\Omega \times [T^-,T]\right).$$

Since both $\mathcal{M}$ and $\Omega \times [T^-, T]$ have the structure of space-times and the time functions and vector fields agree on the overlap, $\widehat{\mathcal{M}}$ inherits the structure of a space-time. Let $G'(t)$, $T^- \leq t \leq T$, be the smooth family of metrics on Ω. The horizontal metrics, G, on $\mathcal{M}$ and this family of metrics on Ω agree on the overlap and hence define a horizontal metric $\widehat{G}$ on $\widehat{\mathcal{M}}$. Clearly, this metric satisfies the Ricci flow equation, so that $(\widehat{\mathcal{M}}, \widehat{G})$ is a generalized Ricci flow extending $(\mathcal{M}, G)$. We call this the *maximal extension of* $(\mathcal{M}, G)$ to time T. Notice that even though the time-slices M_t of $\mathcal{M}$ are compact, it will not necessarily be the case that the time-slice Ω is complete.

At this point we have established the first three of the five conclusions stated in Theorem 11.19. Let us turn to the last two.

3.2. Canonical neighborhoods for $(\widehat{\mathcal{M}}, \widehat{G})$. We continue with the notation and assumptions of the previous subsection. Here we establish the fifth conclusion in Theorem 11.19, namely the existence of strong canonical neighborhoods for $(\widehat{\mathcal{M}}, \widehat{G})$

LEMMA 11.23. *For any $x \in \Omega \times \{T\}$ with $R(x,T) > r_0^{-2}$ one of the following holds:*

(1) (x,T) *is the center of a strong 2ϵ-neck in $(\widehat{\mathcal{M}}, \widehat{G})$.*
(2) *There is a $(2C, 2\epsilon)$-cap in $(\Omega(T), \widehat{G}(T))$ whose core contains (x,T).*
(3) *There is a $2C$-component of $\Omega(T)$ that contains (x,T).*
(4) *There is a 2ϵ-round component of $\Omega(T)$ that contains (x,T).*

PROOF. We fix $x \in \Omega(T)$ with $R(x,T) > r_0^{-2}$. First notice that for all $t < T$ sufficiently close to T we have $R(x,t) > r_0^{-2}$. Thus, for all such t the point (x,t) has a strong (C, ϵ)-canonical neighborhood in $(\mathcal{M}, G) \subset (\widehat{\mathcal{M}}, \widehat{G})$. Furthermore, since $\lim_{t \to T} R(x,t) = R(x,T) < \infty$, for all $t < T$ sufficiently close to T, there is a constant $D < \infty$ such that for any point y contained in a strong (C, ϵ)-canonical neighborhood containing (x,t), we have $D^{-1}R(x,T) \leq R(y,t) \leq DR(x,T)$. Again assuming that $t < T$ is sufficiently close to T, by Lemma 11.2 there is $D' < \infty$ depending only on D, t, and r_0 such that the curvature $R(y,T)$ satisfies $(D')^{-1}R(x,T) \leq R(y,T) \leq D'R(x,T)$. By Lemma 11.21 this implies that there is a compact subset $K \subset \Omega(T)$ containing all the (C, ϵ)-canonical neighborhoods for (x,t). The same lemma implies that the metrics $G(t)|_K$ converge uniformly in the C^∞-topology to $G(T)|_K$. If there is a sequence of t converging to T for which the canonical neighborhood of (y,t) is an ϵ-round component, resp. a C-component, then (y,T) is contained in a 2ϵ-round, resp. a $2C$-component of $\widehat{\Omega}$. If there is a sequence of t_n converging to T so that each (y,t_n) has a canonical neighborhood $\mathcal{C}_n$ which is a (C,ϵ)-cap whose core contains (y,t_n), then by Proposition 9.79, since these caps are all contained in a fixed compact subset K and since the $G(t_n)|_K$ converge uniformly in the C^∞-topology to $G(T)|_K$, it follows that for any n sufficiently large, the metric $G(T)$ restricted to $\mathcal{C}_n$ contains a $(2C,2\epsilon)$-cap $\mathcal{C}$ whose core contains (y,T).

Now we examine the case of strong ϵ-necks.

CLAIM 11.24. *Fix a point $x \in \Omega$. Suppose that there is a sequence $t_n \to T$ such that for every n, the point (x,t_n) is the center of a strong ϵ-neck in $\widehat{\mathcal{M}}$. Then (x,T) is the center of a strong 2ϵ-neck in $\widehat{\mathcal{M}}$.*

PROOF. By an overall rescaling we can assume that $R(x,T) = 1$. For each n let $N_n \subset \Omega$ and let $\psi_n \colon S^2 \times (-\epsilon^{-1}, \epsilon^{-1}) \to N_n \times \{t\}$ be a strong ϵ-neck centered at (x,t_n). Let $B = B(x,T,2\epsilon^{-1}/3)$. Clearly, for all n sufficiently large $B \subset N_n$. Thus, for each point $y \in B$ and each n there is a flow line through y defined on the interval $(t_n - R(x,t_n)^{-1}, t_n]$. Since the $t_n \to T$ and since $R(x,t_n) \to R(x,T) = 1$ as $n \to \infty$, it follows that there is a flow line through y defined on $(T-1, T]$.

Consider the maps

$$\alpha_n \colon B \times (-1, 0] \to \widehat{\mathcal{M}}$$

that send (y,t) to the value at time $t_n - tR(x,t_n)^{-1}$ of the flow line through y. Pulling back the metric $R(x,t_n)\widehat{G}$ by α_n produces the restriction of a strong ϵ-neck structure to B. The maps α_n converge uniformly in the C^∞-topology to the map $\alpha \colon B \times (-1,0] \to \widehat{M}$ defined by sending (y,t) to the value of the flowline through (y,T) at the time $T - t$. Hence, the sequence of metrics $\alpha_n^*(R(x,t_n))\widehat{G}$ on $B \times (-1,0]$ converges uniformly on compact subsets of $B \times (-1, 0]$ in the C^∞-topology to the family $\alpha^*(\widehat{G})$. Then, for all n sufficiently large, the image $\psi_n(S^2 \times (-\epsilon^{-1}/2, \epsilon^{-1}/2))$ is contained in B and has compact closure in B. Since the family of metrics $\psi_n^*\widehat{G}$ on B converge smoothly to $\psi^*\widehat{G}$, it follows that for every n sufficiently large, the restriction of ψ_n to $S^2 \times (-\epsilon^{-1}/2, \epsilon^{-1}/2)$ gives the coordinates showing that the restriction of the family of metrics $\psi^*(\widehat{G})$ to the image $\psi_n(S^2 \times (-\epsilon^{-1}/2, \epsilon^{-1}/2))$ is a strong 2ϵ-neck at time T. □

This completes the proof of the lemma. □

The lemma tells us that every point $x \in \Omega \times \{T\}$ with $R(x) > r_0^{-2}$ has a strong $(2C, 2\epsilon)$-canonical neighborhood. Since, by assumption, points at time before T with scalar curvature at least r_0^{-2} have strong (C,ϵ)-canonical neighborhoods, this completes the proof of the fifth conclusion of Theorem 11.19. It remains to establish the fourth conclusion of that theorem.

3.3. The ends of $(\Omega, g(T))$.

DEFINITION 11.25. A *strong 2ϵ-horn* in $(\Omega, g(T))$ is a submanifold of Ω diffeomorphic to $S^2 \times [0, 1)$ with the following properties:
   (1) The embedding ψ of $S^2 \times [0, 1)$ into Ω is a proper map.
   (2) Every point of the image of this map is the center of a strong 2ϵ-neck in $(\widehat{\mathcal{M}}, \widehat{G})$.
   (3) The image of the boundary $S^2 \times \{0\}$ is the central sphere of a strong 2ϵ-neck.

DEFINITION 11.26. A *strong double 2ϵ-horn* in $(\Omega, g(T))$ is a component of Ω diffeomorphic to $S^2 \times (0, 1)$ with the property that every point of this component is the center of a strong 2ϵ-neck in $\widehat{\mathcal{M}}$. This means that a strong double 2ϵ-horn is a 2ϵ-tube and hence is a component of Ω diffeomorphic to $S^2 \times (-1, 1)$. Notice that each end of a strong double 2ϵ-horn contains a strong 2ϵ-horn.

For any $C' < \infty$, a *C'-capped 2ϵ-horn* in $(\Omega, g(T))$ is a component of Ω that is a the union of a the core of a $(C', 2\epsilon)$-cap and a strong 2ϵ-horn. Such a component is diffeomorphic to an open 3-ball or to a punctured $\mathbb{R}P^3$.

See FIG. 2.

DEFINITION 11.27. Fix any ρ, $0 < \rho < r_0$. We define $\Omega_\rho \subset \Omega$ to be the closed subset of all $x \in \Omega$ for which $R(x, T) \leq \rho^{-2}$. We say that a strong 2ϵ-horn $\psi \colon S^2 \times [0, 1) \to \Omega$ has *boundary contained in Ω_ρ* if its boundary, $\psi(S^2 \times \{0\})$, is contained in Ω_ρ.

LEMMA 11.28. *Suppose that $0 < \rho < r_0$ and that Ω^0 is a component of Ω which contains no point of Ω_ρ. Then one of the following holds:*
   (1) *Ω^0 is a strong double 2ϵ-horn and is diffeomorphic to $S^2 \times \mathbb{R}$.*
   (2) *Ω^0 is a $2C$-capped 2ϵ-horn and is diffeomorphic to $\mathbb{R}^3$ or to a punctured $\mathbb{R}P^3$.*
   (3) *Ω^0 is a compact component and is the union of the cores of two $(2C, 2\epsilon)$-caps and a strong 2ϵ-tube. It is diffeomorphic to S^3, $\mathbb{R}P^3$ or $\mathbb{R}P^3 \# \mathbb{R}P^3$.*
   (4) *Ω^0 is a compact 2ϵ-round component and is diffeomorphic to a compact manifold of constant positive curvature.*
   (5) *Ω^0 is a compact component that fibers over S^1 with fibers S^2.*
   (6) *Ω^0 is a compact $2C$-component and is diffeomorphic to S^3 or to $\mathbb{R}P^3$.*

See FIG.*3*.

PROOF. Let Ω^0 be a component of Ω containing no point of Ω_ρ. Then for every $x \in \Omega^0$, we have $R(x, T) > r_0^{-2}$. Therefore, by Lemma 11.23 (x, T)

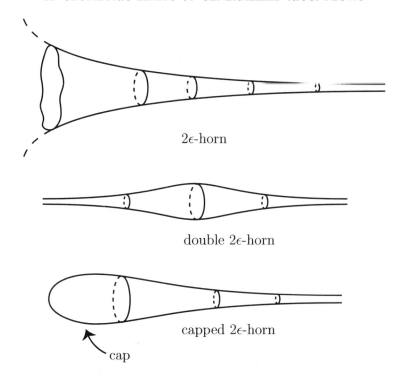

Figure 2. Horns.

has a $(2C, 2\epsilon)$-canonical neighborhood. Of course, this entire neighborhood is contained in $\widehat{\mathcal{M}}$ and hence is contained in Ω^0 (or, more precisely, in the case of strong 2ϵ-necks in the union of maximum backward flow lines ending at points of Ω^0). If the canonical neighborhood of $(x, T) \in \Omega^0$ is a $2C$-component or is a 2ϵ-round component, then of course Ω^0 is that $2C$-component or 2ϵ-round component. Otherwise, each point of Ω^0 is either the center of a strong 2ϵ-neck or contained in the core of a $(2C, 2\epsilon)$-cap. We have chosen 2ϵ sufficiently small so that the result follows from Proposition A.25. □

REMARK 11.29. We do not claim that there are only finitely many such components; in particular, as far as we know there may be infinitely double 2ϵ-horns.

It follows immediately from this lemma that if X is a component of Ω not containing any point of Ω_ρ, then every end of X is contained in a strong 2ϵ-tube. To complete the proof of Theorem 11.19, it remains only to establish the same result for the components of Ω that meet Ω_ρ. That is part of the content of the next lemma.

LEMMA 11.30. *Let $(\mathcal{M}, G)$ be a generalized 3-dimensional Ricci flow defined for $0 \leq t < T < \infty$ satisfying Assumptions 11.18. Fix $0 < \rho < r_0$.*

3. INCOMPLETE SMOOTH LIMITS AT SINGULAR TIMES

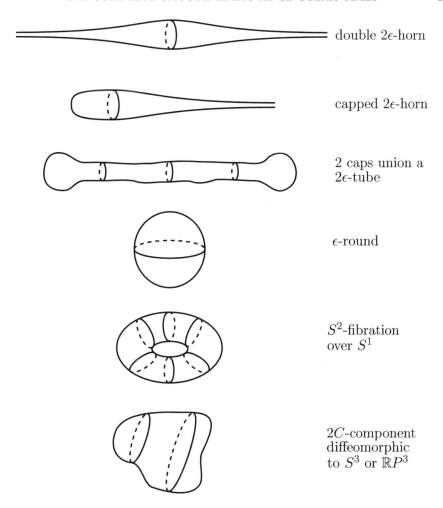

FIGURE 3. Components of Ω disjoint from Ω_ρ.

Let $\Omega^0(\rho)$ be the union of all components of Ω containing points of Ω_ρ. Then $\Omega^0(\rho)$ has finitely many components and is a union of a compact set and finitely many strong 2ϵ-horns each of which is disjoint from Ω_ρ and has its boundary contained in $\Omega_{\rho/2C}$.

PROOF. Since $R\colon \Omega \times \{T\} \to \mathbb{R}$ is a proper function bounded below, Ω_ρ is compact. Hence, there are only finitely many components of Ω containing points of Ω_ρ. Let Ω^0 be a non-compact component of Ω containing a point of Ω_ρ, and let $\mathcal{E}$ be an end of Ω^0. Let

$$X = \{x \in \Omega^0 \big| R(x) \geq 2C^2 \rho^{-2}\}.$$

Then X is a closed set and contains a neighborhood of the end $\mathcal{E}$. Since Ω^0 contains a point of Ω_ρ, $\Omega^0 \setminus X$ is non-empty. Let X_0 be the connected component of X that contains a neighborhood of $\mathcal{E}$. This is a closed, connected set every point of which has a $(2C, 2\epsilon)$-canonical neighborhood. Since X_0 includes an end of Ω^0, no point of X_0 can be contained in an ϵ-round component nor in a C-component. Hence, every point of X_0 is either the center of a strong 2ϵ-neck or is contained in the core of a $(2C, 2\epsilon)$-cap. Since 2ϵ is sufficiently small to invoke Proposition A.21, the latter implies that X_0 is contained either in a 2ϵ-tube which is a union of strong 2ϵ-necks centered at points of X_0 or X_0 is contained in a $2C$-capped 2ϵ-tube where the core of the cap contains a point of X_0. (X_0 cannot be contained in a double capped 2ϵ-tube since the latter is compact.) In the second case, since this capped tube contains an end of Ω^0, it is in fact equal to Ω^0. Since a point of X_0 is contained in the core of the $(2C, 2\epsilon)$-cap, the curvature of this point is at most $2C^2\rho^{-2}$ and hence the curvature at any point of the cap is at least $2C\rho^{-2} > \rho^{-2}$. This implies that the cap is disjoint from Ω_ρ. Of course, any 2ϵ-neck centered at a point of X_0 has curvature at least $C^2\rho^{-2}$ and hence is also disjoint from Ω_ρ. Hence, if Ω^0 is a $2C$-capped 2ϵ-tube and there is a point of X_0 in the core of the cap, then this component is disjoint from Ω_ρ, which is a contradiction. Thus, X_0 is contained in a 2ϵ-tube made up of strong 2ϵ-necks centered at points of X_0.

This proves that X_0 is contained in a strong 2ϵ-tube, Y, every point of which has curvature $\geq C^2\rho^{-2}$. Since X_0 is closed but not the entire component Ω^0, it follows that X_0 has a frontier point y. Of course, $R(y) = 2C^2\rho^{-2}$. Let N be the strong 2ϵ-neck centered at y and let S_N^2 be its central 2-sphere. Clearly, every $y' \in S_N^2$ satisfies $R(y') \leq 4C^2\rho^{-2}$, so that S_N^2 is contained in $\Omega_{\rho/2C}$. Let $Y' \subset Y$ be the complementary component of S_N^2 in Y that contains a neighborhood of the end $\mathcal{E}$. Then the closure of Y' is the required strong 2ϵ-horn containing a neighborhood of $\mathcal{E}$, disjoint from Ω_ρ and with boundary contained in $\Omega_{\rho/2C}$.

The last thing to see is that there are only finitely many such ends in a given component Ω^0. First suppose that the boundary 2-sphere of one of the 2ϵ-horns is homotopically trivial in Ω^0. Then this 2-sphere separates Ω^0 into two components, one of which is compact, and hence Ω^0 has only one boundary component. Thus, we can assume that all the boundary 2-spheres of the 2ϵ-horns are homotopically non-trivial. Suppose that two of these 2ϵ-horns containing different ends of Ω^0 have non-empty intersection. Let N be the 2ϵ-neck whose central 2-sphere is the boundary of one of the 2ϵ-horns. Then the boundary of the other 2ϵ-horn is also contained in N. This means that the union of the two 2ϵ-horns and N is a component of Ω. Clearly, this component has exactly two ends. Thus, we can assume that all the 2ϵ-horns with boundary in $\Omega_{\rho/2C}$ are disjoint. If two of the 2ϵ-horns have boundary components that are topologically parallel in $\Omega^0 \cap \Omega_{\rho/2C}$ (meaning that they are the boundary components of a compact submanifold diffeomorphic to $S^2 \times I$), then Ω^0 is diffeomorphic to $S^2 \times (0, 1)$ and has only two ends. By compactness of $\Omega_{\rho/2C}$, there can only be finitely many disjoint 2ϵ-horns with non-parallel, homotopically non-trivial boundaries in $\Omega^0 \cap \Omega_{\rho/2C}$. This completes the proof of the fact that each component of $\Omega_{\rho/2C}$ has only finitely many ends. □

This completes the proof of Theorem 11.19.

4. Existence of strong δ-necks sufficiently deep in a 2ϵ-horn

We keep the notation and assumptions of the previous section.

THEOREM 11.31. *Fix $\rho > 0$. Then for any $\delta > 0$ there is an $0 < h = h(\delta, \rho) \leq \min(\rho \cdot \delta, \rho/2C)$, implicitly depending on r and (C, ϵ) which are fixed, such that for any generalized Ricci flow $(\mathcal{M}, G)$ defined for $0 \leq t < T < \infty$ satisfying Assumptions 11.18 and for any 2ϵ-horn $\mathcal{H}$ of $(\Omega, g(T))$ with boundary contained in $\Omega_{\rho/2C}$, every point $x \in \mathcal{H}$ with $R(x, T) \geq h^{-2}$ is the center of a strong δ-neck in $(\widehat{\mathcal{M}}, \widehat{G})$ contained in $\mathcal{H}$. Furthermore, there is a point $y \in \mathcal{H}$ with $R(y) = h^{-2}$ with the property that the central 2-sphere of the δ-neck centered at y cuts off an end of the $\mathcal{H}$ disjoint from Ω_ρ. See* FIG. 4.

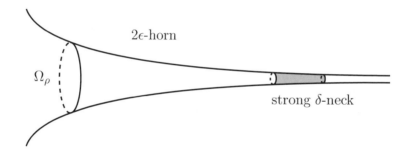

FIGURE 4. δ-necks deep in a 2ϵ-horn.

PROOF. The proof of the first statement is by contradiction. Fix $\rho > 0$ and $\delta > 0$ and suppose that there is no $0 < h \leq \min(\rho \cdot \delta, \rho/2C)$ as required. Then there is a sequence of generalized Ricci flows $(\mathcal{M}_n, G_n)$ defined for $0 \leq t < T_n < \infty$ satisfying Assumptions 11.18 and points $x_n \in \mathcal{M}_n$ with $\mathbf{t}(x_n) = T_n$ contained in 2ϵ-horns $\mathcal{H}_n$ in Ω_n with boundary contained in $(\Omega_n)_{\rho/2C}$ with $Q_n = R(x_n) \to \infty$ as $n \to \infty$ but such that no x_n is the center of a strong δ-neck in $(\mathcal{M}_n, G_n)$. Form the maximal extensions, $(\widehat{\mathcal{M}}_n, \widehat{G}_n)$, to time T of the $(\mathcal{M}_n, G_n)$.

CLAIM 11.32. *The sequence $(\widehat{\mathcal{M}}_n, \widehat{G}_n, x_n)$ satisfies the five hypotheses of Theorem 11.1.*

PROOF. By our assumptions, hypotheses (1) and (3) of Theorem 11.1 hold for this sequence. Also, we are assuming that any point $y \in \mathcal{M}_n$ with

$R(y) \geq r_0^{-2}$ has a strong (C, ϵ)-canonical neighborhood. Since $R(x_n) = Q_n \to \infty$ as $n \to \infty$ this means that for all n sufficiently large, any point $y \in \widehat{\mathcal{M}}_n$ with $R(y) \geq R(x_n)$ has a strong (C, ϵ)-canonical neighborhood. This establishes hypothesis (2) in the statement of Theorem 11.1.

Next, we have:

CLAIM 11.33. *For any $A < \infty$ for all n sufficiently large, the ball $B(x_n, 0, AQ_n^{-1/2})$ is contained in the 2ϵ-horn $\mathcal{H}_n$ and has compact closure in $\mathcal{M}_n$.*

PROOF. Any point $z \in \partial \mathcal{H}_n$ has scalar curvature at most $16C^2\rho^{-2}$ and there is a 2ϵ-neck centered at z. This means that for all y with $d_{G_n}(z, y) < \epsilon^{-1}\rho/2C$ we have $R(y) \leq 32C^2\rho^{-2}$. Hence, for all n sufficiently large, $d_{G_n}(x_n, z) > \epsilon^{-1}\rho/2C$, and thus $d_{Q_n G_n}(x_n, z) > Q_n^{1/2}\epsilon^{-1}\rho/2C$. This implies that, given $A < \infty$, for all n sufficiently large, $z \notin B_{Q_n G_n}(x_n, 0, A)$. Since this is true for all $z \in \partial \mathcal{H}_n$, it follows that for all n sufficiently large $B_{Q_n, G_n}(x_n, 0, A) \subset \mathcal{H}_n$. Next, we must show that, for all n sufficiently large, this ball has compact closure. That is to say, we must show that for every A for all n sufficiently large the distance from x_n to the end of the horn $\mathcal{H}_n$ is greater than $AQ_n^{-1/2}$. If not, then since the curvature at the end of $\mathcal{H}_n$ goes to infinity for each n, this sequence would violate Theorem 10.2. □

Because $B(x_n, 0, AQ_n^{-1/2})$ is contained in a 2ϵ-horn, it is κ-non-collapsed on scales $\leq r$ for a universal $\kappa > 0$ and $r > 0$. Also, because every point in the horn is the center of a strong 2ϵ-neck, for every n sufficiently large and every $y \in B(x_n, 0, AQ_n^{-1/2})$ the flow is defined on an interval through y defined for backward time $R(y)^{-1}$.

This completes the proof that all the hypotheses of Theorem 11.1 hold and establishes Claim 11.32. □

We form a new sequence of generalized Ricci flows from the $(\mathcal{M}_n, G_n)$ by translating by $-T_n$, so that the final time-slice is at $\mathbf{t}_n = 0$, where $\mathbf{t}_n$ is the time function for $\mathcal{M}_n$.

Theorem 11.1 implies that, after passing to a subsequence, there is a limit flow $(M_\infty, g_\infty(t), (x_\infty, 0))$, $t \in [-t_0, 0]$) defined for some $t_0 > 0$ for the sequence $(Q_n \widehat{\mathcal{M}}_n, Q_n \widehat{G}_n, x_n)$. Because of the curvature pinching toward positive assumption, by Theorem 5.33, the limit Ricci flow has non-negative sectional curvature. Of course, $R(x_\infty) = 1$ so that the limit $(M_\infty, g_\infty(0))$ is non-flat.

CLAIM 11.34. *$(M_\infty, g_\infty(0))$ is isometric to the product $(S^2, h) \times (\mathbb{R}, ds^2)$, where h is a metric of non-negative curvature on S^2 and ds^2 is the usual Euclidean metric on the real line.*

PROOF. Because of the fact that the $(\mathcal{M}_n, G_n)$ have curvature pinched toward positive, and since Q_n tend to ∞ as n tends to infinity, it follows that the geometric limit (M_∞, g_∞) has non-negative curvature. In $\mathcal{H}_n$ take a minimizing geodesic ray α_n from x_n to the end of $\mathcal{H}_n$ and a minimizing geodesic β_n from x_n to $\partial \mathcal{H}_n$. As we have seen, the lengths of both α_n and β_n tend to ∞ as $n \to \infty$. By passing to a subsequence, we can assume that the α_n converge to a minimizing geodesic ray α in (M_∞, g_∞) and that the β_n converge to a minimizing geodesic ray β in (M_∞, g_∞). Since, for all n, the union of α_n and β_n forms a piecewise smooth ray in $\mathcal{H}_n$ meeting the central 2-sphere of a 2ϵ-neck centered at x_n in a single point and at this point crossing from one side of this 2-sphere to the other, the union of α and β forms a proper, piecewise smooth map of $\mathbb{R}$ to M_∞ that meets the central 2-sphere of a 2ϵ-neck centered at x_∞ in a single point and crosses from one side to the other at the point. This means that M_∞ has at least two ends. Since (M_∞, g_∞) has non-negative curvature, according to Theorem 2.13, this implies that M_∞ is a product of a surface with $\mathbb{R}$. Since M has non-negative curvature, the surface has non-negative curvature. Since M has positive curvature at least at one point, the surface is diffeomorphic to the 2-sphere. □

According to Theorem 11.1, after passing to a subsequence there is a limit flow defined on some interval of the form $[-t_0, 0]$ for $t_0 > 0$. Suppose that, after passing to a subsequence there is a limit flow defined on $[-T, 0]$ for some $0 < T < \infty$. It follows that for any $t \in [-T, 0]$, the Riemannian manifold $(M_\infty, g_\infty(t))$ is of non-negative curvature and has two ends. Again by Theorem 2.13, this implies that for every $t \in [-T, 0]$ the Riemannian manifold $(M_\infty, g_\infty(t))$ is a Riemannian product of a metric of non-negative curvature on S^2 with $\mathbb{R}$. Thus, by Corollary 4.19 the Ricci flow is a product of a Ricci flow $(S^2, h(t))$ with the trivial flow on $(\mathbb{R}, ds^2)$. It now follows from Corollary 4.14 that for every $t \in (-T, 0]$ the curvature of $g_\infty(t)$ on S^2 is positive.

Let M_n be the zero time-slice of $\mathcal{M}_n$. Since $(M_\infty, g_\infty, x_\infty)$ is the geometric limit of the $(M_n, Q_n G_n(0), x_n)$, there is an exhaustive sequence $x_\infty \in V_1 \subset V_2 \subset \cdots$ of open subsets of M_∞ with compact closure and embeddings $\varphi_n \colon V_n \to M_n$ sending x_∞ to x_n such that $\varphi_n^*(Q_n G_n(0))$ converges in the C^∞-topology, uniformly on compact sets, to g_∞.

CLAIM 11.35. *For any $z \in M_\infty$ for all n sufficiently large, $z \in V_n$, so that $\varphi_n(z)$ is defined. Furthermore, for all n sufficiently large, there is a backward flow line through $\varphi_n(z)$ in the generalized Ricci flow $(Q_n \mathcal{M}_n, Q_n G_n)$ defined on the interval $(-T - (R^{-1}_{Q_n G_n}(\varphi_n(z), 0)/2), 0]$. The scalar curvature is bounded above on this entire flow line by $R(\varphi_n(z), 0)$.*

PROOF. Of course, for any compact subset $K \subset M_\infty$ and any $t' < T$ for all n sufficiently large, $K \subset V_n$, and there is an embedding $\varphi_n(K) \times [-t', 0] \subset Q_n \mathcal{M}_n$ compatible with time and the vector field. The map φ_n defines a map $Q_n^{-1} \varphi_n \colon K \times [-Q_n^{-1} t', 0] \to \mathcal{M}_n$. Since the scalar curvature of the limit is positive, and hence bounded away from zero on the compact set $K \times [-t', 0]$ and since $Q_n \to \infty$ as n tends to infinity, the following is true: For any compact subset $K \subset M$ and any $t' < T$, for all n sufficiently large, the scalar curvature of G_n on the image $Q_n^{-1}\varphi_n(K) \times [-Q_n^{-1}t', 0]$ is greater than r_0^{-2}, and hence for all n sufficiently large, every point in $Q_n^{-1}\varphi_n(K) \times [-Q_n^{-1}t', 0]$ has a strong (C, ϵ)-canonical neighborhood in $\mathcal{M}_n$. Since having a strong (C, ϵ)-canonical neighborhood is invariant under rescaling, it follows that for all n sufficiently large, every point of $\varphi_n(K) \times [-t', 0]$ has a strong (C, ϵ)-canonical neighborhood.

Next we claim that, for all n sufficiently large and for any $t \in [-t', 0]$, the point $(\varphi_n(z), t)$ is the center of a strong ϵ-neck. We have already seen that for all n sufficiently large $(\varphi_n(z), t)$ has a strong (C, ϵ)-canonical neighborhood. Of course, since M_∞ is non-compact, for n sufficiently large, the canonical neighborhood of $(\varphi_n(z), t)$ must either be a (C, ϵ)-cap or a strong ϵ-neck. We shall rule out the possibility of a (C, ϵ)-cap, at least for all n sufficiently large.

To do this, take K to be a neighborhood of $(z, 0)$ in the limit $(M_\infty, g_\infty(0))$ with the topology of $S^2 \times I$ and with the metric being the product of a positively curved metric on S^2 with the Euclidean metric on I. We take K to be sufficiently large to contain the $2C$-ball centered at $(z, 0)$. Because the limit flow is the product of a positively curved flow on S^2 with the trivial flow on $\mathbb{R}$, the flow is distance decreasing. Thus, for every $t \in [-t', 0]$ the submanifold $K \times \{t\}$ contains the ball in $(M_\infty, g_\infty(t))$ centered at (z, t) of radius $2C$. For every n sufficiently large, consider the submanifolds $\varphi_n(K) \times \{t\}$ of $(M_n, Q_nG_n(t))$. Since the metrics $\varphi_n^*Q_nG_n(t)$ are converging uniformly for all $t \in [-t', 0]$ to the product flow on K, for all n sufficiently large and any $t \in [-t', 0]$, this submanifold contains the C-ball centered at $(\varphi_n(z), t)$ in $(M_n, Q_nG_n(t))$. Furthermore, the maximal curvature two-plane at any point of $\varphi_n(K) \times \{t\}$ is almost tangent to the S^2-direction of K. Hence, by Lemma A.2 the central 2-sphere of any ϵ-neck contained $\varphi_n(K) \times \{t\}$ is almost parallel to the S^2-factors in the product structure on K at every point. This implies that the central 2-sphere of any such ϵ-neck is isotopic to the S^2-factor of $\varphi_n(K) \times \{t\}$. Suppose that $(\varphi_n(z), t)$ is contained in the core of a (C, ϵ)-cap $\mathcal{C}$. Then $\mathcal{C}$ is contained in $\varphi_n(K) \times \{t\}$. Consider the ϵ-neck $N \subset \mathcal{C}$ that is the complement of the core of $\mathcal{C}$. Its central 2-sphere, Σ, is isotopic in K to the 2-sphere factor of K, but this is absurd since Σ bounds a 3-ball in the $\mathcal{C}$. This contradiction shows that for all n sufficiently large and all $t \in [-t', 0]$, it is not possible for $(\varphi_n(z), t)$ to be contained in the core of a (C, ϵ)-cap. The only other possibility is that for all n sufficiently large and all $t \in [-t', 0]$ the point $(\varphi_n(z), t)$ is the center of a strong ϵ-neck in $(M_n, Q_nG_n(t))$.

Fix n sufficiently large. Since, for all $t \in [-t', 0]$, the point $(\varphi_n(z), t)$ is the center of a strong ϵ-neck, it follows from Definition 9.78 that for all

4. EXISTENCE OF STRONG δ-NECKS SUFFICIENTLY DEEP IN A 2ϵ-HORN

$t \in [-t', 0]$ we have $R(\varphi_n(z), t) \leq R(\varphi_n(z), 0)$ (this follows from the fact that the partial derivative in the time-direction of the scalar curvature of a strong ϵ-neck of scale one is positive and bounded away from 0). It also follows from Definition 9.78 that the flow near $(\varphi_n(z), -t')$ extends backwards to time

$$-t' - R_{Q_n G_n}^{-1}(\varphi_n(z, t')) < -t' - R_{Q_n G_n}^{-1}(\varphi_n(z, 0)),$$

with the same inequality for scalar curvature holding for all t in this extended interval. Applying this for $t' < T$ but sufficiently close to T establishes the last statement in the claim, and completes the proof of the claim. □

Let Q_0 be the upper bound of the scalar curvature of $(M_\infty, g_\infty(0))$. By the previous claim, Q_0 is also an upper bound for the curvature of $(M_\infty, g_\infty(-t'))$ for any $t' < T$. Applying Theorem 11.1 to the flows

$$(M_n, Q_n G_n(t)), \quad -t' - Q_0^{-1}/2 < t \leq -t',$$

we conclude that there is t_0 depending only on the bound of the scalar curvature of $(M_\infty, g_\infty(-t'))$, and hence depending only on Q_0, such that, after passing to a subsequence the limit flow exists for $t \in [-t' - t_0, -t']$. Since the limit flow already exists on $[-t', 0]$, we conclude that, for this further subsequence, the limit flow exists on $[-t' - t_0, 0]$. Now apply this with $t' = T - t_0/2$. This proves that if, after passing to a subsequence, there is a limit flow defined on $[-T, 0]$, then, after passing to a further subsequence there is a limit flow defined on $[-T - t_0/2, 0]$ where t_0 depends only on Q_0, and in particular, is independent of T. Repeating this argument with $T + (t_0/2)$ replacing T, we pass to a further subsequence so that the limit flow is defined on $[-T - t_0, 0]$. Repeating this inductively, we can find a sequence of subsequences so that for the n subsequence the limit flow is defined on $[-T - nt_0, 0]$. Taking a diagonal subsequence produces a subsequence for which the limit is defined on $(-\infty, 0]$.

The limit flow is the product of a flow on S^2 of positive curvature defined for $t \in (-\infty, 0]$ and the trivial flow on $\mathbb{R}$. Now, invoking Hamilton's result (Corollary 9.50), we see that the ancient solution of positive curvature on S^2 must be a shrinking round S^2. This means that the limit flow is the product of the shrinking round S^2 with $\mathbb{R}$, and implies that for all n sufficiently large there is a strong δ-neck centered at x_n. This contradiction proves the existence of h as required.

Now let us establish the last statement in Theorem 11.31. The subset of $\mathcal{H}$ consisting of all $z \in \mathcal{H}$ with $R(z) \leq \rho^{-2}$ is compact (since R is a proper function), and disjoint from any δ-neck of scale h since $h < \rho/2C$. On the other hand, for any point $z \in \mathcal{H}$ with $R(z) \leq \rho^{-2}$ take a minimal geodesic from z to the end of $\mathcal{H}$. There must be a point y on this geodesic with $R(y) = h^{-2}$. The δ-neck centered at y is disjoint from z (since $h < \rho/2C$) and hence this neck separates z from the end of $\mathcal{H}$. It now follows easily that there is a point $y \in \mathcal{H}$ with $R(y) = h^{-2}$ and such that the central 2-sphere of the δ-neck centered at y divides $\mathcal{H}$ into two pieces with the non-compact piece disjoint from Ω_ρ. □

COROLLARY 11.36. *We can take the function $h(\rho, \delta)$ in the last lemma to be $\leq \delta\rho$, to be a weakly monotone non-decreasing function of δ when ρ is fixed, and to be a weakly monotone non-decreasing function of ρ when δ is held fixed.*

PROOF. If h satisfies the conclusion of Theorem 11.31 for ρ and δ and if $\rho' \geq \rho$ and $\delta' \geq \delta$ then h also satisfies the conclusion of Theorem 11.31 for ρ' and δ'. Also, any $h' \leq h$ also satisfies the conclusion of Theorem 11.31 for δ and ρ. Take a sequence (δ_n, ρ_n) where each of the sequences $\{\delta_n\}$ and $\{\rho_n\}$ is a monotone decreasing sequence with limit 0. Then we choose $h_n = h(\rho_n, \delta_n) \leq \rho_n \delta_n$ as in the statement of Theorem 11.31. We of course can assume that $\{h_n\}_n$ is a non-increasing sequence of positive numbers with limit 0. Then for any (ρ, δ) we take the largest n such that $\rho \geq \rho_n$ and $\delta \geq \delta_n$, and we define $h(\rho, \delta)$ to be h_n for this value of n. This constructs the function $h(\delta, \rho)$ as claimed in the corollary. □

CHAPTER 12

The standard solution

The process of surgery involves making a choice of the metric on a 3-ball to 'glue in'. In order to match approximatively with the metric coming from the flow, the metric we glue in must be asymptotic to the product of a round 2-sphere and an interval near the boundary. There is no natural choice for this metric; yet it is crucial to the argument that we choose an initial metric so that the Ricci flow with these initial conditions has several properties. In this chapter we shall develop the needed conditions on the initial metric and the Ricci flow.

1. The initial metric

Conditions on the initial metric that ensure the required properties for the subsequence flow are contained in the following definition.

DEFINITION 12.1. A *standard initial metric* is a metric g_0 on $\mathbb{R}^3$ with the following properties:
  (i) g_0 is a complete metric.
  (ii) g_0 has non-negative sectional curvature at every point.
  (iii) g_0 is invariant under the usual $SO(3)$-action on $\mathbb{R}^3$.
  (iv) there is a compact ball $B \subset \mathbb{R}^3$ so that the restriction of the metric g_0 to the complement of this ball is isometric to the product $(S^2, h) \times (\mathbb{R}^+, ds^2)$ where h is the round metric of scalar curvature 1 on S^2.
  (v) g_0 has constant sectional curvature $1/4$ near the origin. (This point will be denoted p and is called the *tip* of the initial metric.)
See FIG. 1.

Actually, one can work with an alternative weaker version of the fourth condition, namely:
(iv ') g_0 is asymptotic at infinity in the C^∞-topology to the product of the round metric h_0 on S^2 of scalar curvature 1 with the usual metric ds^2 on the real line. By this we mean that if $x_n \in \mathbb{R}^3$ is any sequence converging to infinity, then the based Riemannian manifolds $(\mathbb{R}^3, g_0, x_n)$ converge smoothly to $(S^2, h_0) \times (\mathbb{R}, ds^2)$. But we shall only use standard initial metrics as given in Definition 12.1.

LEMMA 12.2. *There is a standard initial metric.*

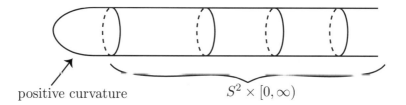

FIGURE 1. A standard initial metric.

PROOF. We construct our Riemannian manifold as follows. Fix Euclidean coordinates (x_0, x_1, x_2, x_3) on $\mathbb{R}^4$. Let $y = f(s)$ be a function defined for $s \geq 0$ and satisfying:

(1) f is C^∞ on $(0, \infty)$.
(2) $f(s) > 0$ for all $s > 0$.
(3) $f''(s) \leq 0$ for all $s > 0$.
(4) There is $s_1 > 0$ such that $f(s) = \sqrt{2}$ for all $s \geq s_1$.
(5) There is $s_0 > 0$ such that $f(s) = \sqrt{4s - s^2}$ for all $s \in [0, s_0]$.

Given such a function f, consider the graph

$$\Gamma = \{(x_0, x_1) \big| x_0 \geq 0 \text{ and } x_1 = f(x_0)\}$$

in the (x_0, x_1)-plane. We define $\Sigma(f)$ by rotating Γ about the x_0-axis in four-space:

$$\Sigma(f) = \{(x_0, x_1, x_2, x_3) \big| x_0 \geq 0 \text{ and } x_1^2 + x_2^2 + x_3^2 = f(x_0)^2\}.$$

Because of the last condition on f, there is a neighborhood of $0 \in \Sigma(f)$ that is isometric to a neighborhood of the north pole in the 3-sphere of radius 2. Because of this and the first item, we see that $\Sigma(f)$ is a smooth submanifold of $\mathbb{R}^4$. Hence, it inherits a Riemannian metric g_0. Because of the fourth item, a neighborhood of infinity of $(\Sigma(f), g_0)$ is isometric to $(S^2, h) \times (0, \infty)$, and in particular, $(\Sigma(f), g_0)$ is complete. Clearly, the rotation action of $S0(3)$ on $\Sigma(f)$, induced by the orthogonal action on the last three coordinates in $\mathbb{R}^4$, is an isometric action with the origin as the only fixed point. It is also clear that $\Sigma(f)$ is diffeomorphic to $\mathbb{R}^3$ by a diffeomorphism that send the $SO(3)$ action to the standard one on $\mathbb{R}^3$.

It remains to compute the sectional curvatures of g_0. Let $q \in \Sigma(f)$ be a point distinct from the fixed point of the $SO(3)$-action. Direct computation shows that the tangent plane to the 2-dimensional $SO(3)$-orbit through q is a principal direction for the curvature, and the sectional curvature on this tangent two-plane is given by

$$\frac{1}{f(q)^2(1 + f'(q)^2)}.$$

On the subspace in $\wedge^2 T_q \Sigma(f)$ perpendicular to the line given by this two-plane, the curvature is constant with eigenvalue
$$\frac{-f''(q)}{f(q)(1+f'(q)^2)^2}.$$
Under our assumptions about f, it is clear that $\Sigma(f)$ has non-negative curvature and has constant sectional curvature $1/4$ near the origin. It remains to choose the function f satisfying items (1) – (5) above.

Consider the function $h(s) = (2-s)/\sqrt{4s-s^2}$. This function is integrable from 0 and the definite integral from zero to s is equal to $\sqrt{4s-s^2}$. Let $\lambda(s)$ be a non-increasing C^∞-function defined on $[0, 1/2]$, with λ identically one near 0 and identically equal to 0 near $1/2$. We extend λ to be identically 1 for $s < 0$ and identically 0 for $s > 1/2$. Clearly,
$$\int_0^2 h(s)\lambda(s-3/2)ds > \int_0^{3/2} h(s)ds > \sqrt{2}$$
and
$$\int_0^2 h(s)\lambda(s)ds < \int_0^{1/2} h(s) < \sqrt{2}.$$
Hence, for some $s_0 \in (1/2, 3/2)$ we have
$$\int_0^2 h(s)\lambda(s-s_0)ds = \sqrt{2}.$$
We define
$$f(s) = \int_0^s h(\sigma)\lambda(\sigma - s_0)d\sigma.$$
It is easy to see that f satisfies all the above conditions. $\square$

The following lemma is clear from the construction.

LEMMA 12.3. *There is $A_0 < \infty$ such that*
$$(\mathbb{R}^3 \setminus B(0, A_0), g(0))$$
is isometric to the product of a round metric on S^2 of scalar curvature 1 with the Euclidean metric on $[0, \infty)$. There is a constant $K < \infty$ such that the volume of $B_{g(0)}(0, A_0)$ is at most K. Furthermore, there is a constant $D < \infty$ so that the scalar curvature of standard initial metric $(\mathbb{R}^3, g(0))$ is bounded above by D and below by D^{-1}.

2. Standard Ricci flows: The statement

Fix once and for all a standard initial metric g_0 on $\mathbb{R}^3$.

DEFINITION 12.4. A *partial standard Ricci flow* is a Ricci flow $(\mathbb{R}^3, g(t))$ defined for $0 \le t < T$, such that $g(0) = g_0$ and such that the curvature is locally bounded in time. We say that a partial standard Ricci flow is a *standard Ricci flow* if it has the property that T is maximal in the sense that there is no extension of the flow to a flow with curvature locally bounded in time defined on an time interval $[0, T')$ with $T' > T$.

Here is the main result of this chapter.

THEOREM 12.5. *There is a standard Ricci flow defined for some positive amount of time. Let $(\mathbb{R}^3, g(t))$, $0 \leq t < T$, be a standard Ricci flow. Then the following hold.*
  (1) **(Uniqueness):** *If $(\mathbb{R}^3, g'(t))$, $0 \leq t < T'$, is a standard Ricci flow, then $T' = T$ and $g'(t) = g(t)$.*
  (2) **(Time Interval):** $T = 1$.
  (3) **(Positive curvature):** *For each $t \in (0, 1)$ the metric $g(t)$ on $\mathbb{R}^3$ is complete of strictly positive curvature.*
  (4) **($SO(3)$-invariance):** *For each $t \in [0, 1)$ the metric $g(t)$ is invariant under the $SO(3)$-action on $\mathbb{R}^3$.*
  (5) **(Asymptotics at ∞):** *For any $t_0 < 1$ and any $\epsilon > 0$ there is a compact subset X of $\mathbb{R}^3$ such that for any $x \in \mathbb{R}^3 \setminus X$ the restriction of the standard flow to an appropriate neighborhood of x for time $t \in [0, t_0]$ is within ϵ in the $C^{[1/\epsilon]}$-topology of the product Ricci flow $(S^2 \times (-\epsilon^{-1}, \epsilon^{-1})), h(t) \times ds^2$, $0 \leq t \leq t_0$, where $h(t)$ is the round metric with scalar curvature $1/(1-t)$ on S^2.*
  (6) **(Non-collapsing):** *There are $r > 0$ and $\kappa > 0$ such that $(\mathbb{R}^3, g(t))$ is κ-non-collapsed on scales less than r for all $0 \leq t < 1$.*

The proof of this result occupies the next few subsections. All the properties except the uniqueness are fairly straightforward to prove. We establish uniqueness by reducing the Ricci flow to the Ricci-DeTurck by establishing the existence of a solution to the harmonic map flow in this case. This technique can be made to work more generally in the case of complete manifolds of bounded curvature, see [**12**], but we preferred to give the more elementary argument that covers this case, where the symmetries allow us to reduce the existence problem for the harmonic map flow to a problem that is the essentially one-dimensional. Also, in the rest of the argument one does not need uniqueness, only a compactness result for the space of all Ricci flows of bounded curvature on each time-slice with the given initial conditions. Kleiner and Lott pointed out to us that this uniqueness can be easily derived from the other properties by arguments similar to those used to establish the compactness of the space of κ-solutions.

3. Existence of a standard flow

For any $R < \infty$, denote by $B_R \subset \mathbb{R}^3$, the ball of radius R about the origin in the metric g_0. For $R \geq A_0 + 1$, a neighborhood of the boundary of this ball is isometric to $(S^2, h) \times ([0, 1], ds^2)$. Thus, in this case, we can double the ball, gluing the boundary to itself by the identity, forming a manifold we denote by S_R^3. The doubled metric will be a smooth Riemannian metric

g_R on S_R^3. Let $p \in S_R^3$ be the image of the origin in the first copy of B_R. Now take a sequence, R_n, tending to infinity to construct based Riemannian manifolds $(S_{R_n}^3, g_{R_n}, p)$ that converge geometrically to $(\mathbb{R}^3, g_0, p)$. For each n, let $(S_{R_n}^3, g_{R_n}(t))$, $0 \le t < T_n$ be maximal Ricci flow with $(S_{R_n}^3, g_{R_n})$ as initial metric. The maximum principle applied to Equation (3.7), $\partial R/\partial t = \triangle R + |\mathrm{Ric}|^2$, then implies by Proposition 2.23 that the maximum of R at time t, $R_{\max}(t)$ obeys the inequality $\partial R_{\max}/\partial t \le R_{\max}(t)^2$, and integrating this inequality (i.e., invoking Lemma 2.22) one finds positive constants t_0 and Q_0 such that for each n, the norm of the scalar curvature of $g_{R_n}(t)$ are bounded by Q_0 on the interval $[0, \max(t_0, T_n))$. By Corollary 4.14, for each n the sectional curvature of the flow $(S_{R_n}^3, g_{R_n}(t))$, $0 \le t < T_n$ is non-negative, and hence the sectional curvature of this flow is also bounded by Q_0 on $[0, \max(t_0, T_n))$. It now follows from Proposition 4.12 and the fact that the T_n are maximal that $T_n > t_0$ for all n. Since the Riemann curvatures of the $(S_{R_n}^3, g_{R_n}(t))$, $0 \le t < t_0$, are bounded independent of n, and since the $(S_{R_n}^3, g_{R_n}, p)$ converge geometrically to $(\mathbb{R}^3, g_0, p)$, it follows from Theorem 5.15 that there is a geometric limiting flow defined on $[0, t_0)$. Since this flow is the geometric limit of flows of uniformly bounded curvature, it has uniformly bounded curvature. Taking a maximal extension of this flow to one of locally bounded curvature gives a standard flow.

4. Completeness, positive curvature, and asymptotic behavior

Let $(\mathbb{R}^3, g(t))$, $t \in [0, T)$, be a partial standard solution. Let $y_i \to \infty$ be the sequence of points in $\mathbb{R}^3$ converging to infinity. From the definition we see that the based Riemannian manifolds $(\mathbb{R}^3, g_0, y_i)$ converge smoothly to $(S^2 \times \mathbb{R}, h(0) \times ds^2)$ where $h(0)$ is the round metric of scalar curvature 1 on S^2.

Let us begin by proving the third item in the statement of Theorem 12.5:

LEMMA 12.6. *For each $t_0 \in [0, T)$ the Riemannian manifold $(\mathbb{R}^3, g(t_0))$ is complete and of positive curvature.*

PROOF. Fix $t_0 \in [0, T)$. By hypothesis $(\mathbb{R}^3, g(t))$, $0 \le t \le t_0$ has bounded curvature. Hence, there is a constant $C < \infty$ such that $g(0) \le Cg(t_0)$, so that for any points $x, y \in \mathbb{R}^3$, we have $d_0(x, y) \le \sqrt{C} d_{t_0}(x, y)$. Since $g(0)$ is complete, this implies that $g(t_0)$ is also complete.

Now let us show that $(M, g(t_0))$ has non-negative curvature. Here, the argument is the analogue of the proof of Corollary 4.14 with one additional step, the use of a function φ to localize the argument. Suppose this is false, i.e., suppose that there is $x \in M$ with $\mathrm{Rm}(x, t_0)$ having an eigenvalue less than zero. Since the restriction of the flow to $[0, t_0]$ is complete and of bounded curvature, according to [**33**] for any constants $C < \infty$ and $\eta > 0$ and any compact subset $K \subset M \times [0, t_0]$ there is $\epsilon > 0$ and a function $\varphi \colon M \times [0, t_0] \to \mathbb{R}$ with the following properties:

(1) $\varphi|_K \leq \eta$.
(2) $\varphi \geq \epsilon$ everywhere.
(3) For each $t \in [0, t_0]$ the restriction of φ to $M \times \{t\}$ goes to infinity at infinity in the sense that for any $\Lambda < \infty$ the pre image $\varphi^{-1}([0, \Lambda] \cap (M \times \{t\})$ is compact.
(4) On all of $M \times [0, t_0]$ we have $\left(\frac{\partial}{\partial t} - \triangle\right) \varphi \geq C\varphi$.

Recall from Section 5 of Chapter 3 is the curvature tensor written with respect to an evolving orthonormal frame $\{F_\alpha\}$ for the tangent bundle. Consider the symmetric, horizontal two-tensor $\widehat{\mathcal{T}} = \mathcal{T} + \varphi g$. Let $\hat{\mu}(x, t)$ denote the smallest eigenvalue of this symmetric two-tensor at (x, t). Clearly, since the curvature is bounded, it follows from the third property of φ that for each $t \in [0, t_0]$ the restriction of $\hat{\mu}$ to $M \times \{t\}$ goes to infinity at infinity in M. In particular, the subset of $(x, t) \in M \times [0, t_0]$ with the property that $\hat{\mu}(x, t) \leq \hat{\mu}(y, t)$ for all $y \in M$ is a compact subset of $M \times [0, t_0]$. It follows from Proposition 2.23 that $f(t) = \min_{x \in M} \hat{\mu}(x, t)$ is a continuous function of t. Choosing $\eta > 0$ sufficiently small and K to include (x, t_0), then $\widehat{\mathcal{T}}$ will have a negative eigenvalue at (x, t_0). Clearly, it has only positive eigenvalues on $M \times \{0\}$. Thus, there is $0 < t_1 < t_0$ so that $\widehat{\mathcal{T}}$ has only positive eigenvalues on $M \times [0, t_1)$ but has a zero eigenvalue at (y, t_1) for some $y \in M$. That is to say, $\mathcal{T} \geq -\varphi g$ on $M \times [0, t_1]$. Diagonalizing $\mathcal{T}$ at any point (x, t) with $t \leq t_1$, all its eigenvalues are at least $-\varphi(x, t_1)$. It follows immediately that on $M \times [0, t_1]$ the smallest eigenvalue of the symmetric form $\mathcal{T}^2 + \mathcal{T}^\#$ is bounded below by 2φ. Thus, choosing $C \geq 4$ we see that for $t \leq t_1$ every eigenvalue of $\mathcal{T}^2 + \mathcal{T}^\#$ is at least $-C\varphi/2$.

We compute the evolution equation using the formula in Lemma 4.13 for the evolution of $\mathcal{T}$ in an evolving orthonormal frame:

$$\frac{\partial \widehat{\mathcal{T}}}{\partial t} = \frac{\partial \mathcal{T}}{\partial t} + \frac{\partial \varphi}{\partial t} g - 2\varphi \text{Ric}(g)$$

$$= \triangle \mathcal{T} + \mathcal{T}^2 + \mathcal{T}^\# + \frac{\partial \varphi}{\partial t} g - 2\varphi \text{Ric}(g)$$

$$= \triangle \widehat{\mathcal{T}} + \mathcal{T}^2 + \mathcal{T}^\# + \left(\frac{\partial \varphi}{\partial t} - \triangle \varphi\right) g - 2\varphi \text{Ric}(g)$$

$$\geq \triangle \widehat{\mathcal{T}} + \mathcal{T}^2 + \mathcal{T}^\# + \varphi \left(Cg - 2\text{Ric}(g)\right).$$

Since every eigenvalue of $\mathcal{T}^2 + \mathcal{T}^\#$ on $M \times [0, t_1]$ is at least $-C\varphi/2$, it follows that on $M \times [0, t_1]$,

$$\frac{\partial \widehat{\mathcal{T}}}{\partial t} \geq \triangle \widehat{\mathcal{T}} + \varphi(Cg/2 - 2\text{Ric}(g)).$$

Once again assuming that C is sufficiently large, we see that for any $t \leq t_1$,

$$\frac{\partial \widehat{\mathcal{T}}}{\partial t} \geq \triangle \widehat{\mathcal{T}}.$$

Thus, at any local minimum $x \in M$ for $\hat{\mu}(\cdot, t)$, we have

$$\frac{\partial \hat{\mu}}{\partial t} \geq 0.$$

This immediately implies by Proposition 2.23 that $\psi(t) = \min_{x \in M} \hat{\mu}(x, t)$ is a non-decreasing function of t. Since its value at $t = 0$ is at least $\epsilon > 0$ and its value at t_1 is zero, this is a contradiction. This establishes that the solution has non-negative curvature everywhere. Indeed, by Corollary 4.20 it has strictly positive curvature for every $t > 0$. $\square$

Now let us turn to the asymptotic behavior of the flow.

Fix $T' < T$. Let y_k be a sequence tending to infinity in $(\mathbb{R}^3, g_0)$. Fix $R < \infty$. Then there is $k_0(R)$ such that for all $k \geq k_0(R)$ there is an isometric embedding $\psi_k \colon (S^2, h) \times (-R, R) \to (\mathbb{R}^3, g_0)$ sending $(x, 0)$ to y_k. These maps realize the product $(S^2, h) \times (\mathbb{R}, ds^2)$ as the geometric limit of the $(\mathbb{R}^3, g_0, y_i)$. Furthermore, for each $R < \infty$ there is a uniform C^∞ point-wise bound to the curvatures of g_0 restricted to the images of the ψ_k for $k \geq k_0(R)$. Since the flow $g(t)$ has bounded curvature on $\mathbb{R}^3 \times [0, T']$, it follows from Theorem 3.29 that there are uniform C^∞ point-wise bounds for the curvatures of $g(t)$ restricted to $\psi_k(S^2 \times (-R, R))$. Thus, by Theorem 5.15, after passing to a subsequence, the flows $\psi_k^* g(t)$ converge to a limiting flow on $S^2 \times \mathbb{R}$. Of course, since the curvature of $g(t)$ is everywhere ≥ 0, the same is true of this limiting flow. Since the time-slices of this flow have two ends, it follows from Theorem 2.13 that every manifold in the flow is a product of a compact surface with $\mathbb{R}$. According to Corollary 4.20 this implies that the flow is the product $(S^2, h(t)) \times (\mathbb{R}, ds^2)$. This means that given $\epsilon > 0$, for all k sufficiently large, the restriction of the flow to the cylinder of length $2R$ centered at y_k is within ϵ in the $C^{[1/\epsilon]}$-topology of the shrinking cylindrical flow on time $[0, T']$. Given $\epsilon > 0$ and $R < \infty$ this statement is true for all y outside a compact ball B centered at the origin.

We have now established the following:

PROPOSITION 12.7. *Given $T' < T$ and $\epsilon > 0$ there is a compact ball B centered at the origin of $\mathbb{R}^3$ with the property that the restriction of the flow $(\mathbb{R}^3 \setminus B, g(t))$, $0 \leq t \leq T'$, is within ϵ in the $C^{[1/\epsilon]}$-topology of the standard evolving cylinder $(S^2, h(t)) \times (\mathbb{R}^+, ds^2)$.*

COROLLARY 12.8. *The maximal time T is ≤ 1.*

PROOF. If $T > 1$, then we can apply the above result to T' with $1 < T' < T$, and see that the solution at infinity is asymptotic to the evolving cylinder $(S^2, h(t)) \times (\mathbb{R}, ds^2)$ on the time interval $[0, T']$. But this is absurd since this evolving cylindrical flow becomes completely singular at time $T = 1$. $\square$

5. Standard solutions are rotationally symmetric

Next, we consider the fourth item in the statement of the theorem. Of course, rotational symmetry would follow immediately from uniqueness. But here we shall use the rotational symmetry to reduce the uniqueness problem to a one-dimensional problem which we then solve. One can also use the general uniqueness theorem for complete, non-compact manifolds due to Chen and Zhu ([**12**]), but we have chosen to present a more elementary, self-contained argument in this special case which we hope will be more accessible.

Let Ric_{ij} be the Ricci tensor and $\text{Ric}_k^i = g^{ij}\text{Ric}_{jk}$ be the dual tensor. Let X be a vector field evolving by

$$\frac{\partial}{\partial t} X = \triangle X + \text{Ric}(X, \cdot)^*. \tag{12.1}$$

In local coordinates $(x^1, \ldots, x^n)$, if $X = X^i \partial_i$, then the equation becomes

$$\frac{\partial}{\partial t} X^i = (\triangle X)^i + \text{Ric}_k^i X^k. \tag{12.2}$$

Let X^* denote the dual one-form to X. In local coordinates we have $X^* = X_i^* dx^i$ with $X_i^* = g_{ij} X^j$. Since the evolution equation for the metric is the Ricci flow, the evolution equation for X^* is

$$\frac{\partial X^*}{\partial t} = \triangle X^* - \text{Ric}(X, \cdot),$$

or in local coordinates

$$\frac{\partial X_i^*}{\partial t} = (\triangle X^*)_i - \text{Ric}_{ij} X^j.$$

LEMMA 12.9. *With X and its dual X^* evolving by the above equations, set $V = \nabla X^*$, so that V is a contravariant two-tensor. In local coordinates we have $V = V_{ij} dx^i \otimes dx^j$ with*

$$V_{ij} = (\nabla_i X)_j = g_{jk}(\nabla_i X)^k.$$

This symmetric two-tensor satisfies

$$\frac{\partial}{\partial t} V = \triangle V - \left(2 R_k{}^{rl}{}_j V_{rl} + \text{Ric}_k^l V_{lj} + \text{Ric}_j^l V_{kl} \right) dx^k \otimes dx^j. \tag{12.3}$$

REMARK 12.10. The covariant derivative acts on one-forms ω in such a way that the following equation holds:

$$\langle \nabla(\omega), \xi \rangle = \langle \omega, \nabla(\xi) \rangle$$

for every vector field ξ. This means that in local coordinates we have

$$\nabla_{\partial_r}(dx^k) = -\Gamma_{rl}^k dx^l.$$

Similarly, the Riemann curvature acts on one-forms ω satisfying

$$\text{Rm}(\xi_1, \xi_2)(\omega)(\xi) = -\omega\left(\text{Rm}(\xi_1, \xi_2)(\xi) \right).$$

Recall that in local coordinates
$$R_{ijkl} = \langle \mathrm{Rm}(\partial_i, \partial_j)(\partial_l), \partial_k \rangle.$$

Thus, we have
$$\mathrm{Rm}(\partial_i, \partial_j)(dx^k) = -g^{ka} R_{ijal} dx^l = -R_{ij}{}^k{}_l dx^l,$$

where as usual we use the inverse metric tensor to raise the index.

Also, notice that $\triangle X_i - \mathrm{Ric}_{ik} X^k = -\triangle_d X_i$, where by $\triangle_d$ we mean the Laplacian associated to the operator d from vector fields to one-forms with values in the vector field. Since

$$-(d\delta + \delta d) X_i = -\nabla_i \left(-\nabla^k X_k\right) - \left(-\nabla^k\right)(\nabla_k X_i - \nabla_i X_k)$$
$$= \nabla_i \nabla^k X_k + \nabla^k \nabla_k X_i - \nabla^k \nabla_i X_k$$
$$= R_i{}^k{}_k{}^j X_j + \nabla^k \nabla_k X_i = \triangle X_i - \mathrm{Ric}_i^j X_j.$$

PROOF. (of Lemma 12.9) The computation is routine, if complicated. We make the computation at a point (p, t) of space-time. We fix local $g(t)$-Gaussian coordinates $(x^1, \ldots, x^n)$ centered at p for space, so that the Christoffel symbols vanish at (p, t).

We compute

(12.4) $\quad \dfrac{\partial}{\partial t} V = \dfrac{\partial}{\partial t}(\nabla X^*) = -\left(\dfrac{\partial}{\partial t}\Gamma^l_{kj}\right) X^*_l dx^k \otimes dx^j + \nabla\left(\dfrac{\partial}{\partial t} X^*\right)$
$$= \left(-\nabla^l \mathrm{Ric}_{kj} + \nabla_k \mathrm{Ric}^l_j + \nabla_j \mathrm{Ric}^l_k\right) X^*_l dx^k \otimes dx^j$$
$$+ \nabla\left(\triangle X^* - \mathrm{Ric}(X, \cdot)\right).$$

We have
$$\nabla(\triangle X^*) = \nabla\left((g^{rs}\left(\nabla_r \nabla_s(X^*) - \Gamma^l_{rs} \nabla_l X^*\right)\right)$$
$$= g^{rs}\left(\nabla\left(\nabla_r \nabla_s(X^*) - \Gamma^l_{rs} \nabla_l X^*\right)\right).$$

Let us recall the formula for commuting ∇ and ∇_r. The following is immediate from the definitions.

CLAIM 12.11. *For any tensor ϕ we have*
$$\nabla(\nabla_r \phi) = \nabla_r(\nabla \phi) + dx^k \otimes \mathrm{Rm}(\partial_k, \partial_r)(\phi) - \nabla_r(dx^l) \otimes \nabla_l(\phi).$$

Applying this to our formula gives
$$\nabla(\triangle X^*) = g^{rs}\bigl(\nabla_r \nabla \nabla_s X^* + dx^k \otimes \mathrm{Rm}(\partial_k, \partial_r)(\nabla_s X^*)$$
$$- \nabla_r(dx^l) \otimes \nabla_l \nabla_s X^* - \nabla(\Gamma^l_{rs} \nabla_l X^*)\bigr).$$

Now we apply the same formula to commute ∇ and ∇_s. The result is

$$\nabla(\triangle X^*) = g^{rs}\Big(\nabla_r\nabla_s\nabla X^* + \nabla_r\big(dx^k \otimes \mathrm{Rm}(\partial_k,\partial_s)X^* - \nabla_s dx^l \otimes \nabla_l X^*\big)$$
$$+ dx^k \otimes \mathrm{Rm}(\partial_k,\partial_r)(\nabla_s X^*)$$
$$- \nabla_r(dx^l) \otimes \nabla_l\nabla_s X^* - \nabla(\Gamma_{rs}^l \nabla_l X^*)\Big).$$

Now we expand

$$\nabla_r\Big(dx^k \otimes \mathrm{Rm}(\partial_k,\partial_s)X^* - \nabla_s dx^l \otimes \nabla_l X^*\Big)$$
$$= \nabla_r(dx^k) \otimes \mathrm{Rm}(\partial_k,\partial_s)X^* + dx^k \otimes \nabla_r(\mathrm{Rm}(\partial_k,\partial_s))X^*$$
$$+ dx^k \otimes \mathrm{Rm}(\partial_k,\partial_s)\nabla_r X^* - \nabla_r\nabla_s dx^l \otimes \nabla_l X^* - \nabla_s dx^l \otimes \nabla_r \nabla_l X^*.$$

Invoking the fact that the Christoffel symbols vanish at the point of space-time where we are making the computation, this above expression simplifies to

$$\nabla_r\Big(dx^k \otimes \mathrm{Rm}(\partial_k,\partial_s)X^* - \nabla_s dx^l \otimes \nabla_l X^*\Big)$$
$$= dx^k \otimes \nabla_r(\mathrm{Rm}(\partial_k,\partial_s))X^* + dx^k \otimes \mathrm{Rm}(\partial_k,\partial_s)\nabla_r X^* - \nabla_r\nabla_s dx^l \otimes \nabla_l X^*.$$

Also, expanding and using the vanishing of the Christoffel symbols we have

$$-\nabla(\Gamma_{rs}^l \nabla_l X^*) = -d\Gamma_{rs}^l \otimes \nabla_l X^* - \Gamma_{rs}^l \nabla\nabla_l X^*$$
$$= -d\Gamma_{rs}^l \otimes \nabla_l X^*.$$

Plugging these computations into the equation above and using once more the vanishing of the Christoffel symbols gives

$$\nabla(\triangle X^*)$$
$$= \triangle(\nabla X^*) + g^{rs}\Big(dx^k \otimes \nabla_r(\mathrm{Rm}(\partial_k,\partial_s))X^* + dx^k \otimes \mathrm{Rm}(\partial_k,\partial_s)\nabla_r X^*$$
$$- \nabla_r\nabla_s dx^l \otimes \nabla_l X^* + dx^k \otimes \mathrm{Rm}(\partial_k,\partial_r)(\nabla_s X^*) - d\Gamma_{rs}^l \otimes \nabla_l X^*\Big).$$

Now by the symmetry of g^{rs} we can amalgamate the second and fourth terms on the right-hand side to give

$$\nabla(\triangle X^*) = \triangle(\nabla X^*) + g^{rs}\Big(dx^k \otimes \nabla_r(\mathrm{Rm}(\partial_k,\partial_s))X^*$$
$$+ 2 dx^k \otimes \mathrm{Rm}(\partial_k,\partial_s)\nabla_r X^* - \nabla_r\nabla_s dx^l \otimes \nabla_l X^*$$
$$- d\Gamma_{rs}^l \otimes \nabla_l X^*\Big).$$

We expand

$$\mathrm{Rm}(\partial_k,\partial_s)\nabla_r X^* = -R_{ks}{}^l{}_j V_{rl} dx^j.$$

5. STANDARD SOLUTIONS ARE ROTATIONALLY SYMMETRIC

Also we have (again using the vanishing of the Christoffel symbols)

$$-\nabla_r \nabla_s dx^l - d\Gamma_{rs}^l = \nabla_r \Gamma_{ks}^l dx^k - \partial_k \Gamma_{rs}^l dx^k$$
$$= R_{rk}{}^l{}_s dx^k.$$

Lastly,

$$\nabla_r (\mathrm{Rm}(\partial_k, \partial_s)) X^* = -(\nabla_r R)_{ks}{}^l{}_j X_l^* dx^j.$$

Plugging all this in and raising indices yields

$$\nabla(\triangle X^*) = \triangle(\nabla X^*) - g^{rs}(\nabla_r R)_{ks}{}^l{}_j X_l^* dx^k \otimes dx^j - 2R_k{}^{rl}{}_j V_{rl} dx^k \otimes dx^j$$
$$+ g^{rs} R_{rk}{}^l{}_s V_{lj} dx^k \otimes dx^j$$
$$= \triangle(\nabla X^*) - g^{rs}(\nabla_r R)_{ks}{}^j{}_l X_j^* dx^k \otimes dx^l - 2R_k{}^{rl}{}_j V_{rl} dx^k \otimes dx^j$$
$$- \mathrm{Ric}_k^l V_{lj} dx^k \otimes dx^j.$$

Thus, we have

$$\nabla(\triangle X^*) - \nabla(\mathrm{Ric}(X, \cdot)^*)$$
$$= \triangle(\nabla X^*) - g^{rs}(\nabla_r R)_{ks}{}^l{}_j X_l^* dx^k \otimes dx^j - 2R_k{}^{rl}{}_j V_{rl} dx^k \otimes dx^j$$
$$- \left(\mathrm{Ric}_k^l V_{lj} + \nabla_k (\mathrm{Ric})_j^l X_l^* + \mathrm{Ric}_j^l V_{kl}\right) dx^k \otimes dx^j,$$

and consequently, plugging back into Equation (12.4), and canceling the two like terms appearing with opposite sign, we have

$$\frac{\partial}{\partial t} V = \left(-\nabla^l \mathrm{Ric}_{kj} + \nabla_j \mathrm{Ric}_k^l\right) X_l^* dx^k \otimes dx^j + \triangle(\nabla X^*)$$
$$- g^{rs}(\nabla_r R)_{ks}{}^l{}_j X_l^* dx^k \otimes dx^j - 2R_k{}^{rl}{}_j V_{rl} dx^k \otimes dx^j$$
$$- \left(\mathrm{Ric}_k^l V_{lj} + \mathrm{Ric}_j^l V_{kl}\right) dx^k \otimes dx^j.$$

The last thing we need to see in order to complete the proof is that

$$-g^{rs}(\nabla_r R)_{ks}{}^l{}_j - \nabla^l \mathrm{Ric}_{kj} + \nabla_j \mathrm{Ric}_k^l = 0.$$

This is obtained by contracting g^{rs} against the Bianchi identity

$$\nabla_r R_{ks}{}^l{}_j + \nabla^l R_{ksjr} + \nabla_j R_{ksr}{}^l = 0.$$

□

Let h_{ij} be defined by $h_{ij} = V_{ij} + V_{ji}$. It follows from (12.3) that

(12.5) $$\frac{\partial}{\partial t} h_{ij} = \triangle_L h_{ij},$$

where by definition $\triangle_L h_{ij} = \triangle h_{ij} + 2R_i{}^{kl}{}_j h_{kl} - \text{Ric}_i^k h_{kj} - \text{Ric}_j^k h_{ki}$ is the **Lichnerowicz Laplacian**. A simple calculation shows that there is a constant $C > 0$ such that

$$\left(\frac{\partial}{\partial t} - \triangle\right) |h_{ij}|^2 = -2|\nabla_k h_{ij}|^2 + 4R^{ijkl} h_{jk} h_{il}, \tag{12.6}$$

$$\frac{\partial}{\partial t} |h_{ij}|^2 \leq \triangle |h_{ij}|^2 - 2|\nabla_k h_{ij}|^2 + C|h_{ij}|^2. \tag{12.7}$$

Note that $X(t)$ is a Killing vector field for $g(t)$ if and only if $h_{ij}(t) = 0$. Since Equation (12.1) is linear and since the curvature is bounded on each time-slice, for any given bounded Killing vector field $X(0)$ for metric $g(0)$, there is a bounded solution $X^i(t)$ of Equation (12.1) for $t \in [0, T]$. Then $|h_{ij}(t)|^2$ is a bounded function satisfying (12.7) and $|h_{ij}|^2(0) = 0$. One can apply the maximum principle to (12.7) to conclude that $h_{ij}(t) = 0$ for all $t \geq 0$. This is done as follows: Let $h(t)$ denote the maximum of $|h_{ij}(x,t)|^2$ on the t time-slice. Note that, for any fixed t the function $|h_{ij}(x,t)|^2$ approaches 0 as x tends to infinity since the metric is asymptotic at infinity to the product of a round metric on S^2 and the standard metric on the line. By virtue of (12.7) and Proposition 2.23, the function $h(t)$ satisfies $dh/dt \leq Ch$ in the sense of forward difference quotients, so that $d(e^{-Ct}h)/dt \leq 0$, also in the sense of forward difference quotients. Thus, by Lemma 2.22, since $h(0) = 0$ and $h \geq 0$, it follows that $e^{-Ct}h(t) = 0$ for all $t \geq 0$, and consequently, $h(t) = 0$ for all $t \geq 0$.

Thus, the evolving vector field $X(t)$ is a Killing vector field for $g(t)$ for all $t \in [0, T)$. The following is a very nice observation of Bennett Chow; we thank him for allowing us to use it. From $h_{ij} = 0$ we have $\nabla_j X^i + \nabla_i X^j = 0$. Taking the ∇_j derivative and summing over j we get

$$\triangle X^i + R_k^i X^k = 0$$

for all t. Hence (12.2) gives $\frac{\partial}{\partial t} X^i = 0$ and $X(t) = X(0)$, i.e., the Killing vector fields are stationary and remain Killing vector fields for the entire flow $g(t)$. Since at $t = 0$ the Lie algebra $so(3)$ of the standard rotation action consists of Killing vector fields, the same is true for all the metrics $g(t)$ in the standard solution. Thus, the rotation group $SO(3)$ of $\mathbb{R}^3$ is contained in the isometry group of $g(t)$ for every $t \in [0, T)$. We have shown:

COROLLARY 12.12. *The standard solution $g(t)$, $t \in [0, T)$, consists of a family of metrics all of which are rotationally symmetric by the standard action of $SO(3)$ on $\mathbb{R}^3$.*

5.1. Non-collapsing.

PROPOSITION 12.13. *For any $r > 0$ sufficiently small, there is a $\kappa > 0$ such that the standard flow is κ-non-collapsed on all scales $\leq r$.*

5. STANDARD SOLUTIONS ARE ROTATIONALLY SYMMETRIC

PROOF. Since the curvature of the standard solution is non-negative, it follows directly that $2|\text{Ric}|^2 \leq R^2$. By Equation (3.7) this gives

$$\frac{\partial R}{\partial t} = \triangle R + 2|\text{Ric}|^2 \leq \triangle R + R^2.$$

Let $C = \max(2, \max_{x \in \mathbb{R}^3} R(x, 0))$. Suppose that $t_0 < T$ and $t_0 < 1/C$.

CLAIM 12.14. *For all $x \in \mathbb{R}^3$ and $t \in [0, t_0]$ we have*

$$R(x, t) \leq \frac{C}{1 - Ct}.$$

PROOF. By the asymptotic condition, there is a compact subset $X \subset \mathbb{R}^3$ such that for any point $p \in \mathbb{R}^3 \setminus X$ and for any $t \leq t_0$ we have $R(p, t) < 2/(1-t)$. Since $C \geq 2$, for all t for which $\sup_{x \in \mathbb{R}^3} R(x, t) \leq 2/(1-t)$, we also have

$$R(x, t) \leq \frac{C}{1 - Ct}.$$

Consider the complementary subset of t, that is to say the subset of $[0, t_0]$ for which there is $x \in \mathbb{R}^3$ with $R(x, t) > C/(1 - Ct)$. This is an open subset of $[0, t_0]$, and hence is a disjoint union of relatively open intervals. Let $\{t_1 < t_2\}$ be the endpoints of one such interval. If $t_1 \neq 0$, then clearly $R_{\max}(t_1) = C/(1 - Ct_1)$. Since $C \geq \sup_{x \in \mathbb{R}^3} R(x, 0)$, this is also true if $t_1 = 0$. For every $t \in [t_1, t_2]$ the maximum of R on the t time-slice is achieved, and the subset of $\mathbb{R}^3 \times [t_1, t_2]$ of all points where maxima are achieved is compact. Furthermore, at any maximum point we have $\partial R/\partial t \leq R^2$. Hence, according to Proposition 2.23 for all $t \in [t_1, t_2]$ we have

$$R_{\max}(t) \leq G(t)$$

where $G'(t) = G^2(t)$ and $G(t_1) = C/(1 - Ct_1)$. It is easy to see that

$$G(t) = \frac{C}{1 - Ct}.$$

This shows that for all $t \in [t_1, t_2]$ we have $R(x, t) \leq \frac{C}{1-Ct}$, completing the proof of the claim. □

This shows that for $t_0 < T$ and $t_0 < 1/C$ the scalar curvature is bounded on $M \times [0, t_0]$ by a constant depending only on C and t_0. Since we are assuming that our flow is maximal, it follows that $T \geq 1/C$.

Since $(\mathbb{R}^3, g_0)$ is asymptotic to $(S^2 \times \mathbb{R}, h(0) \times ds^2)$, by compactness there is $V > 0$ such that for any metric ball $B(x, 0, r)$ on which $|\text{Rm}| \leq r^{-2}$ we have $\text{Vol } B(x, r) \geq Vr^3$. Since there is a uniform bound on the curvature on $[0, 1/2C]$, it follows that there is $V' > 0$ so that any ball $B(q, t, r)$ with $t \leq 1/2C$ on which $|\text{Rm}| \leq r^{-2}$ satisfies $\text{Vol } B(q, t, r) \geq V'r^3$. Set $t_0 = 1/4C$. For any point $x = (p, t)$ with $t \geq 1/2C$ there is a point (q, t_0) such that $l_x(q, t_0) \leq 3/2$; this by Theorem 7.10. Since $B(q, 0, 1/\sqrt{R_{\max}(0)}) \subset \mathbb{R}^3$ has volume at least $V/R_{\max}(0)^{3/2}$, and clearly l_x is bounded above on

$B(q, 0, 1/\sqrt{R_{\max}(0)})$ by a uniform constant, we see that the reduced volume of $B(q, 0, 1/\sqrt{R_{\max}(0)})$ is uniformly bounded from below. It now follows from Theorem 8.1 that there is $\kappa_0 > 0$ such that if $|\mathrm{Rm}|$ is bounded by r^{-2} on the parabolic neighborhood $P(p, t, r, -r^2)$ and $r \le \sqrt{1/4C}$, then the volume of this neighborhood is at least $\kappa_0 r^3$. Putting all this together we see that there is a universal $\kappa > 0$ such that the standard solution is κ-non-collapsed on all scales at most $\sqrt{1/4C}$. $\square$

6. Uniqueness

Now we turn to the proof of uniqueness. The idea is to mimic the proof of uniqueness in the compact case, by replacing the Ricci flow by a strictly parabolic flow. The material we present here is closely related to and derived from the presentation given in [49]. The presentation here is the analogy in the context of the standard solution of DeTurck's argument presented in Section 3 of Chapter 3.

6.1. From Ricci flow to Ricci-DeTurck flow. Let $(M^n, g(t))$, $t \in [t_0, T]$, be a solution of the Ricci flow and let $\psi_t \colon M \to M$, $t \in [t_0, T_1]$ be a solution of the harmonic map flow

$$(12.8) \qquad \frac{\partial \psi_t}{\partial t} = \triangle_{g(t), g(t_0)} \psi_t, \quad \psi_{t_0} = \mathrm{Id}.$$

Here, $\triangle_{g(t), g(t_0)}$ is the Laplacian for maps from the Riemannian manifold $(M, g(t))$ to the Riemannian manifold $(M, g(t_0))$. In local coordinates (x^i) on the domain M and (y^α) on the target M, the harmonic map flow (12.8) can be written as

$$(12.9) \qquad \left(\frac{\partial}{\partial t} - \triangle_{g(t)}\right) \psi^\alpha(x, t) = g^{ij}(x, t) \Gamma^\alpha_{\beta\gamma}(\psi(x, t)) \frac{\partial \psi^\beta(x, t)}{\partial x^i} \frac{\partial \psi^\gamma(x, t)}{\partial x^j}$$

where $\Gamma^\alpha_{\beta\gamma}$ are the Christoffel symbols of $g(t_0)$. Suppose $\psi(x, t)$ is a bounded smooth solution of (12.9) with $\psi_{t_0} = \mathrm{Id}$. Then $\psi(t)$, $t \in [t_0, T_1]$ are diffeomorphisms when $T_1 > t_0$ is sufficiently close to t_0. For any such T_1 and for $t_0 \le t \le T_1$, define $\hat{g}(t) = (\psi_t^{-1})^* g(t)$. Then $\hat{g}(t)$ satisfies the following equation:

$$(12.10) \qquad \frac{\partial}{\partial t} \hat{g}_{ij} = -2\widehat{\mathrm{Ric}}_{ij} + \hat{\nabla}_i W_j(t) + \hat{\nabla}_j W_i(t) \qquad \hat{h}(0) = h(0),$$

where $\widehat{\mathrm{Ric}}_{ij}$ and $\hat{\nabla}_i$ are the Ricci curvature and Levi-Civita connection of $\hat{g}(t)$ respectively and $W(t)$ is the time-dependent 1-form defined by

$$W(t)_j = \hat{g}_{jk}(t) \hat{g}^{pq}(t) \left(\hat{\Gamma}^k_{pq}(t) - \Gamma^k_{pq}(t_0)\right).$$

Here, $\hat{\Gamma}^k_{pq}(t)$ denotes the Christoffel symbols of the metric $\hat{g}(t)$ and $\Gamma^k_{pq}(t_0)$ denotes the Christoffel symbols of the metric $g(t_0)$. (See, for example, ([65] Lemma 2.1).) We call a solution to this flow equation a **Ricci-DeTurck flow** (see [16], or [13] Chapter 3 for details). In local coordinates we have

(12.11)
$$\frac{\partial \hat{g}_{ij}}{\partial t} = \hat{g}^{kl}\nabla_k\nabla_l \hat{g}_{ij} - \hat{g}^{kl}g(t_0)_{ip}\hat{g}^{pq}R_{jkql}(g(t_0)) - \hat{g}^{kl}g(t_0)_{jp}\hat{g}^{pq}R_{ikql}(g(t_0))$$
$$+ \frac{1}{2}\hat{g}^{kl}\hat{g}^{pq}\left[\begin{array}{c}\nabla_i\hat{g}_{pk}\nabla_j\hat{g}_{ql} + 2\nabla_k\hat{g}_{jp}\nabla_q\hat{g}_{il} \\ -2\nabla_k\hat{g}_{jp}\nabla_l\hat{g}_{iq} - 2\nabla_j\hat{g}_{pk}\nabla_l\hat{g}_{iq} - 2\nabla_i\hat{g}_{pk}\nabla_l\hat{g}_{jq}\end{array}\right],$$

where ∇ is the Levi-Civita connection of $g(t_0)$. This is a strictly parabolic equation.

LEMMA 12.15. *Suppose that $g(t)$ solves the Ricci flow equation and suppose that ψ_t solves the harmonic map flow equation, Equation (12.8); then $\hat{g}(t) = (\psi_t^{-1})^*g(t)$ solves the Ricci-DeTurck flow, Equation (12.10) and ψ_t satisfies the following ODE:*

$$\frac{\partial \psi_t}{\partial t} = -\hat{g}^{ij}(t)W(t).$$

PROOF. The first statement follows from the second statement and a standard Lie derivative computation. For the second statement, we need to show

$$\triangle_{g(t),g(0)}\psi^\alpha = -\hat{g}^{pq}\left(\hat{\Gamma}^\alpha_{pq}(t) - \Gamma^\alpha_{pq}(t_0)\right).$$

Notice that this equation is a tensor equation, so that we can choose coordinates in the domain and range so that $\Gamma(t)$ vanishes at the point p in question and $\Gamma(t_0)$ vanishes at $\psi_t(p)$. With these assumptions we need to show

$$g^{pq}(t)\frac{\partial^2\psi^\alpha}{\partial x^p \partial x^q} = -\hat{g}^{pq}(t)\hat{\Gamma}^\alpha_{pq}(t).$$

This is a direct computation using the change of variables formula relating $\hat{\Gamma}$ and Γ. □

COROLLARY 12.16. *Suppose that $(M, g_i(t))$, $t_0 \le t \le T$, are solutions to the Ricci flow equation for $i = 1, 2$. Suppose also that there are solutions*

$$\psi_{1,t}\colon (M, g_1(t)) \to (M, g_1(0))$$

and

$$\psi_{2,t}\colon (M, g_2(t)) \to (M, g_2(0))$$

to the harmonic map equation with $\psi_{1,t_0} = \psi_{2,t_0} = \mathrm{Id}$. Let

$$\hat{g}_1(t) = (\psi_{1,t}^{-1})^*g_1(t) \quad \text{and} \quad \hat{g}_2(t) = (\psi_{2,t}^{-1})^*g_2(t)$$

be the corresponding solutions to the Ricci-DeTurck flow. Suppose that $\hat{g}_1(t) = \hat{g}_2(t)$ for all $t \in [t_0, T]$. Then $g_1(t) = g_2(t)$ for all $t \in [t_0, T]$.

PROOF. Since $\psi_{a,t}$ satisfies the equation

$$\frac{\partial \psi_{a,t}}{\partial t} = -\hat{g}_a^{ij}W(t)_j$$

where the time-dependent vector field $W(t)$ depends only on $\hat{g}_a$, we see that $\psi_{1,t}$ and $\psi_{2,t}$ both solve the same time-dependent ODE and since $\psi_{1,t_0} = \psi_{2,t_0} = \mathrm{Id}$, it follows that $\psi_{1,t} = \psi_{2,t}$ for all $t \in [t_0, T]$. On the other hand, $g_a(t) = \psi_{a,t}^* \hat{g}_u(t)$, so that it follows that $g_1(t) = g_2(t)$ for $t \in [t_0, T]$. □

Our strategy of proof is to begin with a standard solution $g(t)$ and show that there is a solution to the harmonic map equation for this Ricci flow with appropriate decay conditions at infinity. It follows that the solution to the Ricci-DeTurck flow constructed is well-controlled at infinity. Suppose that we have two standard solutions $g_1(t)$ and $g_2(t)$ (with the same initial conditions g_0) that agree on the interval $[0, t_0]$ which is a proper subinterval of the intersection of the intervals of definition of $g_1(t)$ and $g_2(t)$. We construct solutions to the harmonic map flow equation from $g_a(t)$ to $g_a(t_0)$ for $a = 1, 2$. We show that solutions always exist for some amount of time past t_0. The corresponding Ricci-DeTurck flows $\hat{g}_a(t)$ starting at g_{t_0} are well-controlled at infinity. Since the Ricci-DeTurck flow equation is a purely parabolic equation, it has a unique solution with appropriate control at infinity and given initial condition $g_1(t_0) = g_2(t_0)$. This implies that the two Ricci-DeTurck flows we have constructed are in fact equal. Invoking the above corollary, we conclude that $g_1(t)$ and $g_2(t)$ agree on a longer interval extending past t_0. From this it follows easily that $g_1(t)$ and $g_2(t)$ agree on their common domain of definition. Hence, if they are both maximal flows, they must be equal.

7. Solution of the harmonic map flow

In order to pass from a solution to the Ricci flow equation to a solution of the Ricci-DeTurck flow we must prove the existence of a solution of the harmonic map flow associated with the Ricci flow. In this section we study the existence of the harmonic flow (12.8) and its asymptotic behavior at the space infinity when $h(t) = g(t)$ is a standard solution. Here we use in an essential way the rotationally symmetric property and asymptotic property at infinity of $g(t)$. In this argument there is no reason, and no advantage, to restricting to dimension 3, so we shall consider rotationally symmetric complete metrics on $\mathbb{R}^n$, i.e., complete metrics on $\mathbb{R}^n$ invariant under the standard action of $SO(n)$. Let $\theta = (\theta^1, \ldots, \theta^{n-1})$ be local coordinates on the round $(n-1)$-sphere of radius 1, and let $d\sigma$ be the metric on the sphere. We denote by $\hat{r}$ the standard radial coordinate in $\mathbb{R}^n$. Since $g(t)$ is rotationally symmetric and $n \geq 3$, we can write

$$(12.12) \qquad g(t) = dr^2 + f(r,t)^2 d\sigma.$$

Here $r = r(\hat{r}, t)$ is the (time-dependent) radial coordinate on $\mathbb{R}^n$ for the metric $g(t)$.

CLAIM 12.17. *For any fixed t the function $r\colon \mathbb{R}^n \to [0,\infty)$ is a function only of $\hat{r}$. Considered as a function of two variables, $r(\hat{r},t)$ is a smooth function defined for $\hat{r} \geq 0$. It is an odd function of $\hat{r}$. For fixed t it is an increasing function of $\hat{r}$.*

PROOF. Write the metric $g(t) = g_{ij}dx^i dx^j$ and let

(12.13) $\quad x^1 = \hat{r}\cos\theta^1,\ x^2 = \hat{r}\sin\theta^1\cos\theta^2,\ldots,x^n = \sin\theta^1 \cdots \sin\theta^{n-1}.$

We compute $f(r,t)$ by restricting attention to the ray $\hat{r} = x^1$ and $\theta^1 = \cdots = \theta^{n-1} = 0$, i.e., $x^2 = \cdots = x^n = 0$. Then

$$g(t) = g_{11}(\hat{r},0,\ldots,0,t)d\hat{r}^2 + g_{22}(\hat{r},0,\ldots,0,t)\hat{r}^2 d\sigma.$$

Both g_{11} and g_{22} are positive smooth and even in $\hat{r}$. Thus $\sqrt{g_{11}(\hat{r},0,\ldots,0,t)}$ is a positive smooth function defined for all $(\hat{r},t)$ and is invariant under the involution $\hat{r} \mapsto -\hat{r}$. Hence its restriction to $\hat{r} \geq 0$ is an even function. Since

$$r = \int_0^{\hat{r}} \sqrt{g_{11}(\hat{s},0,\ldots,0,t)}\,d\hat{s} = \hat{r}\int_0^1 \sqrt{g_{11}(\hat{r}s,0,\ldots,0,t)}\,ds,$$

we see that $r(\hat{r},t)$ is of the form $\hat{r}\cdot\phi(\hat{r},t)$ where $\phi(\hat{r},t)$ is an even smooth function. This shows that $r(\hat{r},t)$ is an odd function. It is also clear from this formula that $\partial r/\partial \hat{r} > 0$. $\square$

Since, for each t_0, the function $r(\hat{r},t_0)$ is an increasing function of $\hat{r}$, it can be inverted to give a function $\hat{r}(r,t_0)$. In Equation (12.12), we have chosen to write f as a function of r and t, rather than a function of $\hat{r}$ and t. We look for rotationally symmetric solutions to Equation (12.8), i.e., solutions of the form:

(12.14) $\quad \psi(t)\colon \mathbb{R}^n \to \mathbb{R}^n \quad \psi(t)(r,\theta) = (\rho(r,t),\theta) \quad \text{for } t \geq t_0$

$$\psi(r,t_0) = \mathrm{Id}$$

We shall adopt the following conventions: we shall consider functions $f(w,t)$ defined in the closed half-plane $w \geq 0$. When we say that such a function is *smooth* we mean that for each $n,m \geq 0$ we have a continuous function $f_{nm}(w,t)$ defined for all $w \geq 0$ satisfying:

(1)
$$f_{00} = f,$$

(2)
$$\frac{\partial f_{nm}}{\partial t} = f_{n(m+1)}, \quad \text{and}$$

(3)
$$\frac{\partial f_{nm}}{\partial w} = f_{(n+1)m}.$$

In item (3) the partial derivative along the boundary $w = 0$ is a right-handed derivative only. We say such a function is *even* if $f_{(2k+1)m}(0,t) = 0$ for all $k \geq 0$.

We have the following elementary lemma:

LEMMA 12.18. *(a) Suppose that $f(w,t)$ is a smooth function defined for $w \geq 0$. Define $\phi(r,t) = f(r^2, t)$. Then $\phi(r,t)$ is a smooth function defined for all $r \in \mathbb{R}$. Now fix k and let $\hat{r} \colon \mathbb{R}^k \to [0,\infty)$ be the usual radial coordinate. Then we have a smooth family of smooth functions on $\mathbb{R}^k$ defined by*

$$\hat{\phi}(x^1, \ldots, x^k, t) = \phi(\hat{r}(x^1, \ldots, x^k), t) = f(\sum_{i=1}^k (x^i)^2, t).$$

(b) If $\psi(r,t)$ is a smooth function defined for $r \geq 0$ and if it is even in the sense that its Taylor expansion to all orders along the line $r = 0$ involves only even powers of r, then there is a smooth function $f(w,t)$ defined for $w \geq 0$ such that $\psi(r,t) = f(r^2, t)$. In particular, for any $k \geq 2$ the function $\hat{\psi}((x^1, \ldots, x^k), t) = \psi(r(x^1, \ldots, x^k), t)$ is a smooth family of smooth functions on $\mathbb{R}^k$.

PROOF. Item (a) is obvious, and item (b) is obvious away from $r = 0$. We establish item (b) along the line $r = 0$. Consider the Taylor theorem with remainder of order $2N$ in the r-direction for $\psi(r,t)$ at a point $(0, t)$. By hypothesis it takes the form

$$\sum_{i=0} c_i(t) w^{2i} + w^{2N+1} R(w, t).$$

Now we replace w by $\sqrt{r}$ to obtain

$$f(r, t) = \sum_{i=0} c_i r^i + \sqrt{r}^{2N+1} R(\sqrt{r}, t).$$

Applying the usual chain rule and taking limits as $r \to 0^+$ we see that $f(r,t)$ is N times differentiable along the line $r = 0$. Since this is true for every $N < \infty$, the result follows. □

Notice that an even function $f(r,t)$ defined for $r \geq 0$ extends to a smooth function on the entire plane invariant under $r \mapsto -r$. When we say a function $f(r,t)$ defines a smooth family of smooth functions on $\mathbb{R}^n$ we mean that, under the substitution $\hat{f}((x^1, \ldots, x^n), t) = f(r(x^1, \ldots, x^n), t)$, the function $\hat{f}$ is a smooth function on $\mathbb{R}^n$ for each t.

We shall also consider odd functions $f(r,t)$, i.e., smooth functions defined for $r \geq 0$ whose Taylor expansion in the r-direction along the line $r = 0$ involves only odd powers of r. These do not define smooth functions on $\mathbb{R}^n$. On the other hand, by the same argument as above with the Taylor expansion one sees that they can be written as $rg(r,t)$ where g is even, and hence define smoothly varying families of smooth functions on $\mathbb{R}^n$. Notice also that the product of two odd functions $f_1(r,t)f_2(r,t)$ is an even function and hence this product defines a smoothly varying family of smooth function on $\mathbb{R}^n$.

7.1. The properties of r as a function of $\hat{r}$ and t. We shall make a change of variables and write the harmonic map flow equation in terms of r and θ. For this we need some basic properties of r as a function of $\hat{r}$ and t. Recall that we are working on $\mathbb{R}^n$ with its usual Euclidean coordinates $(x^1, \ldots, x^n)$. We shall also employ spherical coordinates $\hat{r}, \theta^1, \ldots, \theta^{n-1}$. (We denote the fixed radial coordinate on $\mathbb{R}^n$ by $\hat{r}$ to distinguish it from the varying radial function $r = r(t)$ that measures the distance from the tip in the metric $g(t)$.)

As a corollary of Claim 12.17 we have:

COROLLARY 12.19. *$r^2(\hat{r}, t)$ is a smoothly varying family of smooth functions on $\mathbb{R}^n$. Also, $\hat{r}$ is a smooth function of (r, t) defined for $r \geq 0$ and odd in r. In particular, any smooth even function of r is a smooth even function of $\hat{r}$ and thus defines a smooth function on $\mathbb{R}^n$. Moreover, there is a smooth function $\xi(w, t)$ such that $d(\log r)/dt = r^{-1}(dr/dt) = \xi(r^2, t)$.*

For future reference we define

$$(12.15) \qquad B(w, t) = \frac{1}{2} \int_0^w \xi(u, t) du.$$

Then $B(r^2, t)$ is a smooth function even in r and hence, as t varies, defines a smoothly varying family of smooth functions on $\mathbb{R}^n$. Notice that

$$\frac{\partial B(r^2, t)}{\partial r} = 2r \frac{\partial B}{\partial w}(w, t)|_{w=r^2} = 2r \left(\frac{1}{2}\xi(r^2, t)\right) = \frac{dr}{dt}.$$

Now let us consider $f(r, t)$.

CLAIM 12.20. *$f(r, t)$ is a smooth function defined for $r \geq 0$. It is an odd function of r.*

PROOF. We have

$$f(r, t) = \hat{r}(r, t)\sqrt{g_{22}(\hat{r}(r,t), 0, \ldots, 0, t)}.$$

Since $\sqrt{g_{22}(\hat{r}, 0, \ldots, 0, t)}$ is a smooth function of $(\hat{r}, t)$ defined for $\hat{r} \geq 0$ and since it is an even function of $\hat{r}$, it follows immediately from the fact that $\hat{r}$ is a smooth odd function of r, that $f(r, t)$ is a smooth odd function of r. $\square$

COROLLARY 12.21. *There is a smooth function $h(w, t)$ defined for $w \geq 0$ so that $f(r, t) = rh(r^2, t)$. In particular, $h(r^2, t)$ defines a smooth function on all of $\mathbb{R}^n$. Clearly, $h(w, t) > 0$ for all $w \geq 0$ and all t.*

We set $\widetilde{h}(w, t) = \log(h(w, t))$, so that $f(r, t) = re^{\widetilde{h}(r^2, t)}$. Notice that $\widetilde{h}(r^2, t)$ defines a smooth function of $\hat{r}^2$ and t and hence is a smoothly varying family of smooth functions on $\mathbb{R}^n$.

7.2. The harmonic map flow equation. Let $\psi(t)\colon \mathbb{R}^n \to \mathbb{R}^n$ be a smoothly varying family of smooth functions as given in Equation (12.14). Using (12.12) and (12.14) it is easy to calculate the energy functional using spherical coordinates with r as the radial coordinate.

$$E(\psi(t)) = \frac{1}{2}\int_{\mathbb{R}^n} |\nabla \psi(t)|^2_{g(t),g_{t_0}} \, dV_{g(t)}$$
$$= \frac{1}{2}\int_{\mathbb{R}^n} \left[\left(\frac{\partial \rho}{\partial r}\right)^2 + (n-1)f^2(\rho, t_0)f^{-2}(r,t)\right] dV_{g(t)}.$$

We set

$$\Xi = (n-1)f(\rho, t_0)\frac{\partial f(\rho, t_0)}{\partial \rho}f^{-2}(r,t).$$

If we have a compactly supported variation $\delta \rho = w$, then letting $d\mathrm{vol}_\sigma$ denote the standard volume element on S^{n-1}, we have

$$\delta E(\psi(t))(w) = \frac{1}{2}\int_{\mathbb{R}^n}\left[2\frac{\partial \rho}{\partial r}\frac{\partial w}{\partial r} + 2\Xi w\right] dV_g(t).$$

Then we can rewrite

$$\delta E(\psi(t))(w) = \int_0^{+\infty}\left[f^{n-1}(r,t)\frac{\partial \rho}{\partial r}\frac{\partial w}{\partial r} + \Xi w f^{n-1}\right] dr \cdot \int_{S^{n-1}} d\mathrm{vol}_\sigma$$
$$= \int_0^{+\infty}\left[-\frac{\partial}{\partial r}\left(f^{n-1}\frac{\partial \rho}{\partial r}\right)w + \Xi w f^{n-1}\right] dr \cdot \int_{S^{n-1}} d\mathrm{vol}_\sigma$$
$$= \int_{\mathbb{R}^n}\left[-f^{1-n}\frac{\partial}{\partial r}\left(\frac{\partial \rho}{\partial r}f^{n-1}\right) + \Xi\right]w\, dV_{g(t)}.$$

The usual argument shows that for a compactly supported variation w we have

$$\delta_w\left(\frac{1}{2}\int_{\mathbb{R}^n}|\nabla_{g(t),g(t_0)}\psi|^2 d\mathrm{vol}\right) = \int_{\mathbb{R}^n}\langle w, -\triangle_{g(t),g(t_0)}\psi\rangle d\mathrm{vol}.$$

Thus,

$$\triangle_{g(t),g(t_0)}\psi = \left[f^{1-n}\frac{\partial}{\partial r}\left(\frac{\partial \rho}{\partial r}f^{n-1}\right) - (n-1)f(\rho,t_0)\frac{\partial f(\rho,t_0)}{\partial \rho}f^{-2}(r,t)\right]\frac{\partial}{\partial r}$$

where we have written this expression using the coordinates (r,θ) on the range $\mathbb{R}^n$ (rather than the fixed coordinates $(\hat{r},\theta)$).

Now let us compute $\partial \psi/\partial t(\hat{r},t)$ in these same coordinates. (We use $\hat{r}$ for the coordinates for ψ in the domain to emphasize that this must be the time derivative at a fixed point in the underlying space.) Of course, by the chain rule,

$$\frac{\partial \psi(\hat{r},t)}{\partial t} = \frac{\partial \psi(r,t)}{\partial r}\frac{\partial r}{\partial t} + \frac{\partial \psi(r,t)}{\partial t}$$
$$= \frac{\partial \rho(r,t)}{\partial r}\frac{\partial r(\hat{r},t)}{\partial t} + \frac{\partial \rho(r,t)}{\partial t}.$$

Consequently, for rotationally symmetric maps as in Equation (12.14) the harmonic map flow equation (12.8) has the following form:

$$\frac{\partial \rho}{\partial t} + \frac{\partial \rho}{\partial r}\frac{\partial r}{\partial t} = \frac{1}{f^{n-1}(r,t)}\frac{\partial}{\partial r}\left(f^{n-1}(r,t)\frac{\partial \rho}{\partial r}\right) - (n-1)f^{-2}(r,t)f(\rho,t_0)\frac{\partial f(\rho,t_0)}{\partial \rho}$$

or equivalently

$$(12.16) \qquad \frac{\partial \rho}{\partial t} = \frac{1}{f^{n-1}(r,t)}\frac{\partial}{\partial r}\left(f^{n-1}(r,t)\frac{\partial \rho}{\partial r}\right)$$
$$- (n-1)f^{-2}(r,t)f(\rho,t_0)\frac{\partial f(\rho,t_0)}{\partial \rho} - \frac{\partial \rho}{\partial r}\frac{\partial r}{\partial t}.$$

The point of rewriting the harmonic map equation in this way is to find an equation for the functions $\rho(r,t), f(r,t)$ defined on $r \geq 0$. Even though the terms in this rewritten equation involve odd functions of r, as we shall see, solutions to these equations will be even in r and hence will produce a smooth solution to the harmonic map flow equation on $\mathbb{R}^n$.

7.3. An equation equivalent to the harmonic map flow equation. We will solve (12.16) for solutions of the form

$$\rho(r,t) = re^{\widetilde{\rho}(r,t)}, \quad t \geq t_0; \quad \widetilde{\rho}(r,t_0) = 0.$$

For ψ as in Equation (12.14) to define a diffeomorphism, it must be the case that $\rho(r,t)$ is a smooth function for $r \geq 0$ which is odd in r. It follows from the above expression that $\widetilde{\rho}(r,t)$ is a smooth function of r and t defined for $r \geq 0$ and even in r, so that it defines a smoothly varying family of smooth functions on $\mathbb{R}^n$. Then some straightforward calculation shows that (12.16) becomes

$$(12.17) \quad \frac{\partial \widetilde{\rho}}{\partial t} = \frac{\partial^2 \widetilde{\rho}}{\partial r^2} + \frac{n+1}{r}\frac{\partial \widetilde{\rho}}{\partial r} + (n-1)\frac{\partial \widetilde{h}(r^2,t)}{\partial r}\frac{\partial \widetilde{\rho}}{\partial r} + \left(\frac{\partial \widetilde{\rho}}{\partial r}\right)^2$$
$$+ \frac{n-1}{r^2}\left[1 - e^{2\widetilde{h}(\rho^2,t_0) - 2\widetilde{h}(r^2,t)}\right] + 2(n-1)\frac{\partial \widetilde{h}}{\partial w}(r^2,t)$$
$$- 2(n-1)e^{2\widetilde{h}(\rho^2,t_0) + 2\widetilde{\rho} - 2\widetilde{h}(r^2,t)}\frac{\partial \widetilde{h}}{\partial w}(\rho^2,t_0) - \frac{2}{r}\frac{\partial r}{\partial t} - \frac{\partial r}{\partial t}\frac{\partial \widetilde{\rho}}{\partial r}.$$

Note that from the definition, $\widetilde{h}(0,t) = 0$, we can write $\widetilde{h}(w,t) = w\widetilde{h}^*(w,t)$ where $\widetilde{h}^*(w,t)$ is a smooth function of $w \geq 0$ and t. So
$$\frac{n-1}{r^2}\left[1 - e^{2\widetilde{h}(\rho^2,t_0) - 2\widetilde{h}(r^2,t)}\right] = \frac{n-1}{r^2}\left[1 - e^{2r^2[e^{2\widetilde{\rho}}\widetilde{h}^*(\rho^2,t_0) - \widetilde{h}^*(r^2,t)]}\right]$$
which is a smooth function of $\widetilde{\rho}, r^2, t$.

Let

(12.18) $\quad G(\widetilde{\rho},w,t) = \dfrac{n-1}{w}\left[1 - e^{2\widetilde{h}(\rho^2,t_0) - 2\widetilde{h}(w,t)}\right] + 2(n-1)\dfrac{\partial \widetilde{h}}{\partial w}(w,t)$

$\qquad\qquad - 2(n-1)e^{2\widetilde{h}(\rho^2,t_0) + 2\widetilde{\rho} - 2\widetilde{h}(w,t)}\dfrac{\partial \widetilde{h}}{\partial w}(\rho^2, t_0) - 2\xi(w,t),$

where ξ is the function from Corollary 12.19. Then $G(\widetilde{\rho},w,t)$ is a smooth function defined for $w \geq 0$. Notice that when r and $\widetilde{\rho}$ are the functions associated with the varying family of metrics $g(t)$ and the solutions to the harmonic map flow, then $G(\widetilde{\rho}, r^2, t)$ defines a smoothly varying family of smooth functions on $\mathbb{R}^n$.

We have the following form of Equation (12.17):

$$\frac{\partial \widetilde{\rho}}{\partial t} = \frac{\partial^2 \widetilde{\rho}}{\partial r^2} + \frac{n+1}{r}\frac{\partial \widetilde{\rho}}{\partial r}$$
$$+ \left[(n-1)\frac{\partial \widetilde{h}}{\partial r} - \frac{\partial B}{\partial r}\right](r^2,t)\frac{\partial \widetilde{\rho}}{\partial r} + \left(\frac{\partial \widetilde{\rho}}{\partial r}\right)^2 + G(\widetilde{\rho}, r^2, t).$$

Now we think of $\widetilde{\rho}$ as a rotationally symmetric function defined on $\mathbb{R}^{n+2}$ and let $\widehat{G}(\widetilde{\rho},(x^1,\ldots,x^{n+2}),t) = G(\widetilde{\rho}, \sum_{i=1}^{n+2}(x^i)^2, t)$ and then the above equation can be written as

(12.19) $\quad \dfrac{\partial \widetilde{\rho}}{\partial t} = \triangle\widetilde{\rho} + \nabla[(n-1)\widetilde{h} - B] \cdot \nabla\widetilde{\rho} + |\nabla\widetilde{\rho}|^2 + G(\widetilde{\rho}, x, t)$

where ∇ and $\triangle$ are the Levi-Civita connection and Laplacian defined by the Euclidean metric on $\mathbb{R}^{n+2}$ respectively and where B is the function defined in Equation (12.15).

REMARK 12.22. The whole purpose of this rewriting of the PDE for $\widetilde{\rho}$ is to present this equation in such a form that all its coefficients represent smooth functions of $\hat{r}$ and t that are even in $\hat{r}$ and hence define smooth functions on Euclidean space of any dimension. We have chosen to work on $\mathbb{R}^{n+2}$ because the expression for the Laplacian in this dimension has the term $((n+1)/r)\partial\widetilde{\rho}/\partial r$.

It is important to understand the asymptotic behavior of our functions at spatial infinity.

CLAIM 12.23. *For any fixed t we have the following asymptotic expansions at spatial infinity.*

(1) $e^{\widetilde{h}(r^2,t)}$ is asymptotic to $\frac{1}{(1-t)r}$.
(2) $\widetilde{h}(r^2,t)$ is asymptotic to $-\log r$.
(3) $\frac{\partial \widetilde{h}}{\partial w}(r^2,t)$ is asymptotic to $-\frac{1}{2r^2}$.
(4) $r^{-1}\frac{\partial r}{\partial t}$ is asymptotic to $\frac{C}{r}$.
(5) $\frac{\partial B(r^2,t)}{\partial r}$ is asymptotic to C.
(6) $|G(\widetilde{\rho},r^2,t)| \leq C_* < \infty$ where $C_* = C_*\left(\sup\{|\widetilde{\rho}|,\widetilde{h}\}\right)$ is a constant depending only on $\sup\{|\widetilde{\rho}|,\widetilde{h}\}$.

PROOF. The first item is immediate from Proposition 12.7. The second and third follow immediately from the first. The fourth is a consequence of the fact that by Proposition 12.7, dr/dt is asymptotic to a constant at infinity on each time-slice. The fifth follows immediately from the fourth and the definition of $B(r^2,t)$. Given all these asymptotic expressions, the last is clear from the expression for G in terms of $\widetilde{\rho}$, r^2, and t. □

7.4. The short time existence. The purpose of this subsection is to prove the following short-time existence theorem for the harmonic map flow equation.

PROPOSITION 12.24. *For any $t_0 \geq 0$ for which there is a standard solution $g(t)$ defined on $[0,T_1]$ with $t_0 < T_1$, there is $T > t_0$ and a solution to Equation (12.19) with initial condition $\widetilde{\rho}(r,t_0) = 0$ defined on the time-interval $[t_0,T]$.*

At this point to simplify the notation we shift time by $-t_0$ so that our initial time is 0, though our initial metric is not g_0 but rather is the t_0 time-slice of the standard solution we are considering, so that now $t_0 = 0$ and our initial condition is $\widetilde{\rho}(r,0) = 0$.

Let $x = (x^1,\ldots,x^{n+2})$ and $y = (y^1,\ldots,y^{n+2})$ be two points in $\mathbb{R}^{n+1}$ and

$$H(x,y,t) = \frac{1}{(4\pi t)^{(n+2)/2}} e^{-\frac{|x-y|^2}{4t}}$$

be the heat kernel. We solve (12.19) by successive approximation [**47**].

Define

$$F(x,\widetilde{\rho},\nabla\widetilde{\rho},t) = \nabla\left[(n-1)\widetilde{h} - B\right]\cdot\nabla\widetilde{\rho} + |\nabla\widetilde{\rho}|^2 + G(\widetilde{\rho},x,t).$$

Let $\widetilde{\rho}_0(x,t) = 0$ and for $i \geq 1$ we define $\widetilde{\rho}_i$ by

(12.20) $$\widetilde{\rho}_i = \int_0^t \int_{\mathbb{R}^{n+2}} H(x,y,t-s)F(y,\widetilde{\rho}_{i-1},\nabla\widetilde{\rho}_{i-1},t)dyds$$

which solves

(12.21) $$\frac{\partial \widetilde{\rho}_i}{\partial t} = \Delta \widetilde{\rho}_i + F(x, \widetilde{\rho}_{i-1}, \nabla \widetilde{\rho}_{i-1}, t) \qquad \widetilde{\rho}_i(x, 0) = 0.$$

To show the existence of $\widetilde{\rho}_i$ by induction, it suffices to prove the following statement: For any $i \geq 1$, if $|\widetilde{\rho}_{i-1}|, |\nabla \widetilde{\rho}_{i-1}|$ are bounded, then $\widetilde{\rho}_i$ exists and $|\widetilde{\rho}_i|, |\nabla \widetilde{\rho}_i|$ are bounded. Assume $|\widetilde{\rho}_{i-1}| \leq C_1, |\nabla \widetilde{\rho}_{i-1}| \leq C_2$ are bounded on $\mathbb{R}^{n+2} \times [0, T]$; then it follows from Claim 12.23 that $G(\widetilde{\rho}_{i-1}, \mathbf{x}, t)$ is bounded on $\mathbb{R}^{n+2} \times [0, T]$,

$$|G(\widetilde{\rho}_{i-1}, x, t)| \leq C_*(C_1, \widetilde{h}),$$

and also because of Claim 12.23 both $|\nabla B|$ and $|\nabla \widetilde{h}|$ are bounded on all of $\mathbb{R}^{n+2} \times [0, T]$, it follows that $F(x, \widetilde{\rho}_{i-1}, \nabla \widetilde{\rho}_{i-1}, t)$ is bounded:

$$|F(x, \widetilde{\rho}_{i-1}, \nabla \widetilde{\rho}_{i-1}, t)| \leq \left[(n-1)\sup|\nabla \widetilde{h}| + \sup|\nabla B|\right] C_2 + C_2^2 + C_*(C_1, \widetilde{h}).$$

Clearly, the last expression is bounded by a constant C_3 depending only on the previous bounds. Hence $\widetilde{\rho}_i$ exists.

The bounds on $|\widetilde{\rho}_i|$ and $|\nabla \widetilde{\rho}_i|$ follow from the estimates

$$|\widetilde{\rho}_i| \leq \int_0^t \int_{\mathbb{R}^{n+2}} H(x, y, t-s) C_3 dy ds \leq C_3 t,$$

and

$$|\nabla \widetilde{\rho}_i| = \left| \int_0^t \int_{\mathbb{R}^{n+2}} [\nabla_x H(x, y, t-s)] F(y, \widetilde{\rho}_{i-1}, \nabla \widetilde{\rho}_{i-1}, t) dy ds \right|$$
$$\leq \int_0^t \int_{\mathbb{R}^{n+2}} |\nabla_x H(x, y, t-s)| C_3 dy ds$$
$$= \int_0^t \int_{\mathbb{R}^{n+2}} \frac{1}{(4\pi(t-s))^{(n+2)/2}} e^{-\frac{|x-y|^2}{4(t-s)}} \frac{|x-y|}{2(t-s)} C_3 dy ds$$
$$\leq \frac{(n+2)C_3}{\sqrt{\pi}} \int_0^t \frac{1}{\sqrt{t-s}} ds = \frac{2(n+2)C_3}{\sqrt{\pi}} \sqrt{t}.$$

Assuming, as we shall, that $T \leq \min\{\frac{C_3}{C_1}, \frac{\pi C_2^2}{4(n+2)^2 C_3^2}\}$, then for $0 \leq t \leq T$ we have for all i,

(12.22) $$|\widetilde{\rho}_i| \leq C_1 \quad \text{and} \quad |\nabla \widetilde{\rho}_i| \leq C_2.$$

We prove the convergence of $\widetilde{\rho}_i$ to a solution of (12.19) via proving that it is a Cauchy sequence in C^1-norm. Note that $\widetilde{\rho}_i - \widetilde{\rho}_{i-1}$ satisfies

(12.23) $$\frac{\partial (\widetilde{\rho}_i - \widetilde{\rho}_{i-1})}{\partial t} = \Delta(\widetilde{\rho}_i - \widetilde{\rho}_{i-1})$$
$$+ F(x, \widetilde{\rho}_{i-1}, \nabla \widetilde{\rho}_{i-1}, t) - F(x, \widetilde{\rho}_{i-2}, \nabla \widetilde{\rho}_{i-2}, t)$$

with

$$(\widetilde{\rho}_i - \widetilde{\rho}_{i-1})(x, 0) = 0,$$

where

$$F(x, \widetilde{\rho}_{i-1}, \nabla\widetilde{\rho}_{i-1}, t) - F(x, \widetilde{\rho}_{i-2}, \nabla\widetilde{\rho}_{i-2}, t)$$
$$= [(n-1)\nabla\widetilde{h} - \nabla B + \nabla(\widetilde{\rho}_{i-1} + \widetilde{\rho}_{i-2})] \cdot \nabla(\widetilde{\rho}_{i-1} - \widetilde{\rho}_{i-2})$$
$$+ G(\widetilde{\rho}_{i-1}, \mathbf{x}, t) - G(\widetilde{\rho}_{i-2}, \mathbf{x}, t).$$

By lengthy but straightforward calculations one can verify the Lipschitz property of $G(\widetilde{\rho}, \mathbf{x}, t)$,

$$|G(\widetilde{\rho}_{i-1}, \mathbf{x}, t) - G(\widetilde{\rho}_{i-2}, \mathbf{x}, t)| \leq C_{\&}(C_1, C_2, \widetilde{f}, \widetilde{f_0}) \cdot |\widetilde{\rho}_{i-1} - \widetilde{\rho}_{i-2}|.$$

This and (12.22) implies

(12.24) $\quad |F(x, \widetilde{\rho}_{i-1}, \nabla\widetilde{\rho}_{i-1}, t) - F(x, \widetilde{\rho}_{i-2}, \nabla\widetilde{\rho}_{i-2}, t)|$
$$\leq C_4 \cdot |\widetilde{\rho}_{i-1} - \widetilde{\rho}_{i-2}| + C_5 \cdot |\nabla\widetilde{\rho}_{i-1} - \nabla\widetilde{\rho}_{i-2}|$$

where $C_4 = C_{\&}(C_1, C_2, \widetilde{f}, \widetilde{f_0})$ and $C_5 = [(n-1)\sup|\nabla\widetilde{f}| + \sup|\nabla B| + 2C_2]$.

Let

$$A_i(t) = \sup_{0 \leq s \leq t, x \in \mathbb{R}^{n+2}} |\widetilde{\rho}_i - \widetilde{\rho}_{i-1}|(x, s),$$
$$B_i(t) = \sup_{0 \leq s \leq t, x \in \mathbb{R}^{n+2}} |\nabla(\widetilde{\rho}_i - \widetilde{\rho}_{i-1})|(x, s).$$

From Equations (12.23) and (12.24) we can estimate $|\widetilde{\rho}_i - \widetilde{\rho}_{i-1}|$ and $|\nabla(\widetilde{\rho}_i - \widetilde{\rho}_{i-1})|$ in the same way as we estimate $|\widetilde{\rho}_i|$ and $|\nabla\widetilde{\rho}_i|$ above; we conclude

$$A_i(t) \leq [C_4 A_{i-1}(t) + C_5 B_{i-1}(t)] \cdot t,$$
$$B_i(t) \leq \frac{2(n+2)[C_4 A_{i-1}(t) + C_5 B_{i-1}(t)]}{\sqrt{\pi}} \cdot \sqrt{t}.$$

Let $C_6 = \max\{C_4, C_5\}$; then we get

$$A_i(t) + B_i(t) \leq \left(C_6 t + \frac{2(n+2)C_6\sqrt{t}}{\sqrt{\pi}}\right) \cdot (A_{i-1}(t) + B_{i-1}(t)).$$

Now suppose that $T \leq T_2$ where T_2 satisfies $C_6 T_2 + \frac{2(n+2)C_6\sqrt{T_2}}{\sqrt{\pi}} = \frac{1}{2}$; then for all $t \leq T$ we have

$$A_i(t) + B_i(t) \leq \frac{1}{2}(A_{i-1}(t) + B_{i-1}(t)).$$

This proves that $\widetilde{\rho}_i$ is a Cauchy sequence in $C^1(\mathbb{R}^{n+2})$. Let $\lim_{i \to +\infty} \widetilde{\rho}_i = \widetilde{\rho}_\infty$. Then $\nabla\widetilde{\rho}_i \to \nabla\widetilde{\rho}_\infty$ and $F(x, \widetilde{\rho}_{i-1}, \nabla\widetilde{\rho}_{i-1}, t) \to F(x, \widetilde{\rho}_\infty, \nabla\widetilde{\rho}_\infty, t)$ uniformly. Hence we get from (12.20),

(12.25) $$\widetilde{\rho}_\infty = \int_0^t \int_{\mathbb{R}^{n+2}} H(x, y, t-s) F(y, \widetilde{\rho}_\infty, \nabla\widetilde{\rho}_\infty, t) dy ds.$$

The next argument is similar to the argument in [**47**], p.21. The function $\widetilde{\rho}_i$ is a smooth solution of (12.21) with $\widetilde{\rho}_i(x,0) = 0$. Also, both $\widetilde{\rho}_i$ and $F(x, \widetilde{\rho}_{i_1}, \nabla \widetilde{\rho}_{i-1}, t)$ are uniformly bounded on $\mathbb{R}^{n+2} \times [0,T]$. Thus, by Theorems 1.11 on p. 211 and 12.1 on p. 223 of [**46**], for any compact $K \subset \mathbb{R}^{n+2}$ and any $0 < t_* < T$, there is C_7 and $\alpha \in (0,1)$ independent of i such that

$$|\nabla \widetilde{\rho}_i(x,t) - \nabla \widetilde{\rho}_i(y,s)| \leq C_7 \cdot \left(|x-y|^\alpha + |t-s|^{\alpha/2}\right)$$

where $x, y \in K$ and $0 \leq t < s \leq t_*$.

Letting $i \to \infty$ we get

$$(12.26) \qquad |\nabla \widetilde{\rho}_\infty(x,t) - \nabla \widetilde{\rho}_\infty(y,s)| \leq C_7 \cdot \left(|x-y|^\alpha + |t-s|^{\alpha/2}\right).$$

Hence $\nabla \widetilde{\rho}_\infty \in C^{\alpha, \alpha/2}$. That is to say $\nabla \widetilde{\rho}_\infty$ is α-Hölder continuous in space and $\alpha/2$-Hölder continuous in time. From (12.25) we conclude that $\widetilde{\rho}_\infty$ is a solution of (12.19) on $\mathbb{R}^{n+2} \times [0,T]$ with $\widetilde{\rho}_\infty(x,0) = 0$.

7.5. The asymptotic behavior of the solutions. In the rest of this subsection we study the asymptotic behavior of solution $\widetilde{\rho}(x,t)$ as $x \to \infty$. First we prove inductively that there is a constant λ and T_3 such that, provided that $T \leq T_3$, for $x \in \mathbb{R}^{n+2}, t \in [0,T]$, we have

$$(12.27) \qquad |\widetilde{\rho}_i(x,t)| \leq \frac{\lambda}{(1+|x|)^2} \quad \text{and} \quad |\nabla \widetilde{\rho}_i(x,t)| \leq \frac{\lambda}{(1+|x|)^2}.$$

Clearly, since $\widetilde{\rho}_0 = 0$, these estimates hold for $i = 0$. It follows from (12.22) and Claim 12.23 that there is a constant C_8 independent of i such that

$$|G(\widetilde{\rho}_i, \mathbf{x}, t)| \leq \frac{C_8}{(1+|x|)^2},$$

$$\left[(n-1)|\nabla \widetilde{h}| + |\nabla B|\right](x,t) \leq C_8.$$

Now we assume these estimates hold for i. Then for $0 \leq t \leq T$ we have

$$|\widetilde{\rho}_i(x,t)|$$
$$\leq \int_0^t \int_{\mathbb{R}^{n+2}} H(x,y,t-s) \left[\frac{C_8 \lambda}{(1+|y|)^2} + \frac{\lambda^2}{(1+|y|)^2} + \frac{C_8}{(1+|y|)^2}\right] dy\, ds$$
$$= \int_0^t \int_{\mathbb{R}^{n+2}} \frac{1}{(4\pi(t-s))^{(n+2)/2}} e^{-\frac{|x-y|^2}{4(t-s)}} \left[\frac{C_8 \lambda + \lambda^2 + C_8}{(1+|y|)^2}\right] dy\, ds$$
$$\leq (C_8 \lambda + \lambda^2 + C_8) \cdot \frac{C(n)t}{(1+|x|)^2}.$$

Also, we have

$$|\nabla \widetilde{\rho}_i(x,t)| \leq \int_0^t \int_{\mathbb{R}^{n+2}} |\nabla_x H(x,y,t-s)| \left[\frac{C_8\lambda + \lambda^2 + C_8}{(1+|y|)^2}\right] dyds$$

$$= \int_0^t \int_{\mathbb{R}^{n+2}} \frac{|x-y|}{2(t-s)} \frac{1}{(4\pi(t-s))^{(n+2)/2}} e^{-\frac{|x-y|^2}{4(t-s)}} \left[\frac{C_8\lambda + \lambda^2 + C_8}{(1+|y|)^2}\right] dyds$$

$$\leq (C_8\lambda + \lambda^2 + C_8) \cdot \frac{C(n)\sqrt{t}}{(1+|x|)^2}.$$

If we choose T_3 such that

$$(C_8\lambda + \lambda^2 + C_8) \cdot C(n)T_3 \leq \lambda \quad \text{and} \quad (C_8\lambda + \lambda^2 + C_8) \cdot C(n)\sqrt{T_3} \leq \lambda,$$

then (12.27) hold for all i. From the definition of $\widetilde{\rho}_\infty$ we conclude that

(12.28) $\quad |\widetilde{\rho}_\infty(x,t)| \leq \dfrac{\lambda}{(1+|x|)^2} \quad \text{and} \quad |\nabla\widetilde{\rho}_\infty(x,t)| \leq \dfrac{\lambda}{(1+|x|)^2}.$

Recall that $\widetilde{\rho}_\infty$ is a solution of the following linear equation (in v):

$$\frac{\partial v}{\partial t} = \Delta v + \nabla[(n-1)\widetilde{h} - B] \cdot \nabla v + G(\widetilde{\rho}_\infty, \mathbf{x}, t),$$
$$v(x,0) = 0.$$

From (12.26) and Claim 12.23 we know that $\nabla[(n-1)\widetilde{h} - B + \widetilde{\rho}_\infty]$ has $C^{\alpha,\alpha/2}$-Hölder-norm bounded (this means α-Hölder norm in space and the $\alpha/2$-Hölder norm in time). By some lengthy calculations we get

$$|G(\widetilde{\rho}_\infty, \mathbf{x}, t)|_{C^{\alpha,\alpha/2}} \leq \frac{C_9}{(1+|x|)^2}.$$

By local Schauder estimates for parabolic equations we conclude

$$|\widetilde{\rho}_\infty|_{C^{2+\alpha,1+\alpha/2}} \leq \frac{C_{10}}{(1+|x|)^2}.$$

Using this estimate one can further show by calculation that

$$|\nabla\nabla[(n-1)\widetilde{f} - B + \widetilde{\rho}_\infty]|_{C^{\alpha,\alpha/2}} \leq C_{11},$$
$$|\nabla G(\widetilde{\rho}_\infty, \mathbf{x}, t)|_{C^{\alpha,\alpha/2}} \leq \frac{C_{12}}{(1+|x|)^2}.$$

By local high order Schauder estimates for parabolic equations we conclude

$$|\nabla\widetilde{\rho}_\infty|_{C^{2+\alpha,1+\alpha/2}} \leq \frac{C_{13}}{(1+|x|)^2}.$$

We have proved the following:

PROPOSITION 12.25. *For a standard solution* $(\mathbb{R}^n, g(t))$, $0 \leq t < T$, *and for any* $t_0 \in [0, T)$ *there is a rotationally symmetric solution* $\psi_t(\mathbf{x}) = x e^{\widetilde{\rho}(\mathbf{x}, t)}$ *to the harmonic map flow*

$$\frac{\partial \psi_t}{\partial t} = \triangle_{g(t), g(t_0)} \psi(t) \qquad \psi(t_0)(\mathbf{x}) = \mathbf{x},$$

and $|\nabla^i \widetilde{\rho}|(\mathbf{x}, t) \leq \frac{C_{14}}{(1+|\mathbf{x}|)^2}$ *for* $0 \leq i \leq 3$ *defined on some non-degenerate interval* $[t_0, T']$.

7.6. The uniqueness for the solutions of Ricci-DeTurck flow. We prove the following general uniqueness result for Ricci-DeTurck flow on open manifolds.

PROPOSITION 12.26. *Let* $\hat{g}_1(t)$ *and* $\hat{g}_2(t)$, $0 \leq t \leq T$, *be two bounded solutions of the Ricci-DeTurck flow on complete and non-compact manifold* M^n *with initial metric* $g_1(t_0) = g_2(t_0) = g$. *Suppose that for some* $1 < C < \infty$ *we have*

$$C^{-1} g \leq \hat{g}_1(t) \leq C g,$$
$$C^{-1} g \leq \hat{g}_2(t) \leq C g.$$

Suppose that in addition we have

$$\|\hat{g}_1(t)\|_{C^2(M), g} \leq C,$$
$$\|\hat{g}_2(t)\|_{C^2(M), g} \leq C.$$

Lastly, suppose there is an exhaustive sequence of compact, smooth submanifolds of $\Omega_k \subset M$, *i.e.,* $\Omega_k \subset \text{int} \Omega_{k+1}$ *and* $\cup \Omega_k = M$ *such that* $\hat{g}_1(t)$ *and* $\hat{g}_2(t)$ *have the same sequential asymptotic behavior at* ∞ *in the sense that for any* $\epsilon > 0$, *there is a* k_0 *arbitrarily large with*

$$|\hat{g}_1(t) - \hat{g}_2(t)|_{C^1(\partial \Omega_{k_0}), g} \leq \epsilon.$$

Then $\hat{g}_1(t) = \hat{g}_2(t)$.

PROOF. Letting $\widetilde{\nabla}$ be the covariant derivative determined by g, then, using the Ricci-DeTurck flow (12.11) for $\hat{g}_1$ and $\hat{g}_2$, we can make the following estimate for an appropriate constant D depending on g. (In these formulas all norms and inner products are with respect to the background metric g.)

$$\frac{\partial}{\partial t}|\hat{g}_1(t) - \hat{g}_2(t)|^2$$
$$= 2\left\langle \frac{\partial}{\partial t}(\hat{g}_1(t) - \hat{g}_2(t)), \hat{g}_1(t) - \hat{g}_2(t) \right\rangle$$
$$\leq 2\left\langle \hat{g}_1^{\alpha\beta}\widetilde{\nabla}_\alpha\widetilde{\nabla}_\beta(\hat{g}_1(t) - \hat{g}_2(t)), (\hat{g}_1(t) - \hat{g}_2(t)) \right\rangle$$
$$+ D|\hat{g}_1(t) - \hat{g}_2(t)|_g^2 + D\left|\widetilde{\nabla}(\hat{g}_1(t) - \hat{g}_2(t))\right||\hat{g}_1(t) - \hat{g}_2(t)|$$
$$\leq \hat{g}_1^{\alpha\beta}\widetilde{\nabla}_\alpha\widetilde{\nabla}_\beta\left(|\hat{g}_1(t) - \hat{g}_2(t)|^2\right)$$
$$- 2\hat{g}_1^{\alpha\beta}\left\langle \widetilde{\nabla}_\beta(\hat{g}_1(t) - \hat{g}_2(t)), \widetilde{\nabla}_\alpha(\hat{g}_1(t) - \hat{g}_2(t)) \right\rangle$$
$$+ D|\hat{g}_1(t) - \hat{g}_2(t)|^2 + D\left|\widetilde{\nabla}(\hat{g}_1(t) - \hat{g}_2(t))\right||\hat{g}_1(t) - \hat{g}_2(t)|$$
$$\leq \hat{g}_1^{\alpha\beta}\widetilde{\nabla}_\alpha\widetilde{\nabla}_\beta\left(|\hat{g}_1(t) - \hat{g}_2(t)|^2\right) - 2C^{-1}\left|\widetilde{\nabla}(\hat{g}_1(t) - \hat{g}_2(t))\right|^2$$
$$+ D|\hat{g}_1(t) - \hat{g}_2(t)|^2 + C^{-1}\left|\widetilde{\nabla}(\hat{g}_1(t) - \hat{g}_2(t))\right|^2 + \frac{D^2}{4C^{-1}}|\hat{g}_1(t) - \hat{g}_2(t)|^2,$$

where the last inequality comes from completing the square to replace the last term in the previous expression. Thus, we have proved

$$(12.29) \qquad \frac{\partial}{\partial t}|\hat{g}_1(t) - \hat{g}_2(t)|_g^2 \leq 2\hat{g}_1^{\alpha\beta}\widetilde{\nabla}_\alpha\widetilde{\nabla}_\beta|\hat{g}_1(t) - \hat{g}_2(t)|_g^2$$
$$+ C_{15}|\hat{g}_1(t) - \hat{g}_2(t)|_g^2$$

pointwise on Ω_k with C_{15} a constant that depends only on n, C and g.

Suppose that $\hat{g}_1(t) \neq \hat{g}_2(t)$ for some t. Then there is a point x_0 such that $|\hat{g}_1(x_0,t) - \hat{g}_2(x_0,t)|_g^2 > \epsilon_0$ for some $\epsilon_0 > 0$.

We choose a k_0 sufficiently large that $x_0 \in \Omega_{k_0}$ and for all $t' \in [t_0,T]$ we have

$$(12.30) \qquad \sup_{x \in \partial\Omega_b} \left|\hat{g}_1(x,t') - \hat{g}_2(x,t')\right|_g^2 \leq \epsilon$$

where $\epsilon > 0$ is a constant to be chosen later.

Recall we have the initial condition $|\hat{g}_1(0) - \hat{g}_2(0)|_g^2 = 0$. Using Equation (12.29) and applying the maximum principle to $|\hat{g}_1(t) - \hat{g}_2(t)|_g^2$) on the domain Ω_{k_0}, we get

$$e^{-C_{15}t}|\hat{g}_1(t) - \hat{g}_2(t)|_g^2(x) \leq \epsilon \qquad \text{for all } x \in \Omega_{k_0}.$$

This is a contradiction if we choose $\epsilon \leq \epsilon_0 e^{-C_{15}T}$. This contradiction establishes the proposition. $\square$

Let $g_1(t)$, $0 \leq t < T_1$, and $g_2(t)$, $0 \leq t < T_2$, be standard solutions that agree on the interval $[0, t_0]$ for some $t_0 \geq 0$. By Proposition 12.25 there are $\psi_1(t)$ and $\psi_2(t)$ which are solutions of the harmonic map flow defined for $t_0 \leq t \leq T$ for some $T > t_0$ for the Ricci flows $g_1(t)$ and $g_2(t)$. Let $\hat{g}_1(t) = (\psi^{-1}(t))^* g_1(t)$ and $\hat{g}_2(t) = (\psi^{-1}(t))^* g_2(t)$. Then $\hat{g}_1(t)$ and $\hat{g}_2(t)$ are two solutions of the Ricci-DeTurck flow with $\hat{g}_1(t_0) = \hat{g}_2(t_0)$. Choose $T' \in (t_0, T]$ such that $\hat{g}_1(t)$ and $\hat{g}_2(t)$ are δ-close to $\hat{g}_1(t_0)$ as required in Proposition 12.26. It follows from Proposition 12.7 and the decay estimate in Proposition 12.25 that $\hat{g}_1(t)$ and $\hat{g}_2(t)$ are bounded solutions and that they have the same sequential asymptotic behavior at infinity. We can apply Proposition 12.26 to conclude $\hat{g}_1(t) = \hat{g}_2(t)$ on $t_0 \leq t \leq T'$. We have proved:

COROLLARY 12.27. *Let $g_1(t)$ and $g_2(t)$ be standard solutions. Suppose that $g_1(t) = g_2(t)$ for all $t \in [0, t_0]$ for some $t_0 \geq 0$. The Ricci-DeTurck solutions $\hat{g}_1(t)$ and $\hat{g}_2(t)$ constructed from standard solutions $g_1(t)$ and $g_2(t)$ with $g_1(t_0) = g_2(t_0)$ exist and satisfy $\hat{g}_1(t) = \hat{g}_1(t)$ for $t \in [t_0, T']$ for some $T' > t_0$.*

8. Completion of the proof of uniqueness

Now we are ready to prove the uniqueness of the standard solution. Let $g_1(t)$, $0 \leq t < T_1$, and $g_2(t)$, $0 \leq t < T_2$, be standard solutions. Consider the maximal interval I (closed or half-open) containing 0 on which g_1 and g_2 agree.

Case 1: $T_1 < T_2$ and $I = [0, T_1)$.

In this case since $g_1(t) = g_2(t)$ for all $t < T_1$ and $g_2(t)$ extends smoothly past time T_1, we see that the curvature of $g_1(t)$ is bounded as t tends to T_1. Hence, $g_1(t)$ extends past time T_1, contradicting the fact that it is a maximal flow.

Case 2: $T_2 < T_1$ and $I = [0, T_2)$.

The argument in this case is the same as the previous one with the roles of $g_1(t)$ and $g_2(t)$ reversed.

There is one more case to rule out.

Case 3: I is a closed interval $I = [0, t_0]$.

In this case, of course, $t_0 < \min(T_1, T_2)$. Hence we apply Proposition 12.25 to construct solutions ψ_1 and ψ_2 to the harmonic map flow for $g_1(t)$ and $g_2(t)$ with ψ_1 and ψ_2 being the identity at time t_0. These solutions will be defined on an interval of the form $[t_0, T]$ for some $T > t_0$. Using these harmonic map flows we construct solutions $\hat{g}_1(t)$ and $\hat{g}_2(t)$ to the Ricci-DeTurck flow defined on the interval $[t_0, T]$. According to Corollary 12.27, there is a uniqueness theorem for these Ricci-DeTurck flows, which implies that $\hat{g}_1(t) = \hat{g}_2(t)$ for all $t \in [t_0, T']$ for some $T' > t_0$. Invoking Corollary 12.16 we conclude that $g_1(t) = g_2(t)$ for all $t \in [0, T']$, contradicting the maximality of the interval I.

If none of these three cases can occur, then the only remaining possibility is that $T_1 = T_2$ and $I = [0, T_1)$, i.e., the flows are the same. This then completes the proof of the uniqueness of the standard flow.

8.1. $T = 1$ and existence of canonical neighborhoods. At this point we have established all the properties claimed in Theorem 12.5 for the standard flow except for the fact that T, the endpoint of the time-interval of definition, is equal to 1. We have shown that $T \leq 1$. In order to establish the opposite inequality, we must show the existence of canonical neighborhoods for the standard solution.

Here is the result about the existence of canonical neighborhoods for the standard solution.

THEOREM 12.28. *Fix $0 < \epsilon < 1$. Then there is $r > 0$ such that for any point (x_0, t_0) in the standard flow with $R(x_0, t_0) \geq r^{-2}$ the following hold.*

(1) *$t_0 > r^2$.*
(2) *(x_0, t_0) has a strong canonical $(C(\epsilon), \epsilon)$-neighborhood. If this canonical neighborhood is a strong ϵ-neck centered at (x_0, t_0), then the strong neck extends to an evolving neck defined for backward rescaled time $(1 + \epsilon)$.*

PROOF. Take an increasing sequence of times t'_n converging to T. Since the curvature of $(\mathbb{R}^3, g(t))$ is locally bounded in time, for each n, there is a bound on the scalar curvature on $\mathbb{R}^3 \times [0, t'_n]$. Hence, there is a finite upper bound R_n on $R(x, t)$ for all points (x, t) with $t \leq t'_n$ for which the conclusion of the theorem does not hold. (There clearly are such points since the conclusion of the theorem fails for all $(x, 0)$.) Pick (x_n, t_n) with $t_n \leq t'_n$, with $R(x_n, t_n) \geq R_n/2$ and such that the conclusion of the theorem does not hold for (x_n, t_n). To prove the theorem we must show that $\overline{\lim}_{n \to \infty} R(x_n, t_n) < \infty$. Suppose the contrary. By passing to a subsequence we can suppose that $\lim_{n \to \infty} R(x_n, t_n) = \infty$. We set $Q_n = R(x_n, t_n)$. We claim that all the hypotheses of Theorem 11.8 apply to the sequence $(\mathbb{R}^3, g(t), (x_n, t_n))$. First, we show that all the hypotheses of Theorem 11.1 (except the last) hold. Since $(\mathbb{R}^3, g(t))$ has non-negative curvature all these flows have curvature pinched toward positive. By Proposition 12.13 there are $r > 0$ and $\kappa > 0$ so that all these flows are κ-non-collapsed on scales $\leq r$. By construction if $t \leq t_n$ and $R(y, t) > 2Q_n \geq R_n$ then the point (y, t) has a strong canonical $(C(\epsilon), \epsilon)$-neighborhood. We are assuming that $Q_n \to \infty$ as $n \to \infty$ in order to achieve the contradiction. Since all time-slices are complete, all balls of finite radius have compact closure.

Lastly, we need to show that the extra hypothesis of Theorem 11.8 (which includes the last hypothesis of Theorem 11.1) is satisfied. This is clear since $t_n \to T$ as $n \to \infty$ and $Q_n \to \infty$ as $n \to \infty$. Applying Theorem 11.8 we conclude that after passing to a subsequence there is a limiting flow which is a κ-solution. Clearly, this and Corollary 9.95 imply that for all sufficiently large n (in the subsequence) the neighborhood as required by the theorem exists. This contradicts our assumption that none of the points (x_n, t_n) have these neighborhoods. This contradiction proves the result. $\square$

8.2. Completion of the proof of Theorem 12.5. The next proposition establishes the last of the conditions claimed in Theorem 12.5.

THEOREM 12.29. *For the standard flow, $T = 1$.*

PROOF. We have already seen in Corollary 12.8 that $T \le 1$. Suppose now that $T < 1$. Take $T_0 < T$ sufficiently close to T. Then according to Proposition 12.7 there is a compact subset $X \subset \mathbb{R}^3$ such that restriction of the flow to $(\mathbb{R}^3 \setminus X) \times [0, T_0]$ is ϵ-close to the standard evolving flow on $S^2 \times (0, \infty), (1-t)h_0 \times ds^2$, where h_0 is the round metric of scalar curvature 1 on S^2. In particular, $R(x, T_0) \le (1+\epsilon)(1-T_0)^{-1}$ for all $x \in \mathbb{R}^3 \setminus X$. Because of Theorem 12.28 and the definition of $(C(\epsilon), \epsilon)$-canonical neighborhoods, it follows that at any point (x, t) with $R(x, t) \ge r^{-2}$, where $r > 0$ is the constant given in Theorem 12.28, we have $\partial R/\partial t(x, t) \le C(\epsilon) R^2(x, t)$. Thus, provided that $T - T_0$ is sufficiently small, there is a uniform bound to $R(x, t)$ for all $x \in \mathbb{R}^3 \setminus X$ and all $t \in [T_0, T)$. Using Theorem 3.29 and the fact that the standard flow is κ-non-collapsed implies that the restrictions of the metrics $g(t)$ to $\mathbb{R}^3 \setminus X$ converge smoothly to a limiting Riemannian metric $g(T)$ on $\mathbb{R}^3 \setminus X$. Fix a non-empty open subset $\Omega \subset \mathbb{R}^3 \setminus X$ with compact closure. For each $t \in [0, T)$ let $V(t)$ be the volume of $(\Omega, g(t)|_\Omega)$. Of course, $\lim_{t \to T} V(t) = \mathrm{Vol}_{g(T)} \Omega > 0$.

Since the limiting metric $g(T)$ exists in a neighborhood of infinity and has bounded curvature there, if the limit metric $g(T)$ exists on all of $\mathbb{R}^3$, then we can extend the flow keeping the curvature bounded. This contradicts the maximality of our flow subject to the condition that the curvature be locally bounded in time. Consequently, there is a point $x \in \mathbb{R}^3$ for which the limit metric $g(T)$ does not exist. This means that $\overline{\lim}_{t \to T} R(x, t) = \infty$. That is to say, there is a sequence of $t_n \to T$ such that setting $Q_n = R(x, t_n)$, we have $Q_n \to \infty$ as n tends to infinity. By Theorem 12.28 the second hypothesis in the statement of Theorem 11.1 holds for the sequence $(\mathbb{R}^3, g(t), (x, t_n))$. All the other hypotheses of this theorem as well as the extra hypothesis in Theorem 11.8 obviously hold for this sequence. Thus, according to Theorem 11.8 the based flows $(\mathbb{R}^3, Q_n g(Q_n^{-1} t' + t_n), (x, 0))$ converge smoothly to a κ-solution. Since the asymptotic volume of any κ-solution is zero (see Theorem 9.59), we see that for all n sufficiently large, the following holds:

CLAIM 12.30. *For any $\epsilon > 0$, there is $A < \infty$ such that for all n sufficiently large we have*
$$\mathrm{Vol} B_{Q_n g}(x, t_n, A) < \epsilon A^3.$$

Rescaling, we see that for all n sufficiently large we have
$$\mathrm{Vol}\, B_g(x, t_n, A/\sqrt{Q_n}) < \epsilon (A/\sqrt{Q_n})^3.$$

Since the curvature of $g(t_n)$ is non-negative and since the Q_n tend to ∞, it follows from the Bishop-Gromov Inequality (Theorem 1.34) that for any $0 < A < \infty$ and any $\epsilon > 0$, for all n sufficiently large we have

$$\operatorname{Vol} B_g(x, t_n, A) < \epsilon A^3.$$

On the other hand, since Ω has compact closure, there is an $A_1 < \infty$ with $\Omega \subset B(x, 0, A_1)$. Since the curvature of $g(t)$ is non-negative for all $t \in [0, T)$, it follows from Lemma 3.14 that the distance is a non-increasing function of t, so that for all $t \in [0, T)$ we have $\Omega \subset B(x, t, A_1)$. Applying the above, for any $\epsilon > 0$ for all n sufficiently large we have

$$\operatorname{Vol}(\Omega, g(t_n)) \leq \operatorname{Vol}_g B(x, t_n, A_1) < \epsilon (A_1)^3.$$

But this contradicts the fact that

$$\lim_{n \to \infty} \operatorname{Vol}(\Omega, g(t_n)) = \operatorname{Vol}(\Omega, g(T)) > 0.$$

This contradiction proves that $T = 1$. □

This completes the proof of Theorem 12.5.

9. Some corollaries

Now let us derive extra properties of the standard solution that will be important in our applications.

PROPOSITION 12.31. *There is a constant $c > 0$ such that for all (p, t) in the standard solution we have*

$$R(p, t) \geq \frac{c}{1 - t}.$$

PROOF. First, let us show that there is not a limiting metric $g(1)$ defined on all of $\mathbb{R}^3$. This does not immediately contradict the maximality of the flow because we are assuming only that the flow is maximal subject to having curvature locally bounded in time. Assume that a limiting metric $(\mathbb{R}^3, g(1))$ exists. First, notice that from the canonical neighborhood assumption and Lemma 11.2 we see that the curvature of $g(T)$ must be unbounded at spatial infinity. On the other hand, by Proposition 9.79 every point of $(\mathbb{R}^3, g(1))$ of curvature greater than R_0 has a $(2C, 2\epsilon)$-canonical neighborhood. Hence, since $(\mathbb{R}^3, g(1))$ has unbounded curvature, it then has 2ϵ-necks of arbitrarily small scale. This contradicts Proposition 2.19. (One can also rule this possibility out by direct computation using the spherical symmetry of the metric.) This means that there is no limiting metric $g(1)$.

The next step is to see that for any $p \in \mathbb{R}^3$ we have $\lim_{t \to 1} R(p, t) = \infty$. Let $\Omega \subset \mathbb{R}^3$ be the subset of $x \in \mathbb{R}^3$ for which $\liminf_{t \to 1} R(x, t) < \infty$. We suppose that $\Omega \neq \emptyset$. According to Theorem 11.19 the subset Ω is open and the metrics $g(t)|_\Omega$ converge smoothly to a limiting metric $g(1)|_\Omega$. On the other hand, we have just seen that there is not a limit metric $g(1)$ defined everywhere. This means that there is $p \in \mathbb{R}^3$ with $\lim_{t \to 1} R(p, t) = \infty$.

Take a sequence t_n converging to 1 and set $Q_n = R(p, t_n)$. By Theorem 11.8 we see that, possibly after passing to a subsequence, the based flows $(\mathbb{R}^3, Q_n g(t' - t_n), (p, 0))$ converge to a κ-solution. Then by Theorem 9.59 for any $\epsilon > 0$ there is $A < \infty$ such that $\operatorname{Vol} B_{Q_n g}(p, t_n, A) < \epsilon A^3$, and hence after rescaling we have $\operatorname{Vol} B_g(p, t_n, A/\sqrt{Q_n}) < \epsilon (A/\sqrt{Q_n})^3$. By the Bishop-Gromov inequality (Theorem 1.34) it follows that for any $0 < A < \infty$, any $\epsilon > 0$ and for all n sufficiently large, we have $\operatorname{Vol} B_g(p, t_n, A) < \epsilon A^3$. Take a non-empty subset $\Omega' \subset \Omega$ with compact closure. Of course, $\operatorname{Vol}(\Omega', g(t))$ converges to $\operatorname{Vol}(\Omega', g(T)) > 0$ as $t \to T$. Then there is $A < \infty$ such that for each n, the subset Ω' is contained in the ball $B(p_0, t_n, A)$. This is a contradiction since it implies that for any $\epsilon > 0$ for all n sufficiently large we have $\operatorname{Vol}(\Omega', g(t)) < \epsilon A^3$. This completes the proof that for every $p \in \mathbb{R}^3$ we have $\lim_{t \to 1} R(p, t) = \infty$.

Fix $\epsilon > 0$ sufficiently small and set $C = C(\epsilon)$. Then for every (p, t) with $R(p, t) \geq r^{-2}$ we have

$$\left| \frac{dR}{dt}(p, t) \right| \leq C R^2(p, t).$$

Fix $t_0 = 1 - 1/2r^2 C$. Since the flow has curvature locally bounded in time, there is $2C \leq C' < \infty$ such that $R(p, t_0) \leq 1/(C'(1 - t_0))$ for all $p \in \mathbb{R}^3$. Since $R(p, t_0) = 1/C'(1 - t_0)$, for all $t \in [t_0, 1)$ we have

$$R(p, t) < \max \left(\left[(C' - C)(1 - t_0) \right]^{-1}, \left[r^{-2} - C(1 - t_0) \right]^{-1} \right).$$

This means that $R(p, t)$ is uniformly bounded as $t \to 1$, contradicting what we just established. This shows that for $t \geq 1 - 1/2r^2 C$ the result holds. For $t \leq 1 - 1/2r^2 C$ there is a positive lower bound on the scalar curvature, and hence the result is immediate for these t as well. □

THEOREM 12.32. *For any $\epsilon > 0$ there is $C'(\epsilon) < \infty$ such that for any point x in the standard solution one of the following holds (see* FIG. *2).*
  (1) *(x, t) is contained in the core of a $(C'(\epsilon), \epsilon)$-cap.*
  (2) *(x, t) is the center of an evolving ϵ-neck N whose initial time-slice is $t = 0$, and this time-slice is disjoint from the surgery cap.*
  (3) *(x, t) is the center of an evolving ϵ-neck defined for rescaled time $1 + \epsilon$.*

REMARK 12.33. At first glance it may seem impossible for a point (x, t) in the standard solution to be the center of an evolving ϵ-neck defined for rescaled time $1 + \epsilon$ since the standard solution itself is only defined for time 1. But this is indeed possible. The reason is because the scale referred to for an evolving neck centered at (x, t) is $R(x, t)^{-1/2}$. As t approaches 1, $R(x, t)$ goes to infinity, so that rescaled time 1 at (x, t) is an arbitrarily small time interval measured in the scale of the standard solution.

9. SOME COROLLARIES

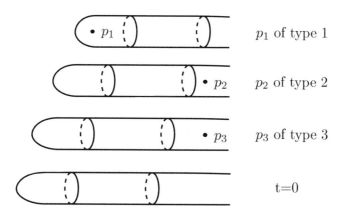

FIGURE 2. Canonical neighborhoods in the standard solution.

PROOF. By Theorem 12.28, there is r_0 such that if $R(x,t) \geq r_0^{-2}$, then (x,t) has a (C,ϵ)-canonical neighborhood and if this canonical neighborhood is a strong ϵ-neck centered at x, then that neck extends to an evolving neck defined for rescaled time $(1+\epsilon)$. By Proposition 12.31, there is $\theta < 1$ such that if $R(x,t) \leq r_0^{-2}$ then $t \leq \theta$. By Proposition 12.7, there is a compact subset $X \subset \mathbb{R}^3$ such that if $t \leq \theta$ and $x \notin X$, then there is an evolving ϵ-neck centered at x whose initial time is zero and whose initial time-slice is at distance at least 1 from the surgery cap. Lastly, by compactness there is $C' < \infty$ such that every (x,t) for $x \in X$ and every $t \leq \theta$ is contained in the core of a (C',ϵ)-cap. □

COROLLARY 12.34. *Fix $\epsilon > 0$. Suppose that (q,t) is a point in the standard solution with $t \leq R(q,t)^{-1}(1+\epsilon))$ and with*

$$(q,0) \in B(p_0, 0, (\epsilon^{-1}/2) + A_0 + 5).$$

Then (q,t) is contained in an $(C'(\epsilon), \epsilon)$-cap.

REMARK 12.35. Recall that p_0 is the origin in $\mathbb{R}^3$ and hence is the tip of the surgery cap. Also, A_0 is defined in Lemma 12.3.

COROLLARY 12.36. *For any $\epsilon > 0$ let $C' = C'(\epsilon)$ be as in Theorem 12.32. Suppose that we have a sequence of generalized Ricci flows $(\mathcal{M}_n, G_n)$, points $x_n \in \mathcal{M}_n$ with $\mathbf{t}(x_n) = 0$, neighborhoods U_n of x_n in the zero time-slice of $\mathcal{M}_n$, and a constant $0 < \theta < 1$. Suppose that there are embeddings $\rho_n \colon U_n \times [0,\theta) \to \mathcal{M}_n$ compatible with time and the vector field so that the Ricci flows $\rho_n^* G_n$ on U_n based at x_n converge geometrically to the restriction of the standard solution to $[0,\theta)$. Then for all n sufficiently large, and any point y_n in the image of ρ_n, one of the following holds:*

(1) *y_n is contained in the core of a $(C'(\epsilon), \epsilon)$-cap.*
(2) *y_n is the center of a strong ϵ-neck.*

(3) y_n is the center of an evolving ϵ-neck whose initial time-slice is at time 0.

PROOF. This follows immediately from Theorem 12.32 and Proposition 9.79. □

There is one property that we shall use later in proving the finite-time extinction of Ricci flows with surgery for manifolds with finite fundamental group (among others). This is a distance decreasing property which we record here.

Notice that for the standard initial metric constructed in Lemma 12.2 we have the following:

LEMMA 12.37. *Let S^2 be the unit sphere in $T_0\mathbb{R}^3$. Equip it with the metric h_0 that is twice the usual metric (so that the scalar curvature of h_0 is 1). We define a map $\rho\colon S^2 \times [0,\infty) \to \mathbb{R}^3$ by sending the point (x,s) to the point at distance s from the origin in the radial direction from 0 given by x (all this measured in the metric g_0). Then $\rho^* g_0 \leq h_0 \times ds^2$.*

PROOF. Clearly, the metric $\rho^* g_0$ is rotationally symmetric and its component in the s-direction is ds^2. On the other hand, since each cross section $\{s\} \times S^2$ maps conformally onto a sphere of radius $\leq \sqrt{2}$ the result follows. □

Part 3

Ricci flow with surgery

CHAPTER 13

Surgery on a δ-neck

1. Notation and the statement of the result

In this chapter we describe the surgery process. For this chapter we fix:

(1) A δ-neck (N, g) centered at a point x_0. We denote by $\rho \colon S^2 \times (-\delta^{-1}, \delta^{-1}) \to N$ the diffeomorphism that gives the δ-neck structure.
(2) Standard initial conditions $(\mathbb{R}^3, g_0)$.

We denote by $h_0 \times ds^2$ the metric on $S^2 \times \mathbb{R}$ which is the product of the round metric h_0 on S^2 of scalar curvature 1 and the Euclidean metric ds^2 on $\mathbb{R}$. We denote by $N^- \subset N$ the image $\rho((-\delta^{-1}, 0] \times S^2)$ and we denote by $s \colon N^- \to (-\delta^{-1}, 0]$ the composition ρ^{-1} followed by the projection to the second factor.

Recall that the standard initial metric $(\mathbb{R}^3, g_0)$ is invariant under the standard $SO(3)$-action on $\mathbb{R}^3$. We let p_0 denote the origin in $\mathbb{R}^3$. It is the fixed point of this action and is called the *tip* of the standard initial metric. Recall from Lemma 12.3 that there are $A_0 > 0$ and an isometry

$$\psi \colon (S^2 \times (-\infty, 4], h_0 \times ds^2) \to (\mathbb{R}^3 \setminus B(p_0, A_0), g_0).$$

The composition of ψ^{-1} followed by projection onto the second factor defines a map $s_1 \colon \mathbb{R}^3 \setminus B(p_0, A_0) \to (-\infty, 4]$. Lastly, there is $0 < r_0 < A_0$ such that on $B(p_0, r_0)$ the metric g_0 is of constant sectional curvature $1/4$. We extend the map s_1 to a continuous map $s_1 \colon \mathbb{R}^3 \to (-\infty, 4 + A_0]$ defined by $s_1(x) = A_0 + 4 - d_{g_0}(p, x)$. This map is an isometry along each radial geodesic ray emanating from p_0. It is smooth except at p_0 and sends p_0 to $4 + A_0$. The pre-images of s_1 on $(-\infty, 4 + A_0)$ are 2-spheres with round metrics of scalar curvature at least 1.

The surgery process is a local one defined on the δ-neck (N, g). The surgery process replaces (N, g) by a smooth Riemannian manifold $(\mathcal{S}, \widetilde{g})$. The underlying smooth manifold $\mathcal{S}$ is obtained by gluing together $\rho(S^2 \times (-\delta^{-1}, 4))$ and $B(p_0, A_0+4)$ by identifying $\rho(x, s)$ with $\psi(x, s)$ for all $x \in S^2$ and all $s \in (0, 4)$. The functions s on N^- and s_1 agree on their overlap and hence together define a function $s \colon \mathcal{S} \to (-\delta^{-1}, 4 + A_0]$, a function smooth except at p_0. In order to define the metric $\widetilde{g}$ we must make some universal choices. We fix once and for all two bump functions $\alpha \colon [1, 2] \to [0, 1]$, which is required to be identically 1 near 1 and identically 0 near 2, and

$\beta\colon [4+A_0-r_0, 4+A_0] \to [0,1]$, which is required to be identically 1 near $4+A_0-r_0$ and identically 0 on $[4+A_0-r_0/2, A_0]$. These functions are chosen once and for all and are independent of δ and (N,g). Next we set $\eta = \sqrt{1-\delta}$. The purpose of this choice is the following:

CLAIM 13.1. *Let* $\xi\colon N \to \mathbb{R}^3$ *be the map that sends* $\rho(S^2 \times [A_0+4, \delta^{-1}))$ *to the origin* $0 \in \mathbb{R}^3$ *(i.e., to the tip of the surgery cap) and for every* $s < A_0+4$ *sends* (x,s) *to the point in* $\mathbb{R}^3$ *in the radial direction* x *from the origin at* g_0-*distance* $A_0 + 4 - s$. *Then* ξ *is a distance decreasing map from* $(N, R(x_0)g)$ *to* $(\mathbb{R}^3, \eta g_0)$.

PROOF. Since $R(x_0)g$ is within δ of $h_0 \times ds^2$, it follows that $R(x_0)g \geq \eta(h_0 \times ds^2)$. But according to Lemma 12.37 the map ξ given in the statement of the claim is a distance non-increasing map from $h_0 \times ds^2$ to g_0. The claim follows immediately. □

The last choices we need to make are of constants $C_0 < \infty$ and $q < \infty$, with $C_0 \gg q$, but both of these are independent of δ. These choices will be made later. Given all these choices, we define a function

$$f(s) = \begin{cases} 0 & s \leq 0, \\ C_0 \delta e^{-q/s} & s > 0, \end{cases}$$

and then we define the metric $\widetilde{g}$ on $\mathcal{S}$ by first defining a metric:

$$\hat{g} = \begin{cases} \exp(-2f(s))R(x_0)\rho^* g, & -\infty < s \leq 1, \\ \exp(-2f(s))\left(\alpha(s)R(x_0)\rho^* g + (1-\alpha(s))\eta g_0\right), & 1 \leq s \leq 2, \\ \exp(-2f(s))\eta g_0, & 2 \leq s \leq A', \\ \left[\beta(s)\exp(-2f(s)) + (1-\beta(s))\exp(-2f(4+A_0))\right]\eta g_0, & A' \leq s \leq A'', \end{cases}$$

where $A' = 4 + A_0 - r_0$ and $A'' = A_0 + 4$. Then we define

$$\widetilde{g} = R(x_0)^{-1}\hat{g}.$$

See FIG. 1.

THEOREM 13.2. *There are constants* $C_0, q, R_0 < \infty$ *and* $\delta'_0 > 0$ *such that the following hold for the result* $(\mathcal{S}, \widetilde{g})$ *of surgery on* (N,g) *provided that* $R(x_0) \geq R_0$, $0 < \delta \leq \delta'_0$. *Define* $f(s)$ *as above with the constants* C_0, δ *and then use f to define surgery on a δ-neck N to produce* $(\mathcal{S}, \widetilde{g})$. *Then the following hold.*

- *Fix* $t \geq 0$. *For any* $p \in N$, *let* $X(p) = \max(0, -\nu(p))$, *where* $\nu(p)$ *is the smallest eigenvalue of* $\mathrm{Rm}(p)$. *Suppose that for all* $p \in N$ *we have:*
  (1) $R(p) \geq \frac{-6}{1+4t}$, *and*
  (2) $R(p) \geq 2X(p)(\log X(p) + \log(1+t) - 3)$, *whenever* $0 < X(p)$.
  Then the curvature of $(\mathcal{S}, \widetilde{g})$ *satisfies the same equations at every point of* $\mathcal{S}$ *with the same value of* t.

1. NOTATION AND THE STATEMENT OF THE RESULT

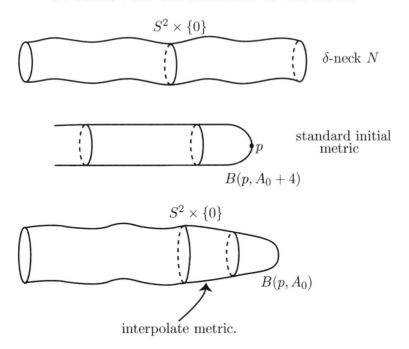

FIGURE 1. Local Surgery

- The restriction of the metric $\widetilde{g}$ to $s^{-1}([1, 4 + A_0])$ has positive sectional curvature.
- Let $\xi \colon N \to \mathcal{S}$ be the map given in Claim 13.1. Then it is a distance decreasing map from g to $\widetilde{g}$.
- For any $\delta'' > 0$ there is $\delta_1' = \delta_1'(\delta'') > 0$ such that if $\delta \leq \min(\delta_1', \delta_0')$, then the restriction of $\hat{g}$ to $B_{\hat{g}}(p_0, (\delta'')^{-1})$ in $(\mathcal{S}, \hat{g})$ is δ''-close in the $C^{[1/\delta'']}$-topology to the restriction of the standard initial metric g_0 to $B_{g_0}(p_0, (\delta'')^{-1})$.

The rest of this chapter is devoted to the proof of this theorem.

Before starting the curvature computations let us make a remark about the surgery cap.

DEFINITION 13.3. The image in $\mathcal{S}$ of $B_{g_0}(p_0, 0, A_0 + 4)$ is called the *surgery cap*.

The following is immediate from the definitions provided that $\delta > 0$ is sufficiently small.

LEMMA 13.4. *The surgery cap in $(\mathcal{S}, \widetilde{g})$ has a metric that differs from the one coming from a rescaled version of the standard solution. Thus, the image of this cap is not necessarily a metric ball. Nevertheless for $\epsilon < 1/200$ the image of this cap will be contained in the metric ball in $\mathcal{S}$ centered at p_0 of radius $R(x_0)^{-1/2}(A_0 + 5)$ and will contain the metric ball centered at p_0*

of radius $R(x_0)^{-1/2}(A_0 + 3)$. Notice also that the complement of the closure of the surgery cap in $\mathcal{S}$ is isometrically identified with N^-.

2. Preliminary computations

We shall compute in a slightly more general setup. Let I be an open interval contained in $(-\delta^{-1}, 4 + A_0)$ and let h be a metric on $S^2 \times I$ within δ in the $C^{[1/\delta]}$-topology of the restriction to this open submanifold of the standard metric $h_0 \times ds^2$. We let $\hat{h} = e^{-2f} h$. Fix local coordinates near a point $y \in S^2 \times I$. We denote by ∇ the covariant derivative for h and by $\widehat{\nabla}$ the covariant derivative for $\hat{h}$. We also denote by (R_{ijkl}) the matrix of the Riemann curvature operator of h in the associated basis of $\wedge^2 T(S^2 \times I)$ and by $(\hat{R}_{ijkl})$ the matrix of the Riemann curvature operator of $\hat{h}$ with respect to the same basis. Recall the formula for the curvature of a conformal change of metric (see, (3.34) on p.51 of [60]):

$$\text{(13.1)} \quad \hat{R}_{ijkl} = e^{-2f} \left(R_{ijkl} - f_j f_k h_{il} + f_j f_l h_{ik} + f_i f_k h_{jl} - f_i f_l h_{jk} \right.$$
$$\left. - (\wedge^2 h)_{ijkl} |\nabla f|^2 - f_{jk} h_{il} + f_{ik} h_{jl} + f_{jl} h_{ik} - f_{il} h_{jk} \right).$$

Here, f_i means $\partial_i f$,

$$f_{ij} = \text{Hess}_{ij}(f) = \partial_i f_j - f_l \Gamma_{ij}^l,$$

and $\wedge^2 h$ is the metric induced by h on $\wedge^2 TN$, so that

$$\wedge^2 h_{ijkl} = h_{ik} h_{jl} - h_{il} h_{jk}.$$

Now we introduce the notation $O(\delta)$. When we say that a quantity is $O(\delta)$ we mean that there is some universal constant C such that, provided that $\delta > 0$ is sufficiently small, the absolute value of the quantity is $\leq C\delta$. The universal constant is allowed to change from inequality to inequality.

In our case we take local coordinates adapted to the δ-neck: (x^0, x^1, x^2) where x^0 agrees with the s-coordinate and (x^1, x^2) are Gaussian local coordinates on the S^2 such that dx^1 and dx^2 are orthonormal at the point in question in the round metric h_0. The function f is a function only of x^0. Hence $f_i = 0$ for $i = 1, 2$. Also, $f_0 = \frac{q}{s^2} f$. It follows that

$$|\nabla f|_h = \frac{q}{s^2} f \cdot (1 + O(\delta)),$$

so that

$$|\nabla f|_h^2 = \frac{q^2}{s^4} f^2 \cdot (1 + O(\delta)).$$

Because the metric h is δ-close to the product $h_0 \times ds^2$, we see that $h_{ij}(y) = (h_0)_{ij}(y) + O(\delta)$ and the Christoffel symbols $\Gamma_{ij}^k(y)$ of h are within δ in the $C^{([1/\delta]-1)}$-topology of those of the product metric $h_0 \times ds^2$. In

2. PRELIMINARY COMPUTATIONS

particular, $\Gamma_{ij}^0 = O(\delta)$ for all ij. The components f_{ij} of the Hessian with respect to h are given by

$$f_{00} = \left(\frac{q^2}{s^4} - \frac{2q}{s^3}\right) f + \frac{q}{s^2} fO(\delta),$$

$$f_{i0} = \frac{q}{s^2} fO(\delta) \quad \text{for } 1 \leq i \leq 2,$$

$$f_{ij} = \frac{q}{s^2} fO(\delta) \quad \text{for } 1 \leq i,j \leq 2.$$

In the following a, b, c, d are indices taking values 1 and 2. Substituting in Equation (13.1) yields

$$\hat{R}_{0a0b} = e^{-2f} \left(R_{0a0b} + \frac{q^2}{s^4} f^2 h_{ab} - h_{ab}(\frac{q^2}{s^4}) f^2 (1 + O(\delta)) + \left(\frac{q^2}{s^4} - \frac{2q}{s^3}\right) fh_{ab} \right.$$
$$\left. + \frac{q}{s^2} fO(\delta) \right)$$
$$= e^{-2f} \left(R_{0a0b} + \left(\frac{q^2}{s^4} - \frac{2q}{s^3}\right) fh_{ab} + h_{ab}(\frac{q^2}{s^4}) f^2 O(\delta) + \frac{q}{s^2} fO(\delta) \right).$$

Also, we have

$$\hat{R}_{ab0c} = e^{-2f} \left(R_{ab0c} - (\wedge^2 h)_{ab0c}(\frac{q^2}{s^4}) f^2 (1 + O(\delta)) + \frac{q}{s^2} fO(\delta) \right)$$
$$= e^{-2f} \left(R_{ab0c} + (\frac{q^2}{s^4}) f^2 O(\delta) + \frac{q}{s^2} fO(\delta) \right).$$

Lastly,

$$\hat{R}_{1212} = e^{-2f} \left(R_{1212} - (\wedge^2 h)_{1212} \frac{q^2}{s^4} f^2 (1 + O(\delta)) + \frac{q}{s^2} fO(\delta) \right)$$
$$= e^{-2f} \left(R_{1212} - \frac{q^2}{s^4} f^2 (1 + O(\delta)) + \frac{q}{s^2} fO(\delta) \right).$$

Now we are ready to fix the constant q. We fix it so that for all $s \in [0, 4 + A_0]$ we have

(13.2) $$q \gg (4 + A_0)^2 \quad \text{and} \quad \frac{q^2}{s^4} e^{-q/s} \ll 1.$$

It follows immediately that $q^2/s^4 \gg q/s^3$ for all $s \in [0, 4 + A_0]$. We are not yet ready to fix the constant C_0, but once we do we shall always require δ to satisfy $\delta \ll C_0^{-1}$ so that for all $s \in [0, 4 + A_0]$ we have

$$\frac{q}{s^2} f^2 \ll \frac{q^2}{s^4} f^2 \ll \frac{q}{s^2} f \ll 1.$$

(These requirements are not circular, since C_0 and q are chosen independent of δ.)

Using these inequalities and putting our computations in matrix form show the following.

COROLLARY 13.5. *There is $\delta_2' > 0$, depending on C_0 and q, such that if $\delta \leq \delta_2'$ then we have*
(13.3)
$$\left(\hat{R}_{ijkl}\right) = e^{-2f}\left[(R_{ijkl}) + \begin{pmatrix} -\frac{q^2}{s^4}f^2 & 0 \\ 0 & \left(\frac{q^2}{s^4} - \frac{2q}{s^3}\right)f\begin{pmatrix} 1 & 0 \\ 0 & 1 \end{pmatrix} \end{pmatrix} + \left(\frac{q}{s^2}fO(\delta)\right)\right].$$

Similarly, we have the equation relating scalar curvatures
$$\hat{R} = e^{2f}\left(R + 4\triangle f - 2|\nabla f|^2\right),$$
and hence
$$\hat{R} = e^{2f}\left(R + 4\left(\frac{q^2}{s^4} - \frac{2q}{s^3}\right)f - 2\frac{q^2}{s^4}f^2 + \frac{q}{s^2}fO(\delta)\right).$$

COROLLARY 13.6. *For any constant $C_0 < \infty$ and any $\delta < \min(\delta_2', C_0^{-1})$ we have $\hat{R} \geq R$.*

PROOF. It follows from the conditions on q that, since $C_0\delta < 1$, we have $f^2 \ll f$ and $q^2/s^4 \gg \max(q/s^3, q/s^2)$. The result then follows immediately from the above formula. □

Now let us compute the eigenvalues of the curvature $R_{ijkl}(y)$ for any $y \in S^2 \times I$.

LEMMA 13.7. *There is a $\delta_3' > 0$ such that the following hold if $\delta \leq \delta_3'$. Let $\{e_0, e_1, e_2\}$ be an orthonormal basis for the tangent space at a point $y \in S^2 \times I$ for the metric $h_0 \times ds^2$ with the property that e_0 points in the I-direction. Then there is a basis $\{f_0, f_1, f_2\}$ for this tangent space so that the following hold:*

(1) *The basis is orthonormal in the metric h.*
(2) *The change of basis matrix expressing the $\{f_0, f_1, f_2\}$ in terms of $\{e_0, e_1, e_2\}$ is of the form $\text{Id} + O(\delta)$.*
(3) *The Riemann curvature of h in the basis $\{f_0 \wedge f_1, f_1 \wedge f_2, f_2 \wedge f_0\}$ of $\wedge^2 T_y(S^2 \times I)$ is*
$$\begin{pmatrix} 1/2 & 0 & 0 \\ 0 & 0 & 0 \\ 0 & 0 & 0 \end{pmatrix} + O(\delta).$$

PROOF. Since h is within δ of $h_0 \times ds^2$ in the $C^{[1/\delta]}$-topology, it follows that the matrix for $h(y)$ in $\{e_0, e_1, e_2\}$ is within $O(\delta)$ of the identity matrix, and the matrix for the curvature of h in the associated basis of $\wedge^2 T_y(S^2 \times I)$ is within $O(\delta)$ of the curvature matrix for $h_0 \times ds^2$, the latter being the diagonal matrix with diagonal entries $\{1/2, 0, 0\}$. Thus, the usual Gram-Schmidt orthonormalization process constructs the basis $\{f_0, f_1, f_2\}$ satisfying the first two items. Let $A = (A^{ab})$ be the change of basis matrix expressing the $\{f_a\}$ in terms of the $\{e_b\}$, so that $A = \text{Id} + O(\delta)$. The curvature of h in this basis is then given by $B^{\text{tr}}(R_{ijkl})B$ where $B = \wedge^2 A$ is the induced change of basis matrix expressing the basis $\{f_0 \wedge f_1, f_1 \wedge f_2, f_2 \wedge f_0\}$ in terms of $\{e_0 \wedge e_1, e_1 \wedge e_2, e_2 \wedge e_0\}$. Hence, in the basis $\{f_0 \wedge f_1, f_1 \wedge f_2,, f_2 \wedge f_0\}$ the curvature matrix for h is within $O(\delta)$ of the same diagonal matrix. For δ sufficiently small, the eigenvalues of the curvature matrix for h are within $O(\delta)$ of $(1/2, 0, 0)$. □

COROLLARY 13.8. *The following holds provided that $\delta \leq \delta'_3$. It is possible to choose the basis $\{f_0, f_1, f_2\}$ satisfying the conclusions of Lemma 13.7 so that in addition the curvature matrix for $(R_{ijkl}(y))$ is of the form*

$$\begin{pmatrix} \lambda & 0 & 0 \\ 0 & \alpha & \beta \\ 0 & \beta & \gamma \end{pmatrix}$$

with $|\lambda - \tfrac{1}{2}| \leq O(\delta)$ and $|\alpha|, |\beta|, |\gamma| \leq O(\delta)$.

PROOF. We have an h-orthonormal basis $\{f_0 \wedge f_1, f_1 \wedge f_2, f_2 \wedge f_0\}$ for $\wedge^2 T_y(S^2 \times \mathbb{R})$ in which the quadratic form $(R_{ijkl}(y))$ is

$$\begin{pmatrix} 1/2 & 0 & 0 \\ 0 & 0 & 0 \\ 0 & 0 & 0 \end{pmatrix} + O(\delta).$$

It follows that the restriction to the h-unit sphere in $\wedge^2 T_y(S^2 \times \mathbb{R})$ of this quadratic form achieves its maximum value at some vector v, which, when written out in this basis, is given by (x, y, z) with $|y|, |z| \leq O(\delta)$ and $|x-1| \leq O(\delta)$. Of course, this maximum value is within $O(\delta)$ of $1/2$. Clearly, on the h-orthogonal subspace to v, the quadratic form is given by a matrix all of whose entries are $O(\delta)$ in absolute value. This gives us a new basis of $\wedge^2 T_y(S^2 \times I)$ within $O(\delta)$ of the given basis in which $(R_{ijkl}(y))$ is diagonal. The corresponding basis for $T_y(S^2 \times \mathbb{R})$ is as required. □

Now we consider the expression $(\hat{R}_{ijkl}(y))$ in this basis.

LEMMA 13.9. *Set $\delta'_4 = \min(\delta'_2, \delta'_3)$. Suppose that $\delta \leq \min(\delta'_4, C_0^{-1})$. Then in the basis $\{f_0, f_1, f_2\}$ for $T_y(S^2 \times I)$ as in Corollary 13.8 we have*

$$(\hat{R}_{ijkl}(y)) - e^{-2f}\left[\begin{pmatrix} \lambda & 0 & 0 \\ 0 & \alpha & \beta \\ 0 & \beta & \gamma \end{pmatrix} + \begin{pmatrix} -\tfrac{q^2}{s^4}f^2 & 0 \\ 0 & \left(\tfrac{q^2}{s^4} - \tfrac{2q}{s^3}\right)f\begin{pmatrix} 1 & 0 \\ 0 & 1 \end{pmatrix} \end{pmatrix}\right]$$

is of the form

$$\left(\frac{q^2}{s^4} f O(\delta)\right)$$

where $\lambda, \alpha, \beta, \gamma$ are the constants in Corollary 13.8 and the first matrix is the expression for $(R_{ijkl}(y))$ in this basis.

PROOF. We simply conjugate the expression in Equation (13.3) by the change of basis matrix and use the fact that by our choice of q and the fact that $C_0\delta < 1$, we have $f \gg f^2$ and $q/s^3 \ll q^2/s^4$. □

COROLLARY 13.10. *Assuming that $\delta \leq \min(\delta'_4, C_0^{-1})$, there is an h-orthonormal basis $\{f_0, f_1, f_2\}$ so that in the associated basis for $\wedge^2 T_y(S^2 \times I)$ the matrix $(R_{ijkl}(y))$ is diagonal and given by*

$$\begin{pmatrix} \lambda & 0 & 0 \\ 0 & \mu & 0 \\ 0 & 0 & \nu \end{pmatrix}$$

with $|\lambda - 1/2| \leq O(\delta)$ and $|\mu|, |\nu| \leq O(\delta)$. Furthermore, in this same basis the matrix $(\hat{R}_{ijkl}(y))$ is

$$e^{-2f}\left[\begin{pmatrix} \lambda & 0 & 0 \\ 0 & \mu & 0 \\ 0 & 0 & \nu \end{pmatrix} + \begin{pmatrix} -\frac{q^2}{s^4}f^2 & 0 \\ 0 & \left(\frac{q^2}{s^4} - \frac{2q}{s^3}\right)f\begin{pmatrix} 1 & 0 \\ 0 & 1 \end{pmatrix} \end{pmatrix} + \frac{q^2}{s^4}fO(\delta)\right].$$

PROOF. To diagonalize the curvature operator, $(R_{ijkl}(y))$, we need only rotate in the $\{f_1 \wedge f_2, f_2 \wedge f_3\}$-plane. Applying this rotation to the expression in Lemma 13.7 gives the result. □

COROLLARY 13.11. *There is a constant $A < \infty$ such that the following holds for the given value of q and any C_0 provided that δ is sufficiently small. Suppose that the eigenvalues for the curvature matrix of h at y are $\lambda \geq \mu \geq \nu$. Then the eigenvalues for the curvature of $\hat{h}$ at the point y are given by λ', μ', ν', where*

$$\left|\lambda' - e^{2f}\left(\lambda - \frac{q^2}{s^4}f^2\right)\right| \leq \frac{q^2}{s^4}fA\delta,$$

$$\left|\mu' - e^{2f}\left(\mu + \left(\frac{q^2}{s^4} - \frac{2q}{s^3}\right)f\right)\right| \leq \frac{q^2}{s^4}fA\delta,$$

$$\left|\nu' - e^{2f}\left(\nu + \left(\frac{q^2}{s^4} - \frac{2q}{s^3}\right)f\right)\right| \leq \frac{q^2}{s^4}fA\delta.$$

In particular, we have

$$\nu' \geq e^{2f}\left(\nu + \frac{q^2}{2s^4}f\right),$$

$$\mu' \geq e^{2f}\left(\mu + \frac{q^2}{2s^4}f\right).$$

PROOF. Let $\{f_0, f_1, f_2\}$ be the h-orthonormal basis as in Corollary 13.10. Then $\{e^f f_0, e^f f_1, e^f f_2\}$ is orthonormal for $\hat{h} = e^{-2f}h$. This change multiplies the curvature matrix by e^{4f}. Since $f \ll 1$, $e^{4f} < 2$ so that the expression for $(\hat{R}_{ijkl}(y))$ in this basis is exactly the same as in Lemma 13.9 except that the factor in front is e^{2f} instead of e^{-2f}. Now, it is easy to see that since $((q^2/s^4)fA\delta)^2 \ll (q^2/s^4)fA\delta$, the eigenvalues will differ from the diagonal entries by at most a constant multiple of $(q^2/s^4)fA\delta$.

The first three inequalities are immediate from the previous corollary. The last two follow since $q^2/s^4 \gg q/s^3$ and $\delta \ll 1$. □

One important consequence of this computation is the following:

COROLLARY 13.12. *For the given value of q and for any C_0, assuming that $\delta > 0$ is sufficiently small, the smallest eigenvalue of $\mathrm{Rm}_{\hat{h}}$ is greater than the smallest eigenvalue of Rm_h at the same point. Consequently, at any point where h has non-negative curvature so does $\hat{h}$.*

PROOF. Since $|\lambda - 1/2|, |\mu|, |\nu|$ are all $O(\delta)$ and since $\frac{q^2}{s^4}f \ll 1$, it follows that the smallest eigenvalue of $(\hat{R}_{ijkl}(y))$ is either μ' or ν'. But it is immediate from the above expressions that $\mu' > \mu$ and $\nu' > \nu$. This completes the proof. □

Now we are ready to fix C_0. There is a universal constant K such that for all $\delta > 0$ sufficiently small and for any δ-neck (N, h) of scale 1, every eigenvalue of Rm_h is at least $-K\delta$. We set

$$C_0 = 2Ke^q.$$

LEMMA 13.13. *With these choices of q and C_0 for any $\delta > 0$ sufficiently small we have $\nu' > 0$ and $\mu' > 0$ for $s \in [1, 4 + A_0]$ and $\lambda' > 1/4$.*

PROOF. Then by the previous result we have

$$\nu' \geq e^{2f}\left(\nu + \frac{q^2}{2s^4}f\right).$$

It is easy to see that since $q \gg (4 + A_0)$ the function $(q^2/2s^4)f$ is an increasing function on $[1, 4 + A_0]$. Its value at $s = 1$ is $(q^2/2)e^{-q}C_0\delta > K\delta$. Hence $\nu + \frac{q^2}{2s^4}f > 0$ for all $s \in [1, 4 + A_0]$ and consequently $\nu' > 0$ on this submanifold. The same argument shows $\mu' > 0$. Since $q^2/s^4 f^2 \ll 1$ and $0 < f$, the statement about λ' is immediate. □

3. The proof of Theorem 13.2

3.1. Proof of the first two items for $s < 4$. We consider the metric in the region $s^{-1}(-\delta^{-1}, 4)$ given by

$$h = \alpha(s)R_g(x_0)\rho^*g + (1 - \alpha(s))\eta g_0.$$

There is a constant $K' < \infty$ (depending on the $C^{[1/\delta]}$-norm of α) such that h is within $K'\delta$ of the product metric in the $C^{[1/(K'\delta)]}$-topology. Thus, if δ is sufficiently small, all of the preceding computations hold with the error term $(q^2/s^4)fAK'\delta$. Thus, provided that δ is sufficiently small, the conclusions about the eigenvalues hold for $e^{-2f}h$ in the region $s^{-1}(-\delta^{-1}, 4)$. But $e^{-2f}h$ is exactly equal to $R(x_0)\widetilde{g}$ in this region. Rescaling, we conclude that on $s^{-1}(-\delta^{-1}, 4)$ the smallest eigenvalue of $\widetilde{g}$ is greater than the smallest eigenvalue of g at the corresponding point and that $R_{\widetilde{g}} \geq R_g$ in this same region.

The first conclusion of Theorem 13.2 follows by applying the above considerations to the case of $h = R_g(x_0)\rho^*g$. Namely, we have:

PROPOSITION 13.14. *Fix $\delta > 0$ sufficiently small. Suppose that for some $t \geq 0$ and every point $p \in N$ the curvature of h satisfies:*

(1) $R(p) \geq \frac{-6}{1+4t}$, *and*
(2) $R(p) \geq 2X(p)(\log X(p) + \log(1+t) - 3)$ *whenever $0 < X(x, t)$.*

Then the curvature $(\mathcal{S}, \widetilde{g})$ satisfies the same equation with the same value of t in the region $s^{-1}(-\delta^{-1}, 4)$. Also, the curvature of $\widetilde{g}$ is positive in the region $s^{-1}[1, 4)$.

PROOF. According to Corollary 13.12, the smallest eigenvalue of $\hat{h}$ at any point p is greater than or equal to the smallest eigenvalue of h at the corresponding point. According to Corollary 13.6, $\hat{R}(p) \geq R(p)$ for every $p \in \mathcal{S}$. Hence, $X_{\hat{h}}(p) \leq X_h(p)$. If $X_h(p) \geq e^3/(1+t)$, then we have

$$\hat{R}(p) \geq R(p)$$
$$\geq 2X_h(p)(\log X_h(p) + \log(1+t) - 3)$$
$$\geq 2X_{\hat{h}}(p)(\log X_{\hat{h}}(p) + \log(1+t) - 3).$$

If $X_h(p) < e^3(1+t)$, then $X_{\hat{h}}(p) < e^3/(1+t)$. Thus, in this case since we are in a δ-neck, provided that δ is sufficiently small, we have $R(p) \geq 0$ and hence

$$\hat{R}(p) \geq R(p) \geq 0 > 2X_{\hat{h}}(p)(\log X_{\hat{h}}(p) + \log(1+t) - 3).$$

This completes the proof in both cases.

This establishes the first item in the conclusion of Theorem 13.2 for $\delta > 0$ sufficiently small on $s^{-1}(-\delta^{-1}, 4)$. As we have seen in Lemma 13.13, the curvature is positive on $s^{-1}[1, 4)$. □

3.2. Proof of the first two items for $s \geq 4$. Now let us show that the curvature on $\widetilde{g}$ is positive in the region $s^{-1}([4, 4+A_0])$. First of all in the preimage of the interval $[4, 4+A_0, -r_0]$ this follows from Corollary 13.12 and the fact that ηg_0 has non-negative curvature. As δ tends to zero, the restriction of $\widetilde{g}$ to the subset $s^{-1}([4+A_0-r_0, 4+A_0])$ tends smoothly to the restriction of the metric g_0 to that subset. The metric g_0 has positive curvature on $s^{-1}([4+A_0-r_0, 4+A_0])$. Thus, for all $\delta > 0$ sufficiently small the metric $\widetilde{g}$ has positive curvature on all of $s^{-1}([4+A_0-r_0, 4+A_0])$. This completes the proof of the first two items.

3.3. Proof of the third item. By construction the restriction of the metric $\widetilde{g}$ to $s^{-1}((-\delta^{-1}, 0])$ is equal to the metric ρ^*g. Hence, in this region the mapping is an isometry. In the region $s^{-1}([0, 4])$ we have $R(x_0)\rho^*g \geq$

ηg_0 so that by construction in this region $\rho^* g \geq \widetilde{g}$. Lastly, in the region $s^{-1}([4, A_0 + 4])$ we have $R(x_0)^{-1} \eta g_0 \geq \widetilde{g}$. On the other hand, it follows from Claim 13.1 that the map from $([0, \delta^{-1}] \times S^2, R(x_0)\rho^* g)$ to $(B(p_0, 4 + A_0), \eta g)$ is distance decreasing. This completes the proof of the third item.

3.4. Completion of the proof. As δ goes to zero, f tends to zero in the C^∞-topology and η limits to 1. From this the fourth item is clear.

This completes the proof of Theorem 13.2.

4. Other properties of the result of surgery

LEMMA 13.15. *Provided that $\delta > 0$ is sufficiently small the following holds. Let (N, g) be a δ-neck and let $(\mathcal{S}, \widetilde{g})$ be the result of surgery along the cental 2-sphere of this neck. Then for any given $0 < D < \infty$ the ball $B_{\widetilde{g}}(p, D + 5 + A_0) \subset \mathcal{S}$ has boundary contained in $s_N^{-1}(-(2D+2), -D/2)$.*

PROOF. The Riemannian manifold $(\mathcal{S}, \widetilde{g})$ is identified by a diffeomorphism with the union of $s_N^{-1}(-\delta^{-1}, 0]$ to $B_{g_0}(p_0, A_0 + 4)$ glued along their boundaries. Thus, we have a natural identification of $\mathcal{S}$ with the ball $B_{g_0}(p, A_0 + 4 + \delta^{-1})$ in the standard solution. This identification pulls back the metric $\widetilde{g}$ to be within 2δ of the standard initial metric. The result then follows immediately for δ sufficiently small. □

CHAPTER 14

Ricci Flow with surgery: the definition

In this chapter we introduce Ricci flows with surgery. These objects are closely related to generalized Ricci flows but they differ slightly. The space-time of a Ricci flow with surgery has an open dense subset that is a manifold, and the restriction of the Ricci flow with surgery to this open subset is a generalized Ricci flow. Still there are other, more singular points allowed in a Ricci flow with surgery.

1. Surgery space-time

DEFINITION 14.1. By a *space-time* we mean a paracompact Hausdorff space $\mathcal{M}$ with a continuous function $\mathbf{t}\colon \mathcal{M} \to \mathbb{R}$, called *time*. We require that the image of $\mathbf{t}$ be an interval I, finite or infinite with or without endpoints, in $\mathbb{R}$. The interval I is called the *time-interval of definition of space-time*. The initial point of I, if there is one, is the *initial time* and the final point of I, if there is one, is the *final time*. The level sets of $\mathbf{t}$ are called the *time-slices* of space-time, and the preimage of the initial (resp., final) point of I is the *initial* (resp., *final*) time-slice.

We are interested in a certain class of space-times, which we call *surgery space-times*. These objects have a 'smooth structure' (even though they are not smooth manifolds). As in the case of a smooth manifold, this smooth structure is given by local coordinate charts with appropriate overlap functions.

1.1. An exotic chart. There is one exotic chart, and we begin with its description. To define this chart we consider the open unit square $(-1, 1) \times (-1, 1)$. We shall define a new topology, denoted by $\mathcal{P}$, on this square. The open subsets of $\mathcal{P}$ are the open subsets of the usual topology on the open square together with open subsets of $(0, 1) \times [0, 1)$. Of course, with this topology the 'identity' map $\iota\colon \mathcal{P} \to (-1, 1) \times (-1, 1)$ is a continuous map. Notice that the restriction of the topology of $\mathcal{P}$ to the complement of the closed subset $[0, 1) \times \{0\}$ is a homeomorphism onto the corresponding subset of the open unit square. Notice that the complement of $(0, 0)$ in $\mathcal{P}$ is a manifold with boundary, the boundary being $(0, 1) \times \{0\}$. (See FIG. 5 in the Introduction.)

Next, we define a 'smooth structure' on $\mathcal{P}$ by defining a sheaf of germs of 'smooth' functions. The restriction of this sheaf of germs of 'smooth functions' to the complement of $(0,1) \times \{0\}$ in $\mathcal{P}$ is the usual sheaf of germs of smooth functions on the corresponding subset of the open unit square. In particular, a function is smooth near $(0,0)$ if and only if its restriction to some neighborhood of $(0,0)$ is the pullback under ι of a usual smooth function on a neighborhood of the origin in the square. Now let us consider the situation near a point of the form $x = (a,0)$ for some $0 < a < 1$. This point has arbitrarily small neighborhoods V_n that are identified under ι with open subsets of $(0,1) \times [0,1)$. We say that a function f defined in a neighborhood of x in $\mathcal{P}$ is *smooth at x* if its restriction to one of these neighborhoods V_n is the pullback via $\iota|_{V_n}$ of a smooth function in the usual sense on the open subset $\iota(V_n)$ of the upper half space. One checks directly that this defines a sheaf of germs of 'smooth' functions on $\mathcal{P}$. Notice that the restriction of this sheaf to the complement of $(0,0)$ is the structure sheaf of smooth functions of a smooth manifold with boundary. Notice that the map $\iota \colon \mathcal{P} \to (-1,1) \times (-1,1)$ is a smooth map in the sense that it pulls back smooth functions on open subsets of the open unit square to smooth functions on the corresponding open subset of $\mathcal{P}$.

Once we have the notion of smooth functions on $\mathcal{P}$, there is the categorical notion of a diffeomorphism between open subsets of $\mathcal{P}$: namely a homeomorphism with the property that it and its inverse pull back smooth functions to smooth functions. Away from the origin, this simply means that the map is a diffeomorphism in the usual sense between manifolds with boundary, and in a neighborhood of $(0,0)$ it factors through a diffeomorphism of neighborhoods of the origin in the square. While $\iota \colon \mathcal{P} \to (-1,1) \times (-1,1)$ is a smooth map, it is not a diffeomorphism.

We define the *tangent bundle* of $\mathcal{P}$ in the usual manner. The tangent space at a point is the vector space of derivations of the germs of smooth functions at that point. Clearly, away from $(0,0)$ this is the usual (2-plane) tangent bundle of the smooth manifold with boundary. The germs of smooth functions at $(0,0)$ are, by definition, the pullbacks under ι of germs of smooth functions at the origin for the unit square, so that the tangent space of $\mathcal{P}$ at $(0,0)$ is identified with the tangent space of the open unit square at the origin. In fact, the map ι induces an isomorphism from the tangent bundle of $\mathcal{P}$ to the pullback under ι of the tangent bundle of the square. In particular, the tangent bundle of $\mathcal{P}$ has a given trivialization from the partial derivatives ∂_x and ∂_y in the coordinate directions on the square. We use this trivialization to induce a smooth structure on the tangent bundle of $\mathcal{P}$: that is to say, a section of $T\mathcal{P}$ is smooth if and only if it can be written as $\alpha \partial_x + \beta \partial_y$ with α and β being smooth functions on $\mathcal{P}$. The smooth structure agrees off of $(0,0) \in \mathcal{P}$ with the usual smooth structure on the tangent bundle of the smooth manifold with boundary. By a *smooth vector field on $\mathcal{P}$* we mean a smooth section of the tangent bundle of $\mathcal{P}$. Smooth vector fields act as derivations on the smooth functions on $\mathcal{P}$.

We let $\mathbf{t}_\mathcal{P}\colon \mathcal{P} \to \mathbb{R}$ be the pullback via ι of the usual projection to the second factor on the unit square. We denote by $\chi_\mathcal{P}$ the smooth vector field $\iota^*\partial_2$. Clearly, $\chi_\mathcal{P}(\mathbf{t}_\mathcal{P}) = 1$. Smooth vector fields on $\mathcal{P}$ can be uniquely integrated locally to smooth integral curves in $\mathcal{P}$. (For a manifold with boundary point, of course only vector fields pointing into the manifold along its boundary can be locally integrated.)

1.2. Coordinate charts for a surgery space-time. Now we are ready to introduce the types of coordinate charts that we shall use in our definition of a surgery space-time. Each coordinate patch comes equipped with a smooth structure (a sheaf of germs of smooth functions) and a tangent bundle with a smooth structure, so that smooth vector fields act as derivations on the algebra of smooth functions. There is also a distinguished smooth function, denoted $\mathbf{t}$, and a smooth vector field, denoted χ, required to satisfy $\chi(\mathbf{t}) = 1$. There are three types of coordinates:

(1) The coordinate patch is an open subset of the strip $\mathbb{R}^n \times I$, where I is an interval, with its usual smooth structure and tangent bundle; the function $\mathbf{t}$ is the projection onto I; and the vector field χ is the unit tangent vector in the positive direction tangent to the foliation with leaves $\{x\} \times I$. The initial point of I, if there is one, is the initial time of the space-time and the final point of I, if there is one, is the final time of the space-time.

(2) The coordinate patch is an open subset of $\mathbb{R}^n \times [a,\infty)$, for some $a \in \mathbb{R}$, with its usual smooth structure as a manifold with boundary and its usual smooth tangent bundle; the function $\mathbf{t}$ is the projection onto the second factor; and the vector field is the coordinate partial derivative associated with the second factor. In this case we require that a not be the initial time of the Ricci flow.

(3) The coordinate patch is a product of $\mathcal{P}$ with an open subset of $\mathbb{R}^{n-1}$ with the smooth structure (i.e., smooth functions and the smooth tangent bundle) being the product of the smooth structure defined above on $\mathcal{P}$ with the usual smooth structure of an open subset of $\mathbb{R}^{n-1}$; the function $\mathbf{t}$ is, up to an additive constant, the pullback of the function $\mathbf{t}_\mathcal{P}$ given above on $\mathcal{P}$; and the vector field χ is the image of the vector field $\chi_\mathcal{P}$ on $\mathcal{P}$, given above, under the product decomposition.

An ordinary Ricci flow is covered by coordinate charts of the first type. The second and third are two extra types of coordinate charts for a Ricci flow with surgery that are not allowed in generalized Ricci flows. Charts of the second kind are smooth manifold-with-boundary charts, where the boundary is contained in a single time-slice, not the initial time-slice, and the flow exists for some positive amount of forward time from this manifold.

All the structure described above for $\mathcal{P}$ — the smooth structure, the tangent bundle with its smooth structure, smooth vector fields acting as derivations on smooth functions — exist for charts of the third type. In addition, the unique local integrability of smooth vector fields hold for coordinate charts of the third type. Analogous results for coordinate charts of the first two types are clear.

Now let us describe the allowable overlap functions between charts. Between charts of the first and second type these are the smooth overlap functions in the usual sense that preserve the functions **t** and the vector fields χ on the patches. Notice that because the boundary points in charts of the second type are required to be at times other than the initial and final times, the overlap of a chart of type one and a chart of type two is disjoint from the boundary points of each. Charts of the first two types are allowed to meet a chart of the third type only in its manifold and manifold-with-boundary points. For overlaps between charts of the first two types with a chart of the third type, the overlap functions are diffeomorphisms between open subsets preserving the local time functions **t** and the local vector fields χ. Thus, all overlap functions are diffeomorphisms in the sense given above.

1.3. Definition and basic properties of surgery space-time.

DEFINITION 14.2. A *surgery space-time* is a space-time $\mathcal{M}$ equipped with a maximal atlas of charts covering $\mathcal{M}$, each chart being of one of the three types listed above, with the overlap functions being diffeomorphisms preserving the functions **t** and the vector fields χ. The points with neighborhoods of the first type are called *smooth points*, those with neighborhoods of the second type but not the first type are called *exposed points*, and all the other points are called *singular points*. Notice that the union of the set of smooth points and the set of exposed points forms a smooth manifold with boundary (possibly disconnected). Each component of the boundary of this manifold is contained in a single time-slice. The union of those components contained in a time distinct from the initial time and the final time is called the *exposed region*. and the boundary points of the closure of the exposed region form the set of the singular points of $\mathcal{M}$. (Technically, the exposed points are singular, but we reserve this word for the most singular points.) An $(n+1)$-dimensional surgery space-time is by definition of homogeneous dimension $n+1$.

By construction, the local smooth functions **t** are compatible on the overlaps and hence fit together to define a global smooth function $\mathbf{t}\colon \mathcal{M} \to \mathbb{R}$, called the *time* function. The level sets of this function are called the *time-slices* of the space-time, and $\mathbf{t}^{-1}(t)$ is denoted M_t. Similarly, the tangent bundles of the various charts are compatible under the overlap diffeomorphisms and hence glue together to give a global smooth tangent bundle on space-time. The smooth sections of this vector bundle, the smooth vector fields on space time, act as derivations on the smooth functions on space-time. The tangent bundle of an $(n+1)$-dimensional surgery space-time is a vector bundle of dimension $n+1$. Also, by construction the local vector fields χ are compatible and hence glue together to define a global vector field, denoted χ. The vector field and time function satisfy

$$\chi(\mathbf{t}) = 1.$$

At the manifold points (including the exposed points) it is a usual vector field. Along the exposed region and the initial time-slice the vector field points into the manifold; along the final time-slice it points out of the manifold.

DEFINITION 14.3. Let $\mathcal{M}$ be a surgery space-time. Given a space K and an interval $J \subset \mathbb{R}$ we say that an embedding $K \times J \to \mathcal{M}$ is *compatible with time and the vector field* if: (i) the restriction of $\mathbf{t}$ to the image agrees with the projection onto the second factor and (ii) for each $x \in X$ the image of $\{x\} \times J$ is the integral curve for the vector field χ. If in addition K is a subset of M_t we require that $t \in J$ and that the map $K \times \{t\} \to M_t$ be the identity. Clearly, by the uniqueness of integral curves for vector fields, two such embeddings agree on their common interval of definition, so that, given $K \subset M_t$ there is a maximal interval J_K containing t such that such an embedding is defined on $K \times J_K$. In the special case when $K = \{x\}$ for a point $x \in M_t$ we say that such an embedding is *the maximal flow line* through x. The embedding of the maximal interval through x compatible with time and the vector field χ is called *the domain of definition* of the flow line through x. For a more general subset $K \subset M_t$ there is an embedding $K \times J$ compatible with time and the vector field χ if and only if, for every $x \in K$, the interval J is contained in the domain of definition of the flow line through x.

DEFINITION 14.4. Let $\mathcal{M}$ be a surgery space-time with I as its time interval of definition. We say that $t \in I$ is a *regular* time if there is an interval $J \subset I$ which is an open neighborhood in I of t, and a diffeomorphism $M_t \times J \to \mathbf{t}^{-1}(J) \subset \mathcal{M}$ compatible with time and the vector field. A time is *singular* if it is not regular. Notice that if all times are regular, then space-time is a product $M_t \times I$ with $\mathbf{t}$ and χ coming from the second factor.

LEMMA 14.5. *Let $\mathcal{M}$ be an $(n+1)$-dimensional surgery space-time, and fix t. The restriction of the smooth structure on $\mathcal{M}$ to the time-slice M_t induces the structure of a smooth n-manifold on this time-slice. That is to say, we have a smooth embedding of $M_t \to \mathcal{M}$. This smooth embedding identifies the tangent bundle of M_t with a codimension-1 subbundle of the restriction of tangent bundle of $\mathcal{M}$ to M_t. This subbundle is complementary to the line field spanned by χ. These codimension-1 sub-bundle along the* *various time-slices fit together to form a smooth, codimension-1 subbundle of the tangent bundle of space-time.*

PROOF. These statements are immediate for any coordinate patch, and hence are true globally. □

DEFINITION 14.6. We call the codimension-1 subbundle of the tangent bundle of $\mathcal{M}$ described in the previous lemma the *horizontal subbundle*, and we denote it $\mathcal{H}T(\mathcal{M})$.

2. The generalized Ricci flow equation

In this section we introduce the Ricci flow equation for surgery space-times, resulting in an object that we call Ricci flow with surgery.

2.1. Horizontal metrics.

DEFINITION 14.7. By a *horizontal metric* G on a surgery space-time $\mathcal{M}$ we mean a C^∞ metric on $\mathcal{H}T\mathcal{M}$. For each t, the horizontal metric G induces a Riemannian metric, denoted $G(t)$, on the time-slice M_t. Associated to a horizontal metric G we have the *horizontal covariant derivative*, denoted ∇. This is a pairing between horizontal vector fields

$$X \otimes Y \mapsto \nabla_X Y.$$

On each time slice M_t it is the usual Levi-Civita connection associated to the Riemannian metric $G(t)$. Given a function F on space-time, by its gradient ∇F we mean its horizontal gradient. The value of this gradient at a point $q \in M_t$ is the usual $G(t)$-gradient of $F|_{M_t}$. In particular, ∇F is a smooth horizontal vector field on space-time. The horizontal metric G on space-time has its (horizontal) curvatures Rm_G. These are smooth symmetric endomorphisms of the second exterior power of $\mathcal{H}T\mathcal{M}$. The value of Rm_G at a point $q \in M_t$ is simply the usual Riemann curvature operator of $G(t)$ at the point q. Similarly, we have the (horizontal) Ricci curvature $\operatorname{Ric} = \operatorname{Ric}_G$, a section of the symmetric square of the horizontal cotangent bundle, and the (horizontal) scalar curvature denoted $R = R_G$. The reason for working in $\mathcal{H}T\mathcal{M}$ rather than individually in each slice is to emphasize the fact that all these horizontal quantities vary smoothly over the surgery space-time.

Suppose that $t \in I$ is not the final time and suppose that $U \subset M_t$ is an open subset with compact closure. Then there is $\epsilon > 0$ and an embedding $i_U \colon U \times [t, t+\epsilon) \subset \mathcal{M}$ compatible with time and the vector field. Of course, two such embeddings agree on their common domain of definition. Notice also that for each $t' \in [t, t+\epsilon)$ the restriction of the map i_U to $U \times \{t'\}$ induces a diffeomorphism from U to an open subset $U_{t'}$ of $M_{t'}$. It follows that the local flow generated by the vector field χ preserves the horizontal subbundle. Hence, the vector field χ acts by Lie derivative on the sections of $\mathcal{H}T(\mathcal{M})$ and on all associated bundles (for example the symmetric square of the dual bundle).

2.2. The equation.

DEFINITION 14.8. A *Ricci flow with surgery* is a pair $(\mathcal{M}, G)$ consisting of a surgery space-time $\mathcal{M}$ and a horizontal metric G on $\mathcal{M}$ such that for every $x \in \mathcal{M}$ we have

$$(14.1) \qquad \mathcal{L}_\chi(G)(x) = -2\mathrm{Ric}_G(x))$$

as sections of the symmetric square of the dual to $\mathcal{H}T(\mathcal{M})$. If space-time is $(n+1)$-dimensional, then we say that the Ricci flow with surgery is n-dimensional (meaning of course that each time-slice is an n-dimensional manifold).

REMARK 14.9. Notice that at an exposed point and at points at the initial and the final time the Lie derivative is a one-sided derivative.

2.3. Examples of Ricci flows with surgery.

EXAMPLE 14.10. One example of a Ricci flow with surgery is $\mathcal{M} = M_0 \times [0, T)$ with time function $\mathbf{t}$ and the vector field χ coming from the second factor. In this case the Lie derivative $\mathcal{L}_\chi$ agrees with the usual partial derivative in the time direction, and hence our generalized Ricci flow equation is the usual Ricci flow equation. This shows that an ordinary Ricci flow is indeed a Ricci flow with surgery.

The next lemma gives an example of a Ricci flow with surgery where the topology of the time-slices changes.

LEMMA 14.11. *Suppose that we have manifolds $M_1 \times (a, b]$ and $M_2 \times [b, c)$ and compact, smooth codimension-0 submanifolds $\Omega_1 \subset M_1$ and $\Omega_2 \subset M_2$ with open neighborhoods $U_1 \subset M_1$ and $U_2 \subset M_2$ respectively. Suppose we have a diffeomorphism $\psi \colon U_1 \to U_2$ carrying Ω_1 onto Ω_2. Let $(M_1 \times (a,b])_0$ be the subset obtained by removing $(M_1 \setminus \Omega_1) \times \{b\}$ from $M_1 \times (a, b]$. Form the topological space*

$$\mathcal{M} = (M_1 \times (a, b])_0 \cup M_2 \times [b, c)$$

where $\Omega_1 \times \{b\}$ in $(M_1 \times (a, b])_0$ is identified with $\Omega_2 \times \{b\}$ using the restriction of ψ to Ω_1. Then $\mathcal{M}$ naturally inherits the structure of a surgery space-time where the time function restricts to $(M_1 \times (a, b])_0$ and to $M_2 \times [b, c)$ to be the projection onto the second factor and the vector field χ agrees with the vector fields coming from the second factor on each of $(M_1 \times (a, b])_0$ and $M_2 \times [b, c)$.

Lastly, given Ricci flows $(M_1, g_1(t))$, $a < t \leq b$, and $(M_2, g_2(t))$, $b \leq t < c$, if $\psi \colon (U_1, g_1(b)) \to (U_2, g_2(b))$ is an isometry, then these families fit together to form a smooth horizontal metric G on $\mathcal{M}$ satisfying the Ricci flow equation, so that $(\mathcal{M}, G)$ is a Ricci flow with surgery.

PROOF. As the union of Hausdorff spaces along closed subsets, $\mathcal{M}$ is a Hausdorff topological space. The time function is the one induced from the projections onto the second factor. For any point outside the b time-slice there is the usual smooth coordinate coming from the smooth manifold $M_1 \times (a, b)$ (if $t < b$) or $M_2 \times (b, c)$ (if $t > b$). At any point of $(M_2 \setminus \Omega_2) \times \{b\}$ there is the smooth manifold with boundary coordinate patch coming from $M_2 \times [b, c)$. For any point in $\text{int}(\Omega_1) \times \{b\}$ we have the smooth manifold structure obtained from gluing $(\text{int}(\Omega_1)) \times (a, b]$ to $\text{int}(\Omega_2) \times [b, c)$ along the b time-slice by ψ. Thus, at all these points we have neighborhoods on which our data determine a smooth manifold structure. Lastly, let us consider a point $x \in \partial \Omega_1 \times \{b\}$. Choose local coordinates $(x^1, \ldots, x^n)$ for a neighborhood V_1 of x such that $\Omega_1 \cap V_1 = \{x^n \leq 0\}$. We can assume that ψ is defined on all of V_1. Let $V_2 = \psi(V_1)$ and take the local coordinates on V_2 induced from the x^i on V_1. Were we to identify $V_1 \times (a, b]$ with $V_2 \times [b, c)$ along the b time-slice using this map, then this union would be a smooth manifold. There is a neighborhood of the point $(x, b) \in \mathcal{M}$ which is obtained from the smooth manifold $V_1 \times (a, b] \cup_\psi V_2 \times [b, c)$ by inducing a new topology where the open subsets are, in addition to the usual ones, any open subset of the form $\{x^n > 0\} \times [b, b')$ where $b < b' \leq c$. This then gives the coordinate charts of the third type near the points of $\partial \Omega_2 \times \{b\}$. Clearly, since the function $\mathbf{t}$ and the vector field $\partial/\partial t$ are smooth on $V_1 \times (a, b] \cup_\psi V_2 \times [b, c)$, we see that these objects glue together to form smooth objects on $\mathcal{M}$.

Given the Ricci flows $g_1(t)$ and $g_2(t)$ as in the statement, they clearly determine a (possibly singular) horizontal metric on $\mathcal{M}$. This horizontal metric is clearly smooth except possibly along the b time-slice. At any point of $(M_2 \setminus \Omega_2) \times \{b\}$ we have a one-sided smooth family, which means that on this set the horizontal metric is smooth. At a point of $\text{int}(\Omega_2) \times \{b\}$, the fact that the metrics fit together smoothly is an immediate consequence of Proposition 3.12. At a point $x \in \partial \Omega_2 \times \{b\}$ we have neighborhoods $V_2 \subset M_2$ of x and $V_1 \subset M_1$ of $\psi^{-1}(x)$ that are isometrically identified by ψ. Hence, again by Proposition 3.12 we see that the Ricci flows fit together to form a smooth family of metrics on $V_1 \times (a, b] \cup_\psi V_2 \times [b, c)$. Hence, the induced horizontal metric on $\mathcal{M}$ is smooth near this point. □

The following is obvious from the definitions.

PROPOSITION 14.12. *Suppose that $(\mathcal{M}, G)$ is a Ricci flow with surgery. Let $\text{int}\,\mathcal{M}$ be the open subset consisting of all smooth $(n+1)$-manifold points, plus all manifold-with-boundary points at the initial time and the final time. This space-time inherits the structure of a smooth manifold with boundary. The restrictions to it of $\mathbf{t}$, of the vector field χ, and of the horizontal metric G, define a generalized Ricci flow whose underlying smooth manifold is $\text{int}\,\mathcal{M}$.*

2. THE GENERALIZED RICCI FLOW EQUATION

2.4. Scaling and translating. Let $(\mathcal{M}, G)$ be a Ricci flow with surgery and let Q be a positive constant. Then we can define a new Ricci flow with surgery by setting $G' = QG$, $\mathbf{t}' = Q\mathbf{t}$ and $\chi' = Q^{-1}\chi$. It is easy to see that the resulting data still satisfies the generalized Ricci flow equation, Equation (14.1). We denote this new Ricci flow with surgery by $(Q\mathcal{M}, QG)$ where the changes in $\mathbf{t}$ and χ are indicated by the factor Q in front of the space-time.

It is also possible to translate a Ricci flow with surgery $(\mathcal{M}, G)$ by replacing the time function $\mathbf{t}$ by $\mathbf{t}' = \mathbf{t} + a$ for any constant a, and leaving χ and G unchanged.

2.5. More basic definitions.

DEFINITION 14.13. Let $(\mathcal{M}, G)$ be a Ricci flow with surgery, and let x be a point of space-time. Set $t = \mathbf{t}(x)$. For any $r > 0$ we define $B(x, t, r) \subset M_t$ to be the metric ball of radius r centered at x in the Riemannian manifold $(M_t, G(t))$.

DEFINITION 14.14. Let $(\mathcal{M}, G)$ be a Ricci flow with surgery, and let x be a point of space-time. Set $t = \mathbf{t}(x)$. For any $r > 0$ and $\Delta t > 0$ we say that the *backward parabolic neighborhood* $P(x, t, r, -\Delta t)$ *exists* in $\mathcal{M}$ if there is an embedding $B(x, t, r) \times (t - \Delta t, t] \to \mathcal{M}$ compatible with time and the vector field. Similarly, we say that the *forward parabolic neighborhood* $P(x, t, r, \Delta t)$ *exists* in $\mathcal{M}$ if there is an embedding $B(x, t, r) \times [t, t + \Delta t) \to \mathcal{M}$ compatible with time and the vector field. A *parabolic neighborhood* is either a forward or backward parabolic neighborhood.

DEFINITION 14.15. Fix $\kappa > 0$ and $r_0 > 0$. We say that a Ricci flow with surgery $(\mathcal{M}, G)$ is κ-*non-collapsed on scales* $\leq r_0$ if the following holds for every point $x \in \mathcal{M}$ and for every $r \leq r_0$. Denote $\mathbf{t}(x)$ by t. If the parabolic neighborhood $P(x, t, r, -r^2)$ exists in $\mathcal{M}$ and if $|\mathrm{Rm}_G| \leq r^{-2}$ on $P(x, t, r, -r^2)$, then $\mathrm{Vol}\, B(x, t, r) \geq \kappa r^3$.

REMARK 14.16. For $\epsilon > 0$ sufficiently small, an ϵ-round component satisfies the first condition in the above definition for some $\kappa > 0$ depending only on the order of the fundamental group of the underlying manifold, but there is no universal $\kappa > 0$ that works for all ϵ-round manifolds. Fixing an integer N let $\mathcal{C}_N$ be the class of closed 3-manifolds with the property that any finite free factor of $\pi_1(M)$ has order at most N. Then any ϵ-round component of any time-slice of any Ricci flow $(\mathcal{M}, G)$ whose initial conditions consist of a manifold in $\mathcal{C}_N$ will have fundamental group of order at most N and hence will satisfy the first condition in the above definition for some $\kappa > 0$ depending only on N.

We also have the notion of the curvature being pinched toward positive, analogous to the notions for Ricci flows and generalized Ricci flows.

DEFINITION 14.17. Let $(\mathcal{M}, G)$ be a 3-dimensional Ricci flow with surgery, whose time domain of definition is contained in $[0, \infty)$. For any $x \in \mathcal{M}$ we denote the eigenvalues of $\mathrm{Rm}(x)$ by $\lambda(x) \geq \mu(x) \geq \nu(x)$ and we set $X(x) = \max(0, \nu(x))$. We say that its *curvature is pinched toward positive* if the following hold for every $x \in \mathcal{M}$:
  (1) $R(x) \geq \frac{-6}{1+4\mathbf{t}(x)}$.
  (2) $R(x) \geq 2X(x)\left(\log X(x) + \log(1+\mathbf{t}(x)) - 3\right)$, whenever $0 < X(x)$.

Let (M, g) be a Riemannian manifold and let $T \geq 0$. We say that (M, g) has curvature *pinched toward positive up to time* T if the above two inequalities hold for all $x \in M$ with $\mathbf{t}(x)$ replaced by T.

Lastly, we extend the notion of canonical neighborhoods to the context of Ricci flows with surgery.

DEFINITION 14.18. Fix constants (C, ϵ) and a constant r. We say that a Ricci flow with surgery $(\mathcal{M}, G)$ *satisfies the strong (C, ϵ)-canonical neighborhood assumption with parameter r* if every point $x \in \mathcal{M}$ with $R(x) \geq r^{-2}$ has a strong (C, ϵ)-canonical neighborhood in $\mathcal{M}$. In all cases except that of the strong ϵ-neck, the strong canonical neighborhood of x is a subset of the time-slice containing x, and the notion of a (C, ϵ)-canonical neighborhood has exactly the same meaning as in the case of an ordinary Ricci flow. In the case of a strong ϵ-neck centered at x this means that there is an embedding $\left(S^2 \times (-\epsilon^{-1}, \epsilon^{-1})\right) \times (\mathbf{t}(x) - R(x)^{-1}, \mathbf{t}(x)] \to \mathcal{M}$, mapping $(q_0, 0)$ to x, where q_0 is the basepoint of S^2, an embedding compatible with time and the vector field, such that the pullback of G is a Ricci flow on $S^2 \times (-\epsilon^{-1}, \epsilon^{-1})$ which, when the time is shifted by $-\mathbf{t}(x)$ and then the flow is rescaled by $R(x)$, is within ϵ in the $C^{[1/\epsilon]}$-topology of the standard evolving round cylinder $\left(S^2 \times (-\epsilon^{-1}, \epsilon^{-1}), h_0(t) \times ds^2\right)$, $-1 < t \leq 0$, where the scalar curvature of the $h_0(t)$ is $1 - t$.

Notice that if x is an exposed point or sufficiently close to an exposed point, then x cannot be the center of a strong ϵ-neck.

CHAPTER 15

Controlled Ricci flows with surgery

We do not wish to consider all Ricci flows with surgery. Rather we shall concentrate on 3-dimensional flows (that is to say 4-dimensional space-times) whose singularities are closely controlled both topologically and geometrically. We introduce the hypotheses that we require these evolutions to satisfy. The main result, which is stated in this chapter and proved in the next two, is that these controlled 3-dimensional Ricci flows with surgery always exist for all time with any compact 3-manifold as initial metric.

0.6. Normalized initial conditions. A compact connected Riemannian 3-manifold $(M, g(0))$ is *normalized* or is *a normalized metric* if it satisfies the following:

(1) $|\text{Rm}(x, 0)| \leq 1$ for all $x \in M$ and
(2) for every $x \in M$ we have $\text{Vol}\, B(x, 0, 1) \geq \omega/2$ where ω is the volume of the unit ball in $\mathbb{R}^3$.

If $(M, g(0))$ is the initial manifold of a Ricci flow with surgery, then we say that it is a *normalized initial metric*, and we shall say that the Ricci flow with surgery *has normalized initial conditions* provided that $(M, g(0))$ is normalized. Of course, given any compact Riemannian 3-manifold $(M, g(0))$ there is a positive constant $Q < \infty$ such that $(M, Qg(0))$ is normalized.

Starting with a normalized initial metric implies that the flow exists and has uniformly bounded curvature for a fixed amount of time. This is the content of the following claim which is an immediate corollary of Theorem 3.11, Proposition 3.12, Theorem 3.28, and Proposition 4.11.

CLAIM 15.1. *There is κ_0 such that the following holds. Let $(M, g(0))$ be a normalized initial metric. Then the solution to the Ricci flow equation with these initial conditions exists for $t \in [0, 2^{-4}]$, and $|R(x,t)| \leq 2$ for all $x \in M$ and all $t \in [0, 2^{-4}]$. Furthermore, for any $t \in [0, 2^{-4}]$ and any $x \in M$ and any $r \leq \epsilon$ we have $\text{Vol}\, B(x, t, r) \geq \kappa_0 r^3$.*

1. Gluing together evolving necks

PROPOSITION 15.2. *There is $0 < \beta < 1/2$ such that the following holds for any $\epsilon < 1$. Let $(N \times [-t_0, 0], g_1(t))$ be an evolving $\beta\epsilon$-neck centered at x with $R(x, 0) = 1$. Let $(N' \times (-t_1, -t_0], g_2(t))$ be a strong $\beta\epsilon/2$-neck. Suppose we have an isometric embedding of $N \times \{-t_0\}$ with $N' \times \{-t_0\}$ and the strong*

$\beta\epsilon/2$-neck structure on $N' \times (-t_1, -t_0]$ is centered at the image of $(x, -t_0]$. Then the union

$$N \times [-t_0, 0] \cup N' \times (-t_1, -t_0]$$

with the induced one-parameter family of metrics contains a strong ϵ-neck centered at $(x, 0)$.

PROOF. Suppose that the result does not hold. Take a sequence of β_n tending to zero and counterexamples

$$(N_n \times [-t_{0,n}, 0], g_{1,n}(t));\ (N'_n \times (-t_{1,n}, -t_{0,n}], g_{2,n}(t)).$$

Pass to a subsequence so that the $t_{0,n}$ tend to a limit $t_{0,\infty} \geq 0$. Since β_n tends to zero, we can take a smooth limit of a subsequence and this limit is an evolving cylinder $(S^2 \times \mathbb{R}, h(t) \times ds^2)$, where $h(t)$ is the round metric of scalar curvature $1/(1-t)$ defined for some amount of backward time. Notice that, for all β sufficiently small, on a $\beta\epsilon$-neck the derivative of the scalar curvature is positive. Thus, $R_{g_{1,n}}(x, -t_{0,n}) < 1$. Since we have a strong neck structure on N'_n centered at $(x, -t_{0,n})$, this implies that $t_{1,n} > 1$ so that the limit is defined for at least time $t \in [0, 1 + t_{0,\infty})$. If $t_{0,\infty} > 0$, then, restricting to the appropriate subset of this limit, a subset with compact closure in space-time, it follows immediately that for all n sufficiently large there is a strong ϵ-neck centered at $(x, 0)$. This contradicts the assumption that we began with a sequence of counterexamples to the proposition.

Let us consider the case when $t_{0,\infty} = 0$. In this case the smooth limit is an evolving round cylinder defined for time $(-1, 0]$. Since $t_{1,n} > 1$ we see that for any $A < \infty$ for all n sufficiently large the ball $B(x_n, 0, A)$ has compact closure in every time-slice and there are uniform bounds to the curvature on $B(x_n, 0, A) \times (-1, 0]$. This means that the limit is uniform for time $(-1, 0]$ on all these balls. Thus, once again for all n sufficiently large we see that $(x, 0)$ is the center of a strong ϵ-neck in the union. In either case we have obtained a contradiction, and hence we have proved the result. See FIG. 1. □

1.1. First assumptions. Choice of C and ϵ: The first thing we need to do is fix for the rest of the argument $C < \infty$ and $\epsilon > 0$. We do this in the following way. We fix $0 < \epsilon \leq \min(1/200, \left(\sqrt{D}(A_0 + 5)\right)^{-1}, \overline{\epsilon}_1/2, \overline{\epsilon}'/2, \epsilon_0)$ where $\overline{\epsilon}_1$ is the constant from Proposition 2.19, $\overline{\epsilon}'$ is the constant from Theorem 9.93, ϵ_0 is the constant from Section 1 of Chapter 10, and A_0 and D are the constants from Lemma 12.3. We fix $\beta < 1/2$, the constant from Proposition 15.2. Then we let C be the maximum of the constant $C(\epsilon)$ as in Corollary 9.94 and $C'(\beta\epsilon/3) + 1$ as in Theorem 12.32.

For all such ϵ, Theorem 10.2 holds for ϵ and Proposition 2.19, Proposition 9.79 and Corollaries 9.94 and 9.95 and Theorems 11.1 and 11.8 hold

1. GLUING TOGETHER EVOLVING NECKS

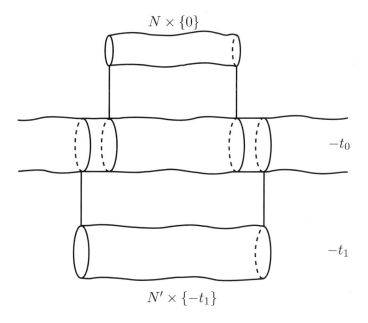

FIGURE 1. Gluing together necks.

for 2ϵ. Also, all the topological results of the Appendix hold for 2ϵ and $\alpha = 10^{-2}$.

Now let us turn to the assumptions we shall make on the Ricci flows with surgery that we shall consider. Let $\mathcal{M}$ be a space-time. Our first set of assumptions are basically topological in nature. They are:

Assumption (1). Compactness and dimension: *Each time-slice M_t of space-time is a compact 3-manifold containing no embedded $\mathbb{R}P^2$ with trivial normal bundle.*

Assumption (2). Discrete singularities: *The set of singular times is a discrete subset of $\mathbb{R}$.*

Assumption (3). Normalized initial conditions: *0 is the initial time of the Ricci flow with surgery and the initial metric $(M_0, G(0))$ is normalized.*

It follows from Assumption (2) that for any time t in the time-interval of definition of a Ricci flow with surgery, with t being distinct from the initial and final times (if these exist), for all $\delta > 0$ sufficiently small, the only possible singular time in $[t - \delta, t + \delta]$ is t. Suppose that t is a singular time. The singular locus at time t is a closed, smooth subsurface $\Sigma_t \subset M_t$. From the local model, near every point of $x \in \Sigma_t$ we see that this surface separates M_t into two pieces:

$$M_t = C_t \cup_{\Sigma_t} E_t,$$

where E_t is the exposed region at time t and C_t is the complement of the interior of E_t in M_t. We call C_t the *continuing region*. It is the maximal subset of M_t for which there is $\delta > 0$ and an embedding

$$C_t \times (t - \delta, t] \to \mathcal{M}$$

compatible with time and the vector field.

Assumption (4). Topology of the exposed regions: *At all singular times t we require that E_t be a finite disjoint union of 3-balls. In particular, Σ_t is a finite disjoint union of 2-spheres.*

The next assumptions are geometric in nature. Suppose that t is a surgery time. Let $\mathcal{M}_{(-\infty, t)}$ be $\mathbf{t}^{-1}((-\infty, t))$ and let $(\widehat{\mathcal{M}}_{(-\infty, t)}, \widehat{G})$ be the maximal extension of $(\mathcal{M}_{(-\infty, t)}, G)$ to time t, as given in Definition 11.22.

Assumption (5). Boundary components of the exposed regions: *There is a surgery control parameter function, $\delta(t) > 0$, a non-increasing function of t, such that each component of $\Sigma_t \subset M_t$ is the central 2-sphere of a strong $\delta(t)$-neck in $(\widehat{\mathcal{M}}_{(-\infty, t)}, \widehat{G})$.*

Suppose that t is a singular time. Then for all $t^- < t$ with t^- sufficiently close to t, the manifolds M_{t^-} are diffeomorphic and are identified under the flow. Applying the flow (backward) to C_t produces a diffeomorphism from C_t onto a compact submanifold with boundary $C_{t^-} \subset M_{t^-}$. Our next assumption concerns the nature of the metrics $G(t^-)$ on the *disappearing region* $D_{t^-} = M_{t^-} \setminus C_{t^-}$. The following holds for every $t^- < t$ sufficiently close to t.

Assumption (6). Control on the disappearing region: *For any singular time t, for all $t^- < t$ sufficiently close to t, each point of $x \in D_{t^-}$ has a strong (C, ϵ)-canonical neighborhood in M_{t^-}.*

Assumption (7). Maximal flow intervals: *Let t be the initial time or a singular time and let t' be the first singular time after t if such exists, otherwise let t' be the least upper bound of the time-interval of definition of the Ricci flow with surgery. Then the restriction of the Ricci flow with surgery to $[t, t')$ is a maximal Ricci flow. That is to say, either $t' = \infty$ or, as $t \to t'$ from below, the curvature of $G(t)$ is unbounded so that this restricted Ricci flow cannot be extended* **as a Ricci flow** *to any larger time.*

From now on C and ϵ have fixed values as described above and all Ricci flows with surgeries are implicitly assumed to satisfy Assumptions (1) – (7).

2. Topological consequences of Assumptions (1) – (7)

Next we show that the topological control that we are imposing on 3-dimensional Ricci flows with surgery is enough to allow us to relate the topology of a time-slice M_T in terms of a later time-slice $M_{T'}$ and topologically standard pieces. This is the result that will be used to establish the topological theorems stated in the introduction.

PROPOSITION 15.3. *Suppose that $(\mathcal{M}, G)$ is a generalized Ricci flow satisfying Assumptions (1) – (7). Let t be a singular time. Then the following holds for any $t^- < t$ sufficiently close to t. The manifold M_{t^-} is diffeomorphic to a manifold obtained in the following way. Take the disjoint union of M_t, finitely many 2-sphere bundles over S^1, and finitely many closed 3-manifolds admitting metrics of constant positive curvature. Then perform connected sum operations between (some subsets of) these components.*

PROOF. Fix $t' < t$ but sufficiently close to t. By Assumption (4) every component of E_t is a 3-ball and hence every component of $\partial E_t = \partial C_t$ is a 2-sphere. Since C_t is diffeomorphic to $C_{t'} \subset M_{t'}$ we see that every component of $\partial C_{t'} = \partial D_{t'}$ is a 2-sphere. Since every component of E_t is a 3-ball, the passage from the smooth manifold $M_{t'}$ to the smooth manifold M_t is effected by removing the interior of $D_{t'}$ from $M_{t'}$ and gluing a 3-ball onto each component of $\partial C_{t'}$ to form M_t.

By Assumption (5) every point of $D_{t'}$ has a strong (C, ϵ)-canonical neighborhood. Since ϵ is sufficiently small it follows from Proposition A.25 that every component of $D_{t'}$ that is also a component of $M_{t'}$ is diffeomorphic either to a manifold admitting a metric of constant positive curvature (a 3-dimensional space-form), to $\mathbb{R}P^3 \# \mathbb{R}P^3$ or to a 2-sphere bundle over S^1. In the passage from $M_{t'}$ to M_t these components are removed.

Now let us consider a component of $D_{t'}$ that is not a component of $M_{t'}$. Such a component is a connected subset of $M_{t'}$ with the property that every point is either contained in the core of a (C, ϵ)-cap or is the center of an ϵ-neck and whose frontier in $M_{t'}$ consists of 2-spheres that are central 2-spheres of ϵ-necks. If every point is the center of an ϵ-neck, then according to Proposition A.19, $D_{t'}$ is an ϵ-tube and in particular is diffeomorphic to $S^2 \times I$. Otherwise $D_{t'}$ is contained in a capped or double capped ϵ-tube. Since the frontier of $D_{t'}$ is non-empty and is the union of central 2-spheres of an ϵ-neck, it follows that either $D_{t'}$ is diffeomorphic to a capped ϵ-tube or to an ϵ-tube. Hence, these components of $D_{t'}$ are diffeomorphic either to $S^2 \times (0,1)$, to D^3, or to $\mathbb{R}P^3 \setminus B^3$. Replacing a 3-ball component of $D_{t'}$ by another 3-ball leaves the topology unchanged. Replacing a component of $D_{t'}$ that is diffeomorphic to $S^2 \times I$ by the disjoint union of two 3-balls has the effect of doing a surgery along the core 2-sphere of the cylinder $S^2 \times I$ in $M_{t'}$. If this 2-sphere separates $M_{t'}$ into two pieces then doing this surgery effects a connected sum decomposition. If this 2-sphere does not separate, then the surgery has the topological effect of doing a connected sum decomposition into two pieces, one of which is diffeomorphic to $S^2 \times S^1$, and then removing that component entirely. Replacing a component of $D_{t'}$ that is diffeomorphic to $\mathbb{R}P^3 \setminus B^3$ by a 3-ball, has the effect of doing a connected sum decomposition on $M_{t'}$ into pieces, one of which is diffeomorphic to $\mathbb{R}P^3$, and then removing that component.

From this description the proposition follows immediately. □

COROLLARY 15.4. *Let $(\mathcal{M}, G)$ be a generalized Ricci flow satisfying Assumptions (1) – (7) with initial conditions $(M, g(0))$. Suppose that for some T the time-slice M_T of this generalized flow satisfies Thurston's Geometrization Conjecture. Then the same is true for the manifold M_t for any $t \leq T$, and in particular M satisfies Thurston's Geometrization Conjecture. In addition:*

(1) *If for some $T > 0$ the manifold M_T is empty, then M is a connected sum of manifolds diffeomorphic to 2-sphere bundles over S^1 and 3-dimensional space-forms, i.e., compact 3-manifolds that admit a metric of constant positive curvature.*
(2) *If for some $T > 0$ the manifold M_T is empty and if M is connected and simply connected, then M is diffeomorphic to S^3.*
(3) *If for some $T > 0$ the manifold M_T is empty and if M has finite fundamental group, then M is a 3-dimensional space-form.*

PROOF. Suppose that M_T satisfies the Thurston Geometrization Conjecture and that t_0 is the largest surgery time $\leq T$. (If there is no such surgery time then M_T is diffeomorphic to M and the result is established.) Let $T' < t_0$ be sufficiently close to t_0 so that t_0 is the only surgery time in the interval $[T', T]$. Then according to the previous proposition $M_{T'}$ is obtained from M_T by first taking the disjoint union of M_T and copies of 2-sphere bundles over S^1 and 3-dimensional space forms. In the Thurston Geometrization Conjecture the first step is to decompose the manifold as a connected sum of prime 3-manifolds and then to treat each prime piece independently. Clearly, the prime decomposition of $M_{T'}$ is obtained from the prime decomposition of M_T by adding a disjoint union with 2-sphere bundles over S^1 and 3-dimensional space forms. By definition any 3-dimensional space-form satisfies Thurston's Geometrization Conjecture. Since any diffeomorphism of S^2 to itself is isotopic to either the identity or to the antipodal map, there are two diffeomorphism types of 2-sphere bundles over S^1: $S^2 \times S^1$ and the non-orientable 2-sphere bundle over S^1. Each is obtained from $S^2 \times I$ by gluing the ends together via an isometry of the round metric on S^2. Hence, each has a homogeneous geometry modeled on $S^2 \times \mathbb{R}$, and hence satisfies Thurston's Geometrization Conjecture. This proves that if M_T satisfies this conjecture, then so does $M_{T'}$. Continuing this way by induction, using the fact that there are only finitely many surgery times completes the proof of the first statement.

Statement (1) is proved analogously. Suppose that M_T is a disjoint union of connected sums of 2-sphere bundles over S^1 and 3-dimensional space-forms. Let t_0 be the largest surgery time $\leq T$ and let $T' < t_0$ be sufficiently close to t_0. (As before, if there is no such t_0 then M_T is diffeomorphic to M and the result is established.) Then it is clear from the previous proposition that $M_{T'}$ is also a disjoint union of connected sums of 3-dimensional space-forms and 2-sphere bundles over S^1. Induction as in the previous case completes the argument for this case.

The last two statements are immediate from this one. □

3. Further conditions on surgery

3.1. The surgery parameters. The process of doing surgery requires fixing the scale h at which one does the surgery. We shall have to allow this scale h to be a function of time, decreasing sufficiently rapidly with t. In fact, the scale is determined by two other functions of time which also decay to zero as time goes to infinity — a canonical neighborhood parameter $r(t)$ determining the curvature threshold above which we have canonical neighborhoods and the surgery control parameter $\overline{\delta}(t)$ determining how close to cylinders (products of the round 2-sphere with an interval) the regions where we do surgery are. In addition to these functions, in order to prove inductively that we can do surgery we need to have a non-collapsing result. The non-collapsing parameter $\kappa > 0$ also decays to zero rapidly as time goes to infinity. Here then are the functions that will play the crucial role in defining the surgery process.

DEFINITION 15.5. We have: (i) a *canonical neighborhood parameter*, $r(t) > 0$, and (ii) a *surgery control parameter* $\overline{\delta}(t) > 0$. We use these to define the surgery scale function $h(t)$. Set $\rho(t) = \overline{\delta}(t)r(t)$. Let $h(t) = h(\rho(t), \overline{\delta}(t)) \leq \rho(t) \cdot \overline{\delta}(t) = \overline{\delta}^2(t)r(t)$ be the function given by Theorem 11.31. We require that $h(0) \leq R_0^{-1/2}$ where R_0 is the constant from Theorem 13.2.

In addition, there is a function $\kappa(t) > 0$ called the *non-collapsing parameter*. All three functions $r(t)$, $\overline{\delta}(t)$ and $\kappa(t)$ are required to be positive, non-increasing functions of t.

We shall consider Ricci flows with surgery $(\mathcal{M}, G)$ that satisfy Assumptions (1) – (7) and also satisfy:
For any singular time t the surgery at time t is performed with control $\overline{\delta}(t)$ and at scale $h(t) = h(\rho(t), \overline{\delta}(t))$, where $\rho(t) = \overline{\delta}(t)r(t)$, in the sense that each boundary component of C_t is the central 2-sphere of a strong $\overline{\delta}(t)$-neck centered at some y with $R(y) = h(t)^{-2}$.

There is quite a bit of freedom in the choice of these parameters. But it is not complete freedom. They must decay rapidly enough as functions of t. We choose to make $r(t)$ and $\kappa(t)$ step functions, and we require $\overline{\delta}(t)$ to be bounded above by a step function of t. Let us fix the step sizes.

DEFINITION 15.6. We set $t_0 = 2^{-5}$, and for any $i \geq 0$ we define $T_i = 2^i t_0$.

The steps we consider are $[0, T_0]$ and then $[T_i, T_{i+1}]$ for every $i \geq 0$. The first step is somewhat special. Suppose that $(\mathcal{M}, G)$ is a Ricci flow with surgery with normalized initial conditions. Then according to Claim 15.1 the flow exists on $[0, T_1]$ and the norm of the Riemann curvature is bounded by 2 on $[0, T_1]$, so that by Assumption (7) there are no surgeries in this time interval. Also, by Claim 15.1 there is a $\kappa_0 > 0$ so that $\operatorname{Vol} B(x, t, r) \leq \kappa_0 r^3$ for every $t \leq T_1$ and $x \in M_t$ and every $r \leq \epsilon$.

DEFINITION 15.7. *Surgery parameter sequences* are sequences

(i) $\mathbf{r} = r_0 \geq r_1 \geq r_2 \geq \cdots > 0$, with $r_0 = \epsilon$,
(ii) $\mathbf{K} = \kappa_0 \geq \kappa_1 \geq \kappa_2 \geq \cdots > 0$ with κ_0 as in Claim 15.1, and
(iii) $\Delta = \delta_0 \geq \delta_1 \geq \delta_2 \geq \cdots > 0$ with $\delta_0 = \min(\beta\epsilon/3, \delta_0', K^{-1}, D^{-1})$ where δ_0' is the constant from Theorem 13.2 and $\beta < 1/2$ is the constant from Proposition 15.2, ϵ is the constant that we have already fixed, and K and D are the constants from Lemma 12.3.

We shall also refer to partial sequences defined for indices $0, \ldots, i$ for some $i > 0$ as surgery parameter sequences if they are positive, non-increasing and if their initial terms satisfy the conditions given above.

We let $r(t)$ be the step function whose value on $[T_i, T_{i+1})$ is r_{i+1} and whose value on $[0, T_0)$ is r_0. We say that a Ricci flow with surgery satisfies the strong (C, ϵ)-canonical neighborhood assumption with parameter $\mathbf{r}$ if it satisfies this condition with respect to the step function $r(t)$ associated with $\mathbf{r}$. This means that any $x \in \mathcal{M}$ with $R(x) \geq r^{-2}(\mathbf{t}(x))$ has a strong (C, ϵ)-canonical neighborhood in $\mathcal{M}$. Let $\kappa(t)$ be the step function whose value on $[T_i, T_{i+1})$ is κ_{i+1} and whose value on $[0, T_0)$ is κ_0. Given $\kappa > 0$, we say that a Ricci flow defined on $[0, t]$ is κ-non-collapsed on scales $\leq \epsilon$ provided that for every point x not contained in a component of its time-slice with positive sectional curvature, if for some $r \leq \epsilon$, the parabolic neighborhood $P(x, \mathbf{t}(x), r, -r^2)$ exists in $\mathcal{M}$ and the norm of the Riemann curvature is bounded on this backward parabolic neighborhood by r^{-2}, then $\operatorname{Vol} B(x, \mathbf{t}(x), r) \geq \kappa r^3$. We say that a Ricci flow with surgery is $\mathbf{K}$-non-collapsed on scales ϵ if for every $t \in [0, \infty)$ the restriction of the flow to $[0, t]$ is $\kappa(t)$-non-collapsed on scales $\leq \epsilon$. Lastly, we fix a non-increasing function $\overline{\delta}(t) > 0$ with $\overline{\delta}(t) \leq \delta_{i+1}$ if $t \in [T_i, T_{i+1})$ for all $i \geq 0$ and $\overline{\delta}(t) \leq \delta_0$ for $t \in [0, T_0)$. We denote the fact that such inequalities hold for all t by saying $\overline{\delta}(t) \leq \Delta$.

Having fixed surgery parameter sequences $\mathbf{K}$, $\mathbf{r}$ and Δ, defined step functions $r(t)$ and $\kappa(t)$, and fixed $\overline{\delta}(t) \leq \Delta$ as above, we shall consider only Ricci flows with surgery where the surgery at time t is defined using the surgery parameter functions $r(t)$ and $\overline{\delta}(t)$. In addition, we require that these Ricci flows with surgery satisfy Assumptions (1) – (7).

What we shall show is that there are surgery parameter sequences **r**, **K** and Δ with the property that for any normalized initial metric and any positive, non-increasing function $\overline{\delta}(t) \leq \Delta$, it is possible to construct a Ricci flow with surgery using the surgery parameters $r(t)$ and $\overline{\delta}(t)$ with the given initial conditions and furthermore that this Ricci flow with surgery satisfies the Assumptions (1) – (7), has curvature pinched toward positive, satisfies the canonical neighborhood assumption, and satisfies the non-collapsing assumption using these parameters.

In fact we shall prove this inductively, constructing the step functions inductively one step at a time. Thus, given surgery parameter sequences indexed by $0, \ldots, i$ we show that there are appropriate choices of r_{i+1}, κ_{i+1} and δ_{i+1} such that the following is true. Given a Ricci flow with surgery defined on time $[0, T_i)$ satisfying all the properties with respect to the first set of data, that Ricci flow with surgery extends to one defined for time $[0, T_{i+1})$ and satisfies Assumptions (1) – (7), the canonical neighborhood assumption and the non-collapsing assumption with respect to the extended surgery parameter sequences, and has curvature pinched toward positive. As stated this is not quite true; there is a slight twist: we must also assume that $\overline{\delta}(t) \leq \delta_{i+1}$ for all $t \in [T_{i-1}, T_{i+1})$. It is for this reason that we consider pairs consisting of sequences Δ and a surgery control parameter $\overline{\delta}(t)$ bounded above by Δ.

4. The process of surgery

We fix surgery parameter sequences $\{r_0, \ldots, r_{i+1}\}$, $\{\kappa_0, \ldots, \kappa_{i+1}\}$ and $\Delta_i = \{\delta_0, \ldots, \delta_i\}$ and also a positive, decreasing function $\overline{\delta}(t) \leq \Delta_i$, defined for $t \leq T_{i+1}$ with $\delta_0 = \min(\alpha\epsilon/3, \delta_0', K^{-1}, D^{-1})$ as above. Suppose that $(\mathcal{M}, G)$ is a Ricci flow with surgery defined for $t \in [0, T)$ that goes singular at time $T \in (T_i, T_{i+1}]$. We suppose that it satisfies Assumptions (1) – (7). Since the flow has normalized initial conditions and goes singular at time T, it follows that $i \geq 1$. We suppose that $(\mathcal{M}, G)$ satisfies the (C, ϵ)-canonical neighborhood assumption with parameter r_{i+1} and that its curvature is pinched toward positive. By Theorem 11.19 we know that there is a maximal extension $(\widehat{\mathcal{M}}, \widehat{G})$ of this generalized flow to time T with the T time-slice being $(\Omega(T), G(T))$. Set $\rho = \overline{\delta}(T)r_{i+1}$, and set $h(T) = h(\rho(T), \overline{\delta}(T))$ as in Theorem 11.31. Since $\overline{\delta}(T) \leq \delta_0 < 1$, we see that $\rho < r_{i+1}$. According to Lemma 11.30 there are finitely many components of $\Omega(T)$ that meet $\Omega_\rho(T)$. Let $\Omega^{\text{big}}(T)$ be the disjoint union of all the components of $\Omega(T)$ that meet $\Omega_\rho(T)$. Lemma 11.30 also tells us that $\Omega^{\text{big}}(T)$ contains a finite collection of disjoint 2ϵ-horns with boundary contained in $\Omega_{\rho/2C}$, and the complement of the union of the interiors of these horns is a compact submanifold with boundary containing Ω_ρ. Let $\mathcal{H}_1, \ldots, \mathcal{H}_j$ be a disjoint union of these 2ϵ-horns. For each i fix a point $y_i \in \mathcal{H}_i$ with $R(y_i) = h^{-2}(T)$. According to Theorem 11.31 for each i there is a strong $\delta(T)$-neck centered at y_i and contained in $\mathcal{H}_i$. We orient the s-direction of the neck so that its positive end lies closer to the end of the horn than its negative end. Let S_i^2 be the center of this strong $\delta(T)$-neck. Let $\mathcal{H}_i^+$ be the unbounded complementary component of S_i^2 in $\mathcal{H}_i$. Let C_T be the complement of $\coprod_{i=1}^j \mathcal{H}_i^+$ in $\Omega^{\text{big}}(T)$. Then we do surgery on these necks as described in Section 1 of Chapter 13, using the constant $q = q_0$ from Theorem 13.2, removing the positive half of the neck, and gluing on the cap from the standard solution. This creates a compact 3-manifold $M_T = C_T \cup_{\coprod_i S_i^2} B_i$, where each B_i is a copy of the metric ball of radius $A_0 + 4$ centered around the tip of the standard solution (with the metric scaled by $h^2(T)$ and then perturbed near the boundary of B_i to match $g(T)$). Notice that in this process we have removed every component of $\Omega(T)$ that does not contain a point of $\Omega_\rho(T)$. The result of this operation is to produce a compact Riemannian 3-manifold (M_T, G_T) which is the T time-slice of our extension of $(\mathcal{M}, G)$. Let $(M_T, G(t))$, $T \leq t < T'$, be the maximal Ricci flow with initial conditions (M_T, G_T) at $t = T$. Our new space-time $(\mathcal{M}', G')$ is the union of $M_T \times [T, T')$ and $(\mathcal{M}, G) \cup C_T \times \{T\}$ along $C_T \times \{T\}$. Here, we view $(\mathcal{M}, G) \cup C_T \times \{T\}$ as a subspace of the maximal extension $(\widehat{\mathcal{M}}, \widehat{G})$. The time functions and vector fields glue to provide analogous data for this new space-time $\mathcal{M}'$. Since the isometric embedding $C_T \subset M_T$ extends to an isometric embedding of a neighborhood of C_T in Ω_T into M_T, according to Lemma 14.11, the horizontal metrics glue together to make a smooth metric on space-time satisfying the generalized Ricci flow equation, and hence defining a Ricci flow with surgery on $(\mathcal{M}', G')$.

Notice that the continuing region at time T is exactly C_T whereas the exposed region is $\coprod_i B_i$, which is a disjoint union of 3-balls. The disappearing region is the complement of the embedding of C_T in $M_{t'}$ for $t' < T$ but sufficiently close to it, the embedding obtained by flowing $C_T \subset \Omega_T$ backward. The disappearing region contains $M_{t'} \setminus \Omega(T)$ and also contains all components of $\Omega(T)$ that do not contains points of $\Omega_\rho(T)$, as well as the ends of those components of $\Omega(T)$ that contain points of $\Omega_\rho(T)$.

DEFINITION 15.8. The operation described in the previous paragraph is the *surgery operation at time T* using the surgery parameters $\overline{\delta}(T)$ and r_{i+1}.

5. Statements about the existence of Ricci flow with surgery

What we shall establish is the existence of surgery satisfying Assumptions (1) – (7) above and also satisfying the curvature pinched toward positive assumption, the strong canonical neighborhood assumption, and the non-collapsing assumption. This requires first of all that we begin with a compact, Riemannian 3-manifold $(M, g(0))$ that is normalized, which we are assuming. It also requires careful choice of upper bounds $\Delta = \{\delta_i\}$ for the surgery control parameter $\overline{\delta}(t)$ and careful choice of the canonical

5. STATEMENTS ABOUT THE EXISTENCE OF RICCI FLOW WITH SURGERY

neighborhood parameter $\mathbf{r} = \{r_i\}$ and of the non-collapsing step function $\mathbf{K} = \{\kappa_i\}$.

Here is the statement that we shall establish.

THEOREM 15.9. *There are surgery parameter sequences*

$$\mathbf{K} = \{\kappa_i\}_{i=1}^{\infty}, \Delta = \{\delta_i\}_{i=1}^{\infty}, \mathbf{r} = \{r_i\}_{i=1}^{\infty}$$

such that the following holds. Let $r(t)$ be the step function whose value on $[T_{i-1}, T_i)$ is r_i. Suppose that $\overline{\delta} \colon [0, \infty) \to \mathbb{R}^+$ is any non-increasing function with $\overline{\delta}(t) \le \delta_i$ whenever $t \in [T_{i-1}, T_i)$. Then the following holds: Suppose that $(\mathcal{M}, G)$ is a Ricci flow with surgery defined for $0 \le t < T$ satisfying Assumptions (1) – (7). In addition, suppose the following conditions hold:

(1) *the generalized flow has curvature pinched toward positive,*
(2) *the flow satisfies the strong (C, ϵ)-canonical neighborhood assumption with parameter $\mathbf{r}$ on $[0, T)$, and*
(3) *the flow is $\mathbf{K}$ non-collapsed on $[0, T)$ on scales $\le \epsilon$.*

Then there is an extension of $(\mathcal{M}, G)$ to a Ricci flow with surgery defined for all $0 \le t < \infty$ and satisfying Assumptions (1) – (7) and the above three conditions.

This of course leads immediately to the existence result for Ricci flows with surgery defined for all time with any normalized initial conditions.

COROLLARY 15.10. *Let $\mathbf{K}$, $\mathbf{r}$ and Δ be surgery parameter sequences provided by the previous theorem. Let $\overline{\delta}(t)$ be a non-increasing positive function with $\overline{\delta}(t) \le \Delta$. Let M be a compact 3-manifold containing no $\mathbb{R}P^2$ with trivial normal bundle. Then there is a Riemannian metric $g(0)$ on M and a Ricci flow with surgery defined for $0 \le t < \infty$ with initial metric $(M, g(0))$. This Ricci flow with surgery satisfies the seven assumptions and is $\mathbf{K}$-non-collapsed on scales $\le \epsilon$. It also satisfies the strong (C, ϵ)-canonical neighborhood assumption with parameter $\mathbf{r}$ and has curvature pinched toward positive. Furthermore, any surgery at a time $t \in [T_i, T_{i+1})$ is done using $\overline{\delta}(t)$ and r_{i+1}.*

PROOF. (Assuming Theorem 15.9) Choose a metric $g(0)$ so that (M, g_0) is normalized. This is possible by beginning with any Riemannian metric on M and scaling it by a sufficiently large positive constant to make it normalized. According to Proposition 4.11 and the definitions of T_i and κ_0 there is a Ricci flow $(M, g(t))$ with these initial conditions defined for $0 \le t \le T_2$ satisfying Assumptions (1) – (7) and the three conditions of the previous theorem. The assumption that M has no embedded $\mathbb{R}P^2$ with trivial normal bundle is needed so that Assumption (1) holds for this Ricci flow. Hence, by the previous theorem we can extend this Ricci flow to a Ricci flow with surgery defined for all $0 \le t < \infty$ satisfying the same conditions. □

Showing that after surgery Assumptions (1) – (7) continue to hold and that the curvature is pinched toward positive is direct and only requires that $\overline{\delta}(t)$ be smaller than some universal positive constant.

LEMMA 15.11. *Suppose that $(\mathcal{M}, G)$ is a Ricci flow with surgery going singular at time $T \in [T_{i-1}, T_i)$. We suppose that $(\mathcal{M}, G)$ satisfies Assumptions (1) - (7), has curvature pinched toward positive, satisfies the strong (C, ϵ)-canonical neighborhood assumption with parameter $\mathbf{r}$ and is $\mathbf{K}$ non-collapsed. Then the result of the surgery operation at time T on $(\mathcal{M}, G)$ is a Ricci flow with surgery defined on $[0, T')$ for some $T' > T$. The resulting Ricci flow with surgery satisfies Assumptions (1) – (7). It also has curvature pinched toward positive.*

PROOF. It is immediate from the construction and Lemma 14.11 that the result of performing the surgery operation at time T on a Ricci flow with surgery produces a new Ricci flow with surgery. Assumptions (1) – (3) clearly hold for the result. and Assumptions (4) and (5) hold because of the way that we do surgery. Let us consider Assumption (6). Fix $t' < T$ so that there are no surgery times in $[t', T)$. By flowing backward using the vector field χ we have an embedding $\psi \colon C_t \times [t', T] \to \widehat{\mathcal{M}}$ compatible with time and the vector field. For any $p \in M_{t'} \setminus \psi(\operatorname{int} C_T \times \{t'\})$ the limit as t tends to T from below of the flow line $p(t)$ at time t through p either lies in $\Omega(T)$ or it does not. In the latter case, by definition we have

$$\lim_{t \to T^-} R(p(t)) = \infty.$$

In the former case, the limit point either is contained in the end of a strong 2ϵ-horn cut off by the central 2-sphere of the strong δ-neck centered at one of the y_i or is contained in a component of $\Omega(T)$ that contains no point of $\Omega_\rho(T)$. Hence, in this case we have

$$\lim_{t \to T^-} R(p(t)) > \rho^{-2} > r_i^{-2}.$$

Since $M_{t'} \setminus \psi(\operatorname{int} C_T \times \{t'\})$ is compact for every t', there is $T_1 < T$ such that $R(p(t)) > r_i^{-2}$ for all $p \in M_{t'} \setminus \psi(\operatorname{int} C_T \times \{t'\})$ and all $t \in [T_1, T)$. Hence, by our assumptions all these points have strong (C, ϵ)-canonical neighborhoods. This establishes that Assumption (6) holds at the singular time T. By hypothesis Assumption (6) holds at all earlier singular times. Clearly, from the construction the Ricci flow on $[T, T')$ is maximal. Hence, Assumption (7) holds for the new Ricci flow with surgery.

From Theorem 13.2 the fact that $\delta(T) \leq \delta_i \leq \delta_0 \leq \delta_0'$ and $h(T) \leq R_0^{-1/2}$ imply that the Riemannian manifold $(M_T, G(T))$ has curvature pinched toward positive for time T. It then follows from Theorem 4.32 that the Ricci flow defined on $[T, T')$ with $(M_T, G(T))$ as initial conditions has curvature pinched toward positive. The inductive hypothesis is that on the time-interval $[0, T)$ the Ricci flow with surgery has curvature pinched toward positive. This completes the proof of the lemma. □

PROPOSITION 15.12. *Suppose that $(\mathcal{M}, G)$ is a Ricci flow with surgery satisfying Assumptions (1) – (7) in Section 1.1. Suppose that T is a surgery time, suppose that the surgery control parameter $\delta(T)$ is less than δ_0 in Definition 15.7, and suppose that the scale of the surgery $h(T)$ is less than $R_0^{-1/2}$ where R_0 is the constant from Theorem 13.2. Fix $t' < T$ sufficiently close to T. Then there is an embedding $\rho\colon M_{t'} \times [t', T) \to \mathcal{M}$ compatible with time and the vector field. Let $X(t')$ be a component of $M_{t'}$ and let $X(T)$ be a component obtained from $X(t')$ by doing surgery at time T. We view $\rho^* G$ as a one-parameter family of metrics $g(t)$ on $X(t')$. There is an open subset $\Omega \subset X(t')$ with the property that $\lim_{t' \to T^-} g(t')|_\Omega$ exists (we denote it by $g(T)|_\Omega$) and with the property that $\rho|_{\Omega \times [t', T)}$ extends to a map $\widehat{\rho}\colon \Omega \times [t', T] \to \mathcal{M}$. This defines a map for $\Omega \subset X(t')$ onto an open subset $\Omega(T)$ of $X(T)$ which is an isometry from the limiting metric $g(T)$ on Ω to $G(T)|_\Omega$. Suppose that all of the 2-spheres along which we do surgery are separating. Then this map extends to a map $X(t') \to X(T)$. For all $t < T$ but sufficiently close to T this extension is a distance decreasing map from $(X(t') \setminus \Omega, g(t))$ to $X(T)$.*

PROOF. This is immediate from the third item in Theorem 13.2. □

REMARK 15.13. If we have a non-separating surgery 2-sphere then there will a component $X(T)$ with surgery caps on both sides of the surgery 2-sphere and hence we cannot extend the map even continuously over all of $X(t')$.

The other two inductive properties in Theorem 15.9 — that the result is **K**-non-collapsed and also that it satisfies the strong (C, ϵ)-canonical neighborhood assumption with parameter **r** — require appropriate inductive choices of the sequences. The arguments establishing these are quite delicate and intricate. They are given in the next two sections.

6. Outline of the proof of Theorem 15.9

Before giving the proof proper of Theorem 15.9 let us outline how the argument goes. We shall construct the surgery parameter sequences **Δ**, **r**, and **K** inductively. Because of Proposition 4.11 we have the beginning of the inductive process. We suppose that we have defined sequences as required up to index i for some $i \geq 1$. Then we shall extend them one more step to sequences defined up to $(i+1)$, though there is a twist: to do this we must redefine δ_i in order to make sure that the extension is possible. In Chapter 16 we establish the non-collapsing result assuming the strong canonical neighborhood result. More precisely, suppose that we have a Ricci flow with surgery $(\mathcal{M}, G)$ defined for time $0 \leq t < T$ with $T \in (T_i, T_{i+1}]$ so that the restriction of this flow to the time-interval $[0, T_i)$ satisfies the inductive hypothesis with respect to the given sequences. Suppose also that the entire Ricci flow with surgery has strong (C, ϵ)-canonical neighborhoods for some $r_{i+1} > 0$. Then there is $\delta(r_{i+1}) > 0$ and $\kappa_{i+1} > 0$ such that, provided that $\overline{\delta}(t) \leq \delta(r_{i+1})$ for all $t \in [T_{i-1}, T)$, the Ricci flow with surgery $(\mathcal{M}, G)$ is κ_{i+1}-non-collapsed on scales $\leq \epsilon$.

In Section 1 of Chapter 17 we show that the strong (C, ϵ)-canonical neighborhood assumption extends for some parameter r_{i+1}, assuming again that $\overline{\delta}(t) \leq \delta(r_{i+1})$ for all $t \in [T_{i-1}, T)$.

Lastly, in Section 2 of Chapter 17 we complete the proof by showing that the number of surgeries possible in $[0, T_{i+1})$ is bounded in terms of the initial conditions and $\overline{\delta}(T)$. The argument for this is a simple volume comparison argument. Namely, under Ricci flow with normalized initial conditions, the volume grows at most at a fixed exponential rate and under each surgery an amount of volume, bounded below by a positive constant depending only on $\overline{\delta}(T_{i+1})$, is removed.

CHAPTER 16

Proof of non-collapsing

The precise statement of the non-collapsing result is given in the next section. Essentially, the proof of non-collapsing in the context of Ricci flow with surgery is the same as the proof in the case of ordinary Ricci flows. Given a point $x \in \mathcal{M}$, one finds a parabolic neighborhood whose size, r', is determined by the constants r_i, C and ϵ, contained in $\mathbf{t}^{-1}([T_{i-1}, T_i))$ and on which the curvature is bounded by $(r')^{-2}$. Hence, by the inductive hypothesis, the final time-slice of this neighborhood is κ_i-non-collapsed. Furthermore, we can choose this neighborhood so that the reduced $\mathcal{L}$-length of its central point from x is bounded by $3/2$. This allows us to produce an open subset at an earlier time whose reduced volume is bounded away from zero. Then using Theorem 8.1 we transfer this conclusion to a non-collapsing conclusion for x. The main issue in this argument is to show that there is a point in each earlier time-slice whose reduced length from x is at most $3/2$. We can argue as in the case of a Ricci flow if we can show that any curve parameterized by backward time starting at x (a point where the hypothesis of κ-non-collapsing holds) that comes close to a surgery cap either from above or below must have large $\mathcal{L}$-length. In establishing the relevant estimates we are forced to require that δ_i be sufficiently small.

1. The statement of the non-collapsing result

Here, we shall assume that after surgery the strong canonical neighborhood assumption holds, and we shall establish the non-collapsing result.

PROPOSITION 16.1. *Suppose that for some $i \geq 0$ we have surgery parameter sequences $\delta_0 \geq \delta_1 \geq \cdots \geq \delta_i > 0$, $\epsilon = r_0 \geq r_1 \geq \cdots \geq r_i > 0$ and $\kappa_0 \geq \kappa_1 \geq \cdots \geq \kappa_i > 0$. Then there is $0 < \kappa \leq \kappa_i$ and for any $0 < r_{i+1} \leq r_i$ there is $0 < \delta(r_{i+1}) \leq \delta_i$ such that the following holds. Suppose that $\overline{\delta} \colon [0, T_{i+1}] \to \mathbb{R}^+$ is a non-increasing function with $\overline{\delta}(t) \leq \delta_j$ for all $t \in [T_j, T_{j+1})$ and $\overline{\delta}(t) \leq \delta(r_{i+1})$ for all $t \in [T_{i-1}, T_{i+1})$. Suppose that $(\mathcal{M}, G)$ is a Ricci flow with surgery defined for $0 \leq t < T$ for some $T \in (T_i, T_{i+1}]$ with surgery control parameter $\overline{\delta}(t)$. Suppose that the restriction of this Ricci flow with surgery to the time-interval $[0, T_i)$ satisfies the hypothesis of Theorem 15.9 with respect to the given sequences. Suppose also that the entire Ricci flow with surgery $(\mathcal{M}, G)$ satisfies Assumptions (1) –*

(7) and the strong (C,ϵ)-canonical neighborhood assumption with parameter r_{i+1}. Then $(\mathcal{M}, G)$ is κ-non-collapsed on all scales $\leq \epsilon$.

REMARK 16.2. Implicitly, κ and $\delta(r_{i+1})$ are also allowed to depend on t_0, ϵ, and C, which are fixed, and also $i+1$. Also recall that the non-collapsing condition allows for two outcomes: if x is a point at which the hypothesis of the non-collapsing holds, then there is a lower bound on the volume of a ball centered at x, or x is contained in a component of its time-slice that has positive sectional curvature.

2. The proof of non-collapsing when $R(x) = r^{-2}$ with $r \leq r_{i+1}$

Let us begin with an easy case of the non-collapsing result, where non-collapsing follows easily from the strong canonical neighborhood assumption, rather than from using $\mathcal{L}$-length and monotonicity along $\mathcal{L}$-geodesics. We suppose that we have a Ricci flow with surgery $(\mathcal{M}, G)$ defined for $0 \leq t < T$ with $T \in [T_i, T_{i+1})$, and a constant $r_{i+1} \leq r_i$, all satisfying the hypothesis of Proposition 16.1. Here is the result that establishes the non-collapsing in this case.

PROPOSITION 16.3. *Let $x \in \mathcal{M}$ with $\mathbf{t}(x) = t$ and with $R(x) = r^{-2} \geq r_{i+1}^{-2}$. Then there is $\kappa > 0$ depending only on C such that $\mathcal{M}$ is κ-non-collapsed at x; i.e., if $R(x) = r^{-2}$ with $r \leq r_{i+1}$, then $\operatorname{Vol} B(x,t,r) \geq \kappa r^3$, or x is contained in a component of M_t with positive sectional curvature.*

PROOF. Since $R(x) \geq r_{i+1}^{-2}$, by assumption any such x has a strong (C, ϵ)-canonical neighborhood. If this neighborhood is a strong ϵ-neck centered at x, then the result is clear for a non-collapsing constant κ which is universal. If the neighborhood is an ϵ-round component containing x, then x is contained in a component of positive sectional curvature. Likewise, if x is contained in a C-component, then by definition it is contained in a component of its time-slice with positive sectional curvature.

Lastly, we suppose that x is contained in the core Y of a (C, ϵ)-cap $\mathcal{C}$. Let $r' > 0$ be such that the supremum of $|\mathrm{Rm}|$ on $B(x,t,r')$ is $(r')^{-2}$. Then, by the definition of a (C, ϵ)-cap, $\operatorname{vol} B(x,t,r') \geq C^{-1}(r')^3$. Clearly, $r' \leq r$ and there is a point $y \in \overline{B(x,t,r')}$ with $R(y) = (r')^{-2}$. On the other hand, by the definition of a (C, ϵ)-cap, we have $R(y)/R(x) \leq C$, so that $r'/r \geq C^{-1/2}$. Thus, the volume of $B(x,t,r)$ is at least $C^{-5/2}r^3$.

This completes an examination of all cases and establishes the proposition. $\square$

3. Minimizing $\mathcal{L}$-geodesics exist when $R(x) \leq r_{i+1}^{-2}$: the statement

The proof of the non-collapsing result when $R(x) = r^{-2}$ with $r_{i+1} < r \leq \epsilon$ is much more delicate. As we indicated above, it is analogous to the proof of non-collapsing for Ricci flows given in Chapter 8. That is to say, in this case the result is proved using the length function on the Ricci flow with surgery and the monotonicity of the reduced volume. Of course, unlike the case of Ricci flows treated in Chapter 8, here not all points of a Ricci flow with surgery $\mathcal{M}$ can be reached by minimizing $\mathcal{L}$-geodesics, or rather more precisely by minimizing $\mathcal{L}$-geodesics contained in the open subset of smooth points of $\mathcal{M}$. (It is only for the latter $\mathcal{L}$-geodesics that the analytic results of Chapter 6 apply.) Thus, the main thing to establish in order to prove non-collapsing is that for any Ricci flow with surgery $(\mathcal{M}, G)$ satisfying the hypothesis of Proposition 16.1, there are minimizing $\mathcal{L}$-geodesics in the open subset of smooth points of $\mathcal{M}$ to 'enough' of $\mathcal{M}$ so that we can run the same reduced volume argument that worked in Chapter 8. Here is the statement that tells us that there are minimizing $\mathcal{L}$-geodesics to 'enough' of $\mathcal{M}$.

PROPOSITION 16.4. *For each r_{i+1} with $0 < r_{i+1} \leq r_i$, there is $\delta = \delta(r_{i+1}) > 0$ (depending implicitly on t_0, C, ϵ, and i) such that if $\overline{\delta}(t) \leq \delta$ for all $t \in [T_{i-1}, T_{i+1}]$ then the following holds. Let $(\mathcal{M}, G)$ be a Ricci flow with surgery satisfying the hypothesis of Proposition 16.1 with respect to the given sequences and r_{i+1}, and let $x \in \mathcal{M}$ have $\mathbf{t}(x) = T$ with $T \in [T_i, T_{i+1})$. Suppose that for some $r \geq r_{i+1}$ the parabolic neighborhood $P(x, r, T, -r^2)$ exists in $\mathcal{M}$ and $|\mathrm{Rm}| \leq r^{-2}$ on this neighborhood. Then there is an open subset U of $\mathbf{t}^{-1}[T_{i-1}, T)$ contained in the open subset of smooth manifold points of $\mathcal{M}$ with the following properties:*

(1) *For every y in U there is a minimizing $\mathcal{L}$-geodesic connecting x to y.*
(2) *$U_t = U \cap \mathbf{t}^{-1}(t)$ is non-empty for every $t \in [T_{i-1}, T)$.*
(3) *For each $t \in [T_{i-1}, T)$ the restriction of $\mathcal{L}$ to U_t achieves its minimum and that minimum is at most $3\sqrt{(T-t)}$.*
(4) *The subset of U consisting of all y with the property that $\mathcal{L}(y) \leq \mathcal{L}(y')$ for all $y' \in \mathbf{t}^{-1}(\mathbf{t}(y))$ has the property that its intersection with $\mathbf{t}^{-1}(I)$ is compact for every compact interval $I \subset [T_{i-1}, T)$.*

The basic idea in proving this result is to show that all paths beginning at x and parameterized by backward time that come close to the exposed regions have large $\mathcal{L}$-length. If we can establish this, then the existence of such paths will not be an impediment to using the analytic estimates from Chapter 6 to show that for each $t \in [T_{i-1}, T)$ there is a point whose $\mathcal{L}$-length from x is at most $3\sqrt{T-t}$, and that the set of points that minimize the $\mathcal{L}$-length from x in a given time-slice form a compact set.

Given Proposition 16.4, arguments from Chapter 8 will be applied to complete the proof of Proposition 16.1.

4. Evolution of neighborhoods of surgery caps

We begin the analysis required to prove Proposition 16.4 by studying the evolution of surgery caps. Proposition 16.5 below is the main result along these lines. Qualitatively, it says that if the surgery control parameter δ is sufficiently small, then as a surgery cap evolves in a Ricci flow with surgery, it stays near the rescaled version of the standard flow for any rescaled time less than 1 unless the entire cap is removed (all at once) by some later surgery. In that case, the evolution of the cap is close to the rescaled version of the standard flow until it is removed. Using this result we will show that if a path parameterized by backward time has final point near a surgery cap and has initial point with scalar curvature not too large, then this path must enter this evolving neighborhood either from the 'top' or 'side' and because of the estimates that we derive in this chapter such a path must have large $\mathcal{L}$-length.

PROPOSITION 16.5. *Given $A < \infty$, $\delta'' > 0$ and $0 < \theta < 1$, there is $\delta_0'' = \delta_0''(A, \theta, \delta'')$ (δ_0'' also depends on r_{i+1}, C, and ϵ, which are all now fixed) such that the following holds. Suppose that $(\mathcal{M}, G)$ is a Ricci flow with surgery defined for $0 \leq t < T$ with surgery control parameter $\overline{\delta}(t)$. Suppose that it satisfies the strong (C, ϵ)-canonical neighborhood assumption at all points x with $R(x) \geq r_{i+1}^{-2}$. Suppose also that $(\mathcal{M}, G)$ has curvature that is pinched toward positive. Suppose that there is a surgery at some time $\overline{t}$ with $T_{i-1} \leq \overline{t} < T$ with $\overline{h}$ as the surgery scale parameter. Set $T' = \min(T, \overline{t} + \theta \overline{h}^2)$. Let $p \in M_{\overline{t}}$ be the tip of the cap of a surgery disk. Then, provided that $\overline{\delta}(\overline{t}) \leq \delta_0''$ one of the following holds:*

(a) *There is an embedding $\rho \colon B(p, \overline{t}, A\overline{h}) \times [\overline{t}, T') \to \mathcal{M}$ compatible with time and the vector field. Let $g'(t)$, $\overline{t} \leq t < T'$, be the one-parameter family of metrics on $B(p, \overline{t}, A\overline{h})$ given by $\rho^* G$. Shifting this family by $-\overline{t}$ to make the initial time 0 and scaling it by $\overline{h}^{-2}$ produces a family of metrics $g(t)$, $0 \leq t < \min((T - \overline{t})\overline{h}^{-2}, \theta)$, on $B_g(p, 0, A)$ that are within δ'' in the $C^{[1/\delta'']}$-topology of the standard flow on the ball of radius A at time 0 centered at the tip of its cap.*

(b) *There is $\overline{t}_+ \in (\overline{t}, T')$ and an embedding $B(p, \overline{t}, A\overline{h}) \times [\overline{t}, \overline{t}_+) \to \mathcal{M}$ compatible with time and the vector field so that the previous item holds with $\overline{t}_+$ replacing T'. Furthermore, for any $t < \overline{t}_+$ but sufficiently close to $\overline{t}_+$ the image of $B(p, \overline{t}, A\overline{h}) \times \{t\}$ is contained in the region $D_t \subset M_t$ that disappears at time $\overline{t}_+$.*

See FIG. 1.

PROOF. The method of proof is to assume that the result is false and take a sequence of counterexamples with surgery control parameters δ_n tending to zero. In order to derive a contradiction we need to be able to take smooth limits of rescaled versions of these Ricci flows with surgery, where the base points are the tips of the surgery caps. This is somewhat delicate since the surgery cap is not the result of moving forward for a fixed amount of time under Ricci flow, and consequently Shi's theorem does not apply.

4. EVOLUTION OF NEIGHBORHOODS OF SURGERY CAPS

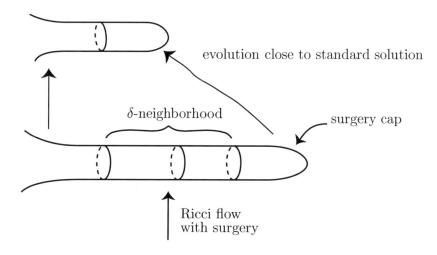

FIGURE 1. Evolution of a surgery cap.

Fortunately, the metrics on the cap are bounded in the C^∞-topology so that Shi's theorem with derivatives does apply. Let us start by examining limits of the sort we need to take.

CLAIM 16.6. *Let (N, g_N) be a strong δ'-neck with N_0 its middle half. Suppose that (S, g) is the result of doing surgery on (the central 2-sphere) of N, adding a surgery cap $\mathcal{C}$ to N^-. Let h be the scale of N. Let $(S_0(N), g')$ be the union of $N_0^- \cup \mathcal{C}$ with its induced metric as given in Section 1 of Chapter 13, and let $(S_0(N), \widehat{g}_0)$ be the result of rescaling g_0 by h^{-2}. Then for every $\ell < \infty$ there is a uniform bound to $|\nabla^\ell \mathrm{Rm}_{\widehat{g}_0}(x)|$ for all $x \in S_0(N)$.*

PROOF. Since (N, g_N) is a strong δ'-neck of scale h, there is a Ricci flow on N defined for backward time h^2. After rescaling by h^{-2} we have a flow defined for backward time 1. Furthermore, the curvature of the rescaled flow is bounded on the interval $(-1, 0]$. Since the closure of N_0 in N is compact, the restriction of $h^{-2} g_N$ to $N_0 \subset N$ at time 0 is uniformly bounded in the C^∞-topology by Shi's theorem (Theorem 3.28). The bound on the kth-derivatives of the curvature depends only on the curvature bound and hence can be taken to be independent of $\delta' > 0$ sufficiently small and also independent of the strong δ'-neck N. Gluing in the cap with a C^∞-metric that converges smoothly to the standard initial metric g_0 as δ' tends to zero using a fixed C^∞-partition of unity produces a family of manifolds uniformly bounded in the C^∞-topology. □

This leads immediately to:

COROLLARY 16.7. *Given a sequence of $\delta'_n \to 0$ and strong δ'_n-necks $(N(n), g_{N(n)})$ of scales h_n and results of surgery $(S_0(N(n)), g(n))$ with tips p_n as in the previous claim, then after passing to a subsequence there is a smooth limit $(S_\infty, g_\infty, p_\infty)$ of a subsequence of the $(S_0(N(n)), h_n^{-2} g_0(n)), p_n)$.*

This limit is the metric from Section 1 of Chapter 13 that gives the standard initial conditions for a surgery cap.

PROOF. That there is a smooth limit of a subsequence is immediate from the previous claim. Since the δ_n tend to zero, it is clear that the limiting metric is the standard initial metric. $\square$

LEMMA 16.8. *Suppose that we have a sequence of 3-dimensional Ricci flows with surgeries $(\mathcal{M}_n, G_n)$ that satisfy the strong (C, ϵ)-canonical neighborhood assumption with parameter r_{i+1}, and have curvature pinched toward positive. Suppose that there are surgeries in $\mathcal{M}_n$ at times t_n with surgery control parameters δ'_n and scales h_n. Let p_n be the tip of a surgery cap for the surgery at time t_n. Also suppose that there is $0 \leq \theta_n < 1$ such that for every $A < \infty$, for all n sufficiently large there are embeddings $B(p_a, t_n, Ah_n) \times [t_n, t_n + h_n^2 \theta_n) \to \mathcal{M}_n$ compatible with time and the vector field. Suppose that $\delta'_n \to 0$ and $\theta_n \to \theta < 1$ as $n \to \infty$. Let $(\mathcal{M}'_n, G'_n, p_n)$ be the Ricci flow with surgery obtained by shifting time by $-t_n$ so that surgery occurs at $\mathbf{t} = 0$ and rescaling by h_n^{-2} so that the scale of the surgery becomes 1. Then, after passing to a subsequence, the sequence converges smoothly to a limiting flow $(M_\infty, g_\infty(t), (p_\infty, 0))$, $0 \leq t < \theta$. This limiting flow is isomorphic to the restriction of the standard flow to time $0 \leq t < \theta$.*

PROOF. Let $Q < \infty$ be an upper bound for the scalar curvature of the standard flow on the time interval $[0, \theta)$. Since $\delta'_n \to 0$, according to the previous corollary, there is a smooth limit at time 0 for a subsequence, and this limit is the standard initial metric. Suppose that, for some $0 \leq \theta' < \theta$, we have established that there is a smooth limiting flow on $[0, \theta']$. Since the initial conditions are the standard solution, it follows from the uniqueness statement in Theorem 12.5 that in fact the limiting flow is isomorphic to the restriction of the standard flow to this time interval. Then the scalar curvature of the limiting flow is bounded by Q. Hence, for any $A < \infty$, for all n sufficiently large, the scalar curvature of the restriction of G'_n to the image of $B_{G'_n}(p_n, 0, 2A) \times [0, \theta']$ is bounded by $2Q$. According to Lemma 11.2 there is an $\eta > 0$ and a constant $Q' < \infty$, each depending only on Q, r_{i+1}, C and ϵ, such that for all n sufficiently large, the scalar curvature of the restriction of G'_n to $B_{G'_n}(p_n, 0, A) \times [0, \min(\theta' + \eta, \theta_n))$ is bounded by Q'. Because of the fact that the curvature is pinched toward positive, this implies that on the same set the sectional curvatures are uniformly bounded. Hence, by Shi's theorem with derivatives (Theorem 3.29), it follows that there are uniform bounds for the curvature in the C^∞-topology. Thus, passing to a subsequence we can extend the smooth limit to the time interval $[0, \theta' + \eta/2]$ unless $\theta' + \eta/2 \geq \theta$. Since η depends on θ (through Q), but is independent of θ', we can repeat this process extending the time-interval of definition of the limiting flow by $\eta/2$ until $\theta' + \eta/2 \geq \theta$. Now suppose that $\theta' + \eta/2 \geq \theta$. Then the argument shows that by passing to a subsequence we can extend the limit to any compact subinterval of $[0, \theta)$. Taking a diagonal sequence allows us to extend it to all of $[0, \theta)$. By the uniqueness of the standard flow, this limit is the standard flow. □

COROLLARY 16.9. *With the notation and assumptions of the previous lemma, for all $A < \infty$, and any $\delta'' > 0$, then for all n sufficiently large, the restriction of G'_n to the image $B_{G'_n}(p_n, 0, A) \times [0, \theta_n)$ is within δ'' in the $C^{[1/\delta'']}$-topology of the restriction of the standard solution to the ball of radius A about the tip for time $0 \le t < \theta_n$.*

PROOF. Let $\eta > 0$ depending on θ (though Q) as well as r_{i+1}, C and ϵ be as in the proof of the previous lemma, and take $0 < \eta' < \eta$. For all n sufficiently large $\theta_n > \theta - \eta'$, and consequently for all n sufficiently large there is an embedding $B_{G_n}(p_n, t_n, Ah_n) \times [t_n, t_n + h_n^2(\theta - \eta')]$ into $\mathcal{M}_n$ compatible with time and with the vector field. For all n sufficiently large, we consider the restriction of G'_n to $B_{G'_n}(p_n, 0, A) \times [0, \theta - \eta']$. These converge smoothly to the restriction of the standard flow to the ball of radius A on the time interval $[0, \theta - \eta']$. In particular, for all n sufficiently large, the restrictions to these time intervals are within δ'' in the $C^{[1/\delta'']}$-topology of the standard flow. Also, for all n sufficiently large, $\theta_n - (\theta - \eta') < \eta$. Thus, by Lemma 11.2, we see that the scalar curvature of G'_n is uniformly bounded (independent of n) on $B_{G'_n}(p_n, 0, A) \times [0, \theta_n)$. By the assumption that the curvature is pinched toward positive, this means that the sectional curvatures of the G'_n are also uniformly bounded on these sets, and hence so are the Ricci curvatures. (Notice that these bounds are independent of $\eta' > 0$.) By Shi's theorem with derivatives (Theorem 3.29), we see that there are uniform bounds on the curvatures in the C^∞-topology on these subsets, and hence bounds in the C^∞-topology on the Ricci curvature. These bounds are independent of both n and η'. Thus, choosing η' sufficiently close to zero, so that $\theta_n - \eta'$ is also close to θ for all n sufficiently large, we see that for all such large n and all $t \in [\theta - \eta', \theta)$, the restriction of G'_n to $B_{G'_n}(p_n, 0, A) \times \{t\}$ is arbitrarily close in the $C^{[1/\delta'']}$-topology to $G'_n(\theta - \eta')$. The same is of course true of the standard flow. This completes the proof of the corollary. □

Now we turn to the proof proper of Proposition 16.5. We fix $A < \infty$, $\delta'' > 0$ and $\theta < 1$. We are free to make A larger so we can assume by Proposition 12.7 that for the standard flow the restriction of the flow to $B(p_0, 0, A) \setminus B(p_0, 0, A/2)$ remains close to a standard evolving $S^2 \times [A/2, A]$ for time $[0, \theta]$. Let $K < \infty$ be a constant with the property that $R(x, t) \le K$ for all $x \in B(p_0, 0, A)$ in the standard flow and all $t \in [0, \theta]$. If there is no $\delta''_0 > 0$ as required, then we can find a sequence $\delta'_n \to 0$ as $n \to \infty$ and Ricci flows with surgery $(\mathcal{M}_n, G_n)$ with surgeries at time t_n with surgery control parameter $\delta_n(t_n) \le \delta'_n$ and surgery scale parameter $h_n = h(r_{i+1}\delta_n(t_n), \delta_n(t_n))$ satisfying the hypothesis of the lemma but not the conclusion. Let T'_n be the final time of $(\mathcal{M}_n, G_n)$. Let $\theta_n \leq \theta$ be maximal subject to the condition that there is an embedding $\rho_n \colon B_{G_n}(x, t_n, Ah_n) \times [t_n, t_n + h_n^2 \theta_n) \to \mathcal{M}_n$ compatible with time and the vector field. Let G'_n be the result of shifting the time by $-t_n$ and scaling the result by h_n^{-2}. According to Corollary 16.9, for all n sufficiently large, the restriction of G'_n to the image of ρ_n is within δ'' in the $C^{[1/\delta'']}$-topology of the standard flow restricted to the ball of radius A about the tip of the standard solution on the time interval $[0, \theta_n)$. If $\theta_n = \min(\theta, (T'_n - t_n)/h_n^2)$, then the first conclusion of Proposition 16.5 holds for $(\mathcal{M}_n, G_n)$ for all n sufficiently large, which contradicts our assumption that the conclusion of this proposition holds for none of the $(\mathcal{M}_n, G_n)$. If on the other hand $\theta_n < \min(\theta, (T'_n - t_n)/h_n^2)$, we need only show that all of $B(x_n, t_n, Ah_n)$ disappears at time $t_n + h_n^2 \theta_n$ in order to show that the second conclusion of Proposition 16.5 holds provided that n is sufficiently large. Again this would contradict the fact that the conclusion of this proposition holds for none of the $(\mathcal{M}_n, G_n)$.

So now let us suppose that $\theta_n < \min(\theta, (T'_n - t_n)/h_n^2)$. Since there is no further extension in forward time for $B(p_n, t_n, Ah_n)$, it must be the case that $t_n + h_n^2 \theta_n$ is a surgery time and there is some flow line starting at a point of $B(p_n, t_n, Ah_n)$ that does not continue to time $t_n + h_n^2 \theta_n$. It remains to show that in this case that for any $t < t_n + h_n^2 \theta_n$ sufficiently close to $t_n + h_n^2 \theta_n$ we have $\rho_n \left(B_{G_n}(x, t_n, Ah_n) \times \{t\} \right) \subset D_t$, the region in M_t that disappears at time $t_n + h_n^2 \theta_n$.

CLAIM 16.10. *Suppose that $\theta_n < \min(\theta, (T'_n - t_n)/h_n^2)$. Let $\Sigma_1, \ldots, \Sigma_k$ be the 2-spheres along which we do surgery at time $t_n + h_n^2 \theta_n$. Then for any $t < t_n + h_n^2 \theta_n$ sufficiently close to $t_n + h_n^2 \theta_n$ the following holds provided that δ'_n is sufficiently small. The image*

$$\rho_n \left(B_{g_n}(x, t_n, Ah_n) \times \{t\} \right)$$

is disjoint from the images $\{\Sigma_i(t)\}$ of the $\{\Sigma_i\}$ under the backward flow to time t of the spheres Σ_i along which we do surgery at time $t_n + h_n^2 \theta_n$.

PROOF. There is a constant $K' < \infty$ depending on θ such that for the standard flow we have $R(x, t) \leq K'$ for all $x \in B(p_0, 0, A)$ and all $t \in [0, \theta)$ for the standard solution. Consider the image

$$\rho_n \left(B(p_n, t_n, Ah_n) \times [t_n, t_n + h_n^2 \theta_n) \right).$$

After time shifting by $-t_n$ and rescaling by h_n^{-2}, the flow G'_n on the image of ρ_n is within δ'' of the standard flow. Thus, we see that for all n sufficiently large and for every point x in the image of ρ_n we have $R_{G'_n}(x) \leq 2K'$ and hence $R_{G_n}(x) \leq 2K' h_n^{-2}$.

Let h'_n be the scale of the surgery at time $t_n + h_n^2 \theta_n$. (Recall that h_n is the scale of the surgery at time t_n.) Suppose that $\rho_n(B(p_n, t_n, Ah_n) \times \{t'\})$ meets one of the surgery 2-spheres $\Sigma_i(t')$ at time t' at a point $y(t')$. Then, for all $t \in [t', t_n + h_n^2 \theta_n)$ we have the image $y(t)$ of $y(t')$ under the flow. All these points $y(t)$ are points of intersection of $\rho_n(B(p, t_n, Ah_n) \times \{t\})$ with $\Sigma_i(t)$. Since $y(t) \in \rho_n(B(p, t_n, Ah_n) \times \{t\})$, we have $R(y(t)) \leq 2K'h_n^{-2}$. On the other hand $R(y(t))(h_n')^2$ is within $O(\delta)$ of 1 as t tends to $t_n + h_n^2\theta_n$. This means that $h_n/h_n' \leq \sqrt{3K'}$ for all n sufficiently large. Since the standard solution has non-negative curvature, the metric is a decreasing function of t, and hence the diameter of $B(p_0, t, A)$ is at most $2A$ in the standard solution. Using Corollary 16.9 we see that for all n sufficiently large, the diameter of $\rho_n(B(p, t_n, Ah_n) \times \{t\})$ is at most $Ah_n \leq 4\sqrt{K'}Ah_n'$. This means that for δ_n' sufficiently small the distance at time t from $\Sigma_i(t)$ to the complement of the t time-slice of the strong $\delta_n(t_n + h_n^2\theta_n)$-neck $N_i(t)$ centered at $\Sigma_i(t)$ (which is at least $(\delta_n')^{-1}h_n'/2$) is much larger than the diameter of

$$\rho_n(B(p_n, t_n, Ah_n) \times \{t\}).$$

Consequently, for all n sufficiently large, the image $\rho_n(B(p_n, t_n, Ah_n) \times \{t\})$ is contained in $N_i(t)$. But by our choice of A, and Corollary 16.9 there is an ϵ-neck of rescaled diameter approximately $Ah_n/2$ contained in $\rho_n(B(p_n, t_n, Ah_n) \times \{t\})$. By Corollary A.3 the spheres coming from the neck structure in

$$\rho_n(B(p_n, t_n, Ah_n) \times \{t\})$$

are isotopic in $N_i(t)$ to the central 2-sphere of this neck. This is a contradiction because in $N_i(t)$ the central 2-sphere is homotopically non-trivial whereas the spheres in $\rho_n(B(p_n, t_n, A\overline{h}_n) \times \{t\})$ clearly bound 3-disks. □

Since $\rho_n(B(p_n, t_n, Ah_n) \times \{t\})$ is disjoint from the backward flow to time t of all the surgery 2-spheres $\Sigma_i(t)$ and since $\rho_n(B(p_n, t_n, Ah_n) \times \{t\})$ is connected, if there is a flow line starting at some point $z \in B(p, t_n, Ah_n)$ that disappears at time $t_n + h_n^2\theta_n$, then the flow from every point of $B(p, t_n, Ah_n)$ disappears at time $t_n + h_n^2\theta_n$. This shows that if $\theta_n < \min(\theta, T_n' - t_n/h_n^2)$, and if there is no extension of ρ_n to an embedding defined at time $t_n + h_n^2\theta_n$, then all forward flow lines beginning at points of $B(p, t_n, Ah_n)$ disappear at time $t_n + h_n^2\theta_n$, which of course means that for all $t < t_n + h_n^2\theta_n$ sufficiently close to $t_n + h_n^2\theta_n$ the entire image $\rho_n(B(p, t_n, Ah_n) \times \{t\})$ is contained in the disappearing region D_t. This shows that for all n sufficiently large, the second conclusion of Proposition 16.5 holds, giving a contradiction.

This completes the proof of Proposition 16.5. □

REMARK 16.11. Notice that it is indeed possible that $B_G(x, t, Ah)$ is removed at some later time, for example as part of a capped ϵ-horn associated to some later surgery.

5. A length estimate

We use the result in the previous section about the evolution of surgery caps to establish the length estimate on paths parameterized by backward time approaching a surgery cap from above.

DEFINITION 16.12. Let $c > 0$ be the constant from Proposition 12.31. Fix $0 < \overline{\delta}_0 < 1/4$ such that if g is within $\overline{\delta}_0$ of g_0 in the $C^{[1/\overline{\delta}]}$-topology then

$$|R_{g'}(x) - R_{g_0}(x)| < c/2 \quad \text{and} \quad |\text{Ric}_{y'} - \text{Ric}_{y_0}| < 1/4.$$

Here is the length estimate.

PROPOSITION 16.13. For any $\ell < \infty$ there is $A_0 = A_0(\ell) < \infty$, $0 < \theta_0 = \theta_0(\ell) < 1$, and for any $A \geq A_0$ for the constant $\delta'' = \delta''(A) = \delta''_0(A, \theta_0, \overline{\delta}_0) > 0$ from Proposition 16.5, the following holds. Suppose that $(\mathcal{M}, G)$ is a Ricci flow with surgery defined for $0 \leq t < T < \infty$. Suppose that it satisfies the strong (C, ϵ)-canonical neighborhood assumption at all points x with $R(x) \geq r_{i+1}^{-2}$. Suppose also that the solution has curvature pinched toward positive. Suppose that there is a surgery at some time $\overline{t}$ with $T_{i-1} \leq \overline{t} < T$ with $\overline{\delta}(\overline{t})$ as the surgery control parameter and with h as the surgery scale parameter. Then the following holds provided that $\overline{\delta}(\overline{t}) \leq \delta''$. Set $T' = \min(T, \overline{t} + h^2\theta_0)$. Let $p \in M_{\overline{t}}$ be the tip of the cap of a surgery disk at time $\overline{t}$. Suppose that $P(p, \overline{t}, Ah, T' - \overline{t})$ exists in $\mathcal{M}$. Suppose that we have $t' \in [\overline{t}, \overline{t} + h^2/2]$ with $t' \leq T'$, and suppose that we have a curve $\gamma(\tau)$ parameterized by backward time $\tau \in [0, T' - t']$ so that $\gamma(\tau) \in M_{T'-\tau}$ for all $\tau \in [0, T' - t']$. Suppose that the image of γ is contained in the closure of $P(p, \overline{t}, Ah, T' - \overline{t}) \subset \mathcal{M}$. Suppose further:

(1) either that $T' = \overline{t} + \theta_0 h^2 \leq T$ or that $\gamma(0) \subset \partial B(p, \overline{t}, Ah) \times \{T'\}$; and
(2) $\gamma(T' - t') \in B(p, \overline{t}, Ah/2) \times t'$.

Then

$$\int_0^{T'-t'} \left(R(\gamma(t)) + |X_\gamma(t)|^2\right) dt > \ell.$$

See FIG. 2.

PROOF. The logic of the proof is as follows. We fix $\ell < \infty$. We shall determine the relevant value of θ_0 and then of A_0 in the course of the argument. Then for any $A \geq A_0$ we define $\delta''(A) = \delta''_0(A, \theta_0, \overline{\delta}_0)$, as in Proposition 16.5.

The integral expression is invariant under time translation and also under rescaling. Thus, we can (and do) assume that $\overline{t} = 0$ and that the scale h of the surgery is 1. We use the embedding of $P(p, 0, A, T') \to \mathcal{M}$ and write the restriction of the flow to this subset as a one-parameter family of metrics $g(t)$, $0 \leq t \leq T'$, on $B(p, 0, A)$. With this renormalization, $0 \leq t' \leq 1/2$, also $T' \leq \theta_0$, and $\tau = T' - t$.

Let us first consider the case when $T' = \theta_0 \leq T$. Consider the standard flow $(\mathbb{R}^3, g_0(t))$, and let p_0 be its tip. According to Proposition 12.31, for all $x \in \mathbb{R}^3$ and all $t \in [0, 1)$ we have $R_{g_0}(x, t) \geq c/(1-t)$. By Proposition 16.5 and since we are assuming that $\overline{\delta}(\overline{t}) \leq \delta'' = \delta''_0(A, \theta_0, \overline{\delta}_0)$, we have that

5. A LENGTH ESTIMATE

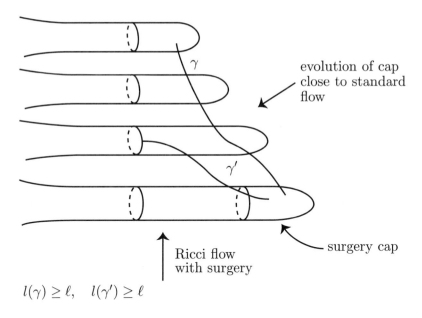

$l(\gamma) \geq \ell, \quad l(\gamma') \geq \ell$

FIGURE 2. Paths in evolving surgery caps are long.

$R(a,t) \geq c/2(1-t)$ for all $a \in B(p,0,A)$ and all $t \in [0,\theta]$. Thus, we have

$$\int_0^{\theta_0 - t'} \left(R(\gamma(\tau)) + |X_\gamma(\tau)|^2 \right) d\tau \geq \int_{t'}^{\theta_0} \frac{c}{2(1-t)} dt$$
$$= \frac{-c}{2} \left(\log(1-\theta_0) - \log(1-t') \right) dt$$
$$\geq \frac{-c}{2} \left(\log(1-\theta_0) + \log(2) \right).$$

Hence, if $\theta_0 < 1$ sufficiently close to 1, the integral will be $> \ell$. This fixes the value of θ_0.

CLAIM 16.14. *There is $A'_0 < \infty$ with the property that for any $A \geq A'_0$ the restriction of the standard solution $g_0(t)$ to $(B(p_0, 0, A) \setminus B(p_0, 0, A/2)) \times [0, \theta_0]$ is close to an evolving family $(S^2 \times [A/2, A], h_0(t) \times ds^2)$. In particular, for any $t \in [0, \theta_0]$, the g_0-distance at time t from $B(p_0, 0, A/2)$ to the complement of $B(p_0, 0, A)$ in the standard solution is more than $A/4$.*

PROOF. This is immediate from Proposition 12.7 and the fact that $\theta_0 < 1$. □

Now fix $A_0 = \max(A'_0, 10\sqrt{\ell})$ and let $A \geq A_0$.

Since $\overline{\delta}_0 < 1/4$ and since $T' \leq \theta_0$, for $\overline{\delta}(\bar{t}) \leq \delta''_0(A, \theta_0, \overline{\delta}_0)$ by Proposition 16.5 the $g(T')$-distance between $B(p, 0, A/2)$ and $\partial B(p, 0, A)$ is at least $A/5$.

Since the flow on $B(p, 0, A) \times [0, T']$ is within $\overline{\delta}_0$ of the standard solution, and since the curvature of the standard solution is non-negative, for any

horizontal tangent vector X at any point of $B(p,0,A) \times [0,T']$ we have that
$$\mathrm{Ric}_g(X,X) \geq -\frac{1}{4}|X|^2_{g_0} \geq -\frac{1}{2}|X|^2_g,$$
and hence
$$\frac{d}{dt}|X|^2_g \leq |X|^2_g.$$
Because $T' \leq 1$, we see that
$$|X|^2_{g(T')} \leq e \cdot |X|^2_{g(t)} < 3|X|^2_{g(t)}$$
for any $t \in [0,T']$.

Now suppose that $\gamma(0) \in \partial B(p,0,A) \times \{T'\}$. Since the image of γ is contained in the closure of $P(p,0,A,T')$, for every $\tau \in [0,T']$ we have $\sqrt{3}|X_\gamma(\tau)|_{g(T'-\tau)} \geq |X_\gamma(\tau)|_{g(T')}$. Since the flow $g(t)$ on $P(p,0,A,T')$ is within $\overline{\delta}_0$ in the $C^{[1/\overline{\delta}_0]}$-topology of the standard flow on the corresponding parabolic neighborhood, $R(\gamma(t)) \geq 0$ for all $t \in [0,T']$. Thus, because of these two estimates we have

(16.1) $$\int_0^{T'-t'} \left(R(\gamma(\tau)) + |X_\gamma(\tau)|^2\right) d\tau \geq \int_0^{T'-t'} \frac{1}{3}|X_\gamma(\tau)|^2_{g(T')} d\tau.$$

Since $\gamma(0) \in \partial B(p,0,A) \times \{T'\}$ and $\gamma(T') \in B(p,0,A/2)$, it follows from Cauchy-Schwarz that

$$(T'-t')^2 \int_0^{T'} |X_\gamma(\tau)|^2_{g(T')} d\tau \geq \left(\int_0^{T'-t'} |X_\gamma(\tau)|_{g(T')} d\tau\right)^2$$
$$\geq \left(d_{g(T')}(B(p,0,A/2), \partial B(p,0,A))\right)^2 \geq \frac{A^2}{25}.$$

Since $T'-t' < 1$, it immediately follows from this and Equation (16.1) that

$$\int_0^{T'-t'} \left(R(\gamma(\tau)) + |X_\gamma(\tau)|^2\right) d\tau \geq \frac{A^2}{75}.$$

Since $A \geq A_0 \geq 10\sqrt{\ell}$, this expression is $> \ell$. This completes the proof of Proposition 16.13. $\square$

5.1. Paths with short $\mathcal{L}_+$-length avoid the surgery caps.

Here we show that a path parameterized by backward time that ends in a surgery cap (or comes close to it) must have long $\mathcal{L}$-length. Let $(\mathcal{M},G)$ be a Ricci flow with surgery, and let $x \in \mathcal{M}$ be a point with $\mathbf{t}(x) = T \in (T_i, T_{i+1}]$. We suppose that these data satisfy the hypothesis of Proposition 16.4 with respect to the given sequences and $r \geq r_{i+1} > 0$. In particular, the parabolic neighborhood $P(x,T,r,-r^2)$ exists in $\mathcal{M}$ and $|\mathrm{Rm}|$ is bounded on this parabolic neighborhood by r^{-2}.

Actually, here we do not work directly with the length function $\mathcal{L}$ defined from x, but rather with a closely related function. We set $R_+(y) = \max(R(y), 0)$.

5. A LENGTH ESTIMATE

LEMMA 16.15. *Given $L_0 < \infty$, there is $\overline{\delta}_1 = \overline{\delta}_1(L_0, r_{i+1}) > 0$, independent of $(\mathcal{M}, G)$ and x, such that if $\overline{\delta}(t) \leq \overline{\delta}_1$ for all $t \in [T_{i-1}, T)$, then for any curve $\gamma(\tau)$, $0 \leq \tau \leq \tau_0$, with $\tau_0 \leq T - T_{i-1}$, parameterized by backward time with $\gamma(0) = x$ and with*

$$\mathcal{L}_+(\gamma) = \int_0^{\tau_0} \sqrt{\tau}\left(R_+(\gamma(\tau)) + |X_\gamma|^2\right) d\tau < L_0,$$

the following two statements hold:

(1) *Set*

$$\tau' = \min\left(\frac{r_{i+1}^4}{(256)L_0^2}, \ln(\sqrt[3]{2})r_{i+1}^2\right).$$

Then for all $\tau \leq \min(\tau', \tau_0)$ we have $\gamma(\tau) \in P(x, T, r/2, -r^2)$.

(2) *Suppose that $\overline{t} \in [T - \tau_0, T)$ is a surgery time with p being the tip of the surgery cap at time $\overline{t}$ and with the scale of the surgery being $\overline{h}$. Suppose $t' \in [\overline{t}, \overline{t} + \overline{h}^2/2]$ is such that there is an embedding*

$$\rho \colon B(p, \overline{t}, (50 + A_0)\overline{h}) \times [\overline{t}, t'] \to \mathcal{M}$$

compatible with time and the vector field. Then the image of ρ is disjoint from the image of γ. See FIG. *3.*

REMARK 16.16. Recall that $(A_0 + 4)\overline{h}$ is the radius of the surgery cap (measured in the rescaled version of the standard initial metric) that is glued in when performing surgery with scale $\overline{h}$.

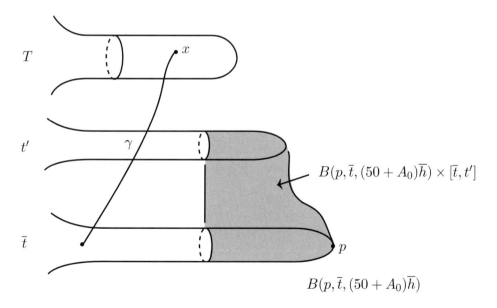

FIGURE 3. Avoiding neighborhoods of surgery caps

PROOF. We define $\ell = L_0/\sqrt{\tau'}$, then define $A = \max(A_0(\ell), 2(50+A_0))$ and $\theta = \theta_0(\ell)$. Here, $A_0(\ell)$ and $\theta_0(\ell)$ are the constants in Proposition 16.13. Lastly, we require $\bar\delta_1 \leq \delta''(A)$ from Proposition 16.13. Notice that, by construction, $\delta''(A) = \delta_0''(A,\theta,\bar\delta_0)$ from Proposition 16.5. Thus, if p is the tip of a surgery cap at time $\bar t$ with the scale of the surgery being $\bar h$, then it follows that for any $\Delta t \leq \theta$, if there is an embedding

$$\rho\colon B(p,\bar t, A\bar h) \times [\bar t, \bar t + \bar h^2 \Delta t] \to \mathcal{M}$$

compatible with time and the vector field, then the induced flow (after time shifting by $-\bar t$ and scaling by $\bar h^{-2}$) is within $\bar\delta_0$ in the $C^{[1/\bar\delta_0]}$-topology of the standard solution. In particular, the scalar curvature at any point of the image of ρ is positive and is within a multiplicative factor of 2 of the scalar curvature at the corresponding point of the standard flow.

Recall that we have $r \geq r_{i+1}$ and that $P(x,T,r,-r^2)$ exists in $\mathcal{M}$ and that $|\text{Rm}| \leq r^{-2}$ on this parabolic neighborhood. We begin by proving by contradiction that there is no $\tau \leq \tau'$ with the property that $\gamma(\tau) \notin P(x,T,r/2,-r^2)$. Suppose there is such a $\tau \leq \tau'$. Notice that by construction $\tau' < r_{i+1}^2 < r^2$. Hence, for the first τ'' with the property that $\gamma(\tau'') \notin P(x,T,r/2,-r^2)$ the point $\gamma(\tau'') \in \partial B(x,T,r/2) \times \{T - \tau''\}$.

CLAIM 16.17. $\int_0^{\tau''} |X_\gamma(\tau)| d\tau > r/2\sqrt{2}$.

PROOF. Since $|\text{Rm}| \leq r^{-2}$ on $P(x,T,r,-r^2)$, it follows that $|\text{Ric}| \leq 2r^{-2}$ on $P(x,T,r,-\tau'')$. Thus, for any tangent vector v at a point of $B(x,T,r)$ we have

$$\left| \frac{d(\langle v,v \rangle_{G(T-\tau)})}{d\tau} \right| \leq 2r^{-2} \langle v,v \rangle_{G(T-\tau)}$$

for all $\tau \in [0, \tau'']$. Integrating gives that for any $\tau \leq \tau''$ we have

$$\exp(-2r^{-2}\tau'')\langle v,v \rangle_{G(T)} \leq \langle v,v \rangle_{G(T-\tau)} \leq \exp(2r^{-2}\tau'')\langle v,v \rangle_{G(T)}.$$

Since $\tau'' \leq \tau'$ and $r \geq r_{i+1}$ by the assumption on τ' we have

$$\exp(2r^{-2}\tau'') \leq \exp(2\sqrt[3]{2}) < 2.$$

This implies that for all $\tau \leq \tau''$ we have

$$\frac{1}{\sqrt{2}}|X_\gamma(\tau)|_{G(T)} < |X_\gamma(\tau)|_{G(T-\tau)} < \sqrt{2}|X_\gamma(\tau)|_{G(T)},$$

and hence

$$\int_0^{\tau''} |X_\gamma(\tau)| d\tau > \frac{1}{\sqrt{2}} \int_0^{\tau''} |X_\gamma(\tau)|_{G(T)} \geq \frac{r}{2\sqrt{2}},$$

where we use the fact that $d_T(\gamma(0), \gamma(\tau'')) = r/2$. □

5. A LENGTH ESTIMATE

Applying Cauchy-Schwarz to $\tau^{1/4}|X_\gamma|$ and $\tau^{-1/4}$ on the interval $[0, \tau'']$ yields

$$\int_0^{\tau''} \sqrt{\tau} \left(R_+(\gamma(\tau)) + |X_\gamma(\tau)|^2 \right) d\tau \geq \int_0^{\tau''} \sqrt{\tau} |X_\gamma(\tau)|^2 d\tau$$

$$\geq \frac{\left(\int_0^{\tau''} |X_\gamma(\tau)| d\tau \right)^2}{\int_0^{\tau''} \tau^{-1/2} d\tau}$$

$$> \frac{r^2}{16\sqrt{\tau''}} \geq L_0.$$

Of course, the integral from 0 to τ'' is less than or equal to the entire integral from 0 to τ_0 since the integrand is non-negative, contradicting the assumption that $\mathcal{L}_+(\gamma) \leq L_0$. This completes the proof of the first numbered statement.

We turn now to the second statement. We impose a further condition on $\overline{\delta}_1$. Namely, require that $\overline{\delta}_1^2 < r_{i+1}/2$. Since $r_i \leq r_0 \leq \epsilon < 1$, we have $\overline{\delta}_1^2 r_i < r_{i+1}/2$. Thus, the scale of the surgery, $\overline{h}$, which is $\leq \overline{\delta}_1^2 r_i$ by definition, will also be less than $r_{i+1}/2$, and hence there is no point of $P(x, T, r, -r^2)$ (where the curvature is bounded by $r^{-2} \leq r_{i+1}^{-2}$) in the image of ρ (where the scalar curvature is greater than $\overline{h}^{-2}/2 > 2r_{i+1}^{-2}$). Thus, if $\tau' \geq \tau_0$ we have completed the proof. Suppose that $\tau' < \tau_0$. It suffices to establish that for every $\tau_1 \in [\tau', \tau_0]$ the point $\gamma(\tau_1)$ is not contained in the image of ρ for any surgery cap and any t' as in the statement. Suppose that in fact there is $\tau_1 \in [\tau', \tau_0]$ with $\gamma(\tau_1)$ contained in the image of $\rho(B(p, \overline{t}, (A_0+50)\overline{h}) \times [\overline{t}, t'])$ where $\overline{t} \leq t' \leq \overline{t} + \overline{h}^2/2$ and where p is the tip of some surgery cap at time $\overline{t}$. We estimate

(16.2)
$$\int_0^{\tau_0} \sqrt{\tau} \left(R_+(\gamma(\tau)) + |X_\gamma(\tau)|^2 \right) d\tau \geq \int_{\tau'}^{\tau_0} \sqrt{\tau} \left(R_+(\gamma(\tau)) + |X_\gamma(\tau)|^2 \right) d\tau$$

$$\geq \sqrt{\tau'} \int_{\tau'}^{\tau_1} \left(R_+(\gamma(\tau)) + |X_\gamma(\tau)|^2 \right) d\tau.$$

Let $\Delta t \leq T - \overline{t}$ be the supremum of the set of s for which there is a parabolic neighborhood $P(p, \overline{t}, A\overline{h}, s)$ embedded in $\mathbf{t}^{-1}((-\infty, T]) \subset \mathcal{M}$. Let $\Delta t_1 = \min(\theta \overline{h}^2, \Delta t)$. We consider $P(p, \overline{t}, A\overline{h}, \Delta t_1)$. First, notice that since $\overline{h} \leq \overline{\delta}_1^2 r_i < r_{i+1}/2$, the scalar curvature on $P(p, \overline{t}, A\overline{h}, \Delta t_1)$ is larger than $\overline{h}^{-2}/2 > r_{i+1}^{-2} \geq r^{-2}$. In particular, the parabolic neighborhood $P(x, T, r, -r^2)$ is disjoint from $P(p, \overline{t}, A\overline{h}, \Delta t_1)$. This means that there is some $\tau'' \geq \tau'$ such that $\gamma(\tau'') \in \partial P(p, \overline{t}, A\overline{h}, \Delta t_1)$ and $\gamma|_{[\tau'', \tau_1]} \subset P(p, \overline{t}, A\overline{h}, \Delta t_1)$. There are two cases to consider. The first is when $\Delta t_1 = \theta \overline{h}^2$, $\tau'' = T - (\overline{t} + \Delta t_1)$ and $\gamma(\tau'') \in B(p, \overline{t}, A\overline{h}) \times \{\overline{t} + \Delta t_1\}$. Then, according to Proposition 16.13,

$$\text{(16.3)} \qquad \int_{\tau''}^{\tau_1} R_+(\gamma(\tau))d\tau > \ell.$$

Now let us consider the other case. If $\Delta t_1 < \theta \bar{h}^2$, this means that either $\bar{t} + \Delta t_1 = T$ or, according to Proposition 16.5, at the time $\bar{t} + \Delta t_1$ there is a surgery that removes all of $B(p, \bar{t}, A\bar{h})$. Hence, under either possibility it must be the case that $\gamma(\tau'') \in \partial B(p, \bar{t}, A\bar{h}) \times \{T - \tau''\}$. Thus, the remaining case to consider is when, whatever Δt_1 is, $\gamma(\tau'') \subset \partial B(p, \bar{t}, A\bar{h}) \times \{T - \tau''\}$. Lemma 16.13 and the fact that $R \geq 0$ on $P(p, \bar{t}, A\bar{h}, \Delta t_1)$ imply that

$$\ell < \int_{\tau''}^{\tau_1} \left(R(\gamma(\tau)) + |X_\gamma(\tau)|^2 \right) d\tau = \int_{\tau''}^{\tau_1} \left(R_+(\gamma(\tau)) + |X_\gamma(\tau)|^2 \right) d\tau.$$

Since $\ell = L_0/\sqrt{\tau'}$ and $\tau'' \geq \tau'$, it follows from Equation (16.2) that in both cases

$$\mathcal{L}_+(\gamma) \geq \int_{\tau''}^{\tau_1} \sqrt{\tau} \left(R_+(\gamma(\tau)) + |X_\gamma(\tau)|^2 \right) d\tau > \ell \sqrt{\tau'} = L_0,$$

which contradicts our hypothesis. This completes the proof of Lemma 16.15. □

5.2. Paths with small energy avoid the disappearing regions.
At this point we have shown that paths of small energy do not approach the surgery caps from above. We also need to rule out that they can be arbitrarily close from below. That is to say, we need to see that paths whose $\mathcal{L}$-length is not too large avoid neighborhoods of the disappearing regions at all times just before the surgery time at which they disappear. Unlike the previous estimates which were universal for all $(\mathcal{M}, G)$ satisfying the hypothesis of Proposition 16.4, in this case the estimates will depend on the Ricci flow with surgery. First, let us fix some notation.

DEFINITION 16.18. Suppose that $\bar{t}$ is a surgery time, that $\tau_1 > 0$, and that there are no other surgery times in the interval $(\bar{t} - \tau_1, \bar{t}]$. Let $\{\Sigma_i(\bar{t})\}_i$ be the 2-spheres on which we do surgery at time $\bar{t}$. Each Σ_i is the central 2-sphere of a strong δ-neck N_i. We can flow the cylinders $J_0(\bar{t}) = \cup_i s_{N_i}^{-1}(-25, 0])$ backward to any time $t \in (\bar{t} - \tau_1, \bar{t}]$. Let $J_0(t)$ be the result. There is an induced function, denoted $\coprod_i s_{N_i}(t)$, on $J_0(t)$. It takes values in $(-25, 0]$. We denote the boundary of $J_0(t)$ by $\coprod_i \Sigma_i(t)$. Of course, this boundary is the result of flowing $\coprod_i \Sigma_i(\bar{t})$ backward to time t. (These backward flows are possible since there are no surgery times in $(\bar{t} - \tau_1, \bar{t})$.) For each $t \in [\bar{t} - \tau_1, \bar{t})$ we also have the region, $D_{\bar{t}}$, that disappears at time $\bar{t}$. It is an open submanifold whose boundary is $\coprod_i \Sigma_i(t)$. Thus, for every $t \in (\bar{t} - \tau_1, \bar{t})$ the subset $J(t) = J_0(t) \cup D_t$ is an open subset of M_t. We define

$$J(\bar{t} - \tau_1, \bar{t}) = \cup_{t \in (\bar{t} - \tau_1, \bar{t})} J(t).$$

Then $J(\bar{t} - \tau_1, \bar{t})$ is an open subset of $\mathcal{M}$. See FIG. 4.

5. A LENGTH ESTIMATE

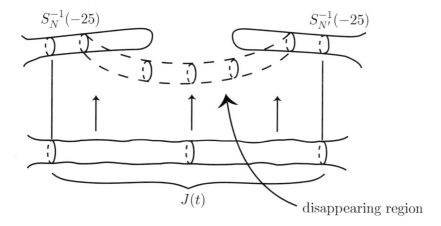

FIGURE 4. Paths of short length avoid disappearing regions.

LEMMA 16.19. *Fix a Ricci flow with surgery $(\mathcal{M}, G)$, a point $x \in \mathcal{M}$ and constants $r \geq r_{i+1} > 0$ as in the statement of Proposition 16.4. For any $1 < \ell < \infty$ the following holds. Suppose that $\bar{t} \in [T_{i-1}, T)$ is a surgery time and that $\gamma(\tau)$ is a path with $\gamma(\tau) \in M_{\bar{t}-\tau}$. Let $\{p_1, \ldots, p_k\}$ be the tips of all the surgery caps at time $\bar{t}$ and let $\bar{h}$ be the scale of surgery at time $\bar{t}$. Suppose that for some $0 < \tau_1 \leq \ell^{-1}\bar{h}^2$ there are no surgery times in the interval $(\bar{t} - \tau_1, \bar{t})$. We identify all M_t for $t \in [\bar{t} - \tau_0, \bar{t})$ with $M_{\bar{t}-\tau_1}$ using the flow. Suppose that $\gamma(0) \in M_{\bar{t}} \setminus \cup_{i=1}^{k} B(p_i, \bar{t}, (50 + A_0)\bar{h})$, and lastly, suppose that*

$$\int_0^{\tau_1} |X_\gamma(\tau)|^2 d\tau \leq \ell.$$

Then γ is disjoint from the open subset $J(\bar{t} - \tau_1, \bar{t}))$ of $\mathcal{M}$.

PROOF. Suppose that the lemma is false and let $\gamma \colon [0, \bar{\tau}] \to \mathcal{M}$ be a path satisfying the hypothesis of the lemma with $\gamma(\bar{\tau}) \in J(\bar{t} - \tau_1, \bar{t})$. Since

$$\gamma(0) \in M_{\bar{t}} \setminus \cup_i B(p_i, \bar{t}, (50 + A_0)\bar{h}),$$

if follows that $\gamma(0)$ is separated from the boundary of $s_{N_i}^{-1}(-25, 0]$ by distance at least $20\bar{h}$. Since the $J_0(t)$ are contained in the disjoint union of strong δ-necks N_i centered at the 2-spheres along which we do surgery, and since $\tau_1 \leq \bar{h}^2/\ell < \bar{h}^2$, it follows that, provided that δ is sufficiently small, for every $t \in [\bar{t} - \tau_1, \bar{t})$, the metric on $J_0(t)$ is at least $1/2$ the metric on $J_0(\bar{t})$. It follows that, for δ sufficiently small, if there is a $\tau \in [0, \tau_1]$ with $\gamma(\tau) \in J(t)$ then $\int_0^{\tau_1} |X_\gamma| d\tau > 10\bar{h}$. Applying Cauchy-Schwarz we see that

$$\int_0^{\tau_1} |X_\gamma(\tau)|^2 d\tau \geq (10\bar{h})^2/\tau_1.$$

Since $\tau_1 \leq \ell^{-1}\bar{h}^2$, we see that

$$\int_0^{\tau'} |X_\gamma(\tau)|^2 d\tau > \ell,$$

contradicting our hypothesis. □

5.3. Limits of a sequence of paths with short $\mathcal{L}$-length. Now using Lemmas 16.15 and 16.19 we show that it is possible to take limits of certain types of sequences of paths parameterized by backward time to create minimizing $\mathcal{L}$-geodesics.

We shall work with a compact subset of $\mathbf{t}^{-1}([T_{i-1}, T])$ that is obtained by removing appropriate open neighborhoods of the exposed regions.

DEFINITION 16.20. Fix $\ell < \infty$. Let $\theta_0 = \theta_0(\ell)$ be as in Proposition 16.13. For each surgery time $\bar{t} \in [T_{i-1}, T]$, let $h(\bar{t})$ be the scale of the surgery. Let $p_1, \ldots, p_k$ be the tips of the surgery caps at time $\bar{t}$. For each $1 \leq j \leq k$, we consider $B_j(\bar{t}) = B(p_j, \bar{t}, (A_0 + 10)h(\bar{t}))$, and we let

$$\Delta t_j \leq \min(\theta_0, (T - \bar{t})/h^2(\bar{t}))$$

be maximal subject to the condition that there is an embedding $\rho_j \colon B_j(\bar{t}) \times [\bar{t}, \bar{t} + h^2(\bar{t})\Delta t_j)$ into $\mathcal{M}$ compatible with time and the vector field. Clearly, $B'_j = B(p_j, \bar{t}, (10 + A_0)h) \cap C_{\bar{t}}$ is contained in $J(\bar{t})$. Let $\bar{t}'$ be the previous surgery time if there is one, otherwise set $\bar{t}' = 0$. Also for each $\bar{t}$ we set $\tau_1(\ell, \bar{t}) = \min\left(h(\bar{t})^2/\ell, \bar{t} - \bar{t}'\right)$. For each $t \in (\bar{t} - \tau_1(\ell, \bar{t}), \bar{t})$ let $\widetilde{J}(t) \subset J(t)$ be the union of D_t, the disappearing region at time t, and $\coprod_i B'_i(t)$, the result of flowing $\coprod_i B'_i$ backward to time t. Then we set

$$\widetilde{J}(\bar{t} - \tau_1(\ell, \bar{t}), \bar{t}) \subset J(\bar{t} - \tau_1(\ell, \bar{t}), \bar{t})$$

equal to the union over $t \in (\bar{t} - \tau_1(\ell, \bar{t}), \bar{t})$ of $\widetilde{J}(t)$.

By construction, for each surgery time $\bar{t}$, the union

$$\nu_{\text{sing}}(\ell, \bar{t}) = \widetilde{J}(\bar{t} - \tau_1(\ell, \bar{t}), \bar{t}) \cup \cup_i B_i \times [\bar{t}, \bar{t} + h^2(\bar{t})\Delta t_i)$$

is an open subset of $\mathcal{M}$ containing all the exposed regions and singular points at time $\bar{t}$.

We define $Y(\ell) \subset \mathbf{t}^{-1}([T_{i-1}, T])$ to be the complement of the $\cup_{\bar{t}} \nu_{\text{sing}}(\ell, \bar{t})$ where the union is over all surgery times $\bar{t} \in [T_{i-1}, T]$. Clearly, $Y(\ell)$ is a closed subset of $\mathbf{t}^{-1}([T_{i-1}, T])$ and hence $Y(\ell)$ is a compact subset contained in the open subset of smooth points of $\mathcal{M}$. (Notice that $Y(\ell)$ depends on ℓ because $\tau_1(\ell, \bar{t})$ and θ_0 depend on ℓ.)

PROPOSITION 16.21. *Fix $0 < L < \infty$. Set*

$$L_0 = L + 4(T_{i+1})^{3/2}.$$

Suppose that for all $t \in [T_{i-1}, T_{i+1}]$, the surgery control parameter $\overline{\delta}(t) \leq \overline{\delta}_1(L_0, r_{i+1})$ where the right-hand side is the constant from Lemma 16.15.

Suppose that γ_n is a sequence of paths in $(\mathcal{M}, G)$ parameterized by backward time $\tau \in [0, \overline{\tau}]$ with $\overline{\tau} \leq T - T_{i-1}$, with $\gamma_n(0) = x$ and with

$$\mathcal{L}(\gamma_n) \leq L$$

for all n. Then:

(1) After passing to a subsequence, there is a limit γ defined on $[0, \overline{\tau}]$. The limit γ is a continuous path and is a uniform limit of the γ_n. The limit is contained in the open subset of smooth points of $\mathcal{M}$ and has finite $\mathcal{L}$-length satisfying

$$\mathcal{L}(\gamma) \leq \liminf_{n \to \infty} \mathcal{L}(\gamma_n).$$

(2) If there is a point $y \in \mathcal{M}_{T-\overline{\tau}}$ such that $\gamma_n(\overline{\tau}) = y$ for all n, and if the γ_n are a sequence of paths parameterized by backward time from x to y with $\lim_{n \to \infty} \mathcal{L}(\gamma_n)$ being no greater than the $\mathcal{L}$-length of any path from x to y, then the limit γ of a subsequence is a minimizing $\mathcal{L}$-geodesic connecting x to y contained in the open subset of smooth points of $\mathcal{M}$.

(3) There is $\ell < \infty$ depending only on L such that any path γ parameterized by backward time from x to a point $y \in \mathbf{t}^{-1}([T_{i-1}, T))$ whose $\mathcal{L}$-length is at most L is contained in the compact subset $Y(\ell)$ given in the previous definition.

PROOF. Given L_0, we set

$$\tau' = \min\left(\frac{r_{i+1}^4}{(256)L_0^2}, \ln(\sqrt[3]{2})r_{i+1}^2\right)$$

as in Lemma 16.15 and then define $\ell = L_0/\sqrt{\tau'}$. We also let

$$A = \min(2(50 + A_0), A_0(\ell)) \quad \text{and} \quad \theta_0 = \theta_0(\ell)$$

be as in Proposition 16.13. Lastly, we let $\overline{\delta}_1(L_0, r_{i+1}) = \delta''(A) = \delta''(A, \theta_0, \overline{\delta}_0)$ from Propositions 16.13 and 16.5. We suppose that $\delta(t) \leq \overline{\delta}_1(L_0, r_{i+1})$ for all $t \in [T_{i-1}, T]$.

Let $\overline{t} \in [T_{i-1}, T]$ be a surgery time, and let $\overline{h}$ be the scale of the surgery at this time. For each surgery cap $\mathcal{C}$ with tip p at a time $\overline{t} \in [T_{i-1}, T]$ let $\Delta t(\mathcal{C})$ be the supremum of those s with $0 \leq s \leq \theta_0 \overline{h}^2$ for which there is an embedding

$$\rho_{\mathcal{C}} \colon B(p, \overline{t}, 2(A_0 + 50)\overline{h}) \times [\overline{t}, \overline{t} + s) \to \mathcal{M}$$

compatible with time and the vector field. We set

$$P_0(\mathcal{C}) = \rho_{\mathcal{C}}\left(B(p, \overline{t}, (A + 50)\overline{h}) \times [\overline{t}, \overline{t} + \min(\overline{h}^2/2, \Delta t(\mathcal{C})))\right).$$

CLAIM 16.22. Any path γ beginning at x and parameterized by backward time misses $P_0(\mathcal{C})$ if $\mathcal{L}(\gamma) < L$.

PROOF. Set $\tau_0 = T - \bar{t}$. Of course, $\tau_0 \leq T - T_{i-1} \leq T_{i+1} - T_{i-1}$. Consider the restriction of γ to $[0, \tau_0]$. We have

$$\int_0^{\tau_0} \sqrt{\tau} \left(R_+(\gamma_n(\tau)) + |X_{\gamma_n}(\tau)|^2\right) d\tau$$
$$\leq \int_0^{T-T_{i-1}} \sqrt{\tau} \left(R_+(\gamma_n(\tau)) + |X_{\gamma_n}(\tau)|^2\right) d\tau$$
$$\leq \int_0^{T-T_{i-1}} \sqrt{\tau} \left(R(\gamma_n(\tau)) + |X_{\gamma_n}(\tau)|^2\right) d\tau + \int_0^{T-T_{i-1}} 6\sqrt{\tau} d\tau$$
$$= \int_0^{T-T_{i-1}} \sqrt{\tau} \left(R(\gamma_n(\tau)) + |X_{\gamma_n}(\tau)|^2\right) d\tau + 4(T - T_{i-1})^{3/2}$$
$$\leq \int_0^{\tau_0} \sqrt{\tau} \left(R(\gamma_n(\tau)) + |X_{\gamma_n}(\tau)|^2\right) d\tau + 4(T_{i+1})^{3/2}.$$

Thus, the hypothesis that $\mathcal{L}(\gamma_n) \leq L$ implies that

(16.4) $$\int_0^{\tau_0} \sqrt{\tau} \left(R_+(\gamma_n(\tau)) + |X_{\gamma_n}(\tau)|^2\right) d\tau \leq L_0.$$

The claim now follows immediately from Lemma 16.15. □

Now set t' equal to the last surgery time before $\bar{t}$ or set $t' = 0$ if $\bar{t}$ is the first surgery time. We set $\tau_1(\bar{t})$ equal to the minimum of $\bar{t} - t'$ and $\bar{h}^2/\ell$.

Assume that $\gamma(0) = x$ and that $\mathcal{L}(\gamma) \leq L$. It follows from Lemma 16.15 that the restriction of the path γ to $[0, \tau']$ lies in a region where the Riemann curvature is bounded above by $r^{-2} \leq r_{i+1}^{-2}$. Hence, since $\bar{h} < \bar{\delta}(t)^2 r_{i+1} \ll r_{i+1}$, this part of the path is disjoint from all strong δ-necks (evolving backward for rescaled time $(-1, 0]$). That is to say, $\gamma|_{[0,\tau']}$ is disjoint from $J_0(t)$ for every $t \in (\bar{t} - \tau_1(\bar{t}), \bar{t})$ for any surgery time $\bar{t} \leq T$. It follows immediately that $\gamma|_{[0,\tau']}$ is disjoint from $J(\bar{t} - \tau_1(\bar{t}), \bar{t})$.

CLAIM 16.23. *For every surgery time $\bar{t} \in [T_{i-1}, T]$, the path γ starting at x with $\mathcal{L}(\gamma) \leq L$ is disjoint from $J(\bar{t}, \bar{t} - \tau_1(\bar{t}))$.*

PROOF. By the remarks above, it suffices to consider surgery times $\bar{t} \leq T - \tau'$. It follows immediately from the previous claim that for any surgery time $\bar{t}$, with the scale of the surgery being $\bar{h}$ and with p being the tip of a surgery cap at this time, we have γ is disjoint from $B(p, \bar{t}, (50 + A_0)\bar{h})$. Also,

$$\int_{T-\bar{t}}^{T-\bar{t}+\tau_1(\bar{t})} \sqrt{\tau}|X_\gamma(\tau)|^2 d\tau \leq \mathcal{L}_+(\gamma) \leq L_0.$$

Since we can assume $T - \bar{t} \geq \tau'$ this implies that

$$\int_{T-\bar{t}}^{T-\bar{t}-\tau_1(\bar{t})} |X_\gamma(\tau)|^2 d\tau \leq L_0/\sqrt{\tau'} = \ell.$$

The claim is now immediate from Lemma 16.19. □

5. A LENGTH ESTIMATE

From these two claims we see immediately that γ is contained in the compact subset $Y(\ell)$ which is contained in the open subset of smooth points of $\mathcal{M}$. This proves the third item in the statement of the proposition. Now let us turn to the limit statements.

Take a sequence of paths γ_n as in the statement of Proposition 16.21. By Lemma 16.15 the restriction of each γ_n to the interval $[0, \min(\overline{\tau}, \tau')]$ is contained in $P(x, T, r/2, -r^2)$. The arguments in the proof of Lemma 7.2 (which involve changing variables to $s = \sqrt{\tau}$) show that, after passing to a subsequence, the restrictions of the γ_n to $[0, \min(\overline{\tau}, \tau')]$ converge uniformly to a path γ defined on the same interval. Furthermore,

$$\int_0^{\min(\overline{\tau},\tau')} \sqrt{\tau} |X_\gamma(\tau)|^2 d\tau \leq \liminf_{n\to\infty} \int_0^{\min(\overline{\tau},\tau')} \sqrt{\tau} |X_{\gamma_n}(\tau)|^2 d\tau,$$

so that

$$(16.5) \quad \int_0^{\min(\overline{\tau},\tau')} \sqrt{\tau} \left(R(\gamma(\tau)) + |X_\gamma(\tau)|^2 \right) d\tau$$

$$\leq \liminf_{n\to\infty} \int_0^{\min(\overline{\tau},\tau')} \sqrt{\tau} \left(R(\gamma_n(\tau)) + |X_{\gamma_n}(\tau)|^2 \right) d\tau.$$

If $\overline{\tau} \leq \tau'$, then we have established the existence of a limit as required. Suppose now that $\overline{\tau} > \tau'$. We turn our attention to the paths $\gamma_n|_{[\tau',\overline{\tau}]}$. Let $T_{i-1} < \overline{t} \leq T - \tau'$ be either a surgery time or $T - \tau'$, and let t' be the maximum of the last surgery time before $\overline{t}$ and T_{i-1}. We consider the restriction of the γ_n to the interval $[T - \overline{t}, T - t']$. As we have seen, these restrictions are disjoint from $J(\overline{t} - \tau_1(\overline{t}), \overline{t})$ and also from the exposed region at time $\overline{t}$, which is denoted $E(\overline{t})$, and from $J_0(\overline{t})$. Let

$$Y = \mathbf{t}^{-1}([T - t', T - \overline{t}]) \setminus \left(J(\overline{t} - \tau_1(\overline{t}), \overline{t}) \cup (E(\overline{t}) \cup J_0(\overline{t})) \right).$$

This is a compact subset with the property that any point $y \in Y$ is connected by a backward flow line lying entirely in Y to a point $y(t')$ contained in $M_{t'}$.

Since Y is compact there is a finite upper bound on the Ricci curvature on Y, and hence to $\mathcal{L}_X(G)$ at any point of Y. Since all backward flow lines from points of Y extend all the way to $M_{t'}$, it follows that there is a constant C' such that

$$|X_{\gamma_n}(\tau)|_{G(t')} \leq C' |X_{\gamma_n}(\tau)|_{G(t)}$$

for all $t \in [t', \overline{t}]$. Our hypothesis that the $\mathcal{L}(\gamma_n)$ are uniformly bounded, the fact that the curvature is pinched toward positive and the fact that there is a uniform bound on the lengths of the τ-intervals imply that the

$$\int_{T-\overline{t}}^{T-t'} \sqrt{\tau} |X_{\gamma_n}(\tau)|^2 d\tau$$

are uniformly bounded. Because $T - \overline{t}$ is at least $\tau' > 0$, it follows that the $\int_{T-\overline{t}}^{T-t'} |X_{\gamma_n}|^2 d\tau$ have a uniform upper bound. This then implies that there

is a constant C_1 such that for all n we have
$$\int_{T-\bar{t}}^{T-t'} |X_{\gamma_n}(\tau)|^2_{G(t')} d\tau \leq C_1.$$

Thus, after passing to a subsequence, the γ_n converge uniformly to a continuous γ defined on $[T-\bar{t}, T-t']$. Furthermore, we can arrange that the convergence is a weak convergence in $W^{1,2}$. This means that γ has a derivative in L^2 and
$$\int_{T-\bar{t}}^{T-t'} |X_\gamma(\tau)|^2 d\tau \leq \liminf_{n\to\infty} \int_{T-\bar{t}}^{T-t'} |X_{\gamma_n}(\tau)|^2 d\tau.$$

Now we do this simultaneously for all $\bar{t} = T - \tau'$ and for all the finite number of surgery times in $[T_{i-1}, T-\tau']$. This gives a limiting path $\gamma\colon [\tau', \bar{\tau}] \to \mathcal{M}$. Putting together the above inequalities we see that the limit satisfies

(16.6) $$\int_{\tau'}^{\bar{\tau}} \sqrt{\tau}\left(R(\gamma(\tau)) + |X_\gamma(\tau)|^2\right) d\tau$$
$$\leq \liminf_{n\to\infty} \int_{\tau'}^{\bar{\tau}} \sqrt{\tau}\left(R(\gamma_n(\tau)) + |X_{\gamma_n}(\tau)|^2\right) d\tau.$$

Since we have already arranged that there is a limit on $[0, \tau']$, this produces a limiting path $\gamma\colon [0, \tau_0] \to \mathcal{M}$. By Inequalities (16.5) and (16.6) we see that
$$\mathcal{L}(\gamma) \leq \liminf_{i\to\infty} \mathcal{L}(\gamma_n).$$
The limit lies in the compact subset $Y(\ell)$ and hence is contained in the open subset of smooth points of $\mathcal{M}$. This completes the proof of the first statement of the proposition.

Now suppose, in addition to the above, that all the γ_n have the same endpoint $y \in M_{T-\tau_0}$ and that $\lim_{n\to\infty} \mathcal{L}(\gamma_n)$ is less than or equal to the $\mathcal{L}$-length of any path parameterized by backward time connecting x to y. Let γ be the limit of a subsequence as constructed in the proof of the first part of this result. Clearly, by what we have just established, γ is a path parameterized by backward time from x to y and $\mathcal{L}(\gamma) \leq \lim_{n\to\infty} \mathcal{L}(\gamma_n)$. This means that γ is a minimizing $\mathcal{L}$-geodesic connecting x to y, an $\mathcal{L}$-geodesic contained in the open subset of smooth points of $\mathcal{M}$.

This completes the proof of the proposition. $\square$

COROLLARY 16.24. *Given $L < \infty$, let $\overline{\delta}_1 = \overline{\delta}_1(L + 4(T_{i+1}^{3/2}, r_{i+1})$ be as given in Lemma 16.15. If $\overline{\delta}(t) \leq \overline{\delta}_1$ for all $t \in [T_{i-1}, T_{i+1}]$, then for any $x \in \mathbf{t}^{-1}([T_i, T_{i+1}))$ and for any $y \in M_{T_{i-1}}$, if there is a path γ parameterized by backward time connecting x to y with $\mathcal{L}(\gamma) \leq L$, then there is a minimizing $\mathcal{L}$-geodesic contained in the open subset of smooth points of $\mathcal{M}$ connecting x to y.*

PROOF. Choose an $\mathcal{L}$-minimizing sequence of paths from x to y and apply the previous proposition. □

5.4. Completion of the proof of Proposition 16.4. Having found a compact subset of the open subset of smooth points of $\mathcal{M}$ that contains all paths parameterized by backward time whose $\mathcal{L}$-length is not too large, we are in a position to prove Proposition 16.4, which states the existence of a minimizing $\mathcal{L}$-geodesics in $\mathcal{M}$ from x and gives estimates on their $\mathcal{L}$-lengths.

PROOF. (of Proposition 16.4). Fix $r \geq r_{i+1} > 0$. Let $(\mathcal{M}, G)$ and $x \in \mathcal{M}$ be as in the statement of Proposition 16.4. We set $L = 8\sqrt{T_{i+1}}(1 + T_{i+1})$, and we set
$$\delta = \min\left(\delta_i, \overline{\delta}_1(L + 4(T_{i+1})^{3/2}, r_{i+1})\right),$$
where $\overline{\delta}_1$ is as given in Lemma 16.15. Suppose that $\overline{\delta}(t) \leq \delta$ for all $t \in [T_{i-1}, T_{i+1})$. We set U equal to the subset of $\mathbf{t}^{-1}([T_{i-1}, T))$ consisting of all points y for which there is a path γ from x to y, parameterized by backward time, with $\mathcal{L}(\gamma) < L$. For each $t \in [T_{i-1}, T)$ we set $U_t = U \cap M_t$. According to Corollary 16.24 for any $y \in U$ there is a minimizing $\mathcal{L}$-geodesic connecting x to y and this geodesic lies in the open subset of $\mathcal{M}$ consisting of all the smooth points of $\mathcal{M}$; in particular, y is a smooth point of $\mathcal{M}$. Let $\mathcal{L}_x \colon U \to \mathbb{R}$ be the function that assigns to each $y \in U$ the $\mathcal{L}$-length of a minimizing $\mathcal{L}$-geodesic from x to y. Of course, $\mathcal{L}_x(y) < L$ for all $y \in U$. Now let us show that the restriction of $\mathcal{L}_x$ to any time-slice $U_t \subset U$ achieves its minimum along a compact set. For this, let $y_n \in U_t$ be a minimizing sequence for $\mathcal{L}_x$ and for each n let γ_n be a minimizing $\mathcal{L}$-geodesic connecting x to y_n. Since $\mathcal{L}(\gamma_n) < L$ for all n, according to Proposition 16.21, we can pass to a subsequence that converges to a limit, γ, connecting x to some point $y \in M_t$ and $\mathcal{L}(\gamma) \leq \inf_n \mathcal{L}(\gamma_n) < L$. Hence, $y \in U_t$, and clearly $\mathcal{L}_x|_{U_t}$ achieves its minimum at y. Exactly the same argument with y_n being a sequence of points at which $\mathcal{L}_x|_{U_t}$ achieves its minimum shows that the subset of U_t at which $\mathcal{L}_x$ achieves its minimum is a compact set.

We set $Z \subset U$ equal to the set of $y \in U$ such that $\mathcal{L}_x(y) \leq \mathcal{L}_x(y')$ for all $y' \in U_{\mathbf{t}(y)}$.

CLAIM 16.25. *The subset $Z' = \{z \in Z | \mathcal{L}_x(z) \leq L/2\}$ has the property that for any compact interval $I \subset [T_{i-1}, T)$ the intersection $\mathbf{t}^{-1}(I) \cap Z'$ is compact.*

PROOF. Fix a compact interval $I \subset [T_{i-1}, T)$. Let $\{z_n\}$ be a sequence in $Z' \cap \mathbf{t}^{-1}(I)$. By passing to a subsequence we can assume that the sequence $\mathbf{t}(z_n) = t_n$ converges to some $t \in I$, and that $\mathcal{L}_x(z_n)$ converges to some $D \leq L/2$. Since the surgery times are discrete, there is a neighborhood J of t in I such that the only possible surgery time in J is t itself. By passing to a further subsequence if necessary, we can assume that $t_n \in J$ for all n. Fix n. First, let us consider the case when $t_n \geq t$. Let γ_n be a minimizing

$\mathcal{L}$-geodesic from x to z_n. Then we form the path $\widehat{\gamma}_n$ which is the union of γ_n followed by the flow line for the vector field χ from the endpoint of γ_n to M_t. (This flowline exists since there is no surgery time in the open interval $(t, t_n]$.) If $t_n < t$, then we set $\widehat{\gamma}_n$ equal to the restriction of γ_n to the interval $[0, T-t]$. In either case let $\hat{y}_n \in M_t$ be the endpoint of $\widehat{\gamma}_n$. Since M_t is compact, by passing to a subsequence we can arrange that the $\hat{y}_n$ converge to a point $y \in M_t$. Clearly, $\lim_{n\to\infty} z_n = y$.

It is also the case that $\lim_{n\to\infty} \mathcal{L}(\widehat{\gamma}_n) = \lim_{n\to\infty} \mathcal{L}(\gamma_n) = D \leq L/2$. This means that $y \in U$ and that $\mathcal{L}_x(y) \leq D \leq L/2$. Hence, the greatest lower bound of the values of $\mathcal{L}_x$ on U_t is at most $D \leq L/2$, and consequently $Z' \cap U_t \neq \emptyset$. Suppose that the minimum value of $\mathcal{L}_x$ on U_t is $D' < D$. Let $z \in U_t$ be a point where this minimum value is realized, and let γ be a minimizing $\mathcal{L}$-geodesic from x to z. Then by restricting γ to subintervals $[0, t-\mu]$ shows that the minimum value of $\mathcal{L}_x$ on $U_{t+\mu} \leq (D'+D)/2$ for all $\mu > 0$ sufficiently small. Also, extending γ by adding a backward vertical flow line from z shows that the minimum value of $\mathcal{L}_x$ on $U_{t-\mu}$ is at most $(D'+D)/2$ for all $\mu > 0$ sufficiently small. (Such a vertical flow line backward in time exists since $z \in U$ and hence z is contained in the smooth part of $\mathcal{M}$.) This contradicts the fact that the minimum values of $\mathcal{L}_x$ on U_{t_n} converge to D as t_n converges to t. This contradiction proves that the minimum value of $\mathcal{L}_x$ on U_t is D, and consequently the point $y \in Z'$. This proves that $Z' \cap \mathbf{t}^{-1}(I)$ is compact, establishing the claim. □

At this point we have established that properties (1), (2), and (4); So it remains only to prove property (3) of Proposition 16.4. To do this we define the reduced length function $l_x \colon U \to \mathbb{R}$ by

$$l_x(q) = \frac{\mathcal{L}_x(q)}{2\sqrt{T-\mathbf{t}(q)}} \quad \text{and} \quad l_x^{\min}(\tau) = \min_{q \in M_t} l_x(q).$$

We consider the subset $\mathcal{S}$ of $\tau' \in (0, T-T_{i-1}]$ with $l_x^{\min}(\tau) \leq L/2$ for all $\tau \leq \tau'$. Recall that by the choice of L, we have $3\sqrt{T-T_{i-1}} < L/2$. Clearly, the minimum value of l_x on $U_{T-\tau}$ converges to 0 as $\tau \to 0$, implying that this set is non-empty. Also, from its definition, $\mathcal{S}$ is an interval with 0 being one endpoint.

LEMMA 16.26. *Let $l_x^{\min}(\tau')$ be the minimum value of l_x on $U_{T-\tau'}$. For any $\tau \in \mathcal{S}$ we have $l_x^{\min}(\tau) \leq 3/2$.*

PROOF. Given that we have already established properties (1), (2), and (4) of Proposition 16.4, this is immediate from Corollary 7.12. □

Now let us establish that $\mathcal{S} = (0, T-T_{i-1}]$. As we remarked above, $\mathcal{S}$ is a non-empty interval with 0 as one endpoint. Suppose that it is of the form $(0, \tau]$ for some $\tau < T-T_{i-1}$. Then by the previous claim, we have $l_x^{\min}(\tau) \leq 3/2$ so that there is an $\mathcal{L}$-geodesic γ from x to a point $y \in M_{T-\tau}$ with $\mathcal{L}(\gamma) \leq 3\sqrt{\tau} < L/2$. This implies that for all $\tau' > \tau$ but sufficiently close to τ, there is a point $y(\tau') \in U_{T-\tau'}$ with $\mathcal{L}_x(y(\tau')) < L/2$. This shows that all τ' greater than and sufficiently close to τ are contained in $\mathcal{S}$. This is a contradiction of the assumption that $\mathcal{S} = (0, \tau]$.

Suppose now that $\mathcal{S}$ is of the form $(0, \tau)$, and set $t = T - \tau$. Let $t_n \to t$ and $z_n \in Z' \cap U_{t'}$. The same argument as above shows that for every n we have $\mathcal{L}_x(z_n) \leq 3\sqrt{T - t_n}$. For all n sufficiently large, there are no surgery times in the interval (t, t_n). Hence, by passing to a subsequence, we can arrange that the z_n converge to a point $z \in M_t$. Clearly,

$$\mathcal{L}_x(z) \leq \limsup_{n \to \infty} \mathcal{L}_x(z_n) \leq 3\sqrt{T - t},$$

so that $\tau \in \mathcal{S}$. This again contradicts the assumption that $\mathcal{S} = (0, \tau)$.

The only other possibility is that the set of τ is $(0, T - T_{i-1}]$ and the minimum value of $\mathcal{L}$ on U_t is at most $3\sqrt{T - t}$ for all $t \in [T_{i-1}, T)$. This is exactly the third property stated in Proposition 16.4. This completes the proof of that proposition. $\square$

6. Completion of the proof of Proposition 16.1

Now we are ready to establish Proposition 16.1, the non-collapsing result. We shall do this by finding a parabolic neighborhood whose size, r', depends only on r_i, C and ϵ, on which the sectional curvature is bounded by $(r')^{-2}$ and so that the $\mathcal{L}$-distance from x to any point of the final time-slice of this parabolic neighborhood is bounded. Recall that in Section 2 we established it when $R(x) = r^{-2}$ with $r \leq r_{i+1} < \epsilon$. Here we assume that $r_{i+1} < r \leq \epsilon$. Fix $\delta = \delta(r_{i+1})$ from Proposition 16.4 and set $L = 8\sqrt{T_{i+1}}(1 + T_{i+1})$.

First of all, in Claim 15.1 we have seen that there is κ_0 such that $\mathbf{t}^{-1}[0, T_1]$ is κ_0-non-collapsed on scales $\leq \epsilon$. Thus, we may assume that $i \geq 1$.

Recall that $\mathbf{t}(x) = T \in (T_i, T_{i+1}]$. Let γ be an $\mathcal{L}$-geodesic contained in the smooth part of $\mathcal{M}$ from x to a point in $M_{T_{i-1}}$ with $\mathcal{L}(\gamma) \leq 3\sqrt{T - T_{i-1}}$. That such a γ exists was proved in Proposition 16.4. We shall find a point y on this curve with $R(y) \leq 2r_i^{-2}$. Then we find a backward parabolic neighborhood centered at y on which $\mathcal{L}$ is bounded and so that the slices have volume bounded from below. Then we can apply the results from Chapter 8 to establish the κ-non-collapsing.

CLAIM 16.27. *There is τ_0 with $\max(\epsilon^2, T - T_i) \leq \tau_0 \leq T - T_{i-1} - \epsilon^2$ such that $R(\gamma(\tau_0)) < r_i^{-2}$.*

PROOF. Let $T' = \max(\epsilon^2, T - T_i)$ and let $T'' = T - T_{i-1} - \epsilon^2$, and suppose that $R(\gamma(\tau)) \geq r_i^{-2}$ for all $\tau \in [T', T'']$. Then we see that

$$\int_{T'}^{T''} \sqrt{\tau} \left(R(\gamma(\tau) + |X_\gamma(\tau)|^2 \right) d\tau \geq \frac{2}{3} r_i^{-2} \left((T'')^{3/2} - (T')^{3/2} \right).$$

Since $R \geq -6$ because the curvature is pinched toward positive, we see that

$$\mathcal{L}(\gamma) \geq \frac{2}{3}r_i^{-2}\left((T'')^{3/2} - (T')^{3/2}\right) - \int_0^{T'} 6\sqrt{\tau}d\tau - \int_{T'''}^{T-T_{i-1}} 6\sqrt{\tau}d\tau$$

$$= \frac{2}{3}r_i^{-2}\left((T'')^{3/2} - (T')^{3/2}\right) - 4(T')^{3/2} - 4\left((T-T_{i-1})^{3/2} - (T'')^{3/2}\right).$$

CLAIM 16.28. *We have the following estimates:*

$$(T'')^{3/2} - (T')^{3/2} \geq \frac{1}{4}(T - T_{i-1})^{3/2},$$

$$4(T')^{3/2} \leq 4(T - T_{i-1})^{3/2},$$

$$4\left((T - T_{i-1})^{3/2} - (T'')^{3/2}\right) \leq \frac{2t_0}{25}\sqrt{(T - T_{i-1})}.$$

PROOF. Since $T_i - T_{i-1} \geq t_0$ and $T \geq T_i$, we see that $T''/(T - T_{i-1}) \geq 0.9$. If $T' = T - T_i$, then since $T < T_{i+1} = 2T_i = 4T_{i-1}$, it follows that $T'/(T - T_{i-1}) \leq 2/3$. If $T' = \epsilon^2$, since $\epsilon^2 \leq t_0/50$, and $T - T_{i-1} \geq t_0$, we see that $T' \leq (T - T_{i-1})/50$. Thus, in both cases we have $T' \leq 2(T - T_{i-1})/3$. Since $(0.9)^{3/2} > 0.85$ and $(2/3)^{3/2} \leq 0.6$, the first inequality follows.

The second inequality is clear since $T' < (T - T_{i-1})$.

The last inequality is clear from the fact that $T'' = T - T_{i-1} - \epsilon^2$ and $\epsilon \leq \sqrt{t_0/50}$. □

Putting these together yields

$$\mathcal{L}(\gamma) \geq \left[\left(\frac{1}{6}r_i^{-2} - 4\right)(T - T_{i-1}) - \frac{2t_0}{25}\right]\sqrt{T - T_{i-1}}.$$

Since

$$r_i^{-2} \geq r_0^{-2} \geq \epsilon^{-2} \geq 50/t_0,$$

and $T - T_{i-1} \geq t_0$ we see that

$$\mathcal{L}(\gamma) \geq \left[\left(\frac{50}{6t_0} - 4\right)t_0 - \frac{2t_0}{25}\right]\sqrt{T - T_{i-1}}$$

$$\geq (8 - 5t_0)\sqrt{T - T_{i-1}}$$

$$\geq 4\sqrt{T - T_{i-1}}.$$

(The last inequality uses the fact that $t_0 = 2^{-5}$.) But this contradicts the fact that $\mathcal{L}(\gamma) \leq 3\sqrt{T - T_{i-1}}$. □

Now fix τ_0 satisfying Claim 16.27. Let γ_1 be the restriction of γ to the subinterval $[0, \tau_0]$, and let $y = \gamma_1(\tau_0)$. Again using the fact that $R(\gamma(\tau)) \geq -6$ for all τ, we see that

(16.7) $\quad \mathcal{L}(\gamma_1) \leq \mathcal{L}(\gamma) + 4(T - T_{i-1})^{3/2} \leq 3(T_{i+1})^{1/2} + 4(T_{i+1})^{3/2}.$

Set $t' = T - \tau_0$. Notice that from the definition we have $t' \leq T_i$. Consider $B = B(y, t', \frac{r_i}{2C})$, and define $\Delta = \min(r_i^2/16C, \epsilon^2)$. According to Lemma 11.2 every point z on a backward flow line starting in B and

6. COMPLETION OF THE PROOF OF PROPOSITION 16.1

defined for time at most Δ has the property that $R(z) \leq 2r_i^{-2}$. For any surgery time $\bar{t}$ in $[t' - \Delta, t') \subset [T_{i-1}, T)$ the scale $\bar{h}$ of the surgery at time $\bar{t}$ is $\leq \delta(\bar{t})^2 r_i$, and hence every point of the surgery cap has scalar curvature at least $D^{-1}\delta(\bar{t})^{-4} r_i^{-2}$. Since $\bar{\delta}(\bar{t}) \leq \bar{\delta} \leq \delta_0 \leq \min(D^{-1}, 1/10)$, it follows that every point of the surgery cap has curvature at least $\delta_0^{-3} r_i^{-2} \geq 1000 r_i^{-2}$. Thus, no point z as above can lie in a surgery cap. This means that the entire backward parabolic neighborhood $P(y, t', \frac{r_i}{2C}, -\Delta)$ exists in $\mathcal{M}$, and the scalar curvature is bounded by $2r_i^{-2}$ on this backward parabolic neighborhood. Because of the curvature pinching toward positive assumption, there is $C' < \infty$ depending only on r_i and such that the Riemann curvature is bounded by C' on $P(y, t', \frac{r_i}{2C}, -\Delta)$.

Consider the one-parameter family of metrics $g(\tau)$, $0 \leq \tau \leq \Delta$, on $B(y, t', \frac{r_i}{2C})$ obtained by restricting the horizontal metric G to the backward parabolic neighborhood. There is $0 < \Delta_1 \leq \Delta/2$ depending only on C' such that for every $\tau \in [0, \Delta_1]$ and every non-zero tangent vector v at a point of $B(y, t', \frac{r_i}{2C})$ we have

$$\frac{1}{2} \leq \frac{|v|^2_{g(\tau)}}{|v|^2_{g(0)}} \leq 2.$$

Set $\hat{r} = \min(\frac{r_i}{32C}, \Delta_1/2)$; it depends only on r_i, C and ϵ. Set $t'' = t' - \Delta_1$. Clearly, $B(y, t'', \hat{r}) \subset B(y, t', \frac{r_i}{2C})$ so that $B(y, t'', \hat{r}) \subset P(y, t', \frac{r_i}{2C}, -\Delta)$. Of course, it then follows that the parabolic neighborhood $P(y, t'', \hat{r}, -\Delta_1)$ exists in $\mathcal{M}$ and

$$P(y, t'', \hat{r}, -\Delta_1) \subset P(y, t', \frac{r_i}{2C}, -\Delta),$$

so that the Riemann curvature is bounded above by C' on the parabolic neighborhood $P(y, t'', \hat{r}, -\Delta_1)$. We set $r' = \min(\hat{r}, (C')^{-1/2}, \sqrt{\Delta_1}/2)$, so that r' depends only on r_i, C, and ϵ. Then the parabolic neighborhood $P(y, t'', r', -(r')^2)$ exists in $\mathcal{M}$ and $|\mathrm{Rm}| \leq (r')^{-2}$ on $P(y, t'', r', -(r')^2)$. Hence, by the inductive non-collapsing assumption either y is contained in a component of $M_{t''}$ of positive sectional curvature or

$$\mathrm{Vol}\, B(y, t'', r') \geq \kappa_i (r')^3.$$

If y is contained in a component of $M_{t''}$ of positive sectional curvature, then by Hamilton's result, Theorem 4.23, under Ricci flow the component of $M_{t''}$ containing y flows forward as a family of components of positive sectional curvature until it disappears. Since there is a path moving backwards in time from x to y, this means that the original point x is contained in a component of its time-slice with positive sectional curvature.

Let us consider the other possibility when $\mathrm{Vol}\, B(y, t'', r') \geq \kappa_i (r')^3$. For each $z \in B(y, t'', r')$ let

$$\mu_z \colon [T - t', T - t''] \to B(y, t', \frac{r_i}{2C})$$

be the $G(t')$-geodesic connecting y to z. Of course

$$|X_{\mu_z}(\tau)|_{G(t')} \leq \frac{r'}{\Delta_1}$$

for every $\tau \in [0, \Delta_1]$. Thus,

$$|X_{\mu_z}(\tau)|_{G(T-\tau)} \leq \frac{\sqrt{2}r'}{\Delta_1}$$

for all $\tau \in [T-t', T-t'']$. Now we let $\widetilde{\mu}_z$ be the resulting path parameterized by backward time on the time-interval $[T-t', T-t'']$. We estimate

$$\mathcal{L}(\widetilde{\mu}_z) = \int_{T-t'}^{T-t''} \sqrt{\tau} \left(R(\widetilde{\mu}_z(\tau)) + |X_{\widetilde{\mu}_z}(\tau)|^2 \right) d\tau$$

$$\leq \sqrt{T-t''} \int_{T-t'}^{T-t''} \left(2r_i^{-2} + \frac{2(r')^2}{\Delta_1^2} \right) d\tau$$

$$\leq \sqrt{T-t''}(2r_i^{-2}\Delta_1 + \frac{1}{2}) \leq \sqrt{T}\left(\frac{1}{16C} + \frac{1}{2} \right).$$

In passing to the last inequality we use the fact, from the definitions, that $r' \leq \sqrt{\Delta_1}/2$ and $\Delta \leq r_i^2/16C$, whereas $\Delta_1 \leq \Delta/2$.

Since $C > 1$, we see that

$$\mathcal{L}(\widetilde{\mu}_z) \leq \sqrt{T}.$$

Putting this together with the estimate, Equation (16.7), for $\mathcal{L}(\gamma_1)$ tells us that for each $z \in B(y, t'', r')$ we have

$$\mathcal{L}(\gamma_1 * \widetilde{\mu}_z) \leq 4(T_{i+1})^{1/2} + 4(T_{i+1})^{3/2} \leq L/2.$$

Hence, by Proposition 16.4 and the choice of L, there is a minimizing $\mathcal{L}$-geodesic from x to each point of $B(y, t'', r')$ of length $\leq L/2$, and these geodesics are contained in the smooth part of $\mathcal{M}$. In fact, by Proposition 16.21 there is a compact subset Y of the open subset of smooth points of $\mathcal{M}$ that contains all the minimizing $\mathcal{L}$-geodesics from x to points of $B(y, t'', r')$.

Then, by Corollary 6.67 (see also, Proposition 6.56), the intersection, B', of $\mathcal{U}_x$ with $B(y, t'', r')$ is an open subset of full measure in $B(y, t'', r')$. Of course, $\mathrm{Vol}\, B' = \mathrm{Vol}\, B(y, t'', r') \geq \kappa_i(r')^3$ and the function l_x is bounded by $L/2$ on B'. It now follows from Theorem 8.1 that there is $\kappa > 0$ depending only on κ_i, r', ϵ and L such that x is κ non-collapsed on scales $\leq \epsilon$. Recall that L depends only on T_{i+1}, and r' depends only on r_i, C, C' and ϵ, whereas C' depends only on r_i. Thus, in the final analysis, κ depends only on κ_i and r_i (and C and ϵ which are fixed). This entire analysis assumed that for all $t \in [T_{i-1}, T_{i+1})$ we have the inequality $\overline{\delta}(t) \leq \overline{\delta}_1(L + 4(t_{i+1})^{3/2}, r_{i+1})$ as in Lemma 16.15. Since L depends only on i and t_0, this shows that the upper bound for δ depends only on r_{i+1} (and on i, t_0, C, and ϵ). This completes the proof of Proposition 16.1.

CHAPTER 17

Completion of the proof of Theorem 15.9

We have established the requisite non-collapsing result assuming the existence of strong canonical neighborhoods. In order to complete the proof of Theorem 15.9 it remains for us to show the existence of strong canonical neighborhoods. This is the result of the next section.

1. Proof of the strong canonical neighborhood assumption

PROPOSITION 17.1. *Suppose that for some $i \geq 0$ we have surgery parameter sequences $\delta_0 \geq \delta_1 \geq \cdots \geq \delta_i > 0$, $\epsilon = r_0 \geq r_1 \geq \cdots \geq r_i > 0$ and $\kappa_0 \geq \kappa_1 \geq \cdots \geq \kappa_i > 0$. For any $r_{i+1} \leq r_i$, let $\delta(r_{i+1}) > 0$ be the constant in Proposition 16.1 associated to these three sequences and to r_{i+1}. Then there are positive constants $r_{i+1} \leq r_i$ and $\delta_{i+1} \leq \delta(r_{i+1})$ such that the following holds. Suppose that $(\mathcal{M}, G)$ is a Ricci flow with surgery defined for $0 \leq t < T$ for some $T \in (T_i, T_{i+1}]$ with surgery control parameter $\bar{\delta}(t)$. Suppose that the restriction of this Ricci flow with surgery to $\mathbf{t}^{-1}([0, T_i))$ satisfies Assumptions (1) – (7) and also the five properties given in the hypothesis of Theorem 15.9 with respect to the given sequences. Suppose also that $\bar{\delta}(t) \leq \delta_{i+1}$ for all $t \in [T_{i-1}, T]$. Then $(\mathcal{M}, G)$ satisfies the strong (C, ϵ)-canonical neighborhood assumption with parameter r_{i+1}.*

PROOF. Suppose that the result does not hold. Then we can take a sequence of $r_a \to 0$ as $a \to \infty$, all less than r_i, and for each a a sequence $\delta_{a,b} \to 0$ as $b \to \infty$ with each $\delta_{a,b} \leq \delta(r_a)$, where $\delta(r_a) \leq \delta_i$ is the constant in Proposition 16.1 associated to the three sequences given in the statement of this proposition and r_a, such that the following holds. For each a, b there is a Ricci flow with surgery $(\mathcal{M}_{(a,b)}, G_{(a,b)})$ defined for $0 \leq t < T_{(a,b)}$ with $T_i < T_{(a,b)} \leq T_{i+1}$ with control parameter $\bar{\delta}_{(a,b)}(t)$ such that the flow satisfies the hypothesis of the proposition with respect to these constants but fails to satisfy the conclusion.

LEMMA 17.2. *For each a and for all b sufficiently large, there is $t_{(a,b)} \in [T_i, T_{(a,b)})$ such that the restriction of $(\mathcal{M}_{(a,b)}, G_{(a,b)})$ to $\mathbf{t}^{-1}\left([0, t_{(a,b)})\right)$ satisfies the strong (C, ϵ)-canonical neighborhood assumption with parameter r_a and such that there is $x \in \mathcal{M}_{(a,b)}$ with $\mathbf{t}(x_{(a,b)}) = t_{(a,b)}$ at which the strong (C, ϵ)-canonical neighborhood assumption with parameter r_a fails.*

PROOF. Fix a. By supposition, for each b there is a point $x \in \mathcal{M}_{(a,b)}$ at which the strong (C, ϵ)-canonical neighborhood assumption fails for the parameter r_a. We call points at which this condition fails *counterexample points*. Of course, since $r_a \leq r_i$ and since the restriction of $(\mathcal{M}_{(a,b)}, g_{(a,b)})$ to $\mathbf{t}^{-1}([0, T_i))$ satisfies the hypothesis of the proposition, we see that any counterexample point x has $\mathbf{t}(x) \geq T_i$. Take a sequence $x_n = x_{n,(a,b)}$ of counterexample points with $\mathbf{t}(x_{n+1}) \leq \mathbf{t}(x_n)$ for all n that minimizes $\mathbf{t}$ among all counterexample points in the sense that for any $\xi > 0$ and for any counterexample point $x \in \mathcal{M}_{(a,b)}$ eventually $\mathbf{t}(x_n) < \mathbf{t}(x) + \xi$. Let $t' = t'_{(a,b)} = \lim_{n \to \infty} \mathbf{t}(x_n)$. Clearly, $t' \in [T_i, T_{(a,b)})$, and by construction the restriction of $(\mathcal{M}_{(a,b)}, G_{(a,b)})$ to $\mathbf{t}^{-1}([0, t'))$ satisfies the (C, ϵ)-canonical neighborhood assumption with parameter r_a. Since the surgery times are discrete, there is $t'' = t''_{(a,b)}$ with $t' < t'' \leq T_{(a,b)}$ and a diffeomorphism $\psi = \psi_{(a,b)} \colon M_{t'} \times [t', t'') \to \mathbf{t}^{-1}([t', t''))$ compatible with time and the vector field. We view $\psi^* G_{(a,b)}$ as a one-parameter family of metrics $g(t) = g_{(a,b)}(t)$ on $M_{t'}$ for $t \in [t', t'')$. By passing to a subsequence we can arrange that $\mathbf{t}(x_n) \in [t', t'')$ for all n. Thus, for each n there are $y_n = y_{n,(a,b)} \in M_{t'}$ and $t_n \in [t', t'')$ with $\psi(y_n, t_n) = x_n$. Since $M_{t'}$ is a compact 3-manifold, by passing to a further subsequence we can assume that $y_n \to x_{(a,b)} \in M_{t'}$. Of course, $t_n \to t'$ as $n \to \infty$ and $\lim_{n \to \infty} x_n = x_{(a,b)}$ in $\mathcal{M}_{(a,b,)}$.

We claim that, for all b sufficiently large, the strong (C, ϵ)-canonical neighborhood assumption with parameter r_a fails at $x_{(a,b)}$. Notice that since $x_{(a,b)}$ is the limit of a sequence where the strong (C, ϵ)-neighborhood assumption fails, the points in the sequence converging to $x_{(a,b)}$ have scalar curvature at least r_a^{-2}. It follows that $R(x_{(a,b)}) \geq r_a^{-2}$. Suppose that $x_{(a,b)}$ satisfies the strong (C, ϵ)-canonical neighborhood assumption with parameter r_a. This means that there is a neighborhood $U = U_{(a,b)}$ of $x_{(a,b)} \in M_{t'}$ which is a strong (C, ϵ)-canonical neighborhood of $x_{(a,b)}$. According to Definition 9.78 there are four possibilities. The first two we consider are that $(U, g(t'))$ is an ϵ-round component or a C-component. In either of these cases, since the defining inequalities given in Definition 9.76 and 9.75 are strong inequalities, all metrics on U sufficiently close to $g(t')$ in the C^∞-topology satisfy these same inequalities. But as n tends to ∞, the metrics $g(t_n)|_U$ converge in the C^∞-topology to $g(t')|_U$. Thus, in these two cases, for all n sufficiently large, the metrics $g(t_n)$ on U are (C, ϵ)-canonical neighborhood metrics of the same type as $g(t'_{(a,b)})|_U$. Hence, in either of these cases, for all n sufficiently large $x_{n,(a,b)}$ has a strong (C, ϵ)-canonical neighborhood of the same type as $x_{(a,b)}$, contrary to our assumption about the sequence $x_{n,(a,b)}$.

Now suppose that there is a (C, ϵ)-cap whose core contains $x_{(a,b)}$. This is to say that $(U, g(t'))$ is a (C, ϵ)-cap whose core contains $x_{(a,b)}$. By Proposition 9.79, for all n sufficiently large, $(U, g(t_n))$ is also a (C, ϵ)-cap with the

1. PROOF OF THE STRONG CANONICAL NEIGHBORHOOD ASSUMPTION

same core. This core contains y_n for all n sufficiently large, showing that x_n is contained in the core of a (C, ϵ)-cap for all n sufficiently large.

Now let us consider the remaining case when $x_{(a,b)}$ is the center of a strong ϵ-neck. In this case we have an embedding

$$\psi_{U_{(a,b)}} \colon U_{(a,b)} \times (t'_{(a,b)} - R^{-1}(x_{(a,b)}), t'_{(a,b)}] \to \mathcal{M}_{(a,b)}$$

compatible with time and the vector field and a diffeomorphism $f_{(a,b)} \colon S^2 \times (-\epsilon^{-1}, \epsilon^{-1}) \to U_{(a,b)}$ such that

$$(f_{(a,b)} \times \mathrm{Id})^* \psi^*_{U_{(a,b)}} (R(x_{(a,b)}) G_{(a,b)})$$

is ϵ-close in the $C^{[1/\epsilon]}$-topology to the evolving product metric

$$h_0(t) \times ds^2, \quad -1 < t \leq 0,$$

where $h_0(t)$ is a round metric of scalar curvature $1/(1-t)$ on S^2 and ds^2 is the Euclidean metric on the interval. Here, there are two subcases to consider.

(i) $\psi_{U_{(a,b)}}$ extends backward past $t'_{(a,b)} - R^{-1}(x_{(a,b)})$.
(ii) There is a flow line through a point $y_{(a,b)} \in U_{(a,b)}$ that is defined on the interval $[t'_{(a,b)} - R^{-1}(x_{(a,b)}), t'_{(a,b)}]$ but with the value of the flow line at $t'_{(a,b)} - R^{-1}(x_{(a,b)})$ an exposed point.

Let us consider the first subcase. The embedding $\psi_{U_{(a,b)}}$ extends forward in time because of the diffeomorphism $\psi_{(a,b)} \colon M_{t'_{(a,b)}} \times [t'_{(a,b)}, t''_{(a,b)}) \to \mathcal{M}_{(a,b)}$ and, by assumption, $\psi_{U_{(a,b)}}$ extends backward in time some amount. Thus, for all n sufficiently large, we can use these extensions of $\psi_{U_{(a,b)}}$ to define an embedding $\psi_{n,(a,b)} \colon U_{(a,b)} \times (\mathbf{t}(x_{n,(a,b)}) - R^{-1}(x_{n,(a,b)}), \mathbf{t}(x_{n,(a,b)})] \to \mathcal{M}_{(a,b)}$ compatible with time and the vector field. Furthermore, since the $\psi_{n,(a,b)}$ converge in the C^∞-topology to $\psi_{U_{(a,b)}}$ as n tends to infinity, the Riemannian metrics $(f_{(a,b)} \times \mathrm{Id})^* \psi^*_{n,(a,b)} (R(x_{n,(a,b)}) G_{a,b})$ converge in the C^∞-topology to the pullback $(f_{(a,b)} \times \mathrm{Id})^* \psi^*_{U_{(a,b)}} (R(x_{(a,b)}) G_{a,b})$. Clearly then, for fixed (a,b) and for all n sufficiently large the pullbacks of the rescalings of these metrics by $R(x_{n,(a,b)})$ are within ϵ in the $C^{[1/\epsilon]}$-topology of the standard evolving flow $h_0(t) \times ds^2, -1 < t \leq 0$, on the product of S^2 with the interval. Under these identifications the points $x_{n,(a,b)}$ correspond to points

$$(p_{n,(a,b)}, s_{n,(a,b)}) \in S^2 \times (-\epsilon^{-1}, \epsilon^{-1})$$

where $\lim_{n \to \infty} s_{n,(a,b)} = 0$. The last thing we do is to choose diffeomorphisms $\varphi_{n,(a,b)} \colon (-\epsilon^{-1}, \epsilon^{-1}) \to (-\epsilon^{-1}, \epsilon^{-1})$ that are the identity near both ends, such that $\varphi_{n,(a,b)}$ carries 0 to $s_{n,(a,b)}$ and such that the $\varphi_{n,(a,b)}$ converge to the identity in the C^∞-topology for fixed (a,b) as n tends to infinity. Then, for all n sufficiently large, the composition

$$S^2 \times (-\epsilon^{-1}, \epsilon^{-1}) \xrightarrow{\mathrm{Id} \times \varphi_{n,(a,b)}} S^2 \times (-\epsilon^{-1}, \epsilon^{-1}) \xrightarrow{f_{(a,b)}} U \xrightarrow{\psi_{n,(a,b)}} \mathcal{M}_{(a,b)}$$

is a strong ϵ-neck centered at $x_{n,(a,b)}$. This shows that for any b for which the first subcase holds, for all n sufficiently large, there is a strong ϵ-neck centered at $x_{n,(a,b)}$.

Now suppose that the second subcase holds for all b. Here, unlike all previous cases, we shall have to let b vary and we shall prove the result only for b sufficiently large. We shall show that for all b sufficiently large, $x_{(a,b)}$ is contained in the core of a (C, ϵ)-cap. This will establish the result by contradiction, for as we showed in the previous case, if $x_{(a,b)}$ is contained in the core of a (C, ϵ)-cap, then the same is true for the x_n for all n sufficiently large, contrary to our assumption.

For the moment fix b. Set $\bar{t}_{(a,b)} = t'_{(a,b)} - R_{G_{(a,b)}}(x_{(a,b)})^{-1}$. Since, by supposition the embedding $\psi_{U_{(a,b)}}$ does not extend backwards past $\bar{t}_{(a,b)}$, it must be the case that $\bar{t}_{(a,b)}$ is a surgery time and furthermore that there is a surgery cap $\mathcal{C}_{(a,b)}$ at this time with the property that there is a point $y_{(a,b)} \in U_{(a,b)}$ such that $\psi_{U_{(a,b)}}(y_{(a,b)}, t)$ converges to a point $z_{(a,b)} \in \mathcal{C}_{(a,b)}$ as t tends to $\bar{t}_{(a,b)}$ from above. (See FIG. 1.) We denote by $p_{(a,b)}$ the tip of $\mathcal{C}_{(a,b)}$, and we denote by $\bar{h}_{(a,b)}$ the scale of the surgery at time $\bar{t}_{(a,b)}$.

Since the statement that $x_{(a,b)}$ is contained in the core of a (C, ϵ)-cap is a scale invariant statement, we are free to replace $(\mathcal{M}_{(a,b)}, G_{(a,b)})$ with $(\widetilde{\mathcal{M}}_{(a,b)}, \widetilde{G}_{(a,b)})$, which has been rescaled to make $\bar{h}_{(a,b)} = 1$ and shifted in time so that $\bar{t}_{(a,b)} = 0$. We denote the new time function by $\widetilde{\mathbf{t}}$. (Notice that this rescaling and time-shifting is different from what we usually do. Normally, when we have a base point like $x_{(a,b)}$ we rescale to make its scalar curvature 1 and we shift time to make it be at time 0. Here we have rescaled based on the scale of the surgery cap rather than $R(x_{(a,b)})$.) We set $\widetilde{Q}_{(a,b)} = R_{\widetilde{G}_{(a,b)}}(x_{(a,b)})$ and we set $\widetilde{t}'_{(a,b)} = \widetilde{\mathbf{t}}(x_{(a,b)})$. Since the initial time of the strong ϵ-neck is zero, $\widetilde{t}'_{(a,b)} = \widetilde{Q}_{(a,b)}^{-1}$. We denote the flow line backward in time from $y_{(a,b)}$ by $y_{(a,b)}(\widetilde{t})$, $0 \leq \widetilde{t} \leq \widetilde{t}'_{(a,b)}$, so that $y_{(a,b)}(\widetilde{t}'_{(a,b)}) = y_{(a,b)}$. Since $U_{(a,b)}$ is a strong ϵ-neck, by our choice of ϵ, it follows from Lemma A.2 and rescaling that $R(\psi(y_{(a,b)}, \widetilde{t}))$ is within $(0.01)\widetilde{Q}_{(a,b)}$ of $\widetilde{Q}_{(a,b)}/(1+\widetilde{Q}_{(a,b)}(\widetilde{t}'_{(a,b)}-\widetilde{t}))$ for all $t \in (0, \widetilde{t}'_{(a,b)}]$. By taking limits as t approaches 0, we see that $R_{\widetilde{G}_{(a,b)}}(z_{(a,b)})$ is within $(0.01)\widetilde{Q}_{(a,b)}$ of $\widetilde{Q}_{(a,b)}/2$. Let D be the universal constant given in Lemma 12.3, so that the scalar curvature at any point of the standard initial metric is at least D^{-1} and at most D. It follows from the third item in Theorem 13.2 that, since we have rescaled to make the surgery scale 1, for all b sufficiently large the scalar curvature on the surgery $\mathcal{C}_{(a,b)}$ is at least $(2D)^{-1}$ and at most $2D$. In particular, for all b sufficiently large

$$(2D)^{-1} \leq R_{\widetilde{G}_{(a,b)}}(z_{(a,b)}) \leq 2D.$$

1. PROOF OF THE STRONG CANONICAL NEIGHBORHOOD ASSUMPTION

Together with the above estimate relating $R_{\widetilde{G}_{(a,b)}}(z_{(a,b)})$ and $\widetilde{Q}_{(a,b)}$, this gives

(17.1) $$(5D)^{-1} \leq \widetilde{Q}_{(a,b)} \leq 5D.$$

Since the flow line from $z_{(a,b)}$ to $y_{(a,b)}$ lies in the closure of a strong ϵ-neck of scale $\widetilde{Q}_{(a,b)}^{-1/2}$, the scalar curvature is less than $6D$ at every point of this flow line. According to Proposition 12.31 there is $\theta_1 < 1$ (depending only on D) such that $R(q,t) \geq 8D$ for all (q,t) in the standard solution with $t \geq \theta_1$.

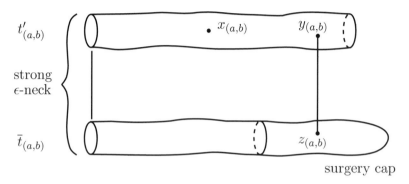

FIGURE 1. A strong neck with initial time in a surgery cap.

By the fifth property of Theorem 12.5 there is $A'(\theta_1) < \infty$ such that in the standard flow, $B(p_0, 0, A)$ contains $B(p_0, \theta_1, A/2)$ for every $A \geq A'(\theta_1)$. We set A equal to the maximum of $A'(\theta_1)$ and

$$3\left((1.2)\sqrt{5D}\epsilon^{-1} + (1.1)(A_0 + 5) + C\sqrt{5D}\right).$$

Now for any $\overline{\delta} > 0$ for all b sufficiently large, we have $\delta_{(a,b)} \leq \delta''(A, \theta_1, \overline{\delta})$, where $\delta''(A, \theta_1, \overline{\delta})$ is the constant given in Proposition 16.13.

CLAIM 17.3. *Suppose b is sufficiently large so that $\delta_{(a,b)} \leq \delta''(A, \theta_1, \overline{\delta}_0)$, where $\overline{\delta}_0$ is the constant given in Definition 16.12. Then $\widetilde{t}'_{(a,b)} \leq \theta_1$.*

PROOF. In this proof we shall fix (a, b), so we drop these indices from the notation. Consider $s \leq \theta_1$ maximal so that there is an embedding

$$\psi = \psi_{(a,b)} \colon B(p_0, 0, A) \times [0, s) \to \widetilde{\mathcal{M}}_{(a,b)}$$

compatible with time and the vector field. First suppose that $s < \theta_1$. Then according to Proposition 16.5 either the entire ball $B(p, 0, A)$ disappears at time s or s is the final time of the time interval of definition for the flow $(\widetilde{\mathcal{M}}_{(a,b)}, \widetilde{G}_{(a,b)})$. Since we have the flow line from $z \in B(p_0, 0, A)$ extending to time $\widetilde{t}' = \widetilde{t}'_{(a,b)}$, in either case this implies that $\widetilde{t}' < s$, proving that $t' < \theta_1$ in this case.

Now suppose that $s = \theta_1$. By the choice of θ_1, for the standard solution the scalar curvature at every (q, θ_1) is at least $8D$. Since $\delta_{(a,b)} \leq \delta''(A, \theta_1, \overline{\delta}_0)$, by the definition of $\overline{\delta}_0$ given in Definition 16.12 and by Proposition 16.5 the scalar curvature of the pullback of the metric under ψ is within a factor of 2 of the scalar curvature of the rescaled standard solution. Hence, the scalar curvature along the flow line (z,t) through z limits to at least $8D$ as t tends to θ_1. Since the scalar curvature on (z,t) for $t \in [0, \widetilde{t}']$ is bounded above by $6D$, it follows that $\widetilde{t}' < \theta_1$ in this case as well. This completes the proof of the claim. $\square$

Thus, we have maps
$$\psi_{(a,b)} \colon B(p_0, 0, A) \times [0, \widetilde{t}'_{(a,b)}] \to \widetilde{\mathcal{M}}_{(a,b)}$$
compatible with time and the vector field, with the property that for each $\delta > 0$, for all b sufficiently large the pullback under this map of $\widetilde{G}_{(a,b)}$ is within δ in the $C^{[1/\delta]}$-topology of the restriction of the standard solution. Let $w_{(a,b)}$ be the result of flowing $x_{(a,b)}$ backward to time 0.

CLAIM 17.4. *For all b sufficiently large, $w_{(a,b)} \in \psi_{(a,b)}(B(p_0, 0, A) \times \{0\})$.*

PROOF. First notice that, by our choice of ϵ, every point in the 0 time-slice of the closure of the strong ϵ-neck centered at $x_{(a,b)}$ is within distance $(1.1)\widetilde{Q}^{-1}_{(a,b)}\epsilon^{-1}$ of $w_{(a,b)}$. In particular,
$$d_{\widetilde{G}_{(a,b)}}(w_{(a,b)}, y_{(a,b)}) < (1.1)\widetilde{Q}^{-1/2}_{(a,b)}\epsilon^{-1}.$$
Since $y_{(a,b)}$ is contained in the surgery cap and the scale of the surgery at this time is 1, $y_{(a,b)}$ is within distance $A_0 + 5$ of $p_{(a,b)}$. Hence, by the triangle inequality and Inequality (17.1), we have
$$d_{\widetilde{G}_{(a,b)}}(w_{(a,b)}, p_{(a,b)}) < (1.1)\widetilde{Q}^{-1/2}_{(a,b)}\epsilon^{-1} + (A_0 + 5)$$
$$< (1.1)\sqrt{5D}\epsilon^{-1} + (A_0 + 5).$$
For b sufficiently large, the image $\psi_{(a,b)}(B(p_0, 0, A))$ contains the ball of radius $(0.95)A$ centered at $p_{(a,b)}$. Since by our choice of A we have $(0.95)A > (1.1)\sqrt{5D}\epsilon^{-1} + (A_0 + 5)$, the claim follows. $\square$

We define $q_{(a,b)} \in B(p_0, 0, A)$ so that $\psi_{(a,b)}(q_{(a,b)}, 0) = w_{(a,b)}$. Of course,
$$\psi_{(a,b)}(q_{(a,b)}, \widetilde{t}'_{(a,b)}) = x_{(a,b)}.$$
If follows from the above computation that for all b sufficiently large we have
$$d_0(q_{(a,b)}, p_0) < (1.15)\widetilde{Q}^{-1/2}_{(a,b)}\epsilon^{-1} + (1.05)(A_0 + 5).$$
Since the standard flow has non-negative curvature, it is a distance non-increasing flow. Therefore,
$$d_{\widetilde{t}'_{(a,b)}}(q_{(a,b)}, p_0) < (1.15)\widetilde{Q}^{-1/2}_{(a,b)}\epsilon^{-1} + (1.05)(A_0 + 5).$$

1. PROOF OF THE STRONG CANONICAL NEIGHBORHOOD ASSUMPTION

Suppose that a point $(q, \widetilde{t}'_{(a,b)})$ in the standard solution were the center of a $\beta\epsilon/3$-neck, where β is the constant from Proposition 15.2. Of course, for all b sufficiently large, $R(q, \widetilde{t}'_{(a,b)}) > (0.99)\widetilde{Q}_{(a,b)}$. Since $\beta < 1/2$ and $\epsilon < \sqrt{5D}(A_0 + 5)/2$ and $\widetilde{Q}_{(a,b)} \leq 5D$, it follows from the above distance estimate that this neck would contain $(p_0, \widetilde{t}'_{(a,b)})$. But this is impossible: since $(p_0, \widetilde{t}'_{(a,b)})$ is an isolated fixed point of an isometric $SO(3)$-action on the standard flow, all the sectional curvatures at $(p_0, \widetilde{t}'_{(a,b)})$ are equal, and this is in contradiction with estimates on the sectional curvatures at any point of an ϵ-neck given in Lemma A.2. We can then conclude from Theorem 12.32 that for all b sufficiently large, the point $(p_0, \widetilde{t}'_{(a,b)})$ is contained in the core of a $(C(\beta\epsilon/3), \beta\epsilon/3)$-cap $Y_{(a,b)}$ in the $\widetilde{t}'_{(a,b)}$ time-slice of the standard solution. Now note that for all b sufficiently large, the scalar curvature of $(q_{(a,b)}, \widetilde{t}'_{(a,b)})$ is at least $(0.99)\widetilde{Q}_{(a,b)}$, since the scalar curvature of $x_{(a,b)}$ is equal to $Q_{(a,b)}$. This implies that the diameter of $Y_{(a,b)}$ is at most

$$(1.01)\widetilde{Q}_{(a,b)}^{-1/2} C(\beta\epsilon/3) < (1.1)\sqrt{5D} C(\beta\epsilon/3).$$

Since $B(p_0, 0, A)$ contains $B(p_0, \widetilde{t}'_{(a,b)}, A/2)$, and since $C > C(\beta\epsilon/3)$, it follows from the definition of A, the above distance estimate, and the triangle inequality that for all b sufficiently large $B(p_0, 0, A) \times \{\widetilde{t}'_{(a,b)}\}$ contains $Y_{(a,b)}$.

Since $C > C(\beta\epsilon/3) + 1$ and since for b sufficiently large $\psi^*_{(a,b)}\widetilde{G}_{(a,b)}$ is arbitrarily close to the restriction of the standard solution metric, it follows from Theorem 12.32 and Proposition 9.79 that for all b sufficiently large, the image $\psi_{(a,b)}(Y_{(a,b)})$ is a (C, ϵ)-cap whose core contains $x_{(a,b)}$. As we have already remarked, this contradicts the assumption that no x_n has a strong (C, ϵ)-canonical neighborhood.

This completes the proof in the last case and establishes Lemma 17.2. $\square$

REMARK 17.5. Notice that even though $x_{(a,b)}$ is the center of a strong ϵ-neck, the canonical neighborhoods of the x_n constructed in the second case are not strong ϵ-necks but rather are (C, ϵ)-caps coming from applying the flow to a neighborhood of the surgery cap $\mathcal{C}$.

Now we return to the proof of Proposition 17.1. For each a, we pass to a subsequence (in b) so that Lemma 17.2 holds for all (a, b). For each (a, b), let $t_{(a,b)}$ be as in that lemma. We fix a point $x_{(a,b)} \in \mathbf{t}^{-1}(t_{(a,b)}) \subset \mathcal{M}_{(a,b)}$ at which the canonical neighborhood assumption with parameter r_a fails. For each a choose $b(a)$ such that $\delta_{b(a)} \to 0$ as $a \to \infty$. For each a we set $(\mathcal{M}_a, G_a) = (\mathcal{M}_{(a,b(a))}, G_{(a,b(a))})$, we set $t_a = t_{(a,b(a))}$, and we let $x_a = x_{(a,b(a))} \in \mathcal{M}_a$. Let $(\widetilde{\mathcal{M}}_a, \widetilde{G}_a)$ be the Ricci flow with surgery obtained from $(\mathcal{M}_a, G_a)$ by shifting t_a to 0 and rescaling the metric and time by $R(x_a)$. We have the points $\widetilde{x}_a$ in the 0 time-slice of $\widetilde{\mathcal{M}}_a$ corresponding to $x_a \in \mathcal{M}_a$. Of course, by construction $R_{\widetilde{G}_a}(\widetilde{x}_a) = 1$ for all a.

We shall take limits of a subsequence of this sequence of based Ricci flows with surgery. Since $r_a \to 0$ and $R(x_a) \geq r_a^{-2}$, it follows that $R(x_a) \to \infty$. By Proposition 16.1, since $\delta_{b(a)} \leq \delta(r_a)$ it follows that the restriction of $(\widetilde{\mathcal{M}}_a, \widetilde{G}_a)$ to $\mathbf{t}^{-1}(-\infty, 0)$ is κ-non-collapsed on scales $\leq \epsilon R_{\widetilde{G}_a}^{1/2}(x_a)$. By passing to a subsequence we arrange that one of the following two possibilities holds:

(i) There is $A < \infty$ and $t' < \infty$ such that, for each a there is a flow line through a point y_a of $B_{\widetilde{G}_a}(\widetilde{x}_a, 0, A)$ that is not defined on all of $[-t', 0]$. (See FIG. 2.)

(ii) For every $A < \infty$ and every $t' < \infty$, for all a sufficiently large all flow lines through points of $B_{\widetilde{G}_a}(\widetilde{x}_a, 0, A)$ are defined on the interval $[-t', 0]$.

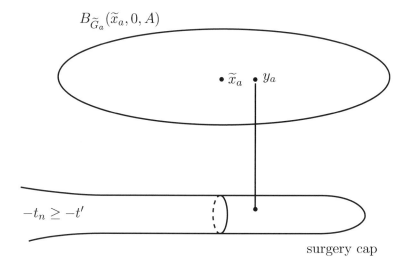

FIGURE 2. Possibility (i).

Let us consider the second case. By Proposition 16.1 these rescaled solutions are κ-non-collapsed on scales $\leq \epsilon R_{G_a}(x_a)^{1/2}$ for all $t < 0$. Since this condition is a closed constraint, the same is true if $t = 0$. Since $R(x_a) \geq r_a^{-2}$, by construction every point $\widetilde{x} \in (\widetilde{\mathcal{M}}_a, \widetilde{G}_a)$ with $R(\widetilde{x}) \geq 1$ and $\mathbf{t}(\widetilde{x}) < 0$ has a strong (C, ϵ)-canonical neighborhood.

CLAIM 17.6. *For all a sufficiently large, every point $\widetilde{x} \in (\widetilde{\mathcal{M}}_a, \widetilde{G}_a)$ with $R(\widetilde{x}) > 1$ and $\mathbf{t}(\widetilde{x}) = 0$ has a $(2C, 2\epsilon)$-canonical neighborhood.*

PROOF. Assume that $\widetilde{x} \in \widetilde{\mathcal{M}}$ has $R(\widetilde{x}) > 1$. Suppose that $\widetilde{x}$ is an exposed point. If a is sufficiently large, then $\delta_{b(a)}$ is arbitrarily close to zero and hence by the last item in Theorem 13.2 and the structure of the standard initial condition, we see that $\widetilde{x}$ is contained in the core of a $(2C, 2\epsilon)$-cap.

Suppose now that $\widetilde{x}$ is not an exposed point. Then we can take a sequence of points $\widetilde{y}_n \in \widetilde{\mathcal{M}}_a$ all lying on the flow line for the vector field through $\widetilde{x}$ converging to $\widetilde{x}$ with $\mathbf{t}(\widetilde{y}_n) < 0$. Of course, for all n sufficiently large $R(\widetilde{y}_n) > 1$, which implies that for all n sufficiently large $\widetilde{y}_n$ has a strong (C, ϵ)-canonical neighborhood. Passing to a subsequence, we can arrange that all of these canonical neighborhoods are of the same type. If they are all ϵ-round components, all C-components, or all (C, ϵ)-caps whose cores contain y_n, then by taking limits and arguing as in the proof of Lemma 11.23 we see that $\widetilde{x}$ has a strong $(2C, 2\epsilon)$-canonical neighborhood of the same type. On the other hand, if $\widetilde{y}_n$ is the center of a strong ϵ-neck for all n, then according to Claim 11.24, the limit point $\widetilde{x}$ is the center of a strong 2ϵ-neck. □

Since we have chosen $\epsilon > 0$ sufficiently small so that Theorem 11.8 applies with ϵ replaced by 2ϵ, applying this theorem shows that we can pass to a subsequence and take a smooth limiting flow of a subsequence of the rescaled flows $(\widetilde{\mathcal{M}}_a, \widetilde{G}_a)$ based at $\widetilde{x}_a$ and defined for all $t \in (-\infty, 0]$. Because the $(\mathcal{M}_a, G_a)$ all have curvature pinched toward positive and since $R(x_a) \to \infty$ as a tends to infinity, this result says that the limiting flow has non-negative, bounded curvature and is κ-non-collapsed on all scales. That is to say, the limiting flow is a κ-solution. By Corollary 9.95 this contradicts the fact that the strong (C, ϵ)-canonical neighborhood assumption fails at x_a for every a. This contradiction shows that in the second case there is a subsequence of the a such that x_a has a strong canonical neighborhood and completes the proof of the second case.

Let us consider the first case. In this case we will arrive at a contradiction by showing that for all a sufficiently large, the point x_a lies in a strong (C, ϵ)-canonical neighborhood coming from a surgery cap. Here is the basic result we use to find that canonical neighborhood.

LEMMA 17.7. *Suppose that there are $A', D', t' < \infty$ such that the following holds for all a sufficiently large. There is a point $y_a \in B_{\widetilde{G}_a}(\widetilde{x}_a, 0, A')$ and a flow line of χ beginning at y_a, defined for backward time and ending at a point z_a in a surgery cap $\mathcal{C}_a$ at time $-t_a$ for some $t_a \leq t'$. We denote this flow line by $y_a(t), -t_a \leq t \leq 0$. Furthermore, suppose that the scalar curvature on the flow line from y_a to z_a is bounded by D'. Then for all a sufficiently large, x_a has a strong (C, ϵ)-canonical neighborhood.*

PROOF. The proof is by contradiction. Suppose the result does not hold. Then there are $A', D', t' < \infty$ and we can pass to a subsequence (in a) such that the hypotheses of the lemma hold for every a but no x_a has a strong (C, ϵ)-canonical neighborhood. The essential point of the argument is to show that **in the units of the surgery scale** the elapsed time between the surgery time and 0 is less than 1 and the distance from the point z_a to the tip of the surgery cap is bounded independent of a.

By Lemma 12.3, the fact that the scalar curvature at z_a is bounded by D' implies that for all a sufficiently large the scale $\bar{h}_a$ of the surgery at time $-t_a$ satisfies

$$(17.2) \qquad h_a^2 \geq (2D'D)^{-1}.$$

(Recall that we are working in the rescaled flow $(\widetilde{\mathcal{M}}_a, \widetilde{G}_a)$.)

Now we are ready to show that the elapsed time is bounded less than 1 in the surgery scale.

CLAIM 17.8. *There is $\theta_1 < 1$, depending on D' and t', such that for all a sufficiently large we have $t_a < \theta_1 \bar{h}_a^2$.*

PROOF. We consider two cases: either $t_a \leq \bar{h}_a^2/2$ or $\bar{h}_a^2/2 < t_a$. In the first case, the claim is obviously true with θ_1 anything greater than $1/2$ and less than 1. In the second case, the curvature everywhere along the flow line is at most $D' < (2t_a D')\bar{h}_a^{-2} \leq (2t' D')\bar{h}_a^{-2}$. Using Proposition 12.31 fix $1/2 < \theta_1 < 1$ so that every point of the standard solution (x, t) with $t \geq (2\theta_1 - 1)$ satisfies $R(x, t) \geq 6t' D'$. Notice that θ_1 depends only on D' and t'. If $t_a < \theta_1 \bar{h}_a^2$, then the claim holds for this value of $\theta_1 < 1$. Suppose $t_a \geq \theta_1 \bar{h}_a^2$, so that $-t_a + (2\theta_1 - 1)\bar{h}_a^2 < 0$. For all a sufficiently large we have $\delta_a \leq \delta_0''(A_0 + 5, \theta_1, \bar{\delta}_0)$ where $\bar{\delta}_0$ is the constant from Definition 16.12 and δ_0'' is the constant from Proposition 16.5. This means that the scalar curvatures at corresponding points of the rescaled standard solution and the evolution of the surgery cap (up to time 0) in $\widetilde{\mathcal{M}}_a$ differ by at most a factor of two. Thus, for these a, we have $R(y_a, (-t_a + (2\theta_1 - 1)\bar{h}_a^2)) \geq 3(t'D')\bar{h}_a^{-2}$ from the definition of $\bar{\delta}_0$ and Proposition 16.5. But this is impossible since $-t_a(2\theta_1 - 1)\bar{h}_a^2 < 0$ and $3t' D'/\bar{h}_a^2 \geq 3t_a D'/\bar{h}_a^2 > D'$ as $t_a > \bar{h}_a^2/2$. Hence, $R(y_a, (-t_a + (2\theta_1 - 1)\bar{h}_a^2)) \leq 2t_a D'\bar{h}_a^{-2} \leq 2t' D'\bar{h}_a^{-2}$. This contradiction shows that if a is sufficiently large then $t_a < \theta_1 \bar{h}_a^2$. □

We pass to a subsequence so that $t_a \bar{h}_a^{-2}$ converges to some $\theta \leq \theta_1$. We define $\widetilde{C}$ to be the maximum of C and $3\epsilon^{-1}\beta^{-1}$. Now, using part 5 of Theorem 12.5 we set $A'' \geq (9\widetilde{C} + 3A')\sqrt{2DD'} + 6(A_0 + 5)$ sufficiently large so that in the standard flow $B(p_0, 0, A'')$ contains $B(p_0, t, A''/2)$ for any $t \leq (\theta_1 + 1)/2$. This constant is chosen only to depend on θ_1, A', and C. As a tends to infinity, δ_a tends to zero which means, by Proposition 16.5, that for all a sufficiently large there is an embedding

$$\rho_a \colon B(p_0, -t_a, A''\bar{h}_a) \times [-t_a, 0] \to \widetilde{\mathcal{M}}_a$$

compatible with time and the vector field such that (after translating by t_a to make the flow start at time 0 and scaling by $\bar{h}_a^{-2}$) the restriction of $\widetilde{G}_a$ to this image is close in the C^∞-topology to the restriction of the standard flow to $B(p_0, 0, A'') \times [0, \bar{h}_a^{-2} t_a]$. The image $\rho_a(p_0, -t_a)$ is the tip p_a of the surgery cap $\mathcal{C}_a$ in $\widetilde{\mathcal{M}}_a$. Thus, for all a sufficiently large the image $\rho_a\left(B(p_0, -t_a, A''\bar{h}_a) \times \{0\}\right)$ contains the $A''\bar{h}_a/3$-neighborhood of the image

1. PROOF OF THE STRONG CANONICAL NEIGHBORHOOD ASSUMPTION

$\rho_a(p_0, 0)$ of the tip of the surgery cap under the flow forward to time 0. By our choice of A'', and Equation (17.2), this means that for all a sufficiently large $\rho_a \left(B(p_0, -t_a, A''\overline{h}_a) \times \{-t_a\} \right)$ contains the $(3\widetilde{C} + A') + 2(A_0 + 5)\overline{h}_a$-neighborhood of $p_a = \rho_a(p_0, -t_a)$. Notice also that, since the standard solution has positive curvature and hence the distance between points is non-increasing in time by Lemma 3.14, the distance at time 0 between $\rho_a(p_0, 0)$ and y_a is less than $2(A_0 + 5)\overline{h}_a$. By the triangle inequality, we conclude that for all a sufficiently large, $\rho_a \left(B(p_0, -t_a, A''\overline{h}_a) \times \{0\} \right)$ contains the $3\widetilde{C}$-neighborhood of x_a. Since the family of metrics on

$$\rho_a \left(B(p_0, -t_a, A''\overline{h}_a) \times [-t_a, 0] \right)$$

(after time-shifting by t_a and rescaling by $\overline{h}_a^{-2}$) are converging smoothly to the ball $B(p_0, 0, A'') \times [0, \theta]$ in the standard flow, for all a sufficiently large, the flow from time $-t_a$ to 0 on the $3\widetilde{C}$-neighborhood of x_a is, after rescaling by $\overline{h}_a^{-2}$, very nearly isometric to the restriction of the standard flow from time 0 to $\overline{h}_a^{-2} t_a$ on the $3\widetilde{C}\overline{h}_a^{-1}$-neighborhood of some point q_a in the standard flow. Of course, since the scalar curvature of x_a is 1, $R(q_a, \overline{h}_a^{-2} t_a)$ in the standard flow is close to $\overline{h}_a^{-2}$. Hence, by Theorem 12.32 there is a neighborhood X of $(q_a, \overline{h}_a^{-2} t_a)$ in the standard solution that either is a (C, ϵ)-cap, or is an evolving $\beta\epsilon/3$-neck centered at $(q_a, \overline{h}_a^{-2} t_a)$. In the latter case either the evolving neck is defined for backward time $(1 + \beta\epsilon/3)$ or its initial time-slice is the zero time-slice and this initial time-slice lies at distance at least 1 from the surgery cap. Of course, X is contained in the $CR(q_a, \overline{h}_a^{-2} t_a)^{-1/2}$-neighborhood of $(q_a, \overline{h}_a^{-2} t_a)$ in the standard solution. Since $\widetilde{C} \geq C$ and $R(q_a, \overline{h}_a^{-2} t_a)$ is close to $\overline{h}_a^{-2}$, the neighborhood X is contained in the $2\widetilde{C}\overline{h}_a^{-1}$-neighborhood of $(q_a, \overline{h}_a^{-2} t_a)$ in the standard solution. Hence, after rescaling, the corresponding neighborhood of x_a is contained in the $3\widetilde{C}$-neighborhood of x_a. If either of the first two cases in Theorem 12.32 occurs for a subsequence of a tending to infinity, then by Lemma 9.79 and the fact that $\widetilde{C} > \max(C, \epsilon^{-1})$, we see that there is a subsequence of a for which x_a either is contained in the core of a (C, ϵ)-cap or is the center of a strong ϵ-neck.

We must examine further the last case. We suppose that for every a this last case holds. Then for all a sufficiently large we have an $\beta\epsilon/3$-neck N_a in the zero time-slice of $\widetilde{\mathcal{M}}_a$ centered at x_a. It is an evolving neck and there is an embedding $\psi \colon N_a \times [-t_a, 0] \to \widetilde{\mathcal{M}}_a$ compatible with time and the vector field so that the initial time-slice $\psi(N_a \times \{-t_a\})$ is in the surgery time-slice M_{-t_a} and is disjoint from the surgery cap, so in fact it is contained in the continuing region at time $-t_a$. As we saw above, the image of the central 2-sphere $\psi(S_a^2 \times \{-t_a\})$ lies at distance at most $A''\overline{h}_a$ from the tip of the surgery cap p_a (where, recall, A'' is a constant independent of a). The 2-sphere, Σ_a, along which we do surgery, creating the surgery cap with p_a as its tip, is the central 2-sphere of a strong $\delta_{b(a)}$-neck. As a tends to infinity the surgery control parameter $\delta_{b(a)}$ tends to zero. Thus, for a sufficiently large this strong $\delta_{b(a)}$-neck will contain a strong $\beta\epsilon/2$- neck N' centered at $\psi(x_a, -t_a)$. Since we know that the continuing region at time $-t_a$ contains a $\beta\epsilon/3$-neck centered at $(x_a, -t_a)$, it follows that N' is also contained in C_{-t_a}. That is to say, N' is contained in the negative half of the $\delta_{b(a)}$-neck centered at Σ_a. Now we are in the situation of Proposition 15.2. Applying this result tells us that x_a is the center of a strong ϵ-neck.

This completes the proof that for all a sufficiently large, x_a has a (C, ϵ)-canonical neighborhood in contradiction to our assumption. This contradiction completes the proof of Lemma 17.7. □

There are several steps in completing the proof of Proposition 17.1. The first step helps us apply the previous claim to find strong (C, ϵ)-canonical neighborhoods.

CLAIM 17.9. *Given any $A < \infty$ there is $D(A) < \infty$ and $\delta(A) > 0$ such that for all a sufficiently large, $|\mathrm{Rm}|$ is bounded by $D(A)$ along all backward flow lines beginning at a point of $B_{\widetilde{G}_a}(\widetilde{x}_a, 0, A)$ and defined for backward time at most $\delta(A)$.*

PROOF. Since all points $y \in (\mathcal{M}_a, G_a)$ with $R_{G_a}(y) \geq r_a^{-2}$ and $\mathbf{t}(y) < \mathbf{t}(x_a)$ have strong (C, ϵ)-canonical neighborhoods, and since $R(x_a) = r_a^{-2}$, we see that all points $y \in (\widetilde{\mathcal{M}}_a, \widetilde{G}_a)$ with $\mathbf{t}(y_a) < 0$ and with $R_{\widetilde{G}_a}(y_a) \geq 1$ have strong (C, ϵ)-canonical neighborhoods. It follows that all points in $(\widetilde{\mathcal{M}}_a, \widetilde{G}_a)$ with $\mathbf{t}(y) \leq 0$ and $R(y) > 1$ have strong $(2C, 2\epsilon)$-canonical neighborhoods. Also, since $\delta_a \leq \delta(r_a)$, where $\delta(r_a)$ is the constant given in Proposition 16.1, and since the condition of being κ-non-collapsed is a closed constraint, it follows from Proposition 16.1 that these Ricci flows with surgery are κ-non-collapsed for a fixed $\kappa > 0$. It is now immediate from Theorem 10.2 that there is a constant $D_0(A)$ such that R is bounded above on $B_{\widetilde{G}_a}(\widetilde{x}_a, 0, A)$ by $D_0(A)$. Since every point $y \in (\mathcal{M}_a, G_a)$ with $R(y) > 1$ with scalar curvature at least 1 has a (C, ϵ)-canonical neighborhood, it follows from the definition that for every such point y we have $|\partial R(y)/\partial t| < CR(y)^2$. Arguing as in Lemma 11.2 we see that there are constants $\delta(A) > 0$ and $D'(A) < \infty$, both depending only on $D_0(A)$, such that the scalar curvature at all points of backward flow lines beginning in $B_{\widetilde{G}_a}(\widetilde{x}_a, 0, A)$ and defined for backward time at most $\delta(A)$ is bounded by $D'(A)$. Since the curvature is pinched toward positive, it follows that there is a $D(A) < \infty$, depending only on $D'(A)$ such that $|\mathrm{Rm}| \leq D(A)$ on the same flow lines. □

CLAIM 17.10. *After passing to a subsequence (in a), either:*
(1) *for each $A < \infty$ there are $D(A) < \infty$ and $t(A) > 0$ such that for all a sufficiently large $P_{\widetilde{G}_a}(\widetilde{x}_a, 0, A, -t(A))$ exists in $\widetilde{\mathcal{M}}_a$ and $|\mathrm{Rm}|$ is bounded by $D(A)$ on this backward parabolic neighborhood, or*
(2) *each x_a has a strong (C, ϵ)-canonical neighborhood.*

PROOF. First notice that if there is $t(A) > 0$ for which the backwards parabolic neighborhood $P = P_{\widetilde{G}_a}(\widetilde{x}_a, 0, A, -t(A))$ exists, then, by Claim 17.9, there are constants $D(A) < \infty$ and $\delta(A) > 0$ such that, replacing $t(A)$ by $\min(t(A), \delta(A))$, $|\mathrm{Rm}|$ is bounded by $D(A)$ on P. Thus, either item (1) holds or passing to a subsequence, we can suppose that there is some $A < \infty$ for which no $t(A) > 0$ as required by item (1) exists. Then, for each a we find a point $y_a \in B_{\widetilde{G}_a}(\widetilde{x}_a, 0, A)$ such that the backwards flow line from y_a meets a surgery cap at a time $-t_a$ where $\lim_{a \to \infty}(t_a) = 0$. Then, by the previous claim, for all a sufficiently large, the sectional curvature along any backward flow line beginning in $B_{\widetilde{G}_a}(\widetilde{x}_a, 0, A)$ and defined for backward time t_a is bounded by a constant $D(A)$ independent of a. Under our assumption this means that for all a sufficiently large, there is a point $y_a \in B_{\widetilde{G}_a}(\widetilde{x}_a, 0, A)$ and a backwards flow line starting at y_a ending at a point z_a of a surgery cap, and the sectional curvature along this entire flow line is bounded by $D(A) < \infty$. Thus, applying Lemma 17.7 produces the strong (C, ϵ)-canonical neighborhood around x_a, proving the claim. □

But we are assuming that no x_a has a strong (C, ϵ)-canonical neighborhood. Thus, the consequence of the previous claim is that for each $A < \infty$ there is a $t(A) > 0$ such that for all a sufficiently large $P_{\widetilde{G}_a}(\widetilde{x}_a, 0, A, -t(A))$ exists in $\widetilde{\mathcal{M}}_a$ and there is a bound, depending only on A for $|\mathrm{Rm}|$ on this backward parabolic neighborhood. Applying Theorem 5.11 we see that, after passing to a subsequence, there is a smooth limit $(M_\infty, g_\infty, x_\infty)$ to the zero time-slices $(\widetilde{\mathcal{M}}_a, \widetilde{G}_a, \widetilde{x}_a)$. Clearly, since the curvatures of the sequence are pinched toward positive, this limit has non-negative curvature.

Lastly, we show that (M_∞, g_∞) has bounded curvature. By part 3 of Proposition 9.79 each point of (M_∞, g_∞) with scalar curvature greater than one has a $(2C, 2\epsilon)$-canonical neighborhood. If a point lies in a 2ϵ-component or in a $2C$-component, then M_∞ is compact, and hence clearly has bounded curvature. Thus, we can assume that each $y \in M_\infty$ with $R(y) > 1$ is either the center of a 2ϵ-neck or is contained in the core of a $(2C, 2\epsilon)$-cap. According to Proposition 2.19 (M_∞, g_∞) does not contain 2ϵ-necks of arbitrarily high curvature. It now follows that there is a bound on the scalar curvature of any 2ϵ-neck and of any $(2C, 2\epsilon)$-cap in (M_∞, g_∞), and hence it follows that (M_∞, g_∞) has bounded curvature.

CLAIM 17.11. *If the constant $t(A) > 0$ cannot be chosen independent of A, then after passing to a subsequence, the x_a have strong (C, ϵ)-canonical neighborhoods.*

PROOF. Let Q be the bound on the scalar curvature of $(M_\infty, g_\infty, x_\infty)$. Then by Lemma 11.2 there is a constant $\Delta t > 0$ such that if $R_{\widetilde{G}_a}(y, 0) \leq 2Q$, then the scalar curvature is bounded by $16Q$ on the backward flow line from y defined for any time $\leq \Delta t$. Suppose that there is $A < \infty$ and a subsequence of a for which there is a flow line beginning at a point $y_a \in B_{\widetilde{G}_a}(\widetilde{x}_a, 0, A)$ defined for backward time at most Δt and ending at a point z_a of a surgery cap. Of course, the fact that the scalar curvature of (M_∞, g_∞) is at most Q implies that for all a sufficiently large, the scalar curvature of $B_{G_a}(\widetilde{x}_a, 0, A)$ is less than $2Q$. This implies that for all a sufficiently large the scalar curvature along the flow line from y_a to z_a in a surgery cap is $\leq 16Q$. Now invoking Lemma 17.7 we see that for all a sufficiently large the point $\widetilde{x}_a$ has a strong (C, ϵ)-canonical neighborhood. This is a contradiction, and this contradiction proves that we can choose $t(A) > 0$ independent of A. □

Since we are assuming that no x_a has a strong (C, ϵ)-canonical neighborhood, this means that it is possible to find a constant $t' > 0$ such that $t(A) \geq t'$ for all $A < \infty$. Now let $0 < T' \leq \infty$ be the maximum possible value for such t'. Then for every A and every $T < T'$ the parabolic neighborhood $P_{\widetilde{G}_a}(\widetilde{x}_a, 0, A, T)$ exists for all a sufficiently large. According to Theorem 11.8, after passing to a subsequence, there is a limiting flow $(M_\infty, g_\infty(t), x_\infty)$, $-T' < t \leq 0$, and this limiting flow has bounded, nonnegative curvature. If $T = \infty$, this limit is a κ-solution, and hence the x_a have strong (C, ϵ)-canonical neighborhoods for all a sufficiently large, which is a contradiction.

Thus, we can assume that $T' < \infty$. Let Q be the bound for the scalar curvature of this flow. Since T' is maximal, for every $t > T'$, after passing to a subsequence, for all a sufficiently large there is $A < \infty$ and a backwards flow line, defined for a time less than t, starting at a point y_a of $B_{\widetilde{G}_a}(\widetilde{x}_a, 0, A)$ and ending at a point z_a of a surgery cap. Invoking Lemma 11.2 again, we see that for all a sufficiently large, the scalar curvature is bounded on the flow line from y_a to z_a by a constant independent of a. Hence, as before, we see that for all a sufficiently large x_a has a strong (C, ϵ)-canonical neighborhood; again this is a contradiction.

Hence, we have now shown that our assumption that the strong (C, ϵ)-canonical neighborhood assumption fails for all r_a and all $\delta_{a,b}$ leads to a contradiction and hence is false.

This completes the proof of Proposition 17.1. □

2. Surgery times don't accumulate

Now we turn to the proof of Theorem 15.9. Given surgery parameter sequences

$$\Delta_i = \{\delta_0, \ldots, \delta_i\},$$
$$\mathbf{r_i} = \{r_0, \ldots, r_i\},$$
$$\mathbf{K_i} = \{\kappa_0, \ldots, \kappa_i\},$$

we let r_{i+1} and δ_{i+1} be as in Proposition 17.1 and then set $\kappa_{i+1} = \kappa(r_{i+1})$ as in Proposition 16.1. Set

$$\mathbf{r_{i+1}} = \{\mathbf{r_i}, r_{i+1}\},$$
$$\mathbf{K_{i+1}} = \{\mathbf{K_i}, \kappa_{i+1}\},$$
$$\Delta_{i+1} = \{\delta_0, \ldots, \delta_{i-1}, \delta_{i+1}, \delta_{i+1}\}.$$

Of course, these are also surgery parameter sequences.

Let $\bar{\delta}\colon [0,T] \to \mathbb{R}^+$ be any non-increasing positive function and let $(\mathcal{M}, G)$ be a Ricci flow with surgery defined on $[0,T)$ for some $T \in [T_i, T_{i+1})$ with surgery control parameter $\bar{\delta}$. Suppose $\bar{\delta} \leq \Delta_{i+1}$ and that this Ricci flow with surgery satisfies the conclusion of Theorem 15.9 with respect to these sequences on its entire interval of definition. We wish to extend this Ricci flow with surgery to one defined on $[0, T')$ for some T' with $T < T' \leq T_{i+1}$ in such a way that $\bar{\delta}$ is the surgery control parameter and the extended Ricci flow with surgery continues to satisfy the conclusions of Theorem 15.9 on its entire interval of definition.

We may as well assume that the Ricci flow $(\mathcal{M}, G)$ becomes singular at time T. Otherwise we would simply extend by Ricci flow to a later time T'. By Proposition 16.1 and Proposition 17.1 this extension will continue to satisfy the conclusions of Theorem 15.9 on its entire interval of definition. If $T \geq T_{i+1}$, then we have extended the Ricci flow with surgery to time T_{i+1} as required and hence completed the inductive step. Thus, we may as well assume that $T < T_{i+1}$.

Consider the maximal extension of $(\mathcal{M}, G)$ to time T. Let T^- be the previous surgery time, if there is one, and otherwise be zero. If the T time-slice, $\Omega(T)$, of this maximal extension is all of M_{T^-}, then the curvature remains bounded as t approaches T from below. According to Proposition 4.12 this means that T is not a surgery time and we can extend the Ricci flow on $(M_{T^-}, g(t))$, $T^- \leq t < T$, to a Ricci flow on $(M_{T^-}, g(t))$, $T^- \leq t < T'$ for the maximal time interval (i.e. so that the flow becomes singular at time T' or $T' = \infty$). But we are assuming that the flow goes singular at T. That is to say, $\Omega(T) \neq M_{T^-}$. Then we can do surgery at time T using $\bar{\delta}(T)$ as the surgery control parameter, setting $\rho(T) = r_{i+1}\delta(T)$. Let $(M_T, G(T))$ be the result of surgery. If $\Omega_{\rho(T)}(T) = \emptyset$, then the surgery process at time T removes all of $M_{T'}$. In this case, the Ricci flow is understood to exist for all time and to be empty for $t \geq T$. In this case we have completed the extension to T_{i+1}, and in fact all the way to $T = \infty$, and hence completed the inductive step in the proof of the proposition.

We may as well assume that $\Omega_{\rho(T)}(T) \neq \emptyset$ so that the result of surgery is a non-empty manifold M_T. Then we use this compact Riemannian 3-manifold as the initial conditions of a Ricci flow beginning at time T. According to Lemma 15.11 the union along $\Omega(T)$ at time T of this Ricci flow with $(\mathcal{M}, G)$ is a Ricci flow with surgery satisfying Assumptions (1) – (7) and whose curvature is pinched toward positive.

Since the surgery control parameter $\overline{\delta}(t)$ is at most $\delta(r_{i+1})$, the constant from Proposition 16.1, for all $t \in [T_{i-1}, T]$, since $T \leq T_{i+1}$, and since the restriction of $(\mathcal{M}, G)$ to $\mathbf{t}^{-1}([0, T_i))$ satisfies Proposition 16.1, we see by Proposition 17.1 that the extended Ricci flow with surgery satisfies the conclusion of Theorem 15.9 on its entire time interval of definition.

Either we can repeatedly apply this process, passing from one surgery time to the next and eventually reach $T \geq T_{i+1}$, which would prove the inductive step, or there is an unbounded number of surgeries in the time interval $[T_i, T_{i+1})$. We must rule out the latter case.

LEMMA 17.12. *Given a Ricci flow with surgery $(\mathcal{M}, G)$ defined on $[0, T)$ with $T \leq T_{i+1}$ with surgery control parameter $\overline{\delta}$ a non-increasing positive function defined on $[0, T_{i+1}]$ satisfying the hypotheses of Theorem 15.9 on its entire time-domain of definition, there is a constant N depending only on the volume of $(M_0, g(0))$, on T_{i+1}, on r_{i+1}, and on $\overline{\delta}(T_{i+1})$ such that this Ricci flow with surgery defined on the interval $[0, T)$ has at most N surgery times.*

PROOF. Let $(M_t, g(t))$ be the t time-slice of $(\mathcal{M}, G)$. If t_0 is not a surgery time, then $\mathrm{Vol}(t) = \mathrm{Vol}(M_t, g(t))$ is a smooth function of t near t_0 and

$$\frac{d\mathrm{Vol}}{dt}(t_0) = -\int_{M_{t_0}} R d\mathrm{vol},$$

so that, because of the curvature pinching toward positive hypothesis, we have $\frac{d\mathrm{Vol}}{dt}(t_0) \leq 6\mathrm{Vol}(t_0)$. If t_0 is a surgery time, then either M_{t_0} has fewer connected components than $M_{t_0^-}$ or we do a surgery in an ϵ-horn of $M_{t_0^-}$. In the latter case we remove the end of the ϵ-horn, which contains the positive half of a $\overline{\delta}(t_0)$-neck of scale $h(t_0)$. We then sew in a ball with volume at most $(1+\epsilon)Kh^3(t_0)$, where $K < \infty$ is the universal constant given in Lemma 12.3. Since $h(t_0) \leq \overline{\delta}^2(t_0)r(t_0) \leq \delta_0^2 r(t_0)$ and since we have chosen $\overline{\delta}(t_0) \leq \delta_0 < K^{-1}$, it follows that this operation lowers volume by at least $\delta^{-1}h^2(t_0)/2$. Since $\overline{\delta}(t_0) \geq \overline{\delta}(T_{i+1}) > 0$ and the canonical neighborhood parameter r at time t_0 is at least $r_{i+1} > 0$, it follows that $h(t_0) \geq h(T_{i+1}) > 0$. Thus, each surgery at time $t_0 \leq T_{i+1}$ along a 2-sphere removes at least a fixed amount of volume depending on $\overline{\delta}(T_{i+1})$ and r_{i+1}. Since under Ricci flow the volume grows at most exponentially, we see that there is a bound depending only on $\overline{\delta}(T_{i+1})$, T_{i+1}, r_{i+1} and $\mathrm{Vol}(M_0, g(0))$ on the number of 2-sphere surgeries that we can do in this time interval. On the other hand, the number of components at any time t is at most $N_0 + S(t) - D(t)$ where N_0 is the number of connected components of M_0, $S(t)$ is the number of 2-sphere surgeries performed in the time interval $[0, t)$ and $D(t)$ is the number of connected components removed by surgeries at times in the interval $[0, t)$. Hence, there is a bound on the number of components in terms of N_0 and

$S(T)$ that can be removed by surgery in the interval $[0, T)$. Since the initial conditions are normalized, N_0 is bounded by the volume of $(M_0, g(0))$. This completes the proof of the result. □

This lemma completes the proof of the fact that for any $T \leq T_{i+1}$, we encounter only a fixed bounded number of surgeries in the Ricci flow with surgery from 0 to T. The bound depends on the volume of the initial manifold as well as the surgery constants up to time T_{i+1}. In particular, for a given initial metric $(M_0, g(0))$ there is a uniform bound, depending only on the surgery constants up to time T_{i+1}, on the number of surgeries in any Ricci flow with surgery defined on a subinterval of $[0, T_{i+1})$. It follows that the surgery times cannot accumulate in any finite interval. This completes the proof of Theorem 15.9.

To sum up, we have sequences Δ, $\mathbf{K}$ and $\mathbf{r}$ as given in Theorem 15.9. Let $\overline{\delta} \colon [0, \infty) \to \mathbb{R}$ be a positive, non-increasing function with $\overline{\delta} \leq \Delta$. Let M be a compact 3-manifold that contains no embedded $\mathbb{R}P^2$ with trivial normal bundle. We have proved that for any normalized initial Riemannian metric (M_0, g_0) there is a Ricci flow with surgery with time-interval of definition $[0, \infty)$ and with (M_0, g_0) as initial conditions. This Ricci flow with surgery is $\mathbf{K}$-non-collapsed and satisfies the strong (C, ϵ)-canonical neighborhood theorem with respect to the parameter $\mathbf{r}$. It also has curvature pinched toward positive. Lastly, for any $T \in [0, \infty)$ if there is a surgery at time T then this surgery is performed using the surgery parameters $\overline{\delta}(T)$ and $r(T)$, where if $T \in [T_i, T_{i+1})$ then $r(T) = r_{i+1}$. In this Ricci flow with surgery, there are only finitely many surgeries on each finite time interval. As far as we know there may be infinitely many surgeries in all.

Part 4

Completion of the proof of the Poincaré Conjecture

CHAPTER 18

Finite-time extinction

Our purpose in this chapter and the next is to prove the following finite-time extinction theorem for certain Ricci flows with surgery which, as we shall show below, when combined with the theorem on the existence of Ricci flows with surgery defined for all $t \in [0, \infty)$ (Theorem 15.9), immediately yields Theorem 0.1, thus completing the proof of the Poincaré Conjecture and the 3-dimensional space-form conjecture.

1. The result

THEOREM 18.1. *Let $(M, g(0))$ be a compact, connected normalized Riemannian 3-manifold. Suppose that the fundamental group of M is a free product of finite groups and infinite cyclic groups. Then M contains no $\mathbb{R}P^2$ with trivial normal bundle. Let $(\mathcal{M}, G)$ be the Ricci flow with surgery defined for all $t \in [0, \infty)$ with $(M, g(0))$ as initial conditions given by Theorem 15.9. This Ricci flow with surgery becomes extinct after a finite time in the sense that the time-slices M_T of $\mathcal{M}$ are empty for all T sufficiently large.*

Let us quickly show how this theorem implies our main result, Theorem 0.1.

PROOF. (of Theorem 0.1 assuming Theorem 18.1). Fix a normalized metric $g(0)$ on M, and let $(\mathcal{M}, G)$ be the Ricci flow with surgery defined for all $t \in [0, \infty)$ produced by Theorem 15.9 with initial conditions $(M, g(0))$. According to Theorem 18.1 there is $T > 0$ for which the time-slice M_T is empty. By Corollary 15.4, if there is T for which M_T is empty, then for any $T' < T$ the manifold $M_{T'}$ is a disjoint union of connected sums of 3-dimensional spherical space forms and 2-sphere bundles over S^1. Thus, the manifold $M = M_0$ is a connected sum of 3-dimensional space-forms and 2-sphere bundles over S^1. This proves Theorem 0.1. In particular, if M is simply connected, then M is diffeomorphic to S^3, which is the statement of the Poincaré Conjecture. Similarly, if $\pi_1(M)$ is finite then M is diffeomorphic to a connected sum of a 3-dimensional spherical space-form and 3-spheres, and hence M is diffeomorphic to a 3-dimensional spherical space-form. □

The rest of this chapter and the next is devoted to the proof of Theorem 18.1, which will then complete the proof of Theorem 0.1.

1.1. History of this approach. The basic idea for proving finite-time extinction is to use a min-max function based on the area (or the closely related energy) of 2-spheres or 2-disks in the manifold. The critical points of the energy functional are harmonic maps and they play a central role in the proof. For a basic reference on harmonic maps see [**59**], [**61**], and [**42**]. Let us sketch the argument. For a compact Riemannian manifold (M, g) every non-zero element $\beta \in \pi_2(M)$ has associated with it an area, denoted $W_2(\beta, g)$, which is the infimum over all maps $S^2 \to M$ in the free homotopy class of β of the energy of the map. We find it convenient to set $W_2(g)$ equal to the minimum over all non-zero homotopy classes β of $W_2(\beta, g)$. In the case of a Ricci flow $g(t)$ there is an estimate (from above) for the forward difference quotient of $W_2(g(t))$ with respect to t. This estimate shows that after a finite time $W_2(g(t))$ must go negative. This is absurd since $W_2(g(t))$ is always non-negative. This means that the Ricci flow cannot exist for all forward time. In fact, using the distance-decreasing property for surgery in Proposition 15.12 we see that, even in a Ricci flow with surgery, the same forward difference quotient estimate holds for as long as π_2 continues to be non-trivial, i.e., is not killed by the surgery. The forward difference quotient estimate means that eventually all of π_2 is killed in a Ricci flow with surgery and we arrive at a time T for which every component of the T time-slice, M_T, has trivial π_2. This result holds for all Ricci flows with surgery satisfying the conclusion of Theorem 15.9.

Now we fix T_0 so that every component of M_{T_0} has trivial π_2. It follows easily from the description of surgery that the same statement holds for all $T \geq T_0$. We wish to show that, under the group-theoretic hypothesis of Theorem 18.1, at some later time $T' > T_0$ the time-slice $M_{T'}$ is empty. The argument here is similar in spirit. There are two approaches. The first approach is due to Perelman [**54**]. Here, one represents a non-trivial element in $\pi_3(M_{T_0}, x_0)$ by a non-trivial element in $\pi_2(\Lambda M, *)$, where ΛM is the free loop space on M and $*$ is the trivial loop at x_0. For any compact family Γ of homotopically trivial loops in M we consider the areas of minimal spanning disks for each of the loops in the family and set $W(\Gamma)$ equal to the maximal area of these minimal spanning disks. For a given element in $\gamma \in \pi_2(\Lambda M)$ we set $W(\gamma)$ equal to the infimum over all representative 2-sphere families Γ for γ of $W(\Gamma)$. Under Ricci flow, the forward difference quotient of this invariant satisfies an inequality and the distance-decreasing property of surgery (Proposition 15.12) says that the inequality remains valid for Ricci flow with surgery. The inequality implies that the value $W(\gamma)$ goes negative in finite time, which is impossible.

The other approach, by Colding-Minicozzi [**15**], is to represent a non-trivial element in $\pi_3(M_T)$ as a non-trivial element in $\pi_1(\text{Maps}(S^2, M))$, and associate to such an element the infimum over all representative families of the maximal energy of the 2-spheres in the family. Again, one shows that under Ricci flow the forward difference quotient of this minimax satisfies an inequality that implies that it goes negative in finite time. As before, the distance-decreasing property of surgery (Proposition 15.12) implies that this inequality is valid for Ricci flows with surgery. This tells us that the manifold must completely disappear in finite time.

Our first reaction was that, of the two approaches, the one considered by Colding-Minicozzi was preferable since it seemed more natural and it had the advantage of avoiding the boundary issues that occupy most of Perelman's analysis in [**54**]. In the Colding-Minicozzi approach one must construct paths of 2-spheres with the property that when the energy of the 2-sphere is close to the maximum value along the path, then the 2-sphere in question represents a point in the space $\text{Maps}(S^2, M)$ that is close to a (usually) non-minimal critical point for the energy functional on this space. Such paths are needed in order to establish the forward difference quotient result alluded to above. In Perelman's approach, one deals only with area-minimizing disks so that one avoids having to deal with non-minimal critical points at the expense of dealing with the technical issues related to the boundary. Since the latter are one-dimensional in nature, they are much easier to handle. In the end we decided to follow Perelman's approach, and that is the one we present here. In [**54**] there were two points that we felt required quite a bit of argument beyond what Perelman presented. In §2.2 on page 4 of [**54**], Perelman asserts that there is a local, pointwise curvature estimate that can be obtained by adapting arguments in the literature; see Lemmas 19.24 and 19.58 for the precise statement. To implement this adaption required further non-trivial arguments. We present these arguments in Section 8 of Chapter 19. In §2.5 on page 5 of [**54**] Perelman asserts that an elementary argument establishes a lower bound on the length of a boundary curve of a minimal annulus; see Proposition 19.35 for a precise statement. While the statement seems intuitively clear, we found the argument, while elementary, was quite intricate. We present this argument in Section 7 of Chapter 19.

The first use of these types of ideas to show that geometric objects must disappear in finite time under Ricci flow is due to Hamilton [**36**]. He was considering a situation where a time-slice $(M, g(t_0))$ of a 3-dimensional Ricci flow had submanifolds on which the metric was close to (a truncated version) of a hyperbolic metric of finite volume. He wished to show that eventually the boundary tori of the truncation were incompressible in the 3-manifold. If not, then there would be an immersed minimal disk in M whose boundary was a non-trivial loop on the torus. He represented this relative homotopy class by a minimal energy disk in $(M, g(t_0))$ and proved the same sort of forward difference quotient estimate for the area of the minimal disk in the relative homotopy class. The same contradiction – the forward difference quotient implies that after a finite time the area would go negative if the disk continued to exist — implies that after a finite amount of time this compressing disk must disappear. Using this he showed that for sufficiently large time all the boundary tori of almost hyperbolic submanifolds in $(M, g(t))$ were incompressible.

In the next section we deal with π_2 and, using W_2, we show that given a Ricci flow with surgery as in Theorem 15.9 there is $T_1 < \infty$ such that for all $T \geq T_1$ every connected component of M_T has trivial π_2. Then in the section after that, by analyzing 3-dimensional analogue W_3, we show that, under the group-theoretic hypothesis of Theorem 18.1, there is a $T_2 < \infty$ such that $M_T = \emptyset$ for all $T \geq T_2$. In both these arguments we need the same type of results – a forward difference inequality for the energy function; the statement that away from surgery times this function is continuous; and lastly, the statement that the value of the energy function at a surgery time is at most the liminf of its values at a sequence of times approaching the surgery time from below.

1.2. Existence of the Ricci flow with surgery. Let $(M, g(0))$ be as in the statement of Theorem 18.1, so that M is a compact, connected 3-manifold whose fundamental group is a free product of finite groups and infinite cyclic groups. By scaling $g(0)$ by a sufficiently large constant, we can assume that $g(0)$ is normalized. Let us show that such a manifold cannot contain an embedded $\mathbb{R}P^2$ with trivial normal bundle. First note that since $\mathbb{R}P^2$ has Euler characteristic 1, it is not the boundary of a compact 3-manifold. Hence, an $\mathbb{R}P^2$ embedded with trivial normal bundle does not separate the connected component of M containing it. Also, any non-trivial loop in $\mathbb{R}P^2$ has non-trivial normal bundle in M so that inclusion of $\mathbb{R}P^2$ into M induces an injection on fundamental groups. Under the fundamental group hypotheses, M decomposes as a connected sum of 3-manifolds with finite fundamental groups and 2-sphere bundles over S^1, see [**39**]. Given an $\mathbb{R}P^2$ with trivial normal bundle embedded in a connected sum, it can be replaced by one contained in one of the connected factors. [Proof: Let $\Sigma = \Sigma_1 \cup \cdots \cup \Sigma_n$ be the spheres giving the connected sum decomposition of M. Deform the $\mathbb{R}P^2$ until it is transverse to Σ and let γ be a circle of intersection of $\mathbb{R}P^2$ with one of the Σ_i that is innermost on Σ_i in the sense that γ bounds a disk D in Σ_i disjoint from all other components of intersection of Σ_i and $\mathbb{R}P^2$. The loop γ also bounds a disk D' in $\mathbb{R}P^2$. Replace D' by D and push D slightly off to the correct side of Σ_i. This will produce a new embedded $\mathbb{R}P^2$ with trivial normal bundle in M and at least one fewer component of intersection with Σ. Continue inductively until all components of intersection with Σ are removed.]

Now suppose that we have an $\mathbb{R}P^2$ with trivial normal bundle embedded disjointly from Σ, and hence embedded in one of the prime factors of M. Since it does not separate this factor, by the Mayer-Vietoris sequence (see

p. 149 of [**38**]) the first homology of the factor in question maps onto $\mathbb{Z}$ and hence the factor in question has infinite fundamental group. But this group also contains the cyclic subgroup of order 2, namely the image of $\pi_1(\mathbb{R}P^2)$ under the map induced by the inclusion. Thus, the fundamental group of this prime factor is not finite and is not infinite cyclic. This is a contradiction. (We have chosen to give a topological argument for this result. There is also an argument using the theory of groups acting on trees which is more elementary in the sense that it uses no 3-manifold topology. Since it is a more complicated, and to us, a less illuminating argument, we decided to present the topological argument.)

Thus, by Theorem 15.9, for any compact 3-manifold M whose fundamental group is a free product of finite groups and infinite cyclic groups and for any normalized metric $g(0)$ on M there is a Ricci flow with surgery $(\mathcal{M}, G)$ defined for all time $t \in [0, \infty)$ satisfying the conclusion of Theorem 15.9 with $(M, g(0))$ as the initial conditions.

DEFINITION 18.2. Let I be an interval (which is allowed to be open or closed at each end and finite or infinite at each end). By a *path of components* of a Ricci flow with surgery $(\mathcal{M}, G)$ defined for all $t \in I$ we mean a connected, open subset $\mathcal{X} \subset \mathbf{t}^{-1}(I)$ with the property that for every $t \in I$ the intersection $\mathcal{X}(t)$ of $\mathcal{X}$ with each time-slice M_t is a connected component of M_t.

Let $\mathcal{X}$ be a path of components in a Ricci flow with surgery $(\mathcal{M}, G)$, a path defined for all $t \in I$. Let I' be a subinterval of I with the property that no point of I' except possibly its initial point is a surgery time. Then the intersection of $\mathcal{X}$ with $\mathbf{t}^{-1}(I')$ is the Ricci flow on the time interval I' applied to $\mathcal{X}(t)$ for any $t \in I'$. Thus, for such intervals I' the intersection, $\mathcal{X}(I')$, of $\mathcal{X}$ with $\mathbf{t}^{-1}(I')$ is determined by the time-slice $\mathcal{X}(t)$ for any $t \in I'$. That is no longer necessarily the case if some point of I' besides its initial point is a surgery time. Let $t \in I$ be a surgery time, distinct from the initial point of I (if there is one), and let $I' \subset I$ be an interval of the form $[t', t)$ for some $t' < t$ sufficiently close to t so that there are no surgery times in $[t', t)$. Then, as we have just seen, $\mathcal{X}(I')$ is a Ricci flow on the connected manifold $\mathcal{X}(t')$. There are several possible outcomes of the result of surgery at time t on this manifold. One possibility is that the surgery leaves this connected component unchanged (affecting only other connected components). In this case, there is no choice for $\mathcal{X}(t)$: it is the continuation to time t of the Ricci flow on $\mathcal{X}(t')$. Another possibility is that $\mathcal{X}(t')$ is completely removed by the surgery at time t. In this case the manifold $\mathcal{X}$ cannot be continued to time t, contradicting the fact that the path of components $\mathcal{X}$ exists for all $t \in I$. The last possibility is that at time t surgery is done on $\mathcal{X}(t')$ using one or more 2-spheres contained in $\mathcal{X}(t')$. In this case the result of surgery on $\mathcal{X}(t')$ results in one or several connected components and $\mathcal{X}(t)$ can be any one of these.

2. Disappearance of components with non-trivial π_2

Let $(\mathcal{M}, G)$ be a Ricci flow with surgery satisfying the conclusions of Theorem 15.9. We make no assumptions about the fundamental group of the initial manifold M_0. In this section we shall show that at some finite time T_1 every connected component of M_{T_1} has trivial π_2 and that this condition persists for all times $T \geq T_1$. There are two steps in this argument. First, we show that there is a finite time T_0 such that after time T_0 every 2-sphere surgery is performed along a homotopically trivial 2-sphere. (Using Kneser's theorem on finiteness of topologically non-trivial families of 2-spheres, one can actually show by the same argument that after some finite time all 2-sphere surgeries are done along 2-spheres that bound 3-balls. But in fact, Kneser's theorem will follow from what we do here.)

After time T_0 the number of components with non-trivial π_2 is a weakly monotone decreasing function of time. The reason is the following. Consider a path of components $\mathcal{X}$ defined for $t \in [T_0, t']$ with the property that each time-slice $\mathcal{X}(t)$ has non-trivial π_2. Using the fact that after time T_0 all the 2-sphere surgeries are along homotopically trivial 2-spheres, one shows easily that $\mathcal{X}$ is determined by its initial time-slice $\mathcal{X}(T_0)$. Also, it is easy to see that if there is a component of $M_{t'}$ with non-trivial π_2, then it is the final time-slice of some path of components defined for $t \in [T_0, t']$ with every time-slice of this path having non-trivial π_2. This then produces an injection from the set of connected components of M_t with non-trivial π_2 into the set of connected components of M_{T_0} with non-trivial π_2.

The second step in the argument is to fix a path $\mathcal{X}(t)$, $T_0 \leq t \leq t'$, of connected components with non-trivial π_2 and to consider the function $W_2 = W_2^{\mathcal{X}}$ that assigns to each $t \in [T_0, t']$ the minimal area of a homotopically non-trivial 2-sphere mapping into $\mathcal{X}(t)$. We show that this function is continuous except at the surgery times. Furthermore, we show that if t is a surgery time, then $W_2(t) \leq \liminf_{t' \to t^-} W_2(t)$. Lastly, we show that at any point $t \geq T_0$ we have

$$\frac{dW_2}{dt}(t) \leq -4\pi - \frac{1}{2} R_{\min}(t) W_2(t),$$

in the sense of forward difference quotients. From the bound $R_{\min}(t) \geq -6/(4t+1)$, it follows that there is $T_1(\mathcal{X})$ such that W_2 with these three properties cannot be non-negative for all $t \in [T_0, T_1(\mathcal{X})]$ and hence $t' < T_1$. Since there are only finitely many components with non-trivial π_2 at time T_0 it follows that there is $T_1 < \infty$ such that every component of M_T has trivial π_2 for every $T \geq T_1$.

2. DISAPPEARANCE OF COMPONENTS WITH NON-TRIVIAL π_2

2.1. A group theory lemma. To bound the number of homotopically non-trivial 2-spheres in a compact 3-manifold we need the following group theory lemma.

LEMMA 18.3. *Suppose that G is a finitely generated group, say generated by k elements. Let $G = G_1 * \cdots * G_\ell$ be a free product decomposition of G with non-trivial free factors, i.e., with $G_i \neq \{1\}$ for each $i = 1, \ldots, \ell$. Then $\ell \leq k$.*

PROOF. This is a consequence of Grushko's theorem [**68**], which says that given a map of a finitely generated free group F onto the free product G, one can decompose the free group as a free product of free groups $F = F_1 * \cdots * F_\ell$ with F_i mapping onto G_i. □

2.2. Homotopically non-trivial families of 2-spheres.

DEFINITION 18.4. Let X be a compact 3-manifold (possibly disconnected). An embedded 2-sphere in X is said to be *homotopically essential* if the inclusion of the 2-sphere into X is not homotopic to a point map of the 2-sphere to X. More generally, let $F = \{\Sigma_1, \ldots, \Sigma_n\}$ be a family of disjointly embedded 2-spheres in X. We say that the family is *homotopically essential* if

   (i) each 2-sphere in the family is homotopically essential, and
   (ii) for any $1 \leq i < j \leq n$, the inclusion of Σ_i into X is not homotopic in X to the inclusion of Σ_j into X.

Notice that if $F = \{\Sigma_1, \ldots, \Sigma_n\}$ is a homotopically essential family of disjointly embedded 2-spheres in X, then any subset F is also homotopically essential.

LEMMA 18.5. *Let X be a compact 3-manifold (possibly disconnected). Then there is a finite upper bound to the number of spheres in any homotopically essential family of disjointly embedded 2-spheres.*

PROOF. Clearly, without loss of generality we can assume that X is connected. If F is a homotopically essential family of 2-spheres in X, then by van Kampen's theorem, see p. 40 of [**38**], there is an induced graph of groups decomposition of $\pi_1(X)$ with all the edge groups being trivial. Since the family is homotopically essential, it follows that the group associated with each vertex of order 1 and each vertex of order 2 is a non-trivial group. The rank of the first homology of the graph underlying the graph of groups, denoted k, is bounded above by the rank of $H_1(X)$. Furthermore, by the theory of graphs of groups there is a free product decomposition of $\pi_1(X)$ with the free factors being the vertex groups and then k infinite cyclic factors. Denote by V_i the number of vertices of order i and by E the number of edges of the graph. The number E is the number of 2-spheres in the family F.

An elementary combinatorial argument shows that
$$2V_1 + V_2 \geq E + 3(1-k).$$
Thus, there is a free product decomposition of $\pi_1(X)$ with at least $E + 3(1-k)$ non-trivial free factors. Since k is bounded by the rank of $H_1(X)$, applying Lemma 18.3 and using the fact that the fundamental group of a compact manifold is finitely presented establishes the result. $\square$

2.3. Two-sphere surgeries are trivial after finite time.

DEFINITION 18.6. Let $(\mathcal{M}, G)$ be a Ricci flow with surgery. We say that a surgery along a 2-sphere $S_0(t)$ at time t in $(\mathcal{M}, G)$ is a *homotopically essential surgery* if, for every $t' < t$ sufficiently close to t, flowing $S_0(t)$ backwards from time t to time t' results in a homotopically essential 2-sphere $S_0(t')$ in $M_{t'}$.

PROPOSITION 18.7. *Let $(\mathcal{M}, G)$ be a Ricci flow with surgery satisfying Assumptions (1) – (7) in Chapter 14. Then there can be only finitely many homotopically essential surgeries along 2-spheres in $(\mathcal{M}, G)$.*

PROOF. Associate to each compact 3-manifold X the invariant $s(X)$ which is the maximal number of spheres in any homotopically essential family of embedded 2-spheres in X. The main step in establishing the corollary is the following:

CLAIM 18.8. *Let $(\mathcal{M}, G)$ be a Ricci flow with surgery and for each t set $s(t) = s(M_t)$. If $t' < t$ then $s(t') \geq s(t)$. If we do surgery at time t along at least one homotopically essential 2-sphere, then $s(t) < s(t')$ for any $t' < t$.*

PROOF. Clearly, for any t_0 we have $s(t) = s(t_0)$ for $t \geq t_0$ sufficiently close to t_0. Also, if t is not a surgery time, then $s(t) = s(t')$ for all $t' < t$ and sufficiently close to t. According to Proposition 15.3, if t is a surgery time then for $t' < t$ but sufficiently close to it, the manifold M_t is obtained from $M_{t'}$ by doing surgery on a finite number of 2-spheres and removing certain components of the result. We divide the operations into three types: (i) surgery along homotopically trivial 2-spheres in $M_{t'}$, (ii) surgery along homotopically non-trivial 2-spheres in $M_{t'}$, (iii) removal of components. Clearly, the first operation does not change the invariant s since it simply creates a manifold that is the disjoint union of a manifold homotopy equivalent to the original manifold with a collection of homotopy 3-spheres. Removal of components will not increase the invariant. The last operation to consider is surgery along a homotopically non-trivial 2-sphere. Let F_t be a homotopically essential family of disjointly embedded 2-spheres in M_t. This family of 2-spheres in M_t can be deformed to miss the 3-disks (the surgery caps) in M_t that we sewed in doing the surgery at time t along a homotopically non-trivial 2-sphere. After deforming the spheres in the family F_t away from the surgery caps, they produce a disjoint family $F'_{t'}$ of 2-spheres in the manifold

$M_{t'}$, for $t' < t$ but t' sufficiently close to t. Each 2-sphere in $F'_{t'}$ is disjoint from the homotopically essential 2-sphere S_0 along which we do surgery at time t. Let $F_{t'}$ be the family $F'_{t'} \cup \{S_0\}$. We claim that $F_{t'}$ is a homotopically essential family in $M_{t'}$.

First, suppose that one of the spheres Σ in $F_{t'}$ is homotopically trivial in $M_{t'}$. Of course, we are in the case when the surgery 2-sphere is homotopically essential, so Σ is not S_0 and hence is the image of one of the 2-spheres in F_t. Since Σ is homotopically trivial, it is the boundary of a homotopy 3-ball B in $M_{t'}$. If B is disjoint from the surgery 2-sphere S_0, then it exists in M_t and hence Σ is homotopically trivial in M_t, which is not possible from the assumption about the family F_t. If B meets the surgery 2-sphere S_0, then since the spheres in the family $F_{t'}$ are disjoint, it follows that B contains the surgery 2-sphere S_0. This is not possible since in this case S_0 would be homotopically trivial in $M_{t'}$, contrary to assumption.

We also claim that no distinct members of $F_{t'}$ are homotopic. For suppose that two of the members Σ and Σ' are homotopic. It cannot be the case that one of Σ or Σ' is the surgery 2-sphere S_0 since, in that case, the other one would be homotopically trivial after surgery, i.e., in M_t. The 2-spheres Σ and Σ' are the boundary components of a submanifold A in $M_{t'}$ homotopy equivalent to $S^2 \times I$. If A is disjoint from the surgery 2-sphere S_0, then A exists in M_t and Σ and Σ' are homotopic in M_t, contrary to assumption. Otherwise, the surgery sphere S_0 must be contained in A. Every 2-sphere in A is either homotopically trivial in A or is homotopic in A to either boundary component. If S_0 is homotopically trivial in A, then it would be homotopically trivial in $M_{t'}$ and this contradicts our assumption. If S_0 is homotopic in A to each of Σ and Σ', then each of Σ and Σ' is homotopically trivial in $M_{t'}$, contrary to assumption. This shows that the family $F_{t'}$ is homotopically essential. It follows immediately that doing surgery on a homotopically non-trivial 2-sphere strictly decreases the invariant s. □

Proposition 18.7 is immediate from this claim and the previous lemma. □

2.4. For all T sufficiently large $\pi_2(M_T) = 0$. We have just established that given any Ricci flow with surgery $(\mathcal{M}, G)$ satisfying the conclusion of Theorem 15.9 there is $T_0 < \infty$, depending on $(\mathcal{M}, G)$, such that all surgeries after time T_0 either are along homotopically trivial 2-spheres or remove entire components of the manifold. Suppose that M_{T_0} has a component $\mathcal{X}(T_0)$ with non-trivial π_2, and suppose that we have a path of components $\mathcal{X}(t)$ defined for $t \in [T_0, T)$ with the property that each time-slice has non-trivial π_2. If T is not a surgery time, then there is a unique extension of $\mathcal{X}$ to a path of components with non-trivial π_2 defined until the first surgery time after T. Suppose that T is a surgery time and let us consider the effect of surgery at time T on $\mathcal{X}(t)$ for $t < T$ but close to it. Since no surgery after time T_0 is done on a homotopically essential 2-sphere there are three possibilities: (i) $\mathcal{X}(t)$ is untouched by the surgery, (ii) surgery is performed on one or more homotopically trivial 2-spheres in $\mathcal{X}(t)$, or (iii) the component $\mathcal{X}(t)$ is completely removed by the surgery. In the second case, the result of the surgery on $\mathcal{X}(t)$ is a disjoint union of components one of which is homotopy equivalent to $\mathcal{X}(t)$, and hence has non-trivial π_2, and all others are homotopy 3-spheres. This implies that there is a unique extension of the path of components preserving the condition that every time-slice has non-trivial π_2, unless the component $\mathcal{X}(t)$ is removed by surgery at time T, in which case there is no extension of the path of components to time T. Thus, there is a unique maximal such path of components starting at $\mathcal{X}(T_0)$ with the property that every time-slice has non-trivial π_2. There are two possibilities for the interval of definition of this maximal path of components with non-trivial π_2. It can be $[T_0, \infty)$ or it is of the form $[T_0, T)$, where the surgery at time T removes the component $\mathcal{X}(t)$ for $t < T$ sufficiently close to it.

PROPOSITION 18.9. *Let $(\mathcal{M}, G)$ be a Ricci flow with surgery satisfying the conclusion of Theorem 15.9. Then there is some time $T_1 < \infty$ such that every component of M_T for any $T \geq T_1$ has trivial π_2. For every $T \geq T_1$, each component of M_T either has finite fundamental group, and hence has a homotopy 3-sphere as universal covering, or has contractible universal covering.*

If M is a connected 3-manifold with $\pi_2(M) = 0$, then the universal covering, $\widetilde{M}$, of M is a 2-connected 3-manifold. The covering $\widetilde{M}$ is compact if and only if $\pi_1(M)$ is finite. In this case $\widetilde{M}$ is a homotopy 3-sphere. If $\widetilde{M}$ is non-compact then $H_3(\widetilde{M}) = 0$, so that all its homology groups and hence, by the Hurewicz theorem, all its homotopy groups vanish. It follows from the Whitehead theorem that $\widetilde{M}$ is contractible in this case. This proves the last assertion in the proposition modulo the first assertion.

The proof of the first assertion of this proposition occupies the rest of this subsection. By the above discussion we see that the proposition holds unless there is a path of components $\mathcal{X}$ defined for all $t \in [T_0, \infty)$ with the property that every time-slice has non-trivial π_2. We must rule out this possibility. To achieve this we introduce the area functional.

LEMMA 18.10. *Let X be a compact Riemannian manifold with $\pi_2(X) \neq 0$. Then there is a positive number $e_0 = e_0(X)$ with the following two properties:*
1. *Any map $f \colon S^2 \to X$ with area less than e_0 is homotopic to a point map.*
2. *There is a minimal 2-sphere $f \colon S^2 \to X$, which is a branched immersion, with the property that the area of $f(S^2) = e_0$ and with the property that f is not homotopic to a point map.*

2. DISAPPEARANCE OF COMPONENTS WITH NON-TRIVIAL π_2

PROOF. The first statement is Theorem 3.3 in [**59**]. As for the second, following Sacks-Uhlenbeck, for any $\alpha > 1$ we consider the perturbed energy E_α given by

$$E_\alpha(s) = \int_{S^2} \left(1 + |ds|^2\right)^\alpha da.$$

According to [**59**] this energy functional is Palais-Smale on the space of $H^{1,2\alpha}$ maps and has an absolute minimum among homotopically non-trivial maps, realized by a map $s_\alpha \colon S^2 \to X$. We consider a decreasing sequence of α tending to 1 and the minimizers s_α among homotopically non-trivial maps. According to [**59**], after passing to a subsequence, there is a weak limit which is a strong limit on the complement of a finite set of points in S^2. This limit extends to a harmonic map of $S^2 \to M$, and its energy is less than or equal to the limit of the α-energies of s_α. If the result is homotopically non-trivial then it realizes a minimum value of the usual energy among all homotopically non-trivial maps, for were there a homotopically non-trivial map of smaller energy, it would have smaller E_α energy than s_α for all α sufficiently close to 1. Of course if the limit is a strong limit, then the map is homotopically non-trivial, and the proof is complete.

We must examine the case when the limit is truly a weak limit. Let s_n be a sequence as above with a weak limit s. If the limit is truly a weak limit, then there is bubbling. Let $x \in S^2$ be a point where the limit s is not a strong limit. Then according to [**59**] pre-composing with a sequence of conformal dilations ρ_n centered at this point leads to a sequence of maps s'_n converging uniformly on compact subsets of $\mathbb{R}^2$ to a non-constant harmonic map s' that extends over the one-point compactification S^2. The energy of this limiting map s' is at most the limit of the α-energies of the s_α. If s' is homotopically non-trivial, then, arguing as before, we see that it realizes the minimum energy among all homotopically non-trivial maps, and once again we have completed the proof. We rule out the possibility that s' is homotopically trivial. Let α be the area, or equivalently the energy, of s'. Let $D \subset \mathbb{R}^2$ be a disk centered at the origin which contains three-quarters of the energy of s' (or equivalently three-quarters of the area of s'), and let D' be the complementary disk to D in S^2. For all n sufficiently large the area of $s'_n|D$ minus the area of $s'_n|D'$ is at least $\alpha/3$. The restrictions of s'_n on ∂D are converging smoothly to $s'|\partial D'$. Let $D_n \subset S^2$ be $\rho_n^{-1}(D)$. Then the area of $s_n|D_n$ equals the area of $s'_n|D$ and hence is at least the area of $s'|D'$ plus $\alpha/4$ for all n sufficiently large. Also, as n tends to infinity the image $s_n(D_n)$ converges smoothly, after reparameterization, to $s'(\partial D)$. Thus, for all n large, we can connect $s_n(\partial D_n)$ to $s'(\partial D')$ by an annulus A_n contained in a small neighborhood of $s'(\partial D')$ and whose area tends to 0 as n goes to infinity. For all n sufficiently large, the resulting 2-sphere Σ_n made out of $s_n|(S^2 \setminus D_n) \cup A_n \cup S'(D')$ is homotopic to $s(S^2)$ since s' is homotopically trivial. Also, for all n sufficiently large, the area of Σ_n is less than the area of s_n minus $\alpha/5$. Reparameterizing this 2-sphere by a conformal map leads to a homotopically non-trivial map of energy less than the area of s_n minus $\alpha/5$. Since as n tends to infinity, the limsup of the areas of the s_n converge to at most c_0, for all n sufficiently large we have constructed a homotopically non-trivial map of energy less than e_0, which contradicts the fact that the minimal α energy for a homotopically non-trivial map tends to e_0 as α tends to 1.

Of course, any minimal energy map of S^2 into M is conformal because there is no non-trivial holomorphic quadratic differential on S^2. It follows that such a map is a branched immersion. □

Now suppose that $\mathcal{X}$ is a path of components defined for all $t \in [T_0, \infty)$ with $\pi_2(\mathcal{X}(t)) \neq 0$ for all $t \in [T_0, \infty)$. For each $t \geq T_0$ we define $W_2(t)$ to be $e_0(\mathcal{X}(t))$, where e_0 is the invariant given in the previous lemma. Our assumption on $\mathcal{X}$ means that $W_2(t)$ is defined and positive for all $t \in [T_0, \infty)$.

LEMMA 18.11.
$$\frac{d}{dt}W_2(t) \leq -4\pi - \frac{1}{2}R_{\min}(t)W_2(t)$$
in the sense of forward difference quotients. If t is not a surgery time, then $W_2(t)$ is continuous at t, and if t is a surgery time, then
$$W_2(t) \leq \liminf_{t' \to t^-} W_2(t').$$

Let us show how this lemma implies Proposition 18.9. Because the curvature is pinched toward positive, we have
$$R_{\min}(t) \geq (-6)/(1+4t).$$
Let $w_2(t)$ be the function satisfying the differential equation
$$\frac{dw_2}{dt} = -4\pi + \frac{3w_2}{1+4t}$$
and $w_2(T_0) = W_2(T_0)$. Then by Lemma 2.22 and Lemma 18.11 we have $W_2(t) \leq w_2(t)$ for all $t \geq T_0$. On the other hand, we can integrate to find
$$w_2(t) = w_2(T_0)\frac{(4t+1)^{3/4}}{(4T_0+1)^{3/4}} + 4\pi(4T_0+1)^{1/4}(4t+1)^{3/4} - 4\pi(4t+1).$$
Thus, for t sufficiently large, $w_2(t) < 0$. This is a contradiction since $W_2(t)$ is always positive, and $W_2(t) \leq w_2(t)$.

This shows that to complete the proof of Proposition 18.9 we need only establish Lemma 18.11.

PROOF. (of Lemma 18.11) Let $f: S^2 \to (X(t_0), g(t_0))$ be a minimal 2-sphere.

CLAIM 18.12.
$$\frac{d\mathrm{Area}_{g(t)}(f(S^2))}{dt}(t_0) \leq -4\pi - \frac{1}{2}R_{\min}(g(t_0))\mathrm{Area}_{g(t_0)}f(S^2).$$

2. DISAPPEARANCE OF COMPONENTS WITH NON-TRIVIAL π_2

PROOF. Recall that, for any immersed surface $f\colon S^2 \to (M, g(t_0))$, we have (see the proof of Theorem 11.1 in [**36**])

$$(18.1) \quad \frac{d}{dt}\mathrm{Area}_{g(t)}(f(S^2))\big|_{t=t_0} = \int_{S^2} \frac{1}{2}\mathrm{Tr}|_{S^2}\left(\frac{\partial g}{\partial t}\right)\bigg|_{t=t_0} da$$

$$= -\int_{S^2}(R - \mathrm{Ric}(\mathbf{n},\mathbf{n}))da$$

where R denotes the scalar curvature of M, Ric is the Ricci curvature of M, and $\mathbf{n}$ is the unit normal vector field of Σ in M. Now suppose that $f(S^2)$ is minimal. We can rewrite this as

$$(18.2) \quad \frac{d}{dt}\mathrm{Area}_{g(t)}(f(S^2))\big|_{t=0} = -\int_{S^2} K_{S^2}da - \frac{1}{2}\int_{S^2}(|A|^2 + R)da,$$

where K_{S^2} is the Gaussian curvature of S^2 and A is the second fundamental form of $f(S^2)$ in M. (Of course, since $f(S^2)$ is minimal, the determinant of its second fundamental form is $-|A|^2/2$.) Even if f is only a branched minimal surface, (18.2) still holds when the integral on the right is replaced by the integral over the immersed part of $f(S^2)$. Then by the Gauss-Bonnet theorem we have

$$(18.3) \quad \frac{d}{dt}\mathrm{Area}_{g(t)}(f(S^2))\big|_{t=t_0} \leq -4\pi - \frac{1}{2}\mathrm{Area}_{g(t_0)}(S^2)\min_{x\in M}\{R_g(x,t_0)\}.$$

$\square$

Since $f(S^2)$ is a homotopically non-trivial sphere in $\mathcal{X}(t)$ for all t sufficiently close to t_0 we see that $W_2(t) \leq \mathrm{Area}_{g(t)}f(S^2)$. Since $\mathrm{Area}_{g(t)}f(S^2))$ is a smooth function of t, the forward difference quotient statement in Lemma 18.11 follows immediately from Claim 18.12.

We turn now to continuity at non-surgery times. Fix $t' \geq T_0$ distinct from all surgery times. We show that the function $e_0(t')$ is continuous at t'. If $f\colon S^2 \to \mathcal{X}(t')$ is the minimal area, homotopically non-trivial sphere, then the area of $f(S^2)$ with respect to a nearby metric $g(t)$ is close to the area of $f(S^2)$ in the metric $g(t')$. Of course, the area of $f(S^2)$ in the metric $g(t)$ is greater than or equal to $W_2(t)$. This proves that $W_2(t)$ is upper semi-continuous at t'. Let us show that it is lower semi-continuous at t'.

CLAIM 18.13. *Let $(M, g(t))$, $t_0 \leq t \leq t_1$, be a Ricci flow on a compact manifold. Suppose that $|\mathrm{Ric}_{g(t)}| \leq D$ for all $t \in [t_0, t_1]$ Let $f\colon S^2 \to (M, g(t_0))$ be a C^1-map. Then*

$$\mathrm{Area}_{g(t_1)}f(S^2) \leq \mathrm{Area}_{g(t_0)}f(S^2)e^{4D(t_1-t_0)}.$$

PROOF. The rate of change of the area of $f(S^2)$ at time t is

$$\int_{f(S^2)} \frac{\partial g}{\partial t}(t)da = -2\int_{f(S^2)} \mathrm{Tr}|_{TS^2}(\mathrm{Ric}_{g(t)})da \leq 4D\mathrm{Area}_{g(t)}f(S^2).$$

Integrating from t_0 to t_1 gives the result. $\square$

Now suppose that we have a family of times t_n converging to a time t' that is not a surgery time. Let $f_n \colon S^2 \to \mathcal{X}(t_n)$ be the minimal area non-homotopically trivial 2-sphere in $\mathcal{X}(t_n)$, so that the area of $f_n(S^2)$ in $\mathcal{X}(t_n)$ is $e_0(t_n)$. Since t' is not a surgery time, for all n sufficiently large we can view the maps f_n as homotopically non-trivial maps of S^2 into $\mathcal{X}(t')$. By the above claim, for any $\delta > 0$ for all n sufficiently large, the area of $f_n(S^2)$ with respect to the metric $g(t')$ is at most the area of $f_n(S^2)$ plus δ. This shows that for any $\delta > 0$ we have $W_2(t') \leq W_2(t_n) + \delta$ for all n sufficiently large, and hence $W_2(t') \leq \liminf_{n \to \infty} W_2(t_n)$. This is the lower semi-continuity.

The last thing to check is the behavior of W_2 near a surgery time t. According to the description of the surgery process given in Section 4 of Chapter 15, we write $\mathcal{X}(t)$ as the union of a compact subset $C(t)$ and a finite number of surgery caps. For every $t' < t$ sufficiently close to t we have an embedding $n_{t'} \colon C(t) \cong C(t') \subset \mathcal{X}(t')$ given by flowing $C(t)$ backward under the flow to time t'. As $t' \to t$ the maps $n_{t'}$ converge in the C^∞-topology to isometries, in the sense that the $n_{t'}^*(g(t'))|_{C(t')}$ converge smoothly to $g(t)|_{C(t)}$. Furthermore, since the 2-spheres along which we do surgery are homotopically trivial they separate $M_{t'}$. Thus, the maps $n_{t'}^{-1} \colon C(t') \to C(t)$ extend to maps $\psi_{t'} \colon \mathcal{X}(t') \to \mathcal{X}(t)$. The image under $\psi_{t'}$ of $\mathcal{X}(t') \setminus C(t')$ is contained in the union of the surgery caps. Clearly, since all the 2-spheres on which we do surgery at time t are homotopically trivial, the maps $\psi_{t'}$ are homotopy equivalences. If follows from Proposition 15.12 that for any $\eta > 0$ for all $t' < t$ sufficiently close to t, the map $\psi_{t'} \colon \mathcal{X}(t') \to \mathcal{X}(t)$ is a homotopy equivalence that is a $(1+\eta)$-Lipschitz map. Thus, given $\eta > 0$ for all $t' < t$ sufficiently close to t, for any minimal 2-sphere $f \colon S^2 \to (\mathcal{X}(t'), g(t'))$ the area of $\psi_{t'} \circ f \colon S^2 \to (\mathcal{X}(t), g(t))$ is at most $(1+\eta)^2$ times the area of $f(S^2)$. Thus, given $\eta > 0$ for all $t' < t$ sufficiently close to t we see that $W_2(t) \leq (1+\eta)^2 W_2(t')$. Since this is true for every $\eta > 0$, it follows that

$$W_2(t) \leq \liminf_{t' \to t^-} W_2(t').$$

This establishes all three statements in Proposition 18.9 and completes the proof of the proposition.

As an immediate corollary of Proposition 18.9, we obtain the sphere theorem for closed 3-manifolds.

COROLLARY 18.14. *Suppose that M is a closed, connected 3-manifold containing no embedded $\mathbb{R}P^2$ with trivial normal bundle, and suppose that $\pi_2(M) \neq 0$. Then either M can be written as a connected sum $M_1 \# M_2$ where neither of the M_i is homotopy equivalent to S^3 or M_1 has a prime factor that is a 2-sphere bundle over S^1. In either case, M contains an embedded 2-sphere which is homotopically non-trivial.*

PROOF. Let M be as in the statement of the corollary. Let g be a normalized metric on M, and let $(\mathcal{M}, G)$ be the Ricci flow with surgery defined for all time with (M, g) as initial conditions. According to Proposition 18.9

there is $T < \infty$ such that every component of M_T has trivial π_2. Thus, by the analysis above, we see that there must be surgeries that kill elements in π_2: either the removal of a component with non-trivial π_2 or surgery along a homotopically non-trivial 2-sphere. We consider the first such surgery in M. The only components with non-trivial π_2 that can be removed by surgery are S^2-bundles over S^1 and $\mathbb{R}P^3 \# \mathbb{R}P^3$. Since each of these has homotopically non-trivially embedded 2-spheres, if the first surgery killing an element in π_2 is removal of such a component, then, because all the earlier 2-sphere surgeries are along homotopically trivial 2-spheres, the homotopically non-trivial embedded 2-sphere in this component deforms back to an embedded, homotopically non-trivial 2-sphere in M. The other possibility is that the first time an element in $\pi_2(M)$ is killed it is by surgery along a homotopically non-trivial 2-sphere. Once again, using the fact that all previous surgeries are along homotopically trivial 2-spheres, deform this 2-sphere back to M producing a homotopically non-trivial 2-sphere in M. □

REMARK 18.15. Notice that it follows from the list of disappearing components that the only ones with non-trivial π_2 are those based on the geometry $S^2 \times \mathbb{R}$; that is to say, 2-sphere bundles over S^1 and $\mathbb{R}P^3 \# \mathbb{R}P^3$. Thus, once we have reached the level T_0 after which all 2-sphere surgeries are performed on homotopically trivial 2-spheres the only components that can have non-trivial π_2 are components of these types. Thus, for example if the original manifold has no $\mathbb{R}P^3$ prime factors and no non-separating 2-spheres, then when we reach time T_0 we have done a connected sum decomposition into components each of which has trivial π_2. Each of these components is either covered by a contractible 3-manifold or by a homotopy 3-sphere, depending on whether its fundamental group has infinite or finite order.

3. Components with non-trivial π_3

Now we assume that the Ricci flow with surgery $(\mathcal{M}, G)$ satisfies the conclusion of Theorem 15.9 and also has initial condition M that is a connected 3-manifold whose fundamental group satisfies the hypothesis of Theorem 18.1. The argument showing that components with non-trivial π_3 disappear after a finite time is, in spirit, very similar to the arguments above, though the technical details are more intricate in this case.

3.1. Forward difference quotient for π_3. Let M be a compact, connected 3-manifold. Fix a base point $x_0 \in M$. Denote by ΛM the free loop space of M. By this we mean the space of C^1-maps of S^1 to M with the C^1-topology. The components of ΛM are the conjugacy classes of elements in $\pi_1(M, x_0)$. The connected component of the identity of ΛM consists of all homotopically trivial loops in M. Let $*$ be the trivial loop at x_0.

CLAIM 18.16. *Suppose that $\pi_2(M, x_0) = 0$. Then $\pi_2(\Lambda M, *) \cong \pi_3(M, x_0)$ and $\pi_2(\Lambda M, *)$ is identified with the free homotopy classes of maps of S^2 to the component of ΛM consisting of homotopically trivial loops.*

PROOF. An element in $\pi_2(\Lambda M, *)$ is represented by a map $S^2 \times S^1 \to M$ that sends $\{pt\} \times S^1$ to x_0. Hence, this map factors through the quotient of $S^2 \times S^1$ obtained by collapsing $\{pt\} \times S^1$ to a point. The resulting quotient space is homotopy equivalent to $S^2 \vee S^3$, and a map of this space into M sending the wedge point to x_0 is, up to homotopy, the same as an element of $\pi_2(M, x_0) \oplus \pi_3(M, x_0)$. But we are assuming that $\pi_2(M, x_0) = 0$. The first statement follows. For the second, notice that since $\pi_2(M, x_0)$ is trivial, $\pi_3(M, x_0)$ is identified with H_3 of the universal covering $\widetilde{M}$ of M. Hence, for any map of S^2 into the component of ΛM containing the trivial loops, the resulting map $S^2 \times S^1 \to M$ lifts to $\widetilde{M}$. The corresponding element in $\pi_3(M, x_0)$ is the image of the fundamental class of $S^2 \times S^1$ in $H_3(\widetilde{M}) = \pi_3(M)$. □

DEFINITION 18.17. Fix a homotopically trivial loop $\gamma \in \Lambda M$. We set $A(\gamma)$ equal to the infimum of the areas of any spanning disks for γ, where by definition a spanning disk is a Lipschitz map $D^2 \to M$ whose boundary is, up to reparameterization, γ. Notice that $A(\gamma)$ is a continuous function of γ in ΛM. Also, notice that $A(\gamma)$ is invariant under reparameterization of the curve γ. Now suppose that $\Gamma \colon S^2 \to \Lambda M$ is given with the image consisting of homotopically trivial loops. We define $W(\Gamma)$ to be equal to the maximum over all $c \in S^2$ of $A(\Gamma(c))$. More generally, given a homotopy class $\xi \in \pi_2(\Lambda M, *)$ we define $W(\xi)$ to be equal to the infimum over all (not necessarily based) maps $\Gamma \colon S^2 \to \Lambda M$ into the component of ΛM consisting of homotopically trivial loops representing ξ of $W(\Gamma)$.

Now let us formulate the analogue of Proposition 18.9 for π_3. Suppose that $\mathcal{X}$ is a path of components of the Ricci flow with surgery $(\mathcal{M}, G)$ defined for $t \in [t_0, t_1]$. Suppose that $\pi_2(\mathcal{X}(t_0), x_0) = 0$ and that $\pi_3(\mathcal{X}(t_0), x_0) \neq 0$. Then, the same two conditions hold for $\mathcal{X}(t)$ for each $t \in [t_0, t_1]$. The reason is that at a surgery time t, since all the 2-spheres in $\mathcal{X}(t')$ ($t' < t$ but sufficiently close to t) along which we are doing surgery are homotopically trivial, the result of surgery is a disjoint union of connected components: one connected component is homotopy equivalent to $\mathcal{X}(t')$ and all other connected components are homotopy 3-spheres. This means that either $\mathcal{X}(t)$ is homotopy equivalent to $\mathcal{X}(t')$ for $t' < t$ or $\mathcal{X}(t)$ is a homotopy 3-sphere. In either case both homotopy group statements hold for $\mathcal{X}(t)$. Even more is true: The distance-decreasing map $\mathcal{X}(t') \to \mathcal{X}(t)$ given by Proposition 15.12 is either a homotopy equivalence or a degree-1 map of $\mathcal{X}(t') \to \mathcal{X}(t)$. In either case, it induces an injection of $\pi_3(\mathcal{X}(t')) \to \pi_3(\mathcal{X}(t))$. In this way a non-zero element in $\xi(t_0) \in \pi_3(\mathcal{X}(t_0))$ produces a family of non-zero elements $\xi(t) \in \pi_3(\mathcal{X}(t))$ with the property that under Ricci flow these elements agree and at a surgery time t the degree-1 map constructed in Proposition 15.12 sends $\xi(t')$ to $\xi(t)$ for all $t' < t$ sufficiently close to it. Since $\pi_2(\mathcal{X}(t))$ is trivial for all t, we identify $\xi(t)$ with a homotopy class of maps of S^2 to $\Lambda \mathcal{X}(t)$. We now define a function $W_\xi(t)$ by associating to each t the invariant $W(\xi(t))$.

Here is the result that is analogous to Lemma 18.11.

PROPOSITION 18.18. *Suppose that $(\mathcal{M}, G)$ is a Ricci flow with surgery as in Theorem 15.9. Let $\mathcal{X}$ be a path of components of $\mathcal{M}$ defined for all $t \in [t_0, t_1]$ with $\pi_2(\mathcal{X}(t_0)) = 0$. Suppose that $\xi \in \pi_3(X(t_0), *)$ is a non-trivial element. Then the function $W_\xi(t)$ satisfies the following inequality in the sense of forward difference quotients:*

$$\frac{dW_\xi(t)}{dt} \leq -2\pi - \frac{1}{2} R_{\min}(t) W_\xi(t).$$

Also, for every $t \in [t_0, t_1]$ that is not a surgery time the function $W_\xi(t)$ is continuous at t. Lastly, if t is a surgery time then

$$W_\xi(t) \leq \liminf\nolimits_{t' \to t^-} W_\xi(t').$$

In the next subsection we assume this result and use it to complete the proof.

3.2. Proof of Theorem 18.1 assuming Proposition 18.18. According to Proposition 18.9 there is T_1 such that every component of M_T has trivial π_2 for every $T \geq T_1$. Suppose that Theorem 18.1 does not hold for this Ricci flow with surgery. We consider a path of components $\mathcal{X}(t)$ of $\mathcal{M}$ defined for $[T_1, T_2]$. We shall show that there is a uniform upper bound to T_2.

CLAIM 18.19. *$\mathcal{X}(T_1)$ has non-trivial π_3.*

PROOF. By hypothesis the fundamental group of M_0 is a free product of infinite cyclic groups and finite groups. This means that the same is true for the fundamental group of each component of M_t for every $t \geq 0$, and in particular it is true for $\mathcal{X}(T_0)$. But we know that $\pi_2(\mathcal{X}(T_0)) = 0$.

CLAIM 18.20. *Let X be a compact 3-manifold. If $\pi_1(X)$ is a non-trivial free product or if $\pi_1(X)$ is isomorphic to $\mathbb{Z}$, then $\pi_2(X) \neq 0$.*

PROOF. See [39], Theorem 5.2 on page 56 (for the case of a copy of $\mathbb{Z}$) and [39] Theorem 7.1 on page 66 (for the case of a free product decomposition). □

Thus, it follows that $\pi_1(\mathcal{X}(T_1))$ is a finite group (possibly trivial). But a 3-manifold with finite fundamental group has a universal covering that is a compact 3-manifold with trivial fundamental group. Of course, by Poincaré duality any simply connected 3-manifold is a homotopy 3-sphere. It follows immediately that $\pi_3(\mathcal{X}(T_1)) \cong \mathbb{Z}$. This completes the proof of the claim. □

Now we can apply Proposition 18.18 to our path of components $\mathcal{X}$ defined for all $t \in [T_1, T_2]$. First recall by Theorem 15.9 that the curvature of $(\mathcal{M}, G)$ is pinched toward positive which implies that $R_{\min}(t) \geq (-6)/(1+4t)$. Let $w(t)$ be the function satisfying the differential equation

$$w'(t) = -2\pi + \frac{3}{1+4t}w(t)$$

with initial condition $w(T_1) = W_\xi(T_1)$. According to Proposition 18.18 and Lemma 2.22 we see that $W_\xi(t) \leq w(t)$ for all $t \in [T_1, T_2]$. But direct integration shows that

$$w(t) = W_\xi(T_1)\frac{(4t+1)^{3/4}}{(4T_1+1)^{3/4}} + 2\pi(4T_0+1)^{1/4}(4t+1)^{3/4} - 2\pi(4t+1).$$

This clearly shows that $w(t)$ becomes negative for t sufficiently large, how large depending only on $W_\xi(T_1)$ and T_1. On the other hand, since $W_\xi(t)$ is the infimum of areas of disks, $W_\xi(t) \geq 0$ for all $t \in [T_1, T_2]$. This proves that T_2 is less than a constant that depends only on T_1 and on the component $\mathcal{X}(T_1)$. Since there are only finitely many connected components of M_{T_1}, this shows that T_2 depends only on T_1 and the Riemannian manifold M_{T_1}. This completes the proof of Theorem 18.1 modulo Proposition 18.18. □

Thus, to complete the argument for Theorem 18.1 it remains only to prove Proposition 18.18.

4. First steps in the proof of Proposition 18.18

In this section we reduce the proof of Proposition 18.18 to a more technical result, Proposition 18.24 below.

4.1. Continuity of $W_\xi(t)$.
In this subsection we establish the two continuity conditions for $W_\xi(t)$ stated in Proposition 18.18.

CLAIM 18.21. *If t is not a surgery time, then $W_\xi(t)$ is continuous at t.*

PROOF. Since t is not a surgery time, a family $\Gamma(t): S^2 \to \Lambda\mathcal{X}(t)$ is also a family $\Gamma(t'): S^2 \to \Lambda\mathcal{X}(t')$ for all nearby t'. The minimal spanning disks for the elements of $\Gamma(t)(x)$ are also spanning disks in the nearby $\mathcal{X}(t')$ and their areas vary continuously with t. But the maximum of the areas of these disks is an upper bound for $W_\xi(t)$. This immediately implies that $W_\xi(t)$ is upper semi-continuous at t.

The result for lower semi-continuity is the same as in the case of 2-spheres. Given a time t distinct from a surgery time and a family $\Gamma: S^2 \to \Lambda\mathcal{X}(t')$ for a time t' near t we can view the family Γ as a map to $\Lambda\mathcal{X}(t)$. The areas of all minimal spanning disks for the loops represented by points Γ measured in $\mathcal{X}(t)$ are at most $(1 + \eta(|t - t'|))$ times their areas measured in $\mathcal{X}(t')$, where $\eta(|t - t'|)$ is a function going to zero as $|t - t'|$ goes to zero.

This immediately implies the lower semi-continuity at the non-surgery time t. □

CLAIM 18.22. *Suppose that t is a surgery time. Then*
$$W_\xi(t) \leq \liminf_{t' \to t^-} W_\xi(t').$$

PROOF. This is immediate from the fact from Proposition 15.12 that for any $\eta > 0$ for every $t' < t$ sufficiently close to t there is a homotopy equivalence $\mathcal{X}(t') \to \mathcal{X}(t)$ which is a $(1+\eta)$-Lipschitz map. □

To prove Proposition 18.18 and hence Theorem 18.1, it remains to prove the forward difference quotient statement for $W_\xi(t)$ given in Proposition 18.18.

4.2. A reduction of Proposition 18.18. Let $\Gamma \colon S^2 \to \Lambda\mathcal{X}(t_0)$ be a family. We must construct an appropriate deformation of the family of loops Γ in order to establish Proposition 18.18. Now we are ready to state the more technical estimate for the evolution of $W(\Gamma)$ under Ricci flow that will imply the forward difference quotient result for $W_\xi(t)$ stated in Proposition 18.18. Here is the result that shows a deformation as required exists.

DEFINITION 18.23. Let $(M, g(t))$, $t_0 \leq t \leq t_1$, be a Ricci flow on a compact 3-manifold. For any a and any $t' \in [t_0, t_1]$ let $w_{a,t'}(t)$ be the solution to the differential equation

(18.4) $$\frac{dw_{a,t'}}{dt} = -2\pi - \frac{1}{2}R_{\min}(t)w_{a,t'}(t)$$

with initial condition $w_{a,t'}(t') = a$. We also denote w_{a,t_0} by w_a.

PROPOSITION 18.24. *Let $(M, g(t))$, $t_0 \leq t \leq t_1$, be a Ricci flow on a compact 3-manifold. Fix a map Γ of S^2 to ΛM whose image consists of homotopically trivial loops and $\zeta > 0$. Then there is a continuous family $\widetilde{\Gamma}(t)$, $t_0 \leq t \leq t_1$, of maps $S^2 \to \Lambda M$ whose image consists of homotopically trivial loops with $[\widetilde{\Gamma}(t_0)] = [\Gamma]$ in $\pi_3(M, *)$ such that for each $c \in S^2$ we have $|A(\widetilde{\Gamma}(t_0)(c)) - A(\Gamma(c))| < \zeta$ and furthermore, one of the following two alternatives holds:*

(i) *The length of $\widetilde{\Gamma}(t_1)(c)$ is less than ζ.*
(ii) $A(\widetilde{\Gamma}(t_1)(c)) \leq w_{A(\widetilde{\Gamma}(t_0)(c))}(t_1) + \zeta$.

Here we shall show that this proposition implies the forward difference quotient result in Proposition 18.18. The next chapter is devoted to proving Proposition 18.24. Let $\mathcal{X}$ be a path of components. Suppose that $\pi_2(\mathcal{X}(t), x_0) = 0$ for all t. Fix t_0 and fix a non-trivial element $\xi \in \pi_3(\mathcal{X}(t_0), x_0)$, which we identify with a non-trivial element in $\xi \in \pi_2(\Lambda\mathcal{X}(t_0), *)$. Fix an interval $[t_0, t_1]$ with the property that there are no surgery times in the interval $(t_0, t_1]$. Restricting to this interval the family $\mathcal{X}(t)$ is a Ricci flow on $\mathcal{X}(t_0)$. In particular, all the $\mathcal{X}(t)$ are identified under the Ricci flow. Let $w(t)$ be the solution to Equation (18.4) with value $w(t_0) = W_\xi(\mathcal{X}, t_0)$. We shall show that $W_\xi(t_1) \leq w(t_1)$. Clearly, once we have this estimate, taking limits as t_1 approaches t_0 establishes the forward difference quotient result at t_0.

DEFINITION 18.25. Let $B(t) = \int_{t'}^{t} \frac{1}{2} R_{\min}(s) ds$.

Direct integration shows the following:

CLAIM 18.26. *We have*

$$w_{a,t'}(t'') = \exp(-B(t'')) \left(a - 2\pi \int_{t'}^{t''} \exp(B(t)) dt \right).$$

If $a' > a$, then for $t_0 \leq t' < t'' \leq t_1$, we have

$$w_{a',t'}(t'') = w_{a,t'}(t'') + (a' - a)\exp(-B(t'')).$$

The next thing to establish is the following.

LEMMA 18.27. *Let (X, g) be a compact Riemannian manifold with trivial π_2. Then there is $\zeta > 0$ such that if $\xi \in \pi_3(\mathcal{X})$ is represented by a family $\Gamma\colon S^2 \to \Lambda X$ with the property that for every $c \in S^2$ the length of the loop $\Gamma(c)$ is less than ζ, then ξ is the trivial homotopy element.*

PROOF. We choose ζ smaller than the injectivity radius of (X, g). Then any pair of points at distance less than ζ apart are joined by a unique geodesic of length less than ζ. Furthermore, the geodesic varies smoothly with the points. Given a map $\Gamma\colon S^2 \to \Lambda X$ such that every loop of the form $\Gamma(c)$ has length at most ζ, we consider the map $f\colon S^2 \to X$ defined by $f(c) = \Gamma(c)(x_0)$, where x_0 is the base point of the circle. Then we can join each point $\Gamma(c)(x)$ to $\Gamma(c)(x_0)$ by a geodesic of length at most ζ to fill out a map of the disk $\hat{\Gamma}(c)\colon D^2 \to X$. This disk is smooth except at the point $\Gamma(c)(x_0)$. The disks $\hat{\Gamma}(c)$ fit together as c varies to make a continuous family of disks parameterized by S^2 or equivalently a map $S^2 \times D^2$ into X whose boundary is the family of loops $\Gamma(c)$. Now shrinking the loops $\Gamma(c)$ across the disks $\hat{\Gamma}(c)$ to $\Gamma(c)(x_0)$ shows that the family Γ is homotopic to a 2-sphere family of constant loops at different points of X. Since we are assuming that $\pi_2(X)$ is trivial, this means the family of loops is in fact trivial as an element of $\pi_2(\Lambda X, *)$, which means that the original element $\xi \in \pi_3(X)$ is trivial. □

Notice that this argument also shows the following:

COROLLARY 18.28. *Let (X, g) be a compact Riemannian manifold. For any $\eta > 0$ there is a $0 < \zeta < \eta/2$ such that any C^1-loop $c\colon S^1 \to X$ of length less than η bounds a disk in X of area less than η.*

4. FIRST STEPS IN THE PROOF OF PROPOSITION 18.18

Now we return to the proof that Proposition 18.24 implies Proposition 18.18. We consider the restriction of the path $\mathcal{X}$ to the time interval $[t_0, t_1]$. As we have already remarked, since there are no surgery times in $(t_0, t_1]$, this restriction is a Ricci flow and all the $\mathcal{X}(t)$ are identified with each other under the flow. Let $w(t)$ be the solution to Equation (18.4) with initial condition $w(t_0) = W_\xi(t_0)$. There are two cases to consider: (i) $w(t_1) \geq 0$ and $w(t_1) < 0$.

Suppose that $w(t_1) \geq 0$. Let $\eta > 0$ be given. Then by Claim 18.26 and Corollary 18.28, there is $0 < \zeta < \eta/2$ such that the following two conditions hold:

(a) Any loop in $\mathcal{X}(t_1)$ of length less than ζ bounds a disk of area less than η.
(b) For every $a \in [0, W_\xi(t_0) + 2\zeta]$ the solution w_a satisfies $w_a(t_1) < w(t_1) + \eta/2$.

Now fix a map $\Gamma\colon S^2 \to \Lambda\mathcal{X}(t_0)$, whose image consists of homotopically trivial loops, with $[\Gamma] = \xi$, and with $W(\Gamma) < W_\xi(t_0) + \zeta$. According to Proposition 18.24 there is a one-parameter family $\widetilde{\Gamma}(t)$, $t_0 \leq t \leq t_1$, of maps $S^2 \to \Lambda\mathcal{X}(t)$, whose images consist of homotopically trivial loops, with $[\widetilde{\Gamma}(t_0)] = [\Gamma] = \xi$ such that for every $c \in S^2$ we have $A(\widetilde{\Gamma}(t_0)(c)) < A(\Gamma(c)) + \zeta$ and one of the following holds:

(i) the length of $\widetilde{\Gamma}(t_1)(c)$ is less than ζ, or
(ii)
$$A(\widetilde{\Gamma}(t_1)(c)) < w_{A(\widetilde{\Gamma}(t_0)(c))}(t_1) + \zeta.$$

Since $A(\widetilde{\Gamma}(t_0)(c)) < A(\Gamma(c)) + \zeta < W_\xi(t_0) + 2\zeta$, it follows from our choice of ζ that for every $c \in S^2$ either

(a) $\widetilde{\Gamma}(t_1)(c)$ has length less than ζ and hence bounds a disk of area less than η, or
(b) $A(\widetilde{\Gamma}(t_1)(c)) < w_{W_\xi(t_0)+2\zeta}(t_1) + \zeta < w(t_1) + \eta/2 + \eta/2 = w(t_1) + \eta$.

Since we are assuming that $w(t_1) \geq 0$, it now follows that for every $c \in S^2$ we have $A(\widetilde{\Gamma}(t_1)(c)) < w(t_1) + \eta$, and hence $W(\widetilde{\Gamma}(t_1)) < w(t_1) + \eta$. This shows that for every $\eta > 0$ we can find a family $\widetilde{\Gamma}(t)$ with $\widetilde{\Gamma}(t_0)$ representing ξ and with $W(\widetilde{\Gamma}(t_1)) < w(t_1) + \eta$. This completes the proof of Proposition 18.24 when $w(t_1) \geq 0$.

Now suppose that $w(t_1) < 0$. In this case, we must derive a contradiction since clearly it must be the case that for any one-parameter family $\widetilde{\Gamma}(t)$ we have $W(\widetilde{\Gamma}(t_1)) \geq 0$. We fix $\eta > 0$ such that $w(t_1) + \eta < 0$. Then using Lemma 18.27 and Claim 18.26, we fix ζ with $0 < \zeta < \eta/2$ such that:

(i) If $\Gamma\colon S^2 \to \Lambda\mathcal{X}(t_1)$ is a family of loops and each loop in the family is of length less than ζ, then the family is homotopically trivial.
(ii) For any $a \in [0, W_\xi(t_0) + 2\zeta]$ we have $w_a(t_1) < w(t_1) + \eta/2$.

We fix a map $\Gamma\colon S^2 \to \mathcal{X}(t_0)$ with $[\Gamma] = \xi$ and with $W(\Gamma) < W_\xi(t_0) + \zeta$. Now according to Proposition 18.24 there is a family of maps $\widetilde{\Gamma}(t)\colon S^2 \to \Lambda\mathcal{X}(t)$ with $[\widetilde{\Gamma}(t_0)] = [\Gamma] = \xi$ and for every $c \in S^2$ we have $A(\widetilde{\Gamma}(t_0)(c)) < A(\Gamma(c)) + \zeta$ and also either $A(\widetilde{\Gamma}(t_1)(c)) \leq w_{A(\widetilde{\Gamma}(t_0)(c))}(t_1) + \zeta$ or the length of $\widetilde{\Gamma}(t_1)(c)$ is less than ζ. It follows that for every $c \in S^2$ we have $A(\widetilde{\Gamma}(t_0)(c)) \leq W(\Gamma) + \zeta < W_\xi(t_0) + 2\zeta$. From the choice of ζ this means that

$$A(\widetilde{\Gamma}(t_1)(c)) < w(t_1) + \eta/2 + \zeta < w(t_1) + \eta < 0$$

if the length of $\widetilde{\Gamma}(t_1)(c)$ is at least ζ. Of course, by definition $A(\widetilde{\Gamma}(t_1)(c)) \geq 0$ for every $c \in S^2$. This implies that for every $c \in S^2$ the loop $\widetilde{\Gamma}(t_1)(c)$ has length less than ζ. By Lemma 18.27 this implies that $\widetilde{\Gamma}(t_1)$ represents the trivial element in $\pi_2(\Lambda\mathcal{X}(t_1))$, which is a contradiction.

At this point, all that remains to do in order to complete the proof of Theorem 18.1 is to establish Proposition 18.24. The next chapter is devoted to doing that.

CHAPTER 19

Completion of the Proof of Proposition 18.24

1. Curve-shrinking

Given Γ, the idea for constructing the one-parameter family $\widetilde{\Gamma}(t)$ required by Proposition 18.24 is to evolve an appropriate approximation $\widetilde{\Gamma}(t_0)$ of Γ by the curve-shrinking flow. Suppose that $(M, g(t))$, $t_0 \le t \le t_1$, is a Ricci flow of compact manifolds and that $c\colon S^1 \times [t_0, t_1] \to (M, g(t_0))$ is a family of parameterized, immersed C^2-curves. We denote by x the parameter on the circle. Let $X(x,t)$ be the tangent vector $\partial c(x,t)/\partial x$ and let $S(x,t) = X(x,t)/(|X(x,t)|_{g(t)})$ be the unit tangent vector to c. We denote by s the arc length parameter on c. We set $H(x,t) = \nabla_{S(x,t)} S(x,t)$, the curvature vector of c with respect to the metric $g(t)$. We define the *curve-shrinking flow* by

$$\frac{\partial c(x,t)}{\partial t} = H(x,t),$$

where $c(x,t)$ is a one-parameter family of curves and $H(x,t)$ is the curvature vector of the curve $c(\cdot, t)$ at the point x with respect to the metric $g(t)$. We denote by $k(x,t)$ the curvature function: $k(x,t) = |H(x,t)|_{g(t)}$. We shall often denote the one-parameter family of curves by $c(\cdot, t)$. Notice that if $c(x,t)$ is a curve-shrinking flow and if $x(y)$ is a reparameterization of the domain circle, then $c'(y,t) = c(x(y), t)$ is also a curve-shrinking flow.

CLAIM 19.1. *For any immersed C^2-curve $c\colon S^1 \to (M, g(t_0))$ there is a curve-shrinking flow $c(x,t)$ defined for $t \in [t_0, t'_1)$ for some $t'_1 > t_0$ with the property that each $c(\cdot, t)$ is an immersion. Either the curve-shrinking flow extends to a curve-shrinking flow that is a family of immersions defined at t'_1 and beyond, or $\max_{x \in S^1} k(x,t)$ blows up as t approaches t'_1 from below.*

For a proof of this result, see Theorem 1.13 in [2].

1.1. The proof of Proposition 18.24 in a simple case. The main technical hurdle to overcome is that in general the curve shrinking flow may not exist if the original curve is not immersed, and even if the original curve is immersed, the curve-shrinking flow can develop singularities, where the curvature of the curve goes to infinity. Thus, we may not be able to define the curve-shrinking flow as a flow defined on the entire interval $[t_0, t_1]$, even though the Ricci flow is defined on this entire interval. But to show the idea of the proof, let us suppose for a moment that the starting curve is embedded and that no singularities develop in the curve-shrinking flow and show how to prove the result.

LEMMA 19.2. *Suppose that $c \in \Lambda M$ is a homotopically trivial, embedded C^2-loop, and suppose that there is a curve-shrinking flow $c(x,t)$ defined for all $t \in [t_0, t_1]$ with each $c(\cdot, t)$ being an embedded smooth curve. Consider the function $A(t)$ which assigns to t the minimal area of a spanning disk for $c(\cdot, t)$. Then $A(t)$ is a continuous function of t and*

$$\frac{dA}{dt}(t) \leq -2\pi - \frac{1}{2} R_{\min}(t) A(t)$$

in the sense of forward difference quotients.

PROOF. According to results of Hildebrandt and Morrey, [**40**] and [**52**], for each $t \in [t_0, t_1]$, there is a smooth minimal disk spanning $c(\cdot, t)$. Fix $t' \in [t_0, t_1)$ and consider a smooth minimal disk $D \to (M, g(t'))$ spanning $c(\cdot, t)$. It is immersed, see [**37**] or [**27**]. The family $c(\cdot, t)$ for t near t' is an isotopy of $c(\cdot, t')$. We can extend this to an ambient isotopy $\varphi_t \colon M \to M$ with $\varphi_{t'} = \mathrm{Id}$. We impose coordinates $\{x_\alpha\}$ on D; we let $h_{\alpha\beta}(t')$ be the metric induced on $\varphi_{t'}(D)$ by $g(t')$, and we let da be the area form induced by the Euclidean coordinates on D. We compute

$$\frac{d}{dt}\Big|_{t=t'} \mathrm{Area}(\varphi_t(D)) = \frac{d}{dt}\Big|_{t=t'} \int_{\varphi_t(D)} \sqrt{\det(h_{\alpha\beta})(t)}\, da.$$

Of course,

$$\frac{d}{dt}\Big|_{t=t'} \int_{\varphi_t(D)} \sqrt{\det(h_{\alpha\beta}(t))}\, da = -\int_{\varphi_{t'}(D)} \left(\mathrm{Tr}\, \mathrm{Ric}^T\right) \sqrt{\det(h_{\alpha\beta}(t))}\, da$$
$$+ \int_{\varphi_{t'}(D)} \mathrm{div}\left(\frac{\partial \varphi_{t'}}{\partial t}\right)^T \sqrt{\det(h_{\alpha\beta}(t))}\, da.$$

Here, Ric^T denotes the restriction of the Ricci curvature of $g(t')$ to the tangent planes of $\varphi_{t'}(D)$ and $\frac{\partial (\varphi_{t'})^T}{\partial t}$ is the component of $\varphi_{t'}$ tangent to $\varphi_{t'}(D)$. Setting $\hat{A}$ equal to the second fundamental form of $\varphi_{t'}(D)$, using the fact that $\varphi_{t'}(D)$ is minimal and arguing as in the proof of Claim 18.12, we have

$$-\int_{\varphi_{t'}(D)} \left(\mathrm{Tr}\, \mathrm{Ric}^T\right) \sqrt{\det(h_{\alpha\beta}(t'))}\, da$$
$$= -\int_{\varphi_{t'}(D)} K_{\varphi_{t'}(D)}\, da - \frac{1}{2} \int_{\varphi_{t'}(D)} (|\hat{A}|^2 + R)\, da$$
$$\leq -\int_{\varphi_{t'}(D)} K_{\varphi_{t'}(D)} \sqrt{\det(h_{\alpha\beta}(t'))}\, da - \frac{1}{2} \mathrm{Area}(\varphi_{t'}(D)) \min_{x \in M}\{R(x, t')\}.$$

Integration by parts shows that

$$\int_{\varphi_{t'}(D)} \operatorname{div}\left(\frac{\partial \varphi_{t'}}{dt}\right)^T \sqrt{\det(h_{\alpha\beta}(t'))} da = -\int_{\varphi_{t'}(\partial D)} \left(\frac{d\varphi_t}{dt}\bigg|_{t=t'}\right) \cdot n\, ds,$$

where n is the inward pointing normal vector to $\varphi_{t'}(D)$ along $\varphi_{t'}(\partial D)$. Of course, by definition, if the variation along the boundary is given by the curve-shrinking flow, then along $\varphi_{t'}(\partial D)$ we have

$$\left(\frac{d\varphi_t}{dt}\bigg|_{t=t'}\right) \cdot n = k_{\text{geod}}.$$

Thus, we have

$$\frac{d}{dt}\bigg|_{t=t'} \int_{\varphi_t(D)} \sqrt{\det(h_{\alpha\beta}(t))}\, da$$
$$\leq -\int_{\varphi_{t'}(D)} K_{\varphi_{t'}(D)}\, da - \int_{\varphi_{t'}(\partial D)} k_{\text{geod}}\, ds - \frac{1}{2} R_{\min}(t')\operatorname{Area}(\varphi_{t'}(D)).$$

Of course, the Gauss-Bonnet theorem allows us to rewrite this as

$$\frac{d}{dt}\bigg|_{t=t'} \int_{\varphi_t(D)} \sqrt{\det(h_{\alpha\beta}(t))}\, da \leq -2\pi - \frac{1}{2} R_{\min}(t')\operatorname{Area}(\varphi_{t'}(D)).$$

□

Let $\psi(t)$ be the solution to the ODE

$$\psi'(t) = -2\pi - \frac{1}{2} R_{\min}(t)\psi(t)$$

with $\psi(t^-) = A(t^-)$. The following is immediate from the previous lemma and Lemma 2.22.

COROLLARY 19.3. *With notation and assumptions as above, if the curve-shrinking flow is defined on the interval $[t^-, t^+]$ and if the curves $c(\cdot, t)$ are embedded for all $t \in [t^-, t^+]$ then*

$$A(t^+) \leq \psi(t^+).$$

Actually, the fact that the loops in the curve-shrinking flow are embedded is not essential in dimensions ≥ 3.

LEMMA 19.4. *Suppose that the dimension of M is at least 3 and that $c(\cdot, t)$ is a C^2-family of homotopically trivial, immersed curves satisfying the curve-shrinking flow defined for $t^- \leq t \leq t^+$. For each t, let $A(t)$ be the infimum of the areas of spanning disks for $c(\cdot, t)$. Then $A(t)$ is a continuous function and, with ψ as above, we have*

$$A(t^+) \leq \psi(t^+).$$

PROOF. We first remark that continuity has already been established. To show the inequality, we begin with a claim.

CLAIM 19.5. *It suffices to prove the following for every $\delta > 0$. There is a C^2-family $\hat{c}(x,t)$ of immersions within δ in the C^2-topology to $c(x,t)$ defined on the interval $[t^-, t^+]$ such that*

$$A(t^+) \leq \psi_{\delta,\hat{c}}(t^+)$$

where $\psi_{\delta,\hat{c}}$ is the solution of the ODE

$$\psi'_{\delta,\hat{c}}(t) = -2\pi + 2\delta L_{\hat{c}}(t) - \frac{1}{2}R_{\min}(t)\psi_{\delta,\hat{c}}(t)$$

with value $A(\hat{c}(t^-))$ at t^-, and where $L_{\hat{c}}(t)$ denotes the length of the loop $\hat{c}(\cdot,t)$.

PROOF. (of the claim) Suppose that for each δ there is such a C^2-family as in the statement of the claim. Take a sequence δ_n tending to zero, and let $\hat{c}_n(\cdot, t)$ be a family as in the claim for δ_n. Then by the continuity of the infimum of areas of the spanning disk in the C^1-topology, we see that

$$\lim_{n\to\infty} A(\hat{c}_n(\cdot, t^{\pm})) = A(c(\cdot, t^{\pm})).$$

Since the $\hat{c}_n(x,t)$ converge in the C^2-topology to $c(x,t)$, the lengths $L(\hat{c}_n(t))$ are uniformly bounded and the $A(\hat{c}_n(t^-))$ converge to $A(c(t))$. Thus, the $\psi_{\delta_n,\hat{c}_n}$ converge uniformly to ψ on $[t^-, t^+]$, and taking limits shows the required inequality for $A(c(t))$, thus proving the claim. □

Now we return to the proof of the lemma. Let $\hat{c}(x,t)$ be a generic C^2-immersion sufficiently close to $c(x,t)$ in the C^2-topology so that the following hold:

(1) the difference of the curvature of $\hat{c}$ and of c at every (x,t) is a vector of length less than δ,
(2) the difference of $\partial \hat{c}/\partial t$ and $\partial c/\partial t$ is a vector of length less than δ,
(3) the ratio of the arc lengths of $\hat{c}$ and c at every (x,t) is between $(1-\delta)$ and $(1+\delta)$.

The generic family $\hat{c}(x,t)$ consists of embedded curves for all but a finite number of $t \in [t^-, t^+]$ and at the exceptional t values the curve is immersed. Let $t_1 < t_2 < \cdots < t_k$ be the values of t for which $\hat{c}(\cdot, t)$ is not embedded. We set $t_0 = t^-$ and $t_{k+1} = t^+$. Notice that it suffices to show that

$$A(\hat{c}(t_{i+1})) - A(\hat{c}(t_i)) \leq \psi_{\delta,\hat{c}}(t_{i+1}) - \psi_{\delta,\hat{c}}(t_i)$$

for $i = 0, \ldots, k$. To establish this inequality for the interval $[t_i, t_{i+1}]$, by continuity it suffices to establish the corresponding inequality for every compact subinterval contained in the interior of this interval. This allows us to assume that the approximating family is a family of embedded curves. Let the endpoints of the parameterizing interval be denoted a and b. Fix $t' \in [a,b]$ and let D be a minimal disk spanning $\hat{c}(\cdot, t')$, and let φ_t be an isotopy as in the argument given the proof of Lemma 19.2. According to this argument we have

$$\frac{d}{dt}A(\hat{c}(t))|_{t=t'} \leq -2\pi - \frac{1}{2}R_{\min}(t')A(c(t'))$$
$$+ \int_{c(x,t')} \left[k_{\text{geod}}(\hat{c}) - \left(\frac{d\varphi_t}{dt}|_{t=t'}\right) \cdot n \right] ds$$

in the sense of forward difference quotients. The restriction of $\frac{d\varphi_t}{dt}|_{t=t'}$ to the boundary of D agrees with $\partial \hat{c}(x,t)/\partial t$. Hence, by our conditions on the approximating family, and since for $c(\cdot, t)$ the corresponding quantities are equal,

$$\left| k_{\text{geod}}(\hat{c}) - \left(\frac{d\varphi_t}{dt}|_{t=t'}\right) \cdot n \right| < 2\delta.$$

Integrating over the circle implies that

$$\frac{d}{dt}A(\hat{c}(t))|_{t=t'} \leq -2\pi - \frac{1}{2}R_{\min}(t')A(c(t')) + 2\delta L_{\hat{c}}(t).$$

The result is then immediate from Lemma 2.22. □

2. Basic estimates for curve-shrinking

Let us establish some elementary formulas. To simplify the formulas we often drop the variables x, t from the notation, though they are understood to be there.

LEMMA 19.6. *Assume that $(M, g(t))$, $t_0 \leq t \leq t_1$, is a Ricci flow and that $c = c(x,t)$ is a solution to the curve-shrinking flow. We have vector fields $X = \partial/\partial x$ and $H = \partial/\partial t$ defined on the domain surface. We denote by $|X|^2_{c^*g}$ the function on the domain surface whose value at (x,t) is $|(X(x,t))|^2_{g(t)}$. We define $S = |X|^{-1}_{c^*g}X$, the unit vector in the x-direction measured in the evolving metric. Then,*

$$\frac{\partial}{\partial t}(|X|^2_{c^*g})(x,t) = -2\text{Ric}_{g(t)}(X(x,t), X(x,t)) - 2k^2|X(x,t)|^2_{g(t)},$$

and

$$[H, S](x,t) = \left(k^2 + \text{Ric}_{g(t)}(S(x,t), S(x,t))\right) S(x,t).$$

PROOF. Notice that as t varies $|X|^2_{c^*g}$ is not the norm of the vector field X with respect to the pullback of a fixed metric $g(t)$. On the other hand, when we compute $\nabla_H X$ at a point (x,t) we are taking a covariant derivative with respect to the pullback of a fixed metric $g(t)$ on the surface. Hence, in computing $H(|X|^2_{c^*g})$ the usual Leibniz rule does not apply. In fact, there are two contributions to $H(|X|^2_{c^*g})$: one, the usual Leibniz rule differentiating in a frozen metric $g(t)$ and the other coming from the effect on $|X|^2_{c^*g}$ of varying the metric with t. Thus, we have

$$H(|X|^2_{c^*g})(x,t) = -2\text{Ric}_{c^*g(t)}(X(x,t), X(x,t)) + 2\langle \nabla_H X(x,t), X(x,t)\rangle_{c^*g(t)}.$$

Since t and x are coordinates on the surface swept out by the family of curves, $\nabla_H X = \nabla_X H$, and hence the second term on the right-hand side of the previous equation can be rewritten as $2\langle \nabla_X H(x,t), X(x,t) \rangle_{c^*g(t)}$. Since $X(x,t)$ and $H(x,t)$ are orthogonal in $c^*g(t)$ and since $X = |X|_{c^*g} S$, computing covariant derivatives in the metric $c^*g(t)$, we have

$$\begin{aligned} 2\langle \nabla_X H, X \rangle_{c^*g(t)} &= -2\langle H, \nabla_X X \rangle_{c^*g(t)} \\ &= -2\langle H, |X|^2_{g(t)} \nabla_S S \rangle_{c^*g(t)} - 2\langle H, |X|_{g(t)} S(|X|_{c^*g}) S \rangle_{c^*g(t)} \\ &= -2\langle H, H \rangle_{c^*g(t)} |X|^2_{g(t)} \\ &= -2k^2 |X|^2_{c^*g}. \end{aligned}$$

This proves the first inequality. As for the second, since X and H commute we have

$$[H, S] = [H, |X|^{-1}_{c^*g} X] = H\left((|X|^2_{c^*g})^{-1/2} \right) X = \frac{-1}{2\left(|X|^2_g\right)^{3/2}} H(|X|_{c^*g})^2) X.$$

According to the first equation, we can rewrite this as

$$[H, S](x, t) = \left(k^2 + \mathrm{Ric}_{c^*g(t)}(S(x,t), S(x,t)) \right) S(x,t).$$

$\square$

Now let us compute the time derivative of k^2. In what follows we drop the dependence on the metric $c^*g(t)$ from all the curvature terms, but it is implicitly there.

LEMMA 19.7.
$$\begin{aligned} \frac{\partial}{\partial t} k^2 &= \frac{\partial^2}{\partial s^2}(k^2) - 2\langle (\nabla_X H)^\perp, (\nabla_S H)^\perp \rangle_{c^*g} + 2k^4 \\ &\quad - 2\mathrm{Ric}(H,H) + 4k^2 \mathrm{Ric}(S,S) + 2\mathrm{Rm}(H,S,H,S), \end{aligned}$$

where the superscript $\perp$ means the image under projection to the orthogonal complement of X.

PROOF. Using the same conventions as above for the function $|H|_{c^*g}$ and leaving the metric implicit, we have

(19.1) $$\frac{\partial}{\partial t} k^2 = \frac{\partial}{\partial t}(|H|^2_{c^*g}) = -2\mathrm{Ric}(H,H) + 2\langle \nabla_H H, H \rangle_{c^*g}.$$

Now we compute (using the second equation from Lemma 19.6)

$$\begin{aligned} \nabla_H H &= \nabla_H \nabla_S S \\ &= \nabla_S \nabla_H S + \nabla_{[H,S]} S + \mathcal{R}(H, S) S \\ &= \nabla_S \nabla_S H + \nabla_S([H,S]) + \nabla_{[H,S]} S + \mathcal{R}(H, S) S \\ &= \nabla_S \nabla_S H + \nabla_S\left((k^2 + \mathrm{Ric}(S,S)) S \right) + (k^2 + \mathrm{Ric}(S,S)) \nabla_S S \\ &\quad + \mathcal{R}(H, S) S \\ &= \nabla_S \nabla_S H + 2(k^2 + \mathrm{Ric}(S,S)) H + S(k^2 + \mathrm{Ric}(S,S)) S + \mathcal{R}(H, S) S. \end{aligned}$$

Using this, and the fact that $\langle H, S \rangle_{c^*g} = 0$, we have
(19.2)
$$2\langle \nabla_H H, H\rangle_{c^*g} = 2g(\nabla_S \nabla_S H, H) + 4k^4 + 4k^2\mathrm{Ric}(S,S)) + 2\mathrm{Rm}(H,S,H,S).$$

On the other hand,

(19.3) $$S(S(\langle H,H\rangle_{c^*g})) = 2\langle \nabla_S \nabla_S H, H\rangle_{c^*g} + 2\langle \nabla_S H, \nabla_S H\rangle_{c^*g}.$$

We write
$$\nabla_S H = (\nabla_S H)^\perp + \langle \nabla_S H, S\rangle_{c^*g} S.$$

Since H and S are orthogonal, we have $\langle \nabla_S H, S\rangle_{c^*g} = -\langle H, \nabla_S S\rangle_{c^*g} = -\langle H, H\rangle$. Thus, we have
$$\nabla_S H = (\nabla_S H)^\perp - \langle H, H\rangle_{c^*g} S.$$

It follows that
$$-2\langle \nabla_S H, \nabla_S H\rangle_{c^*g} = -2\langle (\nabla_S H)^\perp, (\nabla_S H)^\perp\rangle_{c^*g} - 2k^4.$$

Substituting this into Equation (19.3) gives

(19.4) $$2\langle \nabla_S \nabla_S H, H\rangle_{c^*g} = S(S(|H|^2_{c^*g})) - 2\langle (\nabla_S H)^\perp, (\nabla_S H)^\perp\rangle_{c^*g} - 2k^4.$$

Plugging this into Equation (19.2) and using Equation (19.1) yields
$$\frac{\partial}{\partial t}k^2 = -2\mathrm{Ric}(H,H) + S(S\langle H,H\rangle_{c^*g}) - 2\langle (\nabla_S H)^\perp, (\nabla_S H)^\perp\rangle_{c^*g}$$
$$+ 2k^4 + 4k^2\mathrm{Ric}(S,S) + 2\mathrm{Rm}(H,S,H,S).$$

Of course, $S(S(\langle H,H\rangle_{c^*g})) = (k^2)''$ so that this gives the result. □

Grouping together the last three terms in the statement of the previous lemma, we can rewrite the result as

(19.5) $$\frac{\partial}{\partial t}k^2 \leq (k^2)'' - 2\langle (\nabla_S H)^\perp, (\nabla_S H)^\perp\rangle_{c^*g} + 2k^4 + \widehat{C}k^2,$$

where the primes refer to the derivative with respect to arc length along the curve and $\widehat{C}$ is a constant depending only on an upper bound for the norm of the sectional curvatures of the ambient manifolds in the Ricci flow.

CLAIM 19.8. *There is a constant $C_1 < \infty$ depending only on an upper bound for the norm of the sectional curvatures of the ambient manifolds in the Ricci flow $(M, g(t))$, $t_0 \leq t \leq t_1$, such that*
$$\frac{\partial}{\partial t}k \leq k'' + k^3 + C_1 k.$$

PROOF. We set $C_1 = \widehat{C}/2$, where $\widehat{C}$ is as in Inequality (19.5). It follows from Inequality (19.5) that

(19.6) $$2k\frac{\partial k}{\partial t} \leq 2kk'' + 2(k')^2 + 2k^4 - 2\langle (\nabla_S H)^\perp, (\nabla_S H)^\perp\rangle_{c^*g} + \widehat{C}k^2.$$

Since $k^2 = \langle H, H \rangle_{c^*g}$, we see that $(k^2)' = 2\langle \nabla_S H, H \rangle_{c^*g}$. Since H is perpendicular to S, this can be rewritten as $(k^2)' = 2\langle (\nabla_S H)^\perp, H \rangle_{c^*g}$. It follows that
$$k' = \frac{\langle (\nabla_S H)^\perp, H \rangle_{c^*g}}{|H|_{c^*g}}.$$
Hence,
$$(k')^2 \leq \frac{\langle (\nabla_S H)^\perp, H \rangle_{c^*g}^2}{|H|_{c^*g}^2} \leq \langle (\nabla_S H)^\perp, (\nabla_S H)^\perp \rangle_{c^*g}.$$
Plugging this into Equation (19.6) gives
$$\frac{\partial k}{\partial t} \leq k'' + k^3 + C_1 k.$$
□

Now we define the *total length* of the curve $c(x,t)$,
$$L(t) = \int |X|_{c^*g} dx = \int ds.$$
We also define the *total curvature* of the curve $c(x,t)$,
$$\Theta(t) = \int k|X|_{c^*g} dx = \int k\, ds.$$

LEMMA 19.9. *There is a constant $C_2 < \infty$ depending only on an upper bound for the norm of the sectional curvatures of the ambient manifolds in the Ricci flow such that*

(19.7)
$$\frac{d}{dt} L \leq \int (C_2 - k^2) ds$$

and
$$\frac{d}{dt} \Theta \leq C_2 \Theta.$$

PROOF.
$$\frac{d}{dt} L = \int \frac{\partial}{\partial t} \sqrt{|X|_{c^*g}^2}\, dx.$$
By Lemma 19.6 we have
$$\frac{d}{dt} L = \int \frac{1}{2|X|_{c^*g}} \left(-2\mathrm{Ric}(X,X) - 2k^2 |X|_{c^*g}^2 \right) dx.$$
Thus,

(19.8)
$$\frac{d}{dt} L = \int (-\mathrm{Ric}(S,S) - k^2) |X|_{c^*g} dx = \int (-\mathrm{Ric}(S,S) - k^2) ds.$$

The first inequality in the lemma then follows by taking C_2 to be an upper bound for the norm of $\mathrm{Ric}_{g(t)}$.

Now let us consider the second inequality in the statement.
$$\frac{d}{dt} \Theta = \int \frac{\partial}{\partial t} (k|X|_{c^*g}) dx = \int \left(\frac{\partial k}{\partial t} |X|_{c^*g} + k \frac{\partial |X|_{c^*g}}{\partial t} \right) dx.$$

Thus, using Claim 19.8 and the first equation in Lemma 19.6 we have

$$\frac{d}{dt}\Theta \leq \int (k'' + k^3 + C_1 k)ds + \int \frac{k}{2|X|_{c^*g}}(-2\mathrm{Ric}(X,X) - 2k^2|X|^2_{c^*g})dx$$

$$= \int (k'' + k^3 + C_1 k)ds - \int k(\mathrm{Ric}(S,S) + k^2)ds$$

$$= \int (k'' + C_1 k - k\,\mathrm{Ric}(S,S))ds.$$

Since $\int k''ds = 0$ by the fundamental theorem of calculus, we get

$$\frac{d}{dt}\Theta \leq C_2\Theta,$$

for an appropriate constant C_2 depending only on an upper bound for the norm of the sectional curvatures of the ambient family $(M, g(t))$. □

COROLLARY 19.10. *The following holds for the constant C_2 as in the previous lemma. Let $c(x,t)$ be a curve-shrinking flow, let $L(t)$ be the total length of $c(t)$ and let $\Theta(t)$ be the total curvature of $c(t)$. Then for any $t_0 \leq t' < t'' \leq t_1$ we have*

$$L(t'') \leq L(t')e^{C_2(t''-t')},$$
$$\Theta(t'') \leq \Theta(t')e^{C_2(t''-t')}.$$

3. Ramp solutions in $M \times S^1$

As we pointed out in the beginning of Section 1 the main obstacle we must overcome is that the curve-shrinking flow does not always exist for the entire time interval $[t_0, t_1]$. The reason is the following: Even though, as we shall see, it is possible to bound the total curvature of the curve-shrinking flow in terms of the total curvature of the initial curve and the ambient Ricci flow, there is no pointwise estimate on the curvature for the curve-shrinking flow. The idea for dealing with this problem, which goes back to [2], is to replace the original situation of curves in a manifold with graphs by taking the product of the manifold with a circle and using ramps. We shall see that in this context the curve-shrinking flow always exists. The problem then becomes to transfer the information back from the flows of ramps to the original manifold.

Now suppose that the Ricci flow is of the form $(M, g(t)) \times (S^1_\lambda, ds^2)$ where (S^1_λ, ds^2) denotes the circle of length λ. Notice that the sectional curvatures of this product flow depend only on the sectional curvatures of $(M, g(t))$ and, in particular, are independent of λ. Let U denote the vector field made up of unit tangent vectors in the direction of the circle factors. Let $u(x,t) = \langle S, U \rangle_{g(t)}$.

CLAIM 19.11.
$$\frac{\partial u}{\partial t} = u'' + (k^2 + \mathrm{Ric}(S,S))u \geq u'' - C'u,$$
where C' is an upper bound for the norm of the Ricci curvature of $(M, g(t))$.

PROOF. Since U is a constant vector field and hence parallel along all curves and since $\mathrm{Ric}(V, U) = 0$ for all tangent vectors V, by Lemma 19.6 we have
$$\begin{aligned}\frac{\partial}{\partial t}\langle S, U\rangle_{g(t)} &= -2\mathrm{Ric}(S, U) + \langle dc(\nabla_H S), U\rangle_{g(t)} \\ &= \langle dc(\nabla_H S), U\rangle_{g(t)} = \langle dc([H,S] + \nabla_S H), U\rangle_{g(t)} \\ &= (k^2 + \mathrm{Ric}(S,S))u + \langle dc(\nabla_S H), U\rangle_{g(t)} \\ &= (k^2 + \mathrm{Ric}(S,S))u + S(dc(\langle H\rangle), U)_g) \\ &= (k^2 + \mathrm{Ric}(S,S))u + S(\langle dc(\nabla_S S), U\rangle_g) \\ &= (k^2 + \mathrm{Ric}(S,S))u + S(S(u)) = (k^2 + \mathrm{Ric}(S,S))u + u''.\end{aligned}$$
□

DEFINITION 19.12. A curve $c\colon S^1 \to M \times S^1_\lambda$ is said to be a *ramp* if u is strictly positive.

The main results of this section show that the curve-shrinking flow is much better behaved for ramps than for the general smooth curve. First of all, as the next corollary shows, the curve-shrinking flow applied to a ramp produces a one-parameter family of ramps. The main result of this section shows that for any ramp as initial curve, the curve-shrinking flow does not develop singularities as long as the ambient Ricci flow does not.

COROLLARY 19.13. *If $c(x,t)$, $t_0 \leq t < t'_1 < \infty$, is a solution of the curve shrinking flow in $(M, g(t)) \times (S^1_\lambda, ds^2)$ and if $c(t_0)$ a ramp, then $c(t)$ is a ramp for all $t \in [t_0, t'_1)$.*

PROOF. From the equation in Claim 19.11, we see that for C' an upper bound for the norm of the Ricci curvature, we have
$$\frac{\partial}{\partial t}\left(e^{C't}u\right) \geq \left(e^{C't}u\right)''.$$
It now follows from a standard maximum principle argument that the minimum value of $e^{C't}u$ is a non-decreasing function of t. Hence, if $c(t_0)$ is a ramp, then each $c(t)$ is a ramp and in fact $u(x,t)$ is uniformly bounded away from zero in terms of the minimum of $u(x, t_0)$ and the total elapsed time $t_1 - t_0$. □

LEMMA 19.14. *Let $(M, g(t))$, $t_0 \leq t \leq t_1$, be a Ricci flow. Suppose that $c\colon S^1 \to (M \times S^1_\lambda, g(t) \times ds^2)$ is a ramp. Then there is a curve-shrinking flow $c(x,t)$ defined for all $t \in [t_0, t_1]$ with c as the initial condition at time $t = t_0$. The curves $c(\cdot, t)$ are all ramps.*

PROOF. The real issue here is to show that the curve-shrinking flow exists for all $t \in [t_0, t_1]$. Given this, the second part of the statement follows from the previous corollary. If the curve shrinking flow does not exist on all of $[t_0, t_1]$ then by Claim 19.1 there is a $t'_1 \le t_1$ such that the curve-shrinking flow exists on $[t_0, t'_1)$ but k is unbounded on $S^1 \times [t_0, t'_1)$. Thus, to complete the proof we need to see that for any t'_1 for which the curve-shrinking flow is defined on $[t_0, t'_1)$ we have a uniform bound on k on this region.

Using Claim 19.8 and Claim 19.11 we compute

$$\frac{\partial}{\partial t}\left(\frac{k}{u}\right) = \frac{1}{u}\frac{\partial k}{\partial t} - \frac{k}{u^2}\frac{\partial u}{\partial t}$$

$$\le \frac{k'' + k^3 + C_1 k}{u} - \frac{k}{u^2}\left(u'' + (k^2 + \mathrm{Ric}(S,S))u\right)$$

$$= \frac{k''}{u} - \frac{ku''}{u^2} + \frac{C_1 k}{u} - \frac{k}{u}\mathrm{Ric}(S,S).$$

On the other hand,

$$\left(\frac{k}{u}\right)'' = \frac{k''u - u''k}{u^2} - 2\left(\frac{u'}{u}\right)\left(\frac{k'u - u'k}{u^2}\right).$$

Plugging this in, and using the curvature bound on the ambient manifolds we get

$$\frac{\partial}{\partial t}\left(\frac{k}{u}\right) \le \left(\frac{k}{u}\right)'' + \left(\frac{2u'}{u}\right)\left(\frac{k}{u}\right)' + C'\frac{k}{u},$$

for a constant C' depending only on a bound for the norm of the sectional curvature of the ambient Ricci flow. A standard maximum principle argument shows that the maximum of k/u at time t grows at most exponentially rapidly in t. Since u stays bounded away from zero, this implies that for ramp solutions on a finite time interval, the value of k is bounded. □

Next let us turn to the growth rate of the area of a minimal annulus connecting two ramp solutions.

LEMMA 19.15. *Suppose that the dimension, n, of M is at least 3. Let $c_1(x,t)$ and $c_2(x,t)$ be ramp solutions in $(M,g) \times (S^1_\lambda, ds^2)$ with the image under the projection to S^1_λ of each c_i being of degree 1. Let $\mu(t)$ be the infimum of the areas of annuli in $(M \times S^1_\lambda, g(t) \times ds^2)$ with boundary $c_1(x,t) \cup c_2(x,t)$. Then $\mu(t)$ is a continuous function of t and*

$$\frac{d}{dt}\mu(t) \le (2n-1)\mathrm{max}_{x \in M}|\mathrm{Rm}(x,t)|\mu(t),$$

in the sense of forward difference quotients.

PROOF. Fix a time t'. First assume that the loops $c_1(\cdot, t')$ and $c_2(\cdot, t')$ are disjoint. Under Ricci flow the metrics on the manifold immediately become real analytic (see [3]) and furthermore, under the curve-shrinking flow the curves c_1 and c_2 immediately become analytic (see [21]). [Neither of these results is essential for this argument because we could approximate both the metric and the curves by real analytic objects.] Establishing the results for these and taking limits would give the result in general. Since $c_1(\cdot, t')$ and $c_2(\cdot, t')$ are homotopic and are homotopically non trivial, there is an annulus connecting them and there is a positive lower bound to the length of any simple closed curve in any such annulus homotopic to a boundary component. Hence, there is a minimal annulus spanning $c_1(\cdot, t') \coprod c_2(\cdot, t')$ According to results of Hildebrandt ([40]) and Morrey ([52]) any minimal annulus A with boundary the union of these two curves is real analytic up to and including the boundary and is immersed except for finitely many branch points. By shifting the boundary curves slightly within the annulus, we can assume that there are no boundary branch points. Again, if we can prove the result for these perturbed curves, taking limits will give the result for the original ones. Given the deformation vector H on the boundary of the annulus, extend it to a deformation vector $\hat{H}$ on the entire annulus. The first order variation of the area at time t' of the resulting deformed family of annuli is given by

$$\frac{d\text{Area}\, A}{dt}(t') = \int_A (-\text{Tr}(\text{Ric}^T(g(t'))))da + \int_{\partial A} -k_{\text{geod}}ds,$$

where Ric^T is the Ricci curvature in the tangent directions to the annulus. (The first term is the change in the area of the fixed annulus as the metric deforms. The second term is the change in the area of the family of annuli in the fixed metric. There is no contribution from moving the annulus in the normal direction since the original annulus is minimal.) If A is embedded, then by the Gauss-Bonnet theorem, we have

$$\int_{\partial A} -k_{\text{geod}}ds = \int_A K\,da$$

where K is the Gaussian curvature of A. More generally, if A has interior branch points of orders $n_1, \ldots, n_k$ then there is a correction term and the formula is

$$\int_{\partial A} -k_{\text{geod}}ds = \int_A K\,da - \sum_{i=1}^k 2\pi(n_i - 1).$$

Thus, we see

$$\frac{d\text{Area}\, A}{dt}(t') \leq \int_A (-\text{Tr}(\text{Ric}^T(g(t'))) + K)da.$$

On the other hand, since A is a minimal surface, K is at most the sectional curvature of $(M, g(t')) \times (S^1_\lambda, ds^2)$ along the two-plane tangent to the annulus. Of course, the trace of the Ricci curvature along A is at most $2(n-1)|\max_{x \in M} \text{Rm}(x, t')|$. Hence,

$$\frac{d\text{Area}\, A}{dt}(t') \leq (2n-1)|\max_{x \in M} \text{Rm}(x, t')|\mu(t').$$

This computation was done assuming $c_2(\cdot, t')$ is disjoint from $c_1(\cdot, t')$. In general, since the dimension of M is at least 3, given $c_1(\cdot, t')$ and $c_2(\cdot, t')$ we can find $c_3(\cdot, t)$ arbitrarily close to $c_2(\cdot, t')$ in the C^2-sense and disjoint from both $c_1(\cdot, t')$ and $c_2(\cdot, t')$. Let A_3 be a minimal annulus connecting $c_1(\cdot, t')$ to $c_3(\cdot, t')$ and A_2 be a minimal annulus connecting $c_3(\cdot, t')$ to $c_2(\cdot, t')$. We apply the above argument to these annuli to estimate the growth rate of minimal annuli connecting the corresponding curve-shrinking flows. Of course the sum of these areas (as a function of t) is an upper bound for the area of a minimal annulus connecting the curve-shrinking flows starting from $c_1(\cdot, t')$ and $c_2(\cdot, t')$. As we choose $c_3(\cdot, t')$ closer and closer to $c_2(\cdot, t')$, the area of A_2 tends to zero and the area of A_3 tends to the area of a minimal annulus connecting $c_1(\cdot, t')$ and $c_2(\cdot, t')$. This establishes the continuity of $\mu(t)$ at t' and also establishes the forward difference quotient estimate in the general case. □

COROLLARY 19.16. *Given curve-shrinking flows $c_1(\cdot, t)$ and $c_2(\cdot, t)$ for ramps of degree 1 in $(\overline{M}, \overline{g}(t)) \times (S^1_\lambda, ds^2)$ the minimal area of an annulus connecting $c_1(\cdot, t)$ and $c_2(\cdot, t)$ grows at most exponentially with time with an exponent determined by an upper bound on the sectional curvature of the ambient flow, which in particular is independent of λ.*

4. Approximating the original family Γ

Now we are ready to use the curve-shrinking flow for ramps in $M \times S^1_\lambda$ to establish Proposition 18.24 for M. As we indicated above, the reason for replacing the flow $(M, g(t))$ that we are studying with its product with S^1_λ and studying ramps in the product is that the curve-shrinking flow exists for all time $t \in [t_0, t_1]$ for these. By this mechanism we avoid the difficulty of finite time singularities in the curve shrinking flow. On the other hand, we have to translate results for the ramps back to results for the original Ricci flow $(M, g(t))$. This requires careful analysis.

The first step in the proof of Proposition 18.24 is to identify the approximation to the family Γ that we shall use. Here is the lemma that gives the needed approximation together with all the properties we shall use.

Given a loop c in M and $\lambda > 0$ we define a loop c^λ in $M \times S^1_\lambda$. The loop c^λ is obtained by setting $c^\lambda(x) = (c(x), x)$ where we use a standard identification of the domain circle (the unit circle) for the free loop space with S^1_λ, an identification that defines a loop in S^1_λ of constant speed $\lambda/2\pi$.

LEMMA 19.17. *Given a continuous map $\Gamma \colon S^2 \to \Lambda M$ representing an element of $\pi_3(M, *)$ and $0 < \zeta < 1$, there is a continuous map $\widetilde{\Gamma} \colon S^2 \to \Lambda M$ with the following properties:*
  (1) *$[\widetilde{\Gamma}] = [\Gamma]$ in $\pi_3(M, *)$.*
  (2) *For each $c \in S^2$ the loop $\widetilde{\Gamma}(c)$ is a C^2-loop.*
  (3) *For each $c \in S^2$ the length of $\widetilde{\Gamma}(c)$ is within ζ of the length of $\Gamma(c)$.*

(4) For each $c \in S^2$, we have $|A(\widetilde{\Gamma}(c)) - A(\Gamma(c))| < \zeta$.
(5) There is a constant $C_0 < \infty$ depending only on Γ, on the bounds for the norm of the Riemann curvature operator of the ambient Ricci flow, and on ζ such that for each $c \in S^2$ and each $\lambda \in (0,1)$ the total length and the total curvature of the ramp $\widetilde{\Gamma}(c)^\lambda$ are both bounded by C_3.

Before proving this lemma we need some preliminary definitions and constructions.

DEFINITION 19.18. Let $c \colon S^1 \to M$ be a C^1-map. Fix a positive integer n. By a *regular n-polygonal approximation* to c we mean the following. Let $\xi_n = \exp(2\pi i/n)$, and consider the points $p_k = c(\xi_n^k)$ for $k = 1, \ldots, n+1$. For each $1 \leq k \leq n$, let A_k be a minimal geodesic in M from p_k to p_{k+1}. We parameterize A_k by the interval $[\xi_n^k, \xi_n^{k+1}]$ in the circle at constant speed. This gives a piecewise geodesic map $c_n \colon S^1 \to M$.

The following is immediate from the definition.

CLAIM 19.19. *Given $\zeta > 0$ and a C^1-map $c \colon S^1 \to M$ then for all n sufficiently large, the following hold for the n-polygonal approximation c_n of c:*

(a) *The length of c_n is within ζ of the length of c.*
(b) *There is a map of the annulus $S^1 \times I$ to M connecting c_n to c with the property that the image is piecewise smooth and of area less than ζ.*

PROOF. The length of c is the limit of the lengths of the n-polygonal approximations as n goes to infinity. The first item is immediate from this. As to the second, for n sufficiently large, the distance between the maps c and c_n will be arbitrarily small in the C^0-topology, and in particular will be much smaller than the injectivity radius of M. Thus, for each k we can connect A_k to the corresponding part of c by a family of short geodesics. Together, these form an annulus, and it is clear that for n sufficiently large the area of this annulus is arbitrarily small. □

As the next result shows, for $\zeta > 0$, the integer $n(c)$ associated by the previous claim to a C^1-map c can be made uniform as c varies over a compact subset of ΛM.

CLAIM 19.20. *Let $X \subset \Lambda M$ be a compact subset and let $\zeta > 0$ be fixed. Then there is N depending only on X and ζ such the conclusion of the previous claim holds for every $c \in X$ and every $n \geq N$.*

PROOF. Suppose the result is false. Then for each N there is $c_N \in X$ and $n \geq N$ so that the lemma does not hold for c_N and n. Passing to a subsequence, we can suppose that the c_N converge to $c_\infty \in X$. Applying

Claim 19.19 we see that there is N such that the conclusion of Claim 19.19 holds with ζ replaced by $\zeta/2$ for c_∞ and all $n \geq N$. Clearly, then by continuity for all $n \geq N$ the conclusion of Claim 19.19 holds for the n-polygonal approximation for every c_l for all l sufficiently large. This is a contradiction. $\square$

COROLLARY 19.21. *Let $\Gamma\colon S^2 \to \Lambda M$ be a continuous map with the property that $\Gamma(c)$ is homotopically trivial for all $c \in S^2$. Fix $\zeta > 0$. For any n sufficiently large denote by Γ_n the family of loops defined by setting $\Gamma_n(c)$ equal to the n-polygonal approximation to $\Gamma(c)$. There is N such that for all $n \geq N$ we have:*

(1) *Γ_n is a continuous family of n-polygonal loops in M.*
(2) *For each $c \in S^2$, the loop $\Gamma_n(c)$ is a homotopically trivial loop in M and its length is within ζ of the length of $\Gamma(c)$.*
(3) *For each $c \in S^2$, we have $|A(\Gamma_n(c)) - A(\Gamma(c))| < \zeta$.*

PROOF. Given Γ there is a uniform bound over all $c \in S^2$ on the maximal speed of $\Gamma(c)$. Hence, for all n sufficiently large, the lengths of the sides in the n-polygonal approximation to $\Gamma(c)$ will be uniformly small. Once this length is less than the injectivity radius of M, the minimal geodesics between the endpoints are unique and vary continuously with the endpoints. This implies that for n sufficiently large the family Γ_n is uniquely determined and itself forms a continuous family of loops in M. This proves the first item. We have already seen that, for n sufficiently large, for all $c \in S^2$ there is an annulus connecting $\Gamma(c)$ and $\Gamma_n(c)$. Hence, these loops are homotopic in M. The first statement in the second item follows immediately. The last statement in the second item and third item follow immediately from Claim 19.20. $\square$

The next step is to replace these n-polygonal approximations by C^2-curves. We fix, once and for all, a C^∞ function ψ_n from the unit circle to $[0, \infty]$ with the following properties:

(1) ψ_n is non-negative and vanishes to infinite order at the point 1 on the unit circle.
(2) ψ_n is periodic with period $2\pi i/n$.
(3) ψ_n is positive on the interior of the interval $[1, \xi_n]$ on the unit circle, and the restriction of ψ_n to this interval is symmetric about $\exp(\pi i/n)$, and is increasing from 1 to $\exp(\pi i/n)$.
(4) $\int_1^{\xi_n} \psi_n(s)ds = 2\pi/n$.

Now we define a map $\widetilde{\psi}_n\colon S^1 \to S^1$ by

$$\widetilde{\psi}_n(x) = \int_1^x \psi_n(y)dy.$$

It is easy to see that the conditions on ψ_n imply that this defines a C^∞-map from S^1 to S^1 which is a homeomorphism and is a diffeomorphism on the complement of the nth roots of unity.

Now given an n-polygonal loop c_n we define the smoothing $\widetilde{c}_n$ of c_n by $\widetilde{c}_n = c_n \circ \psi_n$. This smoothing $\widetilde{c}_n$ is a C^∞-loop in M with the same length as the original polygonal loop c_n. Notice that the curvature of $\widetilde{c}_n$ is not itself a continuous function: just like the polygonal map it replaces, it has a δ-function at the 'corners' of c_n.

PROOF. (of Lemma 19.17) Given a continuous map $\Gamma \colon S^2 \to \Lambda M$ and $\zeta > 0$ we fix n sufficiently large so that Corollary 19.21 holds for these choices of Γ and ζ. Let $\widetilde{\Gamma} = \widetilde{\Gamma}_n$ be the family of smoothings of the family Γ_n of n-polygonal loops. Since this smoothing operation changes neither the length nor the area of a minimal spanning disk, it follows immediately from the construction and Corollary 19.21 that $\widetilde{\Gamma}$ satisfies the conclusions of Lemma 19.17 except possibly the last one.

To establish the last conclusion we must examine the lengths and total curvatures of the ramps $\widetilde{\Gamma}(c)^\lambda$ associated to this family of C^2-loops. Fix λ with $0 < \lambda < 1$, and consider the product Ricci flow $(M, g(t)) \times (S^1_\lambda, ds^2)$ where the metric on S^1_λ has length λ.

CLAIM 19.22. *For any $0 < \lambda < 1$, the length of the ramp $\widetilde{\Gamma}(c)^\lambda$ is at most λ plus the length of $\Gamma(c)$. The total curvature of $\widetilde{\Gamma}(c)^\lambda$ is at most $n\pi$.*

PROOF. The arc length element for $\widetilde{\Gamma}(c)^\lambda$ is $\sqrt{a(x)^2 + (\lambda/2\pi)^2}dx$, which is at most $(a(x) + \lambda/2\pi)dx$, where $a(x)dx$ is the arc length element for $\widetilde{\Gamma}(c)$. Integrating gives the length estimate.

The total curvature of $\widetilde{\Gamma}(c)^\lambda$ is the sum over the intervals $[\xi_n^k, \xi_n^{k+1}]$ of the total curvature on these intervals. On any one of these intervals we have a curve in a totally geodesic, flat surface: the curve lies in the product of a geodesic arc in M times S^1_λ. Let u and v be unit tangent vectors to this surface, u along the geodesic (in the direction of increasing x) and v along the S^1_λ factor. These are parallel vector fields on the flat surface. The tangent vector $X(x)$ to the restriction of $\widetilde{\Gamma}(c)^\lambda$ to this interval is $L\psi_n(x)u + (\lambda/2\pi)v$, where L is the length of the geodesic segment we are considering. Consider the first-half subinterval $[\xi_n^k, \xi_n^k \cdot \exp(\pi i/n)]$. The tangent vector $X(x)$ is $(\lambda/2\pi)v$ at the initial point of this subinterval and is $L\psi_n(\xi_n^k \cdot \exp(\pi i/2))u + (\lambda/2\pi)v$ at the final point. Throughout this interval the vector is of the form $a(x)u + (\lambda/2\pi)v$ where $a(x)$ is an increasing function of x. Hence, the tangent vector is always turning in the same direction and always lies in the first quadrant (using u and v as the coordinates). Consequently, the total turning (the integral of k against arc-length) over this interval is the absolute value of the difference of the angles at the endpoints. This difference is less than $\pi/2$ and tends to $\pi/2$ as λ tends to zero, unless $L = 0$ in which case there is zero turning for any $\lambda > 0$. By symmetry, the total turning on the second-half subinterval $[\xi_n^k \cdot \exp(\pi i/n), \xi_n^{k+1}]$ is also bounded above by $\pi/2$. Thus, for any $\lambda > 0$, the total turning on one of the segments is bounded above by π. Since there are n segments this gives the upper bound of $n\pi$ on the total turning of $\widetilde{\Gamma}(c)^\lambda$ as required. □

This claim completes the proof of the last property required of $\widetilde{\Gamma} = \widetilde{\Gamma}_n$ and hence completes the proof of Lemma 19.17. □

Having fixed Γ and $\zeta > 0$, we fix n and set $\widetilde{\Gamma} = \widetilde{\Gamma}_n$. We choose n sufficiently large so that $\widetilde{\Gamma}$ satisfies Lemma 19.17. Fix $\lambda \in (0,1)$ and define $\widetilde{\Gamma}^\lambda \colon S^2 \to (M \times S^1_\lambda)$, by setting $\widetilde{\Gamma}^\lambda(c) = \widetilde{\Gamma}(c)^\lambda$.

Fix $c \in S^2$, and let $\widetilde{\Gamma}^\lambda_c(t)$, $t_0 \leq t \leq t_1$, be the curve-shrinking flow given in Lemma 19.14 with initial data the ramp $\widetilde{\Gamma}^\lambda(c)$. As c varies over S^2 these fit together to produce a one-parameter family $\widetilde{\Gamma}^\lambda(t)$ of maps $S^2 \to \Lambda(M \times S^1_\lambda)$. Let p_1 denote the projection of $M \times S^1_\lambda$ to M. Notice that for any λ we have $\widetilde{\Gamma}^\lambda_c(t_0) = \widetilde{\Gamma}^\lambda(c)$, so that $p_1\widetilde{\Gamma}^\lambda_c(t_0) = \widetilde{\Gamma}(c)$. We shall show that for $\lambda > 0$ sufficiently small, the family $p_1\widetilde{\Gamma}^\lambda(t)$ satisfies the conclusion of Proposition 18.24 for the fixed Γ and $\zeta > 0$. We do this in steps. First, we show that fixing one $c \in S^2$, for λ sufficiently small (depending on c) an analogue of Proposition 18.24 holds for the one-parameter family of loops $p_1\widetilde{\Gamma}^\lambda_c(t)$. By this we mean that either $p_1\widetilde{\Gamma}^\lambda_c(t_1)$ has length less than ζ or $A(p_1\widetilde{\Gamma}^\lambda_c(t_1))$ is at most the value $v(t_1) + \zeta$, where v is the solution to Equation (18.4) with initial condition $v(t_0) = A(\widetilde{\Gamma}(c))$. (Actually, we establish a slightly stronger result, see Lemma 19.25.) The next step in the argument is to take a finite subset $\mathcal{S} \subset S^2$ so that for every $c \in S^2$ there is $\hat{c} \in \mathcal{S}$ such that $\widetilde{\Gamma}(c)$ and $\widetilde{\Gamma}(\hat{c})$ are sufficiently close. Then, using the result of a single c, we fix $\lambda > 0$ sufficiently small so that the analogue of Proposition 18.24 for individual curves (or rather the slightly stronger version of it) holds for every $\hat{c} \in \mathcal{S}$. Then we complete the proof of Proposition 18.24 using the fact that for every c the curve $\widetilde{\Gamma}(c)$ is sufficiently close to a curve $\widetilde{\Gamma}(\hat{c})$ associated to an element $\hat{c} \in \mathcal{S}$.

5. The case of a single $c \in S^2$

According to Lemma 19.17, for all $\lambda \in (0,1)$ the lengths and total curvatures of the $\widetilde{\Gamma}^\lambda(c)$ are uniformly bounded for all $c \in S^2$. Hence, by Corollary 19.10 the same is true for $\widetilde{\Gamma}^\lambda_c(t)$ for all $c \in S^2$ and all $t \in [t_0, t_1]$.

CLAIM 19.23. *There is a constant C_4, depending on $t_1 - t_0$, on the curvature bound of the sectional curvature of the Ricci flow $(M, g(t))$, $t_0 \leq t \leq t_1$, on the original family Γ and on ζ such that for any $c \in S^2$ and any $t_0 \leq t' < t'' \leq t_1$ we have*

$$A(p_1\widetilde{\Gamma}^\lambda_c(t'')) - A(p_1\widetilde{\Gamma}^\lambda_c(t')) \leq C_4(t'' - t').$$

PROOF. All the constants in this argument are allowed to depend on $t_1 - t_0$, on the curvature bound of the sectional curvature of the Ricci flow $(M, g(t))$, $t_0 \leq t \leq t_1$, on the original family Γ and on ζ but are independent of λ, $c \in S^2$, and $t' < t''$ with $t_0 \leq t'$ and $t'' \leq t_1$. First, let us consider the surface $S_c^\lambda[t', t'']$ in $M \times S_\lambda^1$ swept out by $c(x, t)$, $t' \leq t \leq t''$. We denote by $\text{Area}(S_c^\lambda[t', t''])$ the area of this surface with respect to the metric $g(t'') \times ds^2$. We compute the derivative of this area for fixed t' as t'' varies. There are two contributions to this derivative: (i) the contribution due to the variation of the metric $g(t'')$ with t'' and (ii) the contribution due to enlarging the surface. The first is

$$\int_{S_c^\lambda[t', t'']} -\text{Tr Ric}^T \, da$$

where Ric^T is the restriction of the Ricci tensor of the ambient metric $g(t'')$ to the tangent planes to the surface and da is the area form of the surface in the metric $g(t'') \times ds^2$. The second contribution is $\int_{c(x,t'')} |H| ds$. According to Lemma 19.9 there is a constant C' (depending only on the curvature bound for the manifold flow, the initial family $\Gamma(t)$ and ζ and $t_1 - t_0$) such that the second term is bounded above by C'. The first term is bounded above by $C'' \text{Area} S_c^\lambda[t', t'']$ where C'' depends only on the bound on the sectional curvatures of the ambient Ricci flow. Integrating we see that there is a constant C'_1 such that the derivative of the area function is at most C'_1. Since its value at t' is zero, we see that $\text{Area } S_c^\lambda[t', t''] \leq C'_1(t'' - t')$. It follows that the area of $p_1 S_c^\lambda[t', t'']$ with respect to the metric $g(t'')$ is at most $C'_1(t'' - t')$.

Now we compute an upper bound for the forward difference quotient of $A(p_1 \widetilde{\Gamma}_c^\lambda(t))$ at $t = t'$. For any $t'' > t'$ we have a spanning disk for $p_1 \widetilde{\Gamma}_c^\lambda(t'')$ defined by taking the union of a minimal spanning disk for $p_1 \widetilde{\Gamma}_c^\lambda(t')$ and the annulus $p_1 S_c^\lambda[t', t'']$. As before, the derivative of the area of this family of disks has two contributions, one coming from the change in the metric over the minimal spanning disk at time t' and the other which we computed above to be at most C'_1. Thus, the derivative is bounded above by $C'_2 A(p_1 \widetilde{\Gamma}_c^\lambda(t')) + C'_1$. This implies that the forward difference quotient of $A(p_1 \widetilde{\Gamma}_c^\lambda(t'))$ is bounded above by the same quantity. It follows immediately that the areas of all the minimal spanning surfaces are bounded by a constant depending only on the areas of the minimal spanning surfaces at time t_0, the sectional curvature of the ambient Ricci flow and $t_1 - t_0$. Hence, there is a constant C_4 such that the forward difference quotient of $A(p_1 \widetilde{\Gamma}_c^\lambda(t))$ is bounded above by C_4. This proves the claim. □

Next, by the uniform bounds on total length of all the curves $\widetilde{\Gamma}_c^\lambda(t)$, it follows from Equation (19.7) that there is a constant C_5 (we take $C_5 > 1$) depending only on the curvature bound of the ambient manifolds and the family Γ such that for any $c \in S^2$ we have

(19.9)
$$\int_{t_0}^{t_1} \int_{\widetilde{\Gamma}_c^\lambda(t)} k^2 ds dt \leq C_5.$$

Thus, for any constant $1 < B < \infty$ there is a subset $I_B(c,\lambda) \subset [t_0, t_1]$ of measure at least $(t_1 - t_0) - C_5 B^{-1}$ such that

$$\int_{\widetilde{\Gamma}_c^\lambda(t)} k^2 ds \leq B$$

for every $t \in I_B(c, \lambda)$. (Later, we shall fix B sufficiently large depending on Γ and ζ.)

Now we need a result for curve-shrinking that in some ways is reminiscent of Shi's theorem for Ricci flows.

LEMMA 19.24. *Let $(M, g(t))$, $t_0 \leq t \leq t_1$, be a Ricci flow. Then there exist constants $\delta > 0$ and $\widetilde{C}_i < \infty$ for $i = 0, 1, 2, \ldots$, depending only on $t_1 - t_0$ and a bound for the norm of the curvature of the Ricci flow, such that the following holds. Let $c(x, t)$ be a curve-shrinking flow that is an immersion for each t. Suppose that at a time t' for some $0 < r < 1$ such that $t' + \delta r^2 < t_1$, the length of $c(\cdot, t')$ is at least r and the total curvature of $c(\cdot, t')$ on any sub-arc of length r is at most δ. Then for every $t \in [t', t' + \delta r^2)$ the curvature k and the higher derivatives satisfy*

$$k^2 \leq \widetilde{C}_0 (t - t')^{-1},$$
$$|\nabla_S H|^2 \leq \widetilde{C}_1 (t - t')^{-2},$$
$$|\nabla_S^i H|^2 \leq \widetilde{C}_i (t - t')^{-(i+1)}.$$

The first statement follows from arguments very similar to those in Section 4 of [**2**]. Once k^2 is bounded by $\widetilde{C}_0/(t - t')$ the higher derivative statements are standard, see [**1**]. For completeness we have included the proof of the first inequality in the last section of this chapter.

We now fix $\delta > 0$ (and also $\delta < 1$) as described in the last lemma for the Ricci flow $(M, g(t))$, $t_0 \leq t \leq t_1$. By Cauchy-Schwarz it follows that for every $t \in I_B(c, \lambda)$, and for any arc J in $\widetilde{\Gamma}_c^\lambda(\cdot, t)$ of length at most $\delta^2 B^{-1}$ we have

$$\int_{J \times \{t\}} k \leq \delta.$$

Applying the previous lemma, for each $a \in I_B(c, \lambda)$ with $a \leq t_1 - B^{-1} - \delta^5 B^{-2}$ we set $J(a) = [a + \delta^5 B^{-2}/2, a + \delta^5 B^{-2}] \subset [t_0, t_1 - B^{-1}]$. Then for all $t \in \cup_{a \in I_B(c,\lambda)} J(a)$ for which the length of $\widetilde{\Gamma}_c^\lambda(\cdot, t)$ is at least $\delta^2 B^{-1}$ we have that k and all the norms of spatial derivatives of H are pointwise uniformly bounded. Since $I_B(c, \lambda)$ covers all of $[t_0, t_1]$ except a subset of measure at most $C_5 B^{-1}$, it follows that the union $\widehat{J}_B(c, \lambda)$ of intervals $J(a)$ for $a \in I_B(c, \lambda) \cap [t_0, t_1 - B^{-1} - \delta^5 B^{-2}]$ cover all of $[t_0, t_1]$ except a subset of measure at most $C_5 B^{-1} + B^{-1} + \delta^5 B^{-1} < 3C_5 B^{-1}$. Now it is straightforward to pass to a finite subset of these intervals $J(a_i)$ that cover all of $[t_0, t_1]$ except a subset of measure at most $3C_5 B^{-1}$. Once we have a finite number of $J(a_i)$, we order them along the interval $[t_0, t_1]$ so that their initial points form an increasing sequence. (Recall that they all have the same length.) Then if we have $J_i \cap J_{i+2} \neq \emptyset$, then J_{i+1} is contained in the union of J_i and J_{i+2} and hence can be removed from the collection without changing the union. In this way we reduce to a finite collection of intervals J_i, with the same union, where every point of $[t_0, t_1]$ is contained in at most two of the intervals in the collection. Once we have arranged this we have a uniform bound, independent of λ and $c \in S^2$, on the number of these intervals. We let $J_B(c, \lambda)$ be the union of these intervals. According to the construction and Lemma 19.24 these sets $J_B(c, \lambda)$ satisfy the following:

(1) $J_B(c, \lambda) \subset [t_0, t_1 - B^{-1}]$ is a union of a bounded number of intervals (the bound being independent of $c \in S^2$ and of λ) of length $\delta^5 B^{-2}/2$.

(2) The measure of $J_B(c, \lambda)$ is at least $t_1 - t_0 - 3C_5 B^{-1}$.

(3) For every $t \in J_B(c, \lambda)$ either the length of $\widetilde{\Gamma}_c^\lambda(t)$ is less than $\delta^2 B^{-1}$ or there are uniform bounds, depending only on the curvature bounds of the ambient Ricci flow and the initial family Γ, on the curvature and its higher spatial derivatives of $\widetilde{\Gamma}_c^\lambda(t)$.

Now we fix $c \in S^2$ and $1 < B < \infty$ and we fix a sequence of λ_n tending to zero. Since the number of intervals in $J_B(c, \lambda)$ is bounded independent of λ, by passing to a subsequence of λ_n we can suppose that the number of intervals in $J_B(c, \lambda_n)$ is independent of n, say this number is N, and that their initial points (and hence the entire intervals since all their lengths are the same) converge as n goes to infinity. Let $\hat{J}_1, \ldots, \hat{J}_N$ be the limit intervals, and for each $i, 1 \leq i \leq N$, let $J_i \subset \hat{J}_i$ be a slightly smaller interval contained in the interior of $\hat{J}_i$. We choose the J_i so that they all have the same length. Let $J_B(c) \subset [t_0, t_1 - B^{-1}]$ be the union of the J_i. Then an appropriate choice of the length of the J_i allows us to arrange the following:

(1) $J_B(c) \subset J_B(c, \lambda_n)$ for all n sufficiently large.

(2) $J_B(c)$ covers all of $[t_0, t_1]$ except a subset of length $4C_5 B^{-1}$.

Now fix one of the intervals J_i making up $J_B(c)$. After passing to a subsequence (of the λ_n), one of the following holds:

(3) there are uniform bounds for the curvature and all its derivatives for the curves $\widetilde{\Gamma}_c^{\lambda_n}(t)$, for all $t \in J_i$ and all n, or

(4) for each n there is $t_n \in J_i$ such that the length of $\widetilde{\Gamma}_c^{\lambda_n}(t_n)$ is less than $\delta^2 B^{-1}$.

By passing to a further subsequence, we arrange that the same one of the Alternatives (3) and (4) holds for every one of the intervals J_i making up $J_B(c)$.

5. THE CASE OF A SINGLE $c \in S^2$

The next claim is the statement that a slightly stronger version of Proposition 18.24 holds for $p_1\widetilde{\Gamma}_c^\lambda(t)$.

LEMMA 19.25. *Given $\zeta > 0$, there is $1 < B < \infty$, with $B > (t_1 - t_0)^{-1}$, depending only on Γ and the curvature bounds on the ambient Ricci flow $(M, g(t))$, $t_0 \leq t \leq t_1$, such that the following holds. Let $t_2 = t_1 - B^{-1}$. Fix $c \in S^2$. Let v_c be the solution to Equation (18.4) with initial condition $v_c(t_0) = A(p_1(\widetilde{\Gamma}(c)))$, so that in our previous notation $v_c = w_{A(p_1(\widetilde{\Gamma}(c)))}$. Then for all $\lambda > 0$ sufficiently small, either $A(p_1\widetilde{\Gamma}_c^\lambda(t_1)) < v_c(t_1) + \zeta/2$ or the length of $\widetilde{\Gamma}_c^\lambda(t)$ is less than $\zeta/2$ for all $t \in [t_2, t_1]$.*

PROOF. In order to establish this lemma we need a couple of claims about functions on $[t_0, t_1]$ that are approximately dominated by solutions to Equation (18.4). In the first claim the function in question is dominated on a finite collection of subintervals by solutions to these equations and the subintervals fill up most of the interval. In the second, we also allow the function to only be approximately dominated by the solutions to Equation (18.4) on these sub-intervals. In both claims the result is that on the entire interval the function is almost dominated by the solution to the equation with the same initial value.

CLAIM 19.26. *Fix C_4 as in Claim 19.23 and fix a constant $\widetilde{A} > 0$. Given $\zeta > 0$ there is $\delta' > 0$ depending on C_4, $t_1 - t_0$, and $\widetilde{A}$ as well as the curvature bound of the ambient Ricci flow such that the following holds. Suppose that $f\colon [t_0, t_1] \to \mathbb{R}$ is a function and suppose that $J \subset [t_0, t_1]$ is a finite union of intervals. Suppose that on each interval $[a, b]$ of J the function f satisfies*

$$f(b) \leq w_{f(a), a}(b).$$

Suppose further that for any $t' < t''$ we have

$$f(t'') \leq f(t') + C_4(t'' - t').$$

Then, provided that the total length of $[t_0, t_1] \setminus J$ is at most δ' and $0 \leq f(t_0) \leq \widetilde{A}$, we have

$$f(t_1) \leq w_{f(t_0), t_0}(t_1) + \zeta/4.$$

PROOF. We write J as a union of disjoint intervals $J_1, \ldots, J_k$ so that $J_i < J_{i+1}$ for every i. Let a_i, resp. b_i, be the initial, resp. final, point of J_i. For each i let δ_i be the length of the interval between J_i and J_{i+1}. (Also, we set $\delta_0 = a_1 - t_0$, and $\delta_k = t_1 - b_k$.) Let $C_6 \geq 0$ be such that $R_{\min}(t) \geq -2C_6$ for all $t \in [t_0, t_1]$. Let $V(a)$ be the maximum value of $|w_{a,t_0}|$ on the interval $[t_0, t_1]$ and let $V = \max_{a \in [0, \widetilde{A}]} V(a)$. Let $C_7 = C_4 + 2\pi + C_6 V$. We shall prove by induction that

$$f(a_i) - w_{f(t_0), t_0}(a_i) \leq \sum_{j=0}^{i-1}\left(C_7 \delta_j \prod_{\ell=j+1}^{i-1} e^{C_6|J_\ell|}\right)$$

and
$$f(b_i) - w_{f(t_0),t_0}(b_i) \leq \sum_{j=0}^{i-1} \left(C_7 \delta_j \prod_{\ell=j+1}^{i} e^{C_6|J_\ell|} \right).$$

We begin the induction by establishing the result at a_1. By hypothesis we know that
$$f(a_1) \leq f(t_0) + C_4 \delta_0.$$
On the other hand, from the defining differential equation for $w_{f(t_0),t_0}$ and the definitions of C_6 and V we have
$$w_{f(t_0),t_0}(a_1) \geq f(t_0) - (C_6 V + 2\pi)\delta_0.$$
Thus,
$$f(a_1) - w_{f(t_0),t_0}(a_1) \leq (C_4 + 2\pi + C_6 V)\delta_0 = C_7 \delta_0,$$
which is exactly the formula given in the case of a_1.

Now suppose that we know the result for a_i and let us establish it for b_i. Let $\alpha_i = f(a_i) - w_{f(t_0),t_0}(a_i)$, and let $\beta_i = f(b_i) - w_{f(t_0),t_0}(b_i)$. Then by Claim 18.26 we have
$$\beta_i \leq e^{C_6|J_i|} \alpha_i.$$
Given the inductive inequality for α_i, we immediately get the one for β_i.

Now suppose that we have the inductive inequality for β_i. Then
$$f(a_{i+1}) \leq f(b_i) + C_4 \delta_i.$$
On the other hand, by the definition of C_6 and V we have
$$w_{f(t_0),t_0}(a_{i+1}) - w_{f(t_0),t_0}(b_i) \geq -(C_6 V + 2\pi)\delta_i.$$
This yields
$$f(a_{i+1}) - w_{f(t_0),t_0}(a_{i+1}) \leq \beta_i + C_7 \delta_i.$$
Hence, the inductive result for β_i implies the result for α_{i+1}. This completes the induction.

Applying this to $a_{k+1} = t_1$ gives
$$f(t_1) - w_{f(t_0),t_0}(t_1) \leq \sum_{j=0}^{k} \left(C_7 \delta_j \prod_{\ell=j+1}^{k} e^{C_6|J_\ell|} \right) \leq C_7 \sum_{j=0}^{k} \delta_j e^{C_6(t_1 - t_0)}.$$

Of course $\sum_{j=1}^{k} \delta_j = t_1 - t_0 - \ell(J) \leq \delta'$, and C_7 only depends on C_6, C_4 and V, while V only depends on $\widetilde{A}$ and C_6 only depends on the sectional curvature bound on the ambient Ricci flow. Thus, given $C_4, \widetilde{A}$ and $t_1 - t_0$ and the bound on the sectional curvature of the ambient Ricci flow, making δ' sufficiently small makes $f(t_1) - w_{f(t_0),t_0}(t_1)$ arbitrarily small. This completes the proof of the claim. □

Here is the second of our claims:

CLAIM 19.27. *Fix $\zeta > 0$, A and C_6, C_4 as in the last claim, and let $\delta' > 0$ be as in the last claim. Suppose that we have $J \subset [t_0, t_1]$ which is a finite disjoint union of intervals with $t_1 - t_0 - |J| \leq \delta'$. Then there is $\delta'' > 0$ (δ'' is allowed to depend on J) such that the following holds. Suppose that we have a function $f \colon [t_0, t_1] \to \mathbb{R}$ such that:*
(1) *For all $t' < t''$ in $[t_0, t_1]$ we have $f(t'') - f(t') \leq C_4(t'' - t')$.*
(2) *For any interval $[a, b] \subset J$ we have $f(b) \leq w_{f(a),a}(b) + \delta''$.*

Then $f(t_1) \leq w_{f(t_0),t_0}(t_1) + \zeta/2$.

PROOF. We define C_7 as in the previous proof. We use the notation $J = J_1 \amalg \cdots \amalg J_k$ with $J_1 < J_2 < \cdots < J_k$ and let δ_i be the length of the interval separating J_{i-1} and J_i. The arguments in the proof of the previous claim work in this context to show that

$$f(a_i) - w_{f(t_0),t_0}(a_i) \leq \sum_{j=0}^{i-1} \left(C_7 \delta_j \prod_{\ell=j+1}^{i-1} (e^{C_6 |J_\ell|} + \delta'') \right).$$

Applying this to a_{k+1} and taking the limit as δ'' tends to zero, the right-hand side tends to a limit smaller than $\zeta/4$. Hence, for δ'' sufficiently small the right-hand side is less than $\zeta/2$. □

Now let us return to the proof of Lemma 19.25. Recall that $c \in S^2$ is fixed. We shall apply the above claims to the curve-shrinking flow $\widetilde{\Gamma}_c^\lambda(t)$ and thus prove Lemma 19.25. Now it is time to fix B. First, we fix $\widetilde{A} = W(\Gamma) + \zeta$, we let C_2 be as in Corollary 19.10, C_4 be as in Claim 19.23, C_5 be as in Equation (19.9), and C_6 be as in the proof of Claim 19.26. Then we have δ' depending on $C_6, C_4, \widetilde{A}$ as in Claim 19.26. We fix B so that:
(1) $B \geq 3C_5(\delta')^{-1}$,
(2) $B \geq 3e^{C_2(t_1 - t_0)} \zeta^{-1}$, and
(3) $B > C_2/(\log 4 - \log 3)$.

The first step in the proof of Lemma 19.25 is the following:

CLAIM 19.28. *After passing to a subsequence of $\{\lambda_n\}$, either:*
(1) *for each n sufficiently large there is $t_n \in J_B(c)$ with the length of $\widetilde{\Gamma}_c^{\lambda_n}(t_n) < \delta^2 B^{-1}$, or*
(2) *for each component $J_i = [t_i^-, t_i^+]$ of $J_B(c)$, after composing $\widetilde{\Gamma}_c^{\lambda_n}(x, t)$ by a reparameterization of the domain circle (fixed in t but a different reparameterization for each n) so that the $\widetilde{\Gamma}_c^{\lambda_n}(t_i^-)$ have constant speed, there is a smooth limiting curve-shrinking flow denoted $\widetilde{\Gamma}_c(t)$, for $t \in J_i$ for the sequence $p_1 \widetilde{\Gamma}_c^{\lambda_n}(t)$, $t_i^- \leq t \leq t_i^+$. The limiting flow consists of immersions.*

PROOF. Suppose that the first case does not hold for any subsequence. Fix a component J_i of $J_B(c)$. Then, by passing to a subsequence, by the fact that $J_B(c) \subset J_B(c, \lambda_n)$ for all n, the curvatures and all the derivatives of the curvatures of $\widetilde{\Gamma}_c^{\lambda_n}(t)$ are uniformly bounded independent of n for all $t \in J_B(c)$. We reparameterize the domain circle so that the $\widetilde{\Gamma}_c^{\lambda_n}(t_i^-)$ have constant speed. By passing to a subsequence we can suppose that the lengths of the $\widetilde{\Gamma}_c^{\lambda_n}(t_i^-)$ converge. The limit is automatically positive since we are assuming that the first case does not hold for any subsequence. Denote by $S_n = S_c^{\lambda_n}(t_i^-)$ the unit tangent vector to $\widetilde{\Gamma}_c^{\lambda_n}(t_i^-)$ and by u_n the inner product $\langle S_n, U \rangle$. Now we have a family of loops with tangent vectors and all higher derivatives bounded. Since u_n is everywhere positive, since $\int u_n ds = \lambda_n$, since the length of the loop $\widetilde{\Gamma}_c^{\lambda_n}(t_i^-)$ is bounded away from 0 independent of n, and since $|(u_n)'| = |\langle \nabla_{S_n} S_n, U \rangle|$ is bounded above independent of n, we see that u_n tends uniformly to zero as n tends to infinity. This means that the $|p_1(S_n)|$ converge uniformly to 1 as n goes to infinity. Since the ambient manifold is compact, passing to a further subsequence we have a smooth limit of the $p_1 \widetilde{\Gamma}_c^{\lambda_n}(t_i^-)$. The result is an immersed curve in $(M, g(t_i^-))$ parameterized at unit speed. Since all the spatial and time derivatives of the $p_1 \widetilde{\Gamma}_c^{\lambda_n}(t)$ are uniformly bounded, by passing to a further subsequence, there is a smooth map $f \colon S^1 \times [t_i^-, t_i^+] \to M$ which is a smooth limit of the sequence $\widetilde{\Gamma}_c^{\lambda_n}(t)$, $t_i^- \leq t \leq t_i^+$. If for some $t \in [t_i^-, t_i^+]$ the curve $f|_{S^1 \times \{t\}}$ is immersed, then this limiting map along this curve agrees to first order with the curve-shrinking flow. Thus, for some $t > t_i^-$ the restriction of f to the interval $[t_i^-, t]$ is a curve-shrinking flow. We claim that f is a curve-shrinking flow on the entire interval $[t_i^-, t_i^+]$. Suppose not. Then there is a first $t' \leq t_i^+$ for which $f|_{S^1 \times \{t'\}}$ is not an immersion. According to Claim 19.1 the maximum of the norms of the curvature of the curves $f(t)$ must tend to infinity as t approaches t' from below. But the curvatures of $f(t)$ are the limits of the curvatures of the family $p_1 \widetilde{\Gamma}_c^{\lambda_n}(t)$ and hence are uniformly bounded on the entire interval $[t_i^-, t_i^+]$. This contradiction shows that the entire limiting surface

$$f \colon S^1 \times [t_i^-, t_i^+] \to (M, g(t))$$

is a curve-shrinking flow of immersions. $\square$

REMARK 19.29. Notice that if the first case holds, then by the choice of B we have a point $t_n \in J_B(c)$ for which the length of $\widetilde{\Gamma}_c^{\lambda_n}(t_n)$ is less than $e^{-C_2(t_1-t_0)}\zeta/3$.

For each n, the family of curves $p_1 \widetilde{\Gamma}_c^{\lambda_n}(t)$ in M all have $p_1 \widetilde{\Gamma}_c^{\lambda_n}(t_0) = \widetilde{\Gamma}(c)$ as their initial member. Thus, these curves are all homotopically trivial. Hence, for each $t \in J_B(c)$ the limiting curve $\widetilde{\Gamma}(c)(t)$ of the $p_1 \widetilde{\Gamma}_c^{\lambda_n}(t)$ is then also homotopically trivial. It now follows from Lemma 19.4, Claim 19.28 and Remark 19.29 that one of the following two conditions holds:

(1) for some $t \in J_B(c)$ the length of $\widetilde{\Gamma}(c)(t)$ is less than or equal to $e^{-C_2(t_1-t_0)}\zeta/3$ or

(2) the function $A(t)$ that assigns to each $t \in J_B(c)$ the area of the minimal spanning disk for $p_1\widetilde{\Gamma}(c)(t)$ satisfies

$$\frac{dA(t)}{dt} \leq -2\pi - \frac{1}{2}R_{\min}(t)A(t)$$

in the sense of forward difference quotients.

By continuity, for any $\delta'' > 0$ then for all n sufficiently large one of the following two conditions holds:

(1) there is $t_n \in J_B(c)$ such that the length of $\widetilde{\Gamma}_c^{\lambda_n}(t_n)$ is less than $e^{-C_2(t_1-t_0)}\zeta/2$, or
(2) for every $t \in J_B(c)$, the areas of the minimal spanning disks for $p_1(\widetilde{\Gamma}_c^{\lambda_n}(t))$ satisfy

$$\frac{dA(p_1\widetilde{\Gamma}_c^{\lambda_n}(t))}{dt} \leq -2\pi - \frac{1}{2}R_{\min}(t)A(p_1\widetilde{\Gamma}_c^{\lambda_n}(t)) + \delta''$$

in the sense of forward difference quotients.

Suppose that for every n sufficiently large, for every $t \in J_B(c)$ the length of $\widetilde{\Gamma}_c^{\lambda_n}(t)$ is at least $e^{-C_2(t_1-t_0)}\zeta/2$. We have already seen in Claim 19.23 that for every $t' < t''$ in $[t_0, t_1]$ the areas satisfy

$$A(p_1(\widetilde{\Gamma}_c^{\lambda_n}(t''))) - A(p_1(\widetilde{\Gamma}_c^{\lambda_n}(t'))) \leq C_4(t''-t').$$

Since the total length of the complement $J_B(c)$ in $[t_0, t_1]$ is at most $3C_5 B^{-1}$, it follows from our choice of B that this total length is at most the constant δ' of Claim 19.26. Invoking Claim 19.27 and the fact that $A(\widetilde{\Gamma}(c)) \leq W(\Gamma(c)) + \zeta \leq W(\Gamma) + \zeta = \widetilde{A}$, we see that for all n sufficiently large we have

$$A(p_1(\widetilde{\Gamma}_{\lambda_n}^c(t_1))) - v_c(t_1) < \zeta/2.$$

The other possibility to consider is that for each n there is $t_n \in J_B(c)$ such that the length of $\widetilde{\Gamma}_c^{\lambda_n}(t_n) < e^{-C_2(t_1-t_0)}\zeta/2$. Since $J_B(c)$ is contained in $[t_0, t_1 - B^{-1}]$, in this case we invoke the first inequality in Corollary 19.10 to see that the length of $\widetilde{\Gamma}_c^{\lambda_n}(t) < \zeta/2$ for every $t \in [t_1 - B^{-1}, t_1]$. This completes the proof of Lemma 19.25. $\square$

6. The completion of the proof of Proposition 18.24

Now we wish to pass from Lemma 19.25 which deals with an individual $c \in S^2$ to a proof of Proposition 18.24 which deals with the entire family $\widetilde{\Gamma}$. Let us introduce the following notation. Suppose that $\omega \subset S^2$ is an arc. Then $\widetilde{\Gamma}(\omega) = \cup_{c \in \omega}\widetilde{\Gamma}(c)$ is an annulus in M and for each $t \in [t_0, t_1]$ we have the annulus $\widetilde{\Gamma}_\omega^\lambda(t)$ in $M \times S_\lambda^1$.

A finite set $\mathcal{S} \subset S^2$ with the property that for $c \in S^2$ there is $\hat{c} \in \mathcal{S}$ and an arc ω in S^2 joining c to $\hat{c}$ so that the area of the annulus $\widetilde{\Gamma}(\omega)$ is less than ν is called a ν-net for $\widetilde{\Gamma}$. Similarly, if for every $c \in S^2$ there is $\hat{c} \in \mathcal{S}$ and an arc ω connecting them for which the area of the annulus $\widetilde{\Gamma}_\omega^\lambda(t_0)$ is less than ν, we say that S is a ν-net for $\widetilde{\Gamma}^\lambda$. Clearly, for any ν there is a subset $S \subset S^2$ that is a ν-net for $\widetilde{\Gamma}$ and for $\widetilde{\Gamma}^\lambda$ for all λ sufficiently small.

LEMMA 19.30. *There is a $\mu > 0$ such that the following holds. Let $c, \hat{c} \in S^2$. Suppose that there is an arc ω in S^2 connecting c to $\hat{c}$ with the area of the annulus $\widetilde{\Gamma}^\lambda_\omega(t_0)$ in $M \times S^1_\lambda$ less than μ. Let $v_{\hat{c}}$, resp., v_c, be the solution to Equation (18.4) with initial condition $v_{\hat{c}}(t_0) = A(\widetilde{\Gamma}(\hat{c}))$, resp., $v_c(t_0) = A(\widetilde{\Gamma}(c))$. If*
$$A(p_1\widetilde{\Gamma}^\lambda_{\hat{c}}(t_1)) \leq v_{\hat{c}} + \zeta/2,$$
then
$$A(p_1(\widetilde{\Gamma}^\lambda_c(t_1)) \leq v_c + \zeta.$$

PROOF. First of all we require that $\mu < e^{-(2n-1)C'(t_1-t_0)}\zeta/4$ where C' is an upper bound for the norm of the Riemann curvature tensor at any point of the ambient Ricci flow. By Lemma 19.15 the fact that the area of the minimal annulus between the ramps $\widetilde{\Gamma}^\lambda_c(t_0)$ and $\widetilde{\Gamma}^\lambda_{\hat{c}}(t_0)$ is less than μ implies that the area of the minimal annulus between the ramps $\widetilde{\Gamma}^\lambda_c(t_1)$ and $\widetilde{\Gamma}^\lambda_{\hat{c}}(t_1)$ is less than $\mu e^{(2n-1)C'(t_1-t_0)} = \zeta/4$. The same estimate also holds for the image under the projection p_1 of this minimal annulus. Thus, with this condition on μ, and for λ sufficiently small, we have
$$\left| A(p_1\widetilde{\Gamma}^\lambda_c(t_1)) - A(p_1\widetilde{\Gamma}^\lambda_{\hat{c}}(t_1)) \right| < \zeta/4.$$

The other condition we impose upon μ is that if $a, \hat{a}$ are positive numbers at most $W(\Gamma) + \zeta$ and if $a < \hat{a} + \mu$ then
$$w_{a,t_0}(t_1) < w_{\hat{a},t_0}(t_1) + \zeta/4.$$
Applying this with $a = A(\widetilde{\Gamma}(c))$ and $\hat{a} = A(\widetilde{\Gamma}(\hat{c}))$ (both of which are at most $W(\widetilde{\Gamma}) < W(\Gamma) + \zeta$), we see that these two conditions on μ together imply the result. □

We must also examine what happens if the second alternative holds for $\widetilde{\Gamma}^\lambda_{\hat{c}}$. We need the following lemma to treat this case.

LEMMA 19.31. *There is $\delta > 0$ such that for any $r > 0$ there is $\overline{\mu} > 0$, depending on r and on the curvature bound for the ambient Ricci flow such that the following holds. Suppose that γ and $\hat{\gamma}$ are ramps in $(M, g(t)) \times S^1_\lambda$. Suppose that the length of γ is at least r and suppose that on any sub-interval I of γ of length r we have*
$$\int_I k ds < \delta.$$
Suppose also that there is an annulus connecting γ and $\hat{\gamma}$ of area less than $\overline{\mu}$. Then the length of $\hat{\gamma}$ is at least $3/4$ the length of γ.

We give a proof of this lemma in the next section. Here we finish the proof of Proposition 18.24 assuming it.

6. THE COMPLETION OF THE PROOF OF PROPOSITION 18.24

CLAIM 19.32. *There is $\mu > 0$ such that the following holds. Suppose that $c, \hat{c} \in S^2$ are such that there is an arc ω in S^2 connecting c and $\hat{c}$ such that the area of the annulus $\widetilde{\Gamma}^\lambda(\omega)$ is at most μ. Set $t_2 = t_1 - B^{-1}$. If the length of $\widetilde{\Gamma}^\lambda_{\hat{c}}(t)$ is less than $\zeta/2$ for all $t \in [t_2, t_1]$, then the length of $p_1 \widetilde{\Gamma}^\lambda_c(t_1)$ is less than ζ.*

PROOF. The proof is by contradiction. Suppose the length of $p_1 \widetilde{\Gamma}^\lambda_c(t_1)$ is at least ζ and the length of $\widetilde{\Gamma}^\lambda_{\hat{c}}(t)$ is less than $\zeta/2$ for all $t \in [t_2, t_1]$. Of course, it follows that the length of $\widetilde{\Gamma}^\lambda_c(t_1)$ is also at least ζ. The third condition on B is equivalent to

$$e^{C_2 B^{-1}} < 4/3.$$

It then follows from Corollary 19.10 that for every $t \in [t_2, t_1]$ the length of $\widetilde{\Gamma}^c_\lambda(t)$ is at least $3\zeta/4$. On the other hand, by hypothesis for every such t, the length of $\widetilde{\Gamma}^\lambda_{\hat{c}}(t)$ is less than $\zeta/2$. It follows from Equation (19.7) that

$$\int_{t_2}^{t_1} \left(\int k^2 ds \right) dt \leq C_2 \left(\int_{t_2}^{t_1} L(\widetilde{\Gamma}^\lambda_c(t)) dt \right) - L(\widetilde{\Gamma}^\lambda_c(t_1)) + L(\widetilde{\Gamma}^\lambda_c(t_2)).$$

(Here L is the length of the curve.) From this and Corollary 19.10 we see that there is a constant C_8 depending on the original family Γ and on the curvature of the ambient Ricci flow such that

$$\int_{t_2}^{t_1} \left(\int_{\widetilde{\Gamma}^\lambda_c(t)} k^2 ds \right) dt \leq C_8.$$

Since $t_1 - t_2 = B^{-1}$, this implies that there is $t' \in [t_2, t_1]$ with

$$\int_{\widetilde{\Gamma}^\lambda_c(t')} k^2 ds \leq C_8 B.$$

By Cauchy-Schwarz, for any subinterval I of length $\leq r$ in $\widetilde{\Gamma}^\lambda_c(t')$ we have

$$\int_I k\, ds \leq \sqrt{C_8 B r}.$$

We choose $0 < r \leq \zeta$ sufficiently small so that $\sqrt{C_8 B r}$ is less than or equal to the constant δ given in Lemma 19.31. Then we set $\overline{\mu}$ equal to the constant given by that lemma for this value of r.

Now suppose that μ is sufficiently small so that the solution to the equation

$$\frac{d\mu(t)}{dt} = (2n-1)|\text{Rm}_{g(t)}|\mu(t)$$

with initial condition $\mu(t_0) \leq \mu$ is less than $\overline{\mu}$ on the entire interval $[t_0, t_1]$. With this condition on μ, Lemma 19.15 implies that for every $t \in [t_0, t_1]$ the ramps $\widetilde{\Gamma}^\lambda_c(t)$ and $\widetilde{\Gamma}^\lambda_{\hat{c}}(t)$ are connected by an annulus of area at most $\overline{\mu}$. In particular, this is true for $\widetilde{\Gamma}_c^\lambda(t')$ and $\widetilde{\Gamma}_{\hat c}^\lambda(t')$. Now we have all the hypotheses of Lemma 19.31 at time t'. Applying this lemma we conclude that

$$L(\widetilde{\Gamma}_{\hat c}^\lambda(t')) \geq \frac{3}{4} L(\widetilde{\Gamma}_c^\lambda(t')).$$

But this is a contradiction since by assumption $L(\widetilde{\Gamma}_{\hat c}^\lambda(t')) < \zeta/2$ and the supposition that $L(p_1\widetilde{\Gamma}_c^\lambda(t_1)) \geq \zeta$ led to the conclusion that $L(\widetilde{\Gamma}_c^\lambda(t')) \geq 3\zeta/4$. This contradiction shows that our supposition that $L(p_1\widetilde{\Gamma}_c^\lambda(t_1)) \geq \zeta$ is false. $\square$

Now we complete the proof of Proposition 18.24.

PROOF. (of Proposition 18.24.) Fix $\mu > 0$ sufficiently small so that Lemma 19.30 and Claim 19.32 hold. Then we choose a $\mu/2$-net X for $\widetilde{\Gamma}$. We take λ sufficiently small so that Lemma 19.25 holds for every $\hat c \in \mathcal{S}$. We also choose λ sufficiently small so that X is a μ-net for $\widetilde{\Gamma}^\lambda$. Let $c \in S^2$. Then there is $\hat c \in \mathcal{S}$ and an arc ω connecting c and $\hat c$ such that the area of $\widetilde{\Gamma}^\lambda(\omega) < \mu$. Let $v_{\hat c}$, resp. v_c, be the solution to Equation (18.4) with initial condition $v_{\hat c}(t_0) = A(\widetilde{\Gamma}(\hat c))$, resp. $v_c(t_0) = A(\widetilde{\Gamma}(c))$. According to Lemma 19.25 either $A(p_1\widetilde{\Gamma}_{\hat c}^\lambda(t_1)) < v_{\hat c}(t_1) + \zeta/2$ or the length of $\widetilde{\Gamma}_{\hat c}^\lambda(t)$ is less than $\zeta/2$ for every $t \in [t_2, t_1]$ where $t_2 = t_1 - B^{-1}$. In the second case, Claim 19.32 implies that the length of $p_1\widetilde{\Gamma}_c^\lambda(t_1)$ is less than ζ. In the first case, Lemma 19.30 tells us that $A(p_1\widetilde{\Gamma}_c^\lambda(t_1)) < v_c(t_1) + \zeta$. This completes the proof of Proposition 18.24. $\square$

7. Proof of Lemma 19.31: annuli of small area

Except for the brief comments that follow, our proof involves geometric analysis that takes place on an abstract annulus with bounds on its area, upper bounds on its Gaussian curvature, and on integrals of the geodesic curvature on the boundary. Proposition 19.35 below gives the precise result along these lines. Before stating that proposition, we show that its hypotheses hold in the situation that arises in Lemma 19.31. Let us recall the situation of Lemma 19.31. We have ramps γ and $\hat\gamma$, which are real analytic embedded curves in the real analytic Riemannian manifold $(M \times S^1_\lambda, g \times ds^2)$. By a slight perturbation we can assume they are disjointly embedded. These curves are connected by an annulus $A_0 \to M \times S^1_\lambda$ of small area, an area bounded above by, say, μ. We take an energy minimizing map of an annulus $\psi \colon A \to M \times S^1_\lambda$ spanning $\gamma \coprod \hat\gamma$. According to [40], ψ is a real analytic map and the only possible singularities (non-immersed points) of the image come from the *branch points* of ψ, i.e., points where $d\psi$ vanishes. There are finitely many branch points. If there are branch points on the boundary, then the restriction of ψ to ∂A will be a homeomorphism rather than a diffeomorphism onto $\gamma \coprod \hat\gamma$. Outside the branch points, ψ is a conformal map onto its image. The image is an area minimizing annulus spanning $\gamma \coprod \hat\gamma$.

7. PROOF OF LEMMA 19.31: ANNULI OF SMALL AREA

Thus, the area of the image is at most μ. According to [**71**] the only branch points on the boundary are false branch points, meaning that a local smooth reparameterization of the map on the interior of A near the boundary branch point removes the branch point. These reparameterizations produce a new smooth structure on A, identified with the original smooth structure on the complement of the boundary branch points. Using this new smooth structure on A the map ψ is an immersion except at finitely many interior branch points. From now on the domain surface A is endowed with this new smooth structure. Notice that, after this change, the domain is no longer real analytic; it is only smooth. Also, the original annular coordinate is not smooth at the finitely many boundary branch points.

The pullback of the metric $g \times ds^2$ is a smooth symmetric two-tensor on A. Off the finite set of interior branch points it is positive definite and hence a Riemannian metric, and in particular, it is a Riemannian metric near the boundary. It vanishes at each interior branch point. Since the geodesic curvature k_{geod} of the boundary of the annulus is given by $k \cdot n$ where n is the unit normal vector along the boundary pointing into A, we see that the restriction of the geodesic curvature to γ, $k_{\text{geod}} \colon \gamma \to \mathbb{R}$ has the property that for any sub-arc I of γ of length r we have

$$\int_I |k_{\text{geod}}| ds < \delta.$$

Lastly, because the map of A into $M \times S^1_\lambda$ is minimal, off the set of interior branch points, the Gaussian curvature of the pulled back metric is bounded above by the upper bound for the sectional curvature of $M \times S^1_\lambda$, which itself is bounded independent of λ and t, by say $C' > 0$.

Next, let us deal with the singularities of the pulled back metric on A caused by the interior branch points. As the next claim shows, it is an easy matter to deform the metric slightly near each branch point without increasing the area much and without changing the upper bound on the Gaussian curvature too much. Here is the result:

CLAIM 19.33. *Let* $\psi \colon A \subset M \times S^1_\lambda$ *be an area-minimizing annulus of area at most* μ *with smoothly embedded boundary as constructed above. Let h be the induced (possibly singular) metric on A induced by pulling back $g \times ds^2$ by ψ, and let $C'' > 0$ be an upper bound on the Gaussian curvature of h (away from the branch points). Then there is a deformation $\tilde{h}$ of h, supported near the interior branch points, to a smooth metric with the property that the area of the deformed smooth metric is at most 2μ and where the upper bound for the curvature of $\tilde{h}$ is $2C''$.*

PROOF. Fix an interior branch point p. Since ψ is smooth and conformal onto its image, there is a disk in A centered at p in which $h = f(z, \bar{z})|dz|^2$ for a smooth function f on the disk. The function f vanishes at the origin and is positive on the complement of the origin. Direct computation shows that the Gaussian curvature $K(h)$ of h in this disk is given by

$$K(h) = \frac{-\Delta f}{2f^2} + \frac{|\nabla f|^2}{2f^3} \leq C,$$

where Δ is the usual Euclidean Laplacian on the disk and

$$|\nabla f|^2 = (\partial f/\partial x)^2 + (\partial f/\partial y)^2.$$

Now consider the metric $(f + \epsilon)|dz|^2$ on the disk. Its Gaussian curvature is

$$\frac{-\Delta f}{2(f+\epsilon)^2} + \frac{|\nabla f|^2}{2(f+\epsilon)^3}.$$

CLAIM 19.34. *For all $\epsilon > 0$ the Gaussian curvature of $(f + \epsilon)|dz|^2$ is at most $2C'''$.*

PROOF. We see that $-\Delta f \leq C'' f^2$, so that

$$\frac{-\Delta f}{(f+\epsilon)^2} + \frac{|\nabla f|^2}{(f+\epsilon)^3} = \frac{(f+\epsilon)(-\Delta f) + |\nabla f|^2}{(f+\epsilon)^3}$$

$$\leq \frac{C'' f^3 - \epsilon \Delta f}{(f+\epsilon)^3} \leq C'' + \frac{\epsilon f^2 C''}{(f+\epsilon)^3} \leq 2C''.$$

□

Now we fix a smooth function $\rho(r)$ which is identically 1 on a subdisk D' of D and vanishes near ∂D and we replace the metric h on the disk by

$$h_\epsilon = (f + \epsilon \rho(r))|dz|^2.$$

The above computation shows that the Gaussian curvature of h_ϵ on D' is bounded above by $2C''$. As ϵ tends to zero the restriction of the metric h_ϵ to $D \setminus D'$ converges uniformly in the C^∞-topology to h. Thus, for all $\epsilon > 0$ sufficiently small the Gaussian curvature of h_ϵ on $D \setminus D'$ will also be bounded by $2C''$. Clearly, as ϵ tends to zero the area of the metric h_ϵ on D tends to the area of h on D.

Performing this construction near each of the finite number of interior branch points and taking ϵ sufficiently small gives the perturbation $\widetilde{h}$ as required. □

Thus, if γ and $\hat{\gamma}$ are ramps as in Lemma 19.31, then replacing $\hat{\gamma}$ by a close C^2 approximation we have an abstract smooth annulus with a Riemannian metric connecting γ and $\hat{\gamma}$. Taking limits shows that establishing the conclusion of Lemma 19.31 for a sequence of better and better approximations to $\hat{\gamma}$ will also establish it for $\hat{\gamma}$. This allows us to assume that γ and $\hat{\gamma}$ are disjoint. The area of this annulus is bounded above by a constant arbitrarily close to μ. The Gaussian curvature of the Riemannian metric is bounded above by a constant depending only on the curvature bounds of the ambient Ricci flow. Finally, the integral of the absolute value of the geodesic curvature over any interval of length r of γ is at most δ.

With all these preliminary remarks, we see that Lemma 19.31 follows from:

PROPOSITION 19.35. *Fix $0 < \delta < 1/100$. For each $0 < r$ and $C'' < \infty$ there is a $\mu > 0$ such that the following holds. Suppose that A is an annulus with boundary components c_0 and c_1. Denote by $l(c_0)$ and $l(c_1)$ the lengths of c_0 and c_1, respectively. Suppose that the Gaussian curvature of A is bounded above by C''. Suppose that $l(c_0) > r$ and that for each sub-interval I of c_0 of length r, the integral of the absolute value of the geodesic curvature along I is less than δ. Suppose that the area of A is less than μ. Then*

$$l(c_1) \geq \frac{3}{4}l(c_0).$$

To us, this statement was intuitively extremely reasonable but we could not find a result along these lines stated in the literature. Also, in the end, the argument we constructed is quite involved, though elementary.

The intuition is that we exponentiate in from the boundary component c_0 using the family of geodesics perpendicular to the boundary. The bounds on the Gaussian curvature and local bounds on the geodesic curvature of c_0 imply that the exponential mapping will be an immersion out to some fixed distance δ or until the geodesics meet the other boundary, whichever comes first. Furthermore, the metric induced by this immersion will be close to the product metric. Thus, if there is not much area, it must be the case that, in the measure sense, most of the geodesics in this family must meet the other boundary before distance δ. One then deduces the length inequality.

There are two main difficulties with this argument that must be dealt with. The first is due to the fact that we do not have a pointwise bound on the geodesic curvature of c_0, only an integral bound of the absolute value over all curves of short length. There may be points of arbitrarily high geodesic curvature. Of course, the length of the boundary where the geodesic curvature is large is very small. On these small intervals the exponential mapping will not be an immersion out to any fixed distance.

We could of course, simply omit these regions from consideration and work on the complement. But these small regions of high geodesic curvature on the boundary can cause focusing (i.e., crossing of the nearby geodesics). We must estimate out to what length along the boundary this happens. Our first impression was that the length along the boundary where focusing occurred would be bounded in terms of the total turning along the arc in c_0. We were not able to establish this. Rather we found a weaker estimate where this focusing length is bounded in terms of the total turning and the area bounded by the triangle cut out by the two geodesics that meet. This is a strong enough result for our application. Since the area is small and the turning on any interval of length r is small, a maximal collection of focusing regions will meet each interval of length r in c_0 in a subset of small total length. Thus, on the complement (which is most of the length of

c_0) the exponential mapping will be an immersion out to length δ and will be an embedding when restricted to each interval of length 1. The second issue to face is to show that the exponential mapping on this set is in fact an embedding, not just an immersion. Here one uses standard arguments invoking the Gauss-Bonnet theorem to rule out various types of pathologies, e.g., that the individual geodesics are not embedded or geodesics that end on c_0 rather than c_1, etc. Once these are ruled out, one has established that the exponential map on this subset is an embedding and the argument finishes as indicated above.

7.1. First reductions. Of course, if the hypothesis of the proposition holds for $r > 0$ then it holds for any $0 < r' < r$. This allows us to assume that $r < \min((C'')^{-1/2}, 1)$. Now let us scale the metric by $4r^{-2}$. The area of A with the rescaled metric is $4r^{-2}$ times the area of A with the original metric. The Gaussian curvature of A with the rescaled metric is less than $r^2 C''/4 \le 1$. Furthermore, in the rescaled metric c_0 has length greater than 2 and the total curvature along any interval of length 1 in c_0 is at most δ. This allows us to assume (as we shall) that $r = 1$, that $C'' \le 1$, and that $l(c_0) \ge 2$. We must find a $\mu > 0$ such that the proposition holds provided that the area of the annulus is less than μ.

The function $k_{\text{geod}} \colon c_0 \to \mathbb{R}$ is smooth. We choose a regular value α for k_{geod} with $1 < \alpha < 1.1$. In this way we divide c_0 into two disjoint subsets, Y where $k_{\text{geod}} > \alpha$, and X where $k_{\text{geod}} \le \alpha$. The subset Y is a union of finitely many disjoint open intervals and X is a disjoint union of finitely many closed intervals.

REMARK 19.36. The condition on k_{geod} implies that for any arc J in c_0 of length 1 the total length of $J \cap Y$ is less than δ.

Fix $\delta' > 0$. For each $x \in X$ there is a geodesic D_x in A whose initial point is x and whose initial direction is orthogonal to c_0. Let $f(x)$ be the minimum of δ and the distance along D_x to the first point (excluding x) of its intersection with ∂A. We set

$$S_X(\delta') = \{(x,t) \in X \times [0, \delta'] \mid t \le f(x)\}.$$

The subset $S_X(\delta')$ inherits a Riemannian metric from the product of the metric on X induced by the embedding $X \subset c_0$ and the standard metric on the interval $[0, \delta']$.

CLAIM 19.37. *There is $\delta' > 0$ such that the following holds. The exponential mapping defines a map* $\exp \colon S_X(\delta') \to A$ *which is a local diffeomorphism and the pullback of the metric on A defines a metric on $S_X(\delta')$ which is at least $(1-\delta)^2$ times the given product metric.*

PROOF. This is a standard computation using the Gaussian curvature upper bound and the geodesic curvature bound. □

7. PROOF OF LEMMA 19.31: ANNULI OF SMALL AREA

Now we fix $0 < \delta' < 1/10$ so that Claim 19.37 holds, and we set $S_X = S_X(\delta')$. We define
$$\partial_+ S_X = \{(x,t) \in S_X \,|\, t = f(x)\}.$$
Then the boundary of S_X is made up of X, the arcs $\{x\} \times [0, f(x)]$ for $x \in \partial X$ and $\partial_+(S_X)$. For any subset $Z \subset X$ we denote by S_Z the intersection $(Z \times [0, \delta]) \cap S_X$, and we denote by $\partial_+(S_Z)$ the intersection of $S_Z \cap \partial_+ S_X$.

Lastly, we fix $\mu > 0$ with $\mu < (1-\delta)^2(\delta')/10$. Notice that this implies that $\mu < 1/100$. We now assume that the area of A is less than this value of μ (and recall that $r = 1$, $C'' = 1$ and $l(c_0) \geq 2$). We must show that $l(c_1) > 3l(c_0)/4$.

7.2. Focusing triangles. By a *focusing triangle* we mean the following. We have distinct points $x, y \in X$ and sub-geodesics $D'_x \subset D_x$ and $D'_y \subset D_y$ that are embedded arcs with x, respectively y, as an endpoint. The intersection $D'_x \cap D'_y$ is a single point which is the other endpoint of each of D'_x and D'_y. Notice that since $D'_x \subset D_x$ and $D'_y \subset D_y$, by construction both D'_x and D'_y have lengths at most δ'. We have an arc ξ in c_0 with endpoints x and y and the loop $\xi * D'_y * (D'_x)^{-1}$ bounds a disk B in A. The arc ξ is called the *base* of the focusing triangle and with, respect to an orientation of c_0, if x is the initial point of ξ then D'_x is called the *left-hand side* of the focusing triangle and D'_y is called its *right-hand side*. See FIG. 1.

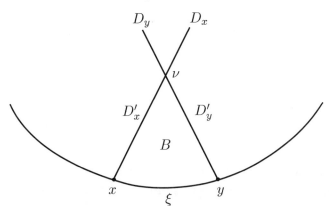

FIGURE 1. Focusing triangle.

Our main goal here is the following lemma which gives an upper bound for the length of the base, ξ, of a focusing triangle in terms of the turning along the base and the area of the region B enclosed by the triangle.

LEMMA 19.38. *Suppose that we have a focusing triangle T with base ξ bounding a disk B in A. Suppose that the length of ξ is at most 1. Then*
$$l(\xi) \leq \left(\int_\xi k_{\text{geod}} ds + \text{Area}(B) \right).$$

PROOF. We begin with a preliminary computation. We denote by $a(B)$ the area of B. We define

$$t_\xi = \int_\xi k_{\text{geod}} ds \quad \text{and} \quad T_\xi = \int_\xi |k_{\text{geod}}| ds.$$

Recall that given a piecewise smooth curve, its *total turning* is the integral of the geodesic curvature over the smooth part of the boundary plus the sum over the break points of π minus the interior angle at the break point. The Gauss-Bonnet theorem tells us that for any compact surface with piecewise smooth boundary the integral of the Gaussian curvature over the interior of the surface plus the total turning around the boundary equals 2π times the Euler characteristic of the surface.

CLAIM 19.39. *The angle θ_B between D'_x and D'_y at the vertex v satisfies*

$$\theta_B \leq t_\xi + a(B)$$

and for any measurable subset $B' \subset B$ we have

$$\theta_B - t_\xi - a(B) \leq \int_{B'} K \, da < a(B).$$

PROOF. Since D'_x and D'_y meet ∂A in right angles, the total turning around the boundary of B is

$$t_\xi + 2\pi - \theta_B.$$

Thus, by Gauss-Bonnet, we have

$$\theta_B = \int_B K \, da + t_\xi.$$

But $K \leq 1$, giving the first stated inequality. On the other hand $\int_B K \, da = \int_B K^+ da + \int_B K^- da$, where $K^+ = \max(K, 0)$ and $K^- = K - K^+$. Since $0 \leq \int_B K^+ da \leq a(B)$ and $\int_B K^- \leq 0$, the second string of inequalities follows. □

In order to make the computation we need to know that this triangle is the image under the exponential mapping of a spray of geodesics out of the vertex v. Establishing that requires some work.

CLAIM 19.40. *Let $a \in \text{int}\,\xi$. There is a shortest path in B from a to v. This shortest path is a geodesic meeting ∂B only in its end points. It has length $\leq (1/2) + \delta'$.*

PROOF. The length estimate is obvious: Since ξ has length at most 1, a path along $\partial A \cap B$ from a to the closest of x and y has length at most $1/2$. The corresponding side has length at most δ'. Thus, there is a path from a to v in B of length at most $(1/2) + \delta'$.

Standard convergence arguments show that there is a shortest path in B from a to v. Fix $a \in \text{int}(\partial A \cap B)$. It is clear that the shortest path cannot

7. PROOF OF LEMMA 19.31: ANNULI OF SMALL AREA

meet either of the 'sides' D'_x and D'_y at any point other than v. If it did, then there would be an angle at this point and a local shortcut, cutting off a small piece of the angle, would provide a shorter path. We must rule out that the shortest path from a to v meets $\partial A \cap B$ in another point. If it does, let a' be the last such point (parameterizing the geodesic starting at a). The shortest path from a then leaves ∂A at a' in the direction tangent at a' to ∂A. (Otherwise, we would have an angle which would allow us to shorten the path just as before.) This means that we have a geodesic γ from v to a' whose interior is contained in the interior of B and which is tangent to ∂A at a'. We label the endpoints of $\partial A \cap B$ so that the union of γ and the interval on $\partial A \cap B$ from a' to y gives a C^1-curve. Consider the disc B' bounded by γ, the arc of ∂A from a' to y, and D'_y. The total turning around the boundary is at most $3\pi/2 + \delta$, and the integral of the Gaussian curvature over B' is at most the area of B, which is less than $\mu < 1/20 < (\pi/4) - \delta$. This contradicts the Gauss-Bonnet theorem. □

CLAIM 19.41. *For any $a \in \partial A \cap B$ there is a unique minimal geodesic in B from a to v.*

PROOF. Suppose not; suppose there are two γ and γ' from v to a. Since they are both minimal in B, each is embedded, and they must be disjoint except for their endpoints. The upper bound on the curvature and the Gauss-Bonnet theorem imply that the angles that they make at each endpoint are less than $\mu < \pi/2$. Thus, there is a spray of geodesics (i.e. geodesics determined by an interval β in the circle of directions at v) coming out of v and moving into B with extremal members of the spray being γ and γ'. The geodesics γ and γ' have length at most $(1/2) + \delta'$, and hence the exponential mapping from v is a local diffeomorphism on all geodesics of length at most the length of γ. Since the angle they make at a is less than $\pi/2$ and since the exponential mapping is a local diffeomorphism near γ, as we move in from the γ end of the spray we find geodesics from v of length less than the length of γ ending on points of γ'. The same Gauss-Bonnet argument shows that the angle that each of these shorter geodesics makes with γ' is at most μ. Consider the subset β' of β which are directions of geodesics in B of length $< (1/2) + \delta'$ that end on points of γ' and make an angle less than μ with γ'. We have just seen that β' contains an open neighborhood of the end of β corresponding to γ. Since the Gaussian curvature is bounded above by 1, and these geodesics all have length at most $1/2 + \delta$, it follows that the exponential map is a local diffeomorphism near all such geodesics. Thus, β' is an open subset of β. On the other, hand if the direction of $\gamma'' \neq \gamma'$ is a point $b'' \in \beta$ which is an endpoint of an open interval β', and if this interval separates b'' from the direction of γ, then the length of γ'' is less than the length of each point in the interval. Hence, the length of γ'' is less than

$(1/2) + \delta'$. Invoking Gauss-Bonnet again we see that the angle between γ'' and γ' is $< \mu$.

This proves that if U is an open interval in β' then the endpoint of U closest to the direction of γ' is also contained in β' (unless that endpoint is the direction of γ'). It is now elementary to see that β' is all of β except the endpoint corresponding to γ'. But this is impossible. Since the exponential mapping is a local diffeomorphism out to distance $(1/2) + \delta'$, and since γ' is embedded, any geodesic from v whose initial direction is sufficiently close to that of γ' and whose length is at most $(1/2) + \delta'$ will not cross γ'. □

See FIG. 2

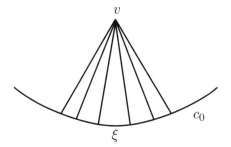

FIGURE 2. Spray of geodesics from v.

REMARK 19.42. The same argument shows that from any $a \in \partial A \cap B$ there is a unique embedded geodesic in B from v to a with length at most $(1/2) + \delta'$. (Such geodesics may cross more than once, but the argument given in the lemma applies to sub-geodesics from v to the first point of intersection along γ.)

Let E be the sub-interval of the circle of tangent directions at v consisting of all tangent directions of geodesics pointing into B at v. The endpoints of E are the tangent directions for D'_x and D'_y. We define a function from ξ to the interval E by assigning to each $a \in \xi$ the direction at v of the unique minimal geodesic in B from v to a. Since the minimal geodesic is unique, this function is continuous and, by the above remark, associates to x and y the endpoints of E. Since geodesics are determined by their initial directions, this function is one-to-one. Hence it is a homeomorphism from ξ to E. That is to say the spray of geodesics coming out of v determined by the interval E produces a diffeomorphism between a wedge-shaped subset of the tangent space at v and B. Each of the geodesics in question ends when it meets ξ.

Now that we have shown that the region enclosed by the triangle is the image under the exponential map from the vertex v of a wedge-shaped region in the tangent space at v, we can make the usual computation relating length and geodesic curvature. To do this we pull back to the tangent space at v, and, using polar coordinates, we write ξ as $\{s = h(\psi); \psi \in E\}$ where s is the radial coordinate and ψ is the angular coordinate. Notice that $h(\psi) \leq (1/2) + \delta'$ for all $\psi \in E$. (In fact, because the angles of intersection at the boundary are all close to $\pi/2$ we can give a much better estimate on h but we do not need it.) We consider the one-parameter family of arcs $\lambda(t)$ defined to be the graph of the function $t \mapsto s(t) = th(\psi)$, for $0 \leq t \leq 1$. We set $l(t)$ equal to the length of $\lambda(t)$.

CLAIM 19.43.
$$\frac{dl}{dt}(t) \leq \max_{\psi \in E} h(\psi) \int_{\lambda(t)} k_{\text{geod}} ds.$$

PROOF. First of all notice that, by construction, the curve ξ, which is defined by $\{s = h(\psi)\}$, is orthogonal to the radial geodesics to the endpoints. As a consequence, $h'(\psi) = 0$ at the endpoints. Thus, each of the curves $\lambda(t)$ is orthogonal to the radial geodesics through its end points. Therefore, as we vary the family $\lambda(t)$ the formula for the derivative of the length is

$$l'(t) = \int_{\lambda(t)} k_{\text{geod}}(\psi) h(\psi) |\cos(\theta(\psi, t))| ds$$

where $\theta(\psi, t)$ is the angle at $(th(\psi), \psi)$ between the curve $s = th(\psi)$ and the radial geodesic. The result follows immediately. □

Next, we must bound the turning of $\lambda(t)$. For this we invoke the Gauss-Bonnet theorem once again. Applying this to the wedge-shaped disk $W(t)$ cut out by $\lambda(t)$ gives

$$\int_{W(t)} K da + \int_{\lambda(t)} k_{\text{geod}} ds = \theta_B.$$

From Claim 19.39 we conclude that

$$\int_{\lambda(t)} k_{\text{geod}} ds \leq t_\xi + a(B).$$

Of course, by Claim 19.40 we have $\max_{\psi \in E} h(\psi) \leq (1/2) + \delta'$. Since $l(0) = 0$, this implies that

$$l(\xi) = l(1) \leq (a(B) + t_B)((1/2) + \delta') < a(B) + t_B.$$

This completes the proof of Lemma 19.38. □

COROLLARY 19.44. *Suppose that T is a focusing triangle with base ξ of length at most 1. Then the length of ξ is at most $\delta + \mu$. More generally, suppose we have a collection of focusing triangles $T_1, \ldots, T_n$ whose bases all lie in a fixed interval of length 1 in c_0. Suppose also that the interiors of disks bounded by these focusing triangles are disjoint. Then the sum of the lengths of the bases is at most $\delta + \mu$.*

PROOF. The first statement is immediate from the previous lemma. The second comes from the fact that the sum of the areas of the disks bounded by the T_i is at most μ and the sum of the total turnings of the ξ_i is at most δ. □

This completes our work on the local focusing issue. It remains to deal with global pathologies that would prevent the exponential mapping from being an embedding out to distance δ'.

7.3. No D_x is an embedded arc with both ends in c_0. One thing that we must show is that the geodesics D_x are embedded. Here is a special case that will serve some of our purposes.

LEMMA 19.45. *For each $x \in X$, there is no non-trivial sub-geodesic of D_x which is a homotopically trivial embedded loop in A.*

PROOF. Were there such a loop, its total turning would be π minus the angle it makes when the endpoints of the arc meet. Since $K \leq 1$ and the area of the disk bounded by this loop is less than the area of A which in turn is less than $\mu < \pi$, one obtains a contradiction to the Gauss-Bonnet theorem. □

Next, we rule out the possibility that one of the geodesics D_x has both endpoints contained in c_0. This is the main result of this section. In a sense, what the argument we give here shows that if there is a D_x with both ends on c_0, then under the assumption of small area, D_x cuts off a thin tentacle of the annulus. But out near the end of this thin tentacle there must be a short arc with large total turning, violating our hypothesis on the integrals of the geodesic curvature over arcs of length at most 1.

LEMMA 19.46. *There is no x for which D_x is an embedded arc with both endpoints on c_0 and otherwise disjoint from ∂A.*

PROOF. Suppose that there were such a D_x. Then D_x separates A into two components, one of which, B, is a topological disk. Let c_0' be the intersection of c_0 with B. We consider two cases: Case (i): $l(c_0') \leq 1$ and Case (ii): $l(c_0') > 1$.

Let us show that the first case is not possible. Since D_x is a geodesic and D_x is perpendicular to c_0 at one end, the total turning around the boundary of B is at most

$$3\pi/2 + \int_{c_0'} k_{\text{geod}} ds < 3\pi/2 + \delta,$$

where the last inequality uses the fact that the length of c_0' is at most 1. On the other hand, $\int_B K da < \mu$, and $\mu < 1/20 < (\pi/2) - \delta$. This contradicts the Gauss-Bonnet theorem.

Now let us consider the second case. Let J be the subinterval of c_0' with one end point being x and with the length of J being 1. We orient J so that x

is its initial point. We set $X_J = J \cap X$. We define $S_{X_J}(B) \subset S_{X_J}$ as follows. For each $y \in X_J$ we let $f_B(y)$ be the minimum of δ' and the distance along D_y to the first point (excluding y) of D_y contained in ∂B and let $D_y(B)$ be the sub-geodesic of D_y of this length starting at y. Then $S_{X_J}(B) \subset S_{X_J}$ is the union over $y \in X_J$ of $[0, f_B(y)]$. Clearly, the exponential mapping defines an immersion of $S_{X_J}(B)$ into B. We need to replace X_J by a slightly smaller subset in order to make the exponential mapping be an embedding. To do this we shall remove bases of a maximal focusing triangles in B.

First notice that for each $y \in X_J$ the exponential mapping is an embedding on $D_y(B)$. The reason is that the image of $D_y(B)$ is a geodesic contained in the ball B. Lemma 19.45 then shows that this geodesic is embedded. This leads to:

CLAIM 19.47. *For any component c of X_J, the restriction of the exponential mapping to $S_c(B) = (c \times [0, \delta')) \cap S_{X_J}(B)$ is an embedding.*

PROOF. Since the geodesics that make up $S_c(B)$ have length at most $\delta' < 1/10$ and since the curvature of the annulus is bounded above by 1, the restriction of the exponential mapping to $S_c(B)$ is a local diffeomorphism. The restriction to each $\{y\} \times [0, f_B(y)]$ is an embedding onto $D_y(B)$. If the restriction of the exponential mapping to $S_c(B)$ is not an embedding, then there are $y \neq y'$ in c such that the geodesics $D_y(B)$ and $D_{y'}(B)$ meet. When they meet, they meet at a positive angle and by the Gauss-Bonnet theorem this angle is less than $\mu + \delta$. Thus, all the geodesics starting at points sufficiently close to y' and between y and y' along c must also meet $D_y(B)$. Of course, if a sequence of $D_{y_i}(B)$ meet $D_y(B)$, then the same is true for the limit. It now follows that $D_{y''}(B)$ meets $D_y(B)$ for all y'' between y and y'. This contradicts the fact that $D_y(B)$ is embedded. □

CLAIM 19.48. *Any focusing triangle for J must contain a component of $J \setminus X_J$. If $\{\mathcal{T}_n\}$ is an infinite sequence of focus triangles for J, then, after passing to a subsequence, there is a limiting focusing triangle for J.*

PROOF. The first statement is immediate from Claim 19.47. Since $X \cap J$ is compact, it is clear that after passing to a subsequence each of the sequence of left-hand sides and the sequence of right-hand sides converge to a geodesic arc orthogonal to J at points of X. Furthermore, these limiting geodesics meet in a point at distance at most δ' from the end of each. The only thing remaining to show is that the limiting left- and right-hand sides do not begin at the same point of X. This is clear since each focusing triangle contains one of the finitely many components of $J \setminus X_J$. □

Using Claim 19.48 we see that if there is a focusing triangle for J there is a first point x_1 in X_J whose associated geodesic contains the left-hand side of a focusing triangle for J. Then since the base length of any focusing triangle is bounded by a fixed constant, invoking again Claim 19.48, that there is a focusing triangle T_1 for J that has left-hand side contained in the geodesic D_{x_1} and has a maximal base among all such focusing triangles, maximal in the sense that the base of this focusing triangle contains the base of any other focusing triangle with left-hand side contained in D_{τ_1}. Denote its base by ξ_1 and denote the right-hand endpoint of ξ_1 by y_1. For the triangle we take the geodesic arcs to the first point of intersection measured along D_{y_1}. Set $J_1 = \overline{J \setminus \xi}$, and repeat the process for J_1. If there is a focusing triangle for J_1 we find the first left-hand side of such focusing triangles and then find the maximal focusing triangle T_2 with this left-hand side.

CLAIM 19.49. *The interior of T_2 is disjoint from the interior of T_1.*

PROOF. Since by construction the interiors of the bases of T_1 and T_2 are disjoint, if the interior of T_2 meets T_1, then one of the sides of T_2 crosses the interior of one of the sides of T_1. But since T_1 is a maximal focusing triangle with its left-hand side, neither of the sides of T_2 can cross the interior of the left-hand side of T_1. If one of the sides of T_2 crosses the interior of the right-hand side of T_1, then the right-hand side of T_1 is the left-hand side of a focusing triangle for J_1. Since by construction the left-hand side of T_2 is the first such, this means that the left-hand side of T_2 is the right-hand side of T_1. This means that the right-hand side of T_2 terminates when it meets the right-hand side of T_1 and hence the right-hand side of T_2 ends the first time that it meets the right-hand side of T_1 and hence does not cross it. □

We continue in this way constructing focusing triangles for J with disjoint interiors. Since each focusing triangle for J contains a component of $J \setminus X_J$, and as there are only finitely many such components, this process must terminate after a finite number of steps. Let $T_1, \ldots, T_k$ be the focusing triangles so constructed, and denote by ξ_i the base of T_i. Let $X'_J = X_J \setminus \cup_{i=1}^k \xi_i$.

DEFINITION 19.50. We call the triangles $T_1, \ldots, T_k$ constructed above, the *maximal set of focusing triangles* for J relative to B.

CLAIM 19.51. *The length of X'_J is at least $1 - 2\delta - \mu$.*

PROOF. Since the interiors of the T_i are disjoint, according to Corollary 19.44, we have $\sum_i l(\xi_i) < \delta + \mu$. We also know by Remark 19.36 that the length of X_J is at least $(1 - \delta)$. Putting these together gives the result. □

We define $S_{X'_J}(B)$ to be the intersection of $S_{X_J}(B)$ with $S_{X'_J}$.

CLAIM 19.52. *The restriction of the exponential mapping to $S_{X'_J}(B)$ is an embedding.*

PROOF. Suppose that we have distinct points x', y' in X'_J such that $D_{x'}(B) \cap D_{y'}(B) \neq \emptyset$. We assume that $x' < y'$ in the orientation on J.

7. PROOF OF LEMMA 19.31: ANNULI OF SMALL AREA

Then there is a focusing triangle for J whose base is the sub-arc of J with endpoints x' and y', and hence the left-hand side of the focusing triangle is contained in $D_{x'}(B)$. Since x' is not a point of $\cup_i \xi_i$ either it lies between two of them, say ξ_j and ξ_{j+1} or it lies between the initial point x of J and the initial point of ξ_1 or it lies between the last ξ_n and the final point of J.

But x' cannot lie before ξ_1, for this would contradict the construction which took as the left-hand endpoint of ξ_1 the first point of J whose geodesic contained the left-hand side of a focusing triangle for J. Similarly, x' cannot lie between ξ_j and ξ_{j+1} for any j since the left-hand endpoint of ξ_{j+1} is the first point at or after the right-hand endpoint of ξ_j whose geodesic contains the left-hand side of a focusing triangle for J. Lastly, x' cannot lie to the right of the last ξ_k, for then we would not have finished the inductive construction. □

We pull back the metric of A to the $S_{X'_J}(B)$ by the exponential mapping. Since this pullback metric is at least $(1-\delta)^2$ times the product of the metric on X'_J induced from c_0 and the usual metric on the interval, and since the map on this subset is an embedding, we see that the area of the region of the annulus which is the image under the exponential mapping of this subset is at least

$$(1-\delta)^2 \int_{X'_J} f_B(x)ds,$$

where s is arc length along X'_J. Of course, the area of this subset is at most μ. This means that, setting Z equal to the subset of $X'_J(\delta')$ given by

$$Z = \{z \in X'_J | f_B(z) < \delta'\},$$

the total length of $X'_J \setminus Z$ satisfies

$$l(X'_J \setminus Z) \leq (1-\delta)^{-2}(\delta')^{-1}\mu < \frac{1}{10},$$

where the last inequality is an immediate consequence of our choice of μ. Thus, the length of Z is at least $(0.9 - 2\delta - \mu) \geq 0.87$. Let $\partial_+ S_Z(B)$ be the union of the final endpoints (as opposed to the initial points) of the $D_x(B)$ as x ranges over Z. Of course, since $f_B(z) < \delta'$ for all $z \in Z$, it must be the case that the exponential mapping embeds $\partial_+ S_Z(B)$ into ∂B. Furthermore, the total length of the image of $\partial_+ S_Z(B)$ is at least $(1-\delta)l(Z) \geq 0.86$. The boundary of B is made up of two pieces: D_x and an arc on c_0. But the length of D_x is at most $\delta' < 1/20$ so that not all of $\partial_+ S_Z(B)$ can be contained in D_x. Thus, there is $z \in Z$, distinct from x such that D_z has both endpoints in c_0. It then follows that all points of Z that are separated (along J) from x by z have the same property. Since the length of Z is at least 0.86, it follows that there is a point $z \in X$ at least distance 0.85 along J from x with the property that D_z has both endpoints in c_0. The complementary component of D_z in A, denoted B', is a disk that is contained in B and the length of $B' \cap c_0$ is at least 0.85 less than the length of $B \cap c_0$.

The length of $B' \cap c_0$ cannot be less than 1, for that gives a contradiction. But if the length of $B' \cap c_0$ is greater than 1, we now repeat this construction replacing B by B'. Continuing in this way we eventually we cut down the length of $B \cap c_0$ to be less than 1 and hence reach a contradiction. $\square$

7.4. For every $x \in X$, the geodesic D_x is embedded. The steps in the above argument, inductively constructing disjoint maximal focusing triangles and showing that their bases have a small length and that off of them the map is an embedding, will be repeated in two other contexts. The next context is to rule out the case when a sub-arc of D_x forms a homotopically non-trivial loop in A.

LEMMA 19.53. *For any $x \in X$ there is no sub-geodesic of D_x that is an embedded loop in A.*

PROOF. We have already treated the case when the loop bounds a disk. Now we need to treat the case when the loop is homotopically non-trivial in A. Let $D'_x \subset D_x$ be the minimal compact sub-geodesic containing x that is not an embedded arc. Let int B be the complementary component of D'_x in A that contains $c_0 \setminus \{x\}$. There is a natural compactification of int B as a disk and an immersion of this disk into A, an immersion that is two-to-one along the shortest sub-geodesic of D'_x from x to the point of intersection of D_x with itself. We do exactly the same construction as before. Take a sub-arc J of length 1 with x as an endpoint and construct $S_{X_J}(B)$ consisting of the union of the sub-geodesics of D_z, for $z \in J \cap X$ that do not cross the boundary of B. We then construct a sequence of maximal focusing triangles along J relative to B just as in the previous case. In this way we construct a subset Z of $X \cap J$ of total length at least 0.87 with the property that for every $z \in Z$ the final end of $D'_z(B)$ lies in ∂B. Furthermore, the length of the arcs that these final ends sweep out is at least 0.86. Hence, since the total length of the part of the boundary of B coming from D'_x is at most $2\delta' < 0.2$, there must be a $z \in Z$ for which $D_z(B)$ has both ends in c_0. This puts us back in the case ruled out in Lemma 19.46. $\square$

7.5. Far apart D_x's don't meet. Now the last thing that can prevent the exponential mapping in the complement of the focusing triangles from being an embedding is that geodesics D_x and D_y might meet even though x and y are far apart along c_0. Our next goal is to rule this out.

LEMMA 19.54. *Let x, y be distinct points of X. Suppose that there are sub-geodesics $D'_x \subset D_x$ and $D'_y \subset D_y$ with a common endpoint. Then the arc $D'_x * (D'_y)^{-1}$ cuts A into two complementary components, exactly one of which is a disk, denoted B. Then it is not possible for $B \cap c_0$ to contain an arc of length 1.*

PROOF. The proof is exactly the same as in Lemma 19.46 except that the part of the boundary of B that one wants to avoid has length at most $2\delta' < 0.2$ instead of δ'. Still, since (in the notation of the proof of Lemma 19.46) the total length of Z is at least 0.87 so that the lengths of the other ends of the D_z as z ranges over Z is at least 0.86, there is $z \in Z$ for which both ends of D_z lie in c_0. Again this puts us back in the case ruled out by Lemma 19.46. □

As a special case of this result we have the following.

COROLLARY 19.55. *Suppose that we have an arc ξ of length at most 1 on c_0. Denote the endpoints of ξ by x and y and suppose that $D_x \cap D_y \neq \emptyset$. Let D'_x and D'_y be sub-geodesics containing x and y respectively ending at the same point, v, and otherwise disjoint. Then the loop $\xi * D'_y * (D'_x)^{-1}$ bounds a disk in A.*

PROOF. If not, then it is homotopically non-trivial in A and replacing ξ by its complement, $c_0 \setminus \mathrm{int}\,\xi$, gives us exactly the situation of the previous lemma. (The length of $c_0 \setminus \mathrm{int}\,\xi$ is at least 1 since the length of c_0 is at least 2.) □

Let us now summarize what we have established so far about the intersections of the geodesics $\{D_x\}_{x \in X}$.

COROLLARY 19.56. *For each $x \in X$, the geodesic D_x is an embedded arc in A. Either it has length δ' or its final point lies on c_1. Suppose there are $x \neq x'$ in X with $D_x \cap D_{x'} \neq \emptyset$. Then there is an arc ξ on c_0 connecting x to x' with the length of ξ at most $\delta + \mu$. Furthermore, for sub-geodesics $D'_x \subset D_x$, containing x, and $D'_{x'} \subset D_{x'}$, containing x', that intersect exactly in an endpoint of each, the loop $\xi * D'_{x'} * (D'_x)^{-1}$ bounds a disk B in A, and the length of ξ is at most the turning of ξ plus the area of B.*

7.6. Completion of the proof. We have now completed all the technical work on focusing and we have also shown that the restriction of the exponential mapping to the complement of the bases of the focusing regions is an embedding. We are now ready to complete the proof of Proposition 19.35.

Let J be an interval of length 1 in c_0. Because of Corollary 19.56 we can construct the maximal focusing triangles for J as follows. Orient J, and begin at the initial point of J. At each step we consider the first x (in the subinterval of J under consideration) which intersects a D_y for some later $y \in J$. If we have such y, then we can construct the sides of the putative triangle for sub-geodesics of D_x and D_y. But we need to know that we have a focusing triangle. This is the content of Corollary 19.55. The same reasoning works when we construct the maximal such focusing triangle with a given left-hand side, and then when we show that in the complement of the focusing triangles the map is an embedding. Thus, as before, for an interval J of length 1, we construct a subset $X'_J \subset X \cap J$ of length at least 0.97 such that the restriction of the exponential mapping to $S_{X'_J}$ is an embedding. Again the area estimate shows that there is a subset $Z \subset X'_J$ whose length is at least 0.87 with the property that for every $z \in Z$ the geodesic D_z has both endpoints in ∂A. By Lemma 19.46, the only possibility for the final endpoints of all these D_z's is that they lie in c_1.

In particular, there are $x \in X$ for which D_x spans from c_0 to c_1. We pick one such, x_0, contained in the interior of X, and use it as the starting point for a construction of maximal focusing triangles all the way around c_0. What we are doing at this point actually is cutting the annulus open along D_{x_0} to obtain a disk and we construct a maximal family of focusing triangles of the interval $[x'_0, x''_0]$ obtained by cutting c_0 open at x_0 relative to this disk. Here x'_0 and x''_0 are the points of the disk that map to x_0 when the disk is identified to form A. Briefly, having constructed a maximal collection of focusing triangles for a subinterval $[x'_0, x]$, we consider the first point y in the complementary interval $[x, x''_0]$ with the property that there is y' in this same interval, further along with $D_y \cap D_{y'} \neq 0$. Then, using Corollary 19.56 we construct the maximal focusing triangle on $[x, x''_0]$ with left-hand side being a sub-geodesic of D_y. We then continue the construction inductively until we reach x''_0. Denote by $\xi_1, \ldots, \xi_k$ the bases of these focusing triangles and let X' be $X \setminus \cup_i \xi_i$.

The arguments above show that the exponential mapping is an embedding of $S_{X'}$ to the annulus.

CLAIM 19.57. *For every subinterval J of length 1 in c_0 the total length of the bases ξ_i that meet J is at most $2\delta + \mu < 0.03$.*

PROOF. Since, by Corollary 19.44, every base of a focusing triangle has length at most $\delta + \mu$, we see that the union of the bases of focusing triangles meeting J is contained in an interval of length $1 + 2(\delta + \mu) < 2$. Hence, the total turning of the bases of these focusing triangles is at most 2δ whereas the sum of their areas is at most μ. The result now follows from Corollary 19.44. □

By hypothesis there is an integer $n \geq 1$ such that the length $l(c_0)$ of c_0 is greater than n but less than or equal to $n + 1$. Then it follows from the above that the total length of the bases of all the focusing triangles in our family is at most

$$(n+1)(2\delta + \mu) < 0.03(n+1) \leq 0.06n \leq 0.06 l(c_0).$$

Since the restriction of the exponential mapping to $S_{X'}$ is an embedding, it follows from Claim 19.37 and the choice of δ' that, for any open subset Z of X', the area of the image under the exponential mapping of S_Z is at least $(1-\delta)^2 \int_Z f(x) ds$, where ds is the arc length along Z. Also, the image under the exponential mapping of $\partial_+(S_Z)$ is an embedded arc in A of length at least $(1-\delta)l(Z)$. Since the length of X' is at least $(0.94)l(c_0)$ and since the area of A is less than $\mu < (1-\delta)^2\delta'/10$, it follows that the subset of X' on which f takes the value δ' has length at most $0.10 < (0.10)l(c_0)$. Hence, there is a subset $X'' \subset X'$ of total length at least $(0.84)l(c_0)$ with the property that $f(x) < \delta'$ for all $x \in X''$. This means that for every $x \in X''$ the geodesic D_x spans from c_0 to c_1, and hence the exponential mapping embeds $\partial_+ S_{X''}$ into c_1. But we have just seen that the length of the image under the exponential mapping of $\partial_+ S_{X''}$ is at least

$$(1-\delta)l(X'') > (0.99)l(X'') > (0.83)l(c_0).$$

It follows that the length of c_1 is at least $(0.83)l(c_0) > 3(l(c_0))/4$.

This completes the proof.

8. Proof of the first inequality in Lemma 19.24

Here is the statement that we wish to establish when the manifold $(W, h(t))$ is the product of $(M, g(t)) \times (S^1_\lambda, ds^2)$.

LEMMA 19.58. *Let $(W, h(t))$, $t_0 \le t \le t_1$, be a Ricci flow and fix $\Theta < \infty$. Then there exist constants $\delta > 0$ and $0 < r_0 \le 1$ depending only on the curvature bound for the ambient Ricci flow and Θ such that the following holds. Let $c(x,t)$, $t_0 \le t \le t_1$, be a curve-shrinking flow with $c(\cdot, t)$ immersed for each $t \in [t_0, t_1]$ and with the total curvature of $c(\cdot, t_0)$ being at most Θ. Suppose that there is $0 < r \le r_0$ and at a time $t' \in [t_0, t_1 - \delta r^2]$ such that the length of $c(\cdot, t')$ is at least r and the total curvature of $c(\cdot, t')$ on any sub-arc of length r is at most δ. Then for every $t \in [t', t' + \delta r^2]$ the curvature k satisfies*

$$k^2 \le \frac{2}{(t-t')}.$$

The rest of this section is devoted to the proof of this lemma. In [**2**] such a local estimate was established when the ambient manifold was Euclidean space and the curve in question is a graph. A related result for hypersurfaces that are graphs appears in [**19**]. The passage from Euclidean space to a general Ricci flow is straightforward, but it is more delicate to use the bound on total curvature on initial sub-arcs of length r to show that in appropriate coordinates the evolving curve can be written as an evolving graph, so that the analysis in [**2**] can be applied.

We fix $\delta > 0$ sufficiently small. We fix $t' \in [t_0, t_1 - \delta r^2]$ for which the hypotheses of the lemma hold. The strategy of the proof is to first restrict to the maximum subinterval of $[t', t_2]$ of $[t', t' + \delta r^2]$ on which k is bounded by $\sqrt{2/(t-t')}$. If $t_2 < t' + \delta r^2$, then k achieves the bound $\sqrt{2/(t-t')}$ at time t_2. We show that in fact on this subinterval k never achieves the bound. The result then follows. To show that k never achieves the bound, we show that on a possibly smaller interval of time $[t', t_3]$ with $t_3 \le t_2$ we can write the restriction of the curve-shrinking flow to any interval whose length at time t' is $(0.9)r$ as a family of graphs in a local coordinate system so that the function f (of arc and time) defining the graph has derivative along the arc bounded in norm by $1/2$. We take $t_3 \le t_2$ maximal with respect to these conditions. Then with both the bound on k and the bound on the derivative of f one shows that the spatial derivative of f never reaches $1/2$ and also that the curves do not move too much so that they always remain in the coordinate patch. The only way that this can happen is that if $t_2 = t_3$, that is to say, on the entire time interval where we have the curvature bound, we also can write the curve-shrinking flow as a flow of graphs with small spatial derivatives. Then it is convenient to replace the curve-shrinking flow by an equivalent flow, introduced in [2], called the graph flow. Applying a simple maximum principle argument to this flow we see that k never achieves the value $\sqrt{2/(t-t')}$ on the time interval $[t', t_2]$ and hence the curvature estimate $k < \sqrt{2/(t-t')}$ holds throughout the interval $(t', t' + \delta r^2]$.

8.1. A bound for $\int kds$. Recall that k is the norm of the curvature vector $\nabla_S S$, and in particular, $k \ge 0$. For any sub-arc $\gamma_{t'}$ of $c(\cdot, t')$ at time t' we let γ_t be the result at time t of applying the curve-shrinking flow to $\gamma_{t'}$. The purpose of this subsection is to show that $\int_{\gamma_t} kds$ is small for all $t \in [t', t' + \delta r^2]$ and all initial arcs $\gamma_{t'}$ of length at most r.

CLAIM 19.59. *There is a constant $D_0 < \infty$, depending only on Θ and the curvature bound of the ambient Ricci flow such that for every $t \in [t', t' + \delta r^2]$ and every sub-arc $\gamma_{t'}$ whose length is at most r, we have $\int_{\gamma_t} kds < D_0$ and $l(\gamma_t) \le D_0 r$, where $l(\gamma_t)$ is the length of γ_t.*

PROOF. This is immediate from Corollary 19.10 applied to all of $c(\cdot, t)$. □

Now we fix $t_2 \le t' + \delta r^2$ maximal subject to the condition that $k(x, t) \le \sqrt{\frac{2}{t-t'}}$ for all x and all $t \in [t', t_2]$. If $t_2 < t' + \delta r^2$ then there is x with $k(x, t_2) = \sqrt{\frac{2}{(t_2-t')}}$.

Now consider a curve $\gamma_{t'}$ of length r. From the integral estimate in the previous claim and the assumed pointwise estimate on k, we see that

$$\int_{\gamma_t} k^2 ds \le \max_{x \in \gamma_t} k(x, t) \int_{\gamma_t} kds < \sqrt{\frac{2}{t-t'}} \cdot D_0.$$

Using Equation (19.8), it follows easily that, provided that $\delta > 0$ is sufficiently small, the length of γ_t is at least $(0.9)r$ for all $t \in [t', t_2]$, and more generally for any subinterval $\gamma'_{t'}$ of $\gamma_{t'}$ and for any $t \in [t', t_2]$ the length of the corresponding interval γ'_t is at least (0.9) times the length of $\gamma'_{t'}$. We introduce a cut-off function on $\gamma_{t'} \times [t', t_2]$ as follows. First, fix a smooth function $\psi \colon [-1/2, 1/2] \to [0, 1]$ which is identically zero on $[-0.50, -0.45]$

and on $[0.45, 0.50]$, and is identically 1 on $[-3/8, 3/8]$. There is a constant D' such that $|\psi'| \leq D'$ and $|\psi''| \leq D'$. Now we fix the midpoint $x_0 \in \gamma_{t'}$ and define the signed distance from (x_0, t), denoted

$$s \colon \gamma_{t'} \times [t', t_2] \to \mathbb{R},$$

as follows:

$$s(x,t) = \int_{x_0}^{x} |X(y,t)| dy.$$

We define the cut-off function

$$\varphi(x,t) = \psi\left(\frac{s(x,t)}{r}\right).$$

CLAIM 19.60. *There is a constant D_1 depending only on the curvature bound for the ambient Ricci flow such that for any sub-arc $\gamma_{t'}$ of length r, defining $\varphi(x,t)$ as above, for all $x \in \gamma_t$ and all $t \in [t', t_2]$ we have*

$$\left|\frac{\partial \varphi(x,t)}{\partial t}\right| \leq \frac{D_1}{r\sqrt{t-t'}} + D_1.$$

PROOF. Clearly,

$$\frac{\partial \varphi(x,t)}{\partial t} = \psi'\left(\frac{s(x,t)}{r}\right) \cdot \frac{1}{r} \frac{\partial s(x,t)}{\partial t}.$$

We know that $|\psi'| \leq D'$ so that

$$\left|\frac{\partial \varphi(x,t)}{\partial t}\right| \leq \frac{D'}{r} \left|\frac{\partial s(x,t)}{\partial t}\right|.$$

On the other hand,

$$s(x,t) = \int_{x_0}^{x} |X(y,t)| dy,$$

so that

$$\left|\frac{\partial s(x,t)}{\partial t}\right| = \left|\int_{x_0}^{x} \frac{\partial |X(y,t)|}{\partial t} dy\right|,$$

By Lemma 19.6 we have

$$\frac{\partial |X(y,t)|}{\partial y} dy = \left(-\mathrm{Ric}(S(y,t), S(y,t)) - k^2(y,t)\right) ds,$$

so that there is a constant D depending only on the bound of the sectional curvatures of the ambient Ricci flow with

$$\left|\frac{\partial s(x,t)}{\partial t}\right| \leq \int_{x_0}^{x} (D + k^2) ds \leq Dl(\gamma_t) + \int_{x_0}^{x} k^2(y,t) ds(y,t),$$

and hence by Claim 19.59

$$\left|\frac{\partial s(x,t)}{\partial t}\right| \leq DD_0 r + \int_{x_0}^{x} k^2(y,t) ds(y,t).$$

Using the fact that $k^2 \leq 2/(t-t')$, we have

$$\int_{x_0}^{x} k^2(y,t) ds(y,t) \leq \sqrt{\frac{2}{t-t'}} \int_{x_0}^{x} k \, ds \leq \frac{\sqrt{2} D_0}{\sqrt{t-t'}}.$$

Putting all this together, we see that there is a constant D_1 such that

$$\left| \frac{\partial \varphi(x,t)}{\partial t} \right| \leq D_1 \left(\frac{1}{r\sqrt{t-t'}} + 1 \right).$$

$\square$

CLAIM 19.61. *There is a constant D_2 depending only on the curvature bound of the ambient Ricci flow and Θ and a constant D_3 depending only on the curvature bound of the ambient Ricci flow, such that for any $t \in [t', t_2]$ and any sub-arc $\gamma_{t'}$ of length r, we have*

$$\left| \frac{d}{dt} \int_{\gamma_t} \varphi k \, ds \right| \leq D_2 \left(1 + \frac{1}{r\sqrt{t-t'}} \right) + \frac{D_2}{r^2} + D_3 \int_{\gamma_t} \varphi k \, ds.$$

PROOF. We have

$$\left| \frac{d}{dt} \int_{\gamma_t} \varphi k \, ds \right| \leq \left| \int_{\gamma_t} \frac{\partial \varphi(x,t)}{\partial t} k \, ds \right| + \left| \int_{\gamma_t} \varphi \frac{\partial (k \, ds)}{\partial t} \right|.$$

Using Claim 19.60 for the first term and Claim 19.8 and arguing as in the proof of Lemma 19.9 for the second term, we have

$$\left| \frac{d}{dt} \int_{\gamma_t} \varphi k \, ds \right| \leq D_1 \left(1 + \frac{1}{r\sqrt{t-t'}} \right) \int_{\gamma_t} k \, ds + \left| \int_{\gamma_t} \varphi k'' \, ds \right| + \int_{\gamma_t} C_1' \varphi k \, ds,$$

where C_1' depends only on the ambient curvature bound. We bound the first term by

$$D_1 D_0 \left(\frac{1}{r\sqrt{t-t'}} + 1 \right),$$

where D_0 is the constant depending on Θ and the ambient curvature bound from Claim 19.59. Since the ends of γ_t are at distance at least $(0.45)r$ from x_0 all $t \in [t', t_2]$, we see that for all $t \in [t', t_2]$,

$$\int_{\gamma_t} \varphi k'' = \int_{c(\cdot, t)} \varphi k''.$$

Integrating by parts we have

$$\int_{c(\cdot, t)} \varphi k'' \, ds = \int_{c(\cdot, t)} \varphi'' k \, ds,$$

where the prime here refers to the derivative along $c(\cdot, t)$ with respect to arc length. Of course $|\varphi''| \leq \frac{D'}{r^2}$. Thus, we see that

$$\left| \int_{\gamma_t} \varphi k'' \, ds \right| \leq \frac{D'}{r^2} \int_{c(\cdot, t)} k \, ds \leq \frac{D' D_0}{r^2}.$$

Putting all this together, we have

$$\left|\frac{d}{dt}\int_{\gamma_t}\varphi k ds\right| \leq D_2\left(1+\frac{1}{r\sqrt{t-t'}}\right) + \frac{D_2}{r^2} + D_3\int_{\gamma_t}\varphi k ds$$

for $D_2 = D_0 \max(D', D_1)$ and $D_3 = C_1'$. This gives the required estimate. □

COROLLARY 19.62. *For any $t \in [t', t_2]$ and any sub-arc $\gamma_{t'}$ of length r we have*

$$\int_{\gamma_t}\varphi k ds \leq D_4\sqrt{\delta}$$

for a constant D_4 that depends only on the sectional curvature bound of the ambient Ricci flow and Θ.

PROOF. This is immediate from the previous result by integrating from t' to $t_2 \leq t' + \delta r^2$, and using the fact that $\delta < 1$ and $r < 1$, and using the fact that

$$\int_{\gamma_{t'}}\varphi k ds \leq \int_{\gamma_{t'}} k ds < \delta$$

since $\gamma_{t'}$ has length at most r. □

This gives:

COROLLARY 19.63. *For $\gamma_{t'} \subset c(\cdot, t')$ a sub-arc of length at most r and for any $t \in [t', t_2]$, we have*

$$\int_{\gamma_t} k ds \leq 2D_4\sqrt{\delta}.$$

For any $t \in [t', t_2]$ and any sub-arc $J \subset c(\cdot, t)$ of length at most $r/2$ with respect to the metric $h(t)$, we have

$$\int_J k(x,t) ds(x,t) \leq 2D_4\sqrt{\delta}.$$

PROOF. We divide an interval $\gamma_{t'} \subset c(\cdot, t')$ of length at most r into two subintervals $\gamma_{t'}'$ and $\gamma_{t'}''$ of lengths at most $r/2$. Let $\hat{\gamma}_{t'}'$ and $\hat{\gamma}_{t'}''$ be intervals of length r containing $\gamma_{t'}'$ and $\gamma_{t'}''$ respectively as middle subintervals. We then apply the previous corollary to $\hat{\gamma}_{t'}'$ and $\hat{\gamma}_{t'}''$ using the fact that $\varphi k \geq 0$ everywhere and $\varphi k = k$ on the middle subintervals of $\hat{\gamma}_{t'}'$ and $\hat{\gamma}_{t'}''$. For an interval $J \subset \gamma_t$ of length $r/2$, according to Lemma 19.10 the length of $\gamma_{t'}|_J$ with respect to the metric $h(t')$ is at most r, and hence this case follows from the previous case. □

8.2. Writing the curve flow as a graph. Now we restrict attention to $[t', t_2]$, the maximal interval in $[t', t' + \delta r^2]$ where $k^2 \le 2/(t-t')$. Let $\gamma_{t'}$ be an arc of length r in $c(\cdot, t')$ and let x_0 be the central point of $\gamma_{t'}$. Denote $\gamma_{t'}(x_0) = p \in W$. We take the $h(t')$-exponential mapping from $T_p W \to W$. This map will be a local diffeomorphism out to a distance determined by the curvature of $h(t')$. For an appropriate choice of the ball (depending on the ambient curvature bound) the metric on the ball induced by pulling back $h(t)$ for all $t \in [t', t_2]$ will be within δ in the C^1-topology to the Euclidean metric $h' = h(t')_p$. By this we mean that

(1) $|\langle X, Y \rangle_{h(t)} - \langle X, Y \rangle_{h'}| < \delta |X|_{h'} |Y|_{h'}$ for all tangent vectors in the coordinate system, and
(2) viewing the connection Γ as a bilinear map on the coordinate space with values in the coordinate space we have $|\Gamma(X,Y)|_{h'} < \delta |X|_{h'} |Y|_{h'}$.

We choose $0 < r_0 \le 1$ so that it is much smaller than this distance, and hence r is also much smaller than this distance. We lift to the ball in $T_p W$.

We fix orthonormal coordinates with respect to the metric h' so that the tangent vector of $\gamma_{t'}(x_0)$ points in the positive x^1-direction. Using these coordinates we decompose the coordinate patch as a product of an interval in the x^1-direction and an open ball, B, spanned by the remaining Euclidean coordinates. From now on we shall work in this coordinate system using this product structure. To simplify the notation in the coming computations, we rename the x^1-coordinate the z-coordinate. Ordinary derivatives of a function α with respect to z are written α_z. When we write norms and inner products without indicating the metric we implicitly mean that the metric is $h(t)$. When we use the Euclidean metric on these coordinates we denote it explicitly. Next, we wish to understand how γ_t moves in the Euclidean coordinates under the curve-shrinking flow. Since we have $|\nabla_S S|_h = k$, it follows that $|\nabla_S S|_{h'} \le \sqrt{1+\delta} k \le 2/\sqrt{t-t'}$, and hence, integrating tells us that for any $x \in \gamma_{t'}$ we have

$$|\gamma_t(x) - \gamma_{t'}(x)|_{h'} \le 4\sqrt{t-t'} \le 4\sqrt{\delta} r.$$

This shows that for every $t \in [t', t_2]$, the curve γ_t is contained in the coordinate patch that we are considering. This computation also implies that the z-coordinate of γ_t changes by at most $4\sqrt{\delta} r$ over this time interval.

Because the total curvature of $\gamma_{t'}$ is small and the metric is close to the Euclidean metric, it follows that the tangent vector at every point of $\gamma_{t'}$ is close to the positive z-direction. This means that we can write $\gamma_{t'}$ as a graph of a function f from a subinterval in the z-line to Y with $|f_z|_{h'} < 2\delta$. By continuity, there is $t_3 \in (t', t_2]$ such that all the curves γ_t are written as graphs of functions (over subintervals of the z-axis that depend on t) with $|f_z|_{h'} \le 1/10$. That is to say, we have an open subset U of the product of the

8. PROOF OF THE FIRST INEQUALITY IN LEMMA 19.24

z-axis with $[t', t_3]$, and the evolving curves define a map $\widetilde{\gamma}$ from U into the coordinate system, where the slices at constant time are graphs $z \mapsto (z, f(z,t))$ and are the curves γ_t. Using the coordinates (z, t) gives a new flow of curves by moving in the t-direction. This new flow is called the *graph flow*. It is a reparameterization of the curve shrinking flow in such a way that the z-coordinate is preserved. We denote by $Z = Z(z,t)$ the image under the differential of the map $\widetilde{\gamma}$ of the tangent vector in the z-direction and by $Y(z,t)$ the image under the differential of $\widetilde{\gamma}$ of the tangent vector in the t-direction. Notice that Z is the tangent vector along the parameterized curves in the graph flow. Since we are now using a different parameterization of the curves from the one determined by the curve-shrinking flow, the tangent vector Z has the same direction but not necessarily the same length as the tangent vector X from the curve-shrinking parameterization. Also, notice that in the Euclidean norm we have $|Z|_{h'}^2 = 1 + |f_z|_{h'}^2$. It follows that on U we have

$$(1-\delta)(1 + |f_z|_{h'}^2) \leq |Z(z,t)|_{h(t)}^2 \leq (1+\delta)(1 + |f_z|_{h'}^2).$$

In particular, because of our restriction to the subset where $|f_z|_{h'} \leq 1/10$ we have $(1-\delta) \leq |Z(z,t)|_{h(t)}^2 \leq (1.01)(1 + \delta)$.

Now we know that $\gamma_{t'}$ is a graph of a function $f(z, t')$ defined on some interval I along the z-axis. Let I' be the subinterval of I centered in I with h'-length (0.9) times the h'-length of I. By the above estimate on $|Z|$ it follows that the restriction of $\gamma_{t'}$ to I' has length between $(0.8)r$ and r, and also that the h'-length of I' is between $(0.8)r$ and r. The above estimate means that, provided that $\delta > 0$ is sufficiently small, for every $t \in [t', t_3]$ there is a subinterval of γ_t that is the graph of a function defined on all of I'. We now restrict attention to the family of curves parameterized by $I' \times [t', t_3]$. For every $t \in [t', t_3]$ the curve $\gamma_t|_{I'}$ has length between $(0.8)r$ and r. The curve-shrinking flow is not defined on this product because under the curve-shrinking flow the z-coordinate of any given point is not constant. But the graph flow defined above, and studied in [2] (in the case of Euclidean background metric), is defined on $I' \times [t', t_3]$ since this flow preserves the z-coordinate. The time partial derivative in the curve-shrinking flow is given by

$$(19.10) \qquad \nabla_S S = \frac{\nabla_Z Z}{|Z|^2} - \frac{1}{|Z|^4} \langle \nabla_Z Z, Z \rangle Z.$$

The time partial derivative in the graph-flow is given by $Y = \partial \widetilde{\gamma} / \partial t$. The tangent vector Y is characterized by being h'-orthogonal to the z-axis and differing from $\nabla_S S$ by a functional multiple of Z.

CLAIM 19.64.
$$Y = \frac{\nabla_Z Z - \langle \Gamma(Z,Z), \partial_z \rangle_{h'} Z}{|Z|^2} = \nabla_S S + \left(\frac{\langle \nabla_Z Z, Z \rangle}{|Z|^4} - \frac{\langle \Gamma(Z,Z), \partial_z \rangle_{h'}}{|Z|^2} \right) Z.$$

PROOF. In our Euclidean coordinates, $Z = (1, f_z)$ so that $\nabla_Z Z = (0, f_{zz}) + \Gamma(Z, Z)$. Thus,
$$\langle \nabla_Z Z, \partial_z \rangle_{h'} = \langle \Gamma(Z, Z), \partial_z \rangle_{h'}.$$
Since $\langle Z, \partial_z \rangle_{h'} = 1$, it follows that
$$\frac{\nabla_Z Z - \langle \Gamma(Z, Z), \partial_z \rangle_{h'} Z}{|Z|^2}$$
is h'-orthogonal to the z-axis and hence is a multiple of Y. Since it differs by a multiple of Z from $\nabla_S S$, it follows that it is Y. This gives the first equation; the second follows from this and Equation (19.10). □

To simplify the notation we set
$$\psi(Z) = \frac{\langle \Gamma(Z, Z), \partial_z \rangle_{h'}}{|Z|^2}.$$
Notice that from the conditions on Γ and h' it follows immediately that $|\psi(Z)| < (1.5)\delta$.

8.3. Proof that $t_3 = t_2$. At this point we have a product coordinate system on which the metric is almost the Euclidean metric in the C^1-sense, and we have the graph flow given by
$$Y = \frac{\partial \widetilde{\gamma}}{\partial t} = \frac{\nabla_Z Z}{|Z|^2} - \psi(Z) Z$$
defined on $[t', t_3]$ with image always contained in the given coordinate patch and written as a graph over a fixed interval I' in the z-axis. For every $t \in [t', t_3]$ the length of $\gamma_{t'}|_{I'}$ in the metric $h(t')$ is between $(0.8)r$ and r. The function $f(z, t)$ whose graphs give the flow satisfies $|f_z|_{h'} \leq 1/10$. Our next goal is to estimate $|f_z|_{h'}$ and show that it is always less than $1/10$ as long as $k^2 \leq 2/(t - t')$ and $t - t' \leq \delta r^2$ for a sufficiently small δ, i.e., for all $t \in [t', t_2]$; that is to say, our next goal is to prove that $t_3 = t_2$. In all the arguments that follow C' is a constant that depends only on the curvature bound for the ambient Ricci flow, but the value of C' is allowed to change from line to line.

The first step in doing this is to consider the angle between $\nabla_Z Z$ and Z.

CLAIM 19.65. *Provided that $\delta > 0$ is sufficiently small, the angle (measured in $h(t)$) between Y and $Z = (1, f_z)$ is greater than $\pi/4$. Also:*

(1)
$$k \leq |Y| < \sqrt{2} k.$$

(2)
$$|\langle \nabla_Z Z, Z \rangle| < (k + 2\delta)|Z|^3.$$

(3)
$$|\nabla_Z Z| < 2(|Y| + \delta).$$

(4)
$$|\langle Y, Z \rangle| \leq |Y||f_z|(1 + 3\delta).$$

PROOF. Under the hypothesis that $|f_z|_{h'} \leq 1/10$, it is easy to see that the Euclidean angle between $(0, f_{zz})$ and $(1, f_z)$ is at most $\pi/2 - \pi/5$. From this, the first statement follows immediately provided that δ is sufficiently small. Since Y is the sum of $\nabla_S S$ and a multiple of Z and since $\nabla_S S$ is $h(t)$-orthogonal to Z, it follows that $|Y| = |\nabla_S S| (\cos(\theta))^{-1}$, where θ is the angle between $\nabla_S S$ and Y. Since Y is a multiple of $(0, f_{zz})$, it follows from the first part of the claim that the $h(t)$-angle between Y and $\nabla_S S$ is less than $\pi/4$. Item (1) of the claim then follows from the fact that by definition $|\nabla_S S| = k$.

Since
$$\frac{\nabla_Z Z}{|Z|^2} = Y + \psi(Z)Z,$$
and $|Z|^2 \leq (1.01)(1 + \delta)$, the third item is immediate. For item (4), since Y is h'-orthogonal to the z-axis, we have
$$|\langle Y, Z \rangle_{h'}| = |\langle Y, (0, f_z) \rangle_{h'}| \leq |Y|_{h'} |f_z|_{h'}.$$

From this and the comparison of $h(t)$ and h', item (4) is immediate. Lastly, let us consider item (2). We have
$$\langle Y, Z \rangle = \frac{\langle \nabla_Z Z, Z \rangle}{|Z|^2} - \langle \Gamma(Z, Z), \partial_z \rangle_{h'}.$$

Thus, from item (4) we have
$$\frac{\langle \nabla_Z Z, Z \rangle}{|Z|^2} \leq |Y||f_z|(1 + 3\delta) + (1.5)\delta.$$

Since $Y < \sqrt{2}k$ and $|f_z| < 1/10$, item (2) follows. □

CLAIM 19.66. *The following hold provided that $\delta > 0$ is sufficiently small:*
(1) $|Z(\psi(Z))| < C'(1 + \delta|Y|)$, *and*
(2) $|Y(\psi(Z))| < C'(|Y| + \delta|\nabla_Z Y|)$.

(Recall that C' is a constant depending only on the curvature bound of the ambient Ricci flow.)

PROOF. For the first item, we write $Z(\psi(Z))$ as a sum of terms where the differentiation by Z acts on the various. When the Z-derivative acts on Γ the resulting term has norm bounded by a constant depending only on the curvature of the ambient Ricci flow. When the Z-derivative acts on one of the Z-terms in $\Gamma(Z, Z)$ the norm of the result is bounded by $2\delta|\nabla_Z Z||Z|$. Action on each of the other Z-terms gives a term bounded in norm by the same expression. Lastly, when the Z-derivative acts on the constant metric h' the norm of the result is bounded by $2\delta^2$. Since $|\nabla_Z Z| \leq 2(|Y| + \delta)$, the first item follows.

We compute $Y(\psi(Z))$ in a similar fashion. When the Y-derivative acts on the Γ, the norm of the result is bounded by $C'|Y|$. When the Y-derivative acts on one of the Z-terms the norm of the result is bounded by $2\delta|\nabla_Y Z|$. Lastly, when the Y-derivative acts on the constant metric h', the norm of the result is bounded by $\delta^2|Y|$. Putting all these terms together establishes the second inequality above. $\square$

Now we wish to compute $\int_{I'\times\{t\}} |Z|^2 dz$. To do this we first note that using the definition of Y, and arguing as in the proof of the first equation in of Lemma 19.6 we have

$$\frac{\partial}{\partial t}|Z|^2 = -2\mathrm{Ric}(Z,Z) + 2\langle \nabla_Y Z, Z\rangle$$
$$= -2\mathrm{Ric}(Z,Z) + 2\langle \nabla_Z Y, Z\rangle.$$

Direct computation shows that

$$2\langle \nabla_Z \left(\frac{\nabla_Z Z}{|Z|^2}\right), Z\rangle = Z\left(\frac{Z(|Z|^2)}{|Z|^2}\right) - 2\frac{|\nabla_Z Z|^2}{|Z|^4}|Z|^2.$$

Thus from the Claim 19.64, we have

(19.11)
$$\frac{\partial}{\partial t}|Z|^2 = 2\langle \nabla_Z Y, Z\rangle - 2\mathrm{Ric}(Z,Z)$$
$$= Z\left(\frac{Z(|Z|^2)}{|Z|^2}\right) - 2\frac{|\nabla_Z Z|^2}{|Z|^4}|Z|^2 - 2\langle \nabla_Z(\psi(Z)Z), Z\rangle - 2\mathrm{Ric}(Z,Z)$$
$$= Z\left(\frac{Z(|Z|^2)}{|Z|^2}\right) - 2|Y|^2|Z|^2 + V,$$

where

$$V = -4|Z|^2\langle Y, \psi(Z)Z\rangle - 2\psi^2(Z)|Z|^4 - 2\langle \nabla_Z(\psi(Z)Z), Z\rangle - 2\mathrm{Ric}(Z,Z).$$

By item (1) in Claim 19.66 and item (4) in Claim 19.65 we have

(19.12) $$|V| < C'(1+\delta|Y|).$$

Using this and the fact that $|Y| \le \sqrt{2}k$ we compute:

$$\frac{d}{dt}\int_{I'\times\{t\}}|Z|^2 dz \le \int_{I'\times\{t\}} Z\left(\frac{Z(|Z|^2)}{|Z|^2}\right) dz + \int_{I'\times\{t\}} \left(C'(1+\delta k)\right) dz$$
$$= \left.\frac{Z(|Z|^2)}{|Z|^2}\right|_0^a + \int_{I'\times\{t\}} \left(C'(1+\delta k)\right) dz$$
$$= 2\left.\frac{\langle \nabla_Z Z, Z\rangle}{|Z|^2}\right|_0^a + \int_{I'\times\{t\}} \left(C'(1+\delta k)\right) dz,$$

where we denote the endpoints of I' by $\{0\}$ and $\{a\}$. By item (2) in Claim 19.65, the first term is at most $2(k+2\delta)\sqrt{(1.01)(1+\delta)}$, which is at most $\frac{8}{\sqrt{t-t'}}$ and the second term is at most $C'(1+\delta k)r$. Now integrating from t' to t we see that for any $t \in [t', t_3]$ we have

$$\int_{I' \times \{t\}} |Z|^2 dz \leq \int_{I' \times \{t'\}} |Z|^2 dz + 16\sqrt{\delta}r + C'\delta r^3 + C'\delta^{3/2}r^2.$$

Since $|f_z(z, t')|_{h'} \leq 2\delta$ and

$$(1-\delta)(1+|f_z|_{h'}^2) \leq |Z|^2 \leq (1+\delta)(1+|f_z|_{h'}^2),$$

we see that $\int_{I' \times \{t'\}} |Z|^2 dz \leq (1+3\delta)\ell_{h'}(I')$. It follows that for any $t \in [t', t_3]$ we have

$$\int_{I' \times \{t\}} |Z|^2 dz \leq (1+3\delta)\ell_{h'}(I') + C'(\sqrt{\delta}r + \delta r^3 + \delta^{3/2}r^2).$$

Since $|Z|^2$ is between $(1-\delta)(1+|f_z|_{h'}^2)$ and $(1+\delta)(1+|f_z|_{h'}^2)$, we see that there is a constant C_1'' depending only on the ambient curvature bound such that for any $t \in [t', t_3]$, denoting by $\ell_{h'}(I')$ the length of I' with respect to h', we have

$$\int_{I' \times \{t\}} |f_z|_{h'}^2 dz \leq 4\delta\ell_{h'}(I') + C_1''(\sqrt{\delta}r + \delta r^3 + \delta^{3/2}r^2).$$

Since $(0.8)r \leq \ell_{h'}(I') \leq r < 1$, we see that provided that δ is sufficiently small, for each $t \in [t', t_3]$ there is $z(t) \in I'$ such $|f_z(z(t), t)|_{h'}^2 \leq 2C_1''\sqrt{\delta}$. If we have chosen δ sufficiently small, this means that for each $t \in [t', t_3]$ there is $z(t)$ such that $|f_z(z(t), t)|_{h'} \leq 1/20$. Since by Corollary 19.63 $\int_{I' \times \{t\}} k ds < 2D_4\sqrt{\delta}$, provided that δ is sufficiently small, it follows that for all $t \in [t', t_3]$ the curve $\gamma_t|_{I'}$ is a graph of (z, t) and $|f_z|_{h'} < 1/10$. But by construction either $t_3 = t_2$ or there is a point in $(z, t_3) \in I' \times \{t_3\}$ with $|f_z(z, t_3)|_{h'} = 1/10$. Hence, it must be the case that $t_3 = t_2$, and thus our graph curve flow is defined for all $t \in [t', t_2]$ and satisfies the derivative bound $|f_z|_{h'} < 1/10$ throughout the interval $[t', t_2]$.

8.4. Proof that $t_2 = t' + \delta r^2$. The last step is to show that the inequality $k^2 < 2/(t-t')$ holds for all $t \in [t', t' + \delta r^2]$.

We fix a point x_0. We continue all the notation, assumptions and results of the previous section. That is to say, we lift the evolving family of curves to the tangent space $T_{x_0}M$ using the exponential mapping, which is a local diffeomorphism. This tangent space is split as the product of the z-axis and B. On this coordinate system we have the evolving family of Riemannian metrics $h(t)$ pulled back from the Ricci flow and also we have the Euclidean metric h' from the metric $h(t')$ on $T_{x_0}M$. We fix an interval I' on the z-axis of h'-length between $(0.8)r$ and r. We choose I' to be centered at x_0 with respect to the z-coordinate. On $I' \times [t', t_2]$ we have the graph-flow which is reparameterization of the pullback of the curve-shrinking flow. The graph-flow is given as the graph of a function f with $|f_z|_{h'} < 1/10$. The vector fields Z and Y are as in the last section.

We follow closely the discussion in Section 4 of [**2**] (pages 293 -294). Since we are not working in a flat background, there are two differences: (i) we take covariant derivatives instead of ordinary derivatives and (ii) there are various correction terms from curvature, from covariant derivatives, and from the fact that Y is not equal to $\nabla_Z Z/|Z|^2$.

Notice that

$$Z\left(\frac{Z(|Z|^2)}{|Z|^2}\right) = \frac{|Z|^2_{zz}}{|Z|^2} - \frac{(|Z|^2_z)^2}{|Z|^4}$$

$$= \frac{|Z|^2_{zz}}{|Z|^2} - 4\langle Y, Z\rangle^2 - 8\langle Y, Z\rangle \psi(Z)|Z|^2 - 4\psi^2(Z)|Z|^4.$$

Thus, it follows from Equation (19.11) that we have

$$\frac{\partial}{\partial t}|Z|^2 = \frac{(|Z|^2)_{zz}}{|Z|^2} - 2|Z|^2|Y|^2 - 4\langle Z, Y\rangle^2 + V,$$

where $|V| \leq C'(1+\delta|Y|)$ for a constant C' depending only on the curvature bound of the ambient flow.

Similar computations show that

$$\frac{\partial}{\partial t}|Y|^2 = \frac{(|Y|^2)_{zz}}{|Z|^2} - \frac{2|\nabla_Z Y|^2}{|Z|^2} - \frac{4}{|Z|^2}\langle \frac{\nabla_Z Z}{|Z|^2}, Y\rangle\langle \nabla_Z Y, Z\rangle$$
$$- 2\mathrm{Ric}(Y,Y) + 2\frac{\mathrm{Rm}(Y, Z, Y, Z)}{|Z|^2} - 2\langle \nabla_Y(\psi(Z)Z), Y\rangle.$$

Of course,

$$\langle \frac{\nabla_Z Z}{|Z|^2}, Y\rangle = |Y|^2 + \psi(Z)\langle Z, Y\rangle.$$

Hence, putting all this together and using Claim 19.66 we have

$$\frac{\partial}{\partial t}|Y|^2 = \frac{(|Y|^2)_{zz}}{|Z|^2} - \frac{2|\nabla_Z Y|^2}{|Z|^2} - \frac{4|Y|^2}{|Z|^2}\langle \nabla_Z Y, Z\rangle + W,$$

where

$$|W| \leq C'|Y|(|Y| + \delta|\nabla_Z Y|).$$

Now let us consider

$$Q = \frac{|Y|^2}{2 - |Z|^2}.$$

Notice that since $|f_z|_{h'} < 1/10$, it follows that $1 - \delta \leq |Z|^2 < (1.01)(1+\delta)$ on all of $[t', t_2]$. We now follow the computations on p. 294 of [**2**], adding in

8. PROOF OF THE FIRST INEQUALITY IN LEMMA 19.24

the error terms.

$$Q_t = \frac{|Y|_t^2}{(2-|Z|^2)} + \frac{|Y|^2|Z|_t^2}{(2-|Z|^2)^2} = \frac{|Y|_{zz}^2}{|Z|^2(2-|Z|^2)} - \frac{2|\nabla_Z Y|^2}{|Z|^2(2-|Z|^2)}$$

$$- \frac{4|Y|^2}{|Z|^2(2-|Z|^2)}\langle \nabla_Z Y, Z\rangle + \frac{W}{(2-|Z|^2)} + \frac{|Y|^2|Z|_{zz}^2}{|Z|^2(2-|Z|^2)^2}$$

$$- \frac{2|Z|^2\|Y\|^4}{(2-|Z|^2)^2} - \frac{4|Y|^2}{(2-|Z|^2)^2}\langle Z, Y\rangle^2 + \frac{|Y|^2}{(2-|Z|^2)^2}V.$$

On the other hand,

$$\frac{Q_{zz}}{|Z|^2} = \frac{|Y|_{zz}^2}{|Z|^2(2-|Z|^2)} + \frac{|Y|^2|Z|_{zz}^2}{|Z|^2(2-|Z|^2)^2} + \frac{2|Y|_z^2|Z|_z^2}{|Z|^2(2-|Z|^2)^2} + \frac{2|Y|^2\left(|Z|_z^2\right)^2}{|Z|^2(2-|Z|^2)^3}.$$

From Claim 19.65 we have

$$|Z|_z^2 = 2\langle \nabla_Z Z, Z\rangle = 2|Z|^2\langle Y, Z\rangle + 2\psi(Z)|Z|^4.$$

Plugging in this expansion gives

$$\frac{Q_{zz}}{|Z|^2} = \frac{|Y|_{zz}^2}{|Z|^2(2-|Z|^2)} + \frac{|Y|^2|Z|_{zz}^2}{|Z|^2(2-|Z|^2)^2}$$

$$+ \frac{8\langle \nabla_Z Y, Y\rangle\langle Y, Z\rangle}{(2-|Z|^2)^2} + \frac{8\psi(z)|Z|^2\langle \nabla_Z Y, Y\rangle}{(2-|Z|^2)^2}$$

$$+ \frac{8|Z|^2|Y|^2\langle Y, Z\rangle^2}{(2-|Z|^2)^3} + \frac{16\psi(Z)|Z|^4|Y|^2\langle Y, Z\rangle}{(2-|Z|^2)^3} + \frac{8\psi^2(Z)|Z|^6|Y|^2}{(2-|Z|^2)^3}.$$

Expanding, we have

$$\frac{Q_{zz}}{|Z|^2} = \frac{|Y|_{zz}^2}{|Z|^2(2-|Z|^2)} + \frac{|Y|^2|Z|_{zz}^2}{|Z|^2(2-|Z|^2)^2} + \frac{8\langle \nabla_Z Y, Y\rangle\langle Y, Z\rangle}{(2-|Z|^2)^2}$$

$$+ \frac{8|Y|^2|Z|^2\langle Y, Z\rangle^2}{(2-|Z|^2)^3} + U,$$

where

$$|U| \leq C'(|Y|^2 + \delta|\nabla_Z Y||Y| + \delta|Y|^3).$$

Comparing the formulas yields

$$Q_t = \frac{Q_{zz}}{|Z|^2} - \frac{8\langle \nabla_Z Y, Y\rangle\langle Y, Z\rangle}{(2-|Z|^2)^2} - \frac{8|Y|^2|Z|^2\langle Y, Z\rangle^2}{(2-|Z|^2)^3}$$

$$- \frac{2|\nabla_Z Y|^2}{|Z|^2(2-|Z|^2)} - \frac{4|Y|^2}{|Z|^2(2-|Z|^2)}\langle \nabla_Z Y, Z\rangle$$

$$- \frac{2|Z|^2\|Y\|^4}{(2-|Z|^2)^2} - \frac{4|Y|^2}{(2-|Z|^2)^2}\langle Z, Y\rangle^2 + A,$$

where

$$|A| \leq C'(|Y|^2 + \delta|\nabla_Z Y||Y| + \delta|Y|^3).$$

Using item (4) of Claim 19.65 this leads to

$$Q_t \leq \frac{Q_{zz}}{|Z|^2} + \frac{8(1+3\delta)|Y|^2|f_z||\nabla_Z Y|}{(2-|Z|^2)^2} - \frac{4|Y|^2\langle \nabla_Z Y, Y\rangle}{|Z|^2(2-|Z|^2)}$$
$$- \frac{2|\nabla_Z Y|^2}{|Z|^2(2-|Z|^2)} - \frac{2|Z|^2|Y|^4}{(2-|Z|^2)^2} + |A|.$$

Next, we have

CLAIM 19.67.

$$|\langle \nabla_Z Y, Z\rangle| \leq (|f_z|(|\nabla_Z Y| + 2\delta|Y|) + \delta|Z|^2|Y|)(1+\delta).$$

PROOF. Since $Y = (0, \phi)$ for some function ϕ, we have $\nabla_Z Y = (0, \phi_z) + \Gamma(Z, Y)$ and hence

$$|\langle \nabla_Z Y, Z\rangle_{h'}| = |\langle \nabla_Z Y, (1, f_z)\rangle_{h'}|$$
$$\leq |\langle f_z, \phi_z\rangle_{h'}| + |\langle \Gamma(Z, Y), Z\rangle_{h'}|$$
$$\leq |f_z|_{h'}|\phi_z|_{h'} + \delta|Z|^2|Y|.$$

On the other hand $\nabla_Z Y = (0, \phi_z) + \Gamma(Z, Y)$ so that $|\phi_z|_{h'} \leq |\nabla_Z Y| + \delta|Z||Y|$. From this the claim follows. □

Now for $\delta > 0$ sufficiently small, using the fact that $1 - \delta < |Z|^2 < (1+\delta)(1.01)$ we can rewrite this as

$$Q_t \leq \frac{Q_{zz}}{|Z|^2} + \frac{8(1+3\delta)|Y|^2|f_z||\nabla_Z Y|}{(2-|Z|^2)^2} + \frac{4|Y|^2(1+\delta)|\nabla_Z Y||f_z|}{|Z|^2(2-|Z|^2)}$$
$$- \frac{2|\nabla_Z Y|^2}{|Z|^2(2-|Z|^2)} - \frac{(1.95)|Y|^4}{(2-|Z|^2)^2} + \widetilde{A},$$

where $\widetilde{A} \leq C'(|Y|^2 + \delta|Y||\nabla_Z Y| + \delta|Y|^3)$. Of course, $|Y||\nabla_Z Y| + |Y|^3 \leq 2|Y|^2 + |\nabla_Z Y|^2 + |Y|^4$. Using this, provided that δ is sufficiently small, we can rewrite this as

$$Q_t \leq \frac{Q_{zz}}{|Z|^2} + \frac{1}{(2-|Z|^2)} \cdot \left[\frac{8(1+3\delta)|\nabla_Z Y||f_z||Y|^2 - (0.9)|Y|^4}{(2-|Z|^2)}\right.$$
$$\left. + \frac{4|Y|^2(1+\delta)|\nabla_Z Y||f_z| - (1.9)|\nabla_Z Y|^2}{|Z|^2}\right] - Q^2 + \widetilde{A}''$$

where $\widetilde{A}'' \leq C'(|Y|^2)$. We denote the quantity within the brackets by B and we estimate

$$B \leq 8(1+3\delta)\frac{|Y|^2|\nabla_Z Y|(1/10)(1+\delta)}{(2-(1.01)(1+\delta))} + \frac{4(1/10)(1+\delta)|Y|^2|\nabla_Z Y|}{(1-\delta)}$$
$$- \frac{(1.9)}{(1.01)(1+\delta)}|\nabla_Z Y|^2 - \frac{(0.9)|Y|^4}{2-(1.01)(1+\delta)}$$
$$\leq (1.6)|Y|^2|\nabla_Z Y| - (0.8)|\nabla_Z Y|^2 - (0.8)|Y|^4$$
$$\leq 0.$$

8. PROOF OF THE FIRST INEQUALITY IN LEMMA 19.24

Therefore,
$$Q_t \leq \frac{Q_{zz}}{|Z|^2} - Q^2 + |\tilde{A}| \leq \frac{Q_{zz}}{|Z|^2} - (Q - C_1')^2 + (C_1')^2,$$

for some constant $C_1' > 1$ depending only on the curvature bound for the ambient Ricci flow.

Denote by l the length of I' under h'. As we have already seen, $(0.8)r \leq l \leq r$. We translate the z-coordinate so that $z = 0$ is one endpoint of I' and $z = l$ is the other endpoint; the point x_0 then corresponds to $z = l/2$. Consider the function $g = l^2/(z^2(l-z)^2)$ on $I' \times [t', t_2]$. Direct computation shows that $g_{zz} \leq 12g^2$. Now set

$$\tilde{Q} = Q - C_1'$$

and

$$h = \frac{1}{t-t'} + \frac{4(1-\delta)^{-1}l^2}{z^2(l-z)^2} + C_1'.$$

Then
$$-h_t + (1-\delta)^{-1}h_{zz} + (C_1')^2 \leq h^2,$$

so that
$$(\tilde{Q} - h)_t \leq \frac{\tilde{Q}_{zz}}{|Z|^2} - \frac{h_{zz}}{1-\delta} - \tilde{Q}^2 + h^2.$$

Since both h and h_{zz} are positive, at any point where $\tilde{Q} - h \geq 0$ and $\tilde{Q}_{zz} < 0$, we have $(\tilde{Q} - h)_t < 0$. At any point where $Q_{zz} \geq 0$, using the fact that $|Z|^2 \geq (1-\delta)$ we have

$$(\tilde{Q} - h)_t \leq (1-\delta)^{-1}(\tilde{Q} - h)_{zz} - \tilde{Q}^2 + h^2.$$

Thus, for any fixed t, at any local maximum for $(\tilde{Q} - h)(\cdot, t)$ at which $(\tilde{Q} - h)$ is ≥ 0 we have $(\tilde{Q} - h)_t \leq 0$. Since $\tilde{Q} - h$ equals $-\infty$ at the end points of I' for all times, there is a continuous function $f(t) = \max_{z \in I'}(\tilde{Q} - h)(z, t)$, defined for all $t \in (t', t_2]$ approaching $-\infty$ uniformly as t approaches t' from above. By the previous discussion, at any point where $f(t) \geq 0$ we have $f'(t) \leq 0$ in the sense of forward difference quotients. It now follows that $f(t) \leq 0$ for all $t \in (t', t_2]$. This means that for all $t \in (t', t_2]$ at the h'-midpoint x_0 of I' (the point where $z = l/2$) we have

$$Q(x_0, t) \leq \frac{1}{t-t'} + \frac{16 \cdot 4(1-\delta)^{-1}}{l^2} + C_1'.$$

Since $l \geq (0.8)r$ and since $t - t' \leq \delta r^2$, we see that provided δ is sufficiently small (depending on the bound of the curvature of the ambient flow) we have

$$Q(x_0, t) < \frac{3}{2(t-t')}$$

for all $t \in [t', t_2]$. Of course, since $|Z|^2 \geq 1 - \delta$ everywhere, this shows that
$$k^2(x_0, t) \leq |Y(x_0, t)|^2 = (2 - |Z(x_0, t)|^2)Q(x_0, t) < \frac{2}{(t - t')}$$
for all $t \subset [t', t_2]$. Since x_0 was an arbitrary point of $c(\cdot, t')$, this shows that $k(x, t) < \sqrt{\frac{2}{t-t'}}$ for all $x \in c(\cdot, t)$ and all $t \in [t', t_2]$. By the definition of t_2 this implies that $t_2 = t' + \delta r^2$ and completes the proof of Lemma 19.58.

APPENDIX

3-manifolds covered by canonical neighborhoods

Recall that an ϵ-neck structure on a Riemannian manifold (N, g) centered at a point $x \in N$ is a diffeomorphism $\psi \colon S^2 \times (-\epsilon^{-1}, \epsilon^{-1}) \to N$ with the property that $x \in \psi(S^2 \times \{0\})$ and the property that $R(x)\psi^*g$ is within ϵ in the $C^{[1/\epsilon]}$-topology of the product metric $h_0 \times ds^2$, where h_0 is the round metric on S^2 of scalar curvature 1 and ds^2 is the Euclidean metric on the interval. Recall that the *scale* of the ϵ-neck is $R(x)^{-1/2}$. We define $s = s_N \colon N \to (-\epsilon^{-1}, \epsilon^{-1})$ as the composition of ψ^{-1} followed by the projection to the second factor.

1. Shortening curves

LEMMA A.1. *The following holds for all $\epsilon > 0$ sufficiently small. Suppose that (M, g) is a Riemannian manifold and that $N \subset M$ is an ϵ-neck centered at x. Let $S(x)$ be the central two-sphere of this neck and suppose that $S(x)$ separates M. Let $y \in M$. Orient s so that y lies in the closure of the positive side of $S(x)$. Let $\gamma \colon [0, a] \to M$ be a rectifiable curve from x to y. If γ contains a point of $s^{-1}(-\epsilon^{-1}, -\epsilon^{-1}/2)$ then there is a rectifiable curve from x to y contained in the closure of the positive side of $S(x)$ whose length is at most the length of γ minus $\frac{1}{2}\epsilon^{-1}R(x)^{-1/2}$.*

PROOF. Since γ contains a point on the negative side of $S(x)$ and it ends on the positive side of $S(x)$, there is a $c \in (0, a)$ such that $\gamma(c) \in S(x)$ and $\gamma|_{(c,a]}$ is disjoint from $S(x)$. Since $\gamma|_{[0,c]}$ has both endpoints in $S(x)$ and also contains a point of $s^{-1}(-\epsilon^{-1}, -\epsilon^{-1}/2)$, it follows that for ϵ sufficiently small, the length of $\gamma|_{[0,c]}$ is at least $3\epsilon^{-1}R(x)^{-1/2}/4$. On the other hand, there is a path μ in $S(x)$ connecting x to $\gamma(c)$ of length at most $2\sqrt{2\pi}(1+\epsilon)$. Thus, if ϵ is sufficiently small, the concatenation of μ followed by $\gamma|_{[c,a]}$ is the required shorter path. $\square$

2. The geometry of an ϵ-neck

LEMMA A.2. *For any $0 < \alpha < 1/8$ there is $\epsilon_1 = \epsilon_1(\alpha) > 0$ such that the following two conditions hold for all $0 < \epsilon \leq \epsilon_1$.*

(1) *If (N, g) is an ϵ-neck centered at x of scale 1 (i.e., with $R(x) = 1$) then the principal sectional curvatures at any point of N are within*

497

$\alpha/6$ of $\{1/2, 0, 0\}$. In particular, for any $y \in N$ we have
$$(1 - \alpha) \leq R(y) \leq (1 + \alpha).$$

(2) There is a unique two-plane of maximal sectional curvature at every point of an ϵ-neck, and the angle between the distribution of two-planes of maximal sectional curvature and the two-plane field tangent to the family of two-spheres of the ϵ-neck structure is everywhere less than α.

PROOF. The principal curvatures and their directions are continuous functions of the metric g in the space of metrics with the C^2-topology. The statements follow immediately. □

COROLLARY A.3. *The following holds for any $\epsilon > 0$ sufficiently small. Suppose that (N, g) is an ϵ-neck and that we have an embedding $f \colon S^2 \to N$ with the property that the restriction of g to the image of this embedding is within ϵ in the $C^{[1/\epsilon]}$-topology to the round metric h_0 of scalar curvature 1 on S^2 and with the norm of the second fundamental form less than ϵ. Then the two-sphere $f(S^2)$ is isotopic in N to any member of the family of two-spheres coming from the ϵ-neck structure on N.*

PROOF. By the previous lemma, if ϵ is sufficiently small for every $n \in N$ there is a unique two-plane, P_n, at each point on which the sectional curvature is maximal. The sectional curvature on this two-plane is close to $1/2$ and the other two eigenvalues of the curvature operator at n are close to zero. Furthermore, P_n makes small g-angles with the tangent planes to the S^2-factors in the neck structure. Under the condition that the restriction of the metric to $f(S^2)$ is close to the round metric h_0 and the norm of the second fundamental form is small, we see that for every $p \in S^2$ the two-plane $df(T_pS^2)$ makes a small g-angle with P_n and hence with the tangent planes to the family of two-spheres coming from the neck structure. Since g is close to the product metric, this means that the angle between $df(T_nS^2)$ and the tangents to the family of two-spheres coming from the neck structure, measured in the product metric, is also small. Hence, the composition of f followed by the projection mapping $N \to S^2$ induced by the neck structure determines a submersion of S^2 onto itself. Since S^2 is compact and simply connected, any submersion of S^2 onto itself is a diffeomorphism. This means that $f(S^2)$ crosses each line $\{x\} \times (-\epsilon^{-1}, \epsilon^{-1})$ transversely and in exactly one point. Clearly then, it is isotopic in N to any two-sphere of the form $S^2 \times \{s\}$. □

LEMMA A.4. *For any $\alpha > 0$ there is $\epsilon_2 = \epsilon_2(\alpha) > 0$ such that the following hold for all $0 < \epsilon \leq \epsilon_2$. Suppose that (N, g) is an ϵ-neck centered at x and $R(x) = 1$. Suppose that γ is a minimal geodesic in N from p to q. We suppose that γ is parameterized by arc length, is of length $\ell > \epsilon^{-1}/100$,*

and that $s(p) < s(q)$. Then for all s in the domain of definition of γ we have
$$|\gamma'(s) - (\partial/\partial s)|_g < \alpha.$$
In particular, the angle between γ' and $\partial/\partial s$ is less than 2α. Also, any member S^2 of the family of two-spheres in the N has intrinsic diameter at most $(1+\alpha)\sqrt{2}\pi$.

PROOF. Let us consider a geodesic μ in the product Riemannian manifold $S^2 \times \mathbb{R}$ with the metric on S^2 being of constant Gaussian curvature $1/2$, i.e., radius $\sqrt{2}$. Its projections, μ_1 and μ_2, to S^2 and to $\mathbb{R}$, respectively, are also geodesics, and $|\mu| = \sqrt{|\mu_1|^2 + |\mu_2|^2}$. For μ to be a minimal geodesic, the same is true of each of its projections. In particular, when μ is minimal, the length of μ_1 is at most $\sqrt{2}\pi$. Hence, for any $\alpha' > 0$, if μ is sufficiently long and if the final endpoint has a larger s-value than the initial point, then the angle between the tangent vectors $\mu'(s)$ and $\partial/\partial s$ is less than α'. This establishes the result for the standard metric on the model for ϵ-necks.

The first statement now follows for all ϵ sufficiently small and all ϵ-necks because minimal geodesics between a pair of points in a manifold vary continuously in the C^1-topology as a function of the space of metrics with the C^k-topology, since $k \geq 2$. The second statement is obvious since the diameter of any member of the family of two-spheres in the standard metric is $\sqrt{2}\pi$. □

COROLLARY A.5. *For any $\alpha > 0$ there is $\epsilon_3 = \epsilon_3(\alpha) > 0$ such that the following hold for any $0 < \epsilon \leq \epsilon_3$ and any ϵ-neck N of scale 1 centered at x.*

(1) *Suppose that p and q are points of N with either $|s(q) - s(p)| \geq \epsilon^{-1}/100$ or $d(p,q) \geq \epsilon^{-1}/100$. Then we have*
$$(1-\alpha)|s(q) - s(p)| \leq d(p,q) \leq (1+\alpha)|s(q) - s(p)|.$$

(2)
$$B(x, (1-\alpha)\epsilon^{-1}) \subset N \subset B(x, (1+\alpha)\epsilon^{-1}).$$

(3) *Any geodesic that exits from both ends of N has length at least $2(1-\alpha)\epsilon^{-1}$.*

COROLLARY A.6. *The following holds for all $\epsilon > 0$ sufficiently small. Let N be an ϵ-neck centered at x. If γ is a shortest geodesic in N between its endpoints and if $|\gamma| > R(x)^{-1/2}\epsilon^{-1}/100$, then γ crosses each two-sphere in the neck structure on N at most once.*

There is a closely related lemma.

LEMMA A.7. *The following holds for every $\epsilon > 0$ sufficiently small. Suppose that (M,g) is a Riemannian manifold and that $N \subset M$ is an ϵ-neck centered at x and suppose that γ is a shortest geodesic in M between its endpoints and that the length of every component of $N \cap |\gamma|$ has* length at least $R(x)^{-1/2}\epsilon^{-1}/8$. Then γ crosses each two-sphere in the neck structure on N at most once; see FIG. 1.

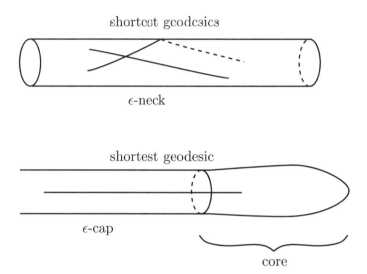

FIGURE 1. Shortest geodesics in necks and caps.

PROOF. We parameterize γ as a map from $[a,b] \to M$. By Corollary A.6, provided that $\epsilon > 0$ is sufficiently small, each component of $\gamma \cap N$ crosses each two-sphere of the neck structure at most once. Suppose that there is some two-sphere $S^2 \times \{x\}$ that is crossed by two different components of γ. Let $c < d$ be two points of intersection of γ with $S^2 \times \{s\}$.

There are two cases to consider. Suppose that the two components of $\gamma \cap N$ cross $S^2 \times \{x\}$ in opposite directions. In this case, since each component of $\gamma \cap N$ has length at least $\epsilon^{-1}/8$, then applying Corollary A.5 we can take the two-sphere that they both cross to be $S^2 \times \{s\}$ for some $s \in (-(0.9)\epsilon^{-1}, (0.9)\epsilon^{-1})$. Applying Corollary A.5 again we see that the distance from this sphere to the complement of N is at least $R(x)^{-1/2}\epsilon^{-1}/20$. Let $c < d$ be the points of intersection. Remove $\gamma([c,d])$ from γ and replace it by a path in $S^2 \times \{s\}$ between $\gamma(c)$ and $\gamma(d)$. If ϵ is sufficiently small, by Lemma A.4 we can choose this path to have length at most 2π, and hence the result will be a shorter path.

The other possibility is that γ crosses $S^2 \times \{s\}$ twice in the same direction. In this case the central two-sphere of N does not separate M and γ makes a circuit transverse to the two-sphere. In particular, by Corollary A.5 the length of $\gamma([c,d])$ is bounded below by $2(1-\alpha)R(x)^{-1/2}\epsilon^{-1}$ where we can take $\alpha > 0$ as close to zero as we want by making ϵ smaller. Clearly, then in this case as well, replacing $\gamma([c,d])$ with a path of length less than $2\pi R(x)^{-1/2}$ on $S^2 \times \{s\}$ will shorten the length of γ. □

2. THE GEOMETRY OF AN ϵ-NECK

COROLLARY A.8. *The following holds for all $\epsilon > 0$ sufficiently small and any $C < \infty$. Let X be an (C, ϵ)-cap in a complete Riemannian manifold (M, g), and let Y be its core and let S be the central two-sphere of the ϵ-neck $N = X - \overline{Y}$. We orient the s-direction in N so that Y lies off the negative end of N. Let $\widehat{Y}$ be the union of $\overline{Y}$ and the closed negative half of N and let S be the boundary of $\widehat{Y}$. Suppose that γ is a minimal geodesic in (M, g) that contains a point of the core Y. Then the intersection of γ with $\widehat{Y}$ is an interval containing an endpoint of γ; see* FIG. 1.

PROOF. If γ is completely contained in $\widehat{Y}$ then the result is clear. Suppose that the path is $\gamma \colon [a, b] \to M$ and $\gamma(d) \in Y$ for some $d \in [a, b]$. Suppose that there are $a' < d < b'$ with $\gamma(a')$ and $\gamma(b')$ contained in S. Then, by Corollary A.5, replacing $\gamma|_{[a',b']}$ with a path on S joining $\gamma(a')$ to $\gamma(b')$ creates a shorter path with the same endpoints. This shows that at least one of the paths $\gamma|_{[a,d]}$ or $\gamma|_{[d,b]}$, let us say $\gamma|_{[a,d]}$, is contained in $\widehat{Y}$. The other path $\gamma|_{[d,b]}$ has an endpoint in Y and exits from $\widehat{Y}$, hence by Corollary A.6 there is a subinterval $[d, b']$ such that either $\gamma(b')$ is contained in the frontier of X or $b = b'$ and furthermore $\gamma([d, b'])$ crosses each two-sphere of the ϵ-neck structure on N at most once. Since γ is not contained in $\widehat{Y}$, there is $b'' \in [d, b']$ such that $\gamma(b'') \in S$. We have constructed a subinterval of the form $[a, b'']$ such that $\gamma([a, b''])$ is contained in $\widehat{Y}$. If $b' = b$, then it follows from the fact that $\gamma|_{[d,b]}$ crosses each two-sphere of N at most once that $\gamma|_{[b'',b]}$ is disjoint from Y. This establishes the result in this case. Suppose that $b' < b$. If there is $c \in [b', b]$ with $\gamma(c) \in \widehat{Y}$ then the length of $\gamma([b'', c])$ is at least twice the distance from S to the frontier of the positive end of N. Thus, we could create a shorter path with the same endpoints by joining $\gamma(b'')$ to $\gamma(c)$ by a path of S. This means that $\gamma|_{[b',b]}$ is disjoint from S and hence from $\widehat{Y}$, proving the result in this case as well. $\square$

We also wish to compare distances from points outside the neck with distances in the neck.

LEMMA A.9. *Given $0 < \alpha < 1$ there is $\epsilon_4 = \epsilon_4(\alpha) > 0$ such that the following holds for any $0 < \epsilon \leq \epsilon_4$. Suppose that N is an ϵ-neck centered at x in a connected manifold M (here we are not assuming that $R(x) = 1$). We suppose that the central 2-sphere of N separates M. Let z be a point outside of the middle two-thirds of N and lying on the negative side of the central 2-sphere of N. (We allow both the case when $z \in N$ and when $z \notin N$.) Let p be a point in the middle half of N. Let $\mu \colon [0, a] \to N$ be a straight line segment (with respect to the standard product metric) in the positive s-direction in N beginning at p and ending at a point q of N. Then*

$$(1 - \alpha)(s(q) - s(p)) \leq d(z, q) - d(z, p) \leq (1 + \alpha)(s(q) - s(p)).$$

PROOF. This statement is clearly true for the product metric on an infinite cylinder, and hence by continuity, for any given α, the result holds for all $\epsilon > 0$ sufficiently small. □

N.B. It is important that the central two-sphere of N separates the ambient manifold M. Otherwise, there may be shorter geodesics from z to q entering the other end of N.

LEMMA A.10. *Given any $\alpha > 0$ there is $\epsilon(\alpha) > 0$ such that the following holds for any $0 < \epsilon \leq \epsilon(\alpha)$. Suppose that N is an ϵ-neck centered at x in a connected manifold M (here we are not assuming that $R(x) = 1$) and that z is a point outside the middle two-thirds of N. We suppose that the central two-sphere of N separates M. Let p be a point in the middle sixth of N at distance d from z. Then the intersection of the boundary of the metric ball $B(z,d)$ with N is a topological 2-sphere contained in the middle quarter of N that maps homeomorphically onto S^2 under the projection mapping $N \to S^2$ determined by the ϵ-neck structure. Furthermore, if $p' \in \partial B(z,d)$ then $|s(p) - s(p')| < \alpha R(x)^{-1/2} \epsilon^{-1}$; see* FIG. *2.*

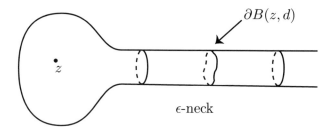

FIGURE 2. Intersection of metric balls and necks.

PROOF. The statement is scale-invariant, so we can assume that $R(x) = 1$. Denote by $S(z,d)$ the boundary of the metric ball $B(z,d)$. We orient s so that z lies to the negative side of the central two-sphere of N. It follows immediately from the previous result that, provided that $\epsilon > 0$ is sufficiently small, $S(z,d)$ intersects any line $y \times (-\epsilon^{-1}/3, \epsilon^{-1}/3)$ in at most one point. To complete the proof we need only show that $S(z,d)$ is contained in $s^{-1}((s(p) - \alpha\epsilon^{-1}, s(p) + \alpha\epsilon^{-1}))$. The distance from d to any point in the two-sphere factor of N containing p is contained in the interval $[d - 2\pi, d + 2\pi]$. Provided that ϵ is sufficiently small depending on α, the result follows immediately from Lemma A.9. □

3. Overlapping ϵ-necks

The subject of this section is the internal geometric properties of ϵ-necks and of intersections of ϵ-necks. We assume that $\epsilon \leq 1/200$.

3. OVERLAPPING ϵ-NECKS

PROPOSITION A.11. *Given $0 < \alpha \le 10^{-2}$, there is $\epsilon_5 = \epsilon_5(\alpha) > 0$ such that the following hold for all $0 < \epsilon \le \epsilon_5$. Let N and N' be ϵ-necks centered at x and x', respectively, in a Riemannian manifold X:*

(1) *If $N \cap N' \ne \emptyset$ then $1 - \alpha < R(x)/R(x') < 1 + \alpha$. In particular, denoting the scales of N and N' by h and h' we have*
$$1 - \alpha < \frac{h}{h'} < 1 + \alpha.$$

(2) *Suppose $y \in N \cap N'$ and S and S' are the two-spheres in the ϵ-neck structures on N and N', respectively, passing through y. Then the angle between TS_y and TS'_y is less than α.*

(3) *Suppose that $y \in N \cap N'$. Denote by $\partial/\partial s_N$ and $\partial/\partial s_{N'}$ the tangent vectors in the ϵ-neck structures of N and N', respectively. Then at the point y, either*
$$|R(x)^{1/2}(\partial/\partial s_N) - R(x')^{1/2}\partial/\partial s_{N'}| < \alpha$$
or
$$|R(x)^{1/2}(\partial/\partial s_N) + R(x')^{1/2}\partial/\partial s_{N'}| < \alpha.$$

(4) *Suppose that one of the two-spheres S' of the ϵ-neck structure on N' is completely contained in N. Then S' is a section of the projection mapping on the first factor*
$$p_1 \colon S^2 \times (-\epsilon^{-1}, \epsilon^{-1}) \to S^2.$$
In particular, S' is isotopic in N to any one of the two-spheres of the ϵ-neck structure on N by an isotopy that moves all points in the interval directions.

(5) *If $N \cap N'$ contains a point y with $(-0.9)\epsilon^{-1} \le s_N(y) \le (0.9)\epsilon^{-1}$, then there is a point $y' \in N \cap N'$ such that*
$$-(0.96)\epsilon^{-1} \le s_N(y') \le (0.96)\epsilon^{-1},$$
$$-(0.96)\epsilon^{-1} \le s_{N'}(y') \le (0.96)\epsilon^{-1}.$$
The two-sphere $S(y')$ in the neck structure on N through y' is contained in N' and the two-sphere $S'(y')$ in the neck structure on N' through y' is contained in N. Furthermore, $S(y')$ and $S'(y')$ are isotopic in $N \cap N'$. Lastly, $N \cap N'$ is diffeomorphic to $S^2 \times (0, 1)$ under a diffeomorphism mapping $S(y)$ to $S^2 \times \{1/2\}$, see FIG. 3.

PROOF. Fix $0 < \epsilon_5(\alpha) \le \min(\epsilon_1(\alpha_1), \epsilon_2(\alpha/3), \epsilon_3(\alpha), \alpha/3)$ sufficiently small so that Corollary A.3 holds. The first two items are then immediate from Lemma A.2. The third statement is immediate from Lemma A.4, and the fourth statement from Corollary A.3. Let us consider the last statement. Let $y \in N \cap N'$ have $-(0.9)\epsilon^{-1} \le s_N(y) \le (0.9)\epsilon^{-1}$. By reversing the s-directions of N and/or N' if necessary, we can assume that $0 \le s_N(y) \le (0.9)\epsilon^{-1}$ and that ∂_{s_N} and $\partial_{s_{N'}}$ almost agree at y. If

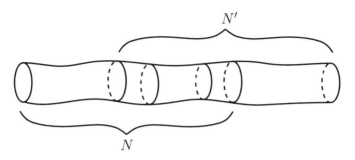

FIGURE 3. Overlapping ϵ-necks.

$-(0.96)\epsilon^{-1} \leq s_{N'}(y) \leq (0.96)\epsilon^{-1}$, we set $y' = y$. Suppose that $s_{N'}(y) > (0.96)\epsilon^{-1}$. We move along the straight line though y in the neck structure on N in the negative direction to a point y' with $(0.96)\epsilon^{-1} = s_{N'}(y')$. According to item (3) of this result we have $-(0.96)\epsilon^{-1} \leq s_N(x') \leq (0.96)\epsilon^{-1}$. There is a similar construction when $s_{N'}(y) < -(0.96)\epsilon^{-1}$. In all cases this allows us to find y' such that both the following hold:

$$-(0.96)\epsilon^{-1} \leq s_N(y') \leq (0.96)\epsilon^{-1},$$
$$-(0.96)\epsilon^{-1} \leq s_{N'}(y') \leq (0.96)\epsilon^{-1}.$$

Let y' be any point satisfying both these inequalities. According to Lemma A.4 and part (1) of this result, the diameter of $S(y')$ is at most $2\pi h$, where h is the scale of N and N'. Since $\epsilon^{-1} \geq 200$, it follows from Corollary A.5 that $S(y')$ is contained in N'. Symmetrically $S'(y')$ is contained in N.

Now consider the intersection of any straight line in the neck structure on N with N'. According to part (3), this intersection is connected. Thus, $N \cap N'$ is a union of open arcs in the s_N-directions through the points of $S(y')$. These arcs can be used to define a diffeomorphism from $N \cap N'$ to $S^2 \times (0,1)$ sending $S(y')$ to $S^2 \times \{1/2\}$. Also, we have the straight line isotopy from $S'(y')$ to $S(y')$ contained in $N \cap N'$. □

4. Regions covered by ϵ-necks and (C, ϵ)-caps

Here we fix $0 < \epsilon \leq 1/200$ sufficiently small so that all the results in the previous two sections hold with $\alpha = 10^{-2}$.

4.1. Chains of ϵ-necks.

DEFINITION A.12. Let (X, g) be a Riemannian manifold. By a *finite chain* of ϵ-necks in (X, g), we mean a sequence $N_a, \ldots, N_b$, of ϵ-necks in (X, g) such that:

(1) for all i, $a \leq i < b$, the intersection $N_i \cap N_{i+1}$ contains the positive-most quarter of N_i and the negative-most quarter of N_{i+1}

and is contained in the positive-most three-quarters of N_i and the negative-most three-quarters of N_{i+1}, and

(2) for all i, $a < i \leq b$, N_i is disjoint from the negative end of N_a.

By an *infinite chain* of ϵ-necks in X we mean a collection $\{N_i\}_{i \in I}$ for some interval $I \subset \mathbb{Z}$, infinite in at least one direction, so that for each finite subinterval J of I the subset of $\{N_i\}_{i \in J}$ is a chain of ϵ-necks.

Notice that in an ϵ-chain $N_i \cap N_j = \emptyset$ if $|i - j| \geq 5$.

LEMMA A.13. *The union U of the N_i in a finite or infinite chain of ϵ-necks is diffeomorphic to $S^2 \times (0,1)$. In particular, it is an ϵ-tube.*

PROOF. Let us first prove the result for finite chains. The proof that U is diffeomorphic to $S^2 \times (0,1)$ is by induction on $b - a + 1$. If $b = a$, then the result is clear. Suppose that we know the result for chains of smaller cardinality. Then $N_a \cup \cdots \cup N_{b-1}$ is diffeomorphic to $S^2 \times (0,1)$. Hence by part (5) of Proposition A.11, U is the union of two manifolds each diffeomorphic to $S^2 \times (0,1)$ meeting in an open subset diffeomorphic to $S^2 \times (0,1)$. Furthermore, by the same result in the intersection there is a two-sphere isotopic to each of the two-sphere factors from the two pieces. It now follows easily that the union is diffeomorphic to $S^2 \times (0,1)$. Now consider an infinite chain. It is an increasing union of finite chains each diffeomorphic to $S^2 \times (0,1)$ and with the two-spheres of one isotopic to the two-spheres of any larger one. It is then immediate that the union is diffeomorphic to $S^2 \times (0,1)$. □

Notice that the frontier of the union of the necks in a finite chain, $U = \cup_{a \leq i \leq b} N_i$, in M is equal to the frontier of the positive end of N_b union the frontier of the negative end of N_a. Thus, we have:

COROLLARY A.14. *Let $\{N_a, \ldots, N_b\}$ be a chain of ϵ-necks. If a connected set Y meets both $U = \cup_{a \leq i \leq b} N_i$ and its complement, then Y either contains points of the frontier of the negative end N_a or of the positive end of N_b.*

The next result shows there is no frontier at an infinite end.

LEMMA A.15. *Suppose that $\{N_0, \cdots\}$ is an infinite chain of ϵ-necks in M. Then the frontier of $U = \cup_{i=0}^{\infty} N_i$ is the frontier of the negative end of N_0.*

PROOF. Suppose that x is a point of the frontier of U. Let $x_i \in U$ be a sequence converging to x. If the x_i were contained in a finite union of the N_k, say $N_0 \cup \cdots \cup N_\ell$, then x would be in the closure of this union and hence by the previous comment would be either in the frontier of the negative end of N_0 or in the frontier of the positive end of N_ℓ. But the frontier of the positive end of N_ℓ is contained in $N_{\ell+1}$ and hence contains no points of the frontier of U. Thus, in this case x is a point of the frontier of the negative end of N_0. If $\{x_i\}$ is not contained in any finite union, then after passing to a subsequence, we can suppose that $x_i \in N_{k(i)}$ where $k(i)$ is an increasing sequence tending to infinity. Clearly $R(x_i)$ converges to $R(x) < \infty$. Hence, there is a uniform lower bound to the scales of the $N_{k(i)}$. For all i sufficiently large, $x_i \notin N_0$. Thus, for such i any path from x_i to x must traverse either N_0 or $N_{k(j)}$ for all $j \geq i+5$. The length of such a path is at least the minimum of the width of N_0 and the width of $N_{k(j)}$ for some j sufficiently large. But we have just seen that there is a positive lower bound to the scales of the $N_{k(j)}$ independent of j, and hence by Corollary A.5 there is a positive lower bound, independent of j, to the widths of the $N_{k(j)}$. This shows that there is a positive lower bound, independent of i, to the distance from x_i to x. This is impossible since x_i converges to x. □

In fact, there is a geometric version of Lemma A.13.

LEMMA A.16. *There is $\epsilon_0 > 0$ such that the following holds for all $0 < \epsilon \leq \epsilon_0$. Suppose that $\{N_j\}_{j \in J}$ is a chain of ϵ-necks in a Riemannian manifold M. Let $U = \cup_{j \in J} N_j$. Then there exist an interval I and a smooth map $p \colon U \to I$ such that every fiber of p is a two-sphere, and if y is in the middle $7/8$'s of N_j then the fiber $p^{-1}(p(y))$ makes a small angle at every point with the family of two-spheres in the ϵ-neck N_j.*

PROOF. Since according to Lemma A.2 the two-spheres for N_j and N_{j+1} almost line up, it is an easy matter to interpolate between the projection maps to the interval to construct a fibration of U by two-spheres with the given property. The interval I is simply the base space of this fibration. □

A finite or infinite chain $\{N_j\}_{j \in J}$ of ϵ-necks is *balanced* provided that for every $j \in J$, not the largest element of J, we have

(A.1) $$(0.99)R(x_j)^{-1/2}\epsilon^{-1} \leq d(x_j, x_{j+1}) \leq (1.01)R(x_j)^{-1/2}\epsilon^{-1},$$

where, for each j, x_j is the central point of N_j.

Notice that in a balanced chain $N_j \cap N_{j'} = \emptyset$ if $|j - j'| \geq 3$.

LEMMA A.17. *There exists $\epsilon_0 > 0$ such that for all $0 < \epsilon \leq \epsilon_0$ the following is true. Suppose that N and N' are ϵ-necks centered at x and x', respectively, in a Riemannian manifold M. Suppose that x' is not contained in N but is contained in the closure of N in M. Suppose also that the two-spheres of the neck structure on N and N' separate M. Then, possibly after reversing the ϵ-neck structures on N and/or N', the pair $\{N, N'\}$ forms a balanced chain.*

PROOF. By Corollary A.5, Inequality (A.1) holds for $d(x, x')$. Once we have this inequality, it follows immediately from the same corollary that, possible after reversing, the s-directions $\{N, N'\}$ makes a balanced chain of ϵ-necks. (It is not possible for the positive end of N_b to meet N_a for this would allow us to create a loop meeting the central two-sphere of N_b

4. REGIONS COVERED BY ϵ-NECKS AND (C,ϵ)-CAPS   507

transversely in a single point, so that this two-sphere would not separate M.) □

LEMMA A.18. *There exists $\epsilon_0 > 0$ such that for all $0 < \epsilon \le \epsilon_0$ the following is true. Suppose that $\{N_a, \ldots, N_b\}$ is a balanced chain in a Riemannian manifold M with $U = \cup_{i=a}^{b} N_i$. Suppose that the two-spheres of the neck structure of N_a separate M. Suppose that x is a point of the frontier of U contained in the closure of the plus end of N_b that is also the center of an ϵ-neck N. Then possibly after reversing the direction of N, we have that $\{N_a, \ldots, N_b, N\}$ is a balanced chain. Similarly, if x is in the closure of the minus end of N_a, then (again after possibly reversing the direction of N) we have that $\{N, N_a, \ldots, N_b\}$ is a balanced ϵ-chain.*

PROOF. The two cases are symmetric; we consider only the first. Since x is contained in the closure of N_b, clearly $N_b \cap N \ne \emptyset$. Also, clearly, provided that $\epsilon > 0$ is sufficiently small, $d(x_b, x)$ satisfies Inequality (A.1) so that Lemma A.17 the pair $\{N_b, N\}$ forms an ϵ-chain, and hence a balanced ϵ-chain. It is not possible for N to meet the negative end of N_a since the central two-sphere of N_a separates M. Hence $\{N_a, \ldots, N_b, N\}$ is a balanced chain of ϵ-necks. □

PROPOSITION A.19. *There exists $\epsilon_0 > 0$ such that for all $0 < \epsilon \le \epsilon_0$ the following is true. Let X be a connected subset of a Riemannian manifold M with the property that every point $x \in X$ is the center of an ϵ-neck $N(x)$ in M. Suppose that the central two-spheres of these necks do not separate M. Then there is a subset $\{x_i\}$ of X such that the necks $N(x_i)$ (possibly after reversing their s-directions) form a balanced chain of ϵ-necks $\{N(x_i)\}$ whose union U contains X. The union U is diffeomorphic to $S^2 \times (0,1)$. It is an ϵ-tube.*

PROOF. According to Lemma A.18 for $\epsilon > 0$ sufficiently small the following holds. Suppose that we have a balanced chain of ϵ-necks $N_a \ldots, N_b$, with N_i centered at $x_i \in X$, whose union U does not contain X. Then one of the following holds:

(1) It is possible to find an ϵ-neck N_{b+1} centered at a point of the intersection of X with the closure of the positive end of N_b so that $N_a, \ldots, N_{b+1}$ is a balanced ϵ-chain.
(2) It is possible to find an ϵ-neck N_{a-1} centered at a point of the intersection of X with the closure of the negative end of N_a so that $N_{a-1}, N_a, \ldots, N_b$ is a balanced ϵ-chain.

Now assume that there is no finite balanced chain of ϵ-necks $N(x_i)$ containing X. Then we can repeatedly lengthen a balanced chain of ϵ-necks centered at points of X by adding necks at one end or the other. Suppose that we have a half-infinite balanced chain $\{N_0, N_1, \ldots, \}$. By Lemma A.15 the frontier of this union is the frontier of the negative end of N_0. Thus, if we can construct a balanced chain which is infinite in both directions, then the union of the necks in this chain is a component of M and hence contains the connected set X. If we can construct a balanced chain that is infinite at one end but not the other that cannot be further extended, then the connected set is disjoint from the frontier of the negative end of the first neck in the chain and, as we have see above, the 'infinite' end of the chain has no frontier. Thus, X is disjoint from the frontier of U in M and hence is contained in U. Thus, in all cases we construct a balanced chain of ϵ-necks containing X. By Lemma A.13 the union of the necks in this chain is diffeomorphic to $S^2 \times (0,1)$ and hence is an ϵ-tube. □

LEMMA A.20. *The following holds for every $\epsilon > 0$ sufficiently small. Let (M, g) be a connected Riemannian manifold. Suppose that every point of M is the center of an ϵ-neck. Then either M is diffeomorphic to $S^2 \times (0,1)$ and is an ϵ-tube, or M is diffeomorphic to an S^2-fibration over S^1.*

PROOF. If the two-spheres of the ϵ-necks do not separate M, then it follows from the previous result that M is an ϵ-tube. If one of the two-spheres does separate, then take the universal covering $\widetilde{M}$ of M. Every point of $\widetilde{M}$ is the center of an ϵ-neck (lifting an ϵ-neck in M) and the two-spheres of these necks separate $\widetilde{M}$. Thus the first case applies, showing that $\widetilde{M}$ is diffeomorphic to $S^2 \times (0,1)$. Every point is the center of an ϵ-neck that is disjoint from all its non-trivial translates under the fundamental group. This means that the quotient is fibered by S^2's over S^1, and the fibers of this fibration are isotopic to the central two-spheres of the ϵ-necks. □

5. Subsets of the union of cores of (C, ϵ)-caps and ϵ-necks.

In this section we fix $0 < \epsilon \leq 1/200$ so that all the results of this section hold with $\alpha = 0.01$.

PROPOSITION A.21. *For any $C < \infty$ the following holds. Suppose that X is a connected subset of a Riemannian three-manifold (M, g). Suppose that every point of X is either the center of an ϵ-neck or is contained in the core of a (C, ϵ)-cap. Then one of the following holds (see* FIG. *4):*

(1) *X is contained in a component of M that is the union of two (C, ϵ)-caps and is diffeomorphic to one of S^3, $\mathbb{R}P^3$ or $\mathbb{R}P^3 \# \mathbb{R}P^3$.*
(2) *X is contained in a component of M that is a double C-capped ϵ-tube. This component is diffeomorphic to S^3, $\mathbb{R}P^3$ or $\mathbb{R}P^3 \# \mathbb{R}P^3$.*
(3) *X is contained in a single (C, ϵ)-cap.*
(4) *X is contained in a C-capped ϵ-tube.*
(5) *X is contained in an ϵ-tube.*
(6) *X is contained in a component of M that is an ϵ-fibration, which itself is a union of ϵ-necks.*

5. SUBSETS OF THE UNION OF CORES OF (C,ϵ)-CAPS AND ϵ-NECKS. 509

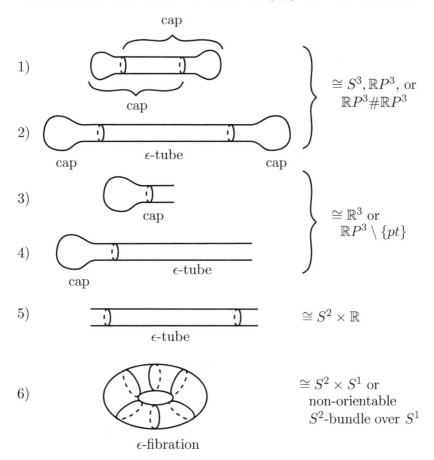

FIGURE 4. Components covered by ϵ-necks and ϵ-caps.

PROOF. We divide the proof into two cases: Case I: There is a point of X contained in the core of a (C, ϵ)-cap. Case II: Every point of X is the center of an ϵ-neck.
Case I: We begin the study of this case with a claim.

CLAIM A.22. *It is not possible to have an infinite chain of (C, ϵ)-caps $C_0 \subset C_1 \subset \cdots$ in M with the property that for each $i \geq 1$, the closure of the core of C_i contains a point of the frontier of C_{i-1}*

PROOF. We argue by contradiction. Suppose there is such an infinite chain. Fix a point $x_0 \in C_0$ and let $Q_0 = R(x_0)$. For each $i \geq 1$ let x_i be a point in the frontier of C_{i-1} that is contained in the closure of the core of C_i. For each i let N_i be the ϵ-neck in C_i that is the complement of the closure of its core. We orient the s_{N_i}-direction so that the core of C_i lies off the negative end of N_i. Let S_i' be the boundary of the core of C_i. It is the central two-sphere of an ϵ-neck N_i' in C_i. We orient the s-direction of N_i' so that the non-compact end of C_i lies off the positive end of N_i'. We denote by h_{i-1} the scale of N_{i-1} and by h'_i the scale of N'_i. By Lemma A.2 the ratio h_{i-1}/h'_i is between 0.99 and 1.01. Suppose that S'_i is disjoint from C_{i-1}. Then one of the complementary components of S'_i in M contains C_{i-i}, and of course, one of the complementary components of S'_i is the core of C_i. These complementary components must be the same, for otherwise C_{i-1} would be disjoint from the core of C_i and hence the intersection of C_{i-1} and C_i would be contained in N_i. This cannot happen since C_{i-1} is contained in C_i. Thus, if S'_i is disjoint from C_{i-1}, then the core of C_i contains C_{i-1}. This means that the distance from x_0 to the complement of C_i is greater than the distance of x_0 to the complement of C_{i-1} by an amount equal to the width of N_i. Since the scale of N_i is at least $C^{-1/2}R(x_0)^{-1/2}$ (see (5) of Definition 9.72), it follows from Corollary A.5 that this width is at least $2(0.99)C^{-1/2}R(x_0)^{-1/2}\epsilon^{-1}$.

Next suppose that S'_i is contained in C_{i-1}. Then one of the complementary components A of S'_i in M has closure contained in C_{i-1}. This component cannot be the core of C_i since the closure of the core of C_i contains a point of the frontier of C_{i-1} in M. Thus, A contains N_i. Of course, $A \neq N_i$ since the frontier of A in M is S'_i whereas N_i has two components to its frontier in M. This means that C_i does not contain A, which is a contradiction since C_i contains C_{i-1} and $A \subset C_{i-1}$.

Lastly, we suppose that S'_i is neither contained in C_{i-1} nor in its complement. Then S'_i must meet N_{i-1}. According to Proposition A.11 the s-directions in N_{i-1} and N'_i either almost agree or are almost opposite. Let $x \in S'_i \cap \partial N_{i-1}$ so that $s_{N'_i}(x) = 0$. Move from x along the $s_{N'_i}$-direction that moves into N_{i-1} to a point x' with $|s_{N_i}(x')| = (0.05)\epsilon^{-1}$. According to Proposition A.11, $(0.94)\epsilon^{-1} < s_{N_{i-1}}(x') < (0.96)\epsilon^{-1}$. Let $S'(x')$ be the two-sphere in the neck structure for N'_i through this point. According to Proposition A.11, $S'(x') \subset N_{i-1}$, and $S'(x')$ is isotopic in N_{i-1} to its central two-sphere. One of the complementary components of $S'(x')$ in C_i, let us call it A', is diffeomorphic to $S^2 \times (0,1)$. Also, one of the complementary components A of $S'(x')$ in M contains the core of C_{i-1}. As before, since $C_{i-1} \subset C_i$, the complementary component A cannot meet C_i in A'. This means that the $s_{N_{i-1}}$- and $s_{N'_i}$-directions almost line up along $S'(x')$. This means that $S'(x') = s_{N'_i}^{-1}(-(0.05)\epsilon^{-1})$. Since the diameter of $S'(x')$ is less than $2\pi h_{i-1}$, and since $s_{N_{i-1}}(x') \geq (0.94)\epsilon^{-1}$, it follows that $S'(x') \subset S_{N_{i-1}}^{-1}((0.9\epsilon^{-1}, \epsilon^{-1}))$. Since the distance from S'_i to the central two-sphere is at least $(0.99)\epsilon^{-1}h'_i$, It follows from Corollary A.5 that the central two-sphere of N_i is disjoint from C_{i-1} and lies off the positive end of N_{i-1}. This implies that the distance from x_0 to the complement of C_i is greater than the distance from x_0 to the complement of C_{i-1} by an amount bounded below by the distance from the central two-sphere of N_i to its positive end. According to Corollary A.5 this distance is at least $(0.99)\epsilon^{-1}h_i$, where h_i is the scale of N_i. But we know that $h_i \geq C^{-1/2}R(x_0)^{-1/2}$.

Thus, all cases either lead to a contradiction or to the conclusion that the distance from x_0 to the complement of C_i is at least a fixed positive amount (independent of i) larger than the distance from x_0 to the complement of C_{i-1}. Since the diameter of any (C,ϵ)-cap is uniformly bounded, this contradicts the existence of an infinite chain $C_0 \subset C_1 \subset \cdots$ contrary to the claim. This completes the proof of the claim. □

Now let us turn to the proof of the proposition. We suppose first that there is a point $x_0 \in X$ that is contained in the core of a (C,ϵ)-cap. Applying the previous claim, we can find a (C,ϵ)-cap C_0 containing x_0 with the property that no point of X contained in the frontier of C_0 is contained in the closure of the core of a (C,ϵ)-cap C_1 that contains C_0.

There are three possibilities to examine:
  (i) X is disjoint from the frontier of C_0.
  (ii) X meets the frontier of C_0 but every point of this intersection is the center of an ϵ-neck.
  (iii) There is a point of the intersection of X with the frontier of C_0 that is contained in the core of (C,ϵ)-cap.

In the first case, since X is connected, it is contained in C_0. In the second case we let N_1 be an ϵ-neck centered at a point of the intersection of X with the frontier of C_0, and we replace C_0 by $C_0 \cup N_1$ and repeat the argument at the frontier of $C_0 \cup N_1$. We continue in this way creating C_0 union a balanced chain of ϵ-necks $C_0 \cup N_1 \cup N_2 \cup \cdots \cup N_k$. At each step it is possible that either there is no point of the frontier containing a point of X, in which case the union, which is a C-capped ϵ-tube, contains X. Another possibility is that we can repeat the process forever creating a C-capped infinite ϵ-tube. By Lemma A.15 this union is a component of M and hence contains X.

We have shown that one of following holds:
  (a) There is a (C,ϵ)-cap that contains X.
  (b) There is a finite or infinite C-capped ϵ-tube that contains X.
  (c) There is a (C,ϵ)-cap or a finite C-capped ϵ-tube $\widetilde{C}$ containing a point of X and there is a point of the intersection of X with the frontier of $\widetilde{C}$ that is contained in the core of a (C,ϵ)-cap.

In the first two cases we have established the proposition. Let us examine the third case in more detail. Let $N_0 \subset C_0$ be the ϵ-neck that is the complement of the closure of the core of C_0. First notice that by Lemma A.20 the union $N_0 \cup N_1 \cup \cdots \cup N_k$ is diffeomorphic to $S^2 \times (0,1)$, with the two-spheres coming from the ϵ-neck structure of each N_i being isotopic to the two-sphere factor in this product structure. It follows immediately that $\widetilde{C}$ is diffeomorphic to C_0. Let C' be a (C,ϵ)-cap whose core contains a point of the intersection of X with the frontier of $\widetilde{C}$. We use the terminology 'the core of $\widetilde{C}$' to mean $\widetilde{C} \setminus N_k$. Notice that if $k = 0$, this is exactly the core of C_0. To complete the proof of the result we must show that the following hold:

CLAIM A.23. *If C' is a (C, ϵ)-cap whose core contains a point of the frontier of $\widetilde{C}$, then $\widetilde{C} \cup C'$ is a component of M containing X.*

PROOF. We suppose that $\widetilde{C}$ is the union of C_0 and a balanced chain $N_0, \ldots, N_k$ of ϵ-necks. We orient this chain so that C_0 lies off the negative end of each of the N_i. Let S' be the boundary of the core of C' and let N' be an ϵ-neck contained in C' whose central two-sphere is S'. We orient the direction $s_{N'}$ so that the positive direction points away from the core of C'. The first step in proving this claim is to establish the following.

CLAIM A.24. *Suppose that there is a two-sphere $\Sigma \subset N'$ contained in the closure of the positive half of N' and also contained in $\widetilde{C}$. Suppose that Σ is isotopic in N' to the central two-sphere S' of N'. Then $\widetilde{C} \cup C'$ is a component of M, a component containing X.*

PROOF. Σ separates $\widetilde{C}$ into two components: A, which has compact closure in $\widetilde{C}$, and B, containing the end of $\widetilde{C}$. The two-sphere Σ also divides C' into two components. Since Σ is isotopic in N' to S', the complementary component A' of Σ in C' with compact closure contains the closure of the core of C'. Of course, the frontier of A in M and the frontier of A' in M are both equal to Σ. If $A = A'$, then the closure of the core of C' is contained in the closure of A and hence is contained in $\widetilde{C}$, contradicting our assumption that C' contains a point of the frontier of $\widetilde{C}$. Thus, A and A' lie on opposite sides of their common frontier. This means that $\overline{A} \cup \overline{A}'$ is a component of M. Clearly, this component is also equal to $\widetilde{C} \cup C'$. Since X is connected and this component contains a point x_0 of X, it contains X. This completes the proof of Claim A.24. □

Now we return to the proof of Claim A.23. We consider three cases.
First Subcase: $S' \subset \widetilde{C}$. In this case we apply Claim A.24 to see that $\widetilde{C} \cup C'$ is a component of M containing X.
Second Subcase: S' **is disjoint from** $\widetilde{C}$. Let A be the complementary component of S' in M containing $\widetilde{C}$. The intersection of A with C' is either the core of C' or is a submanifold of C' diffeomorphic to $S^2 \times (0, 1)$. The first case is not possible since it would imply that the core of C' contains $\widetilde{C}$ and hence contains C_0, contrary to the way we chose C_0. Thus, the core of C' and the complementary component A containing $\widetilde{C}$ both have S' as their frontier and they lie on opposite sides of S'. Since the closure of the core of C' contains a point of the frontier of $\widetilde{C}$, it must be the case that S' also contains a point of this frontier. By Proposition A.11, the neck $N' \subset C'$

meets N_k and there is a two-sphere $\Sigma \subset N' \cap N_k$ isotopic in N' to S' and isotopic in N_k to the central two-sphere of N_k. Because $N_k \subset \widetilde{C}$ and $\widetilde{C}$ is disjoint from the core of C', we see that Σ is contained in the positive half of N'. Applying Claim A.24 we see that $\widetilde{C} \cup C'$ is a component of M containing X.

Third Subcase: $S' \cap \widetilde{C} \neq \emptyset$ **and** $S' \not\subset \widetilde{C}$. Clearly, in this case S' contains a point of the frontier of $\widetilde{C}$ in M, i.e., a point of the frontier of the positive end of N_k in M. Since $N_k \cap N' \neq \emptyset$, by Lemma A.2 the scales of N_k and N' are within 1 ± 0.01 of each other, and hence the diameter of S' is at most 2π times the scale of N_k. Since the central two-sphere S' of N' contains a point in the frontier of the positive end of N_k, it follows from Lemma A.5 that S' is contained on the positive side of the central two-sphere of N_k and that the frontier of the positive end of N_k is contained in N'. By Proposition A.11 there is a two-sphere Σ in the neck structure for N' that is contained in N_k and is isotopic in N_k to the central two-sphere from that neck structure. Let A be the complementary component of Σ in M that contains $\widetilde{C} \setminus N_k$. If the complementary component of Σ that contains $C' \setminus N'$ is not A, then $\widetilde{C} \cup C'$ is a component of M containing X. Suppose that A is also the complementary component of Σ in M that contains $C' \setminus N'$. Of course, A is contained in the core of C'. If $k \geq 1$, we see that A and hence the core of C' contains $\widetilde{C} \setminus N_k$, which in turn contains the core of C_0. This contradicts our choice of C_0. If $k = 0$, then $C_0 = A \cup (N_0 \cap (M \setminus A))$. Of course, $A \subset C'$. Also, the frontier of $N_0 \cap (M \setminus A)$ in M is the union of A and the frontier of the positive end of N_0 in M. But we have already established that the frontier of the positive end of N_0 in M is contained in N'. Since $A \subset C'$, it follows that all of C_0 is contained in C'. On the other hand, there is a point of the frontier of C_0 contained in the closure of the core of C'. This then contradicts our choice of C_0.

This completes the analysis of all the cases and hence completes the proof of Claim A.23. □

The last thing to do in this case in order to prove the proposition in Case I is to show that $\widetilde{C} \cup C'$ is diffeomorphic to S^3, $\mathbb{R}P^3$, or $\mathbb{R}P^3 \# \mathbb{R}P^3$. The reason for this is that $\widetilde{C}$ is diffeomorphic to C_0; hence $\widetilde{C}$ either is diffeomorphic to an open three-ball or to a punctured $\mathbb{R}P^3$. Thus, the frontier of C' in $\widetilde{C}$ is a two-sphere that bounds either a compact three-ball or the complement of an open three-ball in $\mathbb{R}P^3$. Since C' itself is diffeomorphic either to a three-ball or to a punctured $\mathbb{R}P^3$, the result follows.

Case II: Suppose that every point of X is the center of an ϵ-neck. Then if the two-spheres of these necks separate M, it follows from Proposition A.19 that X is contained in an ϵ-tube in M.

It remains to consider the case when the two-spheres of these necks do not separate M. As in the case when the 2-spheres separate, we begin building a balanced chain of ϵ-necks with each neck in the chain centered at a point of X. Either this construction terminates after a finite number of steps in a finite ϵ-chain whose union contains X, or it can be continued infinitely often creating an infinite ϵ chain containing X or at some finite stage (possibly after reversing the indexing and the s-directions of the necks) we have a balanced ϵ-chain $N_a \cup \cdots \cup N_{b-1}$ and a point of the intersection of X with the frontier of the positive end of N_{b-1} that is the center of an ϵ-neck N_b with the property that N_b meets the negative end of N_a. Intuitively, the chain wraps around on itself like a snake eating its tail. If the intersection of $N_a \cap N_b$ contains a point x with $s_{N_a}(x) \geq -(0.9)\epsilon^{-1}$, then according to Proposition A.11 the intersection of $N_a \cap N_b$ is diffeomorphic to $S^2 \times (0,1)$ and the two-sphere in this product structure is isotopic in N_a to the central two-sphere of N_a and is isotopic in N_b to the central two-sphere of N_b. In this case it is clear that $N_a \cup \cdots \cup N_b$ is a component of M that is an ϵ-fibration.

We examine the possibility that the intersection $N_a \cap N_b$ contains some points in the negative end of N_a but is contained in $s_{N_a}^{-1}((-\epsilon^{-1}, -(0.9)\epsilon^{-1}))$. Set $A = s_{N_a}^{-1}((-\epsilon^{-1}, -(0.8)\epsilon^{-1}))$. Notice that since X is connected and X contains a point in the frontier of the positive end of N_a (since we have added at least one neck at this end), it follows that X contains points in $s_{N_a}^{-1}(s)$ for all $s \in [0, \epsilon^{-1})$. If there are no points of X in A, then we replace N_a by an ϵ-neck N_a' centered at a point of $s_{N_a}^{-1}((0.15)\epsilon^{-1}) \cap X$. Clearly, by Lemma A.5 N_a' contains $s_{N_a}^{-1}(-(0.8)\epsilon-1, \epsilon^{-1})$ and is disjoint from $s_{N_a}^{-1}((-\epsilon^{-1}, -(0.9)\epsilon^{-1})$, so that $N_a', N_{a+1}, \ldots, N_b$ is a chain of ϵ-necks containing X. If there is a point of $X \cap A$, then we let N_{b+1} be a neck centered at this point. Clearly, $N_a \cup \cdots \cup N_{b+1}$ is a component, M_0, of M containing X. The preimage in the universal covering of M_0 is a chain of ϵ-necks infinite in both directions. That is to say, the universal covering of M_0 is an ϵ-tube. Furthermore, each point in the universal cover of M_0 is the center of an ϵ-neck that is disjoint from all its non-trivial covering translates. Hence, the quotient M_0 is an ϵ-fibration.

We have now completed the proof of Proposition A.21. $\square$

As an immediate corollary we have:

PROPOSITION A.25. *For all $\epsilon > 0$ sufficiently small the following holds. Suppose that (M, g) is a connected Riemannian manifold such that every point is either contained in the core of a (C, ϵ)-cap in M or is the center of an ϵ-neck in M. Then one of the following holds:*

(1) *M is diffeomorphic to S^3, $\mathbb{R}P^3$ or $\mathbb{R}P^3 \# \mathbb{R}P^3$, and M is either a double C-capped ϵ-tube or is the union of two (C, ϵ)-caps.*
(2) *M is diffeomorphic to $\mathbb{R}^3$ or $\mathbb{R}P^3 \setminus \{\text{point}\}$, and M is either a $(C, \epsilon$-cap or a C-capped ϵ-tube.*
(3) *M is diffeomorphic to $S^2 \times \mathbb{R}$ and is an ϵ-tube.*
(4) *M is diffeomorphic to an S^2-bundle over S^1 and is an ϵ-fibration.*

Bibliography

[1] Steven Altschuler. Singularities of the curve shrinking flow for space curves. *J. Differential Geometry*, 34:491–514, 1991.
[2] Steven Altschuler and Matthew Grayson. Shortening space curves and flow through singularities. *J. Differential Geom.*, 35:283–298, 1992.
[3] Shigetoshi Bando. Real analyticity of solutions of Hamilton's equation. *Math. Z.*, 195(1):93–97, 1987.
[4] Yu. Burago, M. Gromov, and G. Perel'man. A. D. Aleksandrov spaces with curvatures bounded below. *Uspekhi Mat. Nauk*, 47(2(284)):3–51, 222, 1992.
[5] Huai-Dong Cao and Xi-Ping Zhu. A complete proof of the Poincaré and Geometrization conjectures – Application of the Hamilton-Perelman theory of the Ricci flow. *Asian J. of Math*, 10:169–492, 2006.
[6] Jeff Cheeger. Finiteness theorems for Riemannian manifolds. *Amer. J. Math.*, 92:61–74, 1970.
[7] Jeff Cheeger and David Ebin. *Comparison theorems in Riemannian geometry*. North-Holland Publishing Co., Amsterdam, 1975. North-Holland Mathematical Library, Vol. 9.
[8] Jeff Cheeger and Detlef Gromoll. The structure of complete manifolds of nonnegative curvature. *Bull. Amer. Math. Soc.*, 74:1147–1150, 1968.
[9] Jeff Cheeger and Detlef Gromoll. The splitting theorem for manifolds of nonnegative Ricci curvature. *J. Differential Geometry*, 6:119–128, 1971/1972.
[10] Jeff Cheeger and Detlef Gromoll. On the structure of complete manifolds of nonnegative curvature. *Ann. of Math. (2)*, 96:413–433, 1972.
[11] Jeff Cheeger, Mikhail Gromov, and Michael Taylor. Finite propagation speed, kernel estimates for functions of the Laplace operator, and the geometry of complete Riemannian manifolds. *J. Differential Geom.*, 17(1):15–53, 1982.
[12] Bing-Long Chen and Xi-Ping Zhu. Uniqueness of the Ricci flow on complete noncompact manifolds. *J. Differential Geom.*, 74(1):119–154, 2006.
[13] Bennett Chow and Dan Knopf. *The Ricci flow: an introduction*, volume 110 of *Mathematical Surveys and Monographs*. American Mathematical Society, Providence, RI, 2004.
[14] Bennett Chow, Peng Lu, and Lei Ni. *Hamilton's Ricci flow*, volume 77 of *Graduate Studies in Mathematics*. American Mathematical Society, Providence, RI, 2006.
[15] Tobias H. Colding and William P. Minicozzi, II. Estimates for the extinction time for the Ricci flow on certain 3-manifolds and a question of Perelman. *J. Amer. Math. Soc.*, 18(3):561–569 (electronic), 2005.
[16] Dennis M. DeTurck. Deforming metrics in the direction of their Ricci tensors. *J. Differential Geom.*, 18(1):157–162, 1983.
[17] Yu Ding. Notes on Perelman's second paper. Preprint available at URL www.math.lsa.umich.edu/~lott/ricciflow/perelman.html, 2004.
[18] M. P. do Carmo. *Riemannian Geometry*. Birkhäuser, Boston, 1993.
[19] Klaus Ecker and Gerhard Huisken. Interior estimates for hypersurfaces moving by mean curvature. *Invent. Math.*, 105:547–569, 1991.

[20] Lawrence C. Evans and Ronald F. Gariepy. *Measure theory and fine properties of functions*. Studies in Advanced Mathematics. CRC Press, Boca Raton, FL, 1992.

[21] M. Gage and R. S. Hamilton. The heat equation shrinking convex plane curves. *J. Differential Geom.*, 23(1):69–96, 1986.

[22] Sylvestre Gallot, Dominique Hulin, and Jacques Lafontaine. *Riemannian geometry*. Universitext. Springer-Verlag, Berlin, third edition, 2004.

[23] R. E. Greene and H. Wu. Lipschitz convergence of Riemannian manifolds. *Pacific J. Math.*, 131:119–141, 1988.

[24] Detlef Gromoll and Wolfgang Meyer. On complete open manifolds of positive curvature. *Ann. of Math. (2)*, 90:75–90, 1969.

[25] Mikhael Gromov. *Structures métriques pour les variétés riemanniennes*, volume 1 of *Textes mathématiques*. CEDIC/Fernand Nathan, Paris, France, 1981.

[26] Mikhael Gromov and H. Blaine Lawson, Jr. Positive scalar curvature and the Dirac operator on complete Riemannian manifolds. *Inst. Hautes Études Sci. Publ. Math.*, 58:83–196 (1984), 1983.

[27] Robert Gulliver and Frank David Lesley. On boundary branch points of minimizing surfaces. *Arch. Rational Mech. Anal.*, 52:20–25, 1973.

[28] Richard S. Hamilton. The inverse function theorem of Nash and Moser. *Bull. Amer. Math. Soc. (N.S.)*, 7(1):65–222, 1982.

[29] Richard S. Hamilton. Three-manifolds with positive Ricci curvature. *J. Differential Geom.*, 17(2):255–306, 1982.

[30] Richard S. Hamilton. Four-manifolds with positive curvature operator. *J. Differential Geom.*, 24(2):153–179, 1986.

[31] Richard S. Hamilton. The Ricci flow on surfaces. In *Mathematics and general relativity (Santa Cruz, CA, 1986)*, volume 71 of *Contemp. Math.*, pages 237–262. Amer. Math. Soc., Providence, RI, 1988.

[32] Richard S. Hamilton. The Harnack estimate for the Ricci flow. *J. Differential Geom.*, 37(1):225–243, 1993.

[33] Richard S. Hamilton. A compactness property for solutions of the Ricci flow. *Amer. J. Math.*, 117(3):545–572, 1995.

[34] Richard S. Hamilton. The formation of singularities in the Ricci flow. In *Surveys in differential geometry, Vol. II (Cambridge, MA, 1993)*, pages 7–136. Internat. Press, Cambridge, MA, 1995.

[35] Richard S. Hamilton. Four-manifolds with positive isotropic curvature. *Comm. Anal. Geom.*, 5(1):1–92, 1997.

[36] Richard S. Hamilton. Non-singular solutions of the Ricci flow on three-manifolds. *Comm. Anal. Geom.*, 7(4):695–729, 1999.

[37] Robert Hardt and Leon Simon. Boundary regularity and embedded solutions for the oriented Plateau problem. *Ann. of Math. (2)*, 110(3):439–486, 1979.

[38] Allen Hatcher. *Algebraic topology*. Cambridge University Press, Cambridge, 2002.

[39] John Hempel. *3-Manifolds*. Princeton University Press, Princeton, N. J., 1976. Ann. of Math. Studies, No. 86.

[40] Stefan Hildebrandt. Boundary behavior of minimal surfaces. *Arch. Rational Mech. Anal.*, 35:47–82, 1969.

[41] T. Ivey. Ricci solitons on compact three-manifolds. *Diff. Geom. Appl.*, 3:301–307, 1993.

[42] Jürgen Jost. *Two-dimensional geometric variational problems*. Pure and Applied Mathematics (New York). John Wiley & Sons Ltd., Chichester, 1991. A Wiley-Interscience Publication.

[43] Vatali Kapovitch. Perelman's stability theorem. math.DG/0703002, 2007.

[44] Bruce Kleiner and John Lott. Locally collapsed 3-manifolds. In preparation.

[45] Bruce Kleiner and John Lott. Notes on Perelman's papers. math.DG/0605667, 2006.

[46] O. A. Ladyzhenskaja, V. A. Solonnikov, and N. N. Ural′ceva. *Linear and quasilinear equations of parabolic type*. Translated from the Russian by S. Smith. Translations of Mathematical Monographs, Vol. 23. American Mathematical Society, Providence, R.I., 1967.

[47] Peter Li and L-F. Tam. The heat equation and harmonic maps of complete manifolds. *Invent. Math.*, 105:305–320, 1991.

[48] Peter Li and Shing-Tung Yau. On the parabolic kernel of the Schrödinger operator. *Acta Math.*, 156(3-4):153–201, 1986.

[49] Peng Lu and Gang Tian. Uniqueness of standard solutions in the work of Perelman. Available at www.math.lsa.umich.edu/~lott/ricciflow/StanUniqWork2.pdf, 2005.

[50] John Milnor. Towards the Poincaré conjecture and the classification of 3-manifolds. *Notices Amer. Math. Soc.*, 50(10):1226–1233, 2003.

[51] John Morgan and Gang Tian. Completion of Perelman's proof of the Geometrization Conjecture. In preparation.

[52] C. B. Morrey. The problem of Plateau on a Riemannian manifold. *Ann. Math.*, 49:807–851, 1948.

[53] Grisha Perelman. The entropy formula for the Ricci flow and its geometric applications. math.DG/0211159, 2002.

[54] Grisha Perelman. Finite extinction time for the solutions to the Ricci flow on certain three-manifolds. math.DG/0307245, 2003.

[55] Grisha Perelman. Ricci flow with surgery on three-manifolds. math.DG/0303109, 2003.

[56] Stefan Peters. Convergence of Riemannian manifolds. *Compositio Math.*, 62:3–16, 1987.

[57] Peter Petersen. *Riemannian geometry, Second Edition*, volume 171 of *Graduate Texts in Mathematics*. Springer-Verlag, New York, 2006.

[58] Henri Poincaré. Cinquième complément à l'analysis situs. In *Œuvres. Tome VI*, Les Grands Classiques Gauthier-Villars. [Gauthier-Villars Great Classics], pages v+541. Éditions Jacques Gabay, Sceaux, 1996. Reprint of the 1953 edition.

[59] Jonathan Sacks and Karen Uhlenbeck. The existence of minimal immersions of 2-spheres. *Ann. of Math.*, 113:1–24, 1981.

[60] Takashi Sakai. *Riemannian geometry*, volume 149 of *Translations of Mathematical Monographs*. American Mathematical Society, Providence, RI, 1996. Translated from the 1992 Japanese original by the author.

[61] R. Schoen and S.-T. Yau. *Lectures on differential geometry*. Conference Proceedings and Lecture Notes in Geometry and Topology, I. International Press, Cambridge, MA, 1994. Lecture notes prepared by Wei Yue Ding, Kung Ching Chang [Gong Qing Zhang], Jia Qing Zhong and Yi Chao Xu, Translated from the Chinese by Ding and S. Y. Cheng, Preface translated from the Chinese by Kaising Tso.

[62] Richard Schoen and Shing-Tung Yau. The structure of manifolds with positive scalar curvature. In *Directions in partial differential equations (Madison, WI, 1985)*, volume 54 of *Publ. Math. Res. Center Univ. Wisconsin*, pages 235–242. Academic Press, Boston, MA, 1987.

[63] Peter Scott. The geometries of 3-manifolds. *Bull. London Math. Soc.*, 15(5):401–487, 1983.

[64] Natasha Sesum, Gang Tian, and Xiao-Dong Wang. Notes on Perelman's paper on the entropy formula for the Ricci flow and its geometric applications. preprint, 2003.

[65] Wan-Xiong Shi. Deforming the metric on complete Riemannian manifolds. *J. Differential Geom.*, 30(1):223–301, 1989.

[66] Wan-Xiong Shi. Ricci deformation of the metric on complete noncompact Riemannian manifolds. *J. Differential Geom.*, 30(2):303–394, 1989.

[67] Takashi Shioya and Takao Yamaguchi. Volume collapsed three-manifolds with a lower curvature bound. *Math. Ann.*, 333(1):131–155, 2005.

[68] John Stallings. A topological proof of Gruschko's theorem on free products. *Math. Z.*, 90:1–8, 1965.

[69] William P. Thurston. Hyperbolic structures on 3-manifolds. I. Deformation of acylindrical manifolds. *Ann. of Math. (2)*, 124(2):203–246, 1986.

[70] V. Toponogov. Spaces with straight lines. *AMS Translations*, 37:287–290, 1964.

[71] Brian White. Classical area minimizing surfaces with real-analytic boundaries. *Acta Math.*, 179(2):295–305, 1997.

[72] Rugang Ye. On the l function and the reduced volume of Perelman. Available at www.math.ucsb.edu/~yer/reduced.pdf, 2004.

Index

Page numbers in italic refer to definitions.

(C,ϵ)-cap, xviii, *232*, 247
  core of, 250
C-component, xviii, *232*, 247
C_t, 355
D_{t^-}, *356*
E_α, *424*
E_t, 355
H, *437*
$H(X)$, *127*, 128–130, 182
$H(X,Y)$, *123*, 125, 126, 128, 129
$J(a)$, *455*, *456*
$J_B(c)$, *456*
$J_B(c,\lambda)$, *456*
J_i, 456
$K^{\bar\tau}$, *130*
$K^{\bar\tau}_{\tau_1}$, *130*, 130–132
L_x, *117*
L_x^τ, *117*
$W(\Gamma)$, *416*, 430, 433, 435, 462
$W(\gamma)$, *416*
$W(\widetilde{\Gamma})$, 435, 462
$W(\xi)$, *430*
W_2, 420
$W_2(\beta, g)$, *416*
$W_2(g)$, *416*
$W_2(t)$, 426, 427
$W_\xi(t)$, *431*, 433
$\triangle$, *360*, 362
Ω, *279–286*
Ω_ρ, 283
Σ_t, 355
Θ, 444, *444*, 445, 481, 484
χ, *347*
δ-regular point, *83*
$\delta(t)$, *356*, 359
ϵ-cap, xvii
  core of, xviii
ϵ-fibration, *235*
ϵ-neck, xvii, *31*, *232*, 275

  evolving, *37*, *232*
  scale, 31, *499*
  strong, xxii, *37*, *233*, 247
    balanced chain of, 249
ϵ-round component, xviii, *232*, 247
ϵ-tube, xviii, *235*
  capped, *235*, 247
  doubly capped, *235*, 247
  strong, *235*, 280
κ-non-collapsed, xvi, xxvii, 169, *180*, 272
κ-solution, xxviii
  asymptotic soliton, xxviii
  compactness, 225
  qualitative description, 230
$\kappa(t)$, *359*
$\triangle L_x^{\bar\tau}$, 127
$\widehat{\mathcal{M}}$, 281
$\widetilde{L}$, 113
$\widetilde{L}^\tau$, *113*
  gradient, 115
$\widetilde{V}_x$, 140, 170–197
$\widetilde{l}$, 132
$\widetilde{\mathcal{U}}_x$, *116*
$\widetilde{\mathcal{U}}_x(\tau)$, *117*
  monotonicity, 118
k, *437*
l_x, *129*, 149–154, 181–195
$l_{\min}$, *154*
$r(t) > 0$, *359*
$\mathbf{K}$, 360, 362
$\mathbf{r}$, 362
$\mathbf{r}$, 360
$\mathbf{t}$, 346
$\mathcal{HT}(\mathcal{M})$, *348*
$\mathcal{J}(Z,\tau)$, 142
$\mathcal{K}^{\bar\tau}_{\tau_1}$, 127, 131
$\mathcal{L}$, *106*
$\mathcal{L}$-Jacobi equation, 105, *109*

$\mathcal{L}$-Jacobi field, 105, *109*, 114
$\mathcal{L}$-exponential map, 105, *112*
$\mathcal{L}$-geodesic, xxvi, 105, *108*
  minimizing, 109
$\mathcal{L}$-geodesic equation, *108*
$\mathcal{L}$-length, 105, *106*
  Euler-Lagrange equation, 107
  reduced, 129
$\mathcal{L}_+$, *379*
$\mathcal{L}\exp$, *113*
  differential of, *114*
$\mathcal{L}\exp_x^\tau$, *113*
$\mathcal{U}_x$, *117*
$\mathcal{U}_x(\tau)$, *117*, 151
Jac, *109*, 112, 121

ancient solution, xxvii, *180*
asymptotic scalar curvature, 218
asymptotic soliton, *see also* κ-solution
asymptotic volume, *222*

Bianchi identity, 6
Bishop-Gromov Theorem, xvii, *19*
blow-up limit, *99*, 267
Busemann function, *22*

canonical neighborhood, xvii, *232*, 247, 270, 275
  strong, *233*, 267, 268, 274, 280, 281, 323
canonical neighborhood parameter, *359*
compatible with time and the vector field, *347*
cone, 256, 261, 265
conjugate point, 13
continuing region, *355*
converge
  geometrically, xxiv, 84
  Gromov-Hausdorff sense, 93–95, 261, 263
coordinate charts for space-time, 345
curvature, 5
  evolution under Ricci flow, 41
  operator, 7
  pinched toward positive, xxiii, 75–76, 80, *245*, 267, *352*
  Ricci, 7
  Riemann, *see also* Riemann curvature tensor
  scalar, 7
  sectional, 6
curve-shrinking flow, 437, *437*, 439, 441, 445, 446, 449, 455, 481, 482, 487
cut locus, 15

disappearing region, *356*

Einstein manifold, 9, *36*, 81
end, of a manifold, 28
exponential mapping, 14
exposed points, *346*
exposed region, xxxii, *346*

final time, *343*
finite-time extinction, xxxvii
focusing triangle, *469*, *469*, 474, 476, 480
  base, *469*, 475, 480
forward difference quotient, *33*

Gaussian normal coordinates, 14
geodesic, 10
  minimizing, 10
geometric limit, xxiii, 84
  for Ricci flows, 90
    partial, 90
  partial, *84*
gradient shrinking soliton, xx, *38*, 184–185, 198
graph flow, 487, *487*, 488
Gromov-Hausdorff convergence, *see also* converge
Gromov-Hausdorff distance, *92*
Gromov-Hausdorff limit, 219, 264
Grushko's theorem, 421

harmonic map flow, *306*, 308
Hessian, *4*
  of L_x^τ, 122
homotopically essential
  2-sphere, *421*
  family of 2-spheres, *421*
  surgery, *422*
horizontal metric, *348*
horizontal subbundle, *348*
horizontal tangent space, 35
horn, *283*
  capped, 283, *283*
  double, 283, *283*

initial metric, normalized, *353*
initial metric, standard, *293*
initial time, *343*
injectivity domain, *117*
injectivity radius, xvi
injectivity set, *116*

Jacobi field, 12

Kneser's theorem, 420

Laplacian, 5, 16
length space, 95
Levi-Civita connection, *4*
Lichnerowicz Laplacian, 304

maximum principle
　heat equation, 63
　　strong, 69
　scalar curvature, *63–65*
　scalar functions, 29
　tensors, *65–67*

non-collapsing parameter, *359*

parabolic neighborhood, 36, *62*, *351*
parameter, canonical neighborhood, *352*
parameterized by backward time, 106
path of components, 419, *419*, 420, 423, 424, 426, 430, 431, 433
perturbed energy, *424*
Poincaré Conjecture, *ix*
point-picking, 203
polygonal approximation, *450*

ramp, xl, 445, 446, *446*, 447, 449, 450, 453
reduced length function, xxv
　monotonicity of, xxvi
reduced volume, xxvi, 140, 164, 169, 185
　integrand, 142
　monotonicity, 145
Ricci flow, x, *35*
　κ-non-collapsed, *351*
　equation, x, xix, *35*
　generalized, xxi, *60*
　　compatible embedding, 61
　　horizontal metric, *60*
　　space-time, *59*
　　time-slices, *60*
　Harnack inequality for, xxiii, *81*, 181, 273
　initial metric for, 35
　normalized initial conditions, 67
　soliton, 37, 81
　space-time of, xxi, xxxii
　standard metric, xxxi
　standard solution, xvi
　with surgery, xi, xxxi, *349*
　　finite-time extinction, xii
　　long-time existence, xii
　　topological effect, xxxiv
Ricci-DeTurck flow, *306*
Riemann curvature
　tensor, *5*

Riemann curvature operator, 179

Shi's derivative estimates, xxi, *50–52*
Shi's Theorem, *see also* Shi's derivative estimates
shrinking round cylinder, 36
singular points, space-time, *346*
soliton, *see also* Ricci flow
soul, *25*
space-form, 9
　3-dimensional, ix
　conjecture, ix
space-time, *343*
space-time, surgery, *346*
splitting at infinity, 100
splitting result, 204
splitting theorem, xvii, *28*
standard metric, *see also* Ricci flow
standard Ricci flow, *295*
　partial, *295*
strong δ-neck, 287
surgery cap, *333*
　evolution of, *369*, 370
surgery control parameter, *356*, 359
surgery parameter sequence, 360
surgery, on a δ-neck, *331*

Thurston's Geometrization Conjecture, *xv*, 358
time, 346
　regular, *347*
　singular, *347*
time-interval of definition, 343
time-slices, *343*, *346*
tip, of the standard initial metric, *331*
Tits cone, 97, 219
Toponogov theory, xvii, *23–24*
total curvature, *444*, 445, 450, 452, 453, 455, 468
total length, *444*, 445, 450, 454
total turning, *470*, *470*, 471, 474

Uniformization Theorem, 8
upper barrier, 152

volume comparison, *see also* Bishop-Gromov Theorem

DATE DUE

SCI QA 670 .M67 2007

Morgan, John W., 1946-

Ricci flow and the Poincaré conjecture